ADVANCING FRONTIERS OF ECOLOGICAL RESEARCHES IN INDIA

Prof. G.P. Mishra
Commemoration Volume

Editors

Prof. A.K. Kandya
Department of Botany
Dr. H.S. Gour University
Sagar 470003 M.P.

Dr. Asha Gupta
Department of Life Sciences
Manipur University
Imphal 795 003

BISHEN SINGH MAHENDRA PAL SINGH
23-A, New Connaught Place
Dehra Dun - 248 001 (INDIA)
2006

Advancing Frontiers of Ecological Researches in India

ISBN - 81-211-0490-4

Printed by Gajendra Singh Gahlot at Shiva Offset Press, and composed by Doon Phototype Printers, 14, Old Connaught Place, Dehra Dun, India *for* Bishen Singh Mahendra Pal Singh, Dehra Dun, India

PREFACE

The present festschrift "Advancing frontiers of ecological researches in India" is a tribute to Professor Govind Prasad Mishra , F.B.S., F.T.E., Ex Head of the Department of Botany and Dean, Faculty of Life Sciences, Director, Academic Staff College and Executive Councilor, Dr. H.S. Gour University, Sagar, M.P. for his great devotion and outstanding contributions to ecology. This commemoration volume is being brought out at the 70th birthday of Prof. G.P. Mishra as a token of our sincere regards, deep affection and best wishes for him.

This treatise contains 53 invited research articles, most of which are comprehensive reviews on various aspects of ecology by eminent experts of their fields. Each specialist has discussed the present state of art in the subject of his/her specialization, and has also pointed out future challenges in the field for the budding scientists. The topics cover a wide range of aspects of ecology and have been categorized in to ten sections, i.e. Forest Ecology, Grassland Ecology, Wetland Ecology, Cropland Ecology, Microbial Ecology, Weed Ecology, Wildlife Ecology, Pollution Ecology, Ecological Modeling and Applied Ecology etc., in which Prof. Mishra has guided research. Sagar University has been the first centre of ecological studies in India, established by Prof. R. Misra in 1946, that is why he is designated as the Father of Ecology in India. Later on, several of his students spread all over the country and established schools of ecology in various parts of the nation, i.e. Prof. L.P. Mall in Vikram University, Ujjain, Prof. S.C. Pandeya in Surashtra University, Rajkot, Prof. R.R. Das in Jiwaji University, Gwalior, Prof. C.K. Varsheny in Jawahar Lal Nehru University, Delhi and Prof. P.S. Ramakrishnan in North Eastern Hill University, Shillong etc. Prof. R. Misra himself shifted to Banaras Hindu University, Varanasi and established a very strong centre of ecological research there. The Sagar school of ecology continued under the leadership of Prof. G.P. Mishra. With the new generations of ecologists in the country and with the growing realisation of the significance of ecology, several new centres of ecological research like NEERI Nagpur; Institute of Science, Mumbai; Indian Institute of Forest Management (IIFM) Bhopal; M.P. State Forest Research Institute (SFRI) Jabalpur; U.P. Forest Research Centre, Kanpur; G..B. Pant Institute of Himalayan Environment and Development, Almora, Nagpur University, Nagpur; Bhavnagar University, Bhavnagar; Central Pollution Control Board; Agra; Pt. Ravishanker Shukla University, Raipur and Manipur University, Imphal etc. were established in the country in the last fifty years. It has been our serious effort that there must be at least one paper in this festschrift from all the centres of ecology presently working in the country. We are pleased to write that we have been successful in our endeavour.

Prof. Mishra has guided maximum number of Ph.D. students in forest ecology and incidentally we have also received maximum number of papers in this discipline. Prof. Mishra always emphasized to his students for a deeper probe in the forest ecosystem, specially its structure, organic matter dynamics, nutrient cycling and energetics etc. Therefore, there are some excellent papers from the Indian Institute of Forest Management, Bhopal, a central government autonomous organization governed by the Ministry of Environment and Forests. The foremost of them is the review by Dr. Ram Prasad, Ex-Vice Chancellor, Barakatullah University, Bhopal and co-workers on the ecology and forestry with special reference to Central India. This paper summarizes various aspects of ecology like the wasteland development, afforestation in coal mines, re-vegetation of mined out over-burdens, rehabilitation of degraded forests and forestry seed technology etc. These are practical problems tackled by the author personally as Director, MP State Forest Research Institute (SFRI), Jabalpur and Director, Tropical Forestry Research Institute, (TFRI), Govt. of India, Jabalpur.

Another important contribution is by Dr. P.C. Kotwal and his co-worker on the criteria and indicators of sustainable forest management, a new paradigm for monitoring and assessment of forest resources. Dr. P. Bhattacharya has written his experiences from a case study of livelihood generation by tribals of dry tropical forests of Madhya Pradesh. He has emphasized the sustainable management of biodiversity in Joint Forest Management areas. Dr. Manish Mishra has prepared an analytical report from three districts of Orissa on the regeneration and tree density in the natural forests under the impact of harvesting of some selected non-timber forest produce (NTFPs). Such a study indicates the future of natural forests in the country.

Prof. A.K. Kandya from Sagar has given an overview of the forest resources of Madhya Pradesh. Such an inventorisation makes the readers aware of the precious items of the forests called the open treasure of the nature. He has included the salient features of the new forest policy of the state, implemented in 2005 and focussing on the interests of forest dependent communities, not the earning of the revenue.

Participatory carbon management for eco-sustainability of forests and dependent rural communities is a significant aspect dealing with emission of carbon in the form of CO_2 during burning and carbon sequestration through decomposition of organic matter. The management of carbon has been an important point in the Kyoto Protocol and UN Framework of Climate Change (UNFCC). Dr. P.K. Shukla, Director, M.P. State Forest Research Institute, Jabalpur and his colleague have given an authentic document on this important topic. Spatial biodiversity is usually governed by the local inhabitants of an area. Dr. Alka Chaturvedi and her co-workers from Nagpur University have given a significant account of land-use decisions by locals affecting spatial biodiversity. The Convention on Biological Diversity (CBD) also emphasizes that locals must be benefited by biodiversity.

Dr. C.M. Mishra and his group from the UP Forest Research Institute, Kanpur submitted a paper on the recent trends in biodiversity conservation whereas Prof. Sunil Puri from the Department of Forestry, Indira Gandhi Agricultural University, Raipur has prepared an excellent account on the strategies to meet energy challenges through domestication of *Jatropha curcas* called Ratanjot in Hindi, a plant species nowadays used extensively for the commercial production of petroleum.

Prof. S.K. Billore and colleagues from the Institute of Environmental Management and Plant Sciences, Vikram University, Ujjain have described the N_2O source and sink potential in a dry tropical climax forest and in a seral grassland of Central India.

Lot of work on different aspects of ecology has been done in the recent years in the Department of Life Sciences, Manipur University, Imphal. Dr. Asha Gupta and her student have discussed their hydrological studies in a subtropical forest of Manipur, the North-east India. Dr. Manju Sundriyal from G.B. Pant Institute of Himalayan Environment and Development, Almora has presented a very informative review of the ethnobotanical researches with particular reference to wild edible plants. She has given an overall situation of the number of genera and species of the wild plants used in India and the world, their nutritional status and the names of plantrparts which are edible.

Prof. B.R. Pandit, who is a product of Prof. S.C. Pandeya's school of ecology in Saurashtra University, Rajkot and has established a centre of ecological research at Bhavnagar (Gujarat), has contributed a paper on the biodiversity of a dry deciduous forest ecosystem of Gujarat. The investigated site is the part of world famous Gir Forest, the abode of the Asiatic Lions at the peninsular part of Gujarat.

There are three excellent papers in the category of wetland ecology. Prof. R.S. Ambasht from Varanasi has contributed an article on the policies of conservation of wetlands with global and Indian perspectives. This article is of international importance and deals with the extent of wetlands and their conservational strategies in different countries. Dr. Nivanonee and Professor Manihar Sharma from the Deptt. of Life Sciences, Manipur University, Imphal have studied the biomass dynamics and productive status of the macrophytes in Ikoppat lake of Manipur which is a highly eutrophicated fresh water, saucer shaped, natural lake of 13.5 Sq. Km. area and is, at present categorized as threatened. It has 19 species of macrophytes, i.e. *Potamogeton crispus, Jussiaea repens, Azolla pinnata*, *Eichhornia crassipes* and *Echinochloa colonum* etc. which are the characteristic species of the fresh water lakes. In another article from Manipur University, Imphal, Dr. Arun Kumar and co-workers from the Earth Sciences Deptt. have explored recent tectonic activities and have described the results of their studies of Dampi watershed using GIS (Geographic Information System) and remote sensing techniques. This watershed is a tributary of Manipur river basin and its morphometric analysis, neotectonic set up, digital image processing, and digital elevation modeling have been done which are the most modern tools of research.

Three students of Prof. Mishra namely Dr. Archana Shrivastava, Dr. Tara Pathak and Dr. Veenapani Mishra have presented two aspects of weed ecology. Dr. Mrs. V. Mishra (nee Dubey) from C.M.D. College, Bilaspur has discussed the management of weeds in dry seeded rice crops whereas Dr. Mrs. Archana Shrivastava (nee Verma) from Sagar and Dr. Mrs. Tara Pathak (nee Nigam) from Satna have described the biology and adaptive strategies of some weeds. These papers have given an idea of losses in the crop yield done by weeds through the competition for nutrients and water.

Dr. Rajesh Gopal, Director of Project Tiger, Govt. of India, MoEF, New Delhi has discussed a very special wild animal of Kanha National Park of Madhya Pradesh which is also one of the 14 Tiger Reserves of our country. The animal is central Indian Barasingha– *Cervus duvauceli branderi* which is endemic to Kanha only and Dr. Gopal has discussed its population dynamics. One of his colleagues Dr. H.S. Negi has also given a comparative account of wildlife populations in and around Kamha National Park. Dr. Rakesh Shukla has written on the research and monitoring protocol for the adaptive management of the Kanha Tiger Reserve. Dr. B. Prasad from Imphal, Manipur has described a specialized phenomenon of insects called Diapause and also discussed adaptations for it. Similarly, Dr. Prem Kumar Singh, Incharge Wildlife, Forest Deptt. Govt. of Manipur has described the broad based problems and prospects of wildlife of Manipur.

Under the title of Microbial Ecology, we have received seven articles of fundamental as well as practical importance. Prof. B.N. Johri from G. B. Pant University, Pantnagar has contributed his marvelous review on the microbial community dynamics of compost ecosystem, in which he has described the process of composting as well as its structural and functional diversity. Dr. Jaishree Dubey from Sagar, a Ph.D. student of Prof. Mishra has given a comprehensive account on the ecology of terrestrial Blue Green Algae while Dr. S. P. Adhikary from Utkal University, Bhubaneshwar has dealt with an entirely different aspect of Cyanobacteria, i.e. tetrazolium reduction and chlorophyll stability of cyanobacteria in the crust of exposed soil surfaces subjected to heat stress.

Prof. M.L. Naik, Ex- Director General, Chattisgarh Council of Science and Technology and presently from the Deptt. of Biosciences, Pt. Ravishanker Shukla University, Raipur and his group prepared a useful report on the emerging nosocomial pathogen, *Enterobacter* which is not a primary human pathogen but is capable of causing occasional infections in hospitalized patients. Dr. Sarojnalini and co-workers from Manipur University, Imphal have submitted their paper on the microbial flora of two weed fishes. They reported a very high plate count of bacteria (10^7 / g) and 10^4 of fungal counts. Dr. K.V. S. Reddy from L A.D. College for Women, Nagpur and his co-worker have reported a new Blue Green Alga *Anabaena petalonemoides* from the cultivated soils of Yeotmal, Maharashtra and described its morphology and taxonomy for the first time. Dr. S.D. Shankhdarwar and Professor P.B.Nandkar from Nagpur University, Nagpur have discussed the degradation of cyanides by algae. They found that microbial degradation of cyanides was faster and quicker than the chemical degradation.

Nine papers have been categorized under the section Pollution Ecology. These papers have detailed descriptions of pollution and related areas. The most useful of them is the popular article by Dr. S.K. Jain who is the Director of an NGO named Envirotech Associates, based at Indore, Madhya Pradesh. Dr. Jain is a Ph.D. Scholar of Prof. Mishra. In his paper on management of wastes:an ecosystem approach, he has emphasized recycling and reuse of waste and advocated the theme that waste is actually a resource, out of place.

Prof. V.P. Singh from the Institute of Environmental Management and Plant Sciences, Vikram University, Ujjain has contributed a paper on triazoles and discussed them as a tool for ecological research as the derivatives of triazoles are used as fungicides as well as plant growth regulators.

Dr. R.K. Jain and his colleague from the Govt. Model Sciences College, Jabalpur has described a few important aspects of environmental conservation whereas Dr. G.P. Prasain and his co-worker from Manipur University, Imphal have wrote a full length article on environmental auditing. They have emphasized that the users of environmental auditing are companies, customers, banks, general public and almost all the leading Institutions as it is a part of the overall environmental assessment and helps to decrease environmental and public health risks. It also assists in developing environmental management.

Dr. Geet Govind Sinam from the Central Pollution Control Board, Agra has described a preliminary study on source apportionment of respirable suspended particulate matter (RSPM) in the ambient air. He used a range of approaches and combination of models to identify a reliable method for apportion the potential source for RSPM.

Dr. P.R. Choudhary, Assistant Director, EIRA Division, and his group from the National Environmental Engineering Research Institute (NEERI), Nagpur has also discussed a case study on the impact of pulp and paper mill effluent on ground water quality affected by irrigation of sugarcane fields. They detected high level of ground water pollution due to seepage of effluent. Palmer's Pollution Index indicated that pollution was due to inorganic pollutants.

Dr. A.A. Saoji of the Institute of Science, Nagpur has given an excellent account of an important and common aero-allergen–*Aspergillus* which is a major component of the bioaerosol and is also well adapted to air-borne dispersal of its spores. It is an etiologic agent and causes *aspergillosis*.

Dr. S.D. Shankhadarwar and his teacher Professor P.B. Nandkar from the Department of Botany, Nagpur University, Nagpur have contributed a paper on zinc accumulation in two blue green algae–*Chlorococcum humicolum* and *Phormidium foveolarum*. Several Blue-Green Algae, Bacteria and Fungi can bind large amount of metals and the phenomenon is called bio-absorption. In this way, they can be used to remove heavy metals from waste water and industrial effluents. However, high concentration of the metal can be toxic to the organism. The same aspect has been discussed in the two cyano-bacterial species which are resistant to high level of Zn.

Under the section of Grassland Ecology, three papers have received. The first one is by Miss Daisy Angom and Dr. Asha Gupta from Manipur University, Imphal who have discussed the dry matter dynamics in the floating grasslands of Keibul Lamjao National Park, the only abode of endangered deer *Cervus eldi eldi*. The national park has the distinction of being the only floating national park in the world where grasslands grow over the partially decomposed organic matter which keeps on floating over the water surface. In this sense, it has become an unique paper of this festschruft. Second paper is by Dr. Radhapayari Devi and her group from the same University who have described the nitrogen cycling in Canchipur grasslands of Manipur.

In the third paper, Dr. U.N. Singh and his student from D.V.P.G. College, Orai, U.P. have contributed a regional aspect of grasslands, i.e. dry forage production and quality of *Iseilema* grassland community as influenced by legume introduction in Bundelkhand region of Uttar Pradesh. The experiment revealed that the introduction of legumes was advantageous in both the aspects, i.e. quality and quantity of dry forage.

Prof. R.S. Mehrotra from Kurukshetra University, Kurukshetra has been the bosom friend of Prof. G.P. Mishra since his student life. He went to a separate line and developed an excellent working group in plant pathology at Kurukshetra University. He has contributed a paper on some environmentally safe approaches to crop disease control towards sustainable agriculture. In fact, biological control of plant pathogens has not solved the agricultural problems. But, some eco-friendly methods such as soil solarization etc. have a great promise in making our agriculture more sustainable.

Seed technology of tree species has become very important nowadays due to various types of plantation programs to revegetate the wastelands. Mr. Rajeev Baweley and his teacher Dr. U.N. Singh from D.V.P.G. College, Orai. U.P. have contributed a paper on response of pre-sowing treatments on the germination potential of seeds of *Delonix regia* a common avenue tree species being planted in bulk on roadsides. The authors have tested a number of pretreatments and described the most successful ones.

Ecological modeling is the newest development in ecological investigations. Most complicated concepts and interactions can be easily depicted and understood by ecological models. Competition among species, populations and ecosystems for limited resources, stability of ecosystem and predator – prey relationship can be critically learnt by this tool. Dr. M. Ranjit Singh from Manipur University, Imphal has given a good article on the role of mathematics in ecology. This will be a very useful article for the coming generations of ecologists.

Dr. Asha Gupta from the same University has also discussed similar aspect and has described matrix models in analysing population dynamics. Accordingly, better management of species, populations and ecosystems can be understood. The paper not only discusses application of matrix models through some case studies but also describes its disadvantages.

Another article on similar lines also comes from Manipur. Dr. U.C. Laisram of the Deptt. of Science and Technology, Govt. of Manipur, Imphal has discussed application of Global Positioning System (GPS) in ecological studies. GPS is the only system available commercially today to show the exact location on the earth at all times. It is history's most exciting and revolutionary development in location and mapping technology and has been discussed in real world application.

Dr. O.P. Goel, Ex Head of the Geology Deptt. and Dean, Faculty of Science, Kumaon University, Naini Tal, presently Visiting Professor, Manipur University, Imphal has submitted a review on the impact of

Himalayas on modifications of ecosystems of Indian subcontinent and Asia. It is an excellent paper of inter-disciplinary nature and describes a number of processes of geological ages occurred in Asian region.

The understanding of ecological principles behind several natural processes is nowadays being used extensively in several walks of life, i.e. eco-tourism, eco-development of industrial areas, eco-planning of urban colonies, eco-friendly products of various types and eco-management of forests, rivers and other areas, etc. In the section of applied ecology, all such papers have been included which are good examples of application of the knowledge of ecology in our day to day life.

Prof. Y.D. Tiagi from Mohanlal Sukhadia University, Udaipur has contributed a paper on the effect of certain growth regulants on the induction of roots on stem cuttings of *Tylophora indica*, a very important medicinal plant used against Asthama. That is the reason, it was earlier known as *Tylophora asthamatica,* and is extensively used in homoeopathic drugs.

Vermi-culture technology, a modern tool in agriculture, has been described in a nut shell by Dr. B. Narsimha Rao of ICAR Research Campus, Imphal.

Dr. Suprio Chakraborty of the Regional Agricultural Research Station of Assam Agricultural University at Nagaon has contributed an interesting article on the plant genetic resources of north-east India and their sustainable management. He has described the forest trees, cereals, legumes, fruits, vegetables, medicinal plants, orchids, oil seed plants, bamboos and canes and sericulture linked plant genetic resources. He has also discussed the hot spots of plant genetic resources in the NE and reasons for the loss of genetic resources. In this way, it is a very comprehensive account and concludes the ways and means to conserve the plant genetic resources in north-east India.

A very senior pollution ecologist of the country Prof. S.B. Chaphekar from Mumbai has written a very useful document on the eco-management of trees in a poor-rich urban ecosystem. He has described the planning of plantations in a metropolis to keep the environment clean.

Dr. S.R. Gupta, a Ph.D. student of Prof. Mishra, from Bipin Behari College, Jhansi has contributed an important article on increasing production potential of degraded lands in Bundelkhand through agro-forestry. Actually, the hostile climatic conditions, improper land use coupled with low level of soil fertility, low cropping intensity and poor availability of water are the factors responsible for the low productivity in agricultural fields of Bundelkhand. According to this paper, productivity can be enhanced by any kind of agro-forestry system, i.e. agri-silviculture, agri-horticulture, silvi-pasture and agri-sericulture etc.

We are whole heartedly thankful to all the contributors of the papers for this book. Many of them are direct Ph.D. students of Prof. G. P. Mishra but some others are his colleagues, friends and acquaintants. It is heart warming that our call for contributions to the commemoration volume also received an enthusiastic and touching welcome from those well wishers of Prof. Mishra who have not even seen him but have heard of him and known him through his academic contributions. The readiness with which they responded to our request and the quality of the articles reflect the high esteem in which they hold Professor Mishra. We are confident that this treatise will spread the frontiers of knowledge in ecological sciences and would be useful to teachers, research scholars and post-graduate students for getting an up-to-date appraisal of development in the field of plant ecology.

We are thankful to Prof. R.S. Mehrotra, Retd. Head of the Botany Deptt., Kurukshetra University, Kurukshetra for helping us in writing the Biography of Prof. G.P. Mishra and giving us the details of several of his personal experiences shared with Prof. Mishra. We also thank Mr. Niket Kandya for doing lot of our computer work at the cost of his precious study time during his examination days. Last, but not the least, we are deeply indebted to the publisher of this festschrift – M/S Bishen Singh Mahendra Pal Singh and sons who readily accepted our request to publish and market the book to all nooks and corners of the world where ecology is studied. We also take it as a pride in editing this commemoration volume for our esteemed and ideal teacher.

Professor A.K. Kandya and Dr. Asha Gupta
Editors

PROF. G.P. MISHRA
Sagar

PROFESSOR G.P. MISHRA BIOGRAPHY AND CONTRIBUTIONS TO ECOLOGY

Born on 2nd December 1935 in Narsinghpur town of Madhya Pradesh, child Govind Prasad Mishra did his schooling in Gwalior, Hoshangabad and Narsinghpur as his father was a Post Master and transferred from place to place. He passed his Matriculation in 1952 from the Board of Secondary Education of the then Central Province, Nagpur. He did his Intermediate Biology from Holker Science College, Indore in 1954 which was affiliated to the Madhya Bharat School Board. Later, he also did his B.Sc. Biology from the same college in 1956. He came to Sagar for his Post Graduate Education and took admission into the University Teaching Department of Botany for his M.Sc. Degree. He passed his M.Sc. in 1958 from the University of Saugar, Sagar and also took Ph.D. Degree under Prof. L.P. Mall in 1961 as his first Ph.D. student.

The Botany Department of the University had some eminent botanists in its staff at that time. Prof. R. Misra had just left Sagar in 1956 leaving his post of the Head of the Deptt. of Botany in Sagar University to join Banaras Hindu University, Varanasi. He was an eminent ecologist of the country and laid the firm foundation of the Department of Botany at Saugar University and also developed a strong school of ecological studies in it. There were other eminent teachers like Dr. S.B. Saksena (an eminent Mycologist and Plant Pathologist), Dr. T.V. Desikachary (an eminent Algalogist who moved to Madras in 1956), Prof. L.P. Mall (an eminent Ecologist), Dr. Y.D. Tiagi (an eminent Plant Morphologist) and Dr. Y. Sunder Rao (an eminent Cytologist). Two teachers strongly influenced G. P. Mishra during his student life. One was Dr. S. B. Saksena and the other was Dr. L.P. Mall. Ecology was the most liked subject of most of the research scholars in those days and G.P. Mishra had enrolled himself for his Ph.D. Program under Dr. L.P. Mall on a problem of forest ecology, i.e. Ecological studies of some forests of Sagar with special reference to underlying rocks. Prof. R.S. Mehrotra, Ex Head of the Botany Deptt., Kurukshetra University, Kurukshetra, who was also working for his Ph.D. program under the guidance of the great Mycologist and Plant Pathologist of the country, late Prof. S.B. Saksena, FNA, was his closest friend at that time and both were moving on their bicycles to remote areas of Khanpur and Patharia forests for the ecological samplings of forest vegetation. That training of basic principles of ecology later on reflected in the Ph.D. works of the students of Prof. G.P. Mishra who investigated diverse aspects of ecology, i.e. Forest Ecology, Grassland Ecology, Aquatic Ecology, Pollution Ecology, Applied Ecology, Weed Ecology and Algal ecology etc. In fact, there is no branch of ecology which has not been explored or touched by the students of Prof. Mishra..

He joined the Faculty of the Botany Department in the leave vacancy of Dr. L.P. Mall in 1962 and later became Assistant Professor in 1964, Reader in 1971 and Professor in 1978. He served the Department of Botany, Dr. H.S. Gour University for about thirty five years with loyalty as a devoted class-room teacher who never missed his classes, a field expert in identification of plant species, a strict administrator, and an efficient organizer of academic seminars, conferences, symposia, workshops and summer schools etc. Although, looking very simple in his outward behaviour, he had a novel way of teaching his students in M.Sc. Classes wherein he frequently brought into bear his expertise and mastery in field investigations. He has enriched the knowledge of a large number of batches of young students through his stimulating lectures, talks, extempore speeches, curiously a mix of wit and wisdom and possessing a sense of humour. His classroom lectures were very lively. Being a genuine ecologist, he realized the importance of systems ecology and emphasized this aspect to his M.Sc. and Ph.D. students. His lectures on forest ecosystem were simply fascinating. His students were spell bound in his lectures which were full of scientific information. Prof. Mishra has been a scholar of the highest order and has an unique multifaceted personality. By his inborn human nature and sweetness of temperament arising from sympathy and understanding for his students, he won the affection and regard of every one who came in his contact. He has a large number of friends and admirers in India and abroad. Even after his super-annuation, he is still active in academic activities and extends his advises and good wishes to his colleagues, friends, students and any body who wishes to have the benefits of his expertise. He has always been known for his uprightness, integrity and fearlessness which he inherited from his parents and teachers.

It is a fact that vast ecological studies were conducted by the students of Prof. Mishra and a huge amount of data has been collected by them from the Central Indian regions. Late Prof. R. Misra whenever visited Sagar, always used to take keen interest in knowing his work and always appreciated his approaches in tackling diverse ecological problems.

Prof. Mishra surveyed various parts of India leading the ecological tours and botanical excursions which have been acclaimed by his students as most enjoying, rewarding and enriching academic experiences. Prof. Mishra broke new grounds in exploring diverse fields of ecosystem researches in India taking the notice of rich forests and grasslands around Sagar. Due to him, ecology is at cross roads today. He has built several students who are new ecologists and visionaries and think deeply and seriously about the state of affairs in the field with holistic interdisciplinary approach.

The maximum research work done under the guidance of Prof. Mishra at Sagar belongs to the forest ecosystem. In the very beginning, Bhatnagar (1968) assessed the litter production on the forest floor under various tree species. He found that not only quantitative differences were there but also qualitative differences were apparent. The curvature of leaves while the get dried, is specific for each species and depends upon the structure and texture of leaves. Later, Sodhia (1974) analysed many more details of the foliages of several tree species of forest of Sagar. Kandya (1974) investigated the organic matter dynamics in some selected tree species of Sagar and prepared organic matter budget sheets for these species as well as for the tropical dry deciduous mixed teak forests of Sagar. Prasad (1976) performed broad based ecological studies in the forests of Gourjhamar forest Range. With every new Ph.D. student, a new aspect of the forest ecosystem was investigated. Nayak (1977) studied the process of minearl circulation in some forests of Sagar with special reference to *Tectona grandis* (Teak) which is the most dominant tree species of the tropical dry deciduous mixed teak forests of Sagar. Similar kind of work was also done by Johri (1977) in some other forest sites.

Organic layer is the most important and fertile part of the forest floor and Joseph (1978) made comprehensive study of the organic layer in some forests of Sagar. A significant component of the minerals moves out from the plants through stem-flow and through-fall etc. Such an investigation of the hydrological aspects of mineral circulation was done for the first time in forests of Sagar by Yadav (1978) and accepted as a mile-stone in the forest ecosystem research in tropical dry deciduous forests.

Diospyros melanoxylon (Tendu), the Bidi Leaf plant which has also been designated as the Bread Earner tree of Bundelkhand region, has been studied in great details for its natural regeneration in the forests of Sagar by Athya (1980) who pointed out various parameters to indicate leaf quality suitable for Bidi making. During the natural regeneration studies of forest tree species, Prof. Mishra's school obtained a turning point and switched over to the study of forestry seeds, their structure and diversity of germination behaviour. Das (1980) was the pioneer in this field and collected the preliminary database of this aspect.

In between, some students took the topics on various aspects of grassland ecology also. Jain (1971) performed production studies in some grasslands of Sagar. This was also a pioneer work and spelled-out the productivity pattern of various types of grasslands dominated by different grass species. Somewhat similar aspects of grasslands were also investigated by Naik (1973) in the grasslands of Ambikapur, an entirely different locality in a separate edapho-climatic region. Biogeochemical cycle in the grasslands is a very important aspect for the fundamental understanding of the functioning of grassland ecosystem. Dakwale (1975) analysed it for some grasslands of Sagar. At a larger scale, parallel studies were also performed by Trivedi (1977) and Gupta (1977) for the grasslands of Central India. At a very later stage, Mishra (1991) studied the ecology of nitrogen cycling in grasslands. The interests of the students of Prof. Mishra were so varied that Purohit, J. (1977) selected a topic on the ecological study of soil algae of Sagar.

Later on, this work was continued by Jaimini (1987) who investigated the periphyton of Sagar lake. Virtually, Prof. Mishra established new horizons in exploring diverse fields of ecology and ecosystem research. For instance, Pathak (1980) and Shrivastava, A. (1982) initiated deeper probe in ecophysiological studies of some weeds of central India. In a later phase, Dubey (1987) and Rawat (1988) also investigated the weeds of paddy fields of Bilaspur and agricultural weeds of Bundelkhand region respectively.

Prof. Mishra always gave new and newer ideas and thoughts to his students. As a consequence, Khare, D.K. (1981) investigated the ecological effects of forest fire in tropical dry deciduous forests of Sagar whereas Khare, P.K. (1981) studied the environmental biology of forests with special reference to natural regeneration. Shrivastava, V.K. (1982) performed phyto-sociological studies of grasslands of Bundelkhand including productivity and energetics also. Autecological studies of *Echinops echinatus* were carried out by Mahendra (1982) whereas ecological study of *Dalbergia sissoo* in forest plantations of Bundelkhand region with special reference to productive structure, mineral status and accumulation of energy was done by Sengar (1983). Lot of data was generated on forest tree seeds and their seedlings by Mishra (1982) and Yadav (1983).

Spatial and temporal variations in the distribution of vegetation is an inherent property of ecological systems. A large dataset has been made available by the ecological school of Prof. G.P. Mishra for structural and functional aspects of natural, disturbed forests, monoculture plantations and successional woodlands. Choubey (1987) compared the vegetational spectrum and edaphic conditions under natural forests and monocultures of teak whereas Prajapati (1991) worked out the productivity pattern and nutrient cycling in successional forests of western Himalayas.

Seeds play important role in raising plant populations and in turn revegetating the wastelands. Rehabilitation of degraded forests is also achieved by transplanting seedlings of appropriate tree species favourable in particular habitats. Shrivastava (1987), Yadav (1989), Rajput (1992), Purohit, M. (1992), Choubey (1993), Gohlani (1993), Jain (1996) and Balaiya (1997) are those students of the research group of Prof. Mishra who worked out various aspects of seeds, i.e. storability, pre-sowing treatments to hasten germination potential, testing procedures for vitality, viability and vigour of seeds and polymorphism of seeds etc. Most important contrbution in this field was the enhancement in the viability span of the recalcitrant seeds of Sal (*Shorea robusta*) by keeping them in proper storage conditions. This finding change the traditional concept of the field foresters. Not even this, Jaiswal (1990) investigated the pathological aspects and control of diseases of forest tree seedlings in nurseries.

An exceptionally significant and unique aspect was worked out by Bisen (1985) who performed scanning electron microscopic studies of some important bamboos of the Indian sub-continent in relation to their classification. Similarly, Kazmi (1985) carried out analytical studies on another unique aspect, that is, respiration of some plant pathogenic fungi. Bhattacharya (1992) performed investigations on the medicinal plants of Chhindwara District of Madhya Pradesh with special reference to their ethnobotanical aspects. A very important and applied aspect was worked out by Dhuria (1993) who investigated vegetative propagation techniques for some multi-purpose tree species and their field performance. Such a work is very much useful in horticulture as well as silviculture. There have been some more students who were formally not registered for their Ph. D. programs under the supervision of Prof. G.P. Mishra but who planned and completed their works totally under his guidance. Such students are

Pandey (1980), Jain (1981), Pahlariya (1982), Agnihotri (1982) and Gupta (1982). Pandey (1980) worked on an All India Co-ordinated research project sanctioned by BSI to Prof. Mishra on Grasses of Sagar District whereas Jain (1981) carried out eco-physiological studies of forest trees with special reference to *Tectona grandis.* Agnihotri (1982) investigated the productivity of grasslands of Mandla District whereas Gupta (1982) studied the herbaceous flora of Nagpur with special emphasis on the noxious weed *Parthenium hysterophorus.* Pahlariya (1982) investigated entirely new aspect, i.e. pro-cambial system in the embryos of certain members of family Compositae. All the researchers could not be registered with Prof. Mishra as there was no seat vacant under him at that time.

Wildlife management has also been a very specialized field of research which has been undertaken by two students of Prof. Mishra. Kotwal (1987) obtained his prestigious D.Sc. Degree on the evaluation of certain wildlife habitats and their utilization by major mammals in Kanha National Park, whereas Gopal (1996) explored the biology and ecology of hard ground Barasingha in Kanha Tiger Reserve. All these details prove that Prof. Mishra was a genuine ecologist and a multi-faceted dynamic personality.

Prof. A.K. Kandya &
Dr. Asha Gupta
Editors

List of D.Sc. and Ph.D. Students of Prof. G.P. Mishra

D.Sc.

1. Kotwal, P.C. 1987. Ecological studies on evaluation of certain wildlife habitats and their utilization by major mammals in Kanha National Park.

Ph.D.

1. Bhatnagar, S. 1968. Ecological studies of some forests of Sagar with special reference to litter and ground flora.
2. Jain, S.K. 1971. Production studies in some grasslands of Sagar.
3. Naik, M.L. 1973. Ecological studies of some grasslands of Ambikapur.
4. Sodhia, M.K. 1974. Ecological studies of foliage of some forest trees of Sagar.
5. Kandya, A.K. 1974. Organic matter dynamics in four selected forest tree species of Sagar.
6. Dakwale, R.N. 1975. A study of grassland ecosystem with special reference to biogeochemical cycle.
7. Prasad, R. 1976. Ecological studies of some forests of Gourjhamar range.
8. Trivedi, B.K. 1977. Ecological study of some grasslands of central India
9. Gupta, S.R. 1977. An ecological study of *Themeda quadrivalvis* type of grassland of Bundelkhand.
10. Nayak, A.K. 1977. Study of mineral circulation in some forest trees with special reference to teak (*Tectona grandis*) .
11. Purohit, J. 1977. Ecological study of soil algae of Sagar.
12. Johri, R.M. 1977. Mineral circulation studies in some forests of Sagar.
13. Selot, M. 1977. An ecological study of Ratanpur Lake with special reference to physico-chemical properties of water and plankton population.

14. Joseph. R.N. 1978. An ecological study of organic layer of some forests of Sagar.

15. Yadav, A.K. 1978. A study of mineral circulation in some forests of Sagar.

16. Athya, C.D. 1980. Study of natural regeneration of *Diospyros melanoxylon* (Bidi Plant) with special reference to sucker growth and leaf quality.

17. Das, Kamlesh (nee Bains, Kamlesh) 1980. An ecological study of forest tree seeds.

18. Pathak, Tara, (nee Nigam, Tara) 1980. Ecophysiological studies of some weeds.

19. Khare, D.K. 1981. Ecological effects of fire in tropical dry deciduous forests of Sagar.

20. Khare, P.K. 1981. Environmental biology of forests with special reference to natural regeneration.

21. Shrivastava, Archana, (nee Verma, A.) 1982. Ecological studies of some weeds of central India with special reference to *Alysicarpus longifolius*.

22. Shrivastava, V.K. 1982. Phytosociology, productivity and energetics of grasslands of Bundelkhand.

23. Mahendra, O.P. 1982. Autecological study of *Echinops echinatus*.

24. Mishra, Kalpana 1982. Ecophysiological studies of tree seedlings.

25. Sengar, R.B.S. 1983. An ecological study of *Dalbergia sissoo* in forest plantations of Bundekhand region with special reference to productive structure, mineral status and accumulation of energy.

26. Yadav, A.S. 1983. Ecological study of some important forest trees and their seedlings.

27. Kazmi, S.N. 1985. Studies on respiration of some plant pathogenic fungi.

28. Chaubey, O.P. 1987. Comparative studies of vegetational spectrum and edaphic conditions under natural forests and monoculture of teak.

29. Dubey, Veenapani 1987. Ecological studies of paddy fields of Bilaspur.

30. Shrivastava, Mukta (nee M. Khare) 1987. Eco-physiology of seed germination and seedling growth.

31. Jaimini, Ragini (nee R. Gupta) 1987. Ecological study of periphyton of Sagar lake.

32. Rawat, Amita 1988. An ecological study of some agricultural weed plants of Bundelkhand region.

33. Bisen, S.S. 1989. Scanning electron microscopic studies of the epidermis of culms and leaves of some important bamboos of the Indian sub-continent in relation to their classification.

34. Yadav, V. K. 1989. Germination studies of selected forest tree species with special reference to storage condition.

35. Jaiswal, Namita 1990. Ecology and control of diseases of some forest tree species in nurseries.

36. Mishra, Sushma, 1991. Ecology of nitrogen cycling in grasslands.

37. Prajapati, R.K. 1991. Productivity pattern and nutrient cycling in successional forests of the western Himalayas.

38. Rajput, Anjana, 1992. Seed biological studies of some selected tree species of Central India.

39. Bhattacharya, P. 1992. Studies on medicinal plants in forests of Chhindwara District with special reference to ethnobotanical aspects.

40. Purohit, Mamta, 1992. Seed biology of some selected tree species of Central India.

41. Gohlani, Urmila (nee U. Shrawan) 1992. Reproductive biological studies of forest tree seeds.

42. Dhuria, S. S. 1993. Investigation in to the vegetative propagation of some multi-purpose tree species and their field performance.

43. Chaubey, Archana, (nee Sharma, A.) 1993. Studies on pretreatments for hastening seed germination in some forestry seeds.

44. Gopal, R. 1996. The biology and ecology of hard ground Barasingha in Kanha National Park.

45. Jain, Sulekha, 1996. Seed biology of some selected leguminous tree species.

46. Balaiya, Archana, 1997 Seed biology of some selected leguminous tree species of Central India.

The following five persons have worked for their Ph. D. Degree totally under the guidance of Prof. G. P. Mishra, starting from its planning to its completion. But, Prof. G. P. Mishra was not the real supervisor of these candidates. They were registered under other persons. Either a seat was not vacant under Prof. Mishra or there was some other technical reason for this.

47. Asha Gupta 1982. Ecological study of herbaceous flora of Nagpur with special reference to *Parthenium hysterophorus*.

48. G.K. Agnihotri 1982. Productivity of grasslands of Mandla District.

49. J.P. Pahlariya 1982. Study of the pro-cambial system in the embryos of certain members of the family Compositae.

50. H.N. Pandey 1980. Grasses of Sagar District.

51. R.K. Jain 1981. Ecophysiological studies of forest trees with special reference to teak (*Tectona grandis*).

Grassland Ecology: M.L. Naik, S.K. Jain, S.R. Gupta, B.K. Trivedi, R.N. Dakwale, H.N. Pandey, V.K. Shrivastava, G.K. Agnihotri,

Forest Ecolgy: S. Bhatnagar, A.K. Kandya, M.K. Sodhia, Ram Prasad, C.D. Athya, D.K. Khare, P.K. Khare, A.K. Yadav, A.K. Nayak, R.M. Johri, O.P. Choubey, R.B.S. Sengar, R.K. Jain, O.P. Mahendra, R.K. Prajapati, R.N. Joseph, S.S. Dhuria

Weed Ecology: Tara Pathak, Archana Shrivastava, Veenapani Dubey, Amita Rawat, Asha Gupta

Forestry Seed Technology: Kamlesh Das, Kalpana Mishra, V.K. Yadav., Sushma Mishra, Archana Balaiya, Mamta Purohit, A.S. Yadav, Anjana Rajput, Archana Choubey, Sulekha Jain, Urmila Gohlani, Mukta Shrivastava, Namita Jaiswal

Algal Ecology : Jaishri Purohit, Ragini Jaimini

Aquatic Ecology: Manohar Selot

Ethnobotany: Prodyut Bhattacharya

Wildlife Management: P.C. Kotwal, Rajesh Gopal

Scanning Electron Microscopic Study: S.S. Bisen

Fungal Respiration: S.N. Kazmi

Plant Anatomy: J.P. Pahlariya

CONTENTS

Section VI Pollution Ecology

Section X Applied Ecology

SECTION ONE

FOREST ECOLOGY

Advancing Frontiers of Ecological Researches in India, 3-28, 2005
Edited by A.K. Kandya & Asha Gupta
Published by Bishen Singh Mahendra Pal Singh, Dehra Dun, India

Review of researches on various aspects of Ecology and Forestry with special reference to Central India

Ram Prasad, P.C. Kotwal* and Manish Mishra**
Ex-Vice Chancellor, Barkatullah University, Bhopal

Abstract: This review paper is dedicated to a great forest ecologist of Central India, Prof. G P Mishra, who provided insight to a large number of students, who later on followed his path and endeavored to excel in the same field mainly because of his excellent teaching, guidance and encouragement. The present paper reviews more than 100 research publications in national and international journals of Professor Mishra and his students. The paper has been framed under eleven broad headings containing review of each heading together with the relevant reference. The headings are: General/Ecology, Wasteland Development, Forestry Seeds, Forestry Plantations, Grasslands/Grazing, Afforestation in Coal Mines, NTFP including Medicinal Plants, Ecology of Teak, Eucalyptus, Sal, Wildlife management and Sustainable Forest Management. But, these are only a few out of a large number of publications of Professor Mishra and his students covering various aspects of forestry and ecology that if compiled may be quite voluminous. The present paper is a modest attempt to review only some important publications that could become readily available. Several papers dates back to 1960s and all of them are now not readily available. The research works of Prof. Mishra and his students is invaluable contribution to the subject and the region that will be the pathfinder for the new researchers.

INTRODUCTION AND REVIEW

A number of important and useful research works have been done during last two decades on various aspects of forestry in the state of Madhya Pradesh, which continues to be the largest forest state of the country. In terms of bio-diversity also the state has a unique place. The Satpura Ranges act as genetic super highway, which might have witnessed the migration of flora and fauna. As result several elements of flora and fauna found in eastern and Western ghats, some of these are also found in central India. In view of these peculiarities the research results in this part of the country offers valuable insight to the biological research. Some

*, ** Indian Institute of Forest Management, Post Box No. 357, Nehru Nagar, Bhopal - 462003 (M.P.)

of these researches have been published in various journals/books etc. It is worthwhile to review the main findings of these researches. These can be broadly categorized into different headings as given below:

1. General/Ecology
2. Wastelands Development
3. Forestry Seeds
4. Forestry Plantations
5. Grasslands/Grazing
6. Afforestation in Coal Mines
7. NTFP including Medicinal Plants
8. Ecology of Teak
9. Ecology of Eucalyptus
10. Ecology of Sal
11. Wildlife Management
12. Sustainable Forest Management

1. General/Ecology

Prasad and Mishra (1980) studied ecological status of *Diospyros melanoxylon* in dry deciduous Teak Forests of Sagar (M.P.). In this study they have mentioned that in spite of uniform climatic conditions the forest of Sagar are very heterogeneous in composition and quality. The state of vegetation appeared to be the outcome of micro climatic influence resulting from variation in geological formation, physiography and biotic interference. The *D. melanoxylon,* like teak and *Anogeissus latifolia*, out numbers other species in plant density. Its ultimate progression to maturity is one of the lowest. This indicates a high degree of mortality mainly due to biotic factors such as grazing, cutting, leaf plucking and fires. Like other tree species; annual and perennials, *D. melanoxylon* in dry deciduous teak forests have been provided with the phenomenon of dying back to ensure its survival and subsequent establishment. The dying back in this species appears to last upto 10 years of age beyond which normally all round growth takes place and plants are established.

Prasad (1989) studied gregarious flowering pattern in bamboo (*Dendrocalamus strictus*) in Madhya Pradesh. The gregarious flowering in *D. strictus* in Madhya Pradesh has been taking place at an interval of

23 to 47 years. A number of factors are said to affect gregarious flowering. The experience shows that bamboos under moist conditions and where they have been worked properly flower little later than those which are in drier localities and which are constantly hacked are congested and which remain un-worked for long. Study of this physiological cycle is necessary for managing the bamboo forest sustainably. There should be built in mechanism in management system of bamboo forests, which should provide adequate administrative and financial power to the Divisional Forest Officer for providing quick and desired treatment to flowered bamboo areas.

In another study on retrieval of bamboo forests in Madhya Pradesh (Prasad and Chadhar 1988) it was found that with the increase in various development activities the demand for bamboo has increased tremendously. To save the shrinking bamboo forests it is necessary to regulate biotic pressures and improve soil and moisture conservation status of the soil.

Chakravarti and Prasad (1989) in their study on forestry based tribal development mentioned that areas containing a pronounced concentration of tribals also have a comparatively higher proportion of land under forests. Attempted development of forestry resources relegating the tribals to a secondary position of reducing his status to a mere wage earner, wages coming as incidental to the process of forestry development, may have adverse effect on a harmonious balance between forestry and tribals. Intensive forestry based human resources development and actively involving the tribals in the process of regional development would help in maintaining the existing forestry culture in these areas. A number of forestry activities could be identified where such a co-operation can be achieved. Restoration of ecological balance and forest productivity can be further augmented by the active participation of the local people. For the achievement of these goals, the methods of management of forest resources and the utilisation of forest produce may have to be changed in the interests of the tribals.

2. Wasteland Development

The Afforestation of barren and denuded lands within and outside legally constituted forests, has been engaging the attention of foresters for quite a few decades. The book titled 'Technology of Wastelands Development' by Prasad (1988) deals with the generalities of the subject, definition of wastelands, history of formation, nature, classification and extent. The book also attempts to present a broad strategy on integrated package of strategies for bringing wastelands under appropriate uses, especially under forestry systems. The second part of the book is devoted to technology of wasteland development. It also deals with reclamation of mined out areas and over burdens. The third part of the book deals with special technologies, which may broadly be classified under intensive or increased inputs. The chapter on seeds has also been included in this part.

Tissue culture has been identified as one of the major thrust area and this technology is expected to bring fruitful results in the development of wastelands (Prasad and Chadhar 1986). Tree improvement methodology can be applied for obtaining high productive forests by enhancing genetic potential of the species.

Mohammad and Prasad (1986) mentioned that for afforestation and wasteland development use of appropriate artificial cultures of microsymbionts belonging to forest trees such as *Rhizobia, Frankia, mycorrhizal fungi*, free living microbes may be beneficial at the time of seed germination and early growth of seedlings in nursery.

Prasad and Kandya (1989) organised a national seminar at the M.P. State Forest Research Institute, Jabalpur from June 7 to 9, 1989, on a very useful theme–Depletion of soil and forest cover & Technology for control. Large number of participants from all over the country attended the seminar and presented papers in it. Some of them indicated useful indigenous technologies for conservation of forest cover, rehabilitation of degraded forests and plantation in bare lands. It was emphasised by most of the ecologists that active involvement of general public is a must in the protection of vegetal cover on the land.

Prasad and Pandey (1989) observed the effect of closure on rehabilitation of degraded forestland. Since the last few decades, the forestlands particularly near the developing towns have suffered serious depletion and have emerged as ecological Wastelands. Prevailing pressures of human and cattle population and their increasing demand for fuelwood, fodder, timber and diversion of forestlands for cultivation, industrial development, rehabilitation of ever increasing population etc. are the factors responsible for converting productive land into unproductive wastelands near rural habitation. In the study, efforts have been made to understand the revival potentiality of covered wastelands near Bhopal. The study was conducted at three sites viz., Van Vihar National Park, which was protected since 1988, Campus of Indian Institute of Forest Management (IIFM) protected since 1985 and unprotected site, an open site where all biotic factors were in operation. Exclusion of all forms of biotic pressures through fencing has helped to restore natural plant succession and consequent establishment of plant cover. At site II (IIFM Campus) within a protection period of one year, 20 individuals/ha of tree species, 2780 seedlings/saplings of 16 species and 5 species of grasses and herbs were recorded. Detailed studies on the beneficial effects of closure on revival of degraded sites through plant succession and establishment of natural regeneration have been discussed.

3. Forestry Seeds

Attempts have been made to reclaim the Chambal ravines extending over Bhind and Morena districts of M.P. through aerial seeding (Prasad, 1988). Several hectares of ravine land has been reclaimed. About 1000 tonne of seed has been sown from air and another 4000 tonnes of forest seeds were required for the remaining ravines. An initial experience of carrying out systematic sampling has not been found feasible and therefore, an improvised time distance method sampling described was resorted to monitor the success of aerial seeding. The results during the survey show that seedling count is highly variable. On an average 200-300 /ha seedlings were found in seeded areas.

The Manual on Aerial seeding for National Wastelands Development Board (Prasad, 1989) deals with reclamation of wastelands particularly the inaccessible ravines by raising tree plantations. In India and other countries lot of land is of such nature where the approach of air seeding have been attempted with a view to grow plants. The book was prepared with the support of FAO for National Wasteland Development Board, GoI.

Another book on 'Handling of Forestry Seeds in India' is highly useful for foresters, students of seed technology, seed handling organisations and nursery persons (Prasad, and Kandya 1992). The book gives authentic and detailed information about seed collection, processing, storage, treatment and related aspects. Other technical aspects of seeds like purity, dormancy, germination etc. have also been given. The book is a boon for all those interested in raising plantations.

Prasad and Kandya (1989) described the status of seed production of various forest tree species in the state of Madhya Pradesh, and their trading through a Seed Bank established in the M.P. State Forest Research Institute, Jabalpur.

Prasad and Kandya (1991) have given the details of seed development programme run by the government forest department in most of the territorial divisions to obtain high quality seeds of different forest tree species regularly on a sustainable basis, for all the afforestation and reforestation programmes. The four important phases of this programme are described in detail as (1) Selection of Candidate Plus Trees (Seed Bearers, CPTs) or Candidate Elite Trees (CETs) in the natural forests, (2) Development of CPTs into Plus or Elite Trees (3) Development of Seed Stands and (4) Development of Seed Production Areas (SPAs) and Seed Orchards (SOs) , may be that they are Seedling Seed Orcards (SSOs) or Clonal Seed Orechards (CSOs).

Results of the study on effect of sowing depth on the germination of some common forestry seeds indicate that dibbling the seeds at shallow depth has shown better results *in D. sissoo, H. binata, Dendrocalamus L. parviflora* etc. species. In the species whose seeds are small, the plant percent was observed to be higher for seeds sown at shallow depths. Seeds sown at optimum soil depth have rapid rate of germination.

Gradation of seeds into three sizes viz. big, medium, small showed perceptible variation in germination percent. Small size seeds gave more germination %. Upgraded seeds gave less germination while sunk seed gave maximum germination %. Storage of seeds of *L. parviflora* in airtight polythene container indicated better germination % (Prasad, *et al.*, 1988a).

Lagerstroemia parviflora Roxb. is a multipurpose tree species found in central India. Its propagation is difficult on account of poor germination of seeds and plant percent, once established the tree on account of its strong copping power, is difficult to eliminate. The paper is based on the investigations to hasten seed germination.

Prasad, *et al.,* (1993) also observed effect of tree girth on seed germination of *Lagerstromea parviflora.* Seeds were collected from trees of four girth classes. Further the seeds so collected were grouped as (I) un-assorted also designated as Control (C) and (ii) assorted (A) small and thin winged seeds (S). Germination studies with two of seed lots indicated better germination of seeds from higher girth classes. Middle aged tree gave better germination characteristics.

Seeds of *Holoptelia integrifolia* could be assorted on the basis of their colour, size and physical health (Prasad and Jalil 1991). From germination tests it has been observed that reddish brown seeds are healthy (62.65±1.70 viable), yellowish brown seeds gave poor germination while blackish brown seed were found to be sterile. Size also gives good indication of seed germinability. Bigger size recorded germination per cent of 39.50±7.71% followed by 20.00±2.82% in medium size and 18.50 ±4.72% in small sized seeds. Ocular observation can also be relied upon for grading the seeds in "healthy", "unhealthy", 'aborted' and 'empty' categories.

Prasad and Jalil (1987) made observations on the seed viability of Sal (*Shorea robusta*). An experiment to test the viability of freshly collected seeds from trees of different girth classes was undertaken. Sal seeds were stored in airtight polythene bags for about two months to test their viability. It was observed that the deterioration in viability of Sal seeds takes place very fast unless it is properly stored. Severe insect damage to Sal seeds appeared to be responsible for the loss of viability.

Prasad and Chadhar (1988) studied optimum potting mixture for sprouting of root-shoots in some tree species. They reported that root-shoots of *A. auriculiformis, G. arborea, T. arjuna* do not require much nutrients (manure) for their sprouting whereas *A. lebbeck, A. procera and L. leucocephala* need some quantity of organic manures.

Prasad *et al.*, (1992) studied effect of various pre-treatments on seed germination of some forest tree species. Seeds of ten leguminous tree species were subjected to 4 per-sowing treatments. It was observed that mechanical scarification and subsequent soaking in water was found to be better than control, simple soaking in water and acid treatment for inducing seed germination.

Another study (Prasad *et al.*, 1988a) on investigation of the germination behaviour of 4 tree species viz., *Lagerstroemia parvfiflora, Anogeissus latifolia, Anogeissus pendula and Buchanania lanzen* indicated that best results were obtained when seeds were sown in the nursery during the months of June-July. It was also observed that on account of the presence of a large number of infertile seeds in the seed lots, enough seeds should be sown to obtain 1000 seedlings from each seedbed.

Studies were also conducted on emptiness of Teak fruits from different teak forest areas of Madhya Pradesh (Prasad and Jalil, 1986). Teak fruits were collected from different forests of Madhya Pradesh and also from Teak Seed Orchards of Dehradun, Nepanagar, Jabalpur and Seoni. The teak fruits of different origins were compared for their emptiness and germination per cent. The results of this investigation revealed that though the emptiness was almost identical in fruits irrespective of their origin, there was a wide variation in the germination per cent. Generally the seeds from comparatively moist localities gave better germination percentage than the seeds from drier areas.

The coppices of certain tree species were compared with their seeding origin plants for their field survival and growth Thc use of stumps as planting stock on account of being easy to handle can affect economy in transport cost and consequently minimise the cost of plantation.

4. Forestry Plantations

During 1990s, there has been a surge of interest in implementing social forestry programmes in Asia and Pacific. In India social forestry was adopted in a big way during 1980s. Prasad and Bhatnagar, (1995) has compiled book on ' Social Forestry: Experiences Over the Decade'. This book has valuable information about establishment of village wood lots, strip plantations, farm forestry, nursery techniques etc. Participation of people in planning and implementation of plantation programme and other related issues have been discussed in the book. The book particularly deals with historical background, fundamental components of social forestry, and various schemes of social forestry, experiences of different states, involvement of people and Panchayat in social forestry programme.

Effect of pit size on the growth performance of some important species on the overburdens of mined out areas of coal was studied. It was found that irrigation and 60 cm pits were favourably in better survival and plant growth in plantation of *Madhuca latifolia.*

Studies were made on effect of irrigation and soil working on plant growth of forestry plantation raised on Bhata soils of Bilaspur. About 16 litres of water per plant was provided in first and second years and about 30 litres in the third year onward. Observation on plant survival, and height and girth of plants were recorded in December and June every year for four years. The results indicate that irrigation has been only marginally beneficial to tree growth. It appears probable that quantity and frequency of irrigation water applied was far less than the potential water requirement for affecting any substantial increase phyto-mass production.

Establishment of seedlings on wastelands is a problem. Adoption of low input technology (such as bio-fertilizers) and nodulating legumes and non-leguminous trees and shrubs with their appropriate bacterial and fungal symbiont can be useful in establishment of seedlings in such areas. This can provide the base for expansion of forest area and speed up the afforestation programmes (Prasad and Mohammad 1986).

Prasad, and Mohammad (1991) studied soil moisture studies under even-aged plantation of ten broad-leaved tree species. Ten tree species raised as compact block planting of 8-12 years age were studied. Seasonal variations in term of highest moisture content during rainy season followed by winter and summer was observed, No difference was visible due to tree species. Among the study tree species Eucalyptus trees were also included which are accused of higher moisture uptake. No marked change was noticed in moisture content at different soil depth and spacing.

5. Grasslands/Grazing

Some fodder grasses and herbaceous legumes (*Stylosanthus hamata and S. scabra)* were cultivated for assessing the yield of green fodder, dry fodder and seed production. Selling green grass gave gross return of about Rs.10,000/- per hectare. The estimated expenditure for raising one-hectare grass was Rs. 3,262. Dry grass gave less return but with the yield of seed, good return can be obtained. A gross revenue of Rs.10,000 from sale of dry grass and their seeds has been obtained. In view of their capacity to fix atmospheric nitrogen, the overall economics from introducing legume may become attractive. By adopting judicious crop rotation, the soil fertility can also be built up.

Unfortunately, many gazing lands have lost their climax vegetation and are fast turning into wastelands. Prasad and Bhatnagar (1989) studied livestock and fodder position in tribal region of Madhya Pradesh where most of the rural people keep livestock as they are of functional significance to them. The paper analyses the livestock and fodder situation in tribal region and estimates demand supply gap of fodder.

Prasad, and Parihar (1994) made observations on rehabilitation and management of Flowered Bamboo in Mandla-Jabalpur Forest of Madhya Pradesh. They observed that excessive biotic pressures especially the incidence of grazing and fire inhibit clump formation of bamboos. Some clumps get congested posing difficulties in working. As a result the productivity of bamboo clumps get greatly reduced. In the absence of proper working, these clumps dry out and become potential fire hazards.

Biotic pressures from grazing, fire and human interference result in the degradation and consequent reduction in the productivity of forests. These degraded areas contain rooted-stock of plants, which have disappeared on account of excessive gazing, fire, and other forms of land misuse. Once these biotic pressures are relieved, the rootstock of existing trees sprouts with vigorous shoots. In natural teak forest where the bamboo had disappeared after gregarious flowering has re-established as a result of enforcing protection from grazing, fire and other biotic pressure. Investigation brings out the beneficial effect of closure from grazing, fire and other biotic evils on the establishment and growth of *bamboo (Dendrocalamus strictus*) plants (Prasad, 1985).

The effect of grazing on the floristic composition, biomass production and soil properties of a typical grassland of Jabalpur have been assessed by Choubey, *et al.,* (1986). The results show that species diversity at grazed site is lower than the protected grassland. Many of the palatable grass species had disappeared at grazed site. Biomass production, both above ground and below ground was found to be markedly higher at protected site than at the grazed site. Improvement in the soil chemical properties was also observed at the protected site.

6. Afforestation in Coal Mines

Large tracts of rich forests are being cleared for mining in Madhya Pradesh. After mining, the areas are rendered wastelands where site conditions become inhospitable for normal plant growth. However, SFRI Jabalpur has got success in evolving methodologies for the rehabilitation and re-vegetation of such mined out areas. Out of 7 indigenous tree species and one Australian species (*G. pteridifolia)* has found most successful in the Bauxite mined out areas, and produces maximum organic matter (Prasad and Chadhar 1987).

Eight tree species have been planted in Coal Mined areas of SECL at Dhanpuri near Shahdol M.P. All the species have been found successful. However, in terms of height growth and girth increment, Eucalyptus species have been exceptionally good. Maximum tree height of 8.8 M was observed in Eucalyptus hybrid (Prasad and Shrivastava 1991).

The plantations of some leguminous tree species planted on bauxite mined out areas near Amarkantak and coal mine overburden dumps at Dhanpuri near Shahdol were studied for root nodules and microsymbionts. Rhizobia and VAM-fungi association with 5 and 9 year old *Acacia auriculiformis* and *Pongamia pinnata* plantation were studied. On coal mine O.B. dumps different aged plantation of *A. auriculiformis* was found to contain higher nodulation and microsymbionts in comparison to five year old plantation of *P.pinnata* (Prasad and Mohammad 1989).

The activity of mining is invariably associated with digging of earth. The areas taken up for mining are generally those covered with trees i.e. forest. Thus, mining almost in every case results in the destruction of valuable tree-cover. The removal of the plant cover from large areas gives birth to many environmental problems. However, if these areas are soon covered again with the plant cover, then the damage to the environment can be minimised. Technologies of reforestation of mined out sites have been evolved and reported by Prasad (1989).

The impact of protection from livestock grazing (part of natural forest fenced by barbed wires) lead to ameliorating the physical, chemical and biological condition of site. The soil pH under protected areas was found to be lower as compared to open areas (subjected to all type of mal-treatments). The moisture percentage of soil was higher under the closed forest areas as compared to open areas of natural forests. The nitrogen percentage was also found to be higher in soils of protected areas that of open areas. The development of more congenial growing conditions in protected areas of natural forests, responded to better establishment (recruitment rate) of natural regeneration of Sal.

Mining activity, especially the open cast mining creates problems of environmental imbalance. Clearing of forests and digging and heaping the overburdens in the inverted sequence leaves the site desolate and inhospitable. Such sites require special technology and suitable species for bringing back the greenery. Bauxite, coal and dolomite mined out areas and their overburden dumps have been successfully re-vegetated. The successful afforestation trials were carried out on iron-ore mined out areas at Dalli-Rajhara area in Durg District. Among the successful species were *Albizzia spp., Acacia auriculiformis, Dalbergia sissoo, Eucalyptus hybrid, Leucaena leucocephala* etc. However, in terms of biomass the maximum oven dry weight of 32,600 t/ha was recorded in four-year plantation of *L. leucocephala* followed by *Albizzia lebbeck* and *Eucalyptus* sp. (Prasad and Dhuria, 1989).

Bauxite is extracted by open cast mining. In this process the original vegetation is cleared. Reclamation of such area is difficult and requires special techniques as the overburdens are almost devoid of nutrients and moisture retention capacity. Prasad, and Shukla (1985) studied restoration of ecological balance of the bauxite mined areas of Madhya Pradesh. The performance of trial plantation of eight exotic and two indigenous tree species raised by the State Forest Research Institute on bauxite mined areas exploited by BALCO near Amarkantak in Mandla district of Madhya Pradesh were raised in 1979. Biomass productivity was assessed in case of *Eucalyptus camaldulensis, Grevillia pteridifolia, Pinus sp.* by taking biomass measurement separately for wood, bark, leaves and roots of 5 trees for each species. In terms of leaf litter production, *G. pteridifolia* was observed to the most effective producer.

Trial plantation of various tree monocultures was started on bauxite mine overburdens and on coal mine overburdens. In terms of total number of plants, coal mine overburdens appear to have provided more hospitable conditions for natural succession than bauxite mine overburdens. The soil environment appears to have been modified by the addition of huge quantity of litter. The sites are gradually developing to provide more congenial growing environment leading to the ultimate creation of self-sustaining ecosystem.

Prasad and Shukla (1985) found that among the 13 tree species planted in the coal mined areas of Western Coalfield Ltd. at Dhanpuri near Shahdol, Eucalyptus sp. and *Acacia auriculiformis* have been exceptionally good in terms of height growth and diameter. Maximum tree height of 6 m has been observed in Eucalyptus hybrid.

7. NTFP including medicinal plants

Mahua (*Madhuca latifolia*) tree is of immense value to villagers especially the tribals as a source of food, liquor, cattle feed and cash through sale of seeds and flowers. Due to its usefulness it is regarded as tree of poor. Prasad (1992) in his book on Mahua: has rightly described this tree as 'The Tree of Poor'. The book provide useful information on historical, social, economic, silvicultural and medicinal aspects of the tree. It also contains information on conservation and propagation of this important tree species.

Prasad and Bhausar. (1991) have also studied the yield of Mahua flowers and fruits at State Forest Research Institute, Jabalpur from 1987 to 1991. Average sun dried Mahua flower yield of 36.54 kg/tree and fruit 13.0 kg/tree was observed. Trees of 61-90 cm girth at breast height gave lower yield than trees of 180 cm and above girth.

Aonla (*Emblica officinalis*) tree is known for its highly valuable fruits. Population of Aonla trees in the forest is declining due to ruthless felling for fruit collection and firewood purposes. Small trial plantation of this species was raised at Seoni and Jabalpur. Prasad and Danayak (1991) studied the growth and fruiting of Aonla trees.

A large number of tribal communities live in remote and inaccessible parts of the state using an enormous range of wild plants for their food, medicine, fibre, shelter etc. and are still dependent on plants for their daily needs and livelihood. A large number of wild medicinal plants in natural forests of Madhya Pradesh require adequate attention about their conservation and further propagation under natural conditions. There is a need to survey the natural habitat of wild medicinal plants, which are still unknown in respect of their medicinal values. Considering the importance of wild medicinal plants in natural forests of Madhya Pradesh, a list of some valuable wild medicinal plants encountered during survey was prepared by Prasad *et al.*, (1988b) that are being used by local inhabitant tribes around the forests.

Non-wood forest products provide a very important livelihood support to indigenous tribes living on forest fringes. Traditionally the entire economy of these tribes is depended to a large extent on collection activities in the forests. The State has recognised the rights of these fringe dwellers in forest and many programmes have been directed to improve their condition. Prasad and Bhatnagar (1993) discussed the significance and relationship of non-wood forest products to indigenous tribes and plead for a comprehensive policy on non-wood forest products.

Prasad and Pandey (1993) studied wild medicinal plants and their mode of use to cure various ailments by tribal people of Lamni and Achanakmar locality, Chhattisgarh. A total of 113 plant species of 49 families were encountered during the survey.

The vegetational composition of ground flora of Achanakmar is very rich in medicinal point of view. Prasad *et al.,* (1990) studied the ethno-medico- botanical characteristics of indigenous medicinal plants of Lamni and Achanakmar wildlife Sanctuaries. The tribals not only use these plants as food but also to cure their ailments. Ethno-medicinal observations were documented in tabulated form.

Prasad *et al.,* (1990) studied socio-economic and ethno-medico botanical characteristics of plants of Patalkot region M.P. Fifty-five important medicinal plant species were found to be used by the tribals. Out of these 29, were herbs, 12 shrubs, 9 climbers and 5 trees. The percentage of plants having medicinal properties was maximum for herbs (52%), followed by the shrubs (21%), climbers (12%) and trees (9%). From among the plant part of medicinal use, roots were found to be useful in 43% species whereas leaves were found to be medicinally useful in 21% species. Seeds have been found to possess medicinal value only in 7% species, whereas flowers and fruits indicated their medicinal properties in 9% of the studied species. 12% species were recorded to have medicinal property in the entire plant body. Comparative data of Important Value Index (IVI) of herbaceous plants was worked out which show the status of individual plants in the community.

Prasad and Shrivastava (1990) studied the ethno-botany of medicinal plants belonging to family Fabaceae. The work yielded useful information regarding certain wild species of family Fabaceae of central India region, having specific medicinal values as a remedy in different ailments. The study also indicates that the seasonal survey and collection of such plant species is quite useful for identifying medicinally important plants. These studies have also shown that much valuable information has remained unverified and unutilised.

Prasad and Bhatnagar (1989) while studying the economic potential of tree based oilseeds in Madhya Pradesh found that Sal seed collection (a) increases the revenue from forests (b) offers seasonal employment to the tribals and (c) creates greater economic activities in rural areas. The ecological costs in terms of damage to young regeneration, Sal trees and a general bad effect on the Sal forests are heavy. Sal seed collection should be done in a way that sustain and expand the sal forest. Mahua can generate a good oil for both edible and industrial uses. Palas (*Butea monosperma*) is widely available in Madhya Pradesh. Its seeds contain 18-19 per cent oil. It should find good use in soap making. Kusum (*Schlichera oleosa) and* Karanj (*Pongamia pinnata*) also yield oil of industrial application.

In a forest ecosystem the Non-Timber Forest Products (NTFPs) provide maximum inputs into the domestic economy of the rural household. They are vital to the sustenance of the tribals and other weaker sections of the rural population. Looking to the economic and ecological role of the forest in general vis-à-vis the pace of its depletion, sustainable development of the forests and the dependent populations, therefore, assumes greater significance (Prasad, 1995).

The impact of prevalent harvesting practices on Aonla fruits (*Emblica officinalis*) were studied by Prasad *et al.,* (2002). Aonla fruit starts maturing in December but due to great demand by industries and intense competition between collectors, pre mature harvesting of fruits starts in October. Prevalent harvesting practices involve lopping, pollarding of branches and even at some places felling of trees. The study compared the impact of extraction of fruits from Protected Areas as well as from multiple use natural forests.

Uncontrolled harvesting of commercially important medicinal plants like *Chlorophytum spp.* (Safed musli) has virtually reduced the local abundance of this valuable species from many parts of the central Indian forests. un-sustainable harvesting of this species can be gradually eliminated from the natural forests owing to uprooting of the plants on large scale. The export of the product in large quantity and increased demand in the market result in destructive harvest of this species. Observations revealed that this plant species in the state of Madhya Pradesh is literally dying for want of concern from many quarters. Because of increasing demand and unsustainable harvests of the species, an attempt was made by Prasad *et al.,* (2001) to study the ecological impact on the open natural forests and protected areas of Central India.

8. Ecology of Teak

The insecticides, hormones and rallis tracel-2 (micro-nutrient) were used in Teak seed orchards at Betul (M.P) for enhancing the fruiting. The results showed that the growth-promoting chemicals and insecticides were not effective in increasing fruit production. Fruit dropping also does not appear to have been affected by these treatments. Thus in older teak plants these can not been applied. The insect population can be reduced to some extent by the treatment with insecticides, only if it is beyond the tolerable level (Prasad, Ram, Meshram and Jamaluddin 1990).

Inter spaces between Teak trees of clonal seed orchards were planted with *Stylosanthes hamata* for preventing wild grass growth in the area and for building soil fertility of the site. It has been observed that perennial crop of *S. hamata* yield good fodder and also improve soil fertility (Prasad, 1991).

Prasad and Mishra (1985) studied the litter productivity of natural dry deciduous teak forests of Sagar (Madhya Pradesh). It was observed that *T. grandis* alone produced one third of the total litter production of the entire forest stand. The other species, which dominated the litter output in stand, were: *T. tomentosa, D. melanoxylon, Butea monosperma and Miliusa tomentosa*. Small broken pieces of less common species accounted for another one third of the total production. Other important tree species such as *A. latifolia and L. parviflora* in spite of their high plant density contributed only a fraction of the total litter production indicating the immature plant stand of these species. The total leaf litter production in these forests was found to be 4.959 tonnes/ha. In the light of the investigations it was concluded that while relatively younger crop is more efficient in litter production, their distribution pattern is also important. Abundance of plants in juvenile phase instead of seedling and coppice phase appeared to produce maximum foliage. Once the trees cross the juvenile phase, the canopy is closed and thus the size of crown remains constant. It was not due to bigger crown and more leaves but due to thicker branch wood. Besides age the tree density and basal area play equally significant role in leaf litter output. This hypothesis can be explained by the basal area and litter output of *Tectona grandis and Terminalia tomentosa*. In addition to remaining dead parts of the above ground vegetation, leaf litter considerably adds to the forest floor. Herbaceous ground flora also enriches the forest floor with organic matter both directly and indirectly. It traps around its litters and stumps a large quantity of leaf litter, which would have otherwise gone outside the ecosystem.

Choubey *et al.* (1988) also made comparative studies of leaf litter production and nutrient return in teak plantation raised after clear felling of natural forest and in the adjoining natural forests. The leaf litter production was found to be greater in teak plantations than in adjoining natural forests. The per cent concentration of nutrient contents also found to be higher in leaf litter of the plantation area than the leaf litter of natural forest area.

Clear felling of natural forest and subsequent plantation with teak is generally believed to have an adverse ecological impact particularly in respect of floristic composition, species diversity and productivity of ground flora. Choubey *et al. (1988)* studied, total number of ground flora species and stem density (plant/sq. m). These were found to be more pronounced in teak plantation of different ages than adjoining natural forests. The IVI, of those species which were present, both in teak plantation as well as in adjoining natural forest, was found to be higher in plantation areas that the later. The greater improvement in above ground biomass (dry weight in kg/ha) of ground flora was also found in teak plantations than in adjoining natural forests.

Prasad and Jamalludin (1989) made observations on the problem of teak mortality in Central India. Causes of mortality included gregarious flowering in bamboo (which is one of the main associates of teak) followed by its drying and burning which created loss of ground cover, soil desiccation and injury to the roots and standing trees of teak. Measures of protection from such mortality were also described. Proper management of teak forests particularly (1) careful protection of bamboo areas from fire, (2) maintenance of shrubby and herbaceous undergrowth, (3) introduction of fresh stock of teak in the forest were suggested.

A fertiliser experiment using three levels of nitrogen and three levels of phosphorus was laid out in 10 and 20 years old teak plantation in Mandla Forest Division. Observations were made by Prasad, Sah & Bhandari (1986) on the height and diameter which was found significantly increased due to fertiliser dose in both the plantations. The net yield of timber in terms of volume increment was found to be significant in 10 years old plantation.

Prasad and Mishra (1984) carried out studies on standing biomass of various plant parts in selected tree species of dry deciduous teak forest of Gourjhamer Range of Sagar Division in M.P. The teak forests of this Range constituted about 16.4 per cent of the total teak forests and 8.7 per cent of the total mixed forests of Sagar District. The study was conducted on the organic matter productivity of four important tree species viz. *T. grandis, T. tomentosa, A. latifolia and D. melanoxylon* in relation to their girth at breast height (gbh) and age (in *T.grandis*) only. While comparing different species for their net primary productivity it was found that *A. lalifolia* was more efficient in dry matter production followed by *T. tomentosa, T. grandis* and *D. melanoxylon*. At 50-55 cm gbh the total production per tree was 433.50 kg in *A. latifolia*, 184.18 kg in *T. tomentosa*, 157.89 kg in *T. grandis* and 78.65 kg dry matter in *D. melanoxylon*. The mean annual productivity was worked out for *T. grandis* on the basis of prominent annual growth rings whereas in *T. tomentosa, A. latifolia, D. melanoxylon* and other species of dry deciduous forest, due to the obscure growth rings, the assessment of age was not accurate. In such cases, gbh rather than age was found to be useful in comparing the productivity of different species.

9. Ecology of Eucalyptus

An experiment to study the effect of various methods of soil working on survival and height growth of *Euclyptus* hybrid and *E. camaldulensis* was taken up at Jabalpur by Mohammad, and Prasad (1988). There were six soil working treatments, three for ploughing the entire area at three different soil depths and three for ploughing in the plant able strips at three soil depths. Soil working treatment in which the entire area was ploughed given best results both in respect of survival per cent as well as plant height. The detailed results were discussed in the paper.

An experiment was conducted by Prasad, Ram (1988) to evaluate the biodynamic influence of commercially available inoculants of tree living N_2-fixer and p-solubelizer on the seedlings of *Eucalyptus camadulensis* under polypot condition. All the bio-fertilizers containing *Azotobacter chrococoum* strains could not exert a positive biodynamic influence on the seedling production. However, some of them when introduced as mixed inoculant with phosphobacteria resulted in the increase in biological yield. Solid base biofertilizers inocula specially developed for plants used in forestry plantation. The study revealed that bio-fertilisers, if properly placed in the root region of growing seedling can stimulate root and shoot development in nursery and thus help in production of healthy and vigorous planting stocks.

Prasad *et al.*, (1992) studied optimum number of coppice shoots stump of Eucalyptus hybrid. The results indicated that thinning of *Eucalyptus* coppice has practically no effect on growth and yield of coppice shoots. When it is desired to produce small timber poles and fuelwood, retention of all coppice shoots appeared more advantageous.

Prasad, *et al.* (1984) studied dry matter production on 15 year old) and 5 year old plants of *Eucalyptus camaldulensis* raised in campus of Tropical Forest Research Institute, Jabalpur. The entire area of plantation was divided in 5 equal compartments. Five trees from each compartment were randomly selected and felled for above ground and below ground biomass studies. Growth parameters in 15-year-old plantation were 2-4 times that of plants in 5-year-old plantation. The root length was traced to a depth of less than 3 m in all cases. However, the lateral spread was unimaginably very vast. The mean lateral spread of 2.03 m was observed in 5-year-old plants and 9.01 m in 15-year-old plants. The propaganda against Eucalyptus that it is responsible for drying up of underground water reserve is thus disproved from these studies. The vast internal root system may be responsible to utilise even the slightest amount of surface moisture, which may be available just below the surface in any form of precipitation. In terms of mean annual productivity it was 15.92 t/h/year in 15-year-old plants and 6.03 t/h/year in 5-year-old plants. The maximum contribution to the total above ground biomass was observed by stem wood in both plantations.

Sample plot data pertaining to seedling origin crop and first coppice crop of *Eucalyptus hybrid, E tereticornis, E camaldulensis* have been analysed. This study have shown that growth and yield from coppice crop is 2 to 3 times higher than the seedling crop of the same age.

10. Ecology of Sal

Eastern Madhya Pradesh is predominantly a tribal belt and has a substantial area under Sal (*Shorea robusta*) forests. There is a general belief that Sal regions are socio economically backward as compared to non-Sal areas. In order to test this hypothesis, 12 socio-economic indicators of development, were taken to compare the position in tribal region. Land use and cropping patterns were considered. Results show that though this area has a higher rural and tribal population, it is by no means backward in development (Prasad and Bhatanagar 1989).

The investigations and observation relating to Sal mortality revealed a complex interaction between climatic parameters, physical soil attributes (soil compactness) converting the high forest into that of coppice origin and varied biotic factors. These factors have been responsible for all mortality. Their role is only secondary after the complex interactions of climatic, biotic and management factors that have reduced the tree vigour. The causative factors were summarised by Prasad, and Jamaluddin (1985) in their study on Sal mortality in Madhya Pradesh.

11. Wildlife Management:

Kotwal (1987) made extensive ecological studies on evaluation of certain wildlife Habitats and their utilisation by major mammals in Kanha National Park. Kotwal (1982) made studies on evaluation of wildlife habitats- parameters and procedures. Parihar & Kotwal (1989) prepared management plan of Kanha National Park and Project Tiger Kanha for the period 1989-90 to 1989-99 that was considered as ideal plan. Kotwal (1998) made observation on prospects and problems of participatory protected area management in India. Kotwal & Sawarkar (1982) applied remote sensing techniques to wildlife management. Parihar & Kotwal (1986) studied wildlife habitat using high-resolution space photographs in Kanha National Park using results from the joint Indo-Soviet Remote Sensing Experiment TERRA on board Salyut-7. Kotwal & Pandey (1981) studied dynamics of grass production and consumption by wild herbivores in Kanha National Park. Mishra & Kotwal (1987) made detailed observation on plant-herbivore-carnivore system in Kanha National Park. Kotwal (1985) also studied intra -specific fights and cannibalism among tigers in Kanha National Park. Kotwal & Gopal (1993) studied radio-telemetry and field observations on territoriality of tigers in Kanha National Park. Barasingha (*Cervus duvauceli branderi*) which is the State animal of Madhya Pradesh and critically endangered species of deer was studied by Kotwal (1991) and done their re-introduction in alternate natural habitat in Supkhar area of Kanha National Park. Wild buffalo (*Bubalus bubalis* L.) another critically endangered species and State animal of Chattisgarh was studied by Kotwal & Mishra (2003) in Udanti Wildlife Sanctuary.

12. Sustainable Forest Management:

Sustainable Forest Management (SFM) has been recognised as an important element of Sustainable Development at Earth Summit in 1992 at Rio D Janeiro. Because nearly 1/3 of the world population is directly or indirectly dependent on forest resources. Beside the forests provide a number of goods and services that are essential for human survival. This is frontier area for research in the forestry sector. Dr. Prasad was instrumental in initiating the Criteria & Indicators approach for Sustainable Forest Management through "Bhopal-India-Process" (Prasad *et al.* 1999). A number of brain storming sessions and workshops across the country involving a large number of stakeholders (forest officers, scientists, NGOs etc.) have evolved national set of Criteria & Indicators for the country that was also substantiated by a Task Force of Govt. of India thereby accepting this as a Policy matter for countrywide application. The Bhopal-India-Process of SFM also formed the basis of the process for the Sustainable Forest Management of Dry Forest in Asia, which is one of the nine internationally recognised processes of SFM in the world. Nearly 150 countries are following Criteria & Indicators approach for Sustainable Forest Management because it addresses to all the three pillars (social, economic and environmental) of Sustainable Development. Prasad (2000) has reviewed all the important world processes of SFM. Prasad & Kotwal (2001) have reported progress of implementation of Criteria & Indicators approach for Sustainable Forest Management in Asia with special emphasis on Indian initiative. Prasad et al. (2002) have developed a Manual for Operationalising Criteria & Indicators for Sustainable Forest Management at Forest Management Unit level in India.

REFERENCES

Chakravarti, R. and Prasad, Ram (1989). Forestry based tribal development. *J. Trop. Forestry.* **5(1)**:3-11.

Choubey, O.P. and Prasad, Ram (1989). Temperature as a factor hastening germination of *Lagerstroemia parviflora* Seed. *J. Trop. Forestry.* **5(4)**:325-329.

Choubey, O.P.; Mishra, G.P. and Prasad, Ram (1988). Phytosociologial studies of teak plantations and mixed natural forests in Madhya Pradesh. II. Distribution, species diversity, productivity and some quantitative parameters of ground flora, *J. Trop. Forestry.* **4 (11)**:177-187.

Choubey, O.P.; Prasad, Ram and Mishra, G.P. (1988). Litter production and nutrient return in teak plantation and natural forest in M.P. *J. Trop. For.*, **4(3)**:245 -255.

Choubey, OP.; Prasad, Ram and Jamaluddin (1986). Effect of grazing on species diversity, organic matter production and soil properties in a typical grassland of Jabalpur, Madhya Pradesh. *J. Trop. For.*, **2 (iv)**:211-217.

Kotwal, P.C. Re-introduction of Barasingha (*Cervus duvauceli branderi*) in Supkhar (Kanha National Park). *J. Trop. For.* **9(2)**:138-147.

Kotwal, P.C. & Pandey, R.K. (1981). Dynamics of grass production and consumption by wild herbivores in Kanha National Park. Silver Jubilee Symposium of International Society for Tropical Ecology, Varanasi, held at Bhopal pp. 9.

Kotwal, P.C. & Sawarkar, V.B. (1982). Remote Sensing applied to wildlife management. FAO/UNESCO/UNEP International Training course, Rome. pp. 80.

Kotwal, P.C. (1982). Evaluation of wildlife habitats-parameters and procedures. Workshop on techniques in Wildlife research & Management, Kanha. pp. 12.

Kotwal, P.C. (1985). Intra -specific fights and cannibalism in among tigers in Kanha National Park. Cheetal **27**:28-33.

Kotwal, P.C. (1987). Ecological studies on evaluation of certain wildlife Habitats and their utilisation by major mammals in Kanha National Park. D.Sc. thesis Sagar University, India. pp. 395.

Kotwal, P.C. (1998). Participatory Protected Area Management in India: Prospects and Problems. Proceedings of International Seminar on Sustainable Development 31 Aug. – 2 Sept. 1998, Pokhra, Nepal, Pages 334-339.

Kotwal, P.C. and Mishra, Rajendra Prasad (2003). Studies on Wild buffalo (*Bubalus bubalis* L.) in Udanti Wildlife Sanctuary, India. *Tiger Paper*, **Vol. XXX**:No.2, pp 7-11.

Kotwal, P.C.& Gopal, R. (1993). Radio-telemetry and field observations on territoriality of tigers in Kanha National Park. International Symposium on Tiger, New Delhi.

Mishra, G.P. & Kotwal, P.C. (1987). The plant-herbivore-carnivore system in Kanha National Park. IX Symposium of International Society for Tropical Ecology, Varanasi, pp. 17.

Mohammad, G. and Prasad, Ram (1986). Guidelines for application of Bio-Fertilizers in Forest nurseries and plantations. *Vaniki Sandesh* **10(4)**:18-21.

Mohammad, G. and Prasad, Ram (1988). Influence of microbial fertilizers on biomass accumulation in polypotted *Eucalyptus camaldulensis* (Dehn.) seeding. *J. Trop, Forestry.* **4 (1)**:70-77.

Namdeo, R.K., Prasad, Ram and Choubey, O.P. (1989). Impact of grazing closure on some ecological aspects of sal (*Shorea robusta*) natural forest with special reference to its natural regeneration in M.P. *J. Trop. For.*, **5(2&3)**:82-85

Parihar, A.S.& Kotwal, P.C. (1989). Management plan of Kanha National Park and Project Tiger Kanha for the period 1989-90 to 1989-99 pp. 400.

Parihar, J.S. & Kotwal, P.C. (1986). Study of wildlife habitat using high-resolution space photographs - a case study of Kanha National Park. Results from the joint Indo-Soviet Remote Sensing Experiment TERRA on board Salyut-7. I.S.R.O. Bangalore SP-17 pp. 65-82.

Prasad Ram & Kotwal, P.C. (2001). Progress of implementation of Sustainable Management of Dry Forests in Aisa with special emphasis on Indian initiative. Paper presented at UNFF-MAR meeting, Yokohama, 4-8 Nov 2001.

Prasad Ram (1990). Cultivation of fodder grasses and legumes. *Vaniki Sandesh* Vol. No.4, Oct. Dec. pp-1-4.

Prasad Ram (2000). Sustainable Forest Management for Dry Forest of South Asia. Asia Pacific Forestry Commission. Rap Publication 2000/08. FAO Regional Office, Bangkok.

Prasad Ram; Pandey, R.K. and Bajpai, A.K. (1990). Ethnomedocio-botanical studies on indigenous medicinal plants of lamni and Achanakmar wildlife sanctuaries. Paper presented in the national seminar on Medicinal and Aromatic plants, SFRI, Jabalpur (M.P.) April-6-7, 1990.

Prasad, Ram and Muhammad, G. (1986). Role of bio-fertilizers in wasteland development and forest production : A review. *J. Trop. For.*, **2 (IV):**179-200.

Prasad, Ram and Jalil, P. (1986). Emptiness in Teak Fruits from different areas of Madhya Pradesh. *J. Trop. For.*, **2(11):**207-212.

Prasad, Ram and Bhatnagar, Pratibha (1995). Social Forestry: Experiences Over the Decade. International Book Distributors, Dehradun.

Prasad, Ram (1990). Irrigated plantation of common forestry species (1) effect of irrigation and soil working on plant growth. *J. Trop.For.*, **6(III):**195-200.

Prasad Ram; Kotwal, P.C.; Dharmendra Chandurkar; Jadhav, Y.D.; Navin Vivek Horo & Dugaya, D. (2002). Manual for Operationalising Criteria & Indicators for Sustainable Forest Management at Forest Management Unit Level in India. IIFM Publication. ISBN 81-7969-001-6

Prasad Ram; Raghavan, S.; Phukan, B.R. and Bharti, Joshi (1999). Proceedings of national technical workshop on Evolving Criteria & Indicators for Sustainable Forest Management in India. Indian Institute of Forest Management, Bhopal.

Prasad, R. and Bhatanagar, P. (1989). Socio-economic status of tribal region in Madhya Pradesh. *J. Trop. For.*, **5(2&3):**197-202.

Prasad, R. and Bhatnagar, P. (1989). Economic potential of tree based oilseeds in Madhya Pradesh. Paper presented in the National Seminar on Minor Forest Produce and Tribal development, I D F, Jabalpur (M.P.) October 18-20, 1989.

Prasad, Ram & Sah, A.K. & Bhandari, A.S. (1986). Fertilizer trial in ten and twenty years old teak plantation in M.P. *J. Trop. For.*, **2 (1):**47-52.

Prasad, Ram & Mishra, G.P. (1984). Standing biomass for various plant parts in selected tree species of dry deciduous teak forest in M.P., *Ind. For.*, **110(8):**756-782.

Prasad, Ram & Shukla, P.K. (1985). Restoration of Ecological balance of the bauxite mined areas of Madhya Pradesh *J. Trop. For.*, **1(111):**236-245.

Prasad, Ram & Pandey, R.K. (1985). Natural Plant succession in the rehabilitated Bauxite and Coal mine overburdens of Madhya Pradesh *J. Trop. For.*, **1(iv):**309-320.

Prasad, Ram (1985). Effect of grazing closures on the rehabilitation on flowered bamboo areas in Mandla Forests of M.P. *J.Trop For.*, **1(111):**145-154.

Prasad, Ram (1988a). Effect of different methods of soil working on the growth and survival of Eucalyptus hybrid and *E. camaldulensis, J. Trop. For.*, **4(11):**135-140.

Prasad, Ram (1988b). Technology of Wastelands Development, Associated Publishing Company, Karol Bag, New Delhi-110005.

Prasad, Ram (1988c). Effectiveness of Ariel Seeding in Reclamation of Chambal Ravines in Madhya Pradesh. *Indian Forest*, **4(1):**1-18.

Prasad, Ram (1989a). Comparative growth and yield of seedling origin crop and first coppice crop in Eucalyptus hybrid. *Indian For.*, **115(5):**281-285.

Prasad, Ram (1989b). Gregarious flowering pattern in bamboo (*Dendrocalamus strictus*) in Madhya Pradesh. *J. Trop. Forestry.* **6(1):**86-94.

Prasad, Ram (1990). Trials with stump planting of some forestry species. *J. Trop. Forestry.* **6(2):**115-119.

Prasad, Ram (1991). Underplanting of *Stylosanthes* legume crop in Teak seed Orchard. *Vaniki Sandesh* **15(1):**1-5.

Prasad, Ram (1992). Mahua: The Tree of Poor. International Book Distributors, Rajpur Road, Dehradun.

Prasad, Ram (1995). Cluster Approach for socio-economic Development of Tribals through sustained Dev. of Natural Resources in Madhya Pradesh. *J. Trop. Forestry.* **11(3):** 169-176.

Prasad, Ram; Shah, A.K.; Bhandari, A.S. and Choubey, O.P. (1984). Dry matter production by *Eucalyptus camalulensis* Dehn. Plantation in Jabalpur. *Ind.For.*, **110(9): 868-878.**

Prasad, Ram and Parihar, A.S. (1994). Rehabilitation and Management of Flowered Bamboo in Mandla-Jabalpur Forest of Madhya Pradesh. *J. Trop. Forestry* **10(1):**3-7.

Prasad, Ram and Mohammad, G. (1989). Effectiveness of nitrogen fixing trees (NFTs) in improving microbial status of bauxite and coal mined out areas. *J. Trop. Forestry.* **6(1):**86-94.

Prasad, Ram and Mohammad, G. (1991). Soil moisture conservation and utilization pattern under even aged plantations I : Soil moisture studies. *J. Trop. Forestry.* **6(iv).**

Prasad, Ram and Mishra, G.P. (1980). Ecological status of *Diospyros melanoxylon* in dry deciduous Teak Forests of Sagar (M.P.) *Ind.For* **106(1):**41-51.

Prasad, Ram and Mishra, G.P. (1985). Litter productivity of dry deciduous teak forest stands of Sagar, M.P. *Ind. For.*, **111(6):**359-367.

Prasad, Ram and Shrivastava, J.L. (1991). Most successful tree species on coal O.B. dumps. *Vaniki Sandesh,* **15(2):**1-4.

Prasad, Ram and Shrivastava, J.L. (1992). Effect of pit size on the growth performance of some important species on the overburdens of mined out areas of coal. *Vaniki Sandesh,* **16(3):**3-5.

Prasad, Ram and Jamalluddin (1989). Observations on the problem of teak mortality in Central India. Proc. Nat. Seminar on Depletion of Soil and Forest Cover. *J. Trop. Forestry.* **5(2&3):**72-75.

Prasad, Ram and Jamaluddin (1985). Preliminary observations on Sal mortality in Madhya Pradesh. *Ind.For.* **111(5)**:250-271.

Prasad, Ram and Bhausar, K.N. (1991). Yield of Mahua flowers and fruit. *Vaniki Sandesh.* Vol. **XV.** No.1:22-25.

Prasad, Ram and Kandya, A.K. (Eds) (1989). Depletion of soil and forest cover & technology for control. Proc. National Seminar, M.P. State Forest Res. Instt. Jabalpur, June 7–9, *J. Trop. For.* 5 (2+3):1–277.

Prasad, Ram and Kandya, A.K. (1989). Status of commercial production of quality seeds of forestry species in Madhya Pradesh : an over-view. Proc. Natl. Seminar on Forest Seed Technology, Instt. of Forest Genetics and Tree Breeding, Coimbatore, Paper III-1, pp. 1-13.

Prasad, Ram and Kandya, A.K. (1991). Forestry Seed Development Programme : A Forestry Research Institute's Perspective. In–Sen, D.N. and S. Mohammad (Eds.) Marvels of seeds. Proc. Internatl. Seed Symp., Univ. of Jodhpur, Jodhpur. pp. 1-17.

Prasad, Ram and Kandya, A.K. (1992). Handling of forestry seeds in India. Associated Publishing Co., Karol Bagh, New Delhi, p. 421.

Prasad, Ram and Jalil, P. (1987). Observations on the seed viability of Sal (*Shorea robusta*). *Ind. For.*, **113(2)**:89-94.

Prasad, Ram and Jalil, P. (1991). Observation on seed germination on *Holoptelia intergrifolia. Vaniki Sandesh,* **15(2)**:5-9.

Prasad, Ram and Bhatnagar, Pratibha (1989). Livestock and fodder position in tribal region of Madhya Pradesh. *J. Trop. Forestry.* **5(2&3)**:203-205

Prasad, Ram and Bhatnagar, Pratibha (1993). Non-Wood Forest Product and the Indigenous Fringe Dwellers in Madhya Pradesh. *J. Trop. Forestry.* **9(III)**:135-189.

Prasad, Ram and Pandey, R.K. (1989). An observations on the effect of closure for rehabilitation of degraded forest lands : A case study of protected and un-protected wastelands at Bhopal. *J. Trop. Forestry.* **5(2 & 3)**:128-137.

Prasad, Ram and Pandey, R.K. (1993). Ethno-Medico-Botanical studies on indigenous medicinal plants of Lamni and Achanakmar of Bilaspur district of Madhya Pradesh. *J. Trop. Forestry.* **9(1)**:27-40.

Prasad, Ram and Chadhar, S.K. (1986). Relevance of tissue culture technology in wastelands development. *Vaniki Sandesh,* **10(4)**:4-17.

Prasad, Ram and Chadhar, S.K. (1987). *Grevillea pteridifolia (syn.Grevillea banksil*). I-A successful introduction to mined area in Madhya Pradesh-I Botanical characteristics. *Indian For.*, **113(1)**:670-674.

Prasad, Ram and Chadhar, S.K. (1988). Retrieval of bamboo forests in Madhya Pradesh. *Indian For.*, **114(9)**:496-504.

Prasad, Ram and Chadhar, S.K. (1990). Effect of sowing depth on the germination of some common forestry seeds. *Vaniki Sandesh* **14(2)**:1-4.

Prasad, Ram and Dhuria, S.S. (1989). Reclamation of iron ore mined out areas, biomass production efficiency of species. *J. Trop. Forestry.* **5(1)**:51-56.

Prasad, Ram and Chadhar, S.K. (1988). Optimum potting mixture for sprouting of root-shoots in some tree species. Vaniki Sandesh. **12(3)**:7-11.

Prasad, Ram, Jalil, P. and Mishra, Manish (1992). Effect of various pre-treatments on seed germination of some forest tree species. *Vaniki Sandesh.* **16(1)**:6-10.

Prasad, Ram, Pandey, R.K. and Bhattacharya, P. (1990). Socio-economic and ethnomedico botanical studies of Patalkot region: a case study of Bharia tribes. Paper presented in the national seminar on Medicinal and Aromatic plants, SFRI, Jabalpur (M.P.) April 6-7, 1990.

Prasad, Ram; Date, G.P. and Jalil, P. (1988). Observation on storage and germination of seeds of *Largerstroemia parviflora. Vaniki Sandesh*, **12(3)**:2-6.

Prasad, Ram; Kotwal, P.C. & Mishra, M. (2001). "Harvesting Practices of Safed Musli (Chlorophytum spp.) and its ecological and socio-economic impact in the natural forests of Central India. *J. Trop. For.*, **12**:61-65.

Prasad, Ram; Kotwal, P.C. & Mishra, M. (2002). Impact of Harvesting of *Emblica Officinalis* (Aonla) on its Natural Regeneration Health and Vitality of Ecosystem in Central India Forests: *J. Sustainable Forestry*, **Vol. 14** No.4 2002.

Prasad, Ram; Meshram, P.B. and Jamaluddin (1987). Effectiveness of Hormones, Rallis tracel and insecticide on fruiting in Teak Seed Orchards of Madhya Pradesh, *Indian For.*, **113(5)**:323-327.

Prasad, Ram; Pandey, R.K. and Shrivartava, J.L. (1990). Medicinal plants of Fabaceaae (leguminosaea) from Madhya Pradesh- An ethnobotanical approach. Paper presented in the national seminar on Medicinal and Aromatic plants, SFRI, Jabalpur (M.P.) April 6-7, 1990.

Prasad, Ram; Pandey, R.K. and Singh, S.P. (1988). Survey of medicinal wealth of Central India II. Ethnomedico-botanical studies of indigenous plants used by local tribals. *J. Trop. Forestry.* **4(3)**:236-241.

Prasad, Ram; Danayak, S.C. and Goswami, Ashok (1992). Optimum number of coppice shoot on each stool in Eucalyptus hybrid. *Vaniki Sandesh.* **16(1)**:1-5.

Prasad, Ram; Chadhar, S.K. and Jalil, P. (1988). Some observation on germination of four useful but difficult forestry seeds. *J. Trop. Forestry,* **4(4)**:395-398.

Prasad, Ram, Singh, S.P. and Jalil, P. (1993). Effect of tree girth on seed germination of *Lagerstroemia parviflora. J. Trop. Forestry.* **9(1)**:63-69.

Prasad, Ram, Meshram, R.P.B. and Jamaluddin (1990). Possibilities for enhancing the fruiting in Teak Seed Orchards, *Indian For.*, **116(2)**:99-102.

Prasad, R. and Danayak, S.C. (1991). Growth trend in some Aonla plantation in Madhya Pradesh. *Vaniki Sandesh* Vol. **XV.** No 4. 1-4.

Prasad, Ram & Shukla, P.K. (1985). Reclamation & re-vegetation of coal mine overburdens in M.P. *J. Trop. For.*, **1(1)**:79-84.

Advancing Frontiers of Ecological Researches in India, 29-50, 2006
Edited by A.K. Kandya & Asha Gupta
Published by Bishen Singh Mahendra Pal Singh, Dehra Dun, India

The new paradigm for monitoring and assessment of forest resources: Criteria and Indicators for Sustainable Forest Management

P.C. Kotwal and Navin Horo
Indian Institute of Forest Management Bhopal

Abstract: Sustainable forest management maintains the forest, its ecological functions, processes and overall structure in healthy condition, till perpetuity. It also takes care of maintaining the quality of soil and water. Similarly, it does not produce any irreversible consequences or losses to biological diversity, including genes, species, ecosystems, and forests types. That means, there are no extinctions at any level. Such forest management and its planning are based on certain principles and laid objectives. **Criteria** define the essential components of these principles and **Indicators** are ways to measure the status of a criterion. Various organizations and persons have given the C & I for specific types of forests, i.e. ITTO (International Tropical Timber Organizations), African Timber Organization, IIFM (Bhopal, India) and UNCED (UN Committee for Environmental Development) etc. In the present paper, details of all these aspects are discussed.

INTRODUCTION

The forests have long been underscored as a timber resource and various silvicultural practices were innovated to enhance the productivity of timber from the forests. In 1795, G.L. Hartig (1764-1837) introduced the concept of Sustained Yield that was based on the realization that 'wood from the forests is not inexhaustible. It may only be utilized on a scale, which corresponds to its natural rate of growth, if a continuous supply of timber is to be assured.'

Gradually the forestry profession graduated and an understanding of the term "forest management" broadened to accommodate wider environmental issues, such as conservation of biological diversity, social and economic matters and more generally, the concept of sustainability. It was the Brundtland's Report "Our Common Future" *(Brundtland, 1987)* that stressed that the initiatives for development should be people centered and ecologically friendly. This phenomenon on comprehensive management of forests addressing its ecological, economic and socio-cultural functions developed throughout the globe owing to improved understanding of the forest managers and awareness of the general populace.

The global forestry scenario reflects the need for attention and commitment towards Sustainability of forest resources. The total global forest area accounts for approximately 3869.455 million Ha out of which 3682.722 million Ha is natural forests and 186.733 million Ha are Plantations i.e. 29.6% of the total land area, spread in various climatic zones and classified into different forest types. Per capita availability of forests is 0.6 Ha (Global average). Out of this, only 8% of the forest area are protected i.e. approximately 309.56 million Ha. Assessing the loss of the existing forest area, every year almost 9.391 million Ha i.e. 0.2% is lost (every minute approximately 60 acres of forest is lost) owing to irresponsible and illegal practices like fires, encroachments, illegal felling etc. According to current estimates (FAO, 2003), 0.38 per cent of the world's forests were converted to other land uses (i.e. deforested) every year in the 1990s. Besides the loss, the burgeoning demand of forest products especially for wood for paper and timber requirements has been growing multi-fold, which is estimated to be 680 billion broad feet annually; presents alarming statistics. However, the study by World Wildlife Fund, which estimates that the total global requirement of timber and wood can be met just by 20% to 25% of total world's forest resources, through better forestry management practices and reduction of losses.

Sustainable Forest Management

Sustainable Forest Management (SFM) has been recognised as an important aspect of sustainable development during Earth Summit in 1992 at Rio because a large proportion of human population is directly/indirectly dependent on forests. Besides, the forests also provide a number of goods and services to mankind. At the Summit, world leaders adopted the first global policy on Sustainable Forest Management (SFM), known as the "Forest Principles". According to this policy, 'forest resources and lands should be managed sustainably to meet the social, economic, ecological, cultural and spiritual functions and for the maintenance and enhancement of biological diversity'.

Although several definitions have been written and circulated in recent years, most of these have emanated from the forest development sectors. It is time to add another perspective to the debate. To that end, the following is offered:

Sustainable Forest Management is forest management that:

- maintains the forest, its ecological functions, processes and overall structure, in healthy condition, in perpetuity;
- does not degrade soil or water quality;
- does not produce any irreversible consequences or losses to biological diversity, including genes, species, ecosystems, and forest types (i.e. no extinctions);
- applies to the entire forest as an integrated, ecological whole, rather than to any single component or product of the forest;
- can be either active or passive, and does not require the extraction or harvest of a particular product from the forest;
- can be applied to any size or scale of management area, e.g. individual forest management unit or ecosystem, watershed, landscape, forest type, bio-region, nation, etc.; provided, however, that at each different scale, sustainability must be assessed entirely within the boundaries of the defined unit or region; and
- can produce a wide range of environmental, social and economic benefits to society, depending on the scale of the management area and its capabilities and carrying capacity.

Sustainable Forest Management is defined as:

"the process of managing permanent forest land to achieve one or more clearly specified objectives of forest management with regard to the production of a continuous flow of desirable forest products and services without undue reduction of its inherent values and future productivity and without undue undesirable effects on the physical and social environment" (*ITTO 1991*).

Realms of SFM

Ecological security and well-being

India has approximately 2.5% of the total global geographical area and has nearly 1% of forests that is supporting almost 16% of human population and 18% of livestock population. The State of Forests Report (FSI 2001) states that the forest land in India is 23.38% of the total geographical area however only 20.55% of the area is under the forest cover. Compared to the information in the years viz. 1987, 1997 and 1999 the forest cover was 19.49%, 19.27% and 19.39% respectively. When we compare the status of forests in the year 1999 to that in 2001, there is an increase in the dense forest cover by 34580 sq km whereas the open forests have increased by 3665 sq kms. When including the other sources of trees (trees outside forests) the cumulative figure of forest and other tree covers constitutes 23.03% of the total geographic area. Increase in the forest cover involves two aspects viz increase in forests within forest boundary and proliferation of forest trees outside the legal boundary of the forests. The National Forest Policy has envisioned that the total forest cover in the country should be 33% of total geographical area (33% on plains and 66% on slopes). The attempts should hence be towards increasing the forest cover in the country.

India is one among the hot bio-diversity spots of the world. Many of the species are endemic to the forests in India. Hence, it is essential to conserve these resources. India contains a great wealth of biological diversity in its forests, its wetlands and in its marine areas. The stability and progressive enhancement of a biological resource is an efficiently functioning and balanced ecosystem. India has 89,451 of faunal species and almost 49,000 floral species (*GoI, 1999*). This highlights the condition of the forests (in the states) and the health of the associated forest ecosystem. For a resilient and a progressively productive ecosystem all its associated components should work in tandem.

Economic well-being

Forests are an inevitable source of various products and services. The productivity of a forest can be determined, for example, by the amount of biomass it generates over a period of time. Forests produce an array of goods and services. These resources are usable in a variety of ways. But the resources of the forests can be scarce and therefore need optimization in terms of their use as well as reduction in wastage. This is because sustainable use of forest resources is essential to achieve sustainable forest management. There should be no resource utilisation above the maximum allowable removal limits. Over exploitation of the resource would lead to resource depletion, which is detrimental to sustainable forest management. The use of forest resources should not exceed the carrying capacity of the forest nor should forest products be overexploited. Timber and other non-wood forest products and services should be used sustainably but the issues of equitability and optimality should also be given due importance.

There are many factors affecting the productivity of a forest ecosystem viz. soil, moisture regime, micro-flora and fauna, humus content, nutrient recycling mechanism, etc. Minimising the adverse factors such as forest fires, un-regulated grazing of cattle, pests & diseases, extremes of weather conditions such as drought, floods. Forests provide many intangible products that have a bearing on the economic utility of the forests. These products are vital for the stability of a landscape and may also fetch revenue. If forest products and services could be sustained over a period of time the use value of the forest would increase manifold.

2.3 *Socio-cultural functions*

Forests yield many of services including social benefits as well. Therefore the intrinsic value of a forest should be judged on these lines. Many people, including forest dwelling communities, bear allegiance of their ancestry to the forests or tree species therein. People's emotional and spiritual links with forests and its resources have resulted in conserving many of the endangered species, as they are not harmed due only to their spiritual or social significance as the people see them. Sometimes people maintain high spiritual and emotional links to the forestland.

A guiding light for any management regime is the existence and efficient execution of a well-framed policy. This applies to the forest resources too. Under this Criterion a number of indicators that deal with issues linked to the environment under which forest management is practiced are listed. This criterion addresses the specific institutional requirements that are necessary to achieve Sustainable Forest Management in the states.

Importance of SFM:

- To ensure a continuous flow of forest goods and services:

Sustainable forest management in essence aims at meeting the requirements of the existing population without having any affect on availability of resources for the future generations. Hence, advocating optimisation of usage of the forest resources for the current as well the generations to come.

- Contribute towards sustainable development by addressing environmental, economic and social well-being:

As reinstated, majority of the needs of global population, urban and rural, is dependent on forest resources. These dependencies could be in the form for diverse requirements, from wood-based products and minor products (medicinal plants, food plants, oil, tannin, fibers, color etc.) to various intangible products like socio-cultural and environmental needs. More so, at the local conditions, forests serve as major economic resource determining the livelihoods of the population. The relevance and contribution of forest resources to needs of the global population, reiterates it as an important aspect towards sustainable development.

- Monitor progress of forest management at the FMU level (forest divisions)

A progressive management system would ensure the progress of forests towards sustainability. SFM as a holistic paradigm, provides a goal for the management interventions thus, an objectively laid management plan that would effectively contribute to successive well being of population as well the resources. Hence, to know the direction of change, monitoring is essential.

- To fulfill Objective 2000 with states that ' ...the total exports of tropical forest products should come from sustainably managed forests. This means that the any forest product going in the international market will fetch premium price only when it bears the certificate that the product has come from the forest, which is sustainably managed. Without such a certificate there will be difficulty in selling the forest product and it will not fetch appropriate price. This will have an impact on our national economy as well as on the livelihood of the local people.

Besides, SFM being concurrent to sustainable development; it's also important to global economic implications. This is reflected in various international processes and initiatives towards monitoring and assessment of forest management practices contributing to environmental and social well-being.

Monitoring of Sustainable Forest Management

The next question mark that stands ahead is how do we really know whether our initiatives are leading towards sustainability of forests? It's important that there is a system of regular feedback on the management and forest related activities that provide to the decision-makers the information on success and limitations of such practices. Besides providing the information on success or failure, it is also necessary that the extent of success and limitations be well understood to facilitate improved planning and better-adapted practices. In simpler words, there is a strict requirement of a monitoring and assessment system that could provide the periodic information on the status of forest resources as well as the effectivity of management interventions.

Addressing to need of such monitoring and assessment system a hierarchy was developed (also known as Tropenbos framework). The framework has four levels of hierarchy as depicted below.

Principle
↓ Forestry management and planning is based on certain principles or laid objectives

Criteria
↓ Criteria defines the essential components

Indicators
↓ "Indicators" are ways to measure or describe the status of a criterion

Verifiers
Verifiers facilitate in ensuring the information related to indicators

E.g. Criteria and Indicators by African Timber Organisation at Librerville, 1993

***Principle I.* Areas devoted to forestry activities or the permanent forest estate are not declining**

Criterion I.1. Areas devoted to forestry activities or the permanent forest estate are clearly delimited and their boundaries have been well established

Indicators:

I.1.1. There exists a map showing the boundaries of the permanent forest estate.

I.1.2. The boundaries of the permanent forest estate are well marked on the field.

Criteria and Indicators for SFM

The need to reconcile the productive functions with the protective, environmental and social roles, which all types of forests fulfill, was firmly expressed at UNCED. A declaration of 27 guiding principles focusing on the rights and obligations of sovereign states with respect to environment and development was agreed to at UNCED. It included the *Forest Principles* - a Statement of Principles for a global consensus on the management, conservation and Sustainable development of all types of forests and to provide for their multiple and complementary functions and uses. One positive step aimed at developing the *Forest Principles* is that there has been broad international consensus to formulate scientifically sound criteria and indicators for the management, conservation and development of all types of forests.

Criteria and indicators are tools, which can be used to define, implement and monitor sustainable forest management in the broadest sense of the term, nationally and at the forest management unit level.

Criteria: Criteria define the essential factors of forest management against which forest sustainability may be assessed. Each criterion relates to a key management factor, which may be described by one or more qualitative, quantitative or descriptive indicators.

Indicators: Through measurement and monitoring of selected indicators, the effects of forest management action, or inaction, can be assessed and evaluated and action adjusted to ensure that forest management objectives are more likely to be achieved.

The global initiatives for defining the criteria and indicators (C&I) have several approaches. Yet even if no consensus on criteria and indicators can be achieved, SFM is not without importance for forestry. Research results indicate that the main functions of SFM may be found in the emotional and symbolic aspects of the term: its ability to provide a common spirit for the forestry profession the world over, to help provide a guideline for foresters and associated institutions in coping with uncertainty and ignorance regarding management of forests, and of course, its mediating function which serves to bridge the gap between different interests by providing a uniting term so that proper management, monitoring and evaluation of forest resources can be achieved. The C & I approach is an attempt towards arriving at a forest management system so that the direction of change can be ascertained and timely actions to alleviate the foreseeable damages to forest resources can be undertaken. *(Prabhu, 2001).*

Criteria and Indicators often shortened to C & I, is the term used to describe a systematic approach to measuring, monitoring and reporting sustainable forest management. There is no single "best way" to achieve sustainable forest management, but the framework of criteria and indicators is a meaningful approach. In particular, criteria and indicators can be used to encourage more holistic thinking when planning forest management activities, and to bring about greater rigor, openness, transparency and accountability in forest management planning, monitoring and reporting. The C&I indicate the direction of change as regards the forests and also suggest the ways to enhance the process of SFM.

The "Criteria" constitute the essential components of sustainable forest management, i.e. the management of forests as functioning ecosystems. These components include vital functions and attributes of forest ecosystems (such as extent of forest area, biological diversity and forest health), the multiple socio-economic benefits of forests (such as non-wood forest products, timber, recreation and socio-cultural values), and in most cases the policy framework (laws, regulations and economic measures) needed to facilitate sustainable forest management.

The "Indicators" are ways to measure or describe a criterion. The indicators associated with a given criterion serve to define what that criterion is and what it means. For example, "ecosystem and species diversity" are indicators of biological diversity; "area of forest land available for timber production" and "total growing stock" are indicators of forest resource productivity.

Many indicators are quantitative, such as the percent of a country's forest cover. Others are qualitative or descriptive, such as those related to forest planning, public participation and social aspects. Both are important in assessing sustainable forest management at the national level. It is important to note that no single criterion or indicator alone constitutes a measure of sustainability. It is the array of criteria and indicators, which together, when measured over time, that provides the full picture of the state of a country's forests and trends towards sustainable forest management.

International Initiatives on Criteria and Indicators for Sustainable Forest Management

Since UNCED, criteria and indicators for sustainable forest management have been formulated within the framework of several international and national meetings. These include the *Helsinki Process* (for European forests, 1994), the *Montreal Process* (for temperate and boreal forests, 1993 to 1995), the *Tarapoto Proposal* (Amazon Forests 1995), the *Dry Zone Africa Proposal* (1995), the *Near East Proposal* (1996), the *Lepaterique Process of Central America* (1997), Dry Zone Asia Forests (1999) and others. ITTO initiative also has developed a set of 7 criteria and 66 indicators (1992) for National and forest management unit levels in humid tropical forests of member tropical countries (*Johnson, 2001*). The process to develop C & I sets for the dry-zone Asia region (like India) the Regional Initiative for Dry Zone Forest Asia has developed a set of 8 criteria and 49 indicators. The Bhopal-India process is a part of Dry Forest Asia Process that evolved set of C & I for monitoring and assessment (at National levels) for forests in India. Globally there are almost 150 countries that have participated in the process of C & I for SFM and have evolved sets of C&I for monitoring SFM. (The Table-1 depicts summary of all the international processes of SFM and the countries involved) *(Castañeda, 2000; Wijewardana, 1998, 2001)*

The concept of criteria and indicators is a relatively new one and there is no globally agreed set of indicators for a particular criterion, as indicators are to developed or adapted to the ecological, economic, social and institutional conditions and needs of each country in general and an FMU in particular. At the national level, the nature and constitution of a forest policy, rules and regulations, and forest management practices, can be adjusted and gradually improved to take into greater consideration social, economic, environmental, spiritual and cultural requirements as defined by the criteria, to involve and benefit increasingly broad ranges of stakeholder groups. (Prabhu *et al* 1999; Jiang and Jiang, 1997).

Criteria and Indicators for SFM in India

The Indian initiative for development of C&I for SFM was undertaken by Indian Institute of Forest Management (IIFM), Bhopal; a premier education training research and consultancy institution in Asia, in 1998. Through various consultative processes, expert discussions, sessions the process led to development of national level 8 criteria and 51 indicators

Table-1: Brief description of the major international processes on criteria and indicators for SFM

Process	No of criteria	No. of indicators	Applicability	Place of adoption	Date of adoption	No of countries	Participating countries/regions
ITTO Initiative on criteria and indicators	7	66	National and forest management unit levels in humid tropical forests of member tropical countries	Yokohama, Japan	March 1992	12	Cambodia, Fiji, India, Indonesia, Malaysia, Myanmar, Papua New Guinea, Philippines, Thailand, Togo, Trinidad and Tobago, Vanuatu
Dry-Zone Africa Process	7	47	National level	Nairobi, Kenya	November 1995	28	CILSS (9 countries): Burkina Faso, Cape Verde, Chad, Gambia, Guinea-Bissau, Mali, Mauritania, Niger, Senegal IGADD, (7); Dijibouti, Eritrea, Ethiopia, Kenya, Somalia, Sudan, UgandaSADC(12); Angola, Botswana, Lesotho, Malawi, Mauritius, Mozambique, Nambia, South Africa, Swaziland, Tanzania, Zambia Zimbabwe
Pan European Forest Process	6	27 quantitative 101 descriptive	Boreal temperate and Mediterranean type forests in Europe regional and national levels [d]	Helsinki, Finland, Lisbon, Portugal	June 1993 June 1998	37	Albania, Austria, Belarus, Belgium, Bulgaria, Croatia, Czech, Republic, Denmark, Estonia, European Community, Finland, France, Germany, Greece, Hungary, Iceland, Ireland, Italy, Latvia, Liechtenstein, Lithuania, Luxembourg, Monaco, Netherlands, Norway, Poland, Portugal, Romania, Russian Federation, Slovak Republic, Slovenia, Spain, Sweden, Switzerland, Turkey, Ukraine, United Kingdom

Process	No of criteria	No. of indicators	Applicability	Place of adoption	Date of adoption	No of countries	Participating countries/regions
Montreal Process	7 (non-legally binding	67	Temperate and boreal forests in countries outside Europe national level	Santiago, Chile	February 1995	12	Argentina, Australia, Canada, Chile, China, Japan, Republic of Korea, Mexico, New Zealand, Russian Federation, Uruguay, United States
Tarapoto Proposal	1 global 7 national 4 FMU	7 (global) 47 (national) 22 (FMU)	Sponsored by the Amazon Cooperation Treaty	Tarapoto Peru	February 1995	8	Bolivia, Brazil, Colombia, Ecuador, Guyana, Peru, Suriname, Venezuela
Near East Process	7	65	Regional and national levels	Cairo, Egypt	October 1996	30	Afghanistan, Algeria, Azerbaijan, Bahrain, Cyprus, Djibouti, Egypt, Islamic Republic of Iran, Iraq, Jordan, Kuwait, Kyrgyz Republic, Lebanon, Libya, Malta, Mauritania, Morocco, Oman, Pakistan, Qatar, Kingdom of Saudi Arabia, Somalia, Sudan, Syrian Arab Republic Tadjikstan, Tunisia, Turkey, Turkmenistan, United Arab Emirates, Yemen
Lepaterique Process of Central America	4 regional 8 national	40 (regional) 53 (national)	Forest management level criteria and indicators also identified 5 principles and 2 sub principles	Tegucigalpa, Honduras	January 1997	7	Belize, Costa Rica, El Salvador, Guatemala, Hinduras, Nicaragua, Panama
African Timber Organization	28	60	ATO member countries; regional and national levels; also identified 5 principles and 2 sub principles	Libreville, Gabon	January 1993	13	Angola, Cameroon, Central African Republic, Congo, Cote d' lvoire, Democratic Republic of Congo, Equatorial Guinea, Gabon, Ghana, Liberia, Nigeria, Sao Tome and Principe, United Republic of Tanzania

Process	No of criteria	No. of indicators	Applicability	Place of adoption	Date of adoption	No of countries	Participating countries/regions
Regional Initiatives for Dry Forests in Asia	8	49	Dry forests in Asia; national level	Bhopal, India	December 1999	9	Bangladesh, Bhutan, China, India, Mongolia, Myanmar, Nepal, Sri Lanka, Thailand

A. *The totals in this column add up to more than 140 (the total number of countries participating in international process) because many countries are members of more than one process.*

B. *Of the 55 member countries of ITTO only the 12 listed countries participate in ITTO's criteria and indicators process. The rest of the ITTO member countries participate in other process. India, Myanmar and Thailand are also members of the Regional Initiative for Dry Forests in Asia.*

C. *CILSS: Permanent Interstate Committee for Drought Control in the Sahel: SADC: Southern Africa Development Community; IGADD: Intergovernmental Authority on Drought and Development*

D. *Operational Level Guidelines for application at the sub national level have also been developed.*

for SFM in India. Following the processes a pre-project sponsored by ITTO a titled "Sustainable Forest Management with community participation in India" was implemented by India for field testing of the developed C&I at FMU level. The project also initiated a dialogue on SFM in the country through various workshops and seminars most important was the task of sensitizing the communities involved in sustainable management through grassroots level training workshops.

In December 1999, Government of India appointed a National Task Force for reviewing the developed set of C&I. The task force recommendation of a refined set of 8 criteria and 43 indicators was accepted by GoI also designated IIFM as the nodal agency for operationalising SFM in India. Currently the process is in its operationalisation phase. Implementation of C&I as a pilot basis is being undertaken through a project titled "Operational strategy for Sustainable Forestry Development with community participation in India". With respect to the progress in operationalisation, a generic C&I system has been developed and is being implemented in the project areas. Sustainable Forest Management (SFM) is a multi-actor process and informed participation of all actors is a prerequisite for its effective implementation hence, capacity enhancement of forest managers as well as local communities is emphasised under the project along with sensitization of strategic decision-makers to provide conducive policy environment. (*Kotwal and Chandurkar, 2002*).

The B-I Process developed national level criteria and indicators, which were further reviewed by the National Task Force that recommended a national, set of 8 criteria and 43 indicators for monitoring and assessing sustainability of forests in India. The national set of C&I for SFM is listed in table-2.

Bhopal India (B-I) Process

India is a land of diversity both in terms of human as well as natural resources. Due to its diverse eco-regions, India has always been committed to the conservation of forests and biodiversity, with sustainability as the central theme of her Forest Policy (*GoI, 1988*). But in the absence of a pragmatic framework for SFM in India, there was no mechanism to provide feedback on the direction of change taking place. Indian Institute of Forest Management took a timely initiative in 1998 to generate a pool of knowledge for inculcating the idea of SFM in the country. This proactive initiative has been named as the Bhopal-India Process. This initiative found support of a specially constituted Technical Group on SFM and the Government of Madhya Pradesh. This is a pioneering attempt in India c carry forth the cause of SFM. (*Prasad et al, 2001*)

In order to further this cause, a three-day workshop to "Evolving Criteria and Indicators for SFM in India", was organized from January 21-23, 1999, by IIFM in close association with the State Forest Department and the Madhya Pradesh Minor Forest Produce (Trade and Development) Federation, Bhopal (*IIFM, 1999*). The workshop aimed to generate a ground set of national level criteria and indicators to assess SFM and its progress in India. It was decided to adopt a three-tier hierarchical structure for India, involving principles, criteria and indicators, for defining SFM. The discussions at this technical meeting led to the identification of 8 national level criteria and 51 related draft indicators, under the Bhopal-India Process. This set of C & I is relevant not only for the 4 major forest types of India, but also for the Dry forests of South and South East Asia that include, Sri Lanka, Pakistan and Bangladesh as well. These criteria and indicators would serve as a guiding light for similar attempts of evolving and operationalising the C & I system in other South East Asian nations *(Prasad and Kotwal 2001).*

India being one of the few biodiversity rich regions of the sub-continent, has always adopted a planned approach to manage the forests and this is reflected in the continuously evolving Indian Forest Policy. Indian Forest Policies can be broadly classified into pre independent and post independent ones. India's early forest management policies supported agricultural and timber production as highest priority assets, but have gradually evolved to ensure environmental stability and ecological balance over a period of time. The current actual forest cover of India (including the hills) is 63.34 million ha (19.27%). But due to the many issues of concern facing India including deforestation, degradation, increasing demand for forest products, and insufficient financial allocation, an urgent need is being felt to streamline its efforts to ensure its sustainability and maintenance. In spite of the efforts to involve people's participation in forest management (62,890 Joint Forest Management Committees have been constituted managing about 22% (14.25 m ha) of the total forest) (*GoI, 2001*), little has been done to monitor the continuously changing patterns of problems related to forest management like encroachment, grazing, fire, etc. India is a member of ITTO and is committed to the objective, 2000 of ITTO, which states "The objective is that the total exports of tropical timber products should come from sustainably managed forests, by the Year 2000." In order to achieve this objective it is imperative to redefine forest management in the light of the criteria and indicators and operationalise them at all the levels, viz.: local, state and national levels. (*IIFM, 1999*)

Table-2: National level C&I for SFM in India developed under the Bhopal-India Process

Criteria	Indicators
1. Increase in the extent of forest and tree cover	1.1 Area and type of forest cover under - natural forest - man-made forest 1.2 Tree cover outside forest area 1.3 Area of dense and open forest 1.4 Forest area diverted for non- forestry use 1.1 Extent of community managed forest areas
2. Maintenance, conservation and enhancement of bio-diversity	2.1 Area of protected eco-systems 2.2 Area of fragmented eco-systems 2.3 Number of rare, endangered, threatened and endemic species 2.4 Level of species richness, and bio-diversity in selected areas 2.5 Availability of medicinal and aromatic plants in various forest types 2.6 Status of non-destructive harvest of NWFP 2.7 Number of keystone and flagship species in various forest types
3. Maintenance and enhancement of ecosystem function and vitality	3.1 Status of natural regeneration 3.2 Status of secondary forests 3.3 Incidence of: A) Pests and diseases B) Weed infestation C) Grazing and D) Fire
4. Conservation and maintenance of soil and water resources	4.1 Area under watershed treatment 4.2 Soil erosion status 4.3 Area under ravines, saline and alkaline soils 4.4 Ground water table in the vicinity of forest areas
5. Maintenance and enhancement of forest resource productivity	5.1 Growing stock of wood 5.2 Volume of production of identified/ important NWFPs 5.3 Increment in volume of identified species of wood 5.4 Level of financial investment in forestry sector

	5.5 Extent of areas under seed production areas, seedling seed orchards, clonal seed orchards
6. Optimisation of forest resource utilisation.	6.1 Aggregate and per capita consumption of wood and NWFPs 6.2 Import and export of non-wood forest products 6.3 Recorded production of wood and NWFPs 6.4 Direct employment in forestry and forest industries 6.5 Contribution of forests to the income of forest dependent people 6.6 Level of processing and value addition in NWFPs and treatment, seasoning and preservation of wood 6.7 Demand and supply ratio of timber, firewood and fodder
7. Maintenance and enhancement of social cultural and spiritual benefits	7.1 Degree of people's participation: Number of committees and area (s) protected by them 7.2 Use of indigenous technical knowledge : identification, documentation & application 7.3 Quality and extent to which rights and privileges are utilized 7.4 Human development index 7.5 Extent of cultural/sacred–protected landscapes:forests, trees, ponds, streams etc.
8. Adequacy of policy, legal and institutional framework	8.1 Existing policy and legal framework 8.2 Enabling conditions for participation of community, NGO's civil society like JFM resolution, transit rules etc. 8.3 Level of investment in research and development 8.4 Human resource capacity building efforts 8.5 Forest resource accounting 8.6 Monitoring and evaluation mechanism 8.7 Status of information dissemination and utilization

The Way Ahead

Policy Issues

1. *Political commitment*

An enabling political environment is pre-requisite for operationalising sustainable forest management in India. With the changing international scenario, especially related to international trade in forest products, forest certification and reporting to international fora, SFM is more of a necessity for the nation. Congruent to this, policy adjustments and innovations are necessary for promoting and operationalising SFM in India.

2. ***Institutionalisation of C&I***

Institutionalisation of C&I system in the management regime will ensure that C&I becomes the basis of planning, implementation and monitoring of forest management initiatives. The National Working Plan Code is the guiding document for planning forest management in the country. From the National-Working Plan Code, the states develop their own Working Plan Codes. Incorporation of C&I in the National Working Plan Code, accentuating C&I for planning, implementation and reporting, will institutionalize the C&I system at National-level and consequently at the state-level.

3. ***Allocation to forestry sector***

Allocation to forestry sector is currently based on the contribution of the sector to the overall GDP of the nation. The forestry sector is not well represented in the GDP of the nation owing to various accounting distortions. The present system of accounting discounts the environmental, social and cultural values of the forests. Development of a comprehensive Forest Resource Accounting System based on the concept of Total Economic Valuation would be necessary for reflecting the true value and contribution of the sector to the GDP.

4. ***Structural changes***

Along with institutionalization of the C&I system, structural changes in the Forest Department are also necessary. The field level staff which is directly related with operationalisation especially data collection lacks the capabilities and capacities for implementation. Further, with no incentives, promotional avenues and motivation, the efficiency of this stratum is low. Henceforth, structural changes are required especially with respect to fresh recruitments and motivational mechanisms.

5. ***Inter-Sectorial Linkages***

Forestry sector affects other sector of economy and in turn gets affected by the other sector. Therefore, for SFM, development of intersuctorial linkages is necessary. Identification of this linkages and design of appropriate arrangements for strengthening these linkages is imperative for implementing SFM.

Operational issues

1. *Site-specificity*

Forestry programmes in India evolve from and are in cognizance with the National Forest policy objectives. However, the programmes tend to have a tendency of being a broad spectrum for uniform application to all forest areas. This per se discounts the specific ecological, economic and especially the socio cultural environment of the site. Even though implemented with objective considerations, the programme effectiveness may not be as desired.

In C&I for SFM context, this issue is of prime importance. For each site SFM needs to be described through related environmental, economic and social values associated with the specific forests. Further, the indicators also need to be developed for the particular site vis-à-vis the prevalent forestry conditions, relevance and applicability.

2. *Technological inputs*

For operationalising C&I for SFM, innovations would be required for measurement and assessment technologies. While new technologies are to be developed for assessment, the existing technology needs to be upgraded in light of new challenges as posed by C&I assessment. Innovative tools and techniques for collection, compilation, collation and analysis of C&I related data/information would come under the purview.

3. *Cost-effectiveness*

While technological interventions are imperative, the assessment costs is a major area of concern that impedes SFM implementation. Mobilizing financial resources and at the same time developing low-cost technologies is important for operationalising C&I for SFM. Development of simple robust and easy to use indicators is of critical significance, especially for areas under co-management.

4. ***Adequate Capacities***

Assessment of all the indicators is not possible with the existing capacities of the department. For assessing the indicators existing capacities need to be augmented and at the same time new capacities are to be created. Especially for the indicators related to socio economic values apart from enhancement of capacities, support from specialist agencies would also be necessary.

5. ***Peoples participation and collaborative forest management***

The concept participatory regime of Joint Forest Management is envisioned as a vehicle to achieve SFM. With a paradigm shift in management forest, communities living in and around the forest are involved in forest related decision-making and benefit sharing. Hence, involvement of these communities in operationalising SFM is imperative.

Implementations of SFM in JFM areas pose new challenges. First, the communities are to be sensitised about the concepts of C&I. Secondly, site-specific C&I needs to be developed and a functional C&I system needs to be designed. Thirdly and most important is augmenting capabilities of the communities for monitoring, assessment and reporting. Thus, extending SFM to JFM areas is a daunting but essential task.

6. ***Multi-stakeholder synergy***

SFM is a multi-actor process and eliciting informed participation of all actors is quintessential for operationalising SFM. Sensitisation of all stakeholders and development of mechanisms for multi-actor dialogues is imperative. This would also facilitate reconciliation of the often-conflicting interests of the different actors and achieve synergy in action.

REFERENCES

Brundtland, G. (ed.) (1987). "Our common future: The World Commission on Environment and Development", Oxford, Oxford University Press.

Castañeda, F. (2000). Criteria and indicators for sustainable forest management: international processes, current status and the way ahead. *Unasylva* **51**(203).

Food and Agriculture Organisation of the United Nations (2003). State of the World's Forest, 2003, FAO, Rome.

Forest Survey of India (2001). State of Forest Report 2001. pp.130, Ministry of Environment and Forests, Government of India.

GoI (1988). National Forest Policy 1988. Ministry of Environment and Forests; Government of India, New Delhi, India.

GoI (1990). Government resolution on Joint Forest Management. Ministry of Environment and Forests, Government of India, New Delhi, India.

GoI (1999). National Forestry Action Plan (NFAP). Ministry of Environment and Forests, Government of India, Vol I and II, New Delhi.

IIFM (1999). Proceedings of the National Technical Workshop on Evolving Criteria and Indicators for Sustainable Forest Management in India; 21-23 January, 1999. Indian Institute of Forest Management, Bhopal, India.

IIFM (2000). Bhopal-India Process for Sustainable Management of Indian Forests. Sustainable Forest Management, Bhopal, India.

IIFM (2000). Proceedings of National Workshop on National Level Criteria and Indicators for Sustainable Forest Management in India. 4-5 July 2000. Indian Institute of Forest Management. Bhopal, India.

ITTO (1992). Criteria and Indicators for Sustainable Management of Natural Forests, ITTO Policy Development Series No. 7, pp 22.

Jiang YouXu and Jiang, Y.X. (1997). International progress in the study of criteria and indicators for Sustainable Forest Management. *World Forestry Research* **10(2)**:9-14.

Johnson, Steven E. (2001). ITTO's Criteria and Indicators, A Tool for monitoring, assessing and reporting on SFM. International expert meeting on Monitoring, Assessment and Reporting on the Progress towards Sustainable Forest Management. 5-8 November 2001. Yokohama, Japan.

Kotwal, P.C. and Chandurkar, D. (2002). Developing local level Criteria and Indicators. C&I India update, Volume 1. Number 1. pp.14.

Prabhu, R. (2001). Need for criteria and indicators for SFM as a means of monitoring the compliance of the national forest policy in India.

Prabhu, R.; Carol, J. Pierce Colfer, Richard, G. Dudley (1999). Guidelines for developing, testing and selecting criteria and indicators for sustainable forest management: A C & I developer's reference. *CIFOR Publication.* Center for International Forestry Research, Bogor, Indonesia.

Prasad, Ram and Kotwal, P.C. (2001). Progress of implementation of Sustainable Management of Dry forests in Asia with special emphasis on Indian initiative. International expert meeting on Monitoring, Assessment and Reporting on the Progress towards Sustainable Forest Management. 5-8 November 2001. Yokohama, Japan.

Prasad, Ram; Kotwal, P.C.; Chandurkar, Dharmendra; Kharkongar, James T. and Jadhav, Yogesh (1999). Participatory development of C&I for SFM at FMU level for tropical dry deciduous forests of central India. Indian Institute of Forest Management, Bhopal (India).

UNCED (1992). The Earthy Summit, Rio de Janeiro.

Wijewardana, D. (1998). Criteria and indicators for sustainable forest management. ITTO-Tropical Forest Update **8(3)**:4-6.

Wijewardana, Don (2001). Montreal Process Criteria & Indicators for Sustainable Forest Management. International expert meeting on Monitoring, Assessment and Reporting on the Progress towards Sustainable Forest Management. 5-8 November 2001. Yokohama, Japan.

Advancing Frontiers of Ecological Researches in India, 51-76, 2006
Edited by A.K. Kandya & Asha Gupta
Published by Bishen Singh Mahendra Pal Singh, Dehra Dun, India

Lessons in sustainable management of Biodiversity in JFM areas for the livelihood generations of tribals : A case from dry tropical forest of Madhya Pradesh, India

Prodyut Bhattacharya
Indian Institute of Forest Management
Nehru Nagar, P.O. Box 357, Bhopal-462003,
Madhya Pradesh, India
e-mail : prodyut@iifm.org

Abstract: India is the seventh largest country in the world with over 1 billion population, 300 million of which are poor. An estimated 140 million people, most of them tribal depend on forest resources for livelihood and income generation. The majority of forest dwellers are the aboriginal population of tribals, distributed all over the country. Of the 68 million tribal population, 50% directly depend on NTFPs for their livelihood requirements (like food, fodder, medicinal plants, small timber) and income generation from sale of these NTFPs. These tribals are among the most marginal section of population, mostly in state of extreme deprivation with low development index. In this backdrop, the role of forest and NTFPs in developing overall development strategy especially for the tribals assumes great importance.

The paper is an outcome of the 4 years of action research carried out in north-western part of Madhya Pradesh (M.P) where local primitive tribal group Sahariya are a dominant population. A few decades back, Sheopur Forest Division used to be one of the highest revenue-earning Divisions of Madhya Pradesh for producing large quantities of economically valuable NTFPs. This area has tropical dry deciduous forests with 5 months' leafless period. The area presently suffers from extreme scarcity of ground water, erratic and scanty rainfall and limited opportunities for practising agriculture. In such a scenario, Non Timber Forest Products (NTFPs) form the lifeline of around 65,000 rural community living in and around the forests of Sheopur. Sahariya - a primitive tribal group in the region, derives 80% of its livelihood from NTFPs. Gujjars is the second community in the area that has migrated from Rajasthan and has settled here permanently. They are primarily graziers with a livestock-based economy. Both these community groups have frequent NTFP resource-use related conflicts.

The forests of Sheopur are facing rapid degradation. Given the high levels of tribal dependence on NTFPs and low literacy levels, unsustainable and destructive harvest of NTFPs is rampant. Heavy grazing pressure coupled with extreme climatic conditions has reduced natural regeneration of NTFP-yielding species. Though Joint Forest Management Committees have been formed in many villages, these are not functioning effectively owing to pressing economic constraints being faced by the tribals. Distress sale of NTFPs is common due to lack of bargaining power with the NTFP collectors. The abject poverty of Sahariyas residing coupled with the harsh climatic conditions prevailing in the area have, in the recent years, drastically brought down the production levels of NTFPs like Kullu gum (*Sterculia urens*), Aonla (*Emblica officinalis*), Salai gum (*Boswellia serrata*), Bel (*Aegle marmelos*) and Satavar (*Asparagus racemosus*) from the forests of Sheopur and this has directly and adversely impacted local income generation.

In this context, the key areas of concern for bioidiversity conservation in general and sustainable NTFP management in particular, for addressing livelihood concerns in the area are: NTFP resource assessment for calculation of harvest limits; enhancing capacities and introducing incentives for scientific harvest of NTFPs; provision of Microcredit to prevent distress sale of NTFPs; exploring options for processing and value addition of NTFPs to increase the margin of income for the poor tribal NTFP collectors; and market-support services for NTFP-based enterprises.

The paper highlights the learning and experiences from the integrated participatory approach adopted by the Indian Institute of Forest Management, Bhopal for implementation of a 4-year action research project in 8 villages of Sheopur in collaboration with the local Forest Department.

BACKGROUND

Though India has shown remarkable progress in the key development sectors, a large number of rural people still depend on natural resources for livelihood and income generation. Of these an estimated 100 million are forest dwellers who are primarily dependent on forest not only for meeting their multiple household needs but also for their food security. A special category of these forest dwellers is the aboriginal population of tribals, distributed all over the country. Of the 68 million tribal population, 50% directly depend on forest for their livelihood requirements like food, fodder, medicinal plants, small timber and a variety of NTFPs (Non-Timber Forest Products) which they collect from wild. These tribals are among the most marginal section of population, mostly in state of extreme deprivation with very low health, nutrition, and education development index.

In this backdrop, the role of forest in developing overall development strategy especially for the tribals assumes great importance. Studies conducted in many parts of India have indicated the varying but significant dependence of tribals on NTFPs for meeting their livelihood requirements

(Bhattacharya *et al.* 2003, Prasad 2000, Capistrano 2000, Malhotra *et al.* 1992, Rao and Singh 1996). At the national level, it is estimated that the value of NTFPs is nearly US$ 1 billion annually (Poffenberger, 1990). Apart from providing revenue to the State, NTFPs contributes significantly in tribal and rural economy, as about 60% of the product are consumed locally (Rao and Singh 1996) and rest is sold. Most of the NTFPs collected are seasonal in nature and hence their supply is not consistent from forest over the year, but collection of fuelwood and medicinal items from forest is fairly consistent throughout the year. Other estimates suggest that up to 35% of income of tribal household in India come from the collection and sale of unprocessed NTFPs (Giri, Bhattacharya & Santra 2001).

In recent years there has been a great thrust on community oriented forest management in which participation of forest dependent population was envisaged in management and protection of forest. This policy of JFM (Joint Forest Management) as its called has also been reflected in the New State Forest Policy formulated in 1988, which apart from emphasis on community participation, stresses on forest role in meeting local needs of the people in form of fuelwood, fodder, and other NTFPs.

Madhya Pradesh (M.P.) is a centrally located state in India and is the second largest State in terms of area, next only to Rajasthan. M.P. has the largest forest cover among all the States of India with 95221 Sq. Km of forest or 31% of its geographical area. This is around 12.44% of the total forest area of India (MP Forest Department, 2003). The rich diversity of flora and fauna has often placed M.P. as the Guardian State in terms of the biodiversity wealth of India. The entire State is ecologically sensitive and offers unique challenges in conservation and sustenance of natural resources. Of India total forest cover, nearly 28.91% of its forest cover under tropical dry deciduous forest type. These forests are ecologically fragile and in most cases are under extreme pressure mainly due to intense population pressure and high incidence of forest fire. Most of the north-western region of the state (See Map) falls under this forest type.

A few decades back, Sheopur Forest Division used to be one of the highest revenue-earning Divisions of Madhya Pradesh for producing large quantities of economically valuable NTFPs. This area has tropical dry deciduous forests with 5 months' leafless period. The area presently suffers from extreme scarcity of ground water, erratic and scanty rainfall and limited opportunities for practicing agriculture. In such a scenario, Non Timber Forest Products (NTFPs) form the lifeline of around 65,000 rural community living in and around the forests of Sheopur. Sahariya-a primitive tribal group in the region, derives 80% of its livelihood from NTFPs. Gujjars is the second community in the area that has migrated from Rajasthan and has settled here permanently. They are primarily graziers with a livestock-based economy. Both these community groups have frequent NTFP resource-use related conflicts.

The forests of Sheopur are facing rapid degradation. Given the high levels of tribal dependence on NTFPs and low literacy levels, unsustainable and destructive harvest of NTFPs is rampant. Heavy grazing pressure coupled with extreme climatic conditions has reduced natural regeneration of NTFP-yielding species. Though Joint Forest Management Committees have been formed in number of FPCs and VFCs, these are not functioning effectively owing to pressing economic constraints being faced by the tribals. Distress sale of NTFPs is common due to lack of bargaining power with the NTFP collectors. All these factors have in the recent years, drastically brought down the production levels of NTFPs like Kullu gum (*Sterculia urens*), Aonla (*Emblica officinalis*), Salai gum (*Boswellia serrata*), Bel (*Aegle marmelos*) and Satavar (*Asparagus racemosus*) from the forests of Sheopur and this has directly and adversely impacted local income generation.

In this context, the key areas of concern for sustainable management of NTFPs include: NTFP resource assessment for calculation of harvest limits; enhancing capacities and introducing incentives for scientific harvest of NTFPs; provision of Microcredit to prevent distress sale of NTFPs; exploring options for processing and value addition of NTFPs to increase the margin of income for the poor tribal NTFP collectors; and market-support services for NTFP-based enterprises. The paper highlights the learning and experiences from the integrated participatory approach adopted by the Indian Institute of Forest Management, Bhopal for implementation of a 4-year action research project in 8 villages of Sheopur in collaboration with the local Forest Department.

The Study Area

Last three consecutive years of drought forced the household heads of Sironi, Khori, Ameth and Goras villages of Karahal Forest Range in Sheopur to migrate. Only women, children and old-aged people were left behind. As the rainfall was less than 800 mm, the forest condition deteriorated severely; *aonla* and *bel* fruits could not reach maturity and fell in the early, young stage. Whatever fruit survived, it was harvested destructively and prematurely by the drought-affected tribals of the area. Recently, the cattle population has increased three-fold, most of them being the floating population from the neighbouring State of Rajasthan. It is hard to find even a blade of grass in the area's forests. Hundreds of cattle that died due to acute water and fodder scarcity lay scattered throughout the forest areas. In most cases only one tubewell meets the water requirement of about 600-700 inhabitants. Women who are always the most affected due to natural calamities and resource degradation, have been forced to travel upto two kilometers everyday to fetch water from *Kuno* - a seasonal river passing through the forest, that is fast receding as summer progresses.

The situation was not the same 5-6 years ago. The forests of Sheopur used to be the largest producer of NTFPs like *chir (Boswellia serrata)* gum, Aonla (*Embilica officinalis*), Satavar (*Asparagus racemosus*), Bael (*Aegal marmaelos*), Baheda (*Terminalia bellerica*), Mahua (*Madhuca indica*) and Tendu leaf (*Diospyros melanoxylon*). The NTFP traders did a brisk business all over the Division. Though agriculture in the area is totally rainfed, 80% of the local inhabitants - the Sahariyas and Gujjar communities of Sheopur, that are totally dependent on forest products available in the area's forests were able to meet their requirements solely from the forests. Tribals used to enjoy the privilege and rights of collection and sale of most of the NTFPs, except for a few Nationalised ones. In the recent past, *chir* and other gums were also denationalised through a policy change, allowing the tribals to collect and sell these valuable items. Though several Joint Forest Management (JFM) Committees have been formed in the area in the late 1990s, they are not functioning properly due to high levels of forest dependence and low levels of awareness and interest for forest protection among the community.

The forests that are found in this division are mostly of Northern Tropical Dry Deciduous Type. According to the revised classification of forest types of India by Champion and Seth, the forests of the locality can be grouped as follows: -

5A/C1a: Southern Tropical Very Dry Teak Forests

5B/C2: Northern Tropical Dry Mixed Deciduous Forests.

6B: Northern Tropical Thorn Forests.

Madhya Pradesh has 3 kinds of committees :-

Village Forest Committees (VFC) - These committees are formed in the village located in and around 5 km. of forest area which are degraded (forest cover less than 40 % density) with the sole objective of rehabilitating the resource base to a productive level, for the benefit to committee members.

Forest Protection Committees (FPC) - These committees have been formed in the villages falling within 5 kms radius of dense forest area (Forest canopy cover more than 40%) with the objective that these areas where regeneration is adequate and forest condition is good, should be maintained and protected from exploitative use and people living around it should be benefited by it.

Eco-development Committees (EDC) - These committees have been formed in the villages falling within 5 km radius of protected areas (National Parks and Wild Life Sanctuaries). The main objective is to involve people to help in conserving the Bio-diversity rich areas.

S. No.	Type of JFMC	Number	Alloted Forest Area (Sq. Kms)
1	Village Forest Committees	7928	27672
2	Forest Protection Committees	4055	19432
3	Eco-development Committees	675	4896
	Total	12658	52000

Recognizing the need of the local people as the first charge on the forest produce from adjoining area, the JFMC members are entitled for Royalty free Nistar. 'Nistar' is an arrangement made by the Govt. by which the villagers get small timber for their domestic need from the 'Nistar depots' after paying the charges which include 50% cost of the material and its harvesting & transport cost. But in 'Royalty free Nistar' the committee members are entitled to get timber after paying harvesting and transport cost only, which is quite nominal. The JFMC member also get the intermediate yield of materials obtained from thinning and cleaning after paying the harvesting cost. If the final felling coupes of timber is in FPC area then members are entitled to 10% of net profit obtained from the forest produce and if it is in VFC area, then members are entitled to 100% net profit obtained from the forest produce. Similarly in case of felling of Bamboo coupe in JFMC area, the JFMC is entitled to get 20% of the net profit of the forest produce. The amount thus received is further subdivided in three utilization groups.

- 50% goes to JFMC members in cash,
- 30% goes for the Village Resource Development and
- 20% goes for the Forest Development.

In the current year, US $ 1 million has been budgeted for usufruct, share of the JFMC.

Socio Economic Profile

Tribals form the majority population in the Sheopur forest division along with mix of other communities. Due to acute shortage of water in most part of this area, only dry land farming is practiced. Agriculture largely is subsistence and does not contribute significantly to the local economy. For the tribals, goods and services derived from the forest is the mainstay of livelihood (Figure 1). The mixed tropical forest type of the area yields a very rich variety of NTFPs, both in quantum and variety. These include collection of fruits, leaves, bark, stems, gum of trees/plants. The NTFPs are collected, processed and sold at the village level by the tribals. The contribution of NTFPs to the local economy is as high as 80% of the total income of the village (Bhattacharya *et al.,* 1999).

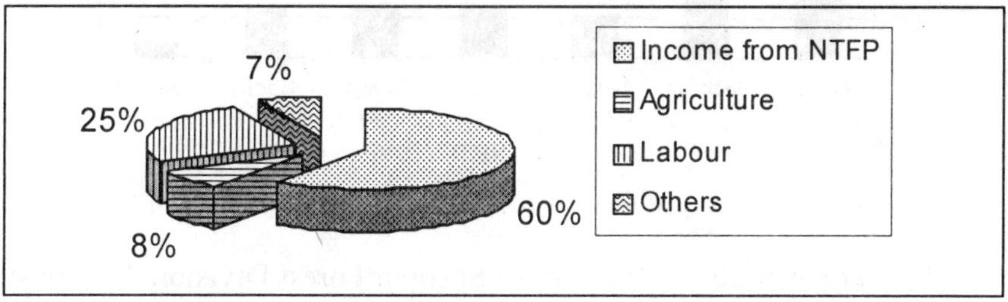

Fig. 1. Main Sources of household income for the tribals in the sampled villages of Sheopur

Implementation of the Project

Looking to the needs and potentiality on NTFP aspect, a project was formulated and being implemented since 1998, keeping in view the tremendous importance of NTFPs in the local forest ecology and in the life and economy of tribals of the area. The area also has myriad of interrelated problems related to forest conservation, dilapidated state of past development initiatives, poor status of development and its impact on forest resources, social dissension and its on forest degradation, and finally breakdown of traditional local level institutions which governed use pattern of forest resources. The tribals inhabiting are "*Saharias*" one of the primitive tribals of India. The Saharias are vagrant tribe. They go under various names such as Saharia, Sour, Sor, Seheriva etc. Another important element in the local demographic composition is the *Gujjars,* the cattle rearing community, which migrated to the District 4-5 decades back from Rajasthan.

The growing stock is dependent upon nature, the depth of the soil and moisture. There is a problem of regeneration and the ground flora is almost absent in most of the patches. However thc mixed forest type is home to large number of NTFP bearing species. An estimated 80% of the tree/ plant species in the forest bear NTFPs of commercial importance to the local people. An estimated 90 different varieties of NTFPs are traded in the local market. There are a few most important ones with their seasonal collection pattern, parts collected and estimated quantity collected are included here (Table-1, Figure-2).

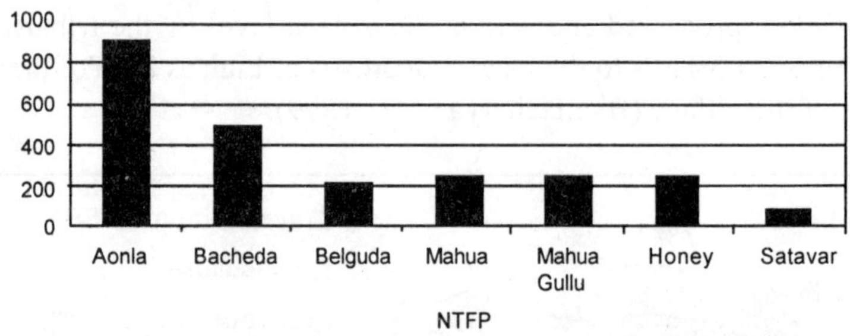

Fig 2. Estimated Quantity of Important NTFPs Produced in the Sheopur Division

There are over 300 villages in the Sheopur Forest Division. In almost all these villages, some or the other variety of NTFPs are collected as the local economy is centered round it. However study done in the area has suggested 60 villages, which are NTFP hot spots. In such villages, NTFPs are critical components for survival of local population. In most of such areas, are also vulnerable from point of view of ecological stability and long-term survival of threatened species.

An estimated 80% of the tree/plant species in the forest bear NTFPs of commercial importance to the local people. An estimated 90 different varieties of NTFPs are traded in the local market. Listed below are the few most important ones with their seasonal collection pattern, parts collected and estimated quantity collected.

Table-1: Collection/harvesting Practice & Ecological impact of NTFPs

S.N	NTFP	Destructive Collection/ harvesting practice	Resulting Ecological Impact
1.	*Emblica officinalis*	Large scale incidence of lopping and cutting of branches and trees for effective maximum collection in shortest time	Very poor regeneration status of Aonla in forest, ripe fruits are almost nil for regeneration
		Collection of premature fruits, not capable of further propagation. Collection takes place as early as September instead of normal fruiting time of November	Due to complete removal of fruits from forest many animals which eat the fruit are deprived Declining trend in Aonla production recorded
2.	Sterculia urens	Deep blazes made for gum collection on the tree trunk often leads to permanent and irreversible damage to the trees	Ecological survey revealed nearly 60% of matured sterculia tree damaged to varying degree of which 20% damaged beyond recovery Decline in gum production due to death of trees adversely affecting livelihood of thousands of gum collectors
3.	Boswellia serrata	Uncontrolled cut for oleo-resin collection leads to damage to tree trunk and may result into less production of oleo-resin, projected tumor like features appearing on bark	Nearly 20% of the *Boswellia* tree is damaged in the forest
4.	Asparagus racemosus	Whole plant is uprooted by collectors Premature harvesting	Adverse impact on regeneration
5.	Honey	Whole bee hive is plucked leading to destruction of habitat and reduction of their availability	Danger of physical injury to collector Collected honey contains impurities Decline in number of bee leads to decline in pollination and hence damage to the forest ecosystem

Source: Vegetation Survey in Sheopur Forest Division, 2002

Methodology

The entire NTFP study is part of an action research project, which is attempting to address the core issues through unique approach involving local community, all primary and secondary stake-holders and all relevant governmental and non - governmental agencies. The primary data is derived from regular visits and meeting with the cross-section of local communities, conducting village PRA exercises, questionnaire and survey, village workshops, personal observation and monitoring exercises for project progress. Ecological data was collected through monitoring of vegetation in permanent sample plots in forest (20m X 20m quadrats), rapid assessment exercises etc. Secondary data was generated through literature review, visit to governmental agencies like local and state level Forest Department offices, Revenue Department offices, Sahariya Vikas Pradhikaran office, local banks etc. Collaboration was also achieved with Central Institute of Agriculture Engineering (CIAE), Bhopal, Indian Institute of Technology, Kharagpur, IGFRI, Jhansi, M.P. Vigyan Sabha and Bharat Gyan Vigyan Samiti, Bhopal, Kovel Foundation, Visakhapattanam and Department of Science and Technology (DST), GoI, New Delhi for extension, S & T based value adition and awareness campaign exercises.

RESULTS AND DISCUSSION

One of the pressing concerns in the dry tropical forests of M.P. is the rampant destruction of NTFP bearing tree/plant species. Apart from illegal and lopping and cutting of the trees, there is widespread damage to the trees/plants due to unsustainable harvesting regime.

Destruction of Habitat

The forest of the region are under intense biotic pressure mainly arising of cattle grazing, illegal lopping and felling and unsustainable collection of NTFPs. Nearly 95% of the nearly 0.4 million cattle of Sheopur graze in open forest (Working Plan of Sheopur Forest Division, 2000), eroding the top soil cover which leads to problem of regeneration and almost complete absence of ground flora in most of the forest patches. Surveys conducted in sampled villages adjoining forest of Karahal Range revealed per household cattle population as high as an average 200 cattle heads. In absence of rotation grazing and stall feeding practices, the entire fodder requirement is met from the forest. For meeting burgeoning fodder requirement, local community often carries out large-scale cutting and lopping of trees. Mere ocular observation in most forest areas show intense lopping carried out in the forest. Encroachment of forest land is also common and at present nearly 0.2% of total forest area of the Division is under encroachment by local people. Encroached forest land is often diverted to agriculture activities.

Ecological Profile of NTFPs

The dry deciduous forest of Sheopur is rich in floral and faunal diversity. The mixed forest patches are particularly rich in variety of trees and plant species bearing large variety of NTFPs including the medicinal and aromatic plants. A quick survey of vegetation will reveal that over three-fourth of forest species yield or have potential to yield commercially valuable NTFPs. The overall canopy cover varies widely between poor 20% to good and dense class of about 70%. Being a large District, the variations in forest vegetation within the District is also significant. Few important NTFPs are occur in specific patches of forest and many species are endemic. The table below presents a brief over view of status of NTFP species in sampled plots in spread over in Sheopur Forest Division (Fig 3).

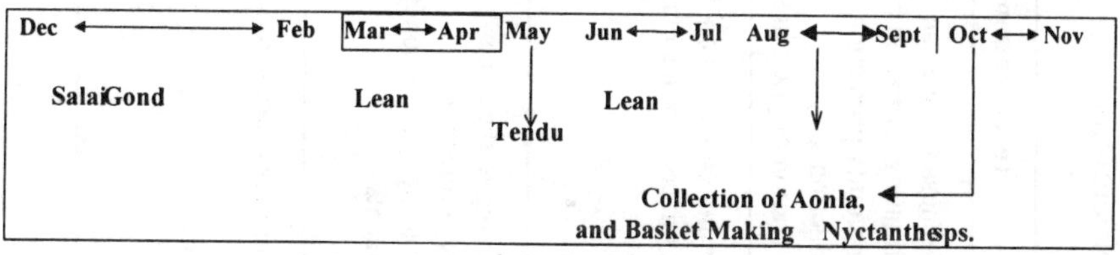

Fig. 3. Availability of NTFPs annually available in Sheopur

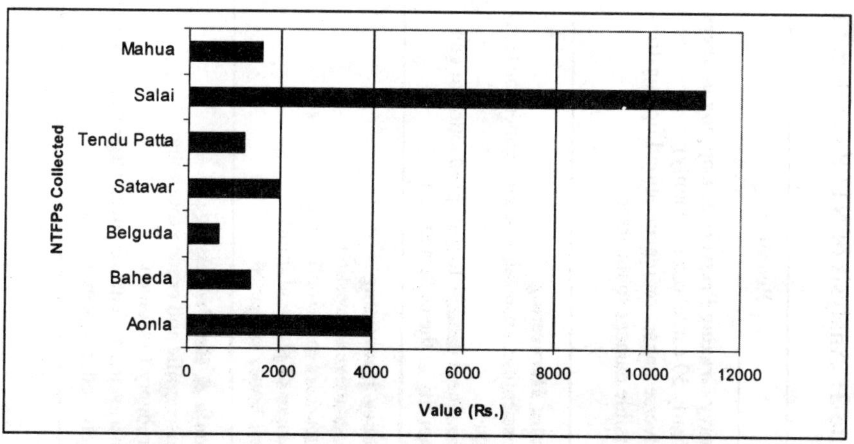

Fig. 4. Estimated Av. Household Income from NTFPs/Year (Rs.)

Table-3: Analysis of NTFP Problems and Interventions

Problem	Intervention	Results
Heavy Grazing Pressure & Lopping of Trees (Nearly 50 cattleholding /family) Adverse impact on regeneration Lopping-Irreversible damage to the trees	Awareness Generation Facilitating formation of VFCs Community protection of large area for grass production. Cultivation of grass in the forest floor	Community vigil on illegal lopping Protection leading to increased grass production used for cattle feed and income through selling
Kullu Harvesting Unscientific, unsustainable gum collection technique Permanent damage to tree, production of gum Impurity in the collected gum	Training collectors on sustainable and scientific um collection technique	Less incidence of damage to the Kullu tree. Better grade of Gum collection
Honey Harvesting Destructive harvesting Impurities in the collection Danger of physical injury Less honey collected	Training to the honey collectors on scientific honey collection Providing special honey collection tools and garments	Increased quantity and better quality of collection Enhanced safety to the collector Halt on destructive harvesting
Aonla & Bel Processing Unscientific and crude processing Drudgery to women Energy inefficient, requires large quantity of fuelwood, pressure on forest	Community based boiler installed on project site Promoting community management of asset	Less drudgery for women for fuelwood collection Increased intra community collaboration Less consumption of fuelwood Better quality of the processed material

Problem	Intervention	Results
Marketing Exploitation of collectors through system of credit disbursement by local traders Largely unorganized selling, high incidence of distress selling Lack of collector's access to market and its information	Formation of NTFP Samiti in 2 villages Organized selling through Samitis Disbursed micro-credit to Samitis toward of lean season	Measurable increase in the income of the collectors from NTFP sale Less incidence of distress selling no credit accepted from the traders, better rate for sale
Destruction of Forest & NTFP bearing trees Organized destruction by outsiders Illegal bark collection of NTFPs, tree cutting and lopping	Organizing the local people for community vigil and protection of forest Influencing policy makers for appropriate legislation for stopping premature harvesting and destruction of NTFPs Rewarding and motivating community leaders for efforts	Better protection of forest, greater awareness Apprehending offenders by local community Legislation by state govt. on Transit Pass regulation to halt premature harvesting of Aonla Facilitating community Leader

Table-4: Major NTFPs available in Sheopur, parts used, and their collection season

Scientific Name	Part (s) used	Collection Season
Emblica officinalis	Fruit and seed	October-December
Asparagus racemosus	Root/tuber	September-October
Aegle marmelos	Fruit pulp	October-November
Teminallia bellerica	Fruit	November-January
Madhuca indica	Flower & Fruits	Flower (May), Fruits (July)
Nyctanthes arbotristis	Stem	November-April
Steculia urens	Gum	December-February
Anogessius latifolia	Gum	December to February
Boswellia serrata	Oleo-resin & Seed	December to May
Dyosporus melanoxylon	leaves	May

Source: Village Survey and PRA Exercises, 2000

Gender and NTFPs

To a limited extent, women also participate in forest protection activities. However, overall the daily activities carried out by the female members of the household are more in number and also more strenuous as compared to those of the male members of the household. Table-6 provides the gender dimension of activities being carried out by the male and female members of the households.

The resource related activity profile of the village community (Table-7) as regards to the work done in nursery, plantation, forestry, agriculture, homestead, off farm labour and labour work in governmental schemes shows that, out of all the activities performed 9 activities are exclusively done by the female members where as they also help in other 22 activities. But on the other hand the male members perform only 6 activities exclusively.

Unsustainable Harvesting

Another pressing concerns in these dry tropical forests is the rampant destruction of NTFP bearing tree. Apart from illegal lopping and cutting of the trees, there is widespread damage to the trees/plants due to unsustainable harvesting regime. Similar problem was noted in other parts

Table-5: Result of Vegetation survey in Sample plot depicting status of NTFPs in Sheopur Forest Division

Vegetation Survey, Sheopur- DST Project Area Tree Species (> 1 m.) - Sample plot size = 20m X 20m.- Square plot				
S.N.	Local name	Botanical name	Number in Sampled area	Density / ha.
1	Kuretha	*Holorrhena antidysentrica*	70	291.9
2	Salai	*Boswelia serrata*	5	20.85
3	Amaltas	*Cassia fistula*	1	4.17
4	Kaker	*Flacourtia ramontchi*	17	70.89
5	Palash	*Butea monosperma*	10	41.7
6	Gurjan	*Lennea coromandelica*	20	83.4
7	Dhawada	*Anogeissus latifolia*	17	70.89
8	Tendu	*Diospyros melanoxylon*	34	141.78
9	Bija	*Pterocarpus marsupium*	2	8.34
10	Khair	*Acacia catechu*	24	100.08
11	Bel	*Aegle marmelos*	9	37.53
12	Jamrasi	*Elaeodendron glaucum*	1	4.17
13	Ghator/ Ghot	*Zyphus xylopyra*	5	20.85
14	Karjoria	*Bridelia retusa*	1	4.17
15	Doodhi / Khini	*Wrinhtia tinctoria*	1	4.17
16	Kanjini	*Mimusops hexandra*	1	4.17
17	Binash	*Limonia crenculata*	3	12.51
18	Mokha	*Schrebera swietenioidas*	1	4.17
19	Amala	*Emblica officinalis*	2	8.34
20	Guter / Ber	*Zizyphus mauratiana*	2	8.34
21	Saheta / Sehra	*Bauhinia retusa*	1	4.17
22	Sahatu	*Morus laevigates*	2	8.34

Source: Vegetation Survey, 2001

of M.P. In a study carried out in to assess the health of *Emblica officinalis* L-"Aonla" tree due to destructive harvesting of its fruit in 20 forest compartments in Bilaspur District of M.P. (Tiwari 1995). In the study it was found that 60% of trees were damaged due to pollarding, lopping. In addition, 9.55 trees out of 65 trees per hectare were felled to effect maximum collection of Aonla fruits. In another study (Prasad, *et al.* 2000) on Aonla, *Buchnania lanzan* L-'Chironji' and *Chlorophytum tuberosum* L-'Safėd Musli', conducted in Central part of M.P., revealed that due to destructive and premature harvesting the regeneration of these species has been adversely impacted and productivity is declining. Murali and Hegde (1996) and Prasad *et al.*, (1999) studied the harvest and regeneration status of Aonla and found that the number of seedlings was less or nil under the trees, which are over harvested for, fruits.

Table-6: Daily Household Chores & other Activities Performed by the Women and Men of Sheopur

S. No.	Time Interval	Gender Activity Women	Men
1	4:00 to 4:30 am	Wake up	Sleep
2	4:30 to 5:30 am	Household & Utensils Cleaning Firing of Chullha Tea Preparation	Wake up Have tea
3	5:30 am to 6:30 am	Cattle tending Fetching water	Child keeping
4	6:30 am to 8:30 am	Cooking and eating	Taking bath and meal
5	8:30 am to 6:30 pm	Collect fuelwood & NTFP from forest Go to market to sell NTFPs/fuelwood	NTFP collection or occasional labor work
6	6:30 am to 8:30 pm	Tending cattle. Milking Washing, taking bath, cooking Caring for the children and aged, Dinner	Milking (some times) Taking bath, gossiping Dinner
7	8:30 pm onwards	Going to bed	Gossiping and listening to radio Sleeping at late night

There are two broad categories of unsustainable harvesting/collection practices of NTFPs. In first case the tree species, as in case of Aonla, is damaged due to large-scale lopping and felling of trees and collection of immature fruits, seeds of which are not viable for further propagation. Such practices have resulted into very poor regeneration status of species in the natural forest and pose threat to their long-term survival (Bhattacharya, *et al.*, 2000). There are hosts of factors both situational and external responsible for this present state. The spurt in the demand for NTFPs and market forces generated by this has contributed to this to a large extent. The local traders of NTFPs in the small towns have traditionally instigated the tribal collectors of NTFPs into destruction of NTFPs trees in the unhealthy competition for making the most of the resource in short-term. At preset, most the NTFPs like fruits of certain trees are harvested completely from the natural forest, leaving nothing for further regeneration

Table 7: Resource-related Activity Profile of Gender in Khori and Shironi Villages

Location	Activities	Gender	Frequency
Nursery Work	Filling poly bags Transplanting seedlings Weeding Clearing, leveling and digging Watering	F F F F F	Occasionally
Plantation in Forest	Digging and pit filling Transplanting seedling Jungle clearing Trench digging Stone wall construction Stump planting	M/F F M M/F M/F M/F	Occasionally
Forestry Work	Collection of NTFPs Fuel wood Shepherding Tree felling Marketing of forest produce Fire protection Patrolling and *chowkidary*	M/F F M M F/M M M	Seasonal
Agricultural work	Ploughing and manuring Sowing Weeding and cleaning Harvesting and harvesting process	M M/F F/M F/M	In rainy season and as and when required
Homestead	Cooking Cleaning & washing Fetching water Tending cattle and goats Milking Caring for the children and aged	F F F F/M F/M F	Daily
Off-farm labor activities	House repairing Stone quarrying Stone digging Stone crushing Stone carrying	F/M F/M F/M F/M	Occasionally
Labor work in government schemes	Road construction Digging pond Construction of check dam Stone wall boundary Building community hall, podium etc.	F/M F/M F/M F/M F/M	Occasionally

Legends

M – Exclusively Male
F - Exclusively Female
M/F – Both Male and Female
M/f – Predominantly Male
F/m – Predominantly Female

or for birds and other animals of the forest. Lack of awareness and training to the collectors is also a major cause. In the second category of unsustainable practices, the collection/harvesting technique adopted by the collectors is putting the species in threat. One major concern is the technique adopted for the collection of gums and resins from trees like, *Boswellia serrata* L- 'Salai', *Sterculia urens* L- 'Kullu' and *Anogeissus pendula* L- 'Dhawda' that are found in large patch of forest in dry deciduous forest of the region. Due to faulty tapping technique employed, there are large numbers of deaths of such trees in the natural forest due to permanent damage to tree trunks. Since regeneration of such species in forest takes long time in natural forest, there is alarming reduction in the tree cover of these group of species (Bhattacharya *et al*, 2002). In case of another NTFP *Asparagus racemosus*, L- Satavar, the whole plant is uprooted to obtain its roots. In the local forest areas of Sheopur, rock bee (*Apis dorsata*) honey is available in abundance. The tribal honey hunters collect them on large-scale but in the process often exposing themselves to grave risk of injury. The method adopted is mostly destructive in which the whole hive is plucked before honey is sieved from it mechanically.

A summary of the destructive collection/harvesting practices of NTFPs and its resulting adverse ecological impact in Table-2.

S&T Intervention for Processing and Value Addition to NTFPs

Value addition and processing of NTFP can play an important role in ensuring remunerative return to the collectors. Even Grading has an overriding influence on product price (Prasad *et al*, 1999). A major thrust of current reform was on improving the current methodology adopted for village level primary processing of various NTFPs before it is actually sold in the market to the local traders. This processing is carried out by the collectors of NTFPs at the household level employing crude, unscientific but simple techniques. The basic steps are boiling, deseeding/depulping and then drying of the raw NTFPs like Aonla (*Emblica officinalis*), Bael (*Aegle marmelos*), Satavar (*Asparagus racemosus*). These NTFPs are also form the bulk of collection for the local people who are almost entirely dependent on them for livelihood and income generation.

The method requires use of household *chulhas*, use of firewood on large scale, which is drawn from the adjoining forest. The process is not energy efficient and results in wastage of fuelwood. This puts additional pressure on the forest. In time heavy demand of fuelwood. The processed product is of inferior quality. Many fruits are not boiled properly making their depulping/deseeding difficult. This also results in incomplete desiccation of fruits, which are more susceptible to rot, and with less shelf life

The drudgery to the women, who are also the main processors, is very high. Frequently children are also engaged in the enterprise.

With the present situation an intervention for improvement was quite imperative. Three main issues were to be addressed-Reducing the pressure on forest, product improvement, and mitigating the drudgery of the collectors in the processing particularly the women and children involved. The answers seem to lie in making S & T intervention through simple, affordable and tested technique. For this a community boiler based on Gasifier was installed on one village site for facilitating scientific boiling of NTFPs like Aonla and Bel fruits. The equipment is basically an energy efficient large capacity oven running on fuelwood chips meant for boiling of forest or agriculture products on community basis.

Micro-credit Support to NTFP Collectors

Distress selling of NTFPs by the tribal collectors is very common in the Sheopur Division. The local NTFP traders provide captive advance to the collectors in the lean income season during May-June. In this period of dry months, the local tribals lack employment opportunity in agriculture and forest activities and are hence forced in the debt trap of NTFP collectors. The debt is returned in form of Aonla collection and the return rate is very low compared to average market rate of the produce during the collection season. Study across sample villages where Aonla is collected reveled that almost three-forth of the household opt for advance every year.

To address the wide spread problem of distress selling, IIFM conceptualized a unique approach of introducing community based micro-credit program for the first time in the forestry area. The main objectives of the program were as follows-

a) To prevent the incident of distress selling of NTFPs

b) To fulfill the subsistence requirement of NTFPs collectors during the lean employment period (May-July) in the year

c) To institutionalize the collective collection/harvesting and marketing of NTFPs through community based

NTFPs institution created for micro-credit management

In this program the NTFP collectors were extended credit support to meet their subsistence need during lean income season. The range of credit support is Rs.200-Rs1000, depending on the pay back capacity of the household opting for credit. The collectors would require paying back in terms of Aonla fruits which is procured by the Samiti at a predetermined rate. The Samiti then arranges to sell the produce in bulk quantity to big traders. The profit in the transaction is then apportioned among the collectors based on the quantity supplied to the Samiti.

The program has proved to be a great success in all the villages. The impact assessment of the program has indicated a measurable and significant enhancement in the income of the NTFP collector from the sale of Aonla fruits. Preliminary result shows an increase of nearly 200% in the sale price of Aonla from the base year largely due to increase in market power of the collectors. The credit recovery rate at nearly 90% is also impressive. The sale of produce in bulk quantity through village level Samiti has resulted into enhanced market leverage for the collectors, resulting in greater returns. In many cases the collectors also resort to selective hoarding of fruits to sell them at opportune time for greater return.

Searching for Solutions

An integrated approach that takes in to account the ecological, financial and social impacts of various alternative actions is required for addressing the key issues, problems and challenges identified for Sheopur Forest Division above. The main reason for the unsustainable collection/harvesting pattern of many of the NTFPs found in the forests of Sheopur is the lack of awareness and capacity among the NTFP collectors to scientifically harvest NTFPs. In this category is included harvest of NTFPs like gum from *Sterculia urens*, honey harvesting and collection of bulbs and tubers. In all such cases there is a valid and strong case for sensitizing the collectors on the impact of destructive harvesting and then transferring appropriate capacity building technique for putting halt to unsustainable collection/harvesting. In case of the remaining few NTFPs, like *Aonla,* the local market forces are responsible for egging on destructive harvest as traders and money lenders often instigate and organize destruction of NTFP bearing tree/plant species in the forest. Due to the very nature and complexity of this problem, it is more difficult to control. For this one need to address larger issue of market reform, equity and ensuring remunerative return to the NTFP collectors.

An Integrated Approach towards Addressing Key Issues

The main reason for the unsustainable collection/harvesting pattern of NTFPs in Sheopur are- Lack of awareness and capacity among the NTFP collectors on sustainable NTFP collection regime and for few NTFPs the reason for destructive harvesting are the local market forces which instigate and organize destruction of NTFP bearing tree/plant species in the forest. In the first category lies case of unsustainable harvesting of most of the NTFPs, like gum extraction from *Sterculia*, destructive honey harvesting, destructive collection of bulbs and tubers from underground. In all such cases the local collectors are not aware of or sensitized of sustainable collection technique for the respective NTFPs. Therefore there is a valid and strong case for sensitizing the collectors on the impact of destructive harvesting and then transferring appropriate capacity building technique for putting halt to unsustainable collection/harvesting. The case of destructive harvesting of Aonla fruits is in contrast due to local market forces and by its very nature complicated and difficult to control. For this one need to address larger issue of market reform, equity and ensuring remunerative return to the NTFP collectors.

The Indian Institute of Forest Management (IIFM), an education and research institution with a wide range of expertise in planning and implementing NTFP development program set on to design an integrated intervention plan for sustainable NTFP management based rural development for Sheopur through an action research project. Initial thrust of the project was on detailed study the project site in terms of its resources, socio-economic profile, community dynamics micro-level assessment of problems, local institutions, its role and constraints. Thereafter opportunities for S&T intervention was sought after eliciting active participation of village community, local Forest Department, S&T technology institutions and other NGOs. To address the core issue of unsustainable harvesting of NTFPs, specific training programmes were organized in field areas for awareness generation and capacity building of NTFP collectors. In this program, training was imparted to the collectors on advanced technique of Kullu gum collection, non-destructive honey harvesting and sustainable Asparagus harvesting. In the all the training program, Frontline Forest Department staffs were actively involved as they are close to the community and are closely associated with protection of forest at grass root level. In course of study of the local condition *vis a vis* the management of NTFPs. It was felt that local traders are indispensable part of the entire marketing reform process and have also a vital role to play in ensuring sustainability of species in the wild through their local influence. Hence the problem of local NTFP traders were also addressed through dialogue and workshops.

Increasing Local Capacities for Scientific NTFP Harvest

Initial thrust of the interventions undertaken in the area was on a detailed study of the site in terms of its resources, socio-economic profile, community dynamics, micro-level assessment of problems, local institutions and their roles and constraints. Thereafter, opportunities for science and technology interventions were sought after eliciting active participation of the village community, local Forest Department, S&T technology institutions and other NGOs. To address the core issue of unsustainable harvesting of NTFPs, specific training programmes were organized in field areas for awareness generation and capacity building of NTFP collectors. In such programs, training was imparted to the collectors on advanced technique of Kullu gum collection, non-destructive honey harvesting and sustainable Asparagus harvesting. In all these programs, the frontline Forest Department staff was actively involved as they are the ones close to the community as well as they are closely associated with protection of forest at the grassroots level. It is clearly understood that if resources have to survive, both men and women must play an equal part in its management. Women of Sheopur who already perform major roles in NTFP collection and processing were, thus, also trained and targeted during awareness programme on sustainable NTFP management. Their existing knowledge of properties and potential uses of various NTFPs also came in handy during the mass mobilization campaigns. During course of earlier studies carried out in Sheopur, it was felt that local traders are indispensable part of the entire marketing reform process and have also a vital role to play in ensuring sustainability of species in the wild through their local influence. Hence the problem of local NTFP traders was also sought to be addressed through dialogue and workshops.

S&T Intervention for Processing and Value Addition to NTFPs

Value addition and processing of NTFP can play an important role in ensuring remunerative return to the collectors. Even Grading has an overriding influence on product price (Prasad *et al.* 1999). A major thrust of current reform was on improving the current methodology adopted for village level primary processing of various NTFPs before it is actually sold in the market to the local traders. IIFM's intervention in this case placed thrust on making simple and appropriate science and technology (S&T) interventions to make processing and value addition quicker and more energy efficient. A community boiler based on Gasifier was installed on one village site for facilitating scientific boiling of NTFPs like Aonla and Bel fruits. The equipment is basically an energy efficient large capacity oven running on fuelwood chips meant for boiling of forest or agriculture products on community basis.

Development of Strong Local Institutions (including marketing linkages) and Inter-Organizational Linkages

One of the organisational development interventions made by IIFM in Sheopur was formation of village level institutions (NTFP collectors' groups) in collaboration with local Forest Department. Efforts were also made to promote collective collection and marketing in village clusters. Efforts are on to craft an appropriate multi-tier platform with focus on local marketing issues, on developing close inter-linkages between collectors and local traders, and organizing local level training and awareness camps for traders, collectors and frontline foresters. IIFM is also developing a local media-based Market Information System (MIS) that would be adapted for local use mainly for the collectors, and strengthen the institutional efforts towards backward flow of market information i.e. from local market to collectors.

Figure 5 below reveals the marketing channel and price flows for important NTFPs in the Sheopur Division. The figure clearly reveals the price increment at the levels of the village market, the local market in the nearest tehsil and the wholesale market in the bigger cities. Given the kind of marginal increments that are visible, it is imperative that the market imperfections are thrashed out before making any headway into the NTFP market.

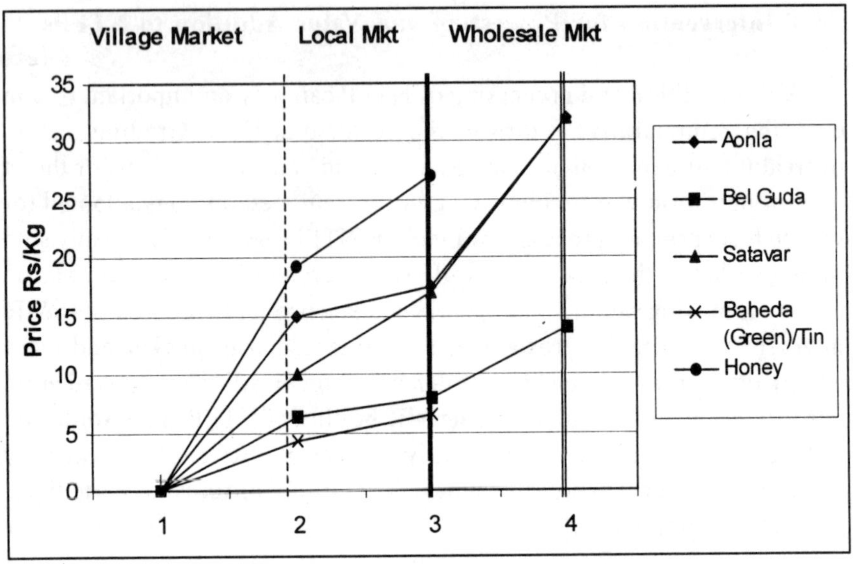

Fig. 5. Marketing Channel & Price Flow of Important NTFPs in the Sheopur Division

Inferences and Future Direction

The NTFP resource in the dry tropical forests of Sheopur is central to the existence of millions of tribals who live close to these forests. In absence of these benefits, the tribal village economy would collapse as alternate sources of income are scarce in these dry areas. The intense pressure on forest from human and cattle and off late the specific threat to many NTFPs due to commercial demands is threatening the very livelihood and existence of aboriginal tribal groups. This case study that is based on the experiences of IIFM's action research and pilot implementation efforts in Sheopur Forest Division of Madhya Pradesh brings home the following points.

A thorough understanding of Sheopur' NTFP resource status and potential, socio economic characteristics of different communities-the Sahariyas and the Gujjars and their dynamics, the present and potential roles of various institutions (JFM Committees, Forest Department, NGOs etc.) and other in-depth baseline an analytical studies have proved crucial to evolving a multi-prong strategy for addressing various problems and challenges in the area.

With the resource in question (i.e., NTFPs) being biological, IIFM benefited from continuous biological monitoring of the forest and of natural NTFP production as this would provide clues to devising an appropriate institutional and market strategy for community-based business development of NTFPs. The efforts made to strengthen community institutions like the NTFP collector groups helped in bringing back to fold, the tribals of Sheopur who were alienated from traditional practices of forest conservation due to socioeconomic compulsions.

Both-the plight and the potential of rural women in Sheopur was assessed in the context of forest protection and NTFP conservation and efforts made to rekindle their concern towards resource conservation benefited the forest-dependent households, the entire village and the area's forests as well. The women among NTFP collectors in Sheopur, who were often sidelined during various developmental programmes earlier, were surprisingly as willing to adopt scientific NTFP-harvesting techniques in the area as the men. The entire community responded to institutional developmental efforts of IIFM as they accepted the need to do so to save their precious forest resource.

Provision of tool kits, dresses and other implements required for scientific harvest of NTFPs to the trained NTFP collectors served further to achieve the objectives of IIFM's capacity building interventions, leading to higher rates of adoption of sustainable NTFP harvesting techniques. Easy access to low cost and timely microcredit to the tribals in Sheopur can make a whole world of difference in the lives of the tribal and on the state of NTFPs and forest. Such micro-credit experiments can be tested and replicated in many more tribal-dominated areas followed by formulation of a suitable credit/loan scheme to be implemented by banks and/or non-banking financial institutions.

Sustainable NTFP management is an important intervention to sustain community interest in forest protection and management. Innovative silvicultural methods need to introduce for special care to NTFPs of the area (Ram Prasad and P. Bhattacharya 2004). Presently almost all NTFPs are harvested by the people, it is essential to keep atleast 40% NTFPs in the forest itself as a thumb rule. With the present exploitative NTFP market regime, especially at the village level, mere transfer of usufruct rights to the collectors has proved to have little meaning for the tribals in Sheopur. The recent changes made by the policy makers related to Transit Pass for *aonla* and its relation to sustainable harvesting of NTFPs also had limited impact as implemented in isolation, such interventions fail to address the entire chain of issues, mainly the market reforms needed to attain a sustainable impact of desired spread and levels. Recently the Forest Department also initiated programs for management of NTFPs like *ex situ* conservation efforts and enforcement of regulations related to harvesting time and technique is yielding positive result. Several such co-ordinated efforts involving multiple agencies-Forest Department and other Government Departments, technical institutes, NGOs and academic institutions like IIFM, are required to tackle the many-faced Dracula of poverty and backwardness afflicting the otherwise resource-rich areas like Sheopur. In the absence of multi-stakeholder collaboration, optimum benefit to collectors is not possible.

For hundreds of villages in the remote and traditionally backward areas of MP, the forest resources are the only means for survival for thousands of people who directly or indirectly dependent on NTFPs for livelihood and income generation. However in past decade or so, the pressure of human and cattle has increased manifold on the forest. Commercialization of many NTFP species which were hitherto collected for household and subsistence purpose led to unsustainable harvesting regime which is threatening the long term availability of many products and will deal a deadly blow to the overall rural economy in tribal areas. To integrate the tribals alienated from traditional practices of forest conservation, it is essential to strengthen the tribal institutions which are in many instances in moribund state. Market reform is key to ensuring remunerative return to collectors and making them empowered for forest conservation. In fact with the present exploitative market regime, mere transfer of usufruct rights to the collectors has little meaning. The recent changes made by the policy makers related to Transit Pass and its relation to sustainable harvesting of NTFPs also had limited impact. This was due to the fact that it failed to address entire chain of issues, mainly the market reforms. Recently the Forest Department also initiated programs for management of NTFPs. These programs like ex-situ conservation efforts and enforcement of regulations related to harvesting time and technique is yielding positive result. In past year IIFM also initiated a DST funded program focussing on Asparagus and its domestication for income generation of tribals in ravine areas of Sheopur Forest Division. It is hoped that the program will promote conservation efforts of Asaparagus and will also bring greater prosperity to hundreds of tribal household.

ACKNOWLEDGMENT

The author acknowledges the financial support received from Sir Dorabjee Tata Trust (SDTT), Mumbai, Department of Science and Technology (DST), GoI, New Delhi and field collaboration from M.P Forest Department.

REFERENCES

Bhattacharya P.; Joshi, B. & Hayat, F. Seyed. (2003). An Improved Method of Tapping Gum from Kullu *Sterculia Urens. Forest Trees and Livelihoods, Vol 13 pp 187-196*

Bhattacharya, P.; Bhagat, N.K. & Hayat, S.F.B. (2000). Science & technology Management Intervention for Value Addition to Select NTFPs for Enhancement of Income of Local People in Madhya Pradesh, DST Sponsored IIFM, Bhopal. Project, Interim Report, 2000 (Unpublished).

Bhattacharya, P.; Bhagat, N.K.; Hayat, S.F. & Srivastava, B. (1999). Science & technology Management Intervention for Value Addition to Select NTFPs for Enhancement of Income of Local People in Madhya Pradesh, DST Sponsored IIFM, Bhopal. Project, Interim Report, 2000 (Unpublished)

Bhattacharya, P.; Joshi, B.; Bhagat. N.K.; Hayat, Faiz (2002). Sustainable Harvesting of Kullu (*Sterculia urens*) Gum. *Indian Institute of Forest Management Newsletter*, Vol **V**, No. 1&2, March & June 2002

Capistrano, D. (2000). NTFP based Enterprise in Forest Conservation and Community Development : India's evolving institutional context. XXI IUFRO World Congress 2000. Kuala Lampur.

Giri, T. K; Bhattacharya, P. & Santra, S.C. (2001). Ecology and Economy of Non-timber Forest Products in village life- A Case from Bankura District in West Bengal. Annals of Forestry, Vol **9**, No. 1. Pp 2.

M.P. Forest Department Official website (2003). http://www.mp.nic.in/forest/figures.html

Malhotra, K.C.; Chandra, S.; Vasulu, T.S.; Majumdar, L. and Kundu, S. (1992). Joint Management of Forest Lands in West Bengal: A Case Study of Jamboni Range in Midnapore District. IBRAD, Calcutta.

Murli, K.S. & Hegde, R. (1996). Sustainable Harvest of NTFPs and Forest Management. In Management of Minor Forest Products for Sustainability. *Economic Botany*, **50**:221-223

Murali K.S.; Shankar, U.; Shaanker, R.U.; Ganeshaiah K.N. and Bawa, K.S. (1996). Extraction of Non-Timber Forest Products in the forests of Biligiri Rangan Hills, India. Economic Botany **50(3):** 252-269

Prasad, Ram (2000). Strategy for Sustainable NTFP Management in India. XXI IUFRO World Congress 2000. Kuala Lampur.

Prasad, R.; Das, S. and Sinha, S. (1999). Value Addition Options for NTFPs at the Primary Collectors Level, *International Forestry review.* **1(1)**

Prasad, R.; Kotwal, P.C. & Mishra, M.(1999). Impact of Harvesting *Emblica officinalis* (Aonla) on its Natural Regeneration Health Vitality of Ecosystem in Central Indian Forests, *Journal of Sustainable Forestry* (under consideration).

Prasad, R.; Kotwal, P.C. & Mishra, M. (2000). Sustainable Harvesting Regime of some NTFPs of Madhya Pradesh. (IIFM) Communicated paper.

Prasad, Ram and Bhattacharya, P. (2004). Silvicultural and Management Issues in Joint Forest management. In Root to Canopy, Ed. Bahuguna *et al.*, Winrock International India and Commonwealth Forestry Association, India Chapter, MoEF, New Delhi, Pp. 255-266.

Rao, A. Ratna and Singh, B.P. (1996). Non wood Forest Products contribution in Tribal Economy. Indian Forester, April 1996.

Tiwari, K.P. (1995). Collection of Aonla (Emblica officinalis) fruits from forest- An Impact Assessment. *Vaniki Sandesh,* 19:(4)

Working plan Sheopur Forest Division (2000).

Advancing Frontiers of Ecological Researches in India, 79-97, 2006
Edited by A.K. Kandya & Asha Gupta
Published by Bishen Singh Mahendra Pal Singh, Dehra Dun, India

Tree Density, Regeneration and Impact of Harvesting of Some Selected NTFPs on the Natural Forests of Three Districts in Orissa State

Manish Mishra, Teki, S.* and R. P. Mishra**
Research Associate,Indian Institute of Forest Management,
P.O Box.357, Nehru Nagar, Bhopal (M.P). India. Email: manishm@iifm.org

Abstract: In a forest ecosystem the Non timber forest products (NTFPs) provide maximum inputs into the domestic economy of the rural household. They are vital to the sustenance of the local tribals and primary collectors. Extraction, processing and marketing of NTFPs is the sources of employment and income to a vast majority of the tribal people. Very often NTFPs are collected haphazardly without looking in to the quality and sustainability of the produce.

In the present paper the tree density of Achar (*Buchanania lanzan)* was found maximum (21.88 tree /ha & IVI 20.95) in Malkangiri district among all the species in the three distrcits. Similarly the density of Mahua tree was also high (16.15 no/ha and IVI 42.35) in Malkangiri while it was observed comparatively very low in other two distrcits. The density of Kullu (*Sterculia urens)* tree is very poor in the natural forests of all the sites (0.63/ha). Similarly poor density & IVI of Siali (*Bauhunia vahlii -climber)* was also observed (3.44 ha & IVI 3.27) in Koraput , followed by other two districts .

The seedlings of *Emblica officinalis* was found more in Koraput district as compared to Malkangiri and Rayagada district. The seedlings of *B. lanzan* and Harra (13.87%/ha.) were recorded highest in Malkangiri than in Rayagada & Koraput districts. In all three districts very few seedlings of Kullu (0.87%) trees were observed. In Rayagada district the maximum damage on *E officinalis* (11.13%) and *B lanzan & B vahlii* (9.50%) plant by felling during fruit/leaves collection was observed in all the damage categories i.e. mildly damage, highly damage. Mahua (97.87%) and Imli (85.75%) trees faced less tree damage (not damaged category) as compared to other selected NTFP species. The local peoples do not fell (almost nil) trees of Mahua, Imli in Malkangiri district of Orissa State.

Key Words: Unsustanable harvesting, Density, Regeneration, Felling, Saplings,Tribals.

---

* Assistant professor, National Institute of Financial Management, Sector-48, Pali Road, Faridabad-121001. Email : teki9@rediffmail.com
** Project Associate, Indian Institute of Forest Management, Nehru Nagar, Bhopal

INTRODUCTION AND REVIEW OF LITERATURE

Few scientific studies have been done on quality and production of Achar fruits as well as seeds. Therefore, it has become necessary to study the phenology, maturity of fruit and seed, fruit and seed yield in order to suggest the optimum level of fruit harvests to maintain ecological sustainability. Rai (1982) observed fruit production of Achar tree girth wise and found that the fruit yield gradually increases according to increase girth size of tree. They have found 6.5 kg fruits in 165 cm girth size tree during the year of 1979 while 7.5 kg was observed during the year of 1980 on same girth of tree. Prasad and Chadhar (1989) studied the fruit and seed yield of *B. lanzan* of different girths and reported that maximum yield is obtained from young to middle sized trees (1.4 to 2.0 kg/trees). They also found that quantity of seed obtained is proportional to that of fruits from which the seed was extracted.

Scientist working at FRI, Dehra Dun also reported fruit production of *B lanzan* (40-50 kg fresh fruits/tree) which yield 8-19 kg after drying and 1.15 kg of finished products (Kernel/seed) per tree (ICFRE, 1996). Chadhar and Sharma (1997) reported variation (6 fold in 30-40 cm girth to 1.5 fold in 91-120 cm girth class) in fruit production of trees of different age class. They reported that the fruit production decreases as increase in tree girth. Chakravesh (1999) studied the distribution of *B lanzan* trees of Chhindwara district in Madhya Pradesh (India). The tree density (Meter^2) and IVI of *B lanzan* 0.0032 & 28.17, *Terminalia chebula* 0.0036 & 33.46, *E. officinalis* 0.002 & 17.48, *T. bellerica* 0.0016 & 34 respectively.

Prasad, Kotwal and Mishra (2000) reported that the average girth and height of *Buchanania lanzan* trees does not vary in the natural forest as well as in Protected Areas (PAs). They have also found flowered trees (5/8 per Ha.) in PAs and 3/4 & 3/5 per hectare found in the natural forest of central India. The average weight of fresh fruit/tree does not found much variation (25 to 35 kg/trees) in both types of forest, irrespective of tree height and girth. They have recorded average fruit yield per tree (6.9 kg/tree). They have also found maximum immature (62/100) fruits during the harvesting period. The floating test was also done to know the maturity of seeds. The ratio of floated and sunken fruit in the PAs was (42-58/100) as compared to (65-35/100) the natural forests.

Mishra *et al*., (2001) have done detail study on the density tree/ha., production, fruit yield, fruit color, fruit quality of Achar (*B. lanzan*) accordingly to girth size of tree in Orissa state. The maximum number of trees fit between 20-50 cm and 51-80 cm girth lasses. Maximum fruit yield per tree was observed in higher GBH classes particularly in Rayagada district (i.e. 141-170 cm). On the other hand least fruit yield per tree was noted in the lower GBH classes (20-50 and 51-80 cm) in Koraput, Malkangiri and Rayagada districts. The maximum flowering trees found between 51-80 girth class while non flowering trees found in lowest girth classes i.e. 20-50 cm. They have also found more mature fruit (black color) in Koraput district than the Malkangiri and Rayagada districts.

Ram Prasad (1993) studied flower and fruit production and marketing of Mahua tree. The lowest yield of mahua flower 6.875 kg/tree and highest yield 144.930 kg/tree was observed. Average yield of 11.780 kg/tree had observed from tree of 61-90 cm girth at breast height and maximum of 90.163 kg/tree of over 180 cm GBH has also observed. The total yield of mahua flower over a period of 5 years (1987-91) from 112 trees was on an average 50.765 kg per tree.

According to Singhal (2000) about 250 to 350 kg Mahua flowers collected/family/year while only 10 kg Achar collected by /family/year. Only 4-5 kg Kullu gum collected annually of local people of Sathar village in Chiklod forest range in Raisen division of M.P. Gera and Baghel (2000) indicate the *B lanzan* production of Orissa State and also inform its important role playing in the livelihood of rural people. The total 4.17 % people involved for seed collection of *B lanzan* in Orissa State.

Patnaik (2001) studies the ecological status of *Buchanania lanzan, Terminalia chebula* and various threats in NTFPs in some forest ranges of Chindwara (M.P.) and Mayurbhanj (Orissa) districts. Study reports that tree density of Achar and Harra was found maximum in agriculture areas while Harra density was substantially lower in both the states. In all the sites of Orissa and M.P. the seedling and sapling density is significantly higher in forest compared to agro forestry and agriculture respectively. The major threat found to be affecting the NTFPs are one epiphytic infection by *Loranthus, Banda* and human impact such as lopping, removal of trees, axe marks, wounds etc.

Prasad *et al.* (2001) reported very deficient regeneration of Aonla (*Emblica officinalis)* tree in the natural forest of Satna, Panna, Sagar and Damoh districts of M.P. They reported causes of Aonla tree depletion mainly due to unsustainable and destructive ways of harvesting like lopping, pollarding, felling etc. The current practice of premature fruit harvesting, damaging trees was found to be rampant in natural forests.

So far, very few scientific studies have been made on the tree density, regeneration and tree damage during harvesting of selected NTFPs. This paper deals with the present method of fruit/leave/gum harvesting by the local peoples and its impact on the natural forests of Orissa State.

MATERIAL AND METHODS

Study area

Orissa the land of lard Jagannath lies on the East Coast in India. Forest cover is the major portion of the state area which through in official statistics stands at 36% (59963 km2) satellite imagery confirms the are

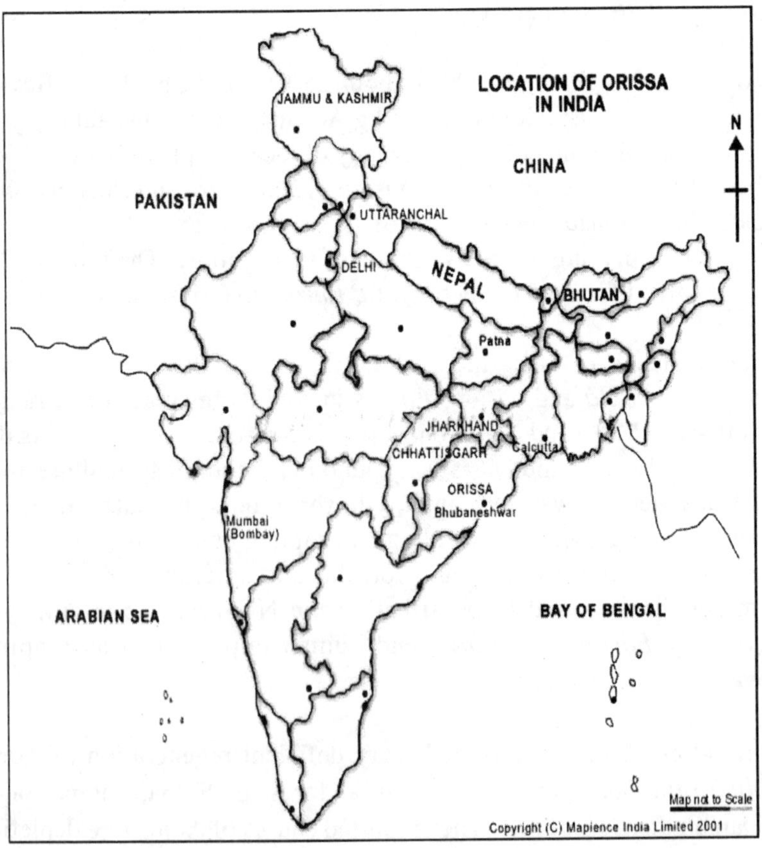

area under forest to be 26% (FSI, 1991). The main types of forest are dry deciduous, which cover approximately 50% and moist deciduous covering 30% of the forest area. The remaining 20% consist of various forest types such as semi evergreen, the littoral and the tidal swamp forests. Predominant vegetation is dry and moist Sal forest with a high tribal population (22%). Forest play a significant role in the socio economic life of the reason populace many of whom depend on NTFP for their subsistence.

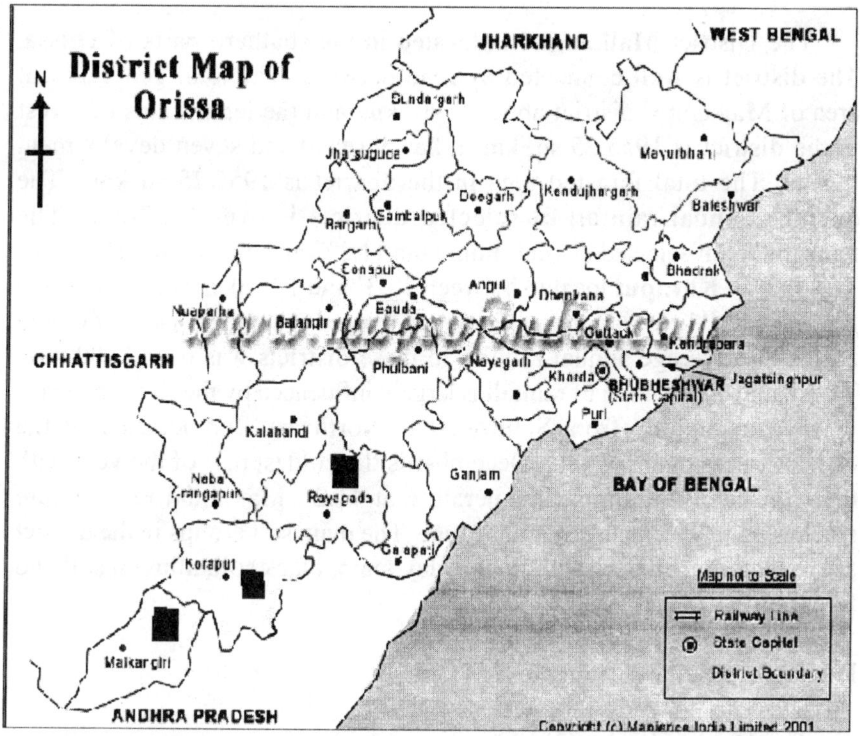

Rayagada district is a place of scientific beauty with hills green with forest and undulating valleys. Total geographical area of the district is 7580 sq km and the forest area of the district constitutes to be 3927.50 sq kms. Tribal populations constitute to be 56.04 per cent of the total population while SC people forms 14.28%. Huge concentration of tribal people show the importance of forest in the districts. There are four ranges in the district. Kashipur block has been selected because of the abundant production of *Tamarind, Harra, Bahera, Achar* and *Kullu gum* in the range.

Table-1: Name of selected Districts and villages in study area.

Districts	Selected Villages
Rayagada	Mandibisi,Kashipur,Shankarda,Godibali,Gumma,Balleshwar, Jimdi, Kailshkota
Malkangiri	Bondaghati, Jharapalli,Keyang, Timurpali, Pedawada, Kabasiguda, Baheraguda, MV-72
Koraput	Ramgiri, Seribeda, Dandawadi, Bhaluguda, Kurali, Sunki, Laudi, Panasput

The District **Malkangiri** is located in the southern parts of Orissa. The district is well connected by road network. The total geographical area of Malkangiri district about 6190^2 km. and the legal status of forest in the district is 1955.25 sq. km. It has 3 Tehsil and seven development blocks. The total forested area in the district is 1955.25 sq kms. The average annual rainfall of selected districts is over 1522-mm. The temperature of selected districts minimum 11.2°C and maximum up to 42°C. The district **Koraput** located between 20°3' and 17° 50' north latitude and 81°71' & 84°1' east longitude. These are located at southern parts of Orissa State. The average annual rainfall in these districts was over 1500 mm. The spatial distribution of rainfall is largely influenced by the Eastern Ghats, which run roughly from Southwest to Northeast. Temperature of the district varies over the year. December is the coldest part of the year with the mean daily minimum temperature at 11.2°. Maximum temperature reaches up to 45°c in the month of June. The main soil groups in the district are red leterite, coarse sandy loam, clay loam, black soil, alluvial soil and forest soil.

Density of selected species:

The 10-sample plot (20x20 m^2) were laid in forest area (near village) on random basis. Quadrat data were used for the analysis of frequency (%), density no/ha and Important Value Index (IVI).

Important Value Index (IVI):

In order to express the dominance and ecological success of any species, which a value, the concept of importance value indexes, has been developed. This index has three characteristics viz. Relative frequency, Relative density and Relative basal area:

$$\text{Relative Frequency} = \frac{\text{Frequency of one species}}{\text{Frequency of all species}} \times 100$$

$$\text{Relative Density} = \frac{\text{Density of one species}}{\text{Density of all species}} \times 100$$

$$\text{Basal Area} = \frac{\text{Basal Area of one species}}{\text{Basal Area of all species}} \times 100$$

IVI = Relative Frequency + Relative Density + Relative Basal area

Regeneration

Young ones of tree species up to height of 0.5m and seedlings of the species were included in this class. Quadrats of 1 x 1 m were laid within each of the 20 x 20 m quadrat. All the seedlings of trees were counted within the quadrat and recorded (Mishra, R. 1968, Kotwal, 1987, Kotwal and Mishra 2000).

Extent of damage during harvesting

Extent of tree damage to the all selected tree species during harvesting of plant parts like fruit, seeds, gum, leaves etc. by lopping, pollarding and felling of trees (by the fruit collectors) ,was observed and recorded. The tree damage was categorised into four types; 1) Not damaged; 2) Moderately damaged; 3) Highly damaged and 4) Completely felled trees. Trees with 25-50% of branches lopped were recorded in the moderately damaged category. Trees with more than 50% of the branches lopped were put in highly damaged category.

Method of fruit/plant parts collection was closely observed at the time of harvesting of each species in the natural forest of study sites. The observations on density, regeneration and tree damage were taken during the year 2000-2001. Collectors used mostly axes for cutting and lopping of trees. However, the damage caused by wild animals and domestic animals was not assessed.

Table-2: Density and IVI of selected NTFPs in study site

Name of species	Districts					
	Malkangiri		Koraput		Rayagada	
	Density no/ha	IVI	Density no/ha	IVI	Density no/ha	IVI
Emblica officinalis	15.37	14.59	12.50	13.62	10.31	11.60
Tamarindus indica	8.23	18.44	4.38	12.06	1.25	2.95
Madhuca latifolia	16.15	42.35	2.19	4.35	4.38	14.
Buchanania lanzan	21.88	20.95	15.63	19.07	14.06	17.06
Sterculia urens	0.63	2.15	0.63	0.81	0.63	1.00
Terminalia chebula	9.90	11.19	9.06	11.79	10.63	13.19
Terminalia bellerica	10.31	11.45	12.19	14.96	9.06	11.79
Bauhinia vahlii	2.19	1.89	3.44	3.27	0.94	1.01
Shorea robusta	6.66	6.15	18.44	27.99	8.44	12.31

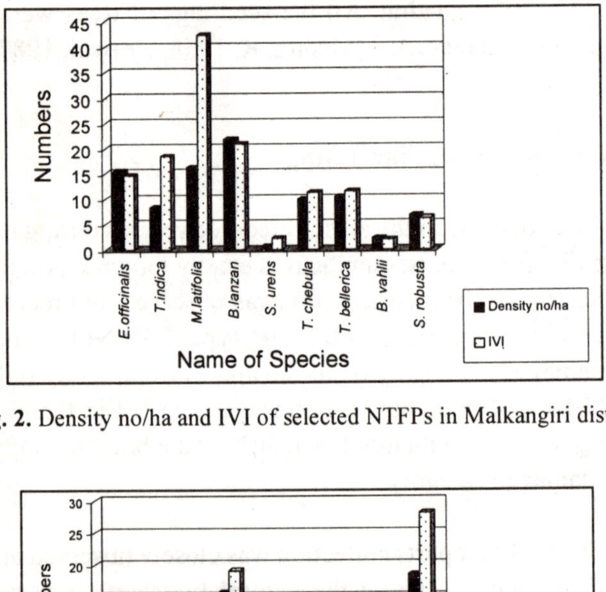

Fig. 2. Density no/ha and IVI of selected NTFPs in Malkangiri district

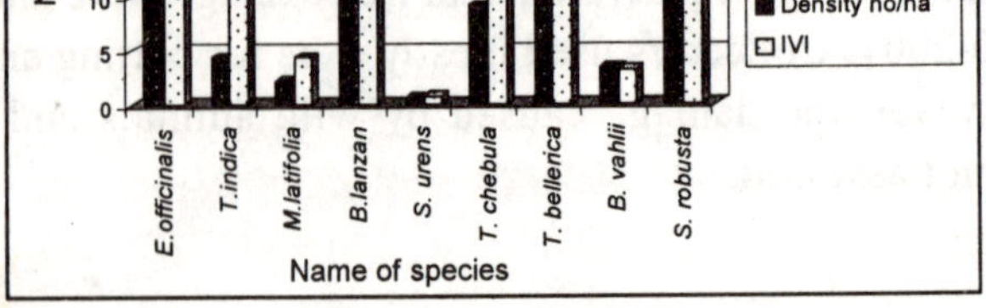

Fig. 3. Density no/ha and IVI of selected NTFPs in Koraput district

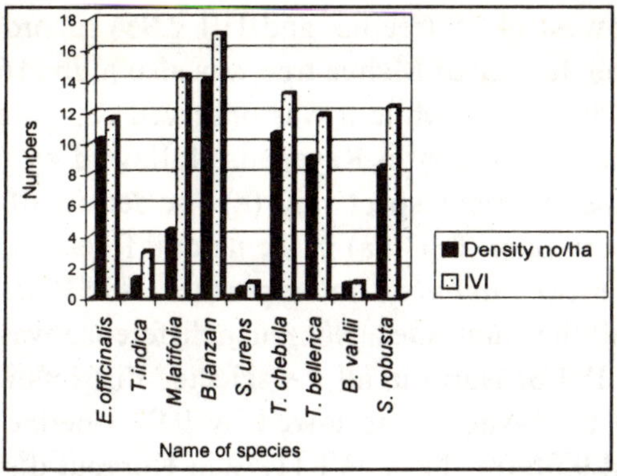

Fig. 4. Density no/ha and IVI of selected NTFPs in Rayagada district

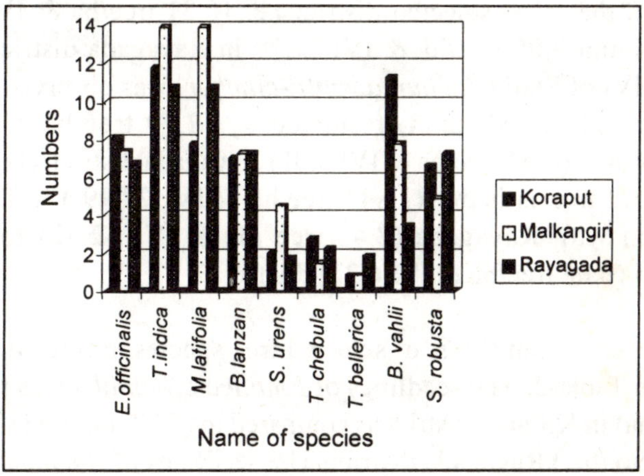

Fig. 5. Regeneration of selected NTFPs in study area

RESULTS

The data presented in Table-2 showed tree density per hectare and IVI of selected NTFPs in three districts of Orissa state. The tree density and IVI of various species are as follows. (a) Achar (*B. lanzan)* is high (21.88 tree /ha & IVI 20.95) in Malkangiri district followed by (15.63 tree/ha. & IVI19.07) Koraput, and (14.06 tree/ha. & IVI 17.06) Rayagada district. (b) high tree density & IVI of Aonla (E *officinalis)* was observed (15.37 tree/ha. and IVI 14.59) in Malkangiri followed by (12.50 tree/ha. and IVI 13.62) Koraput and (10.31 tree/ha. and IVI 11.60) in Rayagada district. (c) the density & IVI of Imli is highest (8.23 tree/ha. and IVI 18.44) in Malkangiri, while low (4.38 tree/ha. and IVI 12.06) in Koraput followed by lowest (1.25 tree/ha. and IVI 2.95) recorded in Rayagada district. (d) The density of Mahua tree was also high (16.15 tree/ha. and IVI 42.35) in Malkangiri while it was observed comparatively very low (4.38 no/ha and IVI 14.30) in Rayagada, followed by (2.19 tree/ha. & IVI 4.35) Koraput district respectively. (e) The density of Kullu (*Sterculia urens)* tree is very poor (0.63/ha) in the natural forests of all the sites. (f) Harra (*Terminalia chebula)* an important NTFP is observed well distributed in all the study sites. Not much difference was observed in the tree density & IVI of Harra in all the selected districts i.e. 10.63 tree/ha and IVI 13.19 in Rayagada, followed by 9.90 tree/ha & IVI 11.19 in Malkangiri and 9.06 tree/ha & IVI 11.79 in Koraput district. (g) Bahera (*Terminalia bellerica*) trees are also common in all the study sites. The tree density & IVI of Bahera is better in Koraput (12.19 tree/ha & IVI 14.96) than the other selected district i.e. 10.31 tree/ha & IVI 11.45 in Malkangiri and 9.06 tree/ha & IVI 11.79 in Rayagada district. (h) poor density & IVI of Siali (*Bauhunia vahlii-climber)* was observed (3.44 tree/ha & IVI 3.27) in Koraput, followed by (2.19 tree/ha & IVI 1.89) Malkangiri and (0.94 tree/ha & IVI 1.01) Rayagada district. (I) the density & IVI of Sal tree is better (18.44 tree/ha & IVI 27.99) in Koraput in comparison with Rayagada (8.44 tree /ha & IVI 12.32) followed by Malkangiri (6.66 tree/ha & IVI 6.25) district.

The regeneration status of selected tree species in three districts was depicted in Table-3. The seedlings of *Emblica officinalis* was found more (8.12 no./ha) in Koraput district as compared to (7.37 no./ha) in Malkangiri and (6.75 no./ha.) Rayagada district. The seedlings of *Buchanania lanzan* was recorded highest (13.87 no./ha) in Malkangiri than in Rayagada (10.75 no./ha.) and (11.75 no./ha) in Koraput district. The seedlings of Harra was recorded high (13.87no./ha) in Malkangiri district followed by (10.75 no./ha.) Rayagada and (7.75 no./ha) in Koraput district. Not much significant difference was observed in the seedlings of *Terminalia bellerica. On an average* 7.16 no./ha was recorded in all the three districts. Very few seedlings of Mahua tree was observed in Malkangiri (4.5 no/ha) followed by (2 no./ha) Koraput and only 1.75 no./ha in Rayagada district. Similarly in case of *T. indica* only few seedlings were observed (2.75 no./ha) in Koraput, (2.25 no./ha) in Rayagada and (1.5 no./ha) in Malkangiri district.

Very few seedlings of Kullu and Imli were recorded in the natural forest of three selected districts. Only 1.87 no./ha of Kullu seedling was recorded in Rayagada district while almost nil (0.87 no./ha) in Malkangiri and Koraput districts. The distribution of *Shorea robusta* was better in

Table-3: Regeneration of selected NTFPs in three district of Orissa state. Quadrat size: 1x1 m2, Seedlings < 0.5 meter

Name of Species	Seedlings tree/ha. in Districts		
	Koraput	Malkangiri	Rayagada
Emblica officinalis	8.12	7.37	6.75
Buchanania lanzan	11.75	13.87	10.75
Terminalia chebula	7.75	13.87	10.75
Terminalia belerica	7.00	7.25	7.25
Madhuca latifolia	2.00	4.50	1.75
Tamarindus indica	2.75	1.50	2.25
Sterculia urens	0.87	0.87	1.87
Shorea robusta	11.25	7.75	3.50
Bauhinia vahlii	6.62	4.87	7.25

Source: Primary data

Koraput (11.25 no./ha) as compared with (7.75 no/ha) Malkangiri and (3.5 no./ha) Rayagada district. A very thin difference was observed in case of Siali seedlings 7.25 no./ha in Rayagada, 6.62 no./ha in Koraput and 4.87 no./ha in Malkangiri district .

Extent of tree damage of selected NTFPs during fruit collection in the study area.

The research team observed that there has been very significant damage occurring to the trees of all selected NTFPs while harvesting by the primary collectors. The team opined that tree damages can be categorized as follows, (a) Not damaged (b) mildly damaged (c) highly damaged and (d) felled trees. The damage materializes mainly in the form of branch lopping, tree felling etc. and also due to unscientific ways of harvest.

Table-4 Shows the extent of damaged trees due to destructive NTFPs harvesting. The data show that the damage on Siali leaves (Climber) plant during collection was maximum (20.87%) in mildly damage and highly damage (5.50%) category, while 9.50% climber were found completely felled in the study area. The data also show that in Rayagada district about 18% Achar trees were damaged mildly, 5.13% has been highly damaged, while 67.37% were found in not damaged category. Local people felled Achar trees (9.50%) to harvest maximum fruits. Similarly in the case of

Aonla , mildly damaged trees were observed more (17.62%) while 66% trees were found not damaged. Maximum felled trees (11.13%) were observed in Aonla in comparison to other NTFP species in the natural forest of the study area. On the other hand most of the Mahua and Imli trees (97.87% and 85.75% respectively) were found undamaged and not a single tree was completely felled (nil) during the harvesting. Very few trees were observed completely felled in case of Kullu (2.88%) and Harra, Bahera (6.00% and 6.88 respectively) during gum and fruit collection.

Table-4: Extent of tree damage of selected NTFPs during fruit collection in Rayagada district

Name of species	Extent of tree damage per hacatre during harvest			
	Not damaged (%)	Mildly damaged (%)	Highly damaged (%)	Felled trees (%)
Emblica officinalis	66.00	17.62	5.25	11.13
Buchanania lanzan	67.37	18.00	5.13	9.50
Terminalia chebula	73.63	14.25	5.24	6.88
Terminalia bellerica	77.87	12.25	3.88	6.00
Madhuca latifolia	97.87	1.51	0.62	Nil
Sterculia urens	81.87	12.15	2.75	2.88
Tamarindus indica	85.75	11.63	2.62	Nil
Bauhinia vahlii	64.13	20.87	5.50	9.50

Source : Primary data

The data presented Table-5, shows by and large less tree damage of selected NTFP species in Malkangiri district in comparison with Rayagada district. Maximum highly damage tree was observed in Aonla (10.86 %) and mildly damaged (12.99%) category in comparison to other NTFP species in the natural forests of Malkangiri district. Completely felled trees were also found more in Aonla (4.58) and Siali climber (4.28%). On the other hand almost no tree damage was observed under completely felled tree category in case of Mahua and Imli. More mildly damage tree observed in Harra, Bahera and Siali climber.

The data presented in Table-6 show extent of damage to the trees by the local people during NTFP collection in the Koraput district. Maximum Aonla trees were felled (10.37%) during fruit collections, while highly damaged (6.26%) and mildly damaged (17.87%) categories were also observed in case of Aonla tree. The second most affected trees was found is Achar, in which local people felled (8.25%) trees during fruit harvesting while highly damaged tree worked to be 5.38%. On the other hand no tree was observed damaged and felled (almost nil) in case of Mahua and Imli tree, in the natural forest of Koraput district. Local people also felled trees of Harra, Bahera and Siali leaves (5.75%, 4.62% and 8.25% respectively) during fruit/leaves collection.

Table-5: Extent of tree damage of selected NTFPs during fruit collectionin Malkangiri district

Name of species	Extent of tree damage during harvesting			
	Not damaged (%)	Mildly damaged (%)	Highly damaged (%)	Felled trees (%)
Emblica officinalis	71.57	12.99	10.86	4.58
Buchanania lanzan	77.14	14.58	4.58	3.70
Terminalia chebula	79.84	13.00	4.85	2.31
Terminalia bellerica	78.43	13.57	5.00	3.00
Madhuca latifolia	99.85	0.15	Nil	Nil
Sterculia urens	90	7.42	1.28	1.3
Tamarindus indica	94.69	4.16	1.15	Nil
Bauhinia vahlii	82.28	8.30	5.14	4.28

Source: primary data

Table-6: Extent of tree damage of selected NTFPs during fruit collection in Koraput district

Name of species	Extent of tree damage during harvesting			
	Not damaged (%)	Mildly damaged (%)	Highly damaged (%)	Felled trees (%)
Emblica officinalis	65.50	17.87	6.26	10.37
Buchanania lanzan	67.00	19.37	5.38	8.25
Terminalia chebula	77.25	13.13	3.87	5.75
Terminalia bellerica	79.87	12.25	3.26	4.62
Madhuca latifolia	98.00	1.25	0.75	Nil
Sterculia urens	77.35	11.25	5.40	6.00
Tamarindus indica	83.75	12.00	4.28	Nil
Bauhinia vahlii	66.26	20.37	5.12	8.25

Source: primary data

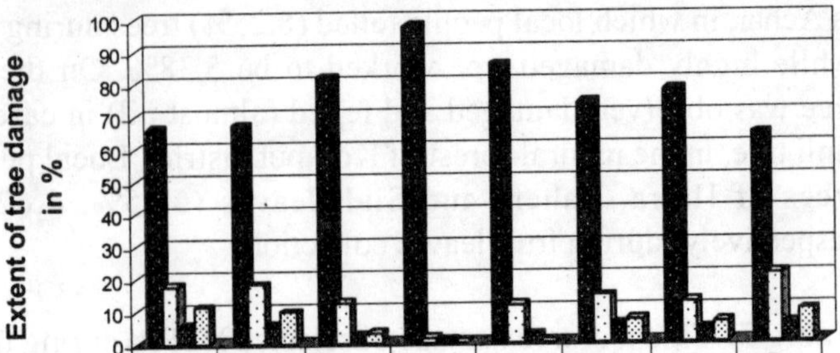

	E.officinali	B.lanzan	S.urens	M.indica	T. indic	T.chebula	T. beller	B. vahlii
Not damaged	66	67.37	81.87	97.87	85.75	73.63	77.87	64.13
Mildly damaged	17.62	18	12.15	1.51	11.63	14.25	12.25	20.87
Highly damaged	5.25	5.13	2.75	0.62	2.62	5.24	3.88	5.5
Completely felled trees	11.13	9.5	2.88	0	0	6.88	6	9.5

Fig. 6. Extent of tree damage of selected NTFP during fruit/leaves/gum collcetion in Rayagada district

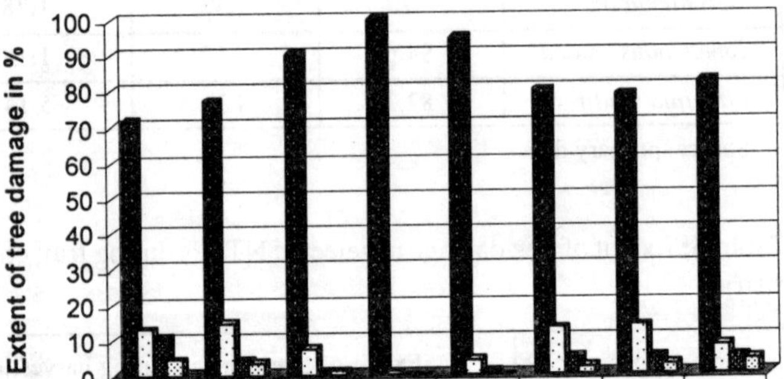

	E.officinalis	B.lanzan	S.urens	M.indica	T. indica	Tchebula	T. bellerica	B. vahlii
Not damaged	71.57	77.14	90	99.85	94.69	79.84	78.43	82.28
Mildly damaged	12.99	14.58	7.42	0.15	4.16	13	13.57	8.3
Highly damaged	10.86	4.58	1.28	0	1.15	4.85	5	5.14
Completely felled trees	4.58	3.7	1.3	0	0	2.31	3	4.28

Fig. 7. Extent of tree damage of selected NTFP during fruit/leaves/gum collcetion in Malkangiri district

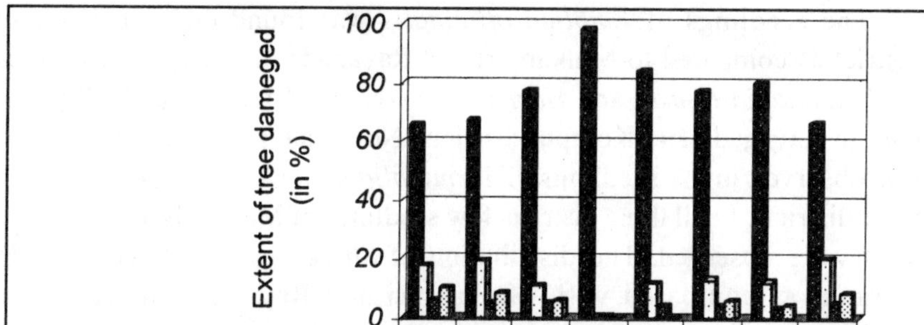

	E.o ffic	B.l an	S.u ren	M.i ndi	T. ind	Tc he	T. bel	B. vah
Not damaged	66	67	77	98	84	77	80	66
Mildly damaged	18	19	11	1.3	12	13	12	20
Highly damaged	6.3	5.4	5.4	0.8	4.3	3.9	3.3	5.1
Completely felled trees	10	8.3	6	0	0	5.8	4.6	8.3

Fig. 8: Extent of tree damage of selected NTFP during fruit/leaves/gum ollcetion in Rayagada district

DISCUSSION AND CONCLUSION

The tree density per hectare and IVI of selected NTFPs in three districts of Orissa state reveals that tree density of Achar (*B. lanzan)* is high in Malkangiri district in comparison with Koraput and Rayagada districts. Similarly high tree density & IVI of Aonla (*E. officinalis*) is observed in Malkangiri district than other two districts. The density & IVI of Imli was found highest in Malkangiri, while low in Koraput followed by lowest in Rayagada district. The density of Mahua tree was also high in Malkangiri while it was observed comparatively very low in Rayagada and Koraput districts. The density of Kullu (*Sterculia urens)* tree is very poor in the natural forests of all the sites. The density & IVI of Kullu was observed high in Malkangiri, while low in Rayagada and Koraput district. Not much difference was observed in the density & IVI of Harra in all the selected districts. Bahera (*Terminalia bellerica*) tree density & IVI is better in Koraput than the other selected district. Poor density & IVI of Siali (*Bauhunia vahlii-climber)* was observed in Koraput, followed by Malkangiri and Rayagada district. The density & IVI of Sal tree is better in Koraput in comparison with Rayagada followed by Malkangiri districts.

The seedlings of *Emblica officinalis* was found more in Koraput district as compared to Malkangiri and Rayagada district. The seedlings of *Buchanania lanzan* and *Harra* were recorded highest in Malkangiri than in Rayagada and Koraput districts. Not much significant difference was observed in the seedlings of *Terminalia bellerica* recorded in all the three districts. In all three districts few seedlings of Mahua, Imli, and Kullu trees were observed. The distribution of *Shorea robusta* was better in Koraput as compared with Malkangiri and Rayagada districts. No significant difference was observed in case of Siali seedlings.

Extent of tree damage during harvesting

In Rayagada district the damage on Siali (Climber) plant during leaf collection was maximum in all the damage categories such as mildly damage, highly damage, while 9.50% climber were found completely felled in the study area. The data further shows that in Rayagada district more (about 23.13%) Achar trees were found damaged mildly as well as highly, while 67.37% were found undamaged. Villagers most often fell Achar trees to harvest maximum fruits. In the case of Aonla it was found that the damage varied substantially than any other selected species in the study area. Number of Aonla tree felled was also very high in comparison to other NTFP species in the natural forest of the study area. Mahua and Imli trees faced less damage than other selected species, not a single tree was found completely felled in all three districts. Very few trees were observed completely felled in case of Kullu, however, the damaged admitted to the kullu was also substantial, sizable number of trees found drying due to unsustainable gum harvest. By and large less tree damage to all selected NTFP species was observed in Malkangiri district in comparison with Rayagada district. With in the admitted damage Aonla faced more mild and high damages during fruit collection the total damage recokned to be about 23.5%. More number of trees of Aonla and Siali were found completely felled in all the districts. Koraput district also experienced lot of damage and complete felling of Aonla trees at the time harvest. The second most affected trees were Achar. The harvester also felled Harra, Bahera trees and Siali climber during harvesting.

Recommendations

- For Aonla, Achar, Harra, Bahera etc. the optimum harvesting time should be notified to the local forest dwellers. In case of Aonla the harvesting period could be after February onward when fruit have attained a coppery red colour and for Achar when the fruit turn black often in June. This will ensure that some fruit fall on the ground and ultimately germinate to become future seedlings and saplings.

- Premature harvesting of Aonla and Achar fruits should be banned or discouraged.
- In the absence of any standards for a sustainable extraction of fruits/ gums harvesting regime, 30-40% of the fruits should be left on trees for future regeneration.
- In view of the importance of Mahua, Imli, Aonla trees it is essential to preserve the existing trees and encourage its planting. Some plus trees of selected species should also be marked at each district from which plucking should not be permitted at all.
- In order to curb the increasing demand Imli, Achar, Aonla and Mahua fruits by industries, the establishment of plantations of these species in village private land, farm lands, community lands need to be encouraged.
- State government without jeopardizing the interest of gatherers need to introduce the system of regulating of harvest and extraction of NTFPs. Regulatory measures like transit pass for Achar, Aonla, Imli, Bahera etc. would ensure that these species are maintained in the forest ecosystem in a viable population at the same it will provide information to foresters as to how much is the potential and how much is being removed from natural forests. This will provide mechanism to regulate optimum extraction.
- End user education is needed to be done by the government agencies.

ACKNOWLEDGMENT

The authors are thankful to Dr. Ram Prasad, IFS, the then Director who gratified the manuscript, providing field facilities and valuable suggestions. The authors are thankful to Dr. P.C. Kotwal, Associate Professor for his valuable guidance. The authors expressed immense gratitude to Regional Centre for National Ecodevelopment board (RCNAEB), Bhopal for funding the project. Thanks are due to the forest department and field staff of the study area of Orissa state for providing cooperation and assistance during the experiment work..

REFERENCES

Chadhar, S.K. & Sharma, M.C. (1997). Trend in fruit production in *Buchanania lanzan* trees. *Vaniki Sandesh, Vol. XXI, No. 3, July to September. SFRI, Jabalpur, India.*

Chakravesh, M.K. (1999). Ecological status of NTFP tree especially in dry deciduous forest of Chhindwada district in Madhya Pradesh. *MRM dissertation No. 36, IIFM, Bhopal, India.*

Champion, H.G. and Seth, S.K. (1968). The forest types of India. New Delhi.

Gera, Mohit and Baghel, L.M.S. (2000). A socio -economic profile of village in Sambalpur with special reference to NTFPs. *Wasteland news, Feb. - April 2000.*

Mishra, M.R. (1996). A study of value addition potential at primary collector's level or Tamarind and Siali leaves in the state of Orissa. *OT report no. 376, submitted at Indian Institute of Forest Management (IIFM), Bhopal, M.P.*

Mishra R. (1968). Ecology work book. *Oxford and IBH Pvt. Ltd. New Delhi.*

Mishra, Manish; Teki ,Surayya and Mishra, R.P. (2001). Observation on the phenology and fruit yield of Achar (*Buchanania lanzan*) tree in the tropical forest of Orissa state. Proc. Symposium on: *The Art and Practice of Conservation Planting under the topic "Tropical Tree Seed Research"* Taiwan. pp. 234-242.

Mishra, Rajendra Prasad & Kotwal, P.C. (2000). Vegetational studies on Indian wild buffalo (Bubalus bubalis L.) habitat in Udanti Wildlife sanctuary. *Journal of Tropical Forestry. October-December 2000. Vol.* 16 (4) pp. 70-77.

Patnaik, Suprava (2000). Comparative study on ecological status of selected NTFP species (Chironji and Harra) in Mayurbhanj district in Orissa and Chhindwara district in M.P. Project report submitted to RCNAEB, Bhopal.

Prasad, Ram and Bhatnagar, P. (1993). Non wood forest products and the indigenous fringe dwellers in M.P. Journal of Tropical Forestry, **9(3): 188-195**.

Prasad, Ram; Kotwal, P.C. and Mishra, Manish (2002). Impact of harvesting of *Emblica officinalis* (Aonla) on its natural regeneration in central Indian forests. Journal of sustainable forestry. Vol. 14, no. 4, 2002.

Prasad, Ram and M. Mishra (2001). Documentation of published and unpublished literature on non-timber forest products in India. *OT report and MRM dissertation on NTFPs from 1989-1999 at IIFM, Bhopal.*

Prasad, R. & Chadhar, S.K. (1989). Results of some field trials of *Buchanania lanzan* in Madhya Pradesh, India. *Vaniki Sandesh, Vol. XXIII, No.2, pp 7-12. SFRI, Jabalpur, India.*

Prasad, R. (1999). Joint forest management in India and the impact of state control over noon wood forest products. *Unsylva. 1998. pp 58-62.*

Prasad, R.; Kotwal, P.C. & Mishra, Manish (2000). Standardizing methodology for sustainable harvest of some important NTFP species in Madhya Pradesh. *Research Project Report submitted at Indian Institute of forest Management, Bhopal, India.*

Rai, Y.C. (1982). *Buchanania lanzan:* A studies on methods of propagation and estimation of fruit yield. *Indian Forester, July 1982.*

Singhal, Rekha (2000). Towards sustainable harvesting of non-wood forest products in India: The role of gender. SEANN *Seminar proceeding on harvesting of non wood forest products October 2-8, 2000 at Nepal.*

Advancing Frontiers of Ecological Researches in India, 99-124, 2006
Edited by A.K. Kandya & Asha Gupta
Published by Bishen Singh Mahendra Pal Singh, Dehra Dun, India

Forest Resources of Madhya Pradesh : An Overview

A. K. Kandya
Department of Botany, Dr. H. S. Gour University, SAGAR, M.P.
e-mail: anand_kandya@rediffmail.com

Abstract: Forests are a huge natural, renewable resource of Madhya Pradesh state. They play a key role in the economy and ecology of the state and support the livelihood of millions of rural people who live in their vicinity. There are several useful products of these forests ranging from timber (Major forest product) to fuel-wood, useful leaves like Tendu Patta, bamboos, wild fruits, fibres, fodder, condiments, dyes, scents, gums, resins, non-edible oils, seeds to a wide variety of medicinal plants (Minor forest products or MFPs). Some of them are nationalized (NMFPs) and others are non-nationalised (NNMFPs). A detailed description of all the non-wood forest products (NWFPs) and non-timber forest products (NTFPs) is given in the present paper. Details of the new State Forest Policy of Madhya Pradesh 2004 enforced by the new Bhartiya Janata Party Government recently are also given.

INTRODUCTION

The whole state of Madhya Pradesh lies in the tropical zone and the temperature in summer exceeds 40°C throughout the state except the hilly region of Pachmarhi. The temperature during winter sometimes touches 0°C in Raisen, Khajuraho or northern parts like Gwalior area, but generally remains between 5 to 10°C. The average annual rainfall ranges between 1500 mm to less than 800 mm in western region.

Presence of the plants of different groups, i.e. algae, fungi, bryophytes, pteridophytes, gymnosperms and angiosperms make the vegetation of a place. But, the presence of only flowering plants of a locality makes the flora of that place. Forests contribute significantly in the biological diversity of a place because there may be representative plants of a large number of species in the forests. But, when any product of the forest is produced at a commercial level, forest is called a resource of it. Timber and wood are the two major products of forests. But, there may be several other, minor products also, like the fruits, leaves, flowers and seeds etc. which are used for different purposes. Lac, honey, horns, hides, Tussar, wax and kosa silk moths are animal products but they are also included in MFPs as they are also a source of revenue to the government and source of livelihood to the poor, rural and tribal population.

Forests occupy 96130 sq. km. area in the state constituting 31.14% of the total geographical area of the sate. Density of forests is good in Mandla, Balaghat, Hoshangabad, Seoni, Betul, Sidhi and Shahdol districts only. Broadly, the forests of Madhya Pradesh are of Tropical type. But, on the basis of geographical distribution, these forests have been categorized in to the following three types :

Tropical Deciduous Forests : They are found in those areas where the average annual rainfall is between 50 to 100 cm. Trees shed all their leaves at a time in the summer season due to scarcity of water. In these forests, good quality timber is produced. Sagon (*Tectona grandis*), Shisham (*Dalbergia latifolia*, Rose wood tree) are the trees found in these forests. These forests are located in Sagar, Jabalpur, Damoh, Chhatarpur, Panna, Betul, Chhindwara, Seoni and Hoshangabad districts etc.

Tropical Semi-deciduous Forests : These forest are found in those areas where the average annual rainfall is from 100 to 150 cm. In these forests, Sal (*Shorea robusta*), Sagon (*Tectona grandis*) trees and bamboos are found more. In these forests, Beeja (*Pterocarpus marsupium*), Mahua (*Madhuca indica*), Harra (*Terminalia chebula*), Tinsa (*Ougenia oogeinensis*), Jamun (*Syzizium cuminii*), Dho (*Anogeissus latifolia*) and Seja (*Lagerstroemia parviflora*) trees are also found. Not all the species of these forests are deciduous in nature. These forests are present in Mandla, Balaghat, Sidhi and Shahdol districts of the state.

Tropical Dry-deciduous Forests : These forests are situated in those areas of the state where the average annual rainfall is 25 to 75 cm. All the species of these forests are deciduous in nature and their height remain between 8 to 20 meters. In these forests Harra (*Terminalia chebula*), Babul (*Acacia nilotica*), Sagon (*Tectona grandis*), Siris (*Albizia odoratissima*), Haldu (*Adina cordifolia*), Palash (*Butea monosperma*), Dho (*Anogeissus latifolia*), Tendu (*Diospuros melanoxylon*), Kika (*Acacia leucophloea*) and Shisham (*Dalbergia latifolia*) species are commonly found. These forests are present in Shivpuri, Ratlam, Neemach, Mandsaur, Tikamgarh, Sheopur, Datia, Gwalior and Khargone districts etc.

Sub-tropical semi-evergreen forest occurs only in higher hills of Pachmarhi. The forests can also be divided into the following two categories on the basis of their economic importance:

Monsoon Forests : In the eastern part of the state where the average annual rainfall is more than 125 cm, these forests are found. Indore, Hoshangabad, Shivpuri, Jabalpur, Rewa and Balaghat are the districts having these forests. They are dominated by Teak trees and therefore earn a huge revenue for the state.

Dry Thorny forests : In those regions of the state where the average annual rainfall is less than 75 cm., these forests are found. In these forests, the number of trees is less. Mostly thorny bushes and grasses are found there. Gwalior, Sidhi, Mandla, Sheopur and Shivpuri are the districts dominated by these forests. *Dalbergia latifolia* (Shisham), *Adina cordifolia* (Haldu), *Madhuca indica* (Mahua), *Acacia arabica* (Babul) and Opuntia are these species found in these forests. Economically, these forest do not earn much revenue for the sate government.

The whole forest area of the state is under the Government Forest Department. For management strategies, it has been categorized in to the following three categories :

Protected Forests : They are looked after by the government but grazing of animals and collection of fuelwood are permitted from them for the native villagers. These forests are protected from felling of trees.

Reserved Forests : In these forests, felling of trees as well as grazing of animals are punishable crimes. They are meant for arresting the soil erosion. From the view point of climate or their utility, these forests are totally banned for cutting of trees.

Unclassified forests : These forests are not yet categorised. In these forests, felling of trees and grazing of animals are permitted.

Forests have also been categorised in the following three categories for administrative purposes :

State Forest : They are completely under the control of state government.

Communal forests : They are under the control of local bodies like the Municipal Committees or Municipal Corporations.

Private forests : They are the personal properties of their owners.

Forests have also been classified according to the specific combinations of plant species in to the following three categories :

Teak Dominated Forests : Tectona grandis is the dominant species in these forests and they are found in a large number of districts of the state, i.e. Jabalpur, Mandla, Seoni, Hoshangabad, Betul, Damoh, Sagar, Tikamgarh, Chhatarpur, Balaghat, Khandwa, Raisen, Panna, Dhar, Jhabua etc. Usually the average annual rainfall in these forests is between 75 to 125 cm. There is normally black cotton soil and a good drainage for water is present. Medicinal plants associated in these forests are : Andrographis paniculata, Abrus precatorius, Adhatoda vasica, Asparagus racemosus, Achyranthus aspera, Cassia fistula, Calotropis gigentea, Chlorphytum tuberosum, Curcuma longa, Curculigo orchioides, Celastris paniculata, Dioscorea hispida, Emblica officinalis, Gardenia lucida, Helectris isora, Holarrhina antidysenterica, Mallotus phillippinesis, Nyctanthes arbotristis, Terminalia bellerica, T. chebula, Elepharispermum subsessile, Tinospora cordifolia etc. Details of these plants are given in later part of the paper.

Sal Dominated Forests : Shorea robusta is the main tree species in these forests which are found in Mandla, Balaghat, Sidhi, Shahdol and Jabalpur districts. Average annual rainfall in these forests is more than 125 cm. therefore these forests are normally dense. The soil is yellow or red. Several species of medicinal plants are found in these forests such as: Anacylius pyrethrum, Allium laptaphyllum, Acorus calamus, Andrographis paniculata, Abrus precatorius, Cassia fistula, Curcuma amada, C. caesia, C. angustifolia, C. longa, Cannabis sativa, Celastris paniculata, Costus speciosus, Dioscorea bulbifera, D. hispida, Emblica officinalis, Gardenia gumifera, Gloriosa superba, Helectris isora, Holarrhina antidysenterica, Hemedesmus indicus, Mallotus phillippinesis, Nyctanthes arbotristis, Prueraria tuberosa, Rauvolfia serpentina, Sida cordifolia, Swertia angustifolia, Urginea indica, Vetiveria zizinoides etc. Details about these plants are given in the later part of the paper separately.

Mixed Miscellaneous Deciduous Forests : Different tree species like Terminalias, Acacias, Cassias, Albizias and others are found in these forests. They are present mainly in Seoni, Jabalpur, Jhabua, Damoh, Sagar, Chhatarpur, Guna, Indore, Datia, Bhopal, Vidisha, Gwalior, Morena, Bhind, Mandsaur, Ratlam, Betul, Tikamgarh, Khandwa, Khargone, Shahdol, Sidhi and Rewa districts. Medicinal plants associated in these forests are : Aegle marmelos, Asparagus racemosus, Abrus precatorius, Achyranthes aspera, Andrographis paniculata, Adiantum lunalatum, Allium cepa, A. sativum, Aloe barbadensis, Cassia fistula, Curcuma amada, C. aromatica, C. longa, C. angustifolia, Chlorophytum tuberosum, Curculigo orchioides, Urginea indica, Cyperus seariosus, Costus speciosus, Dioscorea bulbifera, D. hispida, Celastrus paniculata, Emblica officinalis, Gloriosa superba, Helectris isora, Hemidesmus indicus, Mentha arvensis, Nyctanthes arbotristis, Oscimum vacillicum, Piper longum, Plumbago zeylanica, Psoralea carylifolia, Rauvolfia serpentina, Strichnos nux-vomica, Sida cordifolia, Swertia angustifolia, Terminalia bellerica, T. chebula, Vetivera zizinoides, Withania somniferum etc. Details of these plants are given in later part of the paper.

Botanically, 147 plant families representing 785 genera and 1685 species are found in the state. Out of them, 127 families belong to the group Dicotyledons and remaining 20 families fall under the group Monocotyledons. Following table gives the list of 10 most dominant plant families of India as well as of Madhya Pradesh as per the maximum number of species:

	Plant family	Genera present in M.P	Species present in M.P.	Plant families of India
1.	Fabaceae	65	194	Orchidaceae
2.	Poaceae	78	176	Fabaceae
3.	Asteraceae	55	94	Poaceae
4.	Cyperaceae	13	78	Rubiaceae
5.	Acanthaceae	26	66	Euphorbiaceae
6.	Convolvulaceae	22	61	Acanthaceae
7.	Euphorbiaceae	23	59	Asteraceae
8	Scrophulariaceae	22	49	Cyperaceae
9.	Lamiaceae	23	47	Lamiaceae
10.	Rubiaceae	22	46	Urticaceae

As is clear from the data presented in the above table, except families Urticaceae and Orchidaceae, rest eight plant families are commonly dominant families in the Madhya Pradesh state as well as in the whole country, but their serial numbers are different in both the geographical areas. In place of Urticaceae and Orchidaceae, in Madhya Pradesh, Scrophulariaceae and Convolvulaceae are represented.

Madhya Pradesh is very rich in natural vegetation. The central region of the sate consists of mixed forests in which teak forests occur some times. In the eastern part, moist monsoon forests are present whereas in north western region, thorn forests are present. Following is the detail of the forest areas present in various forest circles of the state :

Forest Circle	Geographical Area (Sq. Km.)	Forest Area (Sq. km.)	% Forest Area
Bhopal	37207	7007	18.83
Indore	46520	10265	22.06
Gwalior	44602	14162	31.75
Balaghat	9245	4055	43.86
Chhindwara	11824	4338	36.69
Betul	10069	3059	39.35
Hoshangabad	10016	3413	34.07
Jabalpur	23421	8858	37.82
Shahdol	13339	5530	41.48
Sagar	31284	8591	26.00
Seoni	13890	4082	29.38
Khandwa	24146	9706	40.19
Rewa	24117	7232	30.00
Total	299680	90298	30.14

There are seven main products of these forests which are nationalized and which the main source of revenue to the government. Madhya Pradesh Minor Forest Produce Federation (MPMFPF) is a government body which controls the marketing management of these MFPs and looks after them during storage. The annual production of these forest products in Madhya Pradesh is given below:

Timber	1.12 Million M^3	Seoni, Chhindwara, Mandla
Fuelwood	2.38 Million M^3	Mandsaur, Hoshangabad, Khandwa, Gwalior
Bamboo	3.00 Million Standard Units	Betul, Balaghat, Khargone, Shahdol, Hoshngabad, Guna
Tendu Patta	4.00 Million Standard Bags	Jabalpur, Rewa, Sagar, Damoh
Sal Seed	1.00 Million Tonnes	(Nationalised Minor Forest Produce, NMFP)
Harra	0.06 Million Tonnes	(NMFP)
Kullu Gum	0.12 Million Tonnes	(NMFP)

Average annual revenue earned from four main mınor forest produce (Timber, Bamboo, Khair and Tendu patta) from different commissionaires of Madhya Pradesh is given in the following table :

Commissionary	Annual revenue earned (Crore Rs) from Timber, Bamboo, Khair and Tendu Patta
Bhopal	61.89
Jabalpur	133.99
Gwalior	3.33
Hoshangabad	30.65
Chambal	2.08
Indore	41.00
Rewa	29.22
Sagar	14.30
Ujjain	3.61

This is clear from the above data that Jabalpur commissionary is very rich in minor forest produce and earns the largest amount of money from them However, other commissionaires are not so rich (Indore, Hoshangabad, Rewa and Sagar). Some of the commissionaires do not produce these MFPs and are absent from the list. Following are the details of these MFPs:

Timber : Most important species are Teak (*Tectona grandis*, Sagon), Sal (*Shorea robusta*), Bija (*Pterocarpus marsupium*), Salai (*Boswellia serreta*), Tinsa (*Ougenia dalbergioides*), Saj (*Terminalia tomentosa*), Haldu (*Adina cordifolia*), Shisham (*Dalbergia latifolia*), and Sandal (*Santalum album*) etc. Teak is the first quality timber whereas other species are used for second category timber. Bamboo occurs mostly as under story in the dry deciduous as well as moist deciduous forests especially on the slopes of the hills, in well drained valleys and heads of streams.

Fuelwood : There is large demand of fuelwood specially by the rural and tribal populations. With rising populations, the demand for wood is going up tremendously. Throughout the country also, there is a pressing problem of scarcity of fuelwood. Deforestation and diversion of forest land for developmental purposes have reduced the fuelwood production. Total ban on cutting healthy, sound trees is also an important reason for reduction. Following are the details of timber yielding species :

Species	Plant Family	Distribution in Madhya Pradesh Districts
Tectona grandis	Verbenaceae	Sagar, Betul, Chhindwara, Hoshangabad, Panna, Dhar, Jhabua, Indore, Jabalpur, Seoni, Dewas, Guna, Sehore and Mandla
Shorea robusta	Dipterocarpaceae	Mandla, Balaghat, Hoshngabad, Jabalpur
Dalbergia sissoo	Papilionaceae	Indore, Guna, Sagar
Dalbergia latifolia	Papilionaceae	Bhopal, Jabalpur, Gwalior, Sagar, Betul, Chhindwara,
Mitragyna parvifolia	Rubiaceae	Damoh, Sagar,
Pterocarpus marsupium	Papilionaceae	Sagar, Panna, Chhatrpur, Raisen, Sehore
Terminalia paniculata	Combretaceae	Indore, Jhabua, Jabalpur, Dewas, Hoshangabd
Anogeissus latifolia	Combretaceae	Damoh, Sagar, Chhatapur, Bhind, Datia,
Anogissus pendula	Combretaceae	Damoh, Sagar, Raisen, Hoshngabad,

Production of timber and fuelwood has been estimated in various districts of the state and the following list has been prepared of those districts which are having surplus production or deficit.

Category	District
Deficit in both timber and fuelwood	Bhopal, Rajgarh, Ujjain, Mandsaur, Datia, Gwalior, Rewa, Tikamgarh, Ratlam, Shajapur, Indore, Dhar, Bhind,
Deficit in fuelwood but surplus in timber	Sehore, Vidisha, Jhabua, Khargone, Khandwa,
Fuelwood supply equals to the demand But timber is surplus	Hoshangabad, Dewas, Damoh
Surplus both in timber and fuelwood	Betul, Raisen, Balaghat, Chhindwara, Mandla, Seoni, Shahdol, Sidhi

There has been a large number of tree species whose wood is used for some very specific purposes, as given below :

For Agricultural Implements: *Anogeissus latifolia, Chloroxylon swetienia, Acacia arabica.*

For Bridge construction: *Acacia arabica, Dipterocarpus.*

For Fencing: *Hopea parviflora, Acacoia arabica.*

For Plough: *Hardwickia binata, Acacia catechu, Anogeissus latifolia.*

For spokes of bullock carts: *Dalbergia latifolia, Tectona grandis, Shorea robusta.*

For wheel of Bullock carts and Tonga: *Dalbergia sissoo, Eugenia jambolana,*

For Plywood Industry: *Bombax malabaricum, Tectona grandis, Mangifera indica, Dalbergia sissoo, D. latifolia.*

Fodder: There are several important grass species found in the meadows (grasslands among forests) of the state. These grasses are nutritious fodder for wild ungulates (herbivorous animals) as well as the domesticated cattle of the nearby villages. The species are : *Andropogon annunnulobus* (Belia grass), *Apluda mutica*, (Kundha), *Bothriocloa pertusa, B. intermedia* (Sandhar), Bromus species (Brome grass), *Cenchrus ciliaris* (Anjan grass or Dhaman grass), *C. setigerus, Chloris gyana, Cynodon dactylon* (Doob ghas), *Dicanthium annulatum, Eragrostis unioloides, Heteropogon contortus, Iselema laxum* (Mushal), Panicum maximum (Guinea grass), *Paspalum dialatum, Pennisetum purpureum* (Elephant grass).

Grasses: There are some grass species which are used for other purposes than grazing like rope making, paper pulp, essential oil extraction and matting etc. They are found throughout the state. Shahdol, Betul, Hoshangabad, Chhindwara, Balaghat and Mandla are good natural resources for them. Details of two of them are given below :

***Vetiveria zizinoides*:** It yields long fibrous roots which are used extensively in the state for making Khus Tatties, or aromatic scented mats which are made wet and hanged during the hot season to keep the rooms cool. This grass is mostly confined to water-logged patches and provide seasonal employment to artisans in Bundelkhand region.

***Eulaliopsis binata* (Sabai grass):** It is a perennial species found on bare slopes and in forest blanks. In recent years, attempts have been made to cultivate this grass for making sabai rope. In M.P. it is required in bulk to tie cut bamboos (industrial and commercial) while exporting them. There are attempts now to propagate this grass on Cattle Proof Trenches (CPTs) and as under-planted species in the afforested areas. According to a study carried out by Forest Development Corporation of M.P. (Jesani, 1979), the activities relating to cultivation, harvesting and rope-making generate an employment of 1000 person-days per year, per hectare of land. Following is the annual production of Sabai grass in some districts of Madhya Pradesh :

District	Approximate Annual Potential (Quintals)
1. Balaghat	600
2. Chhindwara	500
3. Hoshangabad	200
4. Khandwa	500
5. Narsinghpur	5000
6. Satna	500
7. Other districts	4000
Total	11300

Other grasses used by Pulp and Paper Industry : Although Sabai grass is well accepted by pulp and paper industry, there are a number of other grass species which are suitable for paper making. The most important in this category are : *Arundo donax* (nal), *Cymbopogan citrasus* (West Indian lemon), *C. flexuosus* (Lemon grass), *Imperata cylindrical* (siru), *Pennisetum hopenackeri* (Mawai), *Saccharum munja, S. spontaneum* (Kans), *Themeda quadrivalvis* (Gunera), *T. arundinacea* (Ulla), and *Tripsacum loxum* (Gantamala) etc. The natural and man made forests in the state are capable of producing 200,000 to 300,000 tones of useful grasses annually (Prasad and Jalil, 1987).

Tendu Patta : The leaves of Diospyros melanoxylon are a nationalized minor forest produce in the state. Fifty percent of the total production of these leaves in the country occurs in Madhya Pradesh. Thus, it is the state having the largest production of this MFP. About 8 lakh persons of rural and tribal population get their livelihood. Main districts of its production are Sagar, Jabalpur, Mandla, Shahdol, Rewa, etc. Virtually, it is found extensively in almost all districts except one, i.e. Bhind. The main use of the leaves is in making Bidis and Bidi industry is one of the major cottage industries in the state. Annually, about Rs. 3 billion are earned from this MFP to the state government.

Mahul Patta : Leaves of Bauhinia vahlii, a gigantic climber common in central Indian region, are used in making country plates (Pattal) and country bowls (Dona). The fibers from the inner bark are used for making ropes.

Mahua : Madhuca indica is a tree called the bread of the poor. It is found in the drier type of sal forests but also mixed with teak etc. The seeds yield an oil and the flowers are edible. Since the species is indigenous to central Indian region, scattered trees are found in all forests from where corollas and seeds are regularly collected by the local people. Country liquor is prepared from the dried fruits.

Lac : It is the most important animal product found from the forests. It is a resinous secretion formed as a protective covering on the larvae of a minute insect Laccifer lacca. These insect parasites are found mainly on the host trees like *Butea monosperma* (Palash, Chhewala, Dhak), *Ziziphus mauritiana* (Ber), and *Z. xylupyrus* (Ghont). M.P. is the second biggest center of Lac production in the nation after Bihar. Chief districts are Mandla, Jabalpur, Seoni, Shahdol, Hoshangabad, etc. About 85% of the world production of Lac is produced in India.

Sandal wood-Chandan-Trees of Santalum album: This tree is found in Sagar, Damoh and Narsinghpur districts. The oil is one of the most important essential oils produced in India on a commercial scale. It is distilled from the heartwood and roots. When the heartwood is rubbed on a rough stone with little water, it gives a pleasant smell and a cooling effect on the fore-head.

Harra : It is obtained from the fruits of Terminalia chebula and is used for making an agent for tanning and refining the leather. Most of the Harra is sold to other states. Annual production has been measured to be 121.26 metric tons generating an employment of 2.5 lakh man days per year.

Bamboo or Baans : Dendrocalamus strictus plants. Used for making huts as well as paper. Its pulp is used in paper industries of Amlai (Shahdol) and Nepanagar (Khandwa). Balaghat, Umaria and Shahdol districts produce maximum bamboo. But, in Jhabua and Khargone also it is obtained. Bambusa vulgaris and B. arundinacea are also being introduced as they establish fast and their recruitment rate of new culm is relatively more.

Chironji or Achar : Fruits and kernels of Buchanania lanzan, an important, medium sized tree frequently found in the mixed deciduous forests of Seoni, Balaghat and Chhindwara districts of the state. It is almost evergreen because the leaflessness remains for a very short period. This species has its economic importance due to its fruits which are eaten by hill tribal populations. The kernels which have a taste something between Pista and Almond are eaten raw or roasted. Also used in several delicious sweet preparations. Average annual production in the sate is about 8203 quintals and the revenue earned is about Rs. 6,86,78260. Average rate of collection is around Rs.34 per kg whereas the market rate of the kernel is about Rs.180 per Kg.

Aonla-(Euphorbiaceae) : Fruits of the tree *Emblica officinalis*. It is a rich natural resource of vitamin C and therefore is an integral part of a number of ayurvedic formulation. The tree grows as an associate in tropical dry as well as moist deciduous forests. Wood used for agricultural implements, inferior quality furniture. Annual production has been reported to be 462.02 metric tons generating an employment of Rs. 0.25 million man days per year.

Khair : Tree of *Acacia catechu*. The wood of this tree is used in kutch making. There are factories for this in Sagar, Banmore (Morena) and Shivpuri. It is one of the main tree species of the forests of Sagar, Jabalpur, Damoh, Umariya, Morena, Hoshangabad, Sheopur, Guna and Shivpuri districts. It is also used for leather refining and tanning. It is also used for a medicine used in the disorders of digestive tract.

Sal seed oil : It is major, non-edible, tree based fat. It is also used by tribals as edible oil and also for lighting purposes. The seeds are collected before onset of monsoon. In M.P., the trade of Sal seed was nationalized in 1977 and the task of collection and disposal was entrusted to the State Forest Department and MFP Federation. There are several seed units in the state which have been allotted the work of collection and delivery of Sal seeds to various industries, by the MFP Federation. These Seed Units are virtually Registered Primary Co-operative Societies.

Bhilwa : (Anacardiaceae) Obtained from the trees of Semecarpus anacardium, it is used in making ink and paint. There is an industry for this in Chhindwara. The seed is used as a marking nut and has application in Ayurvedic preparations. The seed has come in to lime light for its liquid commercially known as BNSL (Bhilwa Nut Shell Liquid) which is used in varnishes, enamels, paints, water proofing and insulating material. Seoni and Chhindwara are the main districts of it natural distribution in Madhya Pradesh.

Gum : Found from a number of tree species like *Acacia nilotica* (Babul), *A. catechu* (Khair), *Terminalia tomentosa* (Saja), *Anogeissus latifolia* (Dhawra), and *Sterculia urens* etc. Beyond eating, it is used mainly in calico printing, paint industry, medicines and cosmetics etc. Tapping of kullu gum was banned but it was lifted on a trial basis from April 1995 in 3 forest circles having a cumulative production of 8.6 tons per year. Other gums of local importance are obtained from *Bauhinia retusa* (Semla gum), Siris gum from *Albizia lebbek*, Bengal kino from *Butea monosperma, Malabak kino* from *Pterocarpus marsupium* and Katera gum from *Cochlospermum religiosum.*

Tasar : It is also a non-wood minor forest produce (NWFP) or animal origin like lac. Sericulture is very labour intensive operation. Silk is obtained from the cocoons of silkworms. Its production has four phases:1) Cultivation of host plants of silkworms, 2) Rearing silkworms up to cocoon stage, 3) Reeling of cocoons into continuous filaments called Raw Silk and 4) Silk weaving by which filaments are twisted and woven in to fabrics. Tasar worm Antheraca unlike other silk worms is not domesticated. It feeds on trees of *Terminalia arjuna, Shorea robusta, Lagerstroemia parviflora, Syzizium cuminii* and many others. Tasar rearing is practiced in Balaghat and Mandla districts. Raisen, Sidhi, Narsinghpur, Seoni and Chhindwara districts have also made a beginning in this cottage industry as Directorate of Sericulture has established Mulberry Farms and other infrastructure in these districts.

In the state are located several organizations, some belonging to the state government but others belonging to the Central government, but they are actively engaged in several aspects of forests, i.e. scientific management, carrying out research, extending training and education etc. Details are given below :

Madhya Pradesh Rajya Van Vikas Nigam (MP State Forest Development Corporation) : Established in Bhopal on 24th July, 1975 its main function is to exploit the forest products from the viewpoint of revenue, raising plantations of teak and bamboo etc., increasing the productivity of forests, arranging the raw materials for industries requiring paper pulp and production of raw materials for wood based industries, etc. This corporation receives millions of rupees as financial assistance from the world bank for the above works. At present, the capital gain of the Nigam is about 20 crore rupees. The corporation established Narmada Wood Products in 1979 where wood of several species is made stronger and more durable by seasoning and used in wood industries.

Madhya Pradesh State Forest Research Institute (MP Rajya Van Anusandhan Samsthan) : Located in Jabalpur, this organization is responsible for the scientific development and use of forests in the state. It is also used for education and training in several forestry works. It is a state government organization but autonomous in its functioning. The Director of this institute is an IFS officer of the rank of Chief Conservator of Forests (CCF) of M.P. cadre. The institute is doing excellent work for those who are engaged in raising plantations under agro-forestry schemes. Certified seeds of several species are traded by the institute through a SEED BANK and useful information about them is also given to the entrepreneurs. Similarly, nursery technology for cultivating medicinal plants of a wide variety at commercial level is also available here. There can be short term training in seed collection, identification of medicinal plants and tissue culture etc. For collection of forestry seeds, the institute has established 7 seed collection centers in various edapho-climatic zones of the states whose details are given below :

Collection Centre	Species for seed collection	Forest Circles included
SAGAR	*Emblica officinalis, Holoptelia integrifolia, Pongamia pinnata, Cassia siamea, Lagerstroemia parviflora, Albizia procera, Dalbergia sissoo,* and Grasses	Sagar
JABALPUR	*Holoptelia integrifolia, Lagerstroemia parviflora, Acacia catechu, A. nilotica, Gmelina arborea, Dalbergia latifolia, Albizia procera, Pongamia Pinnata,* and Grasses	Rewa, Shahdol and Jabalpur
GWALIOR	Grasses, *Dalbergia latifolia, Acacia catechu, Albizia amara, Prosopis juliflora, Dalbergia Sissoo, Acacia auriculiformis,*	Gwalior
GUNA	*Grasses, Styloxanthes hamata, Acacia Catechu, Albizia amara, A. lebbek, A. procera, Dalbergia sissoo, Acacia nilotica*	Shivpuri
BETUL	*Tectona grandis,* Grasses, *Styloxanthes hamata, Gmelina arborea, Dalbergia latifolia, Albizia lebbek, A. procera, Pongamia pinnata, Bridelia retusa, Lagerstroemia parviflora, Acacia nilotica*	Betul, Hoshangabad
CHHINDWARA	*Tectona grandis,* Grasses, *Gmelina arborea, Dalbergia latifolia, Albizia procera, Terminalia arjuna, Annona squamosa, Acacia nilotica, Albizia lebbek, Pongamia pinnata, Lagerstroemia parviflora,*	Chhindwara, Seoni, Balaghat

BHOPAL	Grasses, *Styloxanthes hamata, Dalbergia latifolia, Albizia lebbek, Cassia siamea,*	Bhopal
INDORE	Grasses, *Styloxanthes hamata, Holoptelia integrifolia, Albizia lebbek, A. procera, Dalbergia sissoo, Acacia auriculiformis, Cassia siamea, Hardwickia binata, Samania saman,*	Indore, Ujjain, Khandwa

For reclamation of various kinds of wastelands, i.e. all afforestation programmes and in the rehabilitation of degraded forests, i.e. all reforestation programmes, the state government has set a plantation target of 2 Lakh hectares of area for every year. Seed required for this plantation programme are detailed below :

Teak	:	6000 Quintals
Bamboo	:	500 Quintals
Grasses and Legumes	:	100 Quintals
Miscellaneous species	:	2100 Quintals
Total	:	8700 Quintals

Out of these, about 700 quintals seeds are collected by the Seed Collection Centres of the SFRI whereas about 8000 quintal seeds are collected by the Territorial Forest Divisions every year.

Tropical Forest Research Institute (TFRI): Also located in Jabalpur, it is one of the 8 research institutes of the Indian Council of Forestry Research and Education (ICFRE). It is Govt. of India funded organisation. It was earlier known by the name Institute of Deciduous Forests (IDF), and even earlier it was a regional center of the Forest Research Institute, Dehra Dun therefore, it was known as Regional Forestry Research Centre (RFRC). This institute caters to the forestry research needs of the states of not only Madhya Pradesh but also of Maharashtra, Orissa and Chhatisgarh. The institute has the mandate of carrying out research on issues like-Non-wood forest products (NWFPs), Eco-restoration of the Vindhyas, Satpura and Amarkantak mountain ranges, Disease and insect pests of deciduous forests and their control measures, Forest fire and forest grazing control, Environmental influences of developmental activities on forest ecosystem, Identification of insect pests on forestry seeds and assessment of damage, Identification of seed micro-flora on teak, neem (*Azadirechta indica*), bamboo, sissoo, casuarina, and siris (*Albizias*).

Biodiversity studies from polluted areas, Contribution of forest to the socio-economic upliftment and well being of tribal and rural masses, Development of land reclamation techniques with the objectives of establishing community forests on mined-out, barren and culturable and un-culturable wastelands.

Centre for Forestry Research and Human Resource Development, Chhindwara: It was also established by ICFRE on 30th March, 1995. But, it works as a satellite center of TFRI since 3rd January, 1996. Its prime object is to develop human resources especially from the tribals through training programmes in Nursery and Plantation Technology and agro-forestry. Its thrust areas include the development of system for sustainable forest management based on sound principles of ecology and genetics, improvement of utilization of forest produce, socio-economic aspects related to resource management and environmental protection, medicinal plant cultivation and eco-restoration of wasteland development. There are Junior and Senior Certificate Courses on these aspects.

Indian Institute of Forest Management : It was established in 1978 with the financial assistance of Swedish International Development Agency (SIDA) with the objective to train the forest officers of different ranks for the new strategies of forest management. Located in the capital of the state-Bhopal, it is the only one organization of its kind in the whole country. Since 1982, it is a fully autonomous institute working directly under the Ministry of Environment and Forests. Its prime object is teaching forest management based on scientific principles. It has courses on Post Graduate Diploma in Forest Management in which all India level open competition is kept for admission. Short term courses on various aspects of forest management are also run regularly for the high ranking forest officers from all over the country. There is also a strong research base for the teaching and training programmes. Consultancy service is also provided by the experts on demand.

State Forest Rangers College : In 1979, a Forest Rangers College was established in Balaghat and in 1980 the other Forest Rangers College was established in Betul so that training of the state forest service officers can be done. Eighty to 100 Forest Ranger Officers are trained every year in these forest colleges. In Shivpuri, Amarkantak and Lakhnadon, three Forest Schools are located for the training of Foresters and Forest Guards. Similarly, training in wildlife protection is given in Kanha National Park.

Medicinal Plants : There are about 500 species of plants belonging to herbs, shrubs and trees which grow in various parts of the state in natural forests as described earlier. But, some of them are also cultivated in specific areas. Following is the list of important species along with their medicinal uses and the average rates at which they are sold by the poor villagers and tribal masses.

Abrus precatorius : Hindi name Ratti, Gumchi - Family Papilionaceae (Fabaceae) : Roots and leaves decoction given in cough and cold, seeds contain abrine, a poisonous principle similar to ricin from Castor seeds, roots diuretic, tonic and emetic, seed administered in affections of nervous system and their paste applied locally in sciatica, stiffness of shoulder joints and paralysis. Abrin suppresses ehrlich ascites tumour growth in mice. Reported from Orchha, Guna, Jabalpur, Katni, Amarkantak, Tikamgarh.

Achyranthes aspera : Hindi name Latjira-Family Amaranthaceae: Decoction of herb diuretic, used in renal dropsis, young leaves are served as spinach.

Adiantum lunulatum : Hindi name Kali jhant - Family Adiantaceae (Fern - Pteridophyte) : Used in blood diseases and epileptic fits, in Bihar the fern is used in cases of rabies, rhizome prescribed for strangury and in fever due to elephantiasis, fronds are burnt in oil and applied to itch. Reported from Pachmarhi, Patalkot, Chanderi, Tilwaraghat, Bargi, Orchha.

Adhatoda vasica - Hindi name Adusa - Family Acanthaceae : Fresh or dried leaves constitute the drug Vasaka used in bronchial troubles and conjunction, leaf juice used also in diarrhoea, dysentery and glandular tumours, powdered leaves used for skin affections, chief principle is vasicine which has also been found to be a promising utero-tonic abortifacient, it may also find use in stopping post-partum hemorrhage, leaves are rich in vitamin C. Reported from Gwalior, Amarkantak, Jabalpur, Chhindwara, Tikamgarh, Chanderi.

Aegle marmelos - Hindi name Bel - Family Rutaceae : Unripe or half ripe fruits astringent, digestive and stomachic, used for diarrhoea and dysentery, Pulp aromatic and cooling, used in form of Sherbat, Marmelosin is the active constituent, it acts as laxative and diuretic, in strong doses a cardiac depressant. Av. Market rate of fruits Rs. 5 to 9 per Kg. Reported from Seoni, Guna, Indore, Amarkantak.

Allium cepa - Hindi name - Pyaz - Family Liliaceae : Onions considered stimulant, diuretic, expectorant, used against flatulence and dysentery, roasted onions are applied as poultice.

Allium sativum - Hindi name Lehsun - Family Liliaceae : Preparations of Garlic are used in pulmonary phthisis, gangrene of lung and whooping cough. Laryngeal tuberculosis, lupus and duodenal ulcers are treated by garlic juices, juice is applied in skin troubles and used as ear drops, cloves also used for flatulence, colic, and a tonic dyspepsia.

Aloe indica - Hindi name Ghrit Kumari - Family Liliaceae : Has been in use for the host of a diseases particularly of the digestive system, medicinally the term aloe is used for dried juice of the leaves, cathartic properties are due to the presence of glycoside called aloin widely used in chronic constipation.

Andrographis paniculata - Hindi Name Kirayat, Kalmegh - Family Acanthaceae : Herb is an ingredient of a medicine commonly used as a bitter tonic and febrifuge, plant is astringent, anodyne, tonic and alexipharmic, used in dysentery, cholera, diabetes, conjunction, influenza, bronchitis, itches and piles. Often substitute for Chirayta. Decoction used for sluggishness of liver and in jaundice. Leaves and roots used as a febrifuge, cholagogue and anthelmintic. Reported from Pachmarhi, Sheopur.

Asparagus racemosus - Hindi name Sataver - Family Liliaceae : Herb tonic, diuretic, and galactagogue , fresh root juice is mixed with honey and given for dyspepsia, roots form a constituent of medicinal oil used for nervous and rheumatic complaints. Av. Market rate of roots Rs. 65 per Kg. Reported from Pachmarhi, Sheopur, Katni, Chhindwara, Bargi, Gwalior, Indore.

Azadirachta indica - Hindi name Neem - Family Meliaceae : Bark used in skin troubles, leaves antiseptic, applied to boils in the form of poultice, decoction given for ulcers and eczema, flowers tonic and stoma chic, Berries purgative, emollient, dried leaves placed in books for keeping away the moths, odour of burning leaves kills insects, seed yield a non-drying oil used for skin affections, neem oil may be mixed with other oils and fats for the manufacture of washing soap, medicated soaps with the odour of neem oil are available. Reported from Jabalpur, Guna, Indore, Sagar Planted on road sides.

Buchanania lanjan-Hindi name Achar, Chironji-Family Anacardiaceae : Seeds eaten , nutritious and palatable, they yield a fatty oil substitute for olive oil and almond oil in indigenous medicines, used also for glandular swelling of the neck, tree yields a gum suitable in diarrhoea and inter-costal pain. Reported from Chanderi, Chhindwara, Guna, Jabalpur, Amarkantak.

Caesalpinia crista - Hindi Name Gataran - Family Caesaloiniaceae : Fruits tonic and anti-pyretic, seeds yield a fatty oil used as a cosmetic and for discharges from the ear, leaves and bark febrifuge, emmenagogue and anthelmintic, seeds employed for colic.

Cassia fistula-Hindi name Amaltas and Kirwara - Family Caesalpiniaceae: Dried fruits used as a purgative, laxative, for habitual constipation, root bark extract found satisfactory as a substitute for Cassia beareana liquidum in the treatment of black water fever. Av. Market Rate of fruits Rs. 5 per Kg. Reported from Chhindwara, Guna, Jabalpur, Amarkantak, Orchha, Datia.

Chlorophytum arundinaceum - Hindi name Safed Musli-Family Liliaceae : Roots used as tonic. Reported from Seoni, Amarkantak, Kanha. Av. Market rate of roots Rs. 650 to 700 per Kg.

Curculigo orchioides - Hindi name Kali Musli - Family Amryllidaceae : Tuberous roots for skin troubles, considered demulcent, diuretic, and tonic. In combination with aromatics and bitters, they are used in piles, diarrhoea, jaundice, and asthama.

Curcuma angustifolia - Hindi name Tikhur-Family Zingiberaceae : Tubers yield starch resembling arrowroot, easily digestible, recommended for invalids and children. Av Market rate of roots and rhizome Rs. 45 to 65 per Kg.

C. aromatica - Hindi name Jangli Haldi - Family Zingiberaceae : Rhizomes enter into compositions used for bruises, contusions and sprains, in medicines they are used as substitute for turmeric (C. domestica) but not as condiment.

C. caesia - Hindi name Kali Haldi - Family Zingiberaceae : Rhizomes used for sprains and bruises, also employed as cosmetic, yield an essential oil which appears to be a good source of Camphor.

Dioscorea hispida - Hindi name Baichandi - Family Dioscoreaceae : Tubers acrid and poisonous, used as food after proper washing only in times of scarcity. They are soaked by the lepchas in running streams for about a week before cooking. Av. Market rate of rhizome Rs. 24 to 30 per Kg.

Embelia ribes - Hindi name Baibirang - Family Myrsinaceae : Fruits used as stomachic, tonic, astringent and anthelmintic against tape worms, active principle being Embelin, dried fruits used as adulterant of black pepper, their decoction used in chest and skin troubles, tender leaves and fruits cooked and eaten.

Emblica offcinalis-Hindi name Aonla -Family Euphorbiaceae - Fruits are astringent, cooling, diuretic, laxative, eaten raw or cooked, also pickled, richest source of vitamin C, contain 20 times as much vitamin C as orange juice. Fruits used in hair dye, dried ones are detergent and used for shampooing hairs, seeds yield a fixed oil, fruits, bark and leaves are rich in tannin, their tannin content being 28%, 8 to 21%, and 22% respectively. Essential component of Trifala (three types of fruits - Harra, Baheda, and Aonla). Av. Market rate of fruits Rs. 18 to 32 per Kg. Reported from Pachmarhi, Guna, Tikamgarh.

Helicteres isora - Hindi name Marorphali - Family Sterculiaceae : Fruits used in intestinal complaints such as colic, diarrohea, chronic dysentery and flatulence, and to improve appetite, juice of roots is beneficial in empyema and stomach affection, also used in diabetes. Reported from Jabalpur, Seoni, Pachmarhi, Betul, Katni and Gwalior. Av. Market arte of fruits Rs. 25 per Kg.

Hemedesmus indicus - Hindi name Anantmool-Family Asclepiadaceae : The dried roots constitute the drug Hemedesmus or Anantmool (Mool = Roots) which have long been used as demulcent , diaphoretic, diuretic and alterative prescribed rheumatism, gravel and other urinary diseases, and in skin troubles. Reported from Bori, Bargi, Patalkot, Katni, Jabalpur, Chanderi.

Holarrhena antidysentrica - Hindi name Kutja, Kurchi - Family Apocynaceae : Bark is astringent, anthelmintic, stomachic, anti-pyretic, tonic and anti-dysentric, used in amoebic dysentery and diarrhoea, contains alkaloid conessine, a gum resin and tannin, Seed also used as an anthelmintic and used for the same purpose as the bark. Av. Market rate Rs. 6 to 9 per Kg. Reported from Jabalpur, Seoni, Tikamgarh, Sagar.

Madhuca latifolia - Hindi name Mahua - Family Sapotaceae : Seeds are a source of Mahua oil used in rheumatism and skin affections and as laxative in case of habitual constipation and piles. Flowers are rich in sugar, used in the preparation of vinegar. Also used for making syrup. They are considered demulcent and tonic. Av. market rate of Flowers Rs. 4 to 5 per Kg.

Nyctanthes arbotristis - Hindi name Harsingar - Family Oleaceae : Leaves anti-bilious and expectorant, used in rheumatism and fever, decoction given in sciatica, juice used as cholagogue, laxative, diaphoretic, diuretic, and anthelmintic powdered seeds used for scurfy affections of the scalp. Bark expectorant, contains two alkaloids, of these the water soluble one stimulates the ciliary movement of oesophagus. Reported from Shivpuri, Guna, Chanderi, Mandla, Ajaygarh, Katni, Chhindwara, Jabalpur, Gwalior.

Rauwolfia serpentina-Hindi name Sarpgandha - Family Apocynaceae : Roots constitute the drug Rauwolfia which has been employed for centuries for relief from nervous disorders including anxiety states, excitement, maniacal behaviour associated with psychosis, schizophrenia, insanity, insomnia, and epilepsy. Extracts of the roots are valued for intestinal troubles. Roots are believed to stimulate uterine contraction and used in case of difficult delivery. Av. Market rate of roots Rs. 85 to 100 per Kg. Reported from Jabalpur, Indore.

Sterculia urens - Hindi name Kullu - Family Sterculiaceae : Pulverised bark given to women to facilitate delivery, leaves and tender shoots yield mucilage when steeped in water, it is used for pleuro-pneumonia in cattle. Tender roots eaten after cooking, leaves are nutritious and used as fodder, rich in vitamin A. Reported from Bori, Sheopur, Jabalpur, Guna, Chhindwara.

Terminalia bellirica - Hindi name Baheda-Family Combretaceae - Ripe fruits used as astringent in combination with chebulic myrobalan obtained from *T. chebula*, half ripe fruits used as a purgative due to presence of an oil having property similar to those of castor oil, which is applied to hair, also applied to rheumatic swelling, mixed with honey fruit pulp employed in ophthalmia. Bark diuretic and gum yielded by the tree demulcent and purgative. Annual production in the state is 5000 metric tons generating an employment of 2.5 lakh man days per year. Av. Market rate of bark Rs. 6 to 8 per Kg. Reported from Indore, Jabalpur, Chanderi.
Terminalia chebula - Hindi name Harra - Family Combretaceae - Fruits

laxative, stomachic, tonic and alterative, form a constituent of trifala. An important ayurvedic medicine, used for host of ailments, laxative principle - a glycoside may be similar to sennoside. Fruit pulp used in dentifrices, coarsely powdered fruit is smoked in asthama, bark diuretic and cardiotonic, kernels yield a fatty oil, tree yields a gum. Av. market rate of small fruits Rs. 30 to 35 per Kg, Large sized Rs. 8 to 13 per Kg, Reported from Patalkot, Pachmarhi, Indore, Jabalpur, Chanderi. Amarkantak.

Tinospora cordifolia - Hindi name Giloy - Family Menispermaceae : Stem is constituent of several ayurvedic preparations, used in general debility, despepsia , fever, and urinary diseases, dried twigs with bark intact constitute the drug, bitter principles present in the drugs show anti-spasmodic, anti-pyretic and anti-inflammatory properties. The drug possesses one fifth of the analgesic effect of sodium salicylate, a kind of starch called giloy ka sat prepared from aqueous extract of dried stem is used as a tonic. Root is a powerful emetic and is used for visceral obstructions. Its watery extract is used in leprosy, a pulverized fruit is used as tonic and also for jaundice and rheumatism. Av. Market rate Rs. 32 per Kg.

Trachyspermum ammi - Hindi name Ajwain - Family Umbelliferae / Apiaceae : Fruits are stimulant, anti-spasmodic, tonic, and carminative, administered in flatulence, a tonic in dyspepsia and diarrhoea and also often recommended in cholera. Ajwain is used for relaxed sore throat and bronchitis and is a common ingredient of cough mixtures. A paste of crushed fruits is applied to the abdomen externally for relief from colic. Reported to be cultivated in Pachmarhi, Mandla.

Withania somnifera - Hindi name Ashwagandha - Family Solanaceae : Roots have long been in use for hiccup, cough, dropsy, rheumatism, and female disorders. And, as a sedative in case of senile debility. The Pharmacological activity is ascribed to the presence of several alkaloids. Leaves used a febrifuge and applied to lesions, painful swellings and sore eyes. Av. Market rate of roots Rs. 50 to 75 per Kg. Reported from open forest areas in Indore, Amarkantak, Jhabua.

WILDLIFE : The state is also very rich in wild animals which are found mostly in the National Parks and Wildlife sanctuaries. Wildlife Protection Act of India (1972) is enforced in the state since 1974. Since 24th October, 1989, hunting of all kinds of wild animals and birds is totally prohibited in the state. The state is known as the Tiger State of India as the maximum number of tigers is present here. There is also a facility available in Kanha National Park for training in wildlife protection measures. Barasingha has been declared as the State Animal on the 25th anniversary of the formation of the state, on 1st Nov. 1981. On the same day, Doodhraj (Paradise fly-catcher) has also been declared the State Bird. It is also called Shah Bulbul and found in plenty in the state. The technical names of these animals are given in the following table:

Zoological names and number of wild animals found in various National Parks, Wildlife Sanctuaries and outside forests of Madhya Pradesh (Courtesy - Chief Wildlife Warden, M.P.)

Animal	Zoological Name	Number in NP	No. outside NP
Chital	Axix axis	45522	4203
Langur (Monkey)	Presbytis entellus	Not reported	NR
Wild pig	Sus scrofa	19103	4440
Samber	Cervus unicolor	9114	2252
Barking deer	Muntiacus muntjak	2166	193
Barasingha	Cervus duvaucelli-branderi	334	-
Gaur	Bos gaurus	2302	1111
Chausingha	Tetracerus quadricornis	201	261
Wild dog	Cuon alpinus	509	415
Jungle cat	Fellus chaus	NR	NR
Tiger	Panthera tigris	284	82
Neelgai (Blue Bull)	Boselephus tragocamelus	5243	6157
Sloth Bear	Melursus ursinus	815	508
Leopard (Panther)	Panthera pardus	234	199
Blackbuck	Antilope cervicapra	37	2835
Mouse Deer	Tragulus meminna	NR	NR
Hyena	Hyaena hyaena	158	190
Python (Ajgar)	Python molurus	NR	NR
Porcupine	Hystrix indica	NR	NR
Peafowl	Pavo cristatus	NR	NR
Jungle fowl	Gallus gallus	NR	NR
Hare	Lepus nigricollis	NR	NR
Mongoose	Herpsetes Edwards	NR	NR

The wild animals exhibit a great diversity because their habitats, i.e. the forests are also of diverse types. Meadows (grasslands) between the dense forests are very much favourable site for the animals which are present in the forests of the state. Under the Protected Area Network (PAN) scheme, the government has established 8 National Parks and 22 Wildlife Sanctuaries where the whole ecosystem has been preserved. Following is the list of the National Parks and Sanctuaries :

Name of the National Park	Area (Sq. Km)	Districts in which located
Bandhavgarh	437	Umariya
Fossil	27	Mandla
Kanha	940	Mandla, Balaghat
Madhav	337	Shivpuri
Panna	543	Panna, Chhatarpur
Pench	293	Seoni, Chhindwara
Satpura	525	Hoshangabad
Van Vihar	445	Bhopal
Name of the Sanctuary		
Bori	518	Hoshangabad
Selana Florican	13	Ratlam
Sardarpur	348	Dhar
Sone	209	Sidhi, Shahdol
Sindhori	288	Raisen
Sanjay (Dubri)	365	Sidhi
Ratapani	689	Raisen
Pench	450	Seoni, Chhindwara
Palpur (Kuno)	345	Morena
Panparha	246	Shahdol
Pachmarhi	462	Hoshangabad
Nauradehi	1035	Sagar
Chambal for Crocodile	320	Morena
Narsinghgarh	57	Rajgarh
Keoni	123	Dewas, Sehore
Ken for Crocodile	45	Chhatarpur, Panna
Karera	202	Shivpuri
Ghatigaon	512	Gwailor
Gandhi Sagar	369	Mandsaur
Fen	111	Mandla
Bagdeera	479	Sidhi
Rala Mandal	not known	Indore

Kanha National Park has also been included under the 9 Project Tiger National Parks of the country where special programme for the protection of Tiger is being run. Similarly, Bandhavgarh National Prak has the highest density of Tigers in the country. The large number of this animal in the forests indicate a very high density of trees and a very sound health of the forest ecosystem.

New State Forest Policy of Madhya Pradesh 2004

Recently, the new Bhartiya Janata Party government in the state has enforced the new State Forest Policy for Madhya Pradesh. The cabinet of Ministers approved it on 4.4.05. The main priority in this Forest Policy is to increase the density of the forest. Since, there is no possibility for enhancing the cover of the forest as no new land area is available for plantation programs, the productivity of the forests can only be increased by enhancing their density. To achieve this target, several attempts will be done which are as under :

1. Any body who plants a plant and maintains it, will be given full authority to cut it. He /she will be in real sense the owner of that tree. At present, one has to seek the permission from the Government Forest Department to cut a green tree even if it has been grown by a person in his own land or in a public place. This will encourage the general masses to plant new seedlings and the green cover over the land will definitely increase.

2. The Bidi Leaf Policy (Tendu Patta Neeti) will also be changed in such a way that the poor leaf collectors get more economic benefit. At present , the leaf plucking and collection is done through co-operative societies.

3. Emphasis will be given on cultivation of medicinai plants (Jadi Booties) so that more raw material for the drug industries is generated and more employment for the marginal poors and tribals can be given.

4. Electricity will be generated from waste and not from coal, by establishing small power generating plants in the forests.

5. Emphasis is given on eco-tourism so that least destruction to the forest ecosystem is faced.

6. The forest area will be declared as Mineral Area so that revenue of the state government will increase.

7. Wasteland development through plantations will be encouraged through investment from private sectors without transferring the land to them.

8. The target of forest management will not be the earning of revenue, but to give priorities for the interests of the forest dependent communities.

9. A Forest Financial Corporation will be established in the state.

10. Special courts of law at district level will be established to settle the forest related crimes early.

11. Forest villages will be converted into revenue villages.

CONCLUSION

It can be easily concluded from the above account that as has been said, Madhya Pradesh is really rich state in all types of forest resources, i.e. timber, fuel, food, fodder, fiber, dyes, wildlife and several other minor forest produce.

REFERENCES

Ambasta, S.P. (1996). The useful plants of India. CSIR Publication, CSIR New Delhi, pp. 918.

Anon. (1988). Madhya Pradesh state of forests. Govt. of Madhya Pradesh, Forest Deptt. State Forest Research Institute, Jabalpur, pp 82.

Jesani, S.C. (1979). Project for cultivation of Sabai grass (Eulaliopsis binata) and Rope Manufacturing Industry. Project Report MPSFDC 44, M.P. Forest Development Corporation, Bhopal.

Lal, J.B. (1998). Policy, Planning and Management issues in NTFPs oriented forestry. Forest Usufructs, 1(1 & 2) : 59-66.

Pandey, R.K. and J.L. Shrivastava (1996). Material for forest flora of Madhya Pradesh. State Forest Research Institute, Jabalpur, pp. 181.

Prasad, R. and Parvez Jalil (1987). Grass production from Man made forests. Vaniki Sandesh, 11: 2-4.

Prasad, R. (1990). Status of medicinal and aromatic plants in Madhya Pradesh. Status Paper, National Seminar on Medicinal and Aromatic Plants, SFRI, Jabalpur. pp. 36.

Prasad, R. and P. Bhatnagar (1991). Socio-economic potential of Minor Forest Produce in Madhya Pradesh. SFRI Bulletin No. 26. M.P. State Forest Research Institute, Jabalpur, pp. 154.

Prasad, R. and Manish Mishra (2001). Non-timber forest products. Indian Institute of Forest Management, Bhopal, pp.76.

Advancing Frontiers of Ecological Researches in India, 125-141, 2006
Edited by A.K. Kandya & Asha Gupta
Published by Bishen Singh Mahendra Pal Singh, Dehra Dun, India

Participatory Carbon Management for Eco-sustainability of Forests and Dependent Rural Communities

P. K. Shukla and R. K. Pandey
State Forest Research Institute, Jabalpur (M.P.)

Abstract: The world has witnessed unprecedented rise in mean annual temperature during the last few decades. This phenomenon is known as global warming. This global warming is ascribed to the green house effect caused by increase in the concentration of green house gases (mainly CO_2) in the atmosphere. A number of global initiatives have been taken to mitigate climate change in recent years; the most significant among them being the adoption of UN framework convention on climate change (UNFCC) and development of Kyoto Protocol. Under this protocol, legally -binding GHG emission reduction targets have been fixed for industrialized countries. The Clean Development Mechanism (CDM) provides cost effective options for industrialized countries to meet their commitment of emission reduction by taking up CDM projects in less developed countries and thereby earning certified emission reductions (CER) credits. Different types of forest carbon projects like forest rehabilitation and regeneration, strict protection of forest areas, multiple use community forestry etc. have a great potential of mitigating climate change. In India, Joint Forest Management (JFM) programme has been very successful in protection and rehabilitation of forests with the participation of local villagers. However, it has been found that JFM has received more success where it has been accompanied with village resource development activities. Infect, it has been established that economic sustainability of dependent rural population is essential for the ecological sustainability of forests. In this paper, the potential of carbon trading as a source of funding for taking up developmental activities in JFM areas has been investigated. It has been estimated that there is a possibility of generation of funds to the tune of about 720 to 1440 billion US dollars per annum which will be more than sufficient to sustain JFM programme for the eco - sustainability of forests and dependent rural communities in India.

Key words: Climate change, global warming, green house effect, carbon emission, carbon sequestration, UNFCC, Kyoto Protocol, Clean Development Mechanism (CDM), LULUCF, Carbon trading, Joint Forest Management (JFM), Forest degradation, deforestation, Sustainable Forest Management (SFM), Eco sustainability, forest carbon projects, participatory carbon management.

INTRODUCTION

In recent decades, there has been growing concern over increase in the average global temperature of the atmosphere near Earth's surface, a phenomenon commonly known as " global warming". Although the mean annual temperature of the Earth has swung up and down by several degrees Celsius over the past 2 million years with intermittent glacial and interglacial periods, the current rise in average global temperature is mainly attributed to human - induced or anthropogenic factors. The global mean annual temperature has risen by approximately 0.5° to 0.7° C during the last 100 years (Botkin and Keller, 1995). A steady rise in temperature has been witnessed especially during the last three decades. Climatic models predict that the global temperature will rise by about 1° - 3.5° C by the year 2100, bringing in its wake disastrous consequences like changed patterns of rainfall, soil moisture and other climatic factors, significant increase in the frequency and intensity of cyclones and hurricanes, rising sea level leading to inundation of coastal areas, just to name a few. It is therefore, imperative to take up effective mitigation measures globally in order to save the humanity from catastrophic consequences of climate change.

Green house effect and green house gases

In order to tackle the problem of global warming, first have to understand what its possible causes are. The sunlight, on reaching the earth, warms both the atmosphere and the surface. The atmosphere then reradiates the heat as infrared radiation. Water vapour and several other gases, namely carbon dioxide (CO_2), methane (CH_4) ozone (O_3) nitrous oxide (N_2O) and chlorofluorocarbons (CFCs), present in the atmosphere, trap some of the solar heat energy. This phenomenon is called green house effect and these gases are, therefore, called "green house gases (GHGs)". In fact, increase in concentration of these GHGs in the atmosphere is responsible for the global warming. Although water vapour and small particles of water in the atmosphere produce about 97% of the total green house warming, we are not particularly worried about the warming caused due to it as it is not significantly increasing in the atmosphere due to anthropogenic pressures. We are worried about the warming caused by the other GHGs viz., CO_2, CFCs, CH_4, N_2O and O_3, all of which have significantly increased in the atmosphere in recent decades, mainly due to the human - induced factors. Among these GHGs, the contribution of CO_2 is most significant. About 50 to 60% of the anthropogenic green house effect is attributed to CO_2 alone and the concentration of this gas has been increasing at an alarming rate of 0.5 % per annum (Abelson, 1990) and if it continues to increase at the same rate, the CO_2 concentration will double in the next 140 years. At present, the concentration of CO_2 in the atmosphere is about 400 ppm and it is likely to reach at 450 ppm by the year 2050.

Carbon emission and carbon sequestration

The main causes of increase in CO_2 concentration in the atmosphere are burning of fossil fuels (coal, petroleum and natural gas) and deforestation. The burning of fossil fuels adds about 5.4 billion metric tons of carbon each year to the atmosphere (Botkin and Keller, 1995). With increase in human population and economic advancement, the consumption of fossil fuels is also increasing resulting in increased CO_2 emission levels. Moreover, these fossil fuel reserves are non-renewable and exhaustible. It is estimated that petroleum, natural gas and coal reserves, at the present level of consumption only, will be exhausted within 45, 70 and 250 years, respectively. It is therefore, necessary to work for the alternative sources of energy like solar, wind and tidal energy etc., which are not only clean but also nearly inexhaustible.

Plants consume atmospheric CO_2 during the process of photosynthesis and trap the carbon in its wood and other parts. This carbon remains trapped till the wood or other plant parts get burnt or are decomposed. This process is called carbon sequestration. After burning or decomposition , the trapped carbon is again released in the atmosphere in the form of CO_2. Forests having trees with long life span, have an important role in carbon sequestration and large scale deforestation is resulting in reduction of carbon sequestration capacity of forests and thereby increase in CO_2 concentration in the atmosphere.

Global initiatives to mitigate climate change

Growing public concern over the possible disastrous consequences of climate change forced the politicians also to think seriously over the problem of global warming and the World Meteorological Organization (WMO) and the United Nations Environment Programme (UNEP) established the Intergovernmental Panel on Climate Change (IPCC) in 1988 to collect and collate scientific information of climate change for policy makers. In the same year, the United Nations General Assembly adopted a resolution on the protection of global climate for the present and future generations of mankind.

The IPCC published its First Assessment Report in the year 1990. This report confirmed that climate change was indeed a grave threat and called for an early global treaty to address the problem. Second World Climate Conference was also held in Geneva in the same year and in this conference also, similar concerns were voiced. In response to these calls, the United Nations General Assembly established an Intergovernmental Negotiating Committee (INC) to conduct negotiations on a frame work convention on climate change in the same year.

The INC adopted the UN Framework Convention on Climate Change (UNFCCC) on 9th May 1992 in its fifth session held at New York. The ultimate objective of the convention was to achieve stabilization of atmospheric concentrations of green house gases at levels that would prevent dangerous anthropogenic interference with the climate system. However, dangerous levels were not defined in the convention and it was open for different interpretations by different parties. The convention embodied the principles of "equity" and "common but differentiated responsibilities" and under these principles, the industrialized countries, having contributed the most to the problem and also having more resources to address it were expected to take the lead by modifying their long term carbon emission trends and also by providing financial and technological resources to help developing countries in tackling the problem. The convention also recognized the linkage of climate change with development as the patterns of energy consumption, land use and demographic growth affect both development and climate change. The convention also emphasized the importance of promoting sustainable development. The convention divides countries into two main groups. Relatively wealthy and industrialized countries were subject to a specific commitment to adopt climate change policies and measures with the non - legally binding aim that they would bring back the GHG emissions to 1990 levels by the year 2000. These countries were also required to provide financial resources to enable developing countries to meet the obligations under the convention and also to take all practicable steps to promote development and transfer of environmentally friendly technologies to developing countries.

The UNFCCC was opened for signature on 4th June, 1992 at the UN Conference on Environment and Development (UNCED), or more popularly known as " Earth Summit" held at Rio de Janeiro, Brazil. The convention entered into force on 21st March,1994 and till now, 186 governments have become parties to the convention. The supreme decision making body of the convention , known as Conference of the Parties (CoP) has been meeting regularly every year to monitor its implementation and to decide the future strategies.

CoP-1, held in Berlin, launched a new round of negotiations to develop a protocol realizing that the commitments of the governments in the UNFCCC were not sufficient to tackle the grave and complex problem of climate change. After two and a half years of intense negotiations, CoP-3 meeting, held in Kyoto, Japan in December, 1997, adopted the Kyoto Protocol to the UNFCCC.

Kyoto Protocol

The Kyoto Protocol provides the framework for action under UNFCCC. The Protocol provides legally binding emission targets. Under the Protocol, the developed countries must put in place domestic policies and measures that cut their GHG emissions. They may also offset their emissions by expanding carbon-sinks. Provisions were also made to address the specific needs and concerns of developing countries. These includes the establishment of a new adaptation fund. Rigorous monitoring procedures were provided to safeguard the integrity of the protocol. A large number of countries, including EU countries, Japan, Canada, Russia, Australia and New Zealand welcomed the Kyoto Protocol. However, the protocol has received a severe setback with the USA., the greatest contributor to GHG emission, under Bush administration reversing its earlier policy decision and indicating its intention not to ratify the Kyoto Protocol.

Commitments

Legally binding emission targets for industrialized countries were fixed under the Kyoto protocol. These targets amount to a total cut among all Annex-I Parties of the Protocol of at least 5 % from 1990 levels by 2008-12, which is the commitment period.

The Protocol's emission targets cover the following six main GHGs- carbon dioxide (CO2), methane (CH_4), nitrous oxide (N_2O), hydrofluorocarbon (HFCs), perfluorocarbons (PFCs) and sulphur hexafluoride (SF_6). These GHGs have different global warming potentials (GWPs).

Besides emission targets, the Protocol contains other general commitments also e.g preparing national climate change mitigation and adaptation programmes, promoting environmental friendly technology transfer, cooperating in scientific research and international climate observation networks and supporting climate change education, training, public awareness and capacity building initiatives.

The land use, land use change and forestry (LULUCF) sector

Parties to the Protocol were allowed to offset their emission by increasing the amount of GHGs removed from the atmosphere by carbon sinks in the LULUCF sector. This sector offers relatively cheaper option to mitigate climate change, either by increasing the removal of GHGs from the atmosphere through carbon sinks or by reducing emissions from this sector. Carbon sinks can be expanded by planting trees, assisting natural regeneration or by accelerating tree growth rates through technological improvements. Emission from this sector can be reduced by curbing deforestation. The eligible activities in LULUCF sector are afforestation, reforestation, averting deforestation, forest management, cropland management and revegetation. Removals of GHGs from sinks generate removal units (RMUs) that an Annex I Party can use to help meet its emission target.

Clean Development Mechanism (CDM)

The clean development mechanism (CDM) provides cost-effective options of climate change mitigation to the industrialised countries by pursuing emission reduction or carbon sink enhancement activities more cheaply abroad than at home . Since the cost of curbing emission or carbon sinks enhancement varies from country to country, it makes economic sense to cut emission or increase removals where it is the cheapest to do so, given that the impact on global warming is the same. CDM projects, taken up in less developed countries through funding from the developed countries generate certified emission reductions (CERs) credits. Thus, through CDM, an international instrument has been created for trading carbon emission off set produced in developing countries.

Deforestation and forest degradation

There are approximately 2.5 billion ha of closed forest and 1.2 billion ha of open woodlands and savannas in the world (Sedjo, 1983). In recent decades, anthropogenic pressures resulting from increasing demand of biological resources have led to high rate of deforestation all over the world. It is estimated that forests covered one quarter of Earth's entire land area in 1950 but only one fifth in 1980 (Botkin and Simpson, 1990). During 1970-80, the world witnessed alarming rates of deforestation in tropical countries. The annual rates of forest loss at the global level during 1980-90 was about 16.3 m ha. Most of this deforestation has occurred in developing countries.

For example, Nepal lost more than half of its forest cover between 1950 and 1980. Land clearing for agriculture is one of the principal causes of deforestation in Nepal. In India also, particularly in tribal belts, considerable forest area has been encroached and cleared for agriculture. Deforestation is the second most important cause of increase in the atmospheric CO2 concentration, after fossil fuels burning. Besides deforestation, the forests of the world, especially the tropical forests are declining at an alarming rate due to various direct / underlying and natural / anthropogenic causes as shown in fig.1.

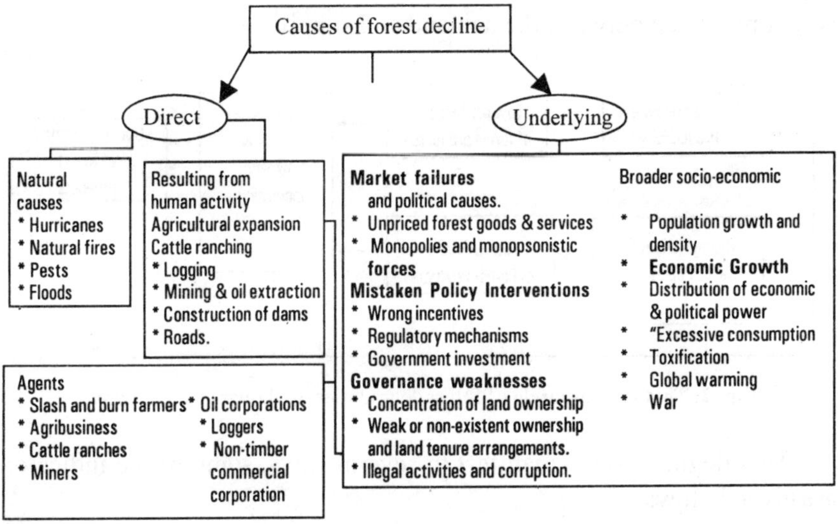

Fig. 1. The causes of forest decline

In India, forest degradation is greater cause of concern than deforestation. With increase in human population, biotic pressures on forests are continuously increasing. The most alarming problem is that of firewood shortage. Whereas firewood provides only less than 1 % of energy in developed countries, it provides one-quarter of the energy in developing countries. Firewood is removed from forests through headloads in quantity much beyond the carrying capacity of the forests. Most of the state governments in India allow free removal of dead, dry and fallen fuel wood from forests. Poor people living inside or near forest areas, collect fuel wood not only for their own consumption but also sell it in nearby towns and it is one of the major source of their livelihood. This practice is leading to over - exploitation of fuel wood in quantity much more than what can be sustainably removed from forests. Another major cause of degradation of forests is over grazing. Population density of unproductive and scrub cattle is the highest in India. These cattle are allowed to graze freely in the forests. Over-grazing leads to soil compaction which in turn results into reduction in soil moisture retention and increase in surface water runoff. The young regeneration is not allowed to establish due to excessive grazing pressure. Other forest products are also being over - exploited. Thus, there is a vicious circle of degradation. The overall productivity of Indian forests is already much less than the global average. Human and cattle population dependent on forests is very high. There is a huge gap between the demand and sustainable production limits of forest products. Therefore, there is over-exploitation of forest products leading to further loss of productivity and widening of gap between the demand and supply.

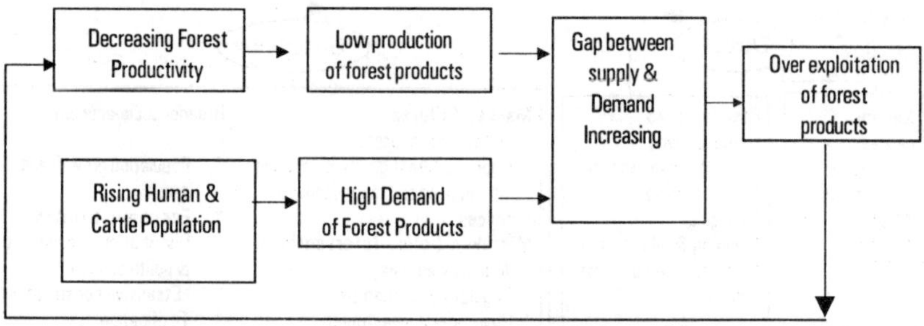

Fig. 2. This vicious circle of forest degradation is shown in fig. 2 above.

This degradation is manifest in several ways, some of the indicators are given below:

(a) Soil: Soil is one of the most valuable natural resources of the Earth but its productivity is going down. Due to excessive collection of leaves, twigs and fuel wood from the forest floor, the whole nutrient cycling is getting disturbed. There is now very little humus and soil organic matter. Over-grazing is leading to soil compaction and reduction in soil porosity. Population of soil micro-organisms is also reducing. In short, there is an overall degradation in physical, chemical and biological properties of soil.

(b) Water: Water is another most vital natural resource without which there can be no life on the earth. Due to over grazing and deforestation, the ground water table is going down and surface run-off is increasing.

(c) Vegetation: There is loss of bio-diversity. Many species have already become extinct and many more have reached at the threshold of extinction. The World Conservation Union has calculated that about 12.5% of the world's 2,70,000 species of plants are threatened by forest decline (WCFSD, 1999). Undesirable weeds are replacing useful species. High forests are gradually being converted into coppice forests. Repeated cuttings are resulting into loss in vigour.

We have come across several instances when trees of particular species suddenly start drying from the top and ultimately die. This phenomenon has been witnessed in sal forests of South Raipur Forest Division (now in Chhattisgarh State) in early nineteen eighties. Quite recently, teak trees have been reported to be dying in forests of Khandwa, Betul and Sheopur districts of Madhya Pradesh. The exact causes of this drying or dying of trees are still not known but it is obvious that it is the consequence of overall ecological degradation.

Sustainable Forest Management (SFM)

From the above discussion, it is clear that we are unable to sustainably manage our forests, especially those located near habitations. There cannot be two opinions that the key to forest management is sustainability. Following the Earth Summit, a number of initiatives were taken up at various levels to support sustainable forest management practices and to promote sustainable forestry development. In India too, criteria and indictors (C & I) for SFM were developed through Bhopal - India Process, embracing ecological, social, economic and institutional considerations.

Ecosystem approach for sustainable forest management

Forests constitute a complex, vital and dynamic ecosystem. It provides habitat and sustenance to a range of living organisms, starting from mammals, reptiles birds and insects to plants and soil micro-organisms, survival of each being dependent on others. Besides this, forests provide invaluable environmental services. Thus, it is essential that forests should be managed in a holistic manner by adopting ecosystem approach. Earlier, forests used to be treated as a factory to produce timber for the state and working plans emphasised merely on the sustainable timber yield, ignoring the importance of non-wood forest products and the environmental services provided by the forests. But now, there is a paradigm shift in the attitude of forest managers and other stakeholders and the view that forest should be managed as an ecosystem in totality, is gradually gaining ground.

Sustainable forest management through people's participation

Rural communities living in and near forests, particularly, the tribal communities, have been traditionally dependent on forests for food, fodder, fuel wood and other requirements and they are the people who have the highest stake in the conservation and sustainable management of forests. Prior to the advent of British rule in India, large tracts of common land, including forest areas, were available and designated for various uses under community vegetation. These common rights were collectively guaranteed by the village community. They were also made secure by customs revered as "sacred". The policy of "reservation" of forests adopted by the British rulers in India led to the alienation of people from the forests. This colonial policy of forest management continued even after Independence till eighties. As a result, forests continued to be subjected to encroachments, illicit felling, rampant forest fires, excessive grazing and unsustainable removal of forest products as participation of local people was not sought in the planning, management, protection or development of forests. It was therefore, realised that the management techniques depending upon the custodial policy had failed to protect the forests in India.

The National Forest Policy of 1988, for the first time, took note of the importance of protecting the livelihood security of the people who live in and around the forests, along with the ecological and environmental objectives of forest management. It was clearly enunciated that the holders of customary rights and concessions in forest areas should be motivated to identify themselves with the protection and development of forests from which they derive benefits. Having regard for the symbiotic relationship between the tribal people and forests, the primary task of all agencies responsible for forest management should be to associate the tribal people closely in the protection, regeneration and development of forests as well as to provide gainful employment to the people living in and around the forests. This led to the concept of "eco-development" where 'eco' denotes to both 'ecological' as well as 'economic'. Thus "eco-development" approach paves the way for the ecological development of forests, along with the economic development of the forest dwellers.

Involvement of forest dependent rural communities as co-managers- The Joint Forest Management (JFM)

In pursuance of the National Forest Policy of 1988, the Government of India, vide Circular No. 6-21/89.FP dated 1st June 1990, issued guidelines to state governments to organise the beneficiaries (forest dependent rural communities) into institutions for ensuring people's participation in forest regeneration and protection. Such beneficiaries would be entitled to the usufruct of all non-wood forest products and a share in proceeds of final tree harvest, subject to successful protection. With this came the concept of 'Joint Forest Management' as a strategy of forestry development with people's participation in the planning and management of forest resources JFM has been defined as the sharing of products, responsibilities, control and decision making authority over forest lands between forest departments and local user groups who are said to be the co-managers of local forest resources.

Table : JFM programme in Madhya Pradesh.

Sn.	Name of the Committees	No. of Committees
1	Village Forest Committees (VFCs)	9035
2	Forest Protection Committees (FPCs)	4201
3	Eco-development Committees (EDCs)	785
	Total	**14021**
4	Total Forest Area under JFM	60.000 sq. km. (63 % of total forest area of the State).
5	Total no. of Families involved in JFM	17 lakhs

Source: M.P. Forest Department (www.mp.nic.in/forests)

In June 1991, the Govt. of India issued a circular supporting joint planning and management of degraded forests in all parts of India. Following the National Forest Policy of 1980 and the subsequent GOI guidelines, most of the states, including Madhya Pradesh, initiated JFM programmes. The Govt. of Madhya Pradesh passed a resolution in 1991 for community participation in preventing illicit felling and rehabilitation of degraded forests. Under the JFM, the user groups have been brought under an institutional structure comprising of Forest Protection Committees (FPCs), Village Forest Committees (VFCs) and Eco-development Committees (EDC's). The (FPCs) are constituted in sensitive forest areas for the protection of 'dense' forests (forests having more than 40 % of canopy cover) VFCs are constituted in open forest areas (forest having less than 40 % canopy cover) and the EDCs in the buffer zones of protected areas i.e. national parks and wildlife sanctuaries. So far, more than 14,000 such committees have been formed in Madhya Pradesh alone by involving 17 lakhs rural forest dependent families to manage 6000 km^2 of forest land, which is 63 % of the total forest area of the State.

The Joint Forest Management Programme has been very successful almost everywhere in enlisting the participation of local communities in the protection and sustainable management of forests. It is now widely accepted that the forests can be conserved and developed only through collaborative efforts with the people. JFM has taken roots in the forestry system and has become more open and responsive to the needs and aspirations of the people.

Village resource development programme (VRDP)

Village resource development has played a key role in the success of JFM in. Under the World Bank assisted Madhya Pradesh Forestry Project, VRDP was the most important component. A number of development works like creation of drinking water and irrigation facilities, construction of village approach roads, school buildings, etc. were taken up in villages. These VRDP works acted as confidence building measures among villagers and helped in removing their alienation and promoting participatory approach.

Eco-Sustainability

The term 'eco-sustainability' refers to both the ecological sustainability of forests as well as the economical sustainability of dependent rural communities. Both are complimentary to each other. Increasing biotic pressures of the burgeoning human population in developing countries of this region, has made the management of forests and other natural resources unsustainable as these resources are being over exploited. Poor rural population residing in and around forests has to depend mainly on forests for their requirements of fuel and fodder. Change in consumption pattern of these forest products is possible only with the economic upliftment of the dependent rural population. It has also been observed that people participated enthusiastically in JFM where village resource development works and income and employment generating activities were undertaken by the forest department and other developmental agencies. It is now evident that conservation and sustainable management of forests is possible only through active participation of dependent rural communities and that in turn, is dependent on simultaneous economic development of villages.

From the above discussion, it follows that sustainable economic development is essential for the success of participatory forest management. Releasing the importance of economic development, most of the state governments initiated entry – point development activities while launching JFM programmes to win the confidence of local people. However, these development activities require sustained funding which is not possible unless some self – sustained mechanism is evolved to generate enough funds for economic development works as a bye - product of sustainable forest management itself. It has been observed during the implementation of JFM programmes in many states that JFM programmes were launched, accompanied with economic development activities under some externally funded forestry projects. Villagers also displayed lot of interest in the protection and development of forests under JFM as there was a massive flow of funds for the economic development of villages also. But with the winding up of the externally funded projects, the funding source dried up and it was difficult to sustain the interest of villagers in JFM. More difficulty was faced in the expansion of JFM programme in new villages in the absence of entry-point development activities.

Sustainable development and CDM

Clean Development Mechanism (CDM) under the Kyoto Protocol provides a potential source of funding to developing countries for projects of sustainable management. Article 12.2 of the Kyoto Protocol specifies that CDM projects should assist host countries in achieving sustainable development.

According to Pearce *et al.* (1988) and Bass *et al* (2000), sustainable development can be viewed as the creation of opportunities and capacities for raising per capita well-being, i.e. the quality of life . Carney (1998) has specified the following five types of assets (or capital) that need to be expanded or made more productive to achieve this end:

1. Natural capital, such as land, forests and water
2. Physical assets, such as farm tools, roads, electricity etc.
3. Financial capital, such as savings
4. Human capital, such as health and education, and
5. Social capital, such as membership in FPCs/VFCs etc.

The concept of sustainable development includes equity which can be achieved by increasing the well being of less advantaged populations. Thus, sustainable development not only promotes opportunities for increasing the above five forms of capital but also facilitates empowerment, ie., strengthening the participation of poor people in local decision making (World Commission on Environment and Development, 1987).

Natural forests and plantations represent more than natural capital. Income from the sale of forest products such as *tendu* leaves, medicinal plants and gums etc. provide financial capital. Forest incomes are a vital economic buffer, particularly for women, children and the poorest households in village communities (Byron and Arnold, 1999). Rehabilitation of degraded forests and eco-development activities under JFM programmes help in the creation of physical assets and human capital. JFM has proved to be a powerful tool for enhancing the social capital also by enlisting the people's participation in forest management. Thus, JFM leads to sustainable development and qualifies for CDM.

Forest-carbon projects and livelihood benefits

Different types of forest carbon projects like forest rehabilitation and regeneration, strict protection of forest areas, multiple-use community forestry (Timber, NTFPs fodder etc.), agro-forestry, pulp wood or timber plantation etc. have potential to mitigate climate change through two major strategies viz., afforestation and reforestation (AR) and averted deforestation (AD). AR helps to mitigate climate change by establishing additional forest area that sequesters carbon as trees grow and then stores carbon for the life of the standing trees. The second strategy AD protects standing forest carbon stocks that would otherwise be lost. While the focus has so far been on strictly protected areas under AD strategy, it is now being realized that multiple-use community forestry can provide greater livelihood benefits. However, AD projects are not presently eligible under CDM (Smith and Sherr, 2002).

In regions where natural forests have been depleted due to intense biotic pressure, the dependent communities can be involved in forest rehabilitation and protection. Techniques such as assisted natural regeneration (ANR) and coppice management can accelerate forest recovery and improve quality (Saxena, 1997).

Participatory Carbon management

JFM had already occupied more than 7 million ha of forest land in India by 1999 (MoEF, 1999) and it has increased to a great extent between 1999 and 2003. FSI report suggests that after JFM coming into operation, the forest cover in India has almost stabilised (Rangachari and Mukherji, 2000). It is obvious that JFM has been extremely successful and it has now come to stay as a viable strategy of sustainable forest management. The key stone for the success of JFM is participatory approach by involving the dependent rural communities, who are the greatest stake holders, in the protection and management of forests. However, as already stated, the interest of these rural communities can be sustained if village development works are also carried out simultaneously.

Now, the question arises from where can we get the funds required for village development sustainably. Emerging global carbon markets can be the answer. Studies carried out in central India revealed that teak and dry deciduous forests under community protection sequester 1 to 3 metric tons of carbon per ha per day and that the best opportunities were in degraded forests with crown cover under 40 % (Poffenberger, 2001). According to the land use /land cover statistics of India generated by remote sensing techniques, the areas under non-forest degraded lands are currently 93.68 million ha and 35.89 million ha (Singh and Lal, 2000). They have estimated that if a conservative managed forest productivity of 5.5 tones per annum could be attained on only 40 million hectares of the available surplus degraded land in India through managed forestry programmes such as short rotation forestry plantations, it is possible to obtain carbon mitigation of about 3.32 Gt in the next 50 years, with an annual reduction of about 0.072 Gt of carbon.

It is therefore, proposed that a 50-year contract be entered into by the village level JFM committees (FPCs/VFCs/EDCs) with the outside financing agencies. The financing organizations would agree to finance carbon sequestered at a fixed rate of US$ 10 per ton per year during the first 25 years of operation, after which the rate would increase to US$ 20 per ton per year (Poffenberger *et al. 2001).* At this rate, there is a possibility of generation of funds to the tune of about 720 billion to 1440 billion US$ annually, which is a gigantic sum to sustain JFM programme for the eco-sustainability of forests and dependent rural communities.

REFERENCES

Abelson, P.H. (1990). Uncertainties about Global Warming. *Science*, 247.

Bass, S., Dubois, O., Moura-costa, P. Pinard, M., Tipper R. and Wilson, C. (2000). Rural livelihood and carbon management. IIED Natural Resource Issues Paper No.1. International Institute for Environment and Development, London.

Botkin, D.B. and Keller, E.A. (1995). *Environmental Science*: Earth as a Living Planet. John Wiley and Sons. Inc. New York.

Botkin, D.B. and Simpson, L. (1990). The first statistically valid estimate of biomass for a large region. *Biogeochemistry*, **9**:161-174.

Byron, N. and Arnold, M. (1999) What future for the people of the tropical forests*? World Development*, **27**:789-805.

Carney, D. (1986). Implementing the sustainable rural livelihood approach In: Carney, D. (ed.) Sustainable rural livelihoods: What contributions can we make? 3-23, DFID, London.

CIFOR (2000). The underlying causes of forest decline Occasional paper No. 30. Centre for International Forestry Research, Indonesia.

FAO, (1997). State of the World's Forests, Food and Agricultural Organization, UN Rome, Italy.

FAO, (1999). United Nations Food and Agricultural Organization. Rome, Italy.

Killmann, W. (2001). Forestry and climate change after COP 6. FAO advisory committee on paper and wood products, FAO, UN Rome Italy.

MoEF (Ministry of Environment and Forests) (1999). National Forestry Action Programme in India. Government of India, New Delhi.

Pearce, D., Bay, B., Newcombe, J., Brunnelo, T. and Bello, T. (1988). The Clean Development Mechanism, benefits of the CDM for developing countries. Unpublished report. CSERGE, University college, London.

Poffenberger, M. (ed) Ravindranath, N.H.: Pandey, D.N., Murthy, I.K., Bist, R. and Jain, D. (2001). Communities and climate change: The Clean Development Mechanism and village based forest restoration in Central India. A case study from Harda Forest Division, Madhya Pradesh, India. Community Forestry International, Inc. and the Indian Institute of Forest Management, Bhopal: Santa Barbara, California, U.S.A.

Rabindranath, N.H., Murali, K.S. and Malhotra, K.C. (2000). Joint forest management and community forestry in India. An ecological and Institutional Assessment. Oxford and Inc. Publishing Co. Pvt. Ltd. pp. 294-297

Rangachari, C.S. and Mukherji, S.D. (2000). Old roots new shoots. A study of joint forest management in Andhra Pradesh, India Winrock International and Ford Foundation, New Delhi.

Saxena, N.C. (1977). The saga of participatory forest management in India. Special Publication. CIFOR, Bogor, Indonesia.

Sedjo, Roger A. (1983). The comparative economics of plantation forestry: a global assessment. Resources for the future. Research paper. John Hopkins Univ. Press. Baltimore, U.S.A.

Sharma, R.C. (2003). Creating awareness among different stakeholders and policy makers regarding the recent technological innovations and research advancement for application in JFM-Chhattisgarh perspective, Proceedings of International workshop on technological innovation and research advancement for application in JFM ICFRE, FAO, pp. 23-40.

Singh, Roma and Lal, M. (2000). Sustainable forestry in India for carbon mitigation. *Current Science*, vol. 78, No. **5**:563-565.

Smith, J. and Scherr, S.J. (2002). Forest carbon and local livelihoods. CIFOR Occasional Paper No.37.

World Commission on Environment and Development (1987). Our common future. Oxford University Press, Oxford, U.K.

World Commission on Forests and Sustainable Development (WCFSD) (1999). Our forests our future. Summer Report of WCFSD, Canada; Winnipeg.

Advancing Frontiers of Ecological Researches in India, 143-160, 2006
Edited by A.K. Kandya & Asha Gupta
Published by Bishen Singh Mahendra Pal Singh, Dehra Dun, India

Land use Decisions by locals affecting Spatial Biodiversity

Alka Chaturvedi, Rucha Ghate[1], and Phani Kumar Garlapati
Department of Botany; Nagpur University, Nagpur

Abstract :Effective management of the bio -resources is a real challenge for the manager, planner or even to appropriators; particularly in the developing countries, where mouths to feed are more than the resources. In India management of the resource by state has been practiced for more than 150 years and has proved to be ineffective. Although the figures of encroachments on the land under forests are soaring, there also are studies indicating existence of communities that are consciously maintaining and managing the forests within their village boundaries, at their own initiatives. Present paper is an effort to analyze how land use decision by locals, affect bio diversity, and what are probable reasons for institutions to emerge and successfully adopt sustainable resource-use strategy.

Biodiversity is an inherent quality of all the living organisms. It is absolutely essential for stability of different ecosystems and proper functioning of their food chains. To a great extent it depends on the people of the region, their culture, social and food habits as well as agriculture priorities. Historically man has lived with diversity but unfortunately man has been the greatest enemy of the diversity even from the prehistoric times. Dependence of the people/society on the forest is forcing countries to think about the quality and quantity of their natural resources.

The unsustainable rate of forest exploitation in tropical regions is a matter of great concern for the governments, environmentalists, and local communities alike. Although deforestation can be attributed to several factors like growing population pressure, technology, government's policy of exclusive management by bureaucracy, commercial logging, poverty etc., the apparent impact is – conversion of forest land into non-forest uses. Forest areas, especially the ones adjoining human habitats, are more likely to get converted into agricultural fields. In recent times, one such important land use change has been encroachment on forest-land for the purpose of cultivation.

---

[1]Department of Economics, Nagpur University, Nagpur

Cultural history of India shows linkage of plants with humans by various ways. Each and every aspect of human life is somehow related to plants. But with the change in values in modern societies, plants remain as a only things to be exploited. At present more than 80 percent of India's forests are under state control. Exclusive ownership and management of the resource by state have been practiced for more than 150 years and have proved to be ineffective. This can be said because neither the target of 33 percent of country's total geographical area under forests (as stipulated in subsequent National Forest Policies) has been achieved nor has it been able to maintain the forest cover that existed at the time of independence. Analysis of the causal factors responsible for the present scenario has brought forth a now well-accepted fact that *de facto* land use decisions of the local populations are more determining than the *de jure* official decisions. Although the figures of encroachments on the land under forests are soaring, there also are studies indicating existence of communities that are consciously maintaining and managing the forests within their village boundaries, at their own initiatives. While some village-communities prefer extending their legal boundaries of cultivable land at the cost of forest-land, discounting future heavily, there are others who subscribe a very low discount rate at the cost of present income and prefer sustainable eco-friendly development practices. A fundamental puzzle that intrigues researchers as well as policy makers is, while some communities organize themselves to solve the problems of institutional supply, why others in similar circumstances do not.

Two aspects discussed in the present paper are: how land use decision by locals, affect bio diversity, and what are probable reasons for institutions to emerge and successfully adopt sustainable resource-use strategy. For this purpose cases of two neighboring villages in Bramhapuri taluka in Chandrpur district of Maharashtra state have been presented. One of these, Saigata, has preferred sustainable ecological use of forest to immediate benefits of higher income by harvesting forest products, or converting the forest land into agriculture land. The other village, Maral Mendha, has heavily encroached upon the forest land for the purpose of cultivation, preferring current income from agriculture. Difference in these two approaches towards land use, reflect on the quality of the resource in terms of diversity of species, regeneration capacity of land etc. Apart from inter comparison between these two villages representing two diverse strategies, intra comparison in case of Saigata, which has different forest management zones, is also attempted to elucidate its impact on density and bio diversity. The paper has been structured in the following manner: brief discussion of institutional studies in common pool resources (CPRs); demographic background of the two villages; results of forest mensuration work in the two villages; institutional analysis; and conclusions.

Theoretical Background of Common Pool Resources

In situations where human being and forests co-exist, sustainability of the resource largely depends upon the suitability of the institution affecting the common-pool resource. A resource, from which it is costly but not impossible to exclude potential consumers and at the same time consumption of the resource is both subtractive and potentially rivalrous. Centralized or public management of forests is an outcome, among other reasons, of the belief reflected in Olson and Hardin's work. Contribution to commons literature has traveled a long way since Olson's (1965) 'zero contribution thesis' through Hardin's 'tragedy of commons' (1968), to many scholars taking cognizance of voluntary associations that are creating and enforcing rules that protect natural resources. Contrary to this, Baland and Platteau (1996) argue that the privatization of common-pool resources or their appropriation and regulation by central authorities tend to eliminate the implicit entitlements and personalized relationships that are characteristics of communal property arrangements. These steps, therefore are likely to impair efficiency, and even more likely to disadvantage traditional users whose rights of use seldom get recognized under privatization or expropriation by the state. Evidence has accumulated that externally imposed rules tend to 'crowd out' endogenous cooperation behavior (Frey, 1994). The greater cooperation that had occurred due to the exogenously created incentive-compatible mechanism appeared to be transient. In a world of no external rules or monitoring, norms can evolve to support cooperation.

While the choice regarding a form of institution for governing the commons depends upon ideological, political, and social background of a given country state, its suitability is determined by the factors important to sustainability, equity, and efficiency of commons. Apart from its effect on forest condition and optimum resource use, shared understanding of social norms plays a crucial role. The writings of scholars of commons, with historical and contemporary evidence, have shown that resource users often create institutional arrangements and management regimes that help them allocate benefit equitably, over long time periods, and with only limited efficiency losses (McKean 1992, Ostrom 1992, Agrawal 1999). Literature published in last two decades provided ample evidence to this. Field research has established today that though temptation to free ride on the provision of collective benefit is a universal problem, self-organized resource governance too has survived in multiple generations (Ostrom, 1990). It is also found that when the users of a common-pool resource organize themselves to devise and enforce some of their own basic rules, they tend to manage local resources more suitably than when rules are externally imposed on them (Tang, 1992; Baland and Platteau, 1996; Wade, 1994).

Among the writings of many scholars, Robert Wade, Elinor Ostrom and Jean-Marie Baland and Jean-Philippe Plateau are significant because of their analyses of sustainable management efforts of common-pool resources. After wide ranging discussions and considerations of many factors, each arrive at a summary set of conditions and conclusions that they believe to be critical to sustainability of commons institutions. Wade's (1988, 1994) important work examines when it is that corporate institutions arise in villages and what accounts for their success in resolving common dilemmas, and provides set of reasons about successful management of commons. Small group size, clear boundaries, and ease in monitoring and enforcement are some of the conditions, according to him, which determine the effectiveness of rules. On the basis of 14 case studies, Ostrom (1990) has designed 8 principles i.e. "essential elements or conditions that help to account for the success of these institutions in sustaining the CPRs and gaining the compliance of generation after generation of appropriators to the rules in use" (1990:90). These are: clearly defined boundaries, congruence, collective-choice arrangements, monitoring, conflict-resolving mechanisms, minimal recognition of rights to organize, and nested enterprises. Later, she further suggested attributes of the resource and attributes of the appropriators that are conducive to an increased likelihood that self-governing associations will form (Ostrom, 1999a). Attributes of the resource are: possibility of feasible improvement of the resource, availability of reliable and valid indicators of the condition of resource system, predictability, and sufficiently small spatial extent for the appropriators to have appropriate knowledge of external boundaries. Attributes of the appropriators are: salience of the resource, common understanding, low discount rate, trust and reciprocity, autonomy, and prior organizational experience and leadership.

Baland and Platteau too have reviewed large number of studies on commons. They suggest that the failure of CPRs in comparison to privatization "rests on the comparison between an idealized fully efficient private property system and the anarchical situation created by open access" (1996:175). They suggest incorporation of cultural and political factors into analysis to know when people cooperate, and when inverted opportunists dominate and make collective action impossible. Small size of user group, location close to the resource, homogeneity among group members, effective enforcement mechanism, and past experience of cooperation are some factors necessary to achieve cooperation. Characteristics of the resource as substantive factor affecting effectiveness of institutions governing the commons too are varied. Volatility and unpredictability in the flow (Wade, 1988), mobility of resource (Naughton-Treves and Sanderson, 1995), stationarity and storage (Blomiquist et al 1994) are some of the aspects in addition to the 'attribute ot the resource' (Ostrom, 1999b) mentioned earlier. Role of technology, population pressures, property rights regimes, heterogeneity, are some of the many other variables studied by different scholars. For the present study we have tried to apply both the 'attributes of appropriators' and 'attributes of the resource', given by Ostrom and see if these are able to answer the fundamental question - why collective action was successful in Saigata but not Maral Mendha.

The Two Case Study Villages

Both Saigata and Maral Mendha are located in Bramhapuri tahsil of Chandrapur district in Eastern Maharashtra, sharing geographical boundaries. The villages are 2 kms. off the Nagpur–Gadchiroli highway (110 kms. from Nagpur). Total population of Saigata is 432 with 88 households. Maral Mendha is slightly a bigger village with population of

Table-1

Sr. No.	Various Aspects	Saigata	Maral Mendha
1	No. of Households	8	125
2	Population	432	700
3	Male	219	250
4	Female	212	350
5	Main Occupation	Cultivation	Cultivation
6	Total forest area	280.13 ha	71.46 ha
7	Per capita forest area	0.65 ha	0.10 ha
8	Legal status of forest	Reserved and Protected forest	Protected forest
9	Concept of wealth in the community	Sufficient food and ownership of cattle	Ownership of agriculture land, house and gold
10	Type of majority houses	Thatched	Concrete and brick
11	Population mix/caste composition	Nav bouddha, Gond, Mana, Mali, Lohar, Dhiwar, Kunbi	Mali Mahar, Teli, Gond, Dhiwar

700 in 125 households. Heterogeneity of population i.e. caste composition is much more in case of Saigata than Maram Mendha. Maral Mendha is dominated by one caste–Mali, which is known as vegetable growers (see Table-1). Main occupation of the villagers in both the villages is farming, and paddy is the main crop grown.

Saigata

In Saigata village collective action has been successful in preserving forest for more than past 20 years. Informally "Krushak Charcha Mandal" was formed in 1973. In Saigata Shri Surybhan Khobragade emerged as the undisputed leader of the people when he started the movement by forming "Nabhovani Shetkari Vachanalaya" (Farmers' Library) and "Nabhovani Shetkari Mandal" (Radio Audience Club). People listened to the agricultural program every day and learn improved farming methods. They also interacted with the experts through this radio program. Shri Khobragade also used traditional art forms (Dandhar) and composed poems, songs, one-act plays to spread the message of forest protection, gradually extending the activity to forest protection. A patch of forest very close to the settlement is known as 'Andhari', meaning 'darkness', because sometime ago it was covered with forest that was so thick that light did not reach the ground. Perturbed by the highly degraded state of the resource then, the community decided to invest in the restoration of the resource. The very first step was barring the outsiders to enter and harvest any forest product. Next step was imposing self-discipline by the community. Gradually the informal, local institution emerged with formation of rules, determining sanctions, and methods of monitoring. A very thoughtful and unique step was determining management zones. A patch of forest was exclusively kept aside for natural generation and nobody was allowed even to enter it. A patch, established for good quality grass, was developed to meet fodder and thatching requirements. Other patches were earmarked for gathering of fuel and small timber and were opened up according to availability. Later, in 1993, when the opportunity of getting associated with Forest Department arose due to Joint Forest Management Program, "Forest Protection Committee" (Van Samrakshan Samiti) was established formally in 1993. With the available departmental funds, plantations were taken up to further improve the quality of forest.

Forest Menstruation

This area was known to have very good forest with the number of dominant species. It was in very good condition till 1970. Later, within 8 years, it was degraded while kept under Revenue and Forest department. Forest map of Saigata (1977-78) shows total forest area of 280.13 ha. and per capita forest was about .65 ha. Present day scenario suggests complete negligence of the area. Although the area is almost flat-land and there is no active or visible erosion of soil, but soil texture, depth, it's profile and thickness of humus layer suggest its poor condition. Development of soil is a very slow process, but at the same time, grazing by livestock, presence of pest in soil and absence of crown cover is also responsible for such type of soil. Absence of epiphytes is clear -cut indication of less ambient humidity.

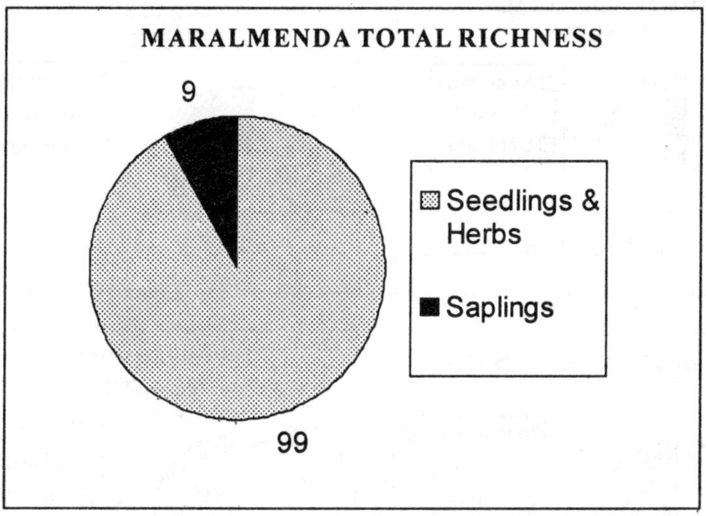

Present observations show increase in the density of shrubs and bushes, but the total area under forest cover is reduced. Reduction in the density of dominant trees are due to anthropogenic influence. There is constant pressure and encroachment from the surrounding villages. But increase in the shrubs and bushes tells us another story of protection and prevailing conditions which are allowing natural growth of these plants. 1979 onwards when local community decided to *manage* this area there own, they distinguished five distinct units for better management and sustainable development. (Figs. 1-6).

Gawatache Beed is kept for the fodder and thatching material. This area shows pressure of grazing animals. Woody plants are not in existence. Only few saplings of *Terminalia* and *Tactona* are visible. Ground vegetation is of annual nature. Dominant plants belonging to family Poaceae, Fabaceae

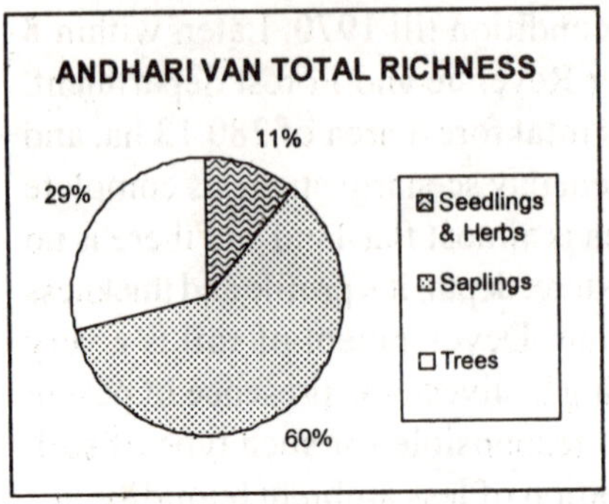

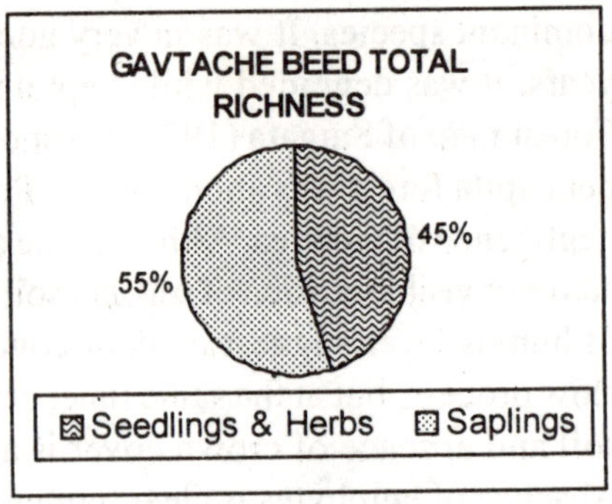

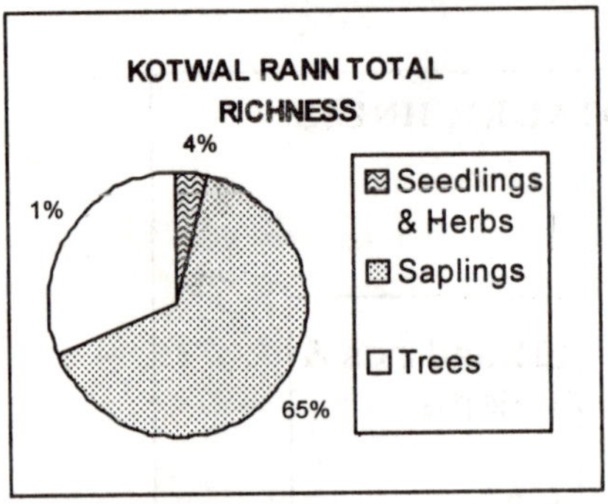

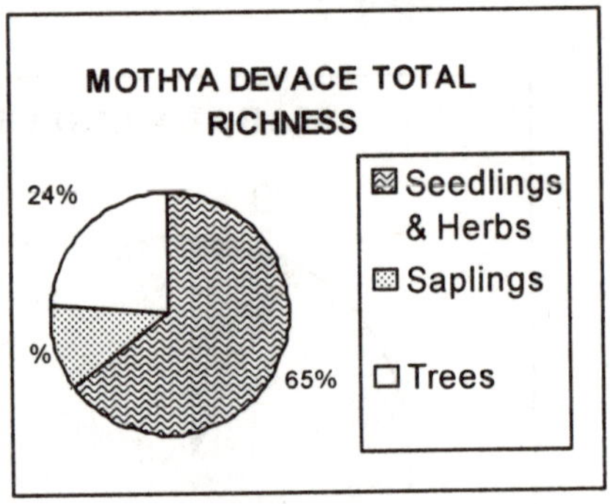

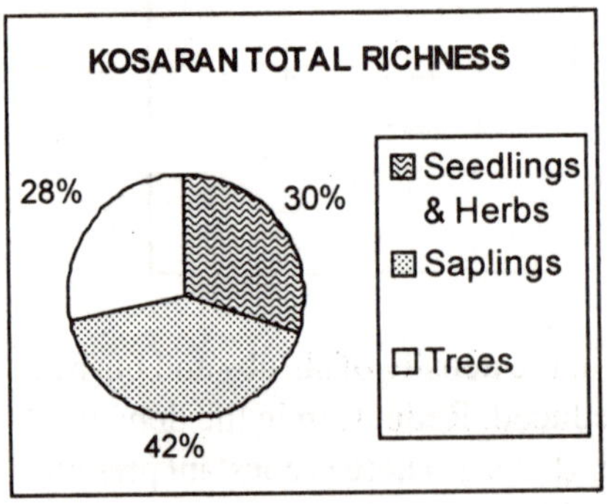

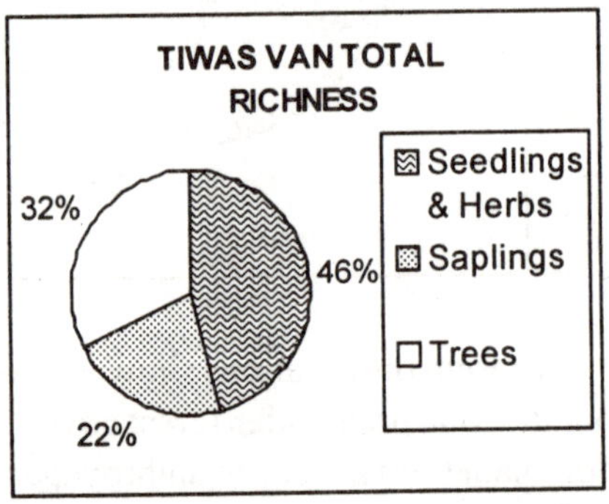

and Acanthaceae. Most common species are *Anotis montholoni, Andrographis echioides, Andropogon pumilus, Ageratum conyzoides, Atylosia scrabroides, Borreria articularis, Blepharis diffusa, Canscora dissecta, Cassia absus , C. tora, Crysopogon sp.,*Crotaleria *ramosissima, Desmodium dichotomum, Grewia hirsuta, Justicia procumbens, Oxalis corniculata, Rungia pectinata, Sida cordifolia,Tridex procumbens, Lepidagathis cristata, Rostellularia diffusa, Indigofera simplicifolia* and *Phyllanthus virgatus*.

Andheri Van is completely prohibited for any fresh wood cutting activity. Only dead, dried wood is allowed to be collected for the purpose of fuel wood. This area shows two storied vegetation. The top strata is dominated by species of family Combrataceae, Meliaceae, Myrtaceae. The average height of the plants in this strata is in between 6-10 meter and DBH between 15-19 centimeter. Most common ones are *Terminalia alata, T. bellerica ,Xeromphis spinosa, Tectona grandis, Cleistanthus collinus, Anogeissus latifolia, Albizia odoratissima, Lagerostroemia parviflora, Butea monosperma.*

Sapling flora is dominated by clumps of *Cleistanthus collinus and Terminalia alata*. Other woody seedling and saplings belong to, *Albizia odoratissima, Xeromphis spinosa, Anogeissus latifolia, Diospyros melanoxylon, Flacourtia indica, Lantana camara, Schelicaria oleosa.* Plants of this region are less than 6 meter in height and below 10 centimeters in diameter.

Calycopteris floribunda and *Mimosa hamata* are prominent woody climbers of the area.

Ground flora is mainly composed of grasses. More frequent species are *Ischaemum pilosum* and *Chrysopogon aciculatus..*

Kotwal Raan is marked for some management practices. It shows restricted felling of Trees, particularly for household needs. Bigger trees are consciously protected.

Ground vegetation is completely absent. Sapling flora is showing high density, *Cleistanthus collinus, Albizia odoratissima, Lagerostroemia parviflora* form dominating saplings. Other species belong to *Butea monosperma, Tectona grandis, Terminalia alata, Mimosa hamata, Ixora arborea, Buchanania lanzan and Leea crispa.*

Woody vegetation presenting majority of the plants below 10 meter in height. Here also *Albizia odoratissima* is dominating species. Co-existent plants are of *Tectona grandis, Terminalia alata, T. bellerica, Pterocarpus marsupium, and Diospyros montana* and *Buchanania lanzan. Tectona grandis* is the only species which is representing plants more than 10 meter in height.

Mothya devache van is likewise barren in ground vegetation. However seedling and sapling form dense vegetation. Saplings of *Terminalia alata, Cleistanthus collinus, Anogeissus latifolia,* show highest rate of density and frequency. Saplings of other species are represented by shrubby *Maytanus emerginata, Grewia hirsuta, Flacourtia indica, Soymida febrifuga, Chloroxylon switiana and Cassia fistula*, in addition to other common species. Trees are below 10 meter in height. Dominating species are *Terminalia alata, and Lagerostroemia parviflora.* But they are having low DBH. Other common species are *Anogeissus latifolia, Buchanania lanzan, Lannea coromondelica Madhuca longifolia, Balanities aegyptica and Zizyphus oenoplia.*

Kosaran is divided in two blocks. In both ground vegetation is rich in species composition but it is sparse and less in frequency. Common species of this area are *Andrographis echioides, Anotis montholony, Ageratum conyzoides, Hemidesmus indicus, Rungia repens, Rostellularia diffusa and Tridex procumbens.* Sapling flora is dominated by *Anogeissus latifolia, Buchanania lanzan* and *Terminalia alata.* Trees are of smaller size, are less than 7 meter in height. Common trees are *Lagerostroemia parviflora, Anogeissus latifolia,Buchanania lanzan, Madhuca longifolia.*

Tiwas Van forms biggest patch ,which is kept for specific forest management practices. From last four years this area is supporting Teak plantation. Ground vegetation is rich in species of family Acanthaceae, Fabaceae and Poaceae. Common species *are Alysicarpus racemosus, Andropogon pumilus. Blepharis repens, Borreria articularis, Cassia tora, Crotoleria juncea, Evolvulus alsinoides, Heteropogon contortus, Indigofera glandulosa , I. linifolia, Lepidagathis cristata, Oxalis corniculata, Peristrophe bicalyculata, Phyllanthus virgatus and Sida cordifolia.*

Sapling flora is dominated by *Anogeissus latifolia Balanities aegyptica*, bunch of *Cleistanthus collinus, Buchanania lanzan, Lagerostroemia parviflora, Diospyros melanoxylon, Mimosa hamata, Zizyphus glaberrima ang Z. oenoplia.* Other species which are present in the area are *Bridelia retusa, Gardenia resinifera, Grewia hirsuta, Ixora arborea, Schleichera oleosa, Terminalia alata and Wrightia tinctoria.* Bunch of *Bambusa arundifolia* are forming common seen here and there.

Trees are of short stature. They are less than 10 meter in height. Dominating species are *Anogeissus latifolia* , *Lagerostroemia parviflora, Terminalia alata.* While other common species are *Cleistanthus collinus, Buchanania lanzan, Lagerostroemia parviflora, Diospyros melanoxylon, Mimosa hamata, Albizia odoratissima, Cassia fistula, Lannea coromandelica and Semecarpus anacardium.*

Analysis of flora and vegetation of **Saigata** shows influence of different management practices in different sectors. **Gawatache Beed**, which is kept for grazing is showing predominance of ground flora. Although some woody saplings are making their appearance. *Xeromphis spinosa* is highest in relative density and relative frequency.

Complete prohibition of wood -cutting is responsible for increase in the density and frequency of woody vegetation in **Andhari van**. Saplings flora is dominated by highest relative frequency (14.28) of *Cleistenthus collinus* and *Terminalia alata*. However *Anogeissus latifolia* is showing heigher density of 27.02%. More frequently distributed trees are *Anogeissus latifolia* and *Tectona grandis. Butea monosperma*, is relatively densely placed but its saplings are completely replaced by *Cleistenthus collinus, Diospyros melanoxylon, Xeromphis spinosa and Mimosa hamata.* Absence of saplings of . *Butea monosperma,* can be attributed due to the over exploitation of young ones, as its leaves are used in making plates and bowls.

Controlled forest clearing in **Kotwal raan** is responsible for high frequency of saplings of *Butea monosperma, Cleistenthus collinus, Lagerostroemia parviflora and Calycopteris floribunda.* Dominant trees are of *Albizia odorattissima, Buchanania lanzan, Tectona grandis and Terminalia alata.* These species are showing less regeneration capacity. Tree flora is comparatively young, as demonstrated by single strata of trees, that too below ten meter in height.

Mothya Devache van is also rich is sapling flora of *Anogeissus latifolia, Cleistenthus collinus, Terminalia alata.* Some new entrance in the area are *of Chloroxylon swietenia, Flacourtia indica, Pterocarpus marsupium and Soymida febrifuga.* Tree species are few in number. They are dominated by *Anogeissus latifolia, Terminalia alata and Lannea coromandelica.*

Kosaran shows frequent clumps of *Bombusa arundifolia. Tectona grandis* is forming dominant sapling flora. *Lagerostroemia parviflora* is the only tree of the area.

Tiwas van is rich in biodiversity of saplings, but comparatively poor in tree vegetation. It is dominated by *Terminalia alata, Anogeissus latifolia*, and *Lagerostroemia parviflora*.

Entire flora of Saigata is showing replacement of older flora by emigration of new species. Plants are of small stature as they are still in sapling stage and small in DBH.

Maralmendha

Maralmendha, sharing its boundary on the west of Saigata, is the village where total deforestation has taken place. Not one full-grown tree is found here though roots of many big trees are seen dug out. Despite being Protected forest under Forest Department's ownership, encroachment on forest land has been going on since long. In 1978 Government of Maharashtra regularized encroachments in the state. This has encouraged people to encroach upon forest land with the hope that their encroachment will also be regularized. The dominant caste of this village being Mali, their main occupation is farming, especially growing vegetables. Destruction of forest took place with the sole objective of extending land under cultivation.

Till 1955 forest land was under the landlords (Malgujars) who collected land cess on agricultural fields and regulated harvesting on forest lands. However, bringing more land under cultivation meant increase in the revenue. Due to this, encroachments on forest lands were neglected to some extent. Later forests were transferred to the Forest Department. Yet, either forest department officials were unable to restrict encroachments or for some obvious reasons, decided to turn a blind eye to the phenomena of illegal occupation. This has resulted in conversion on almost 99 per cent of forest land into cultivated land, owned illegally. Individuals that already owned land have encroached for owning more land and landless have encroached with the ambition to become landowners. In this process two big landowners started brick-kilns that needed lot of fuel. This also contributed to forest degradation. In our group discussions some persons opined that the quality of forest of this area was good as it had valuable trees like teak, but due to restrictions on harvesting, people did not get any returns from the forest. This encouraged illicit felling that destroyed the forest. Due to the ease with which forest land could be encroached, some of the villagers sold out their ancestral land and encroached upon forest land. These are rich and thus influential farmers who have now passed a resolution in Gram Panchayat demanding title deeds for encroached land. It is interesting to note that 90 percent of those who have encroached are landowners and only 10 percent are landless labourers. This is also because substantial expenditure is required for tilling, cleaning and leveling of encroached land, which the poor landless cannot afford. Second important cause of deforestation that came up in the discussions was, heavy pressure from the neighboring villages for fuel and fodder. As the other adjoining villages Chorti and Uchali, do not have any demarcated forest area, they depended on Maral Mendha forests contributing to its destruction.

Prima facie, Maral Mendha appears to be a better off, wealthy village in comparison to Saigata, with brick and concrete houses, large number of cattleheads, bullock-carts, and tractors. Villagers also told the study team that about 30 years back this area was well known for producing onions and sugarcane (both cash crops) in large quantity. But gradually because of insufficient rain & reduced availability of tank-water, it has become difficult to grow these crops. According to the villagers, lack of effective leadership, like the one in Saigata, is the main cause of failure of collective action in protecting their forest.

Maral Mendha is situated on the same land-form and in identical climatic conditions that occur in Saigatta. Topographically it is situated in between N 20 37'18.7', EO 79 47'59.5' and N 20 37'45.8", EO 79 48' 06.4" . Area is almost flat land with 2-5% slop towards north, north-east. It lies in between 251-263 msl altitude.

Forest map of 1977-1978 shows total forest area of 71.46 ha which is divisible into protected forest 70.83 ha and Zudpi forest 00.63 ha, per capita forest was o.12 ha. At that time *Pterocarpus marsupimum, Terminalia alata, Anogeissus latifolia, Lagerostroemia parviflora, Buchanania lanzan, Diospyros melanoxylon, Lannea coromondelica and Boswellia serrata* were dominant species. Deforestation began with the encroachment for fuel, fodder and timber. With the passing of time entire area came under agriculture activity. Present survey shows hardly any woody species. Land is highly degraded. Soil is of poor quality. It shows soil erosion. Impact of heavy grazing and damage by pest is visible. High temperature and exposed surface along with other conditions resulting in poor development of soil.

Soil varies from dark reddish brown, gravel clay loam (2.5 YR,3/6) to sandy loam (5YR, 3/3) and sandy loam clay (7.5 YR, 4/4).

Tree cover is totally absent. Vegetation is in the form of ground flora. Dominant species belong to family Asteraceae, Fabaceae and Poaceae. While thorny shrubs and few saplings are also seen along the boundary of agricultural fields. Shrubby plants belong to *Maytenus emarginata, Ziziphus oenoplia, Diospyros melanoxylon, Chloroxylon swietenia Lagerostroemia parviflora* and *Diospyros montana.* While ground flora is occupied by *Alysicarpus monilifer, Andropogon pumilus, Anotis montana, Ageratum conyzoides, Borreria articularis, B. pusilla, Cassia tora, Chrysopogon fulvus, Desmodium triflorum, Evolvulus alsinoides, Glossocarda boswellia, Hyptis suveolens, Hemidesmus indicus, Indigofera linifolia, Ischaemum pilosum, Pulicaria angustifolia, Spilenthus acmella, Tridex procumbens, Vernonia cinerari, Vicoa indica* and *Zornia diphylla.* Fodder is the only product which can be harvested from the area. Even for fuel villagers have to depend upon neighboring villages(Fig.7).

In Maralmendha, the past record of forests and present observation of the lack of trees manifests negative perception and outlook of certain societies towards natural wealth.

Institutional Analysis

As has been mentioned earlier, although evidence from field research, and also from experimental research challenges the generalizability of the conventional theory, no full-proof alternate theory is yet accepted. However, scholars working on the 'commons' problem more or less agree on a set of variables that enhance the likelihood of appropriators organizing themselves. For the present analysis we have compared the two case studies on the basis of attributes of appropriators as well as the resource suggested by Ostrom. Before that it would be pertinent to see if the two cases are comparable.

Common aspects in Saigata and Maral Mendha are: the geographical location, topography, erstwhile plant species, legal status of forest, main occupation of the people, and the state of forests some 25 years ago. Differences are: degree of heterogeneity, size (though the two differences are not very large), per capita forest land available, difference in priorities taken as proxy to discount rate, also reflected in their concept of wealth (Table-1). It makes a strong case for comparison.

Attributes of the Resource:

1. Feasible improvement: resource conditions are not at a point of deterioration such that it is useless to organize or so underutilization that little advantage results from organization.
2. Indicators: Reliable and valid indicators of the resource system are frequently available at a relatively low cost.
3. Predictability: the flow of resource units is relatively predictable.
4. Spatial extent: the resource system is sufficiently small, given the transportation and communication technology in use, that appropriators can develop accurate knowledge of external boundaries and internal microenvironments.

Table-2

Sr. No.	Attributes of Resource	Saigata	Maral Mendha
1.	Feasible Improvement	Had reached a level of high degradation, organized efforts ensured consistent and feasible improvement	Constant deterioration and encroachment.
2.	Indicators	Yes	Yes
3.	Predictability	Predictable improvement	Predictably reduced availability
4.	Spatial extent	Small, manageable, sufficient for sustenance needs	Too small, insufficient for sustenance needs
5.	Pressure from neighboring communities	Initially low, increased with improved forest quality	Heavy initially, shifted to other forests after total degradation.

The fifth attribute has been added for this study as it has had strong initial impact resulting in deterioration of the resource below the bare minimum. In case of the fourth attribute, it should be noted that though Ostrom mentions the spatial extent in the context of being manageable, another dimension of 'sufficiency' has been added here.

Attributes of the appropriators

1. Salience: Appropriators are dependent on the resource systems for a major portion of their livelihood.
2. Common understanding: appropriators have a shared image of how the resource system operates (attributes 1, 2, 3, and 4) and how their actions affect each other and the resource system.
3. Discount rate: Appropriators use a sufficiently low discount rate in relation to future benefits to be achieved from the resource.
4. Trust: Appropriators trust one another to keep promises and relate to one another with reciprocity.
5. Autonomy: Appropriators are able to determine access and harvesting rules without external authorities countermanding them.
6. Prior organizational experience and local leadership: Appropriators have learnet at least minimal skills of organization through participation in other local associations or learning about ways that neighboring groups have organized.

Table-3

Sr. No.	Attributes of Appropriators	Saigata	Maral Mendha
1.	Salience	High	Low
2.	Common Understanding	Constantly improved	No
3.	Low discount rate	Yes	No
4.	Trust and reciprocity	Yes	No
5.	Autonomy	*De jure* –no, *de facto* - yes	*De jure* –no, *de facto* – yes
6.	Local leadership	Yes	No

Effective leadership which could convince everybody in the village about necessity to protecting forest resource seems to have made a huge impact in the initial success of collective action and emergence of local institution. Since inception there has been no change in the leadership that has ensured that all the decisions would be taken only unanimously and there will be equal participation of all the community members. In addition to this, factors like self-imposed restrictions by the community, monitoring by members within the community, effective patrolling, verifiable improvement in the quality of the resource, improved flow of forest goods like fuel and fodder for which thoughtful management method of zoning was adopted, and importance to social sanctions rather than monetary penalties, are the factors that have contributed to the effective working of the local institution in Saigata.

CONCLUSIONS

Whether communities organize depend on attribute of the resource system and the appropriators themselves that affect the benefits to be achieved and the costs of achieving them. Until recently, the possibility that the appropriators themselves would find ways to organize themselves has not been seriously considered in much of the literature. Field studies in all parts of the world have found that local groups of resource users sometimes themselves and sometimes with the assistance of external authorities, have created a wide diversity of institutional arrangements for coping with common-pool resources. Saigata is one such community.

Success of Saigata is rooted in effective leadership, shared norms, self-imposed discipline, high bequest value of forests, and effective monitoring. As against this, there is no common understanding in Maral Mendha, actions of individuals are based on personal gains, incomes in short run are more important than sustainable development. Effectiveness of forest management in Saigata is well reflected in the data collected through forest mensuration. Same is true for Maral Mendha as well.

ACKNOWLEDGEMENTS

Present paper is the outcome of a pilot study funded by Workshop in Political Theory, Indiana University, Bloomington, US. Authors are grateful to the funding agency and the other research members of the study group.

REFERENCES

Agrawal, Arun (1999). Greener Pastures: Politics, Markets, and Community among a Migrant Pastoral People. Durham, NC: Duke University Press.

Agrawal, Arun. Forthcoming. "Common Resources and Institutional Sustainability", personal communication.

Baland, Jean-Marie and Jean-Philippe Platteau. (1996). *Halting Degradation of Natural Resources: Is There a Role for Rural Communities?* Oxford: Clarendon Press.

Blomquist, William, Edella Schlager, Shui Yan Tang, and Elinor Ostrom (1994). Regularities from the field and possible explanations. In Elinor Ostrom, Roy Gardner, and James Walker. *Rules, Games, and Common-Pool Resources*. Pp. 301-16. Ann Arbor: University of Michigan Press.

FAO (Food and Agriculture Organization) (1999). Status and Progress in the Implementation of National Forest Programmes: Outcomes of an FAO Worldwide Survey. Mimeo. Rome, Italy: FAO.

Frey, Bruno S. (1994). "How Intrinsic Motivation is Crowded Out and In." *Rationality and Society*. 6, pp. 334-52.

Guha, R. (1983). "Forestry in British and Post-British India: A Historical Analysis." *Economic and Political Weekly* 18(4): 1882-96.

Hardin, G. (1968). "The Tragedy of Commons." *Science* **162**:1, 243-48.

Kulkarni, S.D. (1979). "Encroachments on Forest Lands-The Experience in Maharashtra." *Economic and Political Weekly* **19(45)**: 1846-49.

McKean, Margaret (1992). "Management of traditional common lands (*Iriaichi*) in Japan." *In Making the Commons Work: Theory, Practice, and Policy*. (Ed) Daniel W. Bromley, pp. 63-93. San Francisco: Institute for Contemporary Studies Press.

McKean, Margaret, A. (2000). "Common Property: What Is It, What Is It Good For, and What Makes It Work?" In Clark Gibson, Margaret McKean, and Elinor Ostrom, eds. *People and Forests: Communities, Institutions, and Governance*. MIT Press, Cambridge, MA.

Naughton-Treves, Lisa and Steven Sanderson. (1995). Property, politics and wildlife conservation. *World Development* **23(8)**: 1265-75.

Olson, Mancur. (1965). *The Logic of Collective Action: Public Goods and the Theory of Groups*. Cambridge, MA: Harvard University Press.

Ostrom, Elinor. (1990). *Governing the Commons: The Evolution of Institutions for Collective Action*. Cambridge: Cambridge University Press.

Ostrom, Elinor. (1992). *Crafting Institutions for Self-Governing Irrigation Systems*. San Francisco: Institute for Contemporary Studies.

Ostrom, Elinor. (1999a). "Formulating the Commons." In Joanna Burger, Richard Norgaard, Elinor Ostrom, David Plicansky, and Bernard Goldstein, eds. *The Commons Revisited*, Island Press, Washington D.C.

Ostrom, Elinor. (1999b). Self Governance and Forest Resources. Occasional Paper No. 20, Center for International Forestry Research, Bogor, Indonesia.

Tang, Shui Yan. (1992). *Institutions and Collective Action: Self Governance in Irrigation Systems*. San Francisco: ICS Press.

Wade, Robert (1988, 1994). *Village Republics: Economic Conditions for Collective Action in South India*. Oakland: ICS Press.

Advancing Frontiers of Ecological Researches in India, 161-166, 2006
Edited by A.K. Kandya & Asha Gupta
Published by Bishen Singh Mahendra Pal Singh, Dehra Dun, India

Recent Trends in Biodiversity Conservation - A review

C.M. Mishra[1], Bhoomika Srivastava[2] and Veena Mishra[3]

Abstract: Biodiversity is defined variously but the definition to recognize specifically the diversity of ecosystems, species and genes is in common use today. Hence, biodiversity can be classified into genetic, species and ecological diversity.

New genetic variation arises in individuals by gene and chromosome mutations, and in organisms with sexual reproduction it can be spread through the population by recombination. The species level is generally regarded as the most natural to consider whole organism diversity. The ecological importance of a species can have a direct effect on community structure and thus affect the overall biological diversity. It is difficult to make a quantitative assessment of diversity at ecosystem, habitat or community level because they also include soil, parent material and climate. Ecosystem diversity is often evaluated through measures of diversity of the component species.

Human introduction affect biodiversity both directly and indirectly through changes in land and water use. Change in land use are the major causes of habitat destruction and fragmentation. The Teak plantation raised after the conquest of India by the Britishers was the significant cause of erosion of biodiversity over large forested tracts of peninsular India. Invasion of an American weedy plant, *Parthenium* eroded diversity of native biota.

INTRODUCTION

Biological diversity or biodiversity, refers to the universal attribute of all living organism that each individual being is unique that is; no two organism are identical. The biology of biodiversity must include all the aspects of evolutionary and ecological sciences analyzing the origin, changes, and maintenance of the diversity of living organisms.

The multifaced nature of biodiversity is reflected in the many definitions that have been put forward. Perhaps the first definitions to recognize

---

1. Ecologist, Forest Research Institute, U.P. Kanpur, India.
2. Research Scholar, A.N.D. Post Gr. College, Kanpur University, Kanpur.
3. Reader, Zoology Department, A.N.D. Post Gr. College, Kanpur University, Kanpur.

specifically the three components (ecosystems, species and genes) are in common use today.

The contracted form of 'biodiversity' was apparently coined by Walter G. Rosen in 1985 for the first planning meeting of the 'National Forum on Biodiversity' held in Washington DC in September 1986.

Classification of Biodiversity

Classification is an essential process in our daily lives (Davis and Heywood 1963) and a necessary tool for our survival. Biodiversity can also be classified into genetic diversity, species diversity and ecological diversity.

Genetic Diversity

Genetic diversity is reliant on the heritable variation within and between population of organisms. New genetic variation arises in individuals by gene and chromosome mutations, and in organisms with sexual reproduction. It can be spread through the population by recombination. It has been estimated that in humans and fruit flies alike, the number of possible combinations of different forms of each gene sequence exceed the number of atoms in the universe. Other kinds of genetic diversity can be identified at all levels of organization, including the amount of DNA per cell and chromosome structure and number . Only a small fraction (<1%) of genetic material of higher organisms is outwardly expressed in the form and function of the organisms; the purpose of the remaining DNA and the significance of only variation within it are unclear (Thomas *et al.*, 1992).

Each of the estimated 10^9 different genes distributed across the world's biota does not make an identical contribution to overall genetic diversity. In particular, those genes that control fundamental biochemical processes are strongly conserved across different taxa and genetically show little variation although such variation that does exist may exert a strong effect on the viability of the organism; the converse is true of other genes.

Species Diversity

Historically, species are the fundamental descriptive units of the living world and this is why biodiversity is very commonly and incorrectly, used as a synonym of species diversity in particular of "species richness", which is the number of species in a site on habitat.

An estimated 1.7 million species have been described to-date: estimates for the total number of species existing on earth at present vary from 5 million to nearly 100 million. A conservative working estimate suggests there might be around 12.5 million. The species level is generally regarded as the most natural one at which to consider whole organism diversity. While species are also the primary focus of evolutionary mechanisms, and the organization and extinction of species are the principal agents in governing biological diversity, species cannot be recognized and enumerated by systematizes with total precision. The concept of what a species is differ considerably among groups of organisms. It is for this reason among others, that species diversity alone is not a satisfactory basis on which to define biodiversity.

The ecological importance of a species can have a direct effect on community structure, and thus on over all biological diversity. For example a species of tropical rain forest tree that supports an endemic invertebrate fauna of a hundred species makes a greater contribution to the maintenance of global biological diversity than does a European alpine plant that may have no other species wholly dependent on it.

Ecosystem diversity

While it is possible to define what is in principle meant by genetic and species diversity, it is difficult to make a quantitative assessment of diversity at the ecosystem, habitat or community level. There is no unique definition or classification of ecosystems at the global level, and it is difficult in practice to assess ecosystem diversity other than on a local or regional basis, and then only largely in terms of vegetation. Ecosystem are further divorced from genes and species in that they explicitly include abiotic components being partly determined by soil / parent material and climate. To get around this difficulty, ecosystem diversity is often evaluated through measures of the diversity of the component species. This may involve assessment of the relative abundance of different species as well as consideration of the types of species.

Human Introduction

Human introduction affects biodiversity both directly and indirectly through changes in land and water use. In rainfed areas the landuse pattern plays, a major role for deciding the fate of native species (Misra and Dubey 1998). Changes in land use are the major causes of habitat destruction and fragmentation, and these in turn are the major causes of recent extinctions and constitute a major threat to biological diversity (WCMC 1992; Skole and Tuckor 1993).

Human induced changes sometimes increase local species diversity. For example, some grasslands that evolved under low grazing pressure have shown an increase in species diversity as a result of the introduction of livestock and the consequent increase in grazing intensity (Sala *et al.* 1986; Milchunas *et al.* 1988 : Mishra and Mall 1974, 1975, 1976).

This pattern is accounted mainly by the introduction of alien species better adapted to grazing conditions, without the disappearance of native grasses. Further increases in grazing intensity have reduced diversity as introduced grazing tolerant species have become dominant (Chaturvedi and Misra 1985). It is the ability of humans to exchange flora and fauna which has resulted in rapid and major invasions of exotic plant and animal species.

Human impacts on biodiversity

With increased commerce between continents, many insects and pathogens have been introduced into forests without co- evolved biotic controls. For example, virulent pathogens such as Dutch elm disease (*Graphiium ulmi*) and chestunut blight (*Endothia parasitica*) have largely removed their respective host trees from the landscape in the eastern US Gypsy moth (*Lymantria dispar*) and other introduced insects have also caused extensive damage. In addition, invasive plants such as Kudzu (*Pueraria lobata*), a nitrogen fixing vine, have taken over many areas in the Southeastern USA and restricted the regeneration of forests.

Similarly, a climber known as *Teliocora* which has spread in wider areas of sal forests of northern India, has retarded the process of its regeneration (Mishra *et al.* 1994). The reduction in the areas of forests, together with loss of tree species, has an especially hard impact on bird populations. Birds that migrate between temperate and tropical forests are most affected because they are losing habitat in both biomes (Wilcove 1990).

Human and livestock activity have altered plant species composition and vegetation cover in many regions, both by reducing native vegetation and by introducing non-native species either deliberately or accidentally. There is also some evidence that declining bat populations in the

Southwestern state of the USA have led to decreased frequency of effective pollination in some bat pollinated monocarpic species of *Agave* (Howell and Roth 1981). Similarly, in some tree species of eastern Uttar Pradesh, India, the seed formation has been reduced due to poor population of Bat particularly in trees of *Adansonia digitata* (Misra 2001).

Invasion by exotic species

Invasion by exotic species is a serious threat for many wetlands. Neophytes disperse along river course, because the frequent habitat disturbance by water level fluctuations provide good living conditions for ruderal plants. The introduction of Nile perch led to serve reduction and partial extinction of the endemic cichlids in Lake Victoria and associated wet lands (Kaufman 1992; Schofiled and Chapman 1999).

British interest in India's Teak forest, to secure an alternative supply of timber for their merchant ships and navy, was a modern systems of agricultural production create extensive major motivation behind imperial ambitions. The Teak plantation raised after the conquest of India were a significant cause of erosion of biodiversity over large forested tracts of peninsular India (Rawat 1991). More recent times have witnessed the invasion of American weedy plants such as *Parthenium* and *Eupatorium*, all over the tropical world, leading to erosion of diversity of the native biota (Dicastri *et al.* 1989).

When a species enters in an ecosystem in which it previously did not occur, it can either adversely disrupt ecosystem processes or have positive effects.

REFERENCES

Chaturvedi, A.N. and Misra, C.M. (1985). Ecological survey of grasslands at Dudhwa National Park. India. For. **111**(8): 579-582.

Davis, P.H. and Heywood V.H. (1963). Principles of Angiosperm Taxonomy, Oliver and Boyd Edinburg.

Dicastri, F. (1989). History of Biological invasions with special emphasis on the old world. In : Drake, J.A., Mooney, H.A., dicastri, F., Groves, R.H., Kruger, F.J., Rejmanek, M. and Williamson, M.(eds), Biological Invasions: A global Perspective. 1-26 Scope 37, John Wiley, chichester.

Howell, D.J. and Roth, B.S. (1981). Sexual reproduction in agaves : The benefits of bats, the cost of semelparous advertising. Ecology **62**: 1-7.

Kaufman, L. (1992). Catastrophic change in species rich eco systems. Bio. Science **42**: 846-548.

Milchunas, D.G., Sala, O.E. and Lowenroth, W.K. (1988). A generalized model of the effects of grazing by large herbivores on grassland community structure. American Naturalist **132**: 87-106.

Misra, C.M. 2001. Operation Green in Uttar Pradesh, U.P. Forest Department, Lucknow, India.

Misra, C.M. and Dubey, P. 1998. Conservation of Biodiversity in degraded ecosystems, Biodiversity conservation in managed Forests and protected areas. Agro Botanica, Bikaner, India.

Misra, C.M., Srivastava, R.J. and Rai, A.M. 1994. Ecological studies to monitor Biodiversity at Dudhwa and Corbet National parks of Uttar Pradesh, India Reg. conf. Env. & Biod. Abs. (March 7-9, 1994), Kathmandu, Nepal.

Misra, C.M. and Mall, L.P. 1974 . Photosynthetic structure and standing biomass of a grassland community. Trop. Ecol. **15(2)**: 76-80

Misra, C.M. and Mall, L.P. 1975. Architecture and standing biomass of a grassland community. Ann. Arid Zone, **14 (2)**: 119-123.

Misra, C.M. and Mall, L.P. 1976 . Architecture and standing biomass of a grassland Community. App. Ecol. Abst. 2 D6974. IRL, London.

Rawat, A.S. (ed) 1991. History of Forestry in India. Indus Publishing Company, New Delhi.

Sala, O.E., Oesterheld, M., Leon, R.J.C. and Soriano, A 1986. Grazing effects upon plant community structure in sub humid grasslands of Argentina. Vegetatio **67**: 27-32

Schofield, P.J. and Chapman, L.J., 1999. Interactions between Nile perch *Lates niloticus* and other fishes in Lake Nabugabo, Uganda. Environmental Biological of fishes **55**: 343-358.

Skole, D., and Tucker, C. 1993. Tropical deforestation and habitat fragmentation in the Amazon : Satellite data from 1978 to 1988. Science **260**: 1905-1910.

Thomas, C.D., Thomas, J.A. and Warren, M.S. 1992. Distributions of occupied and vacant butterfly habitats in fragmented landscapes. Oecologia. **92**: 563-567.

World Conservation Monitoring Centre, 1992. Global Biodiversity : Status of the earth's living resources. Chapman and Hall, London.

Wilcove, D. 1990. Empty skies, Nature Conservancy Magazine **40**: 4-13.

Advancing Frontiers of Ecological Researches in India, 167-187, 2006
Edited by A.K. Kandya & Asha Gupta
Published by Bishen Singh Mahendra Pal Singh, Dehra Dun, India

Strategies to Meet Energy Challenges Through Domestication of *Jatropha curcas*

Sunil Puri
Department of Forestry, Indira Gandhi Agricultural University, Raipur 492 006 (Chhattisgarh)

Abstract: The present paper deals with the use of jatropha oil as a future energy source. Its use has many positive aspects compared to the use of fossil fuels. There benefits include the reduced use of non renewable fuels, less dependency on foreign fuel, stabilization of income in rural areas, and reduced emission of carbon dioxide and other green house gases. When we compare the carbon dioxide output per unit of electricity produced for several energy sources, coal produces the most, followed by natural gas, solar panels, geothermal, nuclear, hydroelectric, and then in last place, biomass.

Energy Challenges

Energy is the single resource which determines the human carrying capacity on this planet. If the energy input of Antarctica is only from the sun, the energy input is low and the human carrying capacity is near zero. Ecologists view that the carrying capacity of any ecosystem depends upon the amount of energy input into the system. We all know that energy is required to produce food. If energy input is low and food production is low, there will be a low carrying capacity. However, if the energy input is high, and fossil fuels are used to grow food in underground green houses, then the carrying capacity of Antarctica can be temporarily increased above the natural carrying capacity. Therefore, the future human carrying capacity is dependent upon future energy supplies.

Over the next 25 years population growth and economic development could produce a 75% increase in global energy demand. That is amazing and is equivalent to a 2.3% increase in energy demand per year. The basic question arises "Can we really meet the demands of the present without compromising the ability of the future generations to meet their own needs?." Energy will come from the two sources in the future : matter and anti-matter. Anti-matter was detected ten years ago and now researchers can store anti-matter for about 10 days. The reason I am mentioning anti-matter is because it is a new energy source and is based on methane hydrides. There are many different sources of energy like biomass, coal, oil, gas, solar, hydroelectric, nuclear, geothermal, and a surprise energy source like methane hydrides (anti-matter). We have to think about where we want to get our energy in the future. Some want to continue relying on fossil fuels (such as coal and oil) but there are others, like me, who predicts that biomass, solar and wind will be three of the fastest growing source of energy in the future. Wind is economical today and some wind farms exists in western countries. The use of solar-cells might become economical in the near future while the energy from biomass is becoming more economical as the price of oil increases and its stocks are depleted.

The present paper deals with the use of biomass as a future energy source. The use of wood biomass for energy has many positive aspects compared to the use of fossil fuels. There benefits include the reduced use of non renewable fuels, less dependency on foreign fuel, stabilization of income in rural areas, and reduced emission of carbon dioxide and other green house gases. When we compare the carbon dioxide output per unit of electricity produced for several energy sources, coal produces the most, followed by natural gas, solar panels, geothermal, nuclear, hydroelectric, and then in last place, biomass.

For developing countries, like India, the threat of future oil price rise gives maximum cause for concern. Trapped in many cases in circumstances of slow or negative economic growth, foreign indebtedness and current trade deficit, its industrial development prospects are set at a critical disadvantage by any further rise in the cost of fossil fuel imports. When it comes to developing renewable or alternative energy sources, however, India possessed substantial natural advantages, abundant forests, fertile soils and sunlight. It is now apparent that wood and wood-based fuels merit open-minded consideration as technically mature, environmentally sound, economically viable and culturally acceptable options for modern energy development. The most important lesson that remains to be learned is the importance of developing these options in a coordinated way, in keeping with the ecological and socioeconomic realities that prevail from place to place.

An accumulation of FAO research suggests that effective wood energy management requires a coordinated 'systems approach' in which all actions relevant to wood energy production and use support and reinforce one another. Another key point is that there is no standard blueprint for these systems: planning moves should be geared to realities specific to the area in focus.

Among these factors are constraints and limitations always liable to be ranged against wood energy development, including limits on the capacity of the primary resource to support intensified energy-related use. There are, on the other hand, many more situations where scarcity of forests and trees is not a limiting factor and the bioenergy resource is to a certain extent under-utilized. Foremost among other common constraints are lack of the technical capacity, institutional facilities, regulatory mechanisms and educational inputs needed to make the best use of wood energy resources. Lack of political or financial support for wood energy development is another obstacle. It reflects a perception commonly held among policy makers, investors and developers that wood energy is a low-grade or archaic power source by comparison with 'conventional' fuels. This outlook is generally on the retreat as new factors relating to environment enter the picture.

A systems approach to wood energy production and use means giving rein to 'feedback loops' linking forests, trees and people in a mutually supportive and interactive chain of benefits. Development policies and strategies should be aimed at enabling benefits and feedback to flow freely in both directions, to sustainable effect and in measured relation to tropical circumstances.

This paper highlights new opportunities for wood energy use in rural and urban neighbourhoods in commerce and public utility management, without harming the regenerative capacity of trees and forests or reducing their significance as safeguards over soil nutrients, water resources and biodiversity. Wood energy use and production are no panacea for the world's ills. They are often associated with environmental, economic and human problems. Yet, managed sustainably, they can also offer valuable solutions, benefits and safeguards. That promise is the main topic that this paper seeks to explore and affirm.

Substitution and energy mix

In situations where land or water availability and other limiting factors make expanded production or supply of fuel/oil unrealistic, efforts have been made to utilize alternative or substitute energy sources, sucn as processed crop wastes or biofuel from plants. Wherever a choice of energy sources and uses exists, households and businesses will usually improvise the energy 'cocktail' that best suits their needs, tastes and circumstances. This variegated energy mix is often a cornerstone of the informal economic sector on which the functioning of many societies depends.

Fossil-based fuels are predominant sources of energy for economic development in today's world, but abundant and relatively cheap supplies of energy from these sources cannot be expected to grow or last forever. The environmental risks of fossil-based or nuclear fuels also cast doubt on their future viability as staple energy sources. Alternative, renewable sources of energy will in the long run have to be developed, even in relatively wealthy parts of the world. Policies, financial mechanisms and regulations that serve this need already apply in many countries, overturning the conventional economics of commodity supply and demand.

Bioenergy context

The giant potential for world energy from biomass has captured much recent research and development interest. The definition fits traditional fuels such as firewood or animal dung, but in most cases these have not been the main focus of developers attention. Biomass fuels such as wood can also be converted into modern energy 'carriers', such as fuel alcohol or electricity. These can be substituted and mixed in a flexible way for fossil-based fuels, often at a relatively low investment cost.

Biofuels cannot only serve traditional domestic needs such as cooking, heating or lighting but also cater to new energy applications in industry, at a cost competitive with equivalent fossil-based technologies. Because bioenergy production operations are decentralized and labour-intensive, rural employment can be another economic benefit. Production facilities can be developed step-by-step and do not call for the heavy capital investment required by 'hi-tech' energy sources such as hydro-power or nuclear power. Moreover, while 'energy crops' are growing, they fix carbon through photosynthesis and so help mitigate risks of global warming and climate change.

Bioenergy have drawbacks too. Unsustainable exploitation of finite biomass sources can deplete natural forests and woodlands as production of annual crops for energy is often undertaken at the expense of forest clearance. Turning over agricultural land from other uses to 'energy farms' or plantations can also have unwanted social and economic impacts. Even so, with sound planning, consultation, monitoring and implementation, wood energy production and use can avoid these pitfalls.

Converting vegetable oils into fuel oil is another bioenergy process. Raw oil suitable for processing can be obtained from several oil yielding crops and trees, including the fruits of the oil and coconut palms. Diesel engines were originally designed by a German inventor, Rudolf Diesel, to run on vegetable oils. Now the use of modern biodiesel fuels – which can run existing diesel engines without major modification – is promoted in Germany and elsewhere as an environmentally sound alternative to fossil-based diesel. In many tropical countries, as well as in India, new sources of plant-based diesel are under rapid development.

Biodiesel energy from oilseed trees

Use of plant oils in the energy sector is growing, particularly in industrialized countries. Four main technical solutions are being studied, some of which are being piloted commercially:

- Plant oil mixed with diesel for heating systems;
- Purified plant oil for special diesel engines;
- Transterified plant oils for normal diesel engines;
- Transterified plant oil mixed with diesel for normal diesel engines.

These solutions have been widely tested in different countries with excellent results and different field crop species (rapeseed, soya, groundnut, sunflower and others) are being intensively evaluated. Although the technical, environmental and socioeconomic advantages are quite well established, economic viability has to be proven and subsidy is necessary in most cases.

However, many developing countries with plentiful available land and low-cost labour do not face this obstacle. In addition to field crops, established oilseed plantation trees such as oil palms can lend themselves to use as non-wood forest products for energy applications. Novel oilseed tree crops are also being put to this purpose in several tropical countries. For example, Curcas Oil extraction from *Jathropa* seeds is a simple process which can be undertaken in small, local presses. The cost of extraction equipment varies from as low as a few thousand rupees for simple, hand-operated equipment to Rs. 10 lakh for more sophisticated diesel engine powered systems. Pilot units for Curcas Oil production have been successfully launched in Mali, Thailand and elsewhere. In India the initiation has already begin and in due course of time many pilot units will be coming up.

The emerging formula for a 'New Energy Order' has won tacit general acceptance in the international arena but transition from current practices and circumstances to this ideal state of affairs will not be easy or straightforward. Pragmatic guidelines for change are called for, that can be applied and flexibly adapted to real situations that confront the resource planner, manager or user under everyday circumstances.

Given the demand for a continuous and uniform supply of oil and the accelerating depletion of oil reserves, increasing the number of plants species that yield biofuel in cultivation would appear to be an important strategy for meeting a growing demand. In spite of this strategy, few species are cultivated and some species are not cultivated at all. The need is to domesticate and cultivate all those species which have potential to yield biofuel.

Domestication and Commercialization

Domestication is the act or process of converting animals or plants to household use (Leakey & Newton 1994). That often means moving an organism from its natural habitat, introducing it to a modified habitat, and providing the means for it to maintain itself in its new location. This may involve genetic modification and breeding. An example may be removing a oil-producing plant from the wild, planting it near one's house, harvesting the crop, and planting the seeds for a future crop. The domesticated plant or animal may be used for one's own enjoyment or consumption, or it may be domesticated for commercial purposes.

We domesticate because it is amicable to have that particular life form around, because it is easier to deal with at home, or because through cultivation it can be improved. Frequently it is also because it is more economical to produce the product than to continually collect it from the wild. If we do not domesticate, the reasons are probably that there is no need – the organism is plentiful in the wild–or else the habitat requirements are too difficult to duplicate on a farm or in a garden.

In genetic terms, domestication is accelerated and human-induced evolution. The consequences are either loss or gain of genes, altered gene frequencies or modifications to the way genes are packaged (gene complexes). There is much historical evidence of the human role in shaping present-day vegetation patterns (Briggs & Walters 1984). The domestication of trees, then, does not have to occur outside the forest. Many examples of humans changing the species-richness and abundance of desired trees in forests are reported in Asia (Whitmore 1992). Domestication of tree species must therefore be seen as a continuum – from unmolested state, to management of trees in forests, to cultivation of semi-domesticates, to monocultural plantation of advanced generation breeding lines. The level of domestication activity will be dictated by biological, policy, market and social factors.

Domestication of any plant species is concerned with selection and management by humans and is not only about breeding *per se*. Selection can be deliberate or inadvertent. Deliberate choices might be made for fruit characters, growth rate, form, etc., whereas inadvertent selection could be for insect resistance, tolerance of selfing, etc. Of course inadvertent selection could also be carried out for any trait that was genetically linked or correlated to a trait that was being deliberately selected. Traits that confer fitness may be either enhanced or depressed, whichever the form of selection practiced. Management is also linked to the genetics of a plant in that the ability of a plant to be managed in a certain way is often expressed generically. The direction and speed with which domesticated trees diverge from their wild progenitors will depend upon the size of the population, the heritability of traits under selection, the mating system, the intensity of selection and the inherent variability of the traits.

Commercialization can change an area's ecology by favouring or exploiting one species instead of another. This is not necessarily bad, as many creatures modify their environment to suit their needs. We just have to be aware that we are shaping the future by our management activities and plan accordingly. Together, domestication and commercialization may lead to monocultures. While monocultures may not be desirable from a diversity perspective, they may be economically, and perhaps environmentally, acceptable.

In most countries, the trend is towards a greater proportion of obtaining cultivated material. From the perspective of the market, domestication and cultivation provide a number of advantages over wild harvest for production of plant-based oils: 1) while wild collection often offers material adulterated and unwanted, cultivation provides reliable botanical identification; 2) wild harvest volumes are dependent on many factors that cannot be controlled and the irregularity of supply is a common feature. Cultivation guarantees a steady source of raw material; 3) wholesalers can agree on volumes and prices over time with the grower; 4) the selection and development of genotypes with commercially desirable traits from the wild or managed populations may offer opportunities for the economic development of the oil bearing plant species as a crop; 5) cultivation allows controlled post-harvest handling and therefore; 6) quality controls can be assured; and 7) product standards can be adjusted to regulations and consumer preferences; 8) cultivated material can be easily certified organic or biodynamic although certifiers are also presently developing wild crafting standards (Leaman, 2001; Palevitch 1991). In spite of this, domestication of the resource through farming is not always technically feasible. Many species are difficult to cultivate because of certain biological features or ecological requirements.

Sustainable production and harvest

The key to maintaining supply of oil yielding plants is the issue of sustainable resource production and harvest. Of the greatest concern in the development of products based on oil is the ability to ensure sustainability. However, the concept of sustainability is used in a rather cavalier fashion. The truth is that there is little information and knowledge of any production, especially the products from tropical ecosystems. Too little is known about the levels of sustainable production and harvest of many tree species yielding non-timber resources.

Limiting harvest to a sustainable level is complicated by the conflict of interests between use and protection. It requires an effective management system and sound scientific information (Leaman, 2001). In the past few years, field-based methods have been developed for sustainable harvest assessment and monitoring, resulting in the publication of research guidelines and predictive models (Nantel *et al.*, 1996; Schippmann, 1997). Development of a management strategy should include the following elements: 1) assessment of population abundance and distribution; 2) biological studies (growth and regeneration rates, pollination system, seed dispersal, etc.); 3) assessment of tenure and access; 4) potential for confusion with similar species; 5) local knowledge and harvest practices; 6) impact of harvest on viability of individuals; 7) yield and market studies; and 8) assessment of regional and global threat based on all available knowledge and expertise. The management system should include annual harvest quotas; seasonal restrictions; geographical restrictions; restricted harvest of particular plant parts or size classes; and continuous monitoring and evaluation of conservation status (Leman, 2001).

Sustainable harvest is gaining importance as a conservation strategy for most wild-harvested species and their habitats, given their current and potential contributions to local economies and their greater value to harvesters over the long term. The basic idea is that non-destructive harvests and local benefits will maintain population, species and ecosystem diversity (Schippmann *et al.*, 2002).

Potential gains

Trees are predominantly outcrossing. This results in progeny that segregate with respect to parental traits. This wide genetic variation not only provides buffering against differing environments and management practices but also afford the opportunity for selection. Most quantitative traits (height, biomass, etc.) are under low to moderate additive genetic control, which means that only 10-30% of the variation we see for a trait in a population can be captured in individuals following selection. Using a fairly high intensity of selection (say 1 in 50) we could expect the progeny to outyield the group of parents from which it was derived by approximately 15-25%.

Qualitative traits such as fruit shape, taste, tree form, date of first fruiting are usually much more tightly inherited, from selected individuals. The genetic gain we would expect in one cycle of selection for a qualitative trait would be much higher, such that 60% of the progeny might be similar to their parents.

Cloning is the ultimate means of capturing useful genetic variation in that there is no recombination or segregation of genes. There is a need, however, to screen large numbers of trees to identify clones with superior traits. Vegetative propagation, however, is not genetic improvement in itself, and regular testing and introduction of new material is advocated strongly (e.g., Leakey 1991; Leakey and Izac 1996).

Jatropha curcas Production

Jatropha curcas L. (Euphorbiaceae) or jatropha is a large shrub or tree commonly found and utilized throughout most of the tropical and subtropical regions of the world. Several properties of the plant stemming from its hardiness, rapid growth, easy propagation, and wide ranging usefulness have resulted in the spread of jatropha far beyond its original distribution. Foremost, jatropha exhibits a wide environmental tolerance; it is found in both the seasonally dry tropics as well as equatorial regions, and is well adapted for cultivation within the vast areas of marginal and degraded lands in the semi-arid and arid tropics. Propagation is easily accomplished by seed or cuttings (often as simply as pushing a cutting into the ground) and growth is rapid as is implied by its ability to form a thick live hedge nine months after planting. Jatropha probably owes its successful cultivation world-wide to having no known major disease or insect pests (aside from susceptibility to fungi in humid regions) and its non-palatability to livestock so that it will not be grazed out, even during times of drought.

Jatropha's seed yields, up to 50 per cent by weight, a slow-drying oil (known as "curcas oil") which is utilized as an industrial raw material and for manufacturing on a smaller scale. Curcas oil has historically (and currently) been used in the manufacture of candles and particularly, soap. A good quality household soap which lathers heavily is produced using curcas oil and a finer quality soap is produced if curcas oil is combined with palm oil. The Chinese have boiled curcas oil with iron oxide for use as a furniture varnish, although its slow-drying quality makes it less desirable than other materials available for varnish manufacture.

Curcas oil's physical and chemical properties give it potential to be used for a petroleum-substituting fuel. Research on this oil was first initiated during World War II to study the use of curcas oil as a liquid, renewable fuel source to substitute for diesel oil. In recent years, research has been ongoing to further assess curcas oil's application as a fuel substitute; as a crop, jatropha represents a renewable resource with a high energy content and oil yield. Economic feasibility of this technology has yet to be proven, but in the context of a rural community being able to produce energy for their own use, there is great potential.

Rural development

Traditionally, the cultivation of jatropha has been undertaken primarily for protection of crops or pasture lands by serving as a fence to confine livestock or as a hedge for erosion control or as a windbreak. Its secondary uses or products have been exploited mostly for local use, not for significant economic gain. Easily established, jatropha forms a dense hedge within nine months of planting and responds well to cutting and trimming (Banerjee 1989). Although jatropha is exotic to India but has naturalized in different ecological zones of the country. In the eastern parts of India, e.g. in Chhattisgarh and Orissa, rural communities have plants on field bunds and in home gardens as live fence. A survey of 6 main regions of Chhattisgarh revealed that jatropha is present in different types of land use (Table-1). The plant numbers varied in different localities as in more developed areas (like Raipur, capital of Chhattisgarh) it was minimum while in least developed regions (like Pendra Road) it varied from 10-25 plants/ acre. Plants were found in fallows, agricultural fields, forest and homegardens. None of these had been planted and they grew wild and had been retained by farmers on their land. The plant numbers varied significantly in different landuse, e.g., in agricultural fields it was minimum compared to fallow and forest type of lands (Table-1).

Fuel oil

Traditionally the tribals of Chhattisgarh and Orissa have been extracting oil from the seeds of jatropha and the oil is used in lamps. Recently, there has been increasing interest in the use of curcas oil as either a diesel fuel substitute or extender. Research into the use of curcas oil in place of No. 2 diesel fuel has found that curcas oil is a compatible

Table-1: Jatropha plants found in landuse in different regions of Chhattisgarh

Region	Fallow (acre)	Homes	Fields (acre)	Forests (acre)
Pendra Road	12-15	15-25	10-15	15-25
Bilaspur	10-12	10-20	7-10	7-15
Jagdalpur (Bastar)	5-8	6-15	10-15	15-25
Ambikapur	10-15	20-25	7-12	15-25
Raigarh	3-5	7-10	3-5	7-10
Raipur	1	1-3	1-2	3-5

fuel for use in diesel engines (Ishii and Takeuchi 1987). Engine test results have indicated that in order to achieve optimal engine performances, crude curcas oil must first be processed and then blended with diesel oil. A by-product of extraction of curcas oil is the seed cake which can be of value as an organic fertilizer or as a raw material in manufacturing. Thus there is a great potential for use of jatropha oil as a energy mix with diesel. It has been seen that mixing with diesel upto 20% has no negative effects on the engine (Takeda 1982). Presently the yearly consumption of diesel oil is ~ 40 million tones (~ 40% of total petroleum consumption) and India import 70% of its petroleum demand which is 30% of the total import bill. Mixing of curcas oil with diesel will save import bills.

Environmental requirements

Jatropha's adaptability to a wide range of growing conditions is illustrated by its ability to grow well in humid to semi-arid/arid tropical and subtropical environments where it has been reported growing between 480 mm to 2380 mm average annual rainfall. In regions where moisture is not a limiting factor (i.e., continuously wet tropics or under irrigation), jatropha can bloom and produce fruit all year. Although it is found growing in bottomlands and along river banks, jatropha do not survive in saturated soils. A drier climate supposedly improves the oil yield of the seeds, though to withstand times of extreme drought the shrub will shed leaves in an attempt to conserve moisture which results in somewhat decreased growth. Jatropha can survive a light frost and its altitude ranges from sea level to below 1,000 meters.

Jatropha is known to grow well in shallow soils and is commonly found growing in gravel, sandy and clayey soils, although on severely eroded lands its growth form may be stunted. Where it is found growing on arid escarpments, jatropha generally does not exceed 2-3 m in height. Jatropha also is adapted to low fertility sites and alkaline soils, although better yields are obtained on poor quality sites if fertilizers or manures containing small amounts of calcium, magnesium, and sulfur are used. Mycorrhizal associations have been observed with jatropha and are known to aid the plant's growth under conditions where phosphate is limiting.

Propagation

Jatropha is a fast growing plant which can achieve a height of three meters within three years under a variety of growing conditions.

Propagation is not difficult and commonly achieved using either seeds or cuttings. Seed production from plants propagated from seeds, can be expected within three to four years, although in northeastern Thailand, flowering has occurred in plants which had reached a height of one meter only four to five months after planting from seed. Use of branch cuttings for propagation is easy and results in rapid growth; the bush can be expected to start bearing fruit within one year of planting. Again in Thailand, it has been reported that a plant grown from a cutting grew to a height of two meters and yielded 50 fruits eight months after planting. Commonly, initial seed production occurs as early as four to five months after planting of cuttings.

Planting of any type – seeds, seedlings, or cuttings– are best accomplished early in the rainy season. Smaller spacings would be preferred if growing the plant as a hedgerow, for a fence, or for soil conservation purposes. Actual spacing will be determined based on end-use, soil quality / condition, humidity, rainfall, intercropping, etc.

Harvest and seed yield

Oil and seed yields referenced in the literature vary considerably (Table-2). The range of oil and seed yields can be attributed to a variety of factors such as climate, intensity of management, site quality, size of plant, and genetics. The highest seed yields should be expected only under the most favourable growing conditions. Oil yields can be estimated by multiplying the seed yield by the average percent oil yield of 35-40 per cent. Seed production continue for 30-40 years and the plant live for 50 or more years. Flowering of jatropha is associated with the rainy season, both at beginning and end in many locations (September to December in India) thus producing seed crops from one to three times a year. Fruits mature two to four months after flowering and turn yellow when they are mature. Harvesting often takes place during the dry season, so may be less likely to be in conflict with other agricultural labour requirements. The fruits are easily plucked from bushes and manual harvesting is adequate, possibly providing income for women workers.

Variation in oil content of seeds were studied in different geographical regions of India. It is evident from Table-3 that oil content varied from 29 to 55% in the States of Chhattisgarh, Andhra Pradesh, Orissa and Tamil Nadu. It was found to be highest in plants growing in Pendra Road (Chhattisgarh). However, statistically the oil content was at par in those

Table-2: Seed and oil yields in *Jatropha curcas*

Seed yield	Oil yield	Comments	Source
4-6 kg/yr	–	From average size bush	Raina, 1986
4-6 kg/yr	–	5 yr old tree in Thailand	Bhasabutra and Sutiponpeibun, 1982
2-3 kg/tree	–	5 yr old tree in Thailand	Takeda, 1982
4.6 kg/bush	–	Average seed yield	Banerjee, 1989
4-5 kg/tree	–	Madagascar	Martin and Mayeux, 1984
638 kg/ha	1.5-2.0 kg/tree	1x1 m spacing with no fertilizer: NE Thailand	Sukarin, Yamada and Sakaguchi, 1987
400-1200 kg/ha	200-600 kg/ha	Average yield; Cape Verde	Levingston and Zamora, 1983
5000 kg/ha (kernels)	2.5 tonnes/ha	One location in Cape Verde	Martin and Mayeux, 1984
3 tonnes/ha	–	Rainfed 5 years old trees	Wealth of India, 1959
5 ton/acre/yr	–	–	Srivastava, 1984
	1.5-2.3 tonnes/ha	Establishment plantation in Brazil	Calvin, 1985
	3-4 tonnes/ha	3m x 3m spacing; Brazil	Forni-Martins and Diniz de Cruz, 1985

plants growing in Bilaspur and Jagdalpur (Bastar) regions. It is also evident from the Table-3 that oil content was comparatively higher in plants growing

Table-3: Variation in oil content in different geographical regions of *J. curcas*

Region/Provenance	Oil (%)
Chhattisgarh	
Pendra Road	55.2
Bilaspur	54.6
Jagdalpur (Bastar)	52.7
Ambikapur (North Hills)	49.5
Raigarh	45.3
Raipur	48.8
Andhra Pradesh	
Hyderabad	33.8
Orissa	
Bhubneshwar	39.7
Tamil Nadu	
Coimbatore	30.5

N = 50; Seed Coat removed; Seed Weight = 3 g.

in Chhattisgarh followed by those in Orissa and Andhra Pradesh. The lowest oil content was in Tamil Nadu.

Studies were also conducted to know the effect of seed storage on oil content (Table-4). Freshly harvested seeds were stored in January. These were kept at room temperature (25-40°C), in refrigerator (4°C), in desiccator (having silica gel) and in paper bags for a year. Oil was analysed in February, May, August, November and January. Irrespective of the treatment the oil content decreased with time. However, the decrease was only 2%, which is negligible (Table-4). Seeds from the different plants growing in different types of soil were analyzed for oil content. The plant age at which seeds were collected were 5 and 10 years. Variation in oil content within the site due to age was found to be non-significant (Table 5). However, seed oil content varied of the plants growing in different soil types. In moist localities the oil content was higher as compared to dry localities. Among dry localities, the oil content in the seed of the plants growing in vertisol (black soil) was maximum followed by entisol (degraded land) and alfisol (Table-5).

There are several areas which would benefit from further research into the cultivation and utilization of jatropha. Little formal research has

Table-4: Seed storage and oil content (%) in *Jatropha curcas*

Treatment	February	May	August	November	January
Room Temp. (25-40°C)	52.8	52.1	51.8	50.6	50.1
Low 4°C	52.8	52.2	51.0	48.6	48.2
Desicator	52.8	52.2	51.2	49.9	49.4
Paper Bags	52.8	52.1	51.7	49.2	49.1
Fruit	52.8	52.2	51.5	50.3	49.4
Seed	52.8	52.1	51.1	48.9	49.8

Table-5: Oil (%) variation due to soil and moisture in *J. curcas*

Age (Yrs.)	Entisol (Degraded land)		Alfisol (Sandy-loam)		Vertisol (Black soil)
Dry	Moist	Dry	Moist	Dry	
5 +	52.1	57.6	47.5	50.5	54.3
10 +	51.9	56.9	48.3	51.2	55.1

Non-significant variations with age.

been conducted or reported regarding improvement of yields, propagation, or cultivation practices. Trials are needed to select varieties, to breed hardy high-yielding hybrids, and to develop cultivation practices which increases seed yields and oil content (Dehgan 1984).

Requirements for use of Curcas Oil

Curcas oil extraction through pressing is a process well-suited for small or decentralized operations. In many communities, seed oil pressing may already be practiced for extraction of other commonly used oilseeds (e.g., rape-seed, groundnut, and sunflower) although some modification to equipment may be necessary to accommodate new seed types. Extraction and preprocessing of jatropha seeds is similar to that of other oil seeds. Separation of the hull from the kernel is easily done when dried and requires simple preprocessing to maximize oil yield (i.e., grinding, heating, cooling and then pressing).

By-products obtained from oil extraction and processing provide additional locally used and marketable materials. Economic benefits may be realized through the sale or direct use of these by-products (e.g., oil cake from oil extraction and glycerine from transesterification). The sale and use of seed cake as fertilizer can generate and keep revenues within the community, and as a fertilizer may have significant benefits for problematic soils. Glycerine, a raw material used in the production of soap and cosmetics has in some circumstances been found to be worth more than diesel oil so that this by-product can potentially generate significant amounts of revenue. Profits from the sale of by-products could offset the costs associated with curcas oil production, resulting in a more competitively priced fuel.

Benefits of Curcas Oil as a fuel

Curcas oil has shown promise as a source of energy. Seed harvesting, oil extraction and processing can be accomplished in a decentralized, rural environment providing opportunities for business cooperatives and small entrepreneurs. Production of a renewable energy source on a local-scale can lead to self-sufficiency or a decrease in dependence on fuels such as diesel, kerosene, wood, etc. Benefits may include a decrease in dependency on fuelwood gathering (particularly benefiting women) which is often a cause of environmental degradation in many areas (as well as a time consuming task); the establishment of a local renewable oil industry which can provide employment opportunities in the establishment of plantations, harvesting, extraction, and processing; protection of land through revegetation; and production of useful and revenue generating by-products. Furthermore the levels of pollution from engine emissions using curcas oil may be significantly lower or nonexistent under certain engine operating conditions.

Despite the opportunities associated with curcas oil production (or vegetable oils in general), as a fuel it is not viable when compared to global production costs for fossil fuels. However, viewing curcas oil production in the context of its other, previously discussed benefits, may result in economic feasibility under appropriate circumstances.

Though there is a range of village-based technologies currently available for extraction and processing of vegetable oils for small-scale operations, further development and improvement for curcas oil extraction is needed. This may consist of increase use of extension efforts for existing technologies or the development of improved and appropriate technologies for small-scale extraction. Included is the development of simple, inexpensive, and locally manufactured oil extraction equipment to avoid increased dependency on foreign and imported equipment. The development of extraction and processing activities would benefit from the organization of small businesses and provision of management training and technical assistance. Finally, should the use of curcas oil (and other vegetable oils) become extensive, a need would arise for engine designers to work towards the development of design adjustments of more efficient engines.

Mission Mode on *Jatropha*

Studies on the biological variability of jatropha fruit and seeds, its propagation using cheap and simple methods appropriate for mass production needs to be taken up on priority. In the last three years, National Oilseeds and Vegetable Oils Development (NOVOD) Board, Gurgaon, initiated a coordinated initiative across the Chhattisgarh region through Indira Gandhi Agricultural University (IGAU), Raipur, for the domestication and propagation of quality jatropha plants. The prioritization of jatropha in Chhattisgarh is based on farmer preference, product ranking (e.g. oil), and the valuation. The geographic range of this species is throughout Chhattisgarh. In order to capture much of the genetic variation, Germplasm collections are implemented by IGAU. These collections are targeted at visible variation and farmers perception of superiority. It is proposed that in future these collections be made throughout the country, exchanged between the regions and established as living gene banks at three regions (North, South and Central India). Under Mission Mode Programme of NOVOD, each gene bank should have single progenies of collected accessions.

The collected Germplasm be evaluated for growth, phenological traits and oil content. This will be a part of genetic testing, with selection of superior individuals. Molecular studies of the extent of genetic diversity in population will provide a baseline for the domestication programme. Studies related to genetic diversity will be taken up in order to have a fundamentally important basis on which to build a farmer-oriented cultivar selection programme.

The second phase of domestication in the mission mode will identify, select and multiply superior trees. The identified plus-trees will be multiplied through micropropagation and macropropagation techniques. Under this programme Village Nurseries will be established so that the participating farmers are the beneficiaries of the programme. Vegetative propagation techniques will be used for the capture and multiplication of superior phenotypes as putative cultivars. Currently studies are in progress in IGAU to propagate jatropha by cuttings. Simple low technology methods needs to be developed for mature shoots which are more difficult as they become physiologically and ontogenetically old. Research is needed to try to rejuvenate mature crown shoots physiologically, while retaining the benefits of reproductive capacity due to ontogenetic age.

IGAU's tree domestication programme for *J. curcas* is currently focusing much of its resources for mass production of trees that farmers are identifying as superior. There is at present, however, little knowledge about farmers perception of genetic variation or their capacity to identify and capture the potential for improvement through the formation of cultivars based on the propagation of elite trees in a wild population. Evidence from the farmer interviews suggests that there is substantial variation in certain fruit traits, such as fruit size, as well as some variation in the size of the commercially important kernels. Farmers indicate that kernel size is important in the market, although there is yet no indication if this is recognized by the price discrimination. It seems to be generally considered that large kernels will be found in large fruits.

Keeping all these facts in view, it is therefore suggested that NOVOD Board should take up on priority mission mode project on jatropha.

CONCLUSIONS

For developing countries, like India, the threat of future oil price rise gives maximum cause for concern. Trapped in many cases in circumstances of slow or negative economic growth, foreign indebtedness and current trade deficit, its industrial development prospects are set at a critical disadvantage by any further rise in the cost of fossil fuel imports. When it comes to developing renewable or alternative energy sources, however, India possessed substantial natural advantages, abundant forests, fertile soils and sunlight. It is now apparent that wood and wood-based fuels merit open-minded consideration as technically mature, environmentally sound, economically viable and culturally acceptable options for modern energy development. The most important lesson that remains to be learned is the importance of developing these options in a coordinated way, in keeping with the ecological and socioeconomic realities that prevail from place to place.

Curcas Oil has shown promise as a source of energy. Seed harvesting, oil extraction and processing can be accomplished in a decentralized, rural environment providing opportunities for business cooperatives and small entrepreneurs. Production of a renewable energy source on a local-scale can lead to self-sufficiency or a decrease in dependence on fuels such as diesel, kerosene, wood, etc. Benefits may include a decrease in dependency on fuelwood gathering (particularly benefiting women) which is often a cause of environmental degradation in many areas (as well as a time consuming task); the establishment of a local renewable oil industry which can provide employment opportunities in the establishment of plantations, harvesting, extraction, and processing; protection of land through revegetation; and production of useful and revenue generating by-products. Furthermore the levels of pollution from engine emissions using curcas oil may be significantly lower or nonexistent under certain engine operating conditions.

Despite the opportunities associated with curcas oil production (or vegetable oils in general), as a fuel it is not viable when compared to global production costs for fossil fuels. However, viewing curcas oil production in the context of its other, previously discussed benefits, may result in economic feasibility under appropriate circumstances.

Though there is a range of village-based technologies currently available for extraction and processing of vegetable oils for small-scale operations, further development and improvement for curcas oil extraction is needed. This may consist of increase use of extension efforts for existing technologies or the development of improved and appropriate technologies for small-scale extraction. Included is the development of simple, inexpensive, and locally manufactured oil extraction equipment to avoid increased dependency on foreign and imported equipment. The development of extraction and processing activities would benefit from the organization of small businesses and provision of management training and technical assistance. Finally, should the use of curcas oil (and other vegetable oils) become extensive, a need would arise for engine designers to work towards the development of design adjustments of more efficient engines.

REFERENCES

Banerjee, A.K. (1989). Shrubs in tropical forest ecosystems, examples from India. World Bank Technical Paper Number 103.

Briggs, D. & Walters, S.M. (1984). *Plant variation and evolution.* 2nd ed. Cambridge University Press, Cambridge. 412 p.

Dehgan, B. (1984). Phylogenetic significance of interspecific hybridization in jatropha (Euphorbiaceae). Systematic Botany **9(4)**: 467-478.

Leakey R.R.B. (1991). Towards a strategy for clonal forestry: some guidelines based on experience with tropical trees. p 27-42. In: *Tree improvement and breeding,* ed. J.E. Jackson. Royal Forestry Society of England, Wales and Northern Ireland, Tring, England.

Leakey, R.R.B. & Newton, A. (1994). *Domestication of tropical trees for timber and non-timber forest products.* MAB Digest No. 17. UNESCO, Paris. 94 p.

Leakey, R.R.B. & Izac, AMN (1996). Linkages between domestication and commercialization of non-timber forest products: implications for agroforestry. In: Leakey RRB, Temu AB, Melnyk M and Vantomme P (eds), Domestication and Commercialization of Non-timber Forest Products in Agroforestry Systems, pp 1-7. Non-Wood Forest Products No. 9. FAO, Rome, Italy.

Leaman, D. (2001). Conservation, trade, sustainability and exploitation of medicinal plant species. p.1-15 In: P.K. Saxena (ed.). Development of Plant-Based Medicines: Conservation, Efficacy and Safety. Kluwer Academic Publishers, Dordrecht, The Netherlands.

Nantel, P., Gagnon, D. and Nault, A. (1996). Population viability analysis of American Ginseng and Wild Leek harvested in stochastic environments. Conservation Biology **10**: 608-621.

Palevitch, D. (1991). Agronomy applied to medicinal plant conservation. p.168-178 In: O. Akerele, V.H. Heywood and H. Synge (eds.). Conservation of Medicinal Plants. University Press, Cambridge, UK.

Schippmann, U. (1997). Plant uses and species risk. From horticultural to medicinal plant trade. p.161-165 In: J. Newton (ed.). Plants Europaea. Proc. of the 1st European Conference on the conservation of wild plants, Hyeres, France.

Schippmann U., Leaman D.J. and Cunnningham, A.B. (2002). Impact of cultivation and gathering of medicinal plants on biodiversity: global trends and issues. p.1-21 In: Biodiversity and the Ecosystem Approach in Agriculture, Forestry and Fisheries. Ninth Regular Session of the Commission on Genetic Resources for Food and Agriculture. FAO, Rome, Italy.

Whitmore T.C. (1992). *An introduction to tropical rainforests.* Oxford University Press, Oxford. 226 p.

Advancing Frontiers of Ecological Researches in India, 189-200, 2006
Edited by A.K. Kandya & Asha Gupta
Published by Bishen Singh Mahendra Pal Singh, Dehra Dun, India

N_2O Source and Sink Potential in Dry Tropical Climax Forest and Seral Grassland in Central India

S.K. Billore, Vijay P. Singh, K. Kaur and P. Dass
Institute of Environment Management & Plant Sciences
Vikram University, Ujjain (M.P.) 456 010, INDIA

Abstract: Tropical soils are an important natural source of N_2O and contribution from forest and grassland habitats are of much ecological significance in increasing the N_2O pool in the atmosphere. In this study nitrous oxide emissions were measured from *Tectona grandis* L. dominated deciduous forest and grassland soil in the field condition by using closed chamber technique. Different soil parameters like water content, inorganic nitrogen stocks, total nitrogen and organic carbon were measured in soil samples obtained concomitantly while gas flux measurement were made.

Annual mean of N_2O emission from forest and grassland soils were 2.66 and 3.69 mg m^{-2} respectively, with higher emission rates during rainy season. Emissions were positively and significantly correlated with soil moisture and NO_3^--N and negatively correlated with soil temperature in dry deciduous forest. In the grassland soil N_2O fluxes were correlated with NO_3^--N (product of nitrification) and NH_4^+-N.

Results indicate that factors inducing high variability in N_2O production are difficult to predict from soil factors. The complex interplay of the parameters involved, and the heterogeneity of forest–grassland floor emphasized the need for frequent study with large scale integrated measurement techniques to obtain accurate flux estimate of N_2O and to calculate annual emission rate from natural ecosystems.

INTRODUCTION

Nitrous oxide (N_2O) is a relatively long lived trace gas with potent greenhouse properties and a natural species act as reactant in destruction of stratospheric ozone layer (Crutzen, 1979). On a molar basis, N_2O is about 250 times more effective as an absorber of infrared radiation than CO_2 with atmospheric life time of 150 years (Robertson, 1993). As with many greenhouse gases the atmospheric concentration of N_2O has increased from about 285 PPbv (Khalil & Rasmusen, 1988) in the pre–industrial era to about 311.5 PPbv in 1996 (Khalil, 1999). The rate of increase for atmospheric N_2O has been quite variable, but on average it has been 0.3% between 1985 and 1995 (Khalil, 1999). According to an estimate (Crutzen & Ehhalt, 1977) doubling the concentration of N_2O in the atmosphere would cause 10% decrease in the stratospheric ozone layer which would results in a 20% increase in the amount of ultraviolet radiation which reaches the earth surface. According to Davidson (1991) the global budget of source and sinks of atmospheric N_2O is badly out of balance. Due to seasonal and spatial variation in N_2O emission from natural ecosystems are difficult to quantify (Smith *et al.*, 1994).

Soils are an important source of N_2O accounting for 70% of global emission (Conrad, 1996) and produced as result of microbial processes accounting in soil, especially nitrification and denitrification (Billore *et al.*, 1996; Conrad, 1996). The most conspicuous soil variables that operate at the microbial level and influence N_2O production include NO_3^-, NH_4^+ availability, organic matter, pH and O_2 concentration (Firestone & Davidson, 1989), which, in turn, are conditioned by soil water, texture, vegetations and climate (Conrad, 1996).

The tropics are particularly important as sources of nitrogen–containing trace gases because in contrast to most temperate ecosystems, biologically available nitrogen is present in adequate to excess amounts within most intact tropical ecosystems. In tropical conditions natural and managed habitats provide excellent microclimatic conditions for the aerobic nitrification during most part of the year, and for denitrification when the soil becomes completely water saturated immediately after the rains and irrigation practices. In other words tropical soils facing the stressed high temperature condition in summer, and subsequently favourable rainy season boosting trace gas emission under the nitrogen related microbial processes like nitrification and denitrification.

The objective of our research was to estimate the seasonal N_2O fluxes from seral grassland & climax forest and to examine the correlation of the N_2O fluxes with different soil variables.

MATERIALS AND METHODS

2.1 *Study Site*

The selected dry deciduous forest is located 20km. southwards from Indore city (22°43'N latitude and 75°48E longitude, above 573 mean sea level) and moderate grazed grassland situated near the Vikram University Campus, Ujjain (23°11'N latitude and 75°43'E longitude and 491.7m above mean sea level), Central India. The climate of the area is typically monsoonic with hot summer and cool winter. The mean maximum temperature varied from 27.5°C (January) to 46.0°C (May) in forest field and 21.0 (January) to 36.5°C (May) in grassland field, and mean minimum temperature varied from 9.7 (January) to 25.8°C (May) in forest and 13.0 (January) to 29.5°C (May) in grassland field. The year can be divided into three distinct seasons such as Summer (March to Mid June), Rainy (Late June to October) and Winter (November to February). The average annual rainfall is 1012.8 and 1014.0 mm in forest and grassland areas, respectively. 90% of the annual rainfall received during rainy season i.e. mid June to Mid October. About 8 months of the year are dry and 4 months are moist. Major soil characteristics of the sites are summarized in Table-1.

Table-1 : Soil Characteristics for the Surface (15cm) Soil of Study Sites

Parameters	Grassland	Forest
Bulk Density Soil Texture	1.30	1.10
Clay (%)	42.00	36.00
Silt (%)	43.00	30.00
Sand (%)	14.00	17.00
W.H.C.	48.70	46.20
pH	8.10	7.80
Org–C (%)	0.58	0.67
Total–N (%)	0.12	0.10

2.2 Soil Sampling and Analysis

The upper 15 cm of soil, minus litter was sampled adjacent to the chamber used for gas flux measurement on each sampling day. The sample were mixed and pass through 2 mm sieve and stored at 5°C over night before analysis were begun the next day. Soil moisture was determined gravimetrically by oven drying (105°C, 24 hrs.) NH_4–N and NO_3–N were obtained from extracts of field moist soil with 2m KCl (1:5w/v) and analysis by Bremner's distillation method. Total nitrogen (N) was analysed by microkjeldahl method. Organic carbon (C) was analysed by dichromate oxidation and titration with ferrous ammonium sulphate (Walkley, 1947). Soil pH was determined by using a pH meter equipped with glass electrode

(1:2.5 soil : water ratio, w/v). Nitrogen mineralization and nitrification were determined using the aerobic incubation procedure, 20g subsample of field–moist soil was placed in a 100 ml conical flask, which was then closed with a perforated plastic to allow gas exchange while minimizing evaporation loss. Soils were incubated for 21 days at room temperature. NH_4–N and NO_3–N were determined before and after incubation as described above. Net mineralization rates were determined from the difference between inorganic N and net nitrification was from the difference in NO_3–N, at the beginning and end of the incubation and results were expressed on mean daily basis.

2.3 Flux measurements

Two replicate plots were selected to study the N_2O fluxes from each field. Collection of air samples were carried out using closed chamber technique as described by Jonasson & Granat (1984), Parashar *et al.* (1993). To study nitrous oxide fluxes between soil and the atmosphere gas samples were collected for two consecutive years (October 1997 to September 1999) using transparent perspex chamber made from 6mm plexiglass sheet. Aluminium base (54 L x 33 W x 10 H cm) with internal groove size (48 L x 27 W x 2 H cm) were installed manually by digging narrow gully of frame size during the summer and winter seasons in hard and dry soils without much disturbing the soil. During rainy season, the base was installed simply by pushing the aluminium base as soil becomes wet. The base was embedded in the soil two hours in advance to ensure that the ambient soil atmosphere was maintained in a stabilized condition. The perspex chamber (50 L x 30 W x 50 H cm), which fitted into the groove of the aluminium base was put in place at the time of sampling covering an area of 0.1765 m^{-2}. The air inside the chamber was isolated from the outside atmosphere and the system was made air tight by filling the groove in the aluminium base with water. The chamber cover is fitted with a sampling port and a port to which a plastic pump is attached for uniform mixing of air inside the chamber. A sampling port is a hole plugged with a rubber septum, through which a 50 ml glass syringe with hypodermic needle was inserted to collect air samples.

At the time of sampling plants inside chamber affects the actual Box volume. For that plant volume inside the box was calculated in grassland field for the calculation of Actual Box Volume during sampling, while in the forest habitat generally no plants were present inside the chamber.

Air samples for nitrous oxide flux measurements were drawn in pre–evacuated sealed glass vials at 0, 10, 20 and 30 minutes. Each time ambient air sample was also collected near to the chamber to compare with 0 minute flux at the time of calculation. After 30 minutes the chamber was lifted and kept aside. Flux measurements were made in mid morning and mid evening on each sampling day.

The temperature inside the perspex chamber was recorded at the time of sample collection (0, 10, 20, 30 min.) using a thermometer (–10 to 100°C range : Co. Immersion Zeal, England) fixed on inside wall of the chamber for the calculation of box volume. Soil temperature was also measured with a soil thermometer inserted into the soil at 5 cm adjacent the chamber. The collected gas samples in 100 ml glass vials were brought to the laboratory and analyzed for N_2O fluxes in National Physical Laboratory (NPL), New Delhi using a SRI 8610 C gas chromatograph equipped with a 63 Ni electron Capture Detector (ECD) and porapack Q column. The operating temperature of detector, injector and oven was 350°C, 150°C and 80°C, respectively with nitrogen as a carrier gas at the flow rate of 30ml per minute. The gas chromatograph was calibrated, before and after each day of measurement, using 211 and 1242 ppb N_2O as primary standards. 1 ml of air sample was injected in gas chromatograph port with 2 ml glass syringe and separated on a porapack Q column. Peak area of three replicates (almost same peak area was recorded) were averaged for each flux measurement.

RESULTS AND DISCUSSION

3.1 Seasonal Variation

Nitrous oxide emission showed considerable variation sitewise and fluxes varied with seasonal change and this fluctuation in seasonal N_2O pattern may depend on the climatic conditions. Although soil can act as source of and sink for nitrous oxide. Fluxes were high during rainy season in both forest and grassland field (Fig. 1). Higher fluxes during rainy season is supported by Mosier *et al.* (1997), they reports that N_2O emissions peak soon after rainfall. Results are also consistent with findings of Davidson *et al.* (1993) who reported that N_2O fluxes were higher during wet season than in dry season in dry tropical forest of Mexico and same trend was also observed by Verchot *et al.* (1999). While association between N_2O fluxes with wet versus dry season was not observed by Luizao *et al.* (1989).

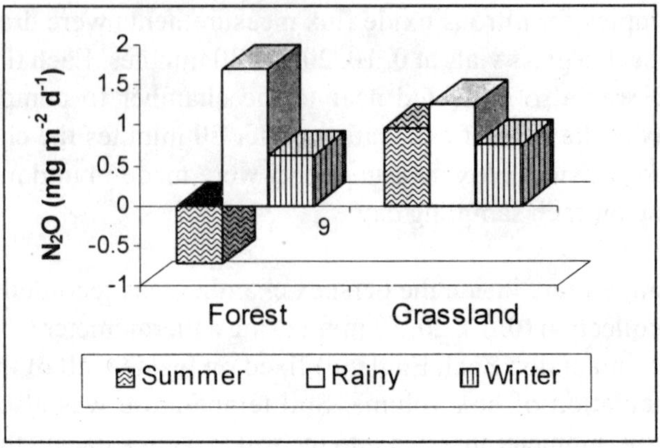

Fig. 1. Seasonal Variation in N_2O Fluxes in Forest and Grassland soils

The negative fluxes of N_2O during the summer season in dry deciduous forest, recorded an unusual trend unlike most of the reported in the literature. This observation that atmospheric N_2O is consumed / decomposed in relatively dry and oxic soils is curious and of ecological significance in the dry tropical habitats, since it is completely unlike that reduction of N_2O via denitrification under these conditions. Soil temperature (38°–46°C) in forest field during summer is greater than temperature optima of nitrification in tropical zone (35°C, Malhi & McGill, 1982) may checked N_2O production via nitrification. Donso *et al.* (1993) reported soil consumption of N_2O during dry season in tropical savannah and Verchot *et al.* (1999) also observed significant negative fluxes of N_2O in tropical cattle pasture during the dry season, but mechanism of negative fluxes is unknown.

Davidson and Cicerone have speculated about possible reduction by minerals such as iron for these negative fluxes, but they have no data to support this idea (E. Davidson, 2001, Personal Communication). In the Russian literature Vedenina & Zavarzin (1977) and Vedenina *et al.* (1980) claiming oxidation of N_2O by soil enzymes such as catalase and peroxidase. According to Ralf Conrad (2001, Personal Communication) enzymatic combustion of N_2O might be a possibility, but it's need further detailed study.

3.2 *Environmental factors influencing N_2O fluxes*

In totality, a strong positive relationship was observed between N_2O flux and year long soil moisture in the dry deciduous forest ($r = 0.89$; $p <$

0.0000; n = 22) but relation was not apparent in grassland field (r = 0.18; p < 0.4164; n =22). Soil moisture was found to be an important regulator of N_2O production by nitrification and denitrification, mainly due to affecting the availability of O_2 (Bollman & Conrad, 1998). The amount of N_2O produced by nitrifiers relatively to NO_2^- and NO_3^- is known to increases as soil moisture increases (Goodroad & Keeney, 1984) or as the partial pressure of oxygen (PO_2) decreases (Goreau, 1980), which is consistent with a positive correlation between soil moisture and N_2O flux. Infect, in grassland field absence of relation between soil moisture and N_2O fluxes does not mean that soil water is an unimportant factor controlling the nature of the N_2O emission from soil. Rather these results suggests that the importance of soil water is apparent when nitrogen is cycling during rainy season, when nitrogen mineralization was highest (Fig. 2) and gas fluxes are more substantial. Several other studies also indicates that interaction among soil warming, moisture, availability of nitrogen cycling and plant physiological activity may control N_2O fluxes (Parton *et al.*, 1988; Matson *et al.*, 1991; Li *et al.*, 1992; Epstein *et al.*, 1998).

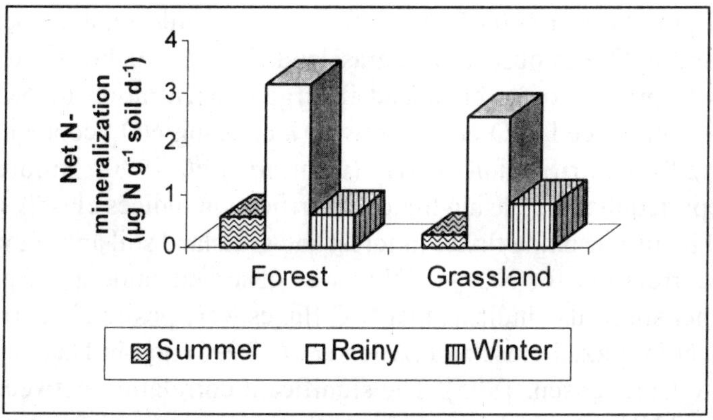

Fig. 2. Seasonal Variation in Net Nitrogen mineralization in Forest and Grassland Ecosystems Under Laboratory Incubation

In grassland field N_2O emissions were not significantly correlated with soil temperature (r = 0.07; p<0.7463; n =22). Ambus & Christensen (1995) also have not observed any correlation with soil temperature in Danish grassland field and Matson *et al.* (1991) in steppe ecosystem, while several studies concur the N_2O emission increases with soil temperature (Mosier *et al.*, 1996; Epstein *et al.*, 1998). While in forest field a strong negative correlation (r = –0.083; p<0.0000; n = 22) was observed with soil temperature & N_2O flux. This is mainly due to significant sink (soil uptake) of N_2O during summer. Maximum flux values grouped near soil temperature 24–30°C and sink of N_2O started when temperature increased from 35–48°C. In contrast, most field studies revealed a positive relationship between N_2O emission and temperature (Slemr *et al.*, 1984; Williams and Fehlenfeld, 1991; Sullivan *et al.*, 1996; Clayton *et al.*, 1997) but in these studies soil temperature never exceeds temperature optima of nitrification and denitrification. While some studies do not observed any relation between soil temperature and N_2O fluxes (Peterjohn *et al.*, 1994; Ambus & Christensen, 1995). Soil temperature is one of the important parameter which changes on dial and seasonal scale & influence the microbial process.

3.3 N-availability and N_2O fluxes

Positive relation between N_2O fluxes and soil nitrate nitrogen indicates possibility that nitrification plays role in N_2O emission from natural terrestrial ecosystems. In the present study positive correlation between N_2O flux and NO_3^-N was observed in both forest ($r = 0.47$; $p <0.0249$; $n =22$) and grassland ($r = 0.65$; $p < 0.05$; $n =21$) while relation between NH_4^+-N and N_2O was observed in grassland field ($r = 0.49$; $p < 0.0500$; $n=21$). N_2O production is enhanced if large concentration of NO_3^- are present and enhanced N_2O emission with increasing NO_3^- concentration originating from nitrification of NH_4^+ (Comfort *et al.*, 1990). Billore *et al.* (1996) reported that increasing trend in nitrification indices closely reflect to pattern in nitrous oxide fluxes in forest and grassland soil and significant positive correlation for NO_3^- Vs N_2O was observed same as in present study. Other study also indicate that N_2O fluxes were positively correlated with NO_3^-N in grazed grassland (Allen *et al.*, 1996) and in Danish forest (Ambus & Christensen, 1995). The significant correlation between N_2O flux and soil NH_4^+–N levels also indicate that nitrification plays a vital role in N_2O emission. Allen *et al.* (1996) also observed a statistically significant relationship between N_2O Vs NH_4^+–N from grazed grassland in U.K. The N_2O flux due to nitrification was found to be strongly correlated with soil NH_4 level (Parton *et al.*, 1988) as well as with soil NO_3 level (Davidson *et al.*, 1986), though a poor relationship has been reported between N_2O flux and size of the soil mineral nitrogen pool (Mosier *et al.*, 1996).

The N_2O emissions in the seral grassland (2.5 kg N_2O–N h^{-1} y^{-1}) was higher than climax forest (1.69 kg N_2O–N ha^{-1} y^{-1}) community. Emission fluxes from seral grassland in present study was much greater than measured in temperate native grassland (0.66 to 1.10 kg N_2O–N ha^{-1} y^{-1})

by Mosier *et al.* (1991) and in Danish coastal grassland (1.24 kg N_2O–N ha^{-1} y^{-1}) by Ambus & Christensen (1995). Two fold greater N_2O fluxes in this study as compare to temperate grassland have provide the evidence for observation of Matson & Vitousek (1990) and Keller *et al.* (1993), that tropical grassland are expected to emit more N_2O as compared to temperate annual rates.

Annual N_2O flux from tropical dry deciduous forest was comparable to 1.9 kg N_2O–N ha^{-1} y^{-1} from Brazilian tropical forest (Luizao *et al.*, 1989) and near to upper range of temperate forest fluxes ranged between 0.1 to 2.25 kg N_2O–N ha^{-1} y^{-1} (Schmidt *et al.*, 1988; Bowden *et al.*, 1990; Struwe & Kjoller, 1994). Fluxes of nitrous oxide appear to be greatly elevated in tropical soil as compared to temperate (Crutzen, 1987; McElory & Wofsy, 1986), while uncertainties are remained due to high variation in daily flux in the present study.

CONCLUSION

Grassland occupying seral ecological status emitting higher N_2O as compared to climax forest soil. These results provide another basis for restoration of climax community and the forest conservation. Crucial habitats factors inducing high variability in N_2O production & emission are difficult to predict and generalize from inter–site soil factors. The probability that any particular site will have a large flux of N_2O is dependent on the coincidence of a number of controlling variables. The overall study leads to the need for large scale integrative measurement techniques to obtain accurate and more frequent flux estimate of N_2O, which may provide a reliable information for extrapolation of N_2O fluxes to larger areas and for predicting emission rates under changing conditions mainly in natural habitats.

ACKNOWLEDGEMENTS

Dr. Prabhat Kumar Gupta, Senior Scientist, Chemistry Division and Dr. D.C. Parashar, Scientist Emeritus, National Physical Laboratory, New Delhi, providing laboratory facilities for N_2O analysis. Department of Science and Technology (DST), Govt. of India, New Delhi, for funding support in the form of Research Project.

REFERENCES

Allen A.G.; Jarvis, S.C.; Headon D.M. (1996). Nitrous oxide emissions from soils due to inputs of nitrogen from excreta return by livestock on grazed grassland in the UK. Soil Biol. Biochem. **28** : 597-607.

Ambus, P. and Christensen, S. (1995). Spatial and seasonal nitrous oxide and methane fluxes in Danish forest, Grassland and Agroecosystems. J. Environ. Qual. **24**: 993-1001.

Billore, S.K.; Numata, M. and Minami, K. (1996). Nitrous oxide emissions from grassland and forest soil through nitrification. Current Sci., 70 **(11)**: 1010-1012.

Bollmann, A. Conrad, R. (1998). Influence of O_2 availability on NO and N_2O release by nitrification and denitrification in soils. Global Change Biol. **4** : 387-396.

Bowden, R.D.; Steudler, P.A.; Meillo, J.M. (1990). Annual nitrous oxide fluxes from temperate forest soils in the Northeastern United States. J. Geophys. Res. **95**: 13997-14005.

Clayton, H.; McTaggart; I.P. Parker, J.; Swan, L.; Smith, K.A. (1997). Nitrous oxide emissions from fertilized grassland a 2 year study of the effects of N fertilizer from and environmental conditions. Biol. Fertil. Soil **25**: 252-226.

Conrad, R. (1996). Metabolism of nitric oxide in soil and soil microorganisms and regulation of flux into the atmosphere. In : Murrell JC, Kelly DP (eds) Microbiology of atmospheric trace gases : sources, sinks and global change processes. Springer, Berlin, Heidelberg New York, pp 167-203.

Comfort, S.D.; Kelling, K.A.; Keeney, D.R.; Converk, J.C. (1990). Nitrous oxide production from injected liquid dairy manure. *Soil Sci. Soc. Am. J.* **54** : 421-427.

Crutzen, P.J. and Ehhalt, D.H. (1977). Effects of nitrogen fertilizers and combustion on the stratospheric ozone layer. Ambio. **6**: 112-117.

Crutzen, P.J. (1979). The role of NO and N_2O in the chemistry of the troposphere and stratosphere. Annu Rev Earth Planet Sci **7**: 443-472.

Crutzen, P.J. (1987). Role of the tropics in atmospheric chemistry. In : Dickinson RE (ed.) The Geophysiology of Amazonia. John Wiley & Sons., New York, pp 107-130.

Davidson, E.A.; Swank, W.T.; Perry, T.O. (1986). Distinguishing between nitrification and denitrification as sources of gaseous nitrogen production in soil. Appl. Environ. Microbiol. **52** : 1280-1286.

Davidson, E.A. (1991). Fluxes of nitrogen oxide and nitric oxide from terrestrial ecosystems. In: Rogers JE, Whitman WB (Eds) Microbial production and consumption of greenhouse gases : Methane, Nitrogen oxide and Halomethanes. American Society for Microbiology, Washington, D.C. pp 219-235.

Davidson, E.A.; Matson, P.A.; Vitousel, P.M.; Riley, R.; Dunkin, K.; Gavcia–Mandez G. Maass, J.M. (1993). Process regulating soil emissions of NO and N_2O in a seasonally dry tropical forest. Ecology **74** : 130-139.

Epstein, H.E.; Burke, I.C.; Mosier, A.R.; Hutchinson, G.L. (1998). Plant functional type effects on trace gas fluxes in the short grass steppe. Biogeochem. **42**: 145-168.

Firestone, M.K. & Davidson, E.A. (1989) Microbiological basis of NO and N_2O production and consumption in soil. In : Exchange of trace gases between terrestrial ecosystems and the atmosphere (eds. MO Andreae and DS Schimal, John Wiley & Sons. Inc. New York, pp 7-21.

Goodroad L.L.; Keeney, D.R. (1984). Nitrous oxide emissions from forest, marsh and prairie ecosystems. J. Environ. Qual. **13**: 448-452.

Goreau, T.J.; Kaplan, W.A.; Wofsy, S.C.; McElory, M.B.; Valois, F.W.; Watson, S.W. (1980). Production of NO_2 and N_2O by nitrifying bacteria at reduced concentrations of oxygen. Appl. Environ. Microbiol. **40**: 526-532.

Jonasson, C. and Granat, L (1984). Emission of nitric oxide from arable land. Tellus 36B : 26-37.

Keller, M.; Veldkam, E. Weltz A.M. and Reiners, W.A. (1993). Effect of pasture age on soil trace–gas emissions from a deforested area of Costa Rica. Nature **365**: 244-246.

Khalil, M.A.K. & Rasmussen, R.A. (1988). Nitrous oxide : Trends and global mass balance over the last 3000 years. Annals of Glaciology **10**: 73-79.

Khalil, M.A.K. (1999). Non–CO_2 Greenhouse gases in the atmosphere. Annu Rev Energy Environ **24**: 645-661.

Li C, Frolking, S. Frolking, T.A. (1992). A model of nitrous oxide evolution from soil drivers by rainfall events, 1, Model structure and sensitivity. J. Geophys. Res. **97**: 9759-9776.

Luizao, F.; Matson, P.; Livingston, G.; Luizao, R.; Vitousek, P.M. (1989). Nitrous oxide flux following tropical land clearing. Global Biogeochem. Cycles. **3**: 281-285.

Malhi, S.S.; McGill, W.B. (1982). Nitrification in three Alberta soils : Effect of temperature, moisture and substrate concentration. Soil Biol. Biochem. **14**: 393-399.

Matson, P.A.; Vitousek, P.M. (1990). Ecosystem approach to a global nitrous oxide budget. Bioscience **40(9)**: 667-672.

Matson, P.A.; Volkmann, C.; Coppinger, K.; Reiners, W.A. (1991). Annual nitrous oxide flux and soil nitrogen characteristics in sagebrush steppe ecosystem. Biogeochem. **14**: 1-12.

McElory, M.B. and Wofsy, S.C. (1986). Tropical forests : Interaction with the atmosphere, In: Tropical Rain Forest and the world atmosphere (Eds.) G.T. Prance, Westview, Boulder Colo. pp 33-60.

Mosier, A.R.; Parton, W.J.; Valentine, D.W.; Ojima, D.S.; Schimel, D.S.; Delgado, J.A. and Kulamala, A. (1996). CH_4 and N_2O fluxes in Colorado shortgrass steppe : I Impact of landscape and nitrogen addition. Global Biogeochem Cycles **10**: 387-399.

Mosier, A.R.; Delgado, J.A.; Cochran, V.L.; Valentine, D.W.; Parton, W.J. (1997). Impact of agriculture on soil consumption of atmospheric CH_4 and a comparison of CH_4 and N_2O flux in ubarctic, temperate and tropical grasslands. Nutrient cycling in agroecosystems **49**: 71–83.

Parashar, D.C.; Gupta, P.K.; Rai, J.; Sharma, R.C. and Singh, N. (1993). Effect of soil temperature on methane emissions from paddy field. Chemosphere **26**: 247-250.

Parton, W.J.; Mosier, A.R.; Schimel, D.S. (1988). Rates and pathways of nitrous oxide production in a shortgrass steppe. Biogeochem. **6**: 45-58.

Peterjohn, W.T.; Melillo, J.M.; Steudler, P.A.; Newkirk, K.M.; Bowles, F.P.; Aber, J.D. (1994). Responses of trace gas fluxes and N availability to experimentally elevated soil temperature. Ecol. Appl. **4**: 617-625.

Robertson, G.P. (1993). Fluxes of nitrous oxide and other nitrogen trace gases from intensively managed landscapes : A global perspective. P 95–108. In : Agricultural ecosystem effects on trace gases and global climate change. ASA Spec Publ 55, ASA, CSSA and SSSA, Madison WI.

Schimdt J, Seiler W and Conrad R (1988) Emission of nitrous oxide from temperate forest soils into the atmosphere. J Atmos Chem **6** : 95-115.

Slemr F.; Conrad, R. Seiler, W. (1984). Nitrous oxide emissions from fertilized and unfertilized soil in a subtropical region (Andalusia, Spain). J. Atmos. Chem. **1**: 159-169.

Smith, K.A.; Clayton, H.; Arah, S.; Christensen, P.; Ambus, P.; Fowler, D.; Hargreaves, K.J. Skiba.; U, Harris, G.W.; Weinhold, F.G.; Klemedtsson, L. and Galle, B. (1994). Micrometerological and chamber methods for measurement of nitrous oxide fluxes between soils and the atmosphere : Overview and conclusion. J Geophys Res **99**: 16541-16548.

Struwe, S.; Kjoller, A. (1994). Potential for N_2O production from beech (*Fragus silvaticus*) forest soils with varying pH. Soil Biol. Biochem. **26**: 1003-1009.

Sullivan, L.J.; Moore, T.C.; Aneja, V.P.; Robarge, W.P.; Pierce, T.E.; Geron, C. Gay, B. (1996). Environmental variables controlling nitric oxide emissions from agricultural soils in the South East United States. Atmos. Environ. **30**: 3573-3582.

Vedenina, I.Y.; Zavarzin, G.A. (1977). Biological removal of nitrous oxide under oxidising conditions. Mikrobiologia **46**: 898-903.

Vedenina, I.Y.; Miller, Y.R.; Kapustin, O.A.; Zavarzin, G.A. (1980). Oxidation of nitrous oxide during decomposition of hydrogen peroxide by catalase. Mikrobiologia

Advancing Frontiers of Ecological Researches in India, 201-208, 2006
Edited by A.K. Kandya & Asha Gupta
Published by Bishen Singh Mahendra Pal Singh, Dehra Dun, India

Hydrological Studies in a Subtropical Forest of Manipur, North-East India

L. Usharani and Asha Gupta
Department of Life Science, Manipur University, Imphal – 795003
E-mail : anjalika_22000@yahoo.co.in

Abstract: The influx of various nutrients (N, P, K, Ca, Mg and Na) in precipitation, throughfall and stem flow was studied in a subtropical forest at Manipur; Leaching account and input-output ratios for various nutrients under evergreen dominated canopy in the forest was investigated. A pronounced modification of precipitation by the canopy was noticed Maximum leachability of K was obtained, it was found that vegetation retains P and Na. Input-output ratio revealed good retention of N.

INTRODUCTION

The flux of nutrients from the atmosphere through rainfall becomes the major net source of nutrients (Kellman *et al.*, 1982). Precipitation reaching and passing through a forest ecosystem follows several pathways before arriving to the forest floor (Caroll, 1980., Rutter and Morton, 1977). Some nutrients may also be absorbed by the Canopy from precipitation (Carroll, 1980, Wittwer and Teubner, 1969). Nutrients move from the forest canopy to the soil primarily through litter fall and through precipitation components (throughfall and stemflow). Franke (1967); Tukey (1970) reported that nutrients are transferred to the forest floor by leaching from the plant and plant tissues. Similarly Miller (1977) assessed that washing from the substances absorbed on the surface of the leaves and twigs also reach the forest floor. Age and physiological condition of leaves are partially responsible in leaching of the minerals from leaf surface (Attiwill, 1966; Curlin, 1968).

The paper reports the influx of various nutrients in precipitation, throughfall and stemflow from a subtropical forest of Manipur, estimate leaching accounts and input-output ratios of various nutrients for understanding the conservation role of forest vegetation.

Study Sites

The study was conducted in Manipur (Latitude 23°83′ N to 25°68′N and longitude 93°21′ E to 94°78′E), a state in the North Eastern corner of India. The state enjoys subtropical climate; January being the coldest month. The average temperature varies from 5° to 29°C. Rainfall is abundant and well distributed.

The study was conducted in a subtropical forest of Manipur. The site is characterized by evergreen trees of *Celtis australis*, *Eugenia praecox*, *Mallotus phillipinensis* The soil is classified as loam.

Rainfall and precipitation flux through forest canopy viz. throughfall and stemflow were measured after Opakunle (1989). Precipitation data from meteorogical department were utilized. A total of 1635.5 mm of gross precipitation was recorded in the hydrological year May 1999-April 2000. A total throughfall of 986.16 mm and stemflow of 76.9 mm were obtained. The net precipitation obtained was 1063.5 mm.

Samples of precipitation, throughfall and stemflow and spring water were collected, brought to laboratory and they were analysed for total Nitrogen, Potassium, Phosphorus, Calcium, Magnessium and Sodium by standard methods. (Trivedy & Goel, 1984).

Average concentration for each nutrient were expressed in mg/l on the basis of mean from three replicate values.

The amount of nutrients in kg ha^{-1} for various months were computed by multiplying the average concentration with the known volume of water in precipitation, throughfall and stemflow for the corresponding month (Opankunle, 1989). The annual amount of various nutrients was computed by summing the nutrient amounts for eight months studied by us. The reason being the four months viz. November, December 1999, January, February 2000 of the hydrological year (1999-2000) experienced only 3.5% of the annual precipitation, thus the amount was found to be insignificant.

Leaching account: The leaching amount for various nutrients were accounted for various nutrients in throughfall and stemflow with that of precipitation.

Input-output ratio: Input-output ratios for all the nutrients were obtained from average value of nutrient in gross precipitation (input) and spring water (output).

RESULTS AND DISCUSSION

The main nutrient concentration (mgl^{-1}) for input, throughfall, stemflow and output for the particular months viz. May, June, July, August, September, October, March and April of the hydrological year (1999-2000) are reflected in Table-1.

Table-1: Mean nutrient concentration (ppm) in precipitation, throughfall, stem flow and output in sub-tropical forest of Manipur.

TOTAL NITROGEN

	Months							
	May '99	Jun '99	Jul '99	Aug '99	Sept '99	Oct '99	Mar '99	Aprl '00
Precipitation	0.42	0.46	0.59	0.66	0.92	1.80	0.38	0.42
Through fall	1.56	1.99	2.13	2.68	3.40	3.80	1.10	2.40
Stemflow	1.58	2.29	2.88	2.93	3.99	4.04	1.56	2.51
Output	0.28	0.43	0.64	0.82	0.97	1.30	0.21	0.28

PHOSPHORUS

	Months							
	May '99	Jun '99	Jul '99	Aug '99	Sept '99	Oct '99	Mar '99	Aprl '00
Precipitation	0.035	0.076	0.094	0.086	0.054	0.036	0.078	0.096
Through fall	0.022	0.026	0.031	0.029	0.025	0.023	0.025	0.027
Stemflow	0.032	0.043	0.068	0.069	0.046	0.036	0.034	0.047
Output	0.020	0.048	0.069	0.063	0.052	0.040	0.030	0.050

POTASSIUM

	Months							
	May '99	Jun '99	Jul '99	Aug '99	Sept '99	Oct '99	Mar '99	Aprl '00
Precipitation	0.03	0.18	0.29	0.38	0.10	0.08	0.04	0.05
Through fall	1.09	2.20	3.42	2.60	1.21	1.06	1.12	1.41
Stemflow	1.80	2.10	3.62	2.70	1.90	1.80	2.60	3.10
Output	0.19	0.24	0.38	0.31	0.26	0.10	0.21	0.39

CALCIUM

	Months							
	May '99	Jun '99	Jul '99	Aug '99	Sept '99	Oct '99	Mar '99	Aprl '00
Precipitation	0.26	0.66	0.91	1.09	1.21	1.60	0.27	0.46
Through fall	0.72	0.94	1.28	1.91	2.63	3.21	0.74	1.11
Stemflow	1.70	2.00	2.60	2.80	3.10	3.40	1.70	2.30
Output	1.09	1.16	1.20	1.35	1.42	1.48	1.04	1.12

MAGNESIUM

	Months							
	May '99	Jun '99	Jul '99	Aug '99	Sept '99	Oct '99	Mar '99	Aprl '00
Precipitation	0.27	0.27	0.28	0.28	0.28	0.28	0.29	0.29
Through fall	0.21	0.91	0.54	0.80	1.26	1.80	0.66	0.80
Stemflow	1.41	1.46	1.58	1.60	2.00	1.26	1.39	1.40
Output	0.36	0.38	0.41	0.43	0.46	0.48	0.53	0.58

SODIUM

	Months							
	May '99	Jun '99	Jul '99	Aug '99	Sept '99	Oct '99	Mar '99	Aprl '00
Precipitation	0.58	0.91	1.23	2.00	2.80	3.00	0.57	1.10
Through fall	0.10	0.13	0.15	0.16	0.18	0.12	0.13	0.14
Stemflow	1.52	1.70	1.80	2.01	2.30	2.41	1.40	2.10
Output	0.78	0.88	0.99	1.09	1.50	1.66	0.70	0.91

The water flux under canopy is related to rainfall-its frequency and intensity and type of vegetation cover which mainly affects the accompanying materials and elements available for transfer (Pathak et al, 1989, Singh and Gupta 1995, Xiao *et al* 2000). The nutrient content of precipitation in Kg ha^{-1} Yr^{-1} were given in the Table-2. Thus it appears that a total of 10.66 Kg ha^{-1} of total nitrogen, 1.06 Kg ha^{-1} of phosphorus, 2.62 Kg ha^{-1} of potassium, 12.67 Kg ha^{-1} of calcium, 4.41 Kg ha^{-1} of magnesium and 21.95 Kg ha^{-1} of sodium are contributed by precipitation. Nutrient concentration in precipitation in various months of the hydrological year (1999-2000) had the following order.

$$Na > Ca > N > Mg > K > P$$

A pronounced modification of precipitation by forest canopy was noticed. The decreasing concentration of phosphorus and sodium with respect to dominance of total nitrogen and potassium is recorded in throughfall, whereas decreasing concentration of phosphorus and magnesium with respect to the dominance of total nitrogen and potassium are observed in stemflow.

Table-2: Annual nutrient input through rainfall and precipitation flux under canopy with leaching account during hydrological year (1999-2000) in a subtropical forest of Manipur (All values in Kg $ha^{-1}yr^{-1}$).

	Total N	P	K	Ca	Mg	Na
Rainfall	10.66	1.06	2.627	12.67	4.41	21.95
Throughfall	22.36	0.32	19.71	14.01	6.66	2.44
Stemflow	1.95	0.115	1.88	1.79	0.84	0.36
Total under canopy	24.31	0.435	21.59	15.8	7.50	2.80
Leaching	+13.65	-0.625	+18.963	+3.13	+3.09	-19.15

The nutrient concentration of throughfall in the studied hydrological year had the following order.

$$N > K > Ca > Mg > Na > P$$

Similarly the nutrient concentration in stemflow water revealed the following order.

$$N > K > Ca > Na > Mg > P$$

The nutrient in precipitation accounted for 19.97 % of Total N, 1.98 % of P, 4.92 % of K, 23.74 % of Ca, 8.26 % of Mg and 41.12 % of Na of the total nutrient amounting to 53.37 Kg ha^{-1} Yr^{-1} of nutrient input.

In the study site throughfall accounted for 34.14 % of total N, 0.48 % of P, 30.09 % of K, 21.38 % of Ca, 10.16 % of Mg and 3.72 % of Na amounting to 65.5 Kg ha^{-1} Yr^{-1} of total nutrient, whereas stemflow accounted for 28.11 % of N, 1.65 % of P, 27.10 % of K, 25.81 % of Ca, 12.11 % of Mg and 5.19 % of Na amounting to 6.935 Kg ha^{-1} Yr^{-1}.

Preliminary analysis of stemflow waters of the oaks indicated that although the quantities of water flowing down the stems were small when expressed on an area basis (Carlisle *et. al.* 1965) the concentrations of some elements (particularly, calcium, magnesium and potassium) were relatively high. It was possible, therefore, that small but significant amounts of these elements could be reaching the soil from this source. Our observation is tallying in case of K and Ca.

The stemflow water contained relatively higher concentration of nutrients compared to throughfall water.

72.43 kg ha^{-1} of total nutrients were received under canopy accounting for 33.56 % of total nitrogen, 0.60 % of phosphorus, 29.80 % of potassium, 21.81 % of calcium, 10.35 % of magnesium and 3.86 % of sodium respectively.

Leaching of nutrient was initiated at the end of dry season and went on increasing till the month of August in P and K and then decreased in September, October, the amount became low after the winter season, whereas for other nutrient, the leachability increased with the arrival of rain and continuous increase in total N, Ca, Na and Mg is observed, again after winter it declined. Over all data was found to be similar to those found in other tropical forest. Amount of nutrient in throughfall showed 2 typical trends:

1. A continuous increase for N, Ca, Mg and Na and
2. Increase followed by decrease in P and K.

Leaching of P is negligible. Maximum leachability of K is observed. Our observation is similar to Reversat (1975).

Concentration of nutrients in stemflow water is higher than that of throughfall. The nutrients increase in order of N > K> Ca > Na > Mg > P. However as the nutrients in throughfall water are added from leaves, Mg is more than Na. Similar results have been reported by Negi *et al*, 1993 in their study in *Pinus caribaea* stand. Ratio of potassium in throughfall to rainfall is slightly lower in tropical forests suggesting the nutrients such as potassium may be conserved at leaf surface in tropical forests.

To reduce the loss of nutrients through leaching it appears that the leaves have evolved nutrient conservation mechanisms. The epiphyllae like algae, lichens, fungi, and various animals, form a dense mat on the leaf surface by colonizing the leaves of tropical trees. Witkamp (1970) has shown that nuclide retention on tropical forest leaves in Puerto Rico with epiphyllae was 1.7 to 20 times greater than those from which the epiphyllae were removed. Thus epiphyllae reduce the leaching of nutrients from the atmosphere and constitute a sort of ecological mechanism for nutrient conservation.

By calculating input-output ratio we wanted to see whether the ratio reflected any trend in ability to conserve the nutrient under the forest canopy. Our ratios showed variation as to the effectiveness of ionic conservation mechanism.

Input/Output ratio revealed good retention of nitrogen, poor retention of potassium, calcium and magnesium. Thus, these are the nutrient passing through the rooting zone to spring water. Comparing of spring water with throughfall water showed that spring water had low nutrient concentration. Thus, soil-root system efficiently retains the nutrient brought to it in litter and through throughfall. In output, it was found that in case of cation this efficiency increased in order of calcium, potassium and magnesium (Table-3). The highest output concentration was for sodium, calcium and nitrogen, these are also the nutrients with highest concentration in the input.

Table-3: Input/Output ratio for nutrients in a subtropical forest site, Manipur.

Parameter	Input/Output Ratio
Total N	1.15
P	1.50
K	0.54
Ca	0.65
Mg	0.62
Na	1.43

ACKNOWLEDGEMENT

One of us (L. Usharani) is thankful to Co-ordinator, WWF, India, CCV programme for selecting her as conservation corps volunteer for the period 1998-2000 and granting her fellowship for the project "Studies on keystone species in sacred groves of Manipur".

REFERENCES

Attiwill, P.M. (1966). The chemical composition of rain water in relation to cycling of nutrients in mature Encalyptus forest. Plant and Soil. **24**: 390-406.

Carlisle, A. Brown A.H.F. and White, E.J. (1965). The interception of precipitation by Oak (Quercus petracea) on a high rainfall site. Q.J. For. 59, 140-143.

Carroll, G.C. (1980). Forest canopies and independent subsystems. In R.H. Waring (ed.) Forests : fresh perspectives from ecosystem analysis Oregon state Univ. Press, Oregon. pp 87-108.

Curlin, J.W. (1968). Nutrient cycling as a factor in site productivity and forest fertilization. In tree growth and forest soils (C.T. Youngberg and C.B. Davy. (eds.) p. 313-325. Proc. Third North Am. Forest Soils Cong. Oreg. State Univ. Press, Corvallis.

Franke, W. (1967). Mechanisms of foliar penetration of solutions. Annu. Rev. Plant Physiol. **18**: 281-300.

Kellman, M.; Hudson, J. and Sanmugadas, K. (1982). Temporal variability in atmospheric nutrient influx to a tropical ecosystem. Biotropica. **14**:1-19.

Miller, H.C. (1977). Assessing the contribution of crown leaching, rain and dry deposition to the chemistry of throughfall and stemflow. Unpublished paper distributed at workshop meeting to consider methods involved in studies of acid precipitation to forest ecosystem, Edinburgh, U.K. 7 p.

Negi J.D.S.; Sharma, S.C. and Taneja P.L. (1993). Interception and chemical composition of precipitation in Pinus caribaca stand. Indian Forester **119(1)**: 25-32.

Opankunle, J.S. (1989). Throughfall, Stemflow and rainfall interception in a cacoa plantation in South western Nigeria. Tropical Ecology **30(2)**: 244-252.

Pathak, P.C.; Rawat, Y.S. and Singh, R.P. (1989). Rainfall interception by leaf litter in Central Himalayan forests. Ind. For. **115(11)**: 120-123.

Reversat, F.B. (1975). Nutrients in throughfall and their quantitative importance in rainforest minerals cycles. In Ecological studies, Analysis and Synthesis edited by E. Medina and F.B. Golley. Springer Verlag New York. Inc. : 153-159.

Rutter and Morton, A.J. (1977). A predictive model of rainfall interception in forest (III). Sensitivity of the model of stand parameters and meteorogical variables. J. Appl. Ecol. **14**:567-588.

Singh, R.P. and Gupta, M.K. (1995). Measurements of interception, rainfall stem-flow and throughfall in *Quercus leucotrichosphora*. (*A. camus* Ex. Bahadur) Forest J. Tropical Forestry. **11(11)**: 83-90.

Trivedy, R.K. and Goel, P.K. (1984). Chemical and biological methods for water pollution studies. Published by Environmental Publications, Karad, India.

Tukey, H.B. Jr. (1970). Leaching of metabolites from foliage and its implication in the tropical rain forest. In : H.T. Odum and R.F. Pigeon (Editors). A tropical Rain Forest. A study of Irradiation and Ecology at E 1 Verde, Puerto Rico. U.S. Atomic Energy Commission, Washington, D.C. pp. H 155-H160.

Witkamp, M. (1966). Decomposition of leaf litter in relation to environmental microflora and microbial population. Ecology **47**: 194-201.

Witkamp, M. (1970). Mineral retention by epiphyllic organisms. In : H.T. Odum and R.F. Pigeon (Editors), A Tropical Rain Forest, A Study of Irradiation and Ecology at El Verde, Puerto Rico. U.S. Atomic Energy commission, Washington D.C. pp. H177-H179.

Wittwer, S.H. and Teubner, F.G. (1969). Foliar absorption of mineral nutrients. Annu. Rev Plant Physiol. **10**:13-32.

Xiao, I.E. (2000). Winter rainfall interception by two mature open-grown trees in Davis California. Hydrol. Process. **14**: 763-784.

Advancing Frontiers of Ecological Researches in India, 209-232, 2006
Edited by A.K. Kandya & Asha Gupta
Published by Bishen Singh Mahendra Pal Singh, Dehra Dun, India

An Overview of Ethnobiological Researches with particular reference to Wild Edible Plants

Manju Sundriyal
G. B. Pant institute of Himalayan Environment and Development
Kosi-Katarmal, Almora, Uttranchal
manjusundriyal@yahoo.co.in

Abstract : There has been an increasing interest on documentation of species that are of considerable interests to human societies. Since last two decades, considerable literatures have accumulated in ethnobotany, particularly on wild edible and medicinal plants. The present paper attempts to compile the emerging trends on ethnobotany, particularly wild edible plants along with information on species diversity, methodological issues, nutritive status and propagation of selected species. It also provides information on the International and National scenarios on ethnobotanical aspects. It is expected that this review will be useful for researchers and development workers to study on this area and develop suitable strategies for promotion and conservation of these species.

INTRODUCTION

Ethnobotany has been a subject of interest since time immemorial. People use plants for meeting their basic requirements of food, fibre, construction materials, fuelwood, medicines, dyes and various other purposes (Harshberger 1896, Janaki Ammal 1955). In the course of human civilisation some of these plants were domesticated and cultivated in large scale by human beings. However, still a huge diversity of plant exists in wilder habitats. Expansion of agriculture for meeting human needs of food has put pressure on the wild habitats by converting land into agricultural purposes. Still a great chunk of population specially in underdeveloped and poor countries rely on wild edible plants (Jain 1995). World wide the traditional local communities are extremely knowledgeable about local plants and other natural resources on which they directly depend (Martin 1995). Tribal communities have screened large variety of species for edible purposes through trials and errors (Arora and Pandey 1996). There has been a growing interest in the scientific researches on wild edible plants mainly for enumeration (taxonomic) purpose and subsequently nutritional studies were also added. Ethnobiology is a potential subject for understanding development processes, for sustainable livelihood of mountain communities and natural resources management (Pei Shengji 1995). The present paper provides, in brief, some emerging trend on ethnobiological researches and its possible implications in near future.

Ethnobotany and international scenario

Internationally there has been a few important reporting on the wild plants. A document on Wild Fruits of Great Britain (Cameron 1917) and Food Plants of Philippines (Wester 1925) were prepared in first quarter of the 20th century. Saunders (1934) has reported on the useful plants of United States and Canada. Burkill (1935) has compiled "A Dictionary of Economic Products" for Malaya. A list of the useful plants of tropical west Africa was prepared by Dalzell (1937), while a compilation of "Edible Roots and Berries of Northern Canada" was given by Porsild (1937). Hill (1939) has compiled information on wild fruits of Britain. Subsequently country wise compilation of wild edible plants was done and published for China (Porterfied 1951, Chang 1965) USA and Canada (Medsger 1943, Fernald and Kensy 1943, Harington and Matsumura 1967), for Union of Soviet Russia (Koleshikov 1943), for Philippines (Brown 1951-54), for South Pacific Islands (Barrau 1959), for Nepal (Malla *et al.* 1982), and for India (Singh and Arora 1978, Arora and Pandey 1996). Tanaka (1976) has given "Cyclopedia of Edible Plants of the World". Ramis (1993) has reported on the "Wild Fruits in Puerto Rico". "A Methodological Manual for Ethnobotany" was compiled very recently (Martin 1995). Thus there are numerous records world over to describe the wide variety of foods which have been part of local diets not only of hunter-gathers, but also of agro-pastoralists (Grivetti *et al.* 1987). Some other most prominent early studies include the report by the International Tropical Timber Organisation in 1988 (Panayotou and Ashton 1992) and the monograph on non-timber forest products by de Beer and McDermott (1989) which argued that the sensible extraction and marketing of plants and animals could increase the rural income up to some extent. Since the late 1980s, enough studies on the human use of wild plants and animals in the tropical forest have accumulated (Plotkin and Famolare 1992 and references their in). Salick *et al.* (1995) has reported non-timber forest products integrated with natural forest management in Rio San Juan. Few studies were also concentrated on the economic value of non-timber tropical forest products (Godoy and Feaw 1989, Schwartzhan 1989). A few reports are available on conservation of tropical moist forests, their genetic resources, and valuing the conservation of genetic resources

(Myers 1980, Oldfield 1981-84). Alcorn (1981) has given noncrop resources management and some factors influencing botanical resource perception amongst the Huastec community in Mexico. Britta (1990) has given dietary use of wild plants in rural Swaziland, and illustrated the need for systematic incorporation of information on current use of wild food resources in agricultural research, agroforestry, forestry and agricultural programmes, and in other activities dealing with sustained food security for the peoples of Africa.

Some weeds are also considered as a scarcely used food resource (Rapoport *et al.* 1995) and among the 18 most obnoxious weeds of the world (Holm *et al.* 1977), 16 species are included in Kunkel's (1984) list of 12,560 plants for human consumption. In the Sanelian zone from Senegal to Kenya, 800 food plants have been reported (Becker 1986). North-American Indians used at least 1112 species of food plants (Duke 1992), which represent 6% of the 18956 species of their flowering plants. Campbell (1986) has given the use of wild food plants in drought for Botswana. Most of the above studies and other reports revealed that by and large the plights of Non-Timber Forest Products (NTFPs) are not much better with few if any forests under management for sustainable production (Hall and Bawa 1993). In most of the cases, the poor management is caused due to lack of incentives from Government sides (Repetto and Gillis 1988).

In the last few years however there is a noticeable change in attitude and increasing attention is given to the role and potential importance of wild food resources in many developing countries. Thus for example, increased farming systems approach is observed now in ethnobiological research (Sundriyal 1999). Much more consideration is given to entire food systems, including procurement methods such as gathering wild plants (Frankenberger 1985), promotion of wild food resources in various programmes on traditional and under-utilised foods in several regions of Africa (FAO 1985), and edible wild plants by the forestry sector in programmes on management of natural forests and in community or social forestry actions (Hoskins 1985, Ogle 1988). The importance of wild foods is more often noted in famine relief and emergency situation (Longhurst 1986, Sperling 1987).

Indian status on ethnobotany

Researches of Schultes (1962) in Amazon and Barrau (1966) and Conclin (1962) in Asia greatly prompted Indian workers into Ethnobotany.

Janaki Ammal (1955) and Jain (1963, 1965) initiated intensive studies on ethnobotany in India. Subsequently various institutes (viz. Botanical Survey of India, ICAR Centres, NBPGR Centres, Universities) and other organisations started work on the discipline. CSIR compils work on Plant Wealth of India, which is updated at time to time (Anonymous 1948-72). Most of the ethnobotanical studies in India centred around the reporting of medicinal plants and wild edible plants from diverse habitat (Jain 1994). In the early 1980s, the Department of Environment, Government of India funded an All India Co-ordinated Research Project on Ethnobiology (DOE 1984). In last four decades of ethnobiological research, over 400 research papers have been published in various national/international journals and periodicals, and nearly 20 books have also been published. The Society of Ethnobotanists was established in 1982 considering the importance of thc subject in the country, and a journal *Ethnobotany* is being published since 1989, thus has put the discipline into firm footing in India.

Numerous ethnobotanical studies have been done throughout the length and breadth of the country (Jain and Tarafder 1970, Pal and Srivastava 1976, Jain 1981, 1984, Manilal 1981, Vishnu Mittree 1981, Ramachandran and Nair 1981, Saxena *et al.*1981, Janaki Ammal and Nagendra Prasad 1984, Pushpagandhan and Atal 1984). High genetic diversity mainly belongs to the family Poaceae, Orchidaceae, Fabaceae, Caesalpiniaceae, Cyperaceae, etc. in the world as well as in India (Table-1). Efforts were also made to compile literature and update knowledge on the ethnobotany from time to time (Mudgal 1984, Jain 1994). Though the important contribution to the subject of ethnobotany came from medicinal plants, substantial literature has been generated on wild edible plants.

In India nearly 15000-20000 plant species have good medicinal value. However the the tribal communities use about only 7000 plant species for their medicinal utility. The Ayurveda system of medicine uses 700 plant species with over 360740 practitioners throughout the country. Similarly the Unani system also uses 700 plants by 29701 practitioners. All these three systems are evolved from the traditional knowledge base. As per UNDP report nearly 130 pure chemical substances are extracted from 100 species of higher plants, which are used throughout the world. India has over 150 very useful medicinal plants. CSIR laboratories have identified 400 useful plants for their further screening for medicinal value. The annual value of medicinal plants derived from developing countries is about US $ 32 billion cover Rs. 100,000 crore). There are 47 major modern plant based drug on world market and there is a predicted 328 more potential drugs with an estimated value of US $ 147 billion. Government of India is eyeing a potential Rs. 400 crore market for herbal medicine in current five years plan and leading Indian research centres are already working to seek patentable genes using the latest genetic engineering technique and plant based drugs from countries vast collections.

Table-1: The number of genera and species occurring in India and North Eastern region in respect of some dominant families of higher plants of the world (Source - Arora and Pandey, 1996).

Name of family		No. of genera / species		
		N.E. Region	India	World
1.	Poaceae	160/500	240/1100	620/10000
2.	Orchidaceae	104/700	200/1500	735/20000
3.	Fabaceae	50/200	100/750	482/12000
4.	Caesalpiniaceae	11/42	23/80	152/2800
5.	Mimosaceae	10/35	15/75	56/2800
6.	Asteraceae	25/70	135/710	900/1300
7.	Cyperaceae	14/ 175	21/350	90/4000
8.	Lamiaceae	30/95	65/380	180/3500
9.	Scrophulariaceae	15/35	60/350	220/3000
10.	Acanthaceae	25/125	70/340	250/2500
11.	Euphorbiaceae	55/160	65/340	300/5000
12.	Rubiaceae.	50/170	80/280	500/6000
13.	Utricaceae	15/45	25/114	45/550
14.	Zingiberaceae	18/73	20/115	46/850

In India some workers have reported on wild edibles and under exploited economically useful plants (Khan 1980, Thothathri *et al.* 1990). 225 plant species of ethnobotanical importance were recorded from the Santhal Pargana district of Bihar by Goel *et al.* (1984). Maharashtra came to ethnobotanical prominence due to the study carried out on folklore and traditional beliefs associated with sacred groves (Vartak and Gadgil 1981). 120 wild edible plants were recorded from hilly regions of Maharashtra and Goa by Vartak (1981). Gunjatkar and Vartak (1982) have reported wild edible legumes and other 33 species of Pune district. Jain (1984) has studied the ethnobotany of Morni and Kabser Hills in Ambala district of Haryana, and reported 40 plant species used by the local inhabitants, of these 26 species were used as medicine and remaining 14 as fodder and edible fruits. Saxena and Dutta (1975) have reported 81 plant species of Orissa used by the rural folk for antifertility and as medicine, fibre and food. Singh and Maheshwari (1985) studied the forest life and economy of the tribals of Varanasi district and recorded 77 wild plants used for food, fibre, medicinal, oil, house making, cordage baskets, broom making, and fuel. Throughout the length and breadth of the country 647 wild edible species produce edible fruits, which varied from 357 genera and 112 families. Out of that 107 species are domesticated/semi-domesticated. Nearly 521 species are used as leafy vegetables and potherbs, of which 72 are domesticated/semi-domesticated. Nearly 145 wild edible species are consumed as roots and tubers which vary from 92 genera and 42 families. Nearly 33 species are adopted to be grown in the field.

The importance of folk claims is gaining attention of scientists in India and abroad (Brandis *et al.*1989, Gangwar and Ramakrishnan 1990, Jain 1981, Edington 1990, Singh and Khan 1989). Janaki Ammal (1984) studied substitute food plants of certain tribals of South India. About 25 plant species are reported which are used on various occasions by tribal women as herbal cosmetics and natural make-up (Vishwa Vihari 1995). Indo-Amazonian ethnobotanic connections through similar uses of some common plants are given by Jain *et al.* (1995). Fernandes *et al.* (1988) have compiled extensive data on many NTFPs harvested by indigenous societies in Orissa. Parmar and Kaushal (1982) has reported wild fruits, while genetic resources of under-exploited fruits has been given by Pareek and Sharma (1993). Wild germplasm of temperate fruits has been reported by Kishore and Randhawa (1993). They have reported that the different fruit species found growing wild in the Himalaya can be broadly classified as pome fruits, stone fruits and berries, an attempt has been made to highlight some of the important fruit species under each of these classes. Uma Shankar *et al.* (1996) has reported sustainable extraction of non-timber forest product Amla *(Phyllanthus emblica)* in the forest of Biligiri Rangan Hills, India. The native tribe of Biligiri Rangan hills extracts a wide array of non-timber forest products from tropical forests (Murali *et al.*1996). It is also used as a source of dyes and tannins and for making pickles, jellies and preserves (Krishnamurthy 1993). Some literature on ethnobotany pertaining to medicinal and wild edible plants is also compiled in the form of books. 'Glimpses of Indian Ethnobotany' consists the compilation of the research work done in 1980s (Jain 1981). Later on the work was revised and modified as 'Contribution to Ethnobotany of India' (Jain 1991) with the purpose that more and more researches and institutions will take up ethnobotanical studies considering the rich plant folklore in India. 'A Manual of Ethnobotany' was edited to help fieldwork in ethnobotany subject and to provide the basic framework for persons interested in the subject (Jain 1987). Some compilation has been done in the form of special issues for some journals and published as books, e.g. Ethnobotany in India (Maheshwari et al. 1993) is based on compilation of research articles for the Journal of Economic and Taxonomic Botany.

Himalayan and North Eastern states

In the Himalaya several floristic reports are available which also include little notes on economic aspects of the plants (Duthie 1906, Osmaston 1927, Smythe 1938, Babu 1977, Bhattacharya and Goel 1982, Semwal and Gaur 1981). Atkinson (1882) in his Himalayan Gazetteer has noted the uses of local plants by the natives. Some 450 wild edible plant species of the Himalaya have been described by Badhwar and Fernandez (1964), and details are provided about plant description, distribution and uses. Economic Botany of the Himalayan region deals with food of the people which also includes wild edible plants (Atkinson 1989). Sporadic information on wild relatives and related species of crop plants is also available on various floristic accounts and floras (Hooker 1896, Collett 1971, Kanjilal *et al.* 1940). A compilation of the wild edible plant diversity of the Himalayan region is done by Samant & Dhar (1997).

High diversity of wild relatives occurs in tropical, sub-tropical and temperate fruit types in different phyto-geographical zones of the country. This genetic diversity mainly belongs to the Rutaceae, Moraceae, Rosaceae, Musaceae, Anacardiaceae, Rhamnaceae, Myrtaceae, Guttiferae, Euphorbiaceace and Vitaceae. Among the tropical/sub-tropical types, maximum concentrations of wild species are in *Citrus* and *Musa,* which occurs in the north-eastern region (Bhattacharya and Dutta 1951, Chakravorty 1951). Various studies revealed that some temperate wild fruits such as *Pyrus, Prunus, Rubus, Ribes* and *Sorbus* are prominantly found in the Himalaya. Some wild species like *Prunus prostrata* and *P. tomentosa* are found to western Himalaya, while *P. jenkinsii* occurs in the eastern belt. Some species are distributed in the whole Himalayan zone, viz., *P. cornuta, P. nepalensis* allied to *P. pandus, P. cornata* and *P. acuminata.* Likewise in case of *Pyrus* species, *P. communis* occurs largely in Kashmir, *P. kumaoni* is also confined to western Himalaya, however, *P. pashia* and *P. baccata* of the western Himalaya are widely distributed eastwards to the sub-tropical zone of the north-eastern hills (Chakravorty 1951, Arora and Nayar 1983). Information on diversity of wild related taxa of crops of agriculture and horticulture importance occurring in different phytogeographical regions of India have been compiled by Arora and Nayar (1983). It is reported that wild crop relatives of *Alocasia, Amomum,*

Alpinia, Abelmoschus, Brassica, Colocasia, Chorchorus, Curcuma, Camellia, Citrus, Cucumis, Dioscorea, Docynia, Digitaria, Eurya, Erianthus, Hedychium, Hibiscus, Mucuna, Morus, Musa, Mangifera, Momordica, Prunus, Rubus, Sorbus, Setaria, Trichosanthes and *Vitis* mainly occur in the eastern and north-eastern region, while wild crop relatives of *Avena, Agropyron, Amaranthus, Allium, Cucumis, Carum, Eremopyrum, Elymus, Fagopyrum, Ficus, Hordeum, Juglans, Malus, Pyrus, Prunus, Punica, Rubus, Ribes,* and *Sorbus* are more confined to the western Himalayan region. Some wild relatives of crop species represent a much wider distribution viz. *Abelmoschus, Amaranthus, Chenopodium, Coccinia, Eleusine, Saccharum* and *Solanum* (Arora and Nayar 1983). In subtropical/tropical and humid/sub-humid areas there is a high diversity of wild edible species and other related taxa of rice, wheat barley, coix, finger millet and fox-tail millet (Table-2). Ethnobotanical investigations in the north west Himalaya is presented by numerous workers (Kachroo *et al.* 1977, Dhar and Siddique 1993, Virjee *et al.* 1984, Sharma and Singh 1989, Navchoo *et al.* 1990). Some traditional beverages, narcotics, wild edible plants and foods have also been reported for Ladakh (Navchoo *et al.* 1993).

Table-2: Diversity of crop plants concentrated in the warm humid tropical and sub-tropical regions of North-Eastern India (Arora and Pandey, 1996).

Phytogeographical zone	No. of species of wild relatives of crop plants
1. North-East region (including Eastern Himalaya)	189
2. Malabar	142
3. Western Himalaya	113
4. Deccan	96
5. Gangetic plains	73
6. North-Western plains	49

Numerous studies are available on the medicinal, wild edible and other useful plants in the central Himalaya, which includes Garhwal and Kumaun regions of Uttar Pradesh hills (Juyal and Uniyal 1960, Uniyal 1977, Gaur *et al.* 1977, Upreti 1983, Balodi 1988, Paliwal and Badoni 1990, Negi and Pant 1990, Negi *et al.* 1993, Samant and Pangtey 1995). Pangtey (1989) has given ethnobotanical notes on the Bhotia tribes of Kumaon. Negi and Gaur (1990) has presented distribution and usage of seabuckthorn

(*Hippophae*). Little known endemic wild *Allium* species in the central Himalaya was also reported by Negi and Gaur (1991). Little known wild edible fruits from Garhwal were recorded by Gaur and Semwal (1983). Rawat and Pangtey (1989) have reported 148 plant species of various uses from alpine region of Kumaon. *Utrica diocia* has been widely used by the Bhotias of Chamoli Garhwal for food, fodder, fiber, fuel, medicine and dye (Harish Singh 1989). 200 plant species uses by Tharus mainly for food, fibre, medicine oil, handicraft and fuelwood reported by Singh and Maheswari (1990). Ethnobotanical notes are presented on Bhoxa tribe of Bijnor and Pauri Garhwal (Maheshweri and Singh 1984). It can be said that the prosperity of the local people is inextricably linked with food plants resources therefore the management planning should consider maximum benefits out of the inventory of wild edible plants.

Lessor-known crops of food value in Garhwal Himalaya and a strategy to conserve them has given by Maikhuri *et al.* (1992). Underutilized edible plants of the Sikkim Himalaya are reported by Sundriyal & Sundriyal (2003). The native people should be involved directly in raising, tending, managing, harvesting, transportation and marketing the major and minor wild food plants so that they could produce economic incentives to locals and also satisfy their nutritional requirement (Negi and Gaur 1994).

North-east India is predominantly covered under the humid tropics, exhibits maximum diversity of plant-wealth perhaps in the whole of Indian sub-continent and considered as the centre of origin for a large number plants (Vavilov 1951). The germplasm resource potential of the area not only consist flora (approximately 7000 species) and fauna (160 species), but also consider ethnic races cover (100 tribal communities), physio-geographical as well as agro-ecological conditions (Sharma and Hore 1993). The region has about 190 species of wild relatives of cultivated plants thus exhibits the maximum genetic variability (Rao and Murti 1990). Out of about 315 families of flowering plants in India more than 200 families are represented in this region alone. Similarly, out of about 800 species of food-plants of India about 300 species are reported from this region (Table-3) (Watt 1989-99, Rao 1972, Kharsati 1973, Joseph and Murti 1981, Singh *et al.* 1983, Arora and Pandey 1996). Native food plants of the north eastern tribals are given by Arora (1981). Pal (1984) has recorded 129 plant species used by Subansiri tribals in Arunachal Pradesh. Forest types of Arunachal Pradesh have been given by Kaul (1987). Haridason *et al.* (1990) has reported use of 90 wild edible plants from Arunachal Pradesh. The first ethnobotanical report on the Mao Nagas of Manipur is reported by Mao

(1993). Pandey *et al.* (1993) has reported indigenous minor fruit genetic resources and their marketing status in north-eastern hills of India. Minor and less known fruits of Assam have been reported by Neog and Mohan (1994). A report on survey of edible plants in the bazars (markets) of Meghalaya and Manipur was presented by some workers (Singh and Singh 1985). *Mangifera sylvatica, Elaeocarpus floribundus, Myrica esculenta, Docynia indica* and *D. hookeriana* wild fruits widely occur in north-eastern region. In *Musa* also maximum variability occurs in the north-eastern region in the wild types, *M. balbisiana* and *M. acuminata* and in related types. *Musa itinerans* and *Musa flaviflora* (*M. thomsoni*) is confined to Manipur and Meghalaya, whereas *M. sikkimensis* is widely distributed in North Bengal, Sikkim, Khasi hills and Manipur (Chakravorty 1951). Marketing status of selected wild edible plants of Sikkim Himalaya is also assessed Hajra & Chakravarty (1981) and Sundriyal & Sundriyal (2004).

Table-3: Distribution of different wild edible species used for various purposes in different phytogeographical regions of India (after Arora and Pandey 1996)

Plant part used	West Himalaya	East Himalaya	N-E Himalaya	Gangetic plains	Indus plain	Eastern Ghats	Western Ghats	Total
Root/Rhizome /Tubers	82	24	71	39	18	53	76	100
Leafy vegetables	203	35	222	178	100	216	234	500
Buds and flowers	47	2	57	31	17	40	48	100
Fruits	209	42	286	170	70	219	282	600
Seeds and nuts	35	5	50	47	13	48	61	100
Miscellaneous								40

Most of the papers emphasised that the potential diversity resources are threatened because of modernisation of agriculture changing local culture and various developmental activities, and there is a need of future management of wild relatives of various crops and should consider the urgency of conservation of plant resources in north eastern hills of India (Ramakrishnan 1984). Some information related to ethnobotanical observations is also available for Sikkim Himalaya (Bennet 1983, Srivastava *et al.* 1987, Kumar *et al.* 1994, Sundriyal & Rai 1996, Rai *et al.* 1994). A list of plants used by natives of North Bengal and Sikkim for medicinal purpose was recorded by Basu (1990). Wild plants used as a source of food by different ethnic group particularly by Lepchas, Bhutias and Nepalese of Sikkim are reported by Raju and Krishna (1990) and Singh (1995). Sundriyal *et al.* (1999) have recorded wild edible and other useful plants of the Sikkim Himalaya. Hajra and Chakraborty (1981) have enumerated 35 wild species of plants and their marketing potential of Sikkim. Raju and Mandal (1990) has reported Papri (*Podophyllum hexandrum*), an important medicinal plant from Sikkim. Yonzone *et al.* (1981) has reported a contribution to the ethnobotany of Darjeeling hills while Bhujel *et al.* (1985) described wild edible plants of the same area. Traditional fermented foods and beverages of Darjeeling and Sikkim was reported by Tamang and Sarkar (1988). About 29 prominent plant species are used for medicinal purpose in folk life of Darjeeling and Sikkim (Chettri *et al.* 1992). Information on forest structure, phenology, fruit yield, regeneration and marketing status of some prominent ethnobiological plant species of the Sikkim Himalaya, is also presented by (Sundriyal & Sundriyal 2004). Recently information on status of medicinal plants in Sikkim Himalaya (Rai and Sharma 1994) and strategies for their cultivation (Sundriyal and Sharma 1995) was also made available.

Nutritive values of wild edible plants

A few reports are available from different parts of the world on the nutritional status of some wild edible plants (Aykroyd 1956, Gopalan et al. 1971, Massefield 1973, Berris 1975, Singh and Arora 1978, Wolfgang 1982, Atal *et al.* 1984, Tandon 1984, Kaul *et al.* 1985). Various workers have reported nutritional status of wild leafy vegetables (Munro and Bassir 1969, Bassir 1979, Smith 1983, Akpanyong *et al.* 1995) and fruits (Zennie and Ogzewalla 1977, Ifon and Bassir 1979, Bajracharya 1980a, b, Mbiyangandu 1985, Uthaiah 1988, Roy *et al.* 1989, Mazza and Miniati 1993, Saka *et al.* 1992). These papers suggest the need to meet nutrient requirements through an increased use of the cheap wild edible food items with good nutritive potential. Detailed nutrient analysis of wild edible plant is presented by Sundriyal & Sundriyal (2001).

Pal and Banerjee (1971) surveyed the nutritional property of the less known plants used by the tribals in the Andhra Pradesh. Chemical composition of some Himalayan fruits (*Myrica, Rubus* and *Morus*) was presented by Tiwari *et al.* (1979). Bajracharya (1980 a, b) has analysed nutrients of 40 edible wild fruits of Nepal and reported that most of them were comparable to cultivated fruits in terms of minerals and recommended them for cultivation. Two wild edible fruits i.e. *Baccaurea sapida* and *Flacourtia jangomas* are considered very rich in vitamin C, proteins and minerals (Mohan 1987). Wesche-Ebeling (1995) has reported on the contribution to the botany and nutritional value of some wild *Amaranthus* species (viz. *A. retroflexus*, *A. viridis*, *A. palmeri* and *A. blitoides)* as potential sources of vegetable greens. Nutritional value of some lesser known wild plants (viz. *Alpinia malaccensis*, *Angiopteris evecta*, *Calamus tennis*, *Cyathea gigentea*, *Dendrocalamus hamiltonii*, *Dioscorea bulbifera*, *Pinaga gracillis*, *Sphenoclea zeylanica* and *Wallichia densiflora*) used by Nishi tribe of Arunachal Pradesh in north east India was reported by Maikhuri (1991). Negi (1991) has given nutrient composition and chemical analysis of *Ficus* species.

Approaches, methodologies and value addition

Ethnobotanical studies can help to understand relationship between plants and human beings, and also to show local communities to define their need for plant resources more clearly, thus help identify conservation issues. The basic concept and methods in the ethnobotany was given by Jain (1989). More recently methods for the economic evaluation of non-timber tropical forest products were presented (Ricardo *et al*. 1993). Information on approaches for developing sustainable extraction systems for tropical forests was given by Boot and Gullison (1995). Recent high rates of tropical deforestation have resulted in a search for uses of these forests that can provide economic incentives for their long term retention (Anderson 1990). Pamella and Baba (1993) presented methods to assess the impact of extraction of non-timber tropical products on plant population. The economic value and sustainable harvest of plants and animals from the tropical forest, and methods are given by Ricardo and Bawa (1993). Murali *et al*. (1996) have also reported impact of NTFPs extraction on regeneration, population structure and species composition in the forest. There are many examples of depletion of populations as result of over exploitation in the tropics (Browder 1992, Homma 1992, Mepstad *et al*. 1992). A manual on methods useful for working with communities to learn about their knowledge and uses of the plant world was compiled by Martin (1995).

Most of the wild edible plant parts have low keeping quality and therefore available for short periods, even sold at very less prices. Cursory information are available on the efforts made for value addition to wild edible plants. A dictionary of Indian raw materials and industrial products was presented in the form of Wealth of India, which presents some information on by-products of wild plants (Anonymous 1948-1972). Fruits of seabuckthorn are used for the wide range of by-products and have been recommended for plantation in fragile mountains (Lu and Rongson 1992, Negi *et al.* 1990). Fruits of *Ficus* tree species are good for jelly and jam production and thus has potential of income generation (Dhyani and Khali 1993). Similarly *Myrica* species is also considered for juice and jam production (Dhyani and Dhar 1993). Maikhuri *et. al.* (1994) has reported the cost-benefit analysis of the by-products made from some wild fruits and thus recommended them for sustainable rural development in the Garhwal Himalaya.

Germination, propagation and growth strategies

Some information is available on seed germination, propagation and growth pattern of some economically important plant species (Kandya and Kandya 1988, Nautiyal *et al.* 1987, Todaria and Negi 1992, Brand *et al.* 1994, Fox *et al.* 1994, Tiwari and Dhar 1996). Tiwari and Dhar (1997) has shown that rooting ability of *Aisandra butyracea* stem cuttings caused significant benefit in mass clonal propagation of the species. Nautiyal and Purohit (1986) has reported effect of auxin on seasonal rooting response of stem cuttings of *Berberis* species from different altitudes. Dhyani and Khali (1993) have reported rooting response of branch cuttings of two multipurpose *Ficus* tree species. Sundriyal & Sundriyal (2001) have recorded seed germination and stem-cutting response to hormonal treatment in six wild edible fruit species of the Sikkim Himalaya. A large number of species are yet to be covered for studying their germination, propagation and growth strategies that will havw significant benefit to adopt them in traditional agroforestry systems.

Future Implications

With the foregoing discussion it is reported that nearly 3000 plant species have been used as food at some time during human civilisation world over, out of which 150 species were ever cultivated and less than 20 species are meeting 90% of the world food need (N.R.C. 1982). In fact, just three species (viz. wheat, rice and maize) is meeting over half of the human food demand (Wilkes 1981). Dependence on such a small number of plants carries high risk for monoculture, catastrophic failure such as diseases and climatic stresses. Therefore it is wise to find out new varieties of edible species and lessor known crops with the rapid depletion of wild habitats and genetic diversity. There has been growing interest to find out new crops and wild edible plants so that some of these plants could be harvested and exploited for large scale cultivation. There is a greater need to conserve and do value addition for large number of ethnobiological species. Gene-pool banks are need to be maintained at different locations. Value addition may bring high returns to local communities through product diversification and quality control (Sundriyal & Sundriyal 2004). The nutritional role of various edible species is need to be investigated further. Since these species are found in diverse areas, they may play an important role in nutritional balance of local communities. A few selected species, depending on local choice, may be targeted for domestication and researches are need to be focussed on this direction. If properly planned along with due policy attention to ethnobiological sector, a large number of species could be used to eradicate poverty of rural communities in diverse places.

ACKNOWLEDGEMENTS

I thank Director, G.B. pant Institute of Himalayan Environment & Development for providing necessary facilities to undertake the present research work.

REFERENCES

Alcorn, J.B. (1981a). Huastec non crop resource management. *Human Ecology* **9**: 395-417

Anonymous (1948-72 & 1970-82). *Wealth of India: Raw materials*, Vol. 1-12. Council of Scientific and Industrial Research (Reprinted), Delhi.

Arora, R.K. (1981). Native food plants of north-eastern tribals, pp. 91-106, In: *Glimpses of Indian Ethnobotany* (Jain, S.K., ed). Oxford & IBH.New Delhi.

Arora, R.K. and E.R. Nayar (1983). Wild relatives and related species of crop plants in India- Their diversity and distribution. *Bull. Botanical Survey of India* **25(1-4)**: 35-45.

Arora, R.K. and Pandey, A. (1996). *Wild edible plants of India : Conservation and use*. Indian Council of Agricultural Research. National Bureau of Plant Genetic Resources, New Delhi.

Atkinson, E.T. (1982). *The Himalayan Gazetter*. Vol. I Part II. Reprinted by Cosmo Publications, India (1973).

Atkinson, E.T. (1989). *The Economic Botany of the Himalayas*. (Ed.). Cosmo Publications New Delhi, India.

Babu, C.R. (1977). *Herbaceous Flora of Dehradun*.. Council of Scientific and Industrial Research. New Delhi, P. 721.

Badhwar, R.L and R.R. Fernandes (1964). *Edible Wild Plants of Himalaya*. Government Publication, Delhi.

Balodi, B. (1988). Introductory note on the ethnobotany of Gori valley. *J. Econ. Taxo. Bot.* **12(2)**: 453-455.

Barrau, J. (1959). The Sago plants and other food plants of March dwellers in the South pacific. *Economic Botany* **13**: 151-162.

Basu, P.K. (1990). Less known medicinal plants and non-conventional food plants in the hills of North Bengal and Sikkim. In: *Economic Plants and Microbes*, (Ed. R.P. Purkayastha), pp. 101-105, Today and Tomorrow's Printers and Publishers, New Delhi.

Becker, B. (1986). Wild plants for human nutrition in the Sahelian zone. *J. of Arid Environments* **11**: 61-64.

Bennet, S.S.R. (1983). Ethnobotanical studies in Sikkim. *Indian Forester* **109(7)**: 477-481.

Bhattacharya, S.C. and S.Dutta (1951). Citrus varicties of Assam. *Indian J. Genet. & Pl. Br.* **11(1)**: 57-62.

Bhujel, R.B.; Tamang, K.K. and G.S.Yonzone (1985). Edible wild plants of Darjeeling district. *J. Bengal Nat. Hist. Soci., New Series* **3(1)**:76-83.

Brand, J.E.; Effendi, M. and Fox, J.E.D. (1994). Seed size and germination of *Santalum album* from Timor. Mulga Research Centre, School of Environmental Biology. Curtin University of Technology GPO Box 1987.

Britta, A. Ogle (1990). Dietary use of wild plant resources in rural Swaziland. *Mitt. Inst. Allg. Bot. Hamberg Band* 23b S 895-910.

Brown, W.H. (1951-54). *Useful plants of the Philippines*. 3 Vols Manila.

Browder, J.O. (1992). Social and economic constraints on the development of market. Oriented extractive reserves in Amazon rain forests, pp. 33-42, *In*: (D. C. Nepstad and S.,Schwartzman, eds.) *Non-timber products from tropical forests: evaluation of a conservation and development strategy*. Advances in Economic Botany I. The New York Botanical Garden. NY.

Burkill, I.H.(1935). *Dictionary of Economic Products of Malaya Peninsula*. 2 vols. London.

Cameron, L.C.R. (1917). *The wild Foods of Great Britain*. London.

Campbell, A. (1986). The use of wild food plants, and drought in Botswana. *Journal of Arid Environments* **11**: 81-91.

Chadha, K.L. and O.P. Pareek (1993). Fruit research in India- History Infrastructure and Achievements. *Advances in Horticulture I - Fruit Crops Part I*, pp. 1-41, Malhotra publi. House, New Delhi.

Chakravorty, A.K. (1951). Origin of cultivated banana of southeast Asia. *Indian Journal of Genetics* **11**: 34-46.

Chang, W.T. (1965). Utilization of wild plants in communist China. *Economic Botany* **19**: 3-15.

Chauhan, V. and Chauhan, N.S. (1988). Ethnobotany of Trans-Giri area of Sirmour district of Himachal Pradesh. *Bull. Medico Ethnobot. Res.* **9(3-4)**: 106-122.

Chettri, R.; Rai, B. and Khawas, D.B. (1992). Certain medicinal plants in the folklore and folk life of Darjeeling and Sikkim Hills, India. *J. Econ. and Taxo. Bot. (*Addl. Ser.*)* **10**: 393-398.

Collett, H. (1971). *Flora Simlensis* (Reprint ed.). Bishan Singh and Mahendra Pal Singh, Dehradun.

Conclin, H.C. (1962). Ethnobotanical problems in comparative study of folk taxonomy. Proc. Ninth Pacific Science Congress, Pacific Science Assoc. **4**: (Botany)

Dalzell, J.M. (1937). *The Useful Plants of Tropical West Africa,* London.

De Beer, J.H. and M.J. Mc Dermott. (1989). Economic value of non-timber forest Products in South-east Asia. Council for the International Union of the Conservation of Nature, The Netherlands.

Dhar, U. and Siddque, M.A.A. (1993). Ethnobotanical studies and biological diversity of Suru valley (Zanskar). *In:* Glimpses in plant research vol. X, pp 31-39. Medicinal plants new vistas of research (part I) (Govil, J.N. ,V.K, Singh and Shamami Hashmi, eds.). Today and Tomorrow's Printer & Publishers, New Delhi.

Dhyani, P.P. and Dhar, U. (1993). *Box Myrtle (Myrica esculenta* Buch. Ham.*): A promising tree crop of Himalaya*. GBPIHED. Almora.

Dhyani, P. P. and Khali, M.P. (1993). Rooting Response of branch Cuttings of two multi-Purpose *Ficus* tree crops. *The International Tree Crops Journal* 7: 241-247.

Duke, J.A. (1992). *Handbook of edible weeds*. CRC Press, Bocaraton, Florida, USA.

Duthie, J. F. (1906). Catalogue of the plants of Kumaon and of the adjacent portion of Garhwal and Tibet based on the collection made by Strachcy and Winterbottom during the years 1846-49. London.

FAO, (1985). *The Role of Minor crops in Nutrition and Food Security*. The Committee of Agriculture, Eighth Session, March 1985.

Fernald, M.L. and Kensy, A.C. (1943). *Edible Wild Plants of Eastern North America*. New York, USA.

Fernandes, W.; Menon, G. and Viegas, P. (1988). *Forests, environment and tribal economy: Deforestation impoverishment and marginalisation in Orissa*. Indian Social institute, Tribes of India Series 2. Indian Social Institute, New Delhi. pp.363.

Fox, J.E.D.; Barrett, D.R.; Brand, J.E. and Effendi, M. (1994). Germination in *Santalum album*. l. Recent research in Western Australia and a protocol for Timor, Indonesia. *International Journal of Ecology and Environmental Sciences* **20**: 345-356.

Frankenburger, T. (1985*). Adding a Food Consumption Perspectives to Farming System Research.* US Department of Agriculture, Office of Nutrition. Bureau for Science and Technology, AID Project, p. 931-1171.

Gangwar, A.K. and Ramakrishnan, P.S. (1990). Ethnobotanical notes on some tribes of Arunachal Pradesh, northeastern India. *Economic Botany* **44(1)**: 94-105.

Gaur, R.D. and Semwal, J.K. (1983). Some little known wild edibles of Garhwal Himalaya. *Man & Environment* 7: 161-165.

Goel, A.K.; Sahoo, A.K. and Mudgal, V. (1984). *A contribution to the ethnobotany of Santhal Pargana, Bihar*. Botanical Survey of India, Howrah.

Godoy, R. and Feaw, T.C. (1989). The profitablility of smallholder rattan cultivation in central Borneo. *Human Ecology* **7**: 397-420.

Grivetti, L.E.; Frentzel, C.J.; Ginsberg, K.E.; Howell, K. E. and Ogle, B.M. (1987). Bush foods and edible weeds of agriculture: Perspectives on dietary use of wild plants in Africa, their role in maintaining human nutritional status and implications for agricultural development. *In*: (Rais, A., ed.), *Health and Disease in Tropical Africa.* Geographical Viewpoints, London.

Hajra, P.K. and Chakraborty (1981). A survey of wild plants sold in the Lal market of Gangtok. *Indian Journal of Forestry* **4(3)**: 217-220.

Hall, P. and Bawa, K.S. (1993). Methods to assess the impact of extraction of non-timber forest products on plant populations. *Economic Botany* **47**: 234-247.

Harish, S. (1989). Ethnobotanical studies on *Urtica dioica* among the Bhotias of Chamoli Garhwal, Uttar Pradesh. *J. Econ. Taxo. Bot.* **13(8)**: 719-724.

Haridasan, K.; Bhuyan, L.R. and Deori, M.L. (1990). Wild edible plants of Arunachal Pradesh. *Arunachal Forest News* **18 (1&2)**: 1-8.

Harrington, H.D. and Matsumura, V. (1967). *Edible Native Plants of the Rocky Mountains.* Univ. New Hempshire, Mexico.

Harshberger, J.W. (1896). The purpose of ethno-botany. *The Botanical Gazette* **21**: 146-154.

Hill, J. (1939). *Wild Foods of Britain*, London.

Holm, L.G.; Plucknett, D.L.; Pancho, J.V. and Herberger, J.P. (1977). *The world's worst weeds.* University of Nawaii, Honolulu, Hawaii, USA.

Homma, A. K.O. (1992). The dynamics of extraction in Amazonia: a historical perspective, pp. 23-31. *In:* (D.C. Nepstad and S. Schwartzman, eds.), *Non-timber products from tropical forests: Evaluation of a conservation and development strategy.* Advances in Economic Botany I. The NewYork Botanical Garden. Bronx, NewYork.

Hoskins, M. (1985). The promise in trees. *Food and Nutrition* **11(2)**: 44-46.

Hynniewta, T.M. and Wadhwa, B.M. (1988-89). *Ethnobiology of Arunachal Pradesh.* Phase I. Botanical Survey of India, Shillong.

Jain, S.K. (1963). Magico-Religious beliefs about plants among the tribals of Bastar, Madhya Pradesh. *Q. J. Mythic Soc.* **54**: 73-94.

Jain, S.K. (1965a). Wooden musical instruments of the Gonds of Central India. *Ethnomusicology* **9**: 39-42.

Jain, S.K. (Ed.) (1981). *Glimpses of Indian Ethnobotany.* Oxford and IBH Publishing Co. New Delhi.

Jain, S.K. (1984). Ethnòbotany of Morni and Kabsar (district Ambala, Haryana). *J. Econ. and Taxon. Botany* **5**: 809-813.

Jain, S.K. (1987). *A Manual of Ethnobotany.* Scientific Publishers. Jodhpur.

Jain, S.K. (1994). Ethnobotany and research on medicinal plants in India. Ethnobotany and the search for new drugs. Wiley, Chichester (Ciba Foundation Sumposium 185) p. 153-168.

Jain, S.K. and Rao, R.R. (1977). *A Handbook of Field and Herbarium Methods: Planning, Preparation and Publications of Scientific Papers*. Today & Tomorrows Publications, New Delhi.

Jain, S.K. and Tarafdar, C.R. (1970). Medicinal plantlore of the Santals: A revival of P.O. Boddings work. *Economic Botany* **24**: 241-278.

Jain, S.K.; Fernandes, V.F.; Snehlata and Ayub, A. (1995). Indo-Amazonian ethnobotanic connections- Similar uses of some common plants. *Ethnobotany* **7**: 29-37.

Jamir, N.S. and Rao, R.R. (1990). Fifty new or interesting medicinal plants used by the Zealangs of Nagaland, India. *Ethnobotany* **2(1-2)**: 11-18.

Janaki Ammal, E.K. (1955). An introduction to the subsistence economy of India. Background Paper No. 10. Wenner-Gren Foundation. International Symosium. *In*: (William L.T. Jr., ed.) *Man's role in changing the face of the earth,* pp.324-335. University of Chicago Press, Chicago.

Janaki Ammal, E.K. and Nagendra Prasad, P. (1984). Ethnobotanical findings on *Costus speciosus* (Koen.) Sm. among the Knanikkars of Tamil Nadu. *J. Econ. Taxo. Bot.* **5**: 129-133.

Joseph, J.S. and S.K. Murty (1981). Plant resources of north-east India. *In:* Seminar on resource potential of North-East India. April 29-May 1, Meghalaya Science Society, Shillong.

Juyal, S.P. and M.R. Uniyal (1960). Medicinal Plants of commercial and traditional importance in Bhillangana Valley of Tehri-Garhwal. *Nagarjun* **10**: 26-36.

Kachroo, P.; Sapru, B.L. and Dhar, U. (1977). *Flora of Ladakh*. Bishen Singh Mahendra Pal Singh, Dehradun, India.

Kandya, A.K. and Kandya, S. (1988). A review of seed research on *Shorea robusta* in India. *International Journal of Ecology and Environmental Sciences* **14**: 245-258.

Kanjilal, U.N.; Kanjilal and Das, A. (1934-1940). *Flora of Assam*. 11 vols. Govt. of Assam, Shillong.

Kapur, S.K. and Sarin, Y.K. (1982). Economically useful plants of Trikuta hills. (Jammu & Kashmir). *Indian Journal of Forestry* **5(2)**: 105-110.

Kaul, R.N. and Haridasan, K. (1987). Forest types of Arunachal Pradesh a Preliminary study. *J. Econ. Taxo. Bot.* **9(2)**: 379-389.

Khan, S.S. (1980). Medicibotany of aquatic macrophytes of Bhopal. *Nagarjun* **23(10)**: 217-219.

Kharsati, U.M. (1973). The wild flora of Meghalaya and their economic importance relating to horticulture and arboriculture *In*: *Horticulture Development in India*, Council of India Publ., Government of Meghalaya, Shillong.

Kishore, D.K. and Randhawa, S.S. (1993). Wild germplasm of temperate fruits. *Advances in Horticulture*. Vol. 1-Fruit crops Part 1: (K.L. Chadha and Pareek, eds.), Malhotra Publishing House, New Delhi.

Kolesnikov, B. P. (1943). *Edible wild Plants*. Moscow (in Russian).

Krishnamurthy, T. (1993). *Minor forest products of India*. Oxford & IBH Publishing Co., Pvt. Ltd, New Delhi.

Kumar, S.; Singh, P. and Singh, V. (1994). Ethno-medico aspects of some arboreal and arborescent taxa of Sikkim. *Higher Plants of Indian Subcontinent* **3**: 161-166. Bishen Singh and Mahendra Pal Singh Publisher, Dehradun.

Kunkel, G. (1984). *Plants for human consumption*. Koeltz Scientific Books, Kowningsten, Germany.

Longhurst, R. (1986). Household food strategies in response to seasonality and famine. *IDS Bulletin* **17**: 3.

Lu and Rongson (1992). Sea buckthorn: A multipurpose plant species for fragile mountains. ICIMOD Occasional Paper No. 20. ICIMOD, Kathamandu, Nepal.

Maheshwari, J.K. and Singh, J.P. (1984). Contribution to the Ethnobotany of Bhoxa tribe of Bijnor and Pauri Garhwal districts, U. P. *J. Econ. Taxo. Bot.* **5(2)**: 251-259.

Maheshwari, J.K.; Kunkel, G.; Bhandari, M.M. and Duke, D.A. (1993). *Ethnobotany in India*. Scientific Publishers, Jodhpur.

Maikhuri, R.K.; Nautiyal, M.C. and Khali, M.P. (1992). Lesser known crops of food value in Garhwal Himalaya and a strategy to conserve them. *FAO/IBPGR Plant Genetic Resources Newsletter* **86**: 33-36.

Maikhuri, R.K.; Semwal, R.L.; Singh, A. and Nautiyal, M.C. (1994). Wild fruits as a contribution to sustainable rural development: A case study from the Garhwal Himalaya. *Int. J. Sustain. Dev. & World Ecol.* **1**: 56-68.

Malla, S.B.; Rajbhandari, S.B.; Shrestha, T.B.; Adhikari, P.M. and Adhikari, S.R. (1982). *Wild Edible Plants of Nepal*. Department of Medicinal Plants, Bulletin No. 9. Royal Governament of Nepal, Kathmandu, Nepal.

Manilal, K.S. (1981). *Hortus malabaricus*, Indian Ethnobotany cermelite Missionaries. *In*: (John, K.J., ed.) *The Christian heritage of Karala*. Fr. George Velipara mbail, Cochin.

Mao, A.A. (1993). Preliminary report on the folklore botany of Mao Nagas of Manipur (India). *Ethnobotany* **5**: 143-147.

Martin, Gary J. (1995). *Ethnobotany- A Methods Manual*. Chapman and Hall, London.

Medsger, O.P. (1943). *Edible Wild Plants*. New York.

Megoneitso and Rao, R.R. (1983). Ethnobotanical studies in Nagaland-62 medicinal plants used by Angami Nagas. *J. Econ. Taxo. Bot.* **4**: 167-172.

Mudgal, V.; Pal, D.C. and Jain, S.K. (1984). On botanical and ethnobotanical aspects of betel vine. *In*: (Roy. S. K. *et al.*, eds.), pp. 17-21, Proc. National workshop on Betelvine 1982. The Agricultural Soc. of India, Calcutta.

Murali, K. S.; Uma Shanker, R.; Uma Shaanker, K.; Ganeshaiah, N. and Bawa, K.S. (1996). Extraction of non-timber forest products in the forests of Bilgiri Rangan Hills. India: Impact of NTFP extraction on regeneration, population structure and species composition. *Economic Botany* **50(2)**: 252-269.

Nautiyal, P.C. and Purohit, A.N. (1986). Effect of auxin on seasonal rooting response of stem cutting of *Berberis* species from different altitudes. *Indian Journal of Plant Physiology* **29(3)**: 286-290.

Nautiyal, M.C.; Rawat, A.S.; Bhadula, S.K. and Purohit, A.N. (1987). Seed germination in *Podophyllum hexandrum. Seed Research* **15(2)**: 206-209.

Negi, K.S. and Gaur, R.D. (1994). Principal wild food plants of western Himalaya Uttar Pradesh. India. p. 1-78, *In:* (B.K. Gupta, ed.), *Higher Plants of Indian Subcontinent.* Bishan Singh Mahendra Pal Singh, Dehradun.

Negi, K.S. and Pant, K.C. (1990). Notes on ethnobotany of the Gangwal- A tribe of Garhwal Himalaya. *Ethbobotany* **2**: 81-85.

Negi, K.S.; Pant, K.C.; and Gaur, R.D. (1990). Seabuckthorn, *Hippophae*-A pickle plant from Central Himalaya. *Acta Botanica Indica* **18**: 274-275.

Negi, K.S.; Tiwari, J.K.; Gaur, R.D. and Pant, K.C. (1993). Notes on ethnobotany of five districts of Garhwal Himalaya. Uttar Pradesh, India. *Ethnobotany* **5**: 73-81.

Negi, Y.S. (1991). *Nutrient composition and chemical analysis of Ficus species*. D.Phil. Thesis. H.N.B. Garhwal University, Srinagar, India.

Neog, M. and Mohan, N.K. (1994). Minor and less-Known fruits of Assam. *Indian Horticiture.* July-September, 1994.

Neogi, B., N.V. Prasad and R.R. Rao (1989). Ethnobotany of some weeds of Khasi and Garo Hills, Meghalaya, Northeastern India. *Economic Botany* **43(4)**: 471-479.

N.R.C. (National Research Council) (1982). *Ecological aspects of development in the humid tropics*. National Acad. of Sciences, Washington. DC, 297 p.

Ogle, B.M. (1988). Nutrition in the Forest. Trees and People Project in Zambia. Report on a consultancy for GCP/ZAM/040/SWE. International Rural Development Centre, Swedish University of Agricultural Sciences.

Osmaston, A.E. (1927). *Forest Flora of Kumaon.* Government Press, Allahabad.

Pal, D.C. and J.N. Srivastava, (1976). Preliminary notes on ethnobotany of Singhbum district, Bihar. *Bull. Bot. Sur. India* **18**: 247-250.

Pal, G.D. (1984). Observations on ethbnobotany of tribals of Subansiri Arunachal Pradesh. *Bull. Bot. Sur. India* **26(1-2)**: 26-37.

Paliwal, G.S. and Badoni, A.K. (1990). Ethbnobotany of the Hill tribes of Uttarkashi. *J. Econ. Taxo. Bot.* **14(2)**: 421-442.

Panayotou, T. and Ashton, P. (1992). *Not by timber alone: The case for multiple use management of tropical forests*. Island Press, CA, USA.

Pandey, G.; Sharma, B.D. Hore, D.K. and Rao, N.V. (1993). Indigenous minor fruits genetic resources and their marketing status in north-eastern hill of India. *Journal of Hill Research* **6(I)**: 1-4.

Pangtey, Y.P.S.; Samant, S.S. and Rawat, G.S. (1989). Ethnobotanical notes on the Bhotia tribes of Kumaon. *Indian Journal of Forestry* **12(3)**: 191-196.

Pareek, O.P. and Sharma, S. (1993). Under-utilized fruits. *Indian Horticulture*, April-June, 1993.

Parmar, C. and Kaushal, M.K. (1982). *Wild Fruits*. Kalyani Publishers, Ludhiana, India.

Pei Shengji (1995). Sustainable livelihood for mountain communities: tradition and transition natural resources management. International preparatory seminar-cum workshop to develop projects on resource management and sustainable livelihood for traditional societies in south and central Asia (I. A. Khan, ed.) 10-14 September, University of Agriculture, Faisalabad, Pakistan.

Plotkin, M. and L. Famolare, (ed) (1992). *Sustainable harvest and marketing of rain forest Products.* Island Press, Washington DC.

Porterfied, Jr. M.M. (1951). Principal Chinese vegetable foods and food plants of China town markets. *Economic Botany* **5**: 3-37.

Porsild, A.E. (1937). *Edible Roots and Berries of Northern Canada.* Nat. Mus., Ottawa, Canada.

Pushpangadan, P. and Atal, C.K. (1984). Ethno-medico-botanical investigations in Kerala-I. Some primitive tribals of western ghats and their herbal medicine. *J.Ethno-pharm.* **11**: 59-77.

Rai, L.K. and Sharma, E. (1994). *Medicinal Plants of the Sikkim Himalaya: Status, Usage and Potential.* Himavikas Occasional Publication No. 5. Bishen Singh Mahendra Pal Singh, Dehradun.

Rai, Y.K.; Rai, L.K. and Sundriyal, Manju (1994). On the availability and marketing aspects of wild edible plants in Mamlay watershed: A Pilot Report. *SSS Newsletter* **7**: 4.

Raju, D.C.S. and Krishna, B. (1990). Less known edible Plants of Sikkim. *In*: *Economic plants and microbes.* (R. P. Porkayastna, ed.), pp. 83-86. Today & Tomorrow's Print. & Publi., New Delhi.

Raju, D.C.S. and Mandal, N.R. (1990). Papri (*Podophyllum hexandrum* Royle)-an important medicinal plant. *J. Hill Res* **3**: 176-178.

Ramachandran, V.S. and Nair, V.J. (1981). Ethno-botanical studies in Cannanore district, Kerala State (India). *J. Econ. Taxo. Bot.* **2**: 65-72.

Ramakrishnan, P.S. (1984). The need to conserve silent valley and tropical rain forest ecosystems in India. *Environment and Conservation* **11**: 170-171.

Rao, M.K.U. and Shanpru, R. (1981). Some plants in the life of the Garos of Meghalaya. In: (S.K. Jain, ed.) pp. 153-160, *Glimpses of Indian Ethnobotany*. Oxford & IBM Publishing Co., New Delhi.

Rao, R.R. and Neogi, B. (1980). Observation on the ethbobotany of the Khasi and Garo tribes in Meghalaya (India). *J. Econ. Taxo. Bot.* **1**: 157-162.

Rao, R.R. and Murti, S.K. (1990). North-East India- A major centre for plant diversity in India. *Indian Journal of Forestry* **13(3)**: 214-222.

Rao, R.S. (1972). Botanical Survey work in Arunachal Pradesh. Arunachal Research Bulletin, pp. 49-79. Silver Jubilee Volume (ed. B. Shastri), Itanagar.

Rapoport, E.H.; Raffale, E.; Ghermandi, L. and Margutti, L. (1995). Edible weeds: A scarcely used resource. *Bulletin of the Ecological Society of America* **76(3)**: 163-166.

Rawat, G.S. and Y.P.S. Pangtey (1989). A contribution to the ethnobotany of alpine region of Kumaon. *J. Econ. Taxon. Bot.* **11**(1):139-148.

Repetto, R. and Gillis, M. (1988). *Public Policies and the misuse of forest resources.* Cambridge University Press, New York, USA.

Salick, J.A. Mejia and Anderson, T. (1995). Non-timber forest products integrated with natural forest management, Rio San Juan, Nicaragua. *Ecological Applications* **5(4)**: 878-895.

Samant, S.S. and Dhar, U. (1997). Diversity, endemism and economic potential of wild edible plants of Indian Himalaya. International journal of sustainable Development and World Ecology **4**: 179-191.

Samant, S.S. and Pangtey, Y.P.S. (1995). Diversity of Ichthyotoxic plants of Kumaon Himalaya. *Indian Journal of Forestry* **18(I)**: 80-86.

Saunders, C.F. (1934). *Useful Plants of United States and Canada*, New York, USA.

Saxena, A.P. and Vyas, K.N. (1981). Ethbobotanical records on infectious diseases from tribals of Banda dist. *J. Econ. Taxo. Bot.* **2**: 191-194.

Saxena, H.O. and Dutta, P.K. (1975). Studies on the ethnobotany of Orissa. *Bull. Bot. Sur. India* **17**: 124-131.

Sharma, P.K. and Singh, V. (1989). Ethnobotanical studies in Northwest and Tran-Himalaya. Ethnoveterinary medicinal plants used in Jammu and Kashmir, India. *J. Ethnopharm.* **27(1-2)**: 63-70.

Singh, H.B. and Arora, R.K. (1978). *Wild Edible Plants of India* (Ist Ed.), ICAR Publication, New Delhi. 88 p.

Singh, K.K. and Maheshwari, J.K. (1985). Forest in the life and economy of the tribals of Varanasi District, Uttar Pradesh, *J. Econ. Taxo. Bot.* **6(I)**: 109-116.

Singh, S.R.; Kumar, P.; Singh, N.Z. and Singh, L.J. (1988). Ethnobotanical Studies on wild plants in the market of Manipur. II. *J. Eco. and Taxo. Bot.* **12(1)**: 113.

Singh, S. R. and Singh, N.I. (1985). A Preliminary ethnobotanical study on wild edible Plants in the markets of Manipur. I. *J. Econ. Taxo. Bot.* **6**: 699-703.

Singh, Virendra (1995). Lessor Known wild edibles of Sikkim Himalaya. *J. Econ. Taxo. Bot.* **19(2)**: 385-390.

Singh, V.K. and Khan, A.M. (1989). Medicinal plants folklores: A strategy towards conquest of human ailments. *In: Glimpses of Plant Research* Vol. X (J. N. Govil, ed.). Today and Tomorrows Printers and Publishers, New Delhi.

Smythe, F.S. (1938). *The Valley of Flowers*. London.

Sperling, L. (1987). Food acquisition during the African drought of 1983-1984: A shift of Kenyan Herders. *Disasters* **11**: 263-271.

Srivastava, T.N.; Kapahi, B.K. and Atal, C.K. (1987). Ethnomedico botanical investigation in Sikkim. *J. Econ. Taxo. Bot.* **11(2)**: 413-421.

Sundriyal, R.C. and Sharma, E. (1996). Anthropogenic pressure on tree structure and biomass in the temperate forest of Mamlay watershed in Sikkim. *For. Ecol. Manag.* **81**: 113-134.

Sundriyal, Manju and Rai, L.K. (1996). Wild edible plants of the Sikkim Himalaya. *Journal of Hill Research* **9(2)**: 267-278.

Sundriyal, Manju and Sundriyal, R.C. (1998). Wild plants of food value in Sikkim Himalaya: Aspects of diversity and utilization. *In: Perspectives for Planning and Development in North East India* (Eds. R.C. Sundriyal, Uma Shankar and T.C. Upreti), pp. 138-149. G.B. Pant Institute of Himalayan Environment & Development, *Himavikas Occasional Publication No. 11*.

Sundriyal Manju (1999). Distribution, propagation and nutritive values of some wild edible plants of the Sikkim Himalaya. (Ph.D. thesis). HNB Garhwal University, Srinagar, Garhwal, Uttranchal.

Sundriyal, Manju, R.C. Sundriyal, E. Sharma and A.N. Purohit (1999). Wild edibles and other useful plants from the Sikkim Himalaya, India. *Oecologia Montana*, **7**: 43-54, Slovakia.

Sundriyal, Manju, and Sundriyal, R.C. (2000a). Potential of wild edible plants in the Sikkim Himalaya: Conservation concerns. . *Non-timber Forest Produce* **7(3/4)**: 253-262.

Sundriyal, Manju and Sundriyal, R.C. (2000b). Seed germination and stem-cuttings response to hormonal treatment in six wild edible fruit species of Sikkim Himalaya. *Indian Forester* **127(6)**: 695-706.

Sundriyal, Manju and Sundriyal, R.C. (2001). Wild edible plants of the Sikkim Himalaya: Nutritive values. *Economic Botany*, **55(3)**: 313-334, USA.

Sundriyal Manju and Sundriyal, R.C. (2003). Underutilized edible plants of the Sikkim Himalaya: Need for domestication. *Current Science* **85(6)**: 731-733.

Sundriyal, Manju and Sundriyal, R.C. (2004a). Structure, phenology, fruit yield, regeneration and marketing status of some prominent ethnobiological plant species of the Sikkim Himalaya, India. *Journal of Ethnobiology* **24(1)**: (USA) (in press).

Sundriyal, Manju and Sundriyal, R.C. (2004b). Wild edible plants of the Sikkim Himalaya: Marketing , value addition and Implication for Management. *Economic Botany*, **58(2)**: 300-315.

Tamang, J.P.; Sarkar, P.K. and Clifford W. Hesseltine (1988). Traditional fermented foods and beverages of Derjeeling and Sikkim: A review. *J. Sci. Food. Agric.* **44**: 375-385.

Tewari, A. and Dhar, U. (1996). An investigation on seed germination of Indian butter tree *Aisandra butyracea. Seed Science & Technology* **24**: 211-218.

Thothathri, K.; Sen, R. and Pal, D.C. (1990). Ethnobotany of less known oil yielding plants. *Indian J. Forestry* **13(3)**: 268-270.

Uma Shankar, K.S. Murali, R. Umashaanker, K.N. Ganeshaiah and Bawa, K.S. (1996). Extraction of non-timber forest products in the forest of Biligiri Rangan Hills India. 3. Productivity, extraction and prospects of sustainable harvest of Amla (*Phyllanthus emblica*). *Economic Botany* **50(3)**: 270-279.

Uma Shankar, K.S. Murali, R. Umashaanker, K.N.Ganeshaiah and K.S. Bawa (1998). Extraction of non timber forest products of Biligiri Rangan Hills,India. 4.Impact on Floristic diversity and population structure in a thorn scrub forest. *Economic Botany* **53(3)**: 302-315.

Uniyal, M. R. (1977). *Uttarkhand Vanaushadi Darshika*. Central Council for Research in Indian medicine and Homeopathy, New Delhi.

Upreti, M.C. (1983). Utilization of Pine needles. *Himalayan Research and Development* **2**: 41-42.

Vartak, V.D. and M. Gadgil (1981). Studies on Sacred groves along the western Gahts from Maharashtra and Goa: Role of beliefs and folk-lores. In: (Jain, S.K., ed.), pp. 272-278, *Glimpses of Indian Ethnobotany*. Oxford and IBH, New Delhi.

Vavilov, N.I. (1950). The origin, variation, immunity and breeding of cultivated plants. *Chron. Bot.* **13**: 364.

Vihari, V. (1995). Ethnobotany of Cosmetics of Indo-Nepal Border. *Ethbobotany* **7**: 89-94.

Virjee, Dar, G.H.; P. Kachroo, and G.M. Bhat (1984). Taxo-ethbobotanical studies of the rural areas in district Rajouri (Jammu). *J. Eco. Taxo. Bot.* **5(4)**: 831-838.

Vishnu Mittre, (1981). Wild plants in Indian folk life: A historical perspective. *In:* (Jain, S.K., ed.), *Glimpses of Indian Ethnobotany*. Oxford and IBH, Co., New Delhi.

Wester, P.J. (1925). *The Food Plants of the Philippines*. Dept. Agri. & Nat. Res. Bur. Agri. Bull. 39, Philippines.

Wilkes, H.G. (1981). New and potential crops, or what to anticipate for the future. Paper presented for annual meeting of the American Association for the advancement of Science, January 1981, Toronto, Canada.

Advancing Frontiers of Ecological Researches in India, 233-244, 2006
Edited by A.K. Kandya & Asha Gupta
Published by Bishen Singh Mahendra Pal Singh, Dehra Dun, India

Biodiversity of Tropical Dry Deciduous Forest Ecosystem

B.R. Pandit
Department of Life Sciences,
Bhavnagar University, Bhavnagar-364 002

Abstract: The Gir Forest, famous as the abode of Asiatic lions is situated in the Saurashtra region of Peninsular Gujarat (20^0 41' N and 21^0 50' N, 70^0 50' E and 71^0 50' E). The total forest area is 1412 km^2. The area is hilly terrain elevation varies from 152 to 531 m above msl. The climate is very hot and dry. The rains are highly erratic, both in intensity and frequency; with average annual rainfall is 700 mm. The climate is supporting typically dry deciduous forest ecosystem. The forest is enriched with a faunal diversity including 31 species of mammals and comprising the floral diversity of 431 species of plants.

The present study enumerated 431 species, which belonged to 294 genera and 94 families of flowering plants, out of which, Dicotyledons contributed 371 plant species belonging to 246 genera and 80 families, which is quite higher than that of Monocotyledons. The Monocotyledons contributed only 60 species belonging to 48 genera and 14 families. Leaving Poaceae and Cyperaceae, Monocotyledons are poorly represented. Proportion of the Monocotyledons to Dicotyledons was recorded 1:5.71 of families, 1:5.12 of genera and 1:6.18 of species. The ratio of the total number of genera to species was recorded 1:1.46. 42 families were found monotypic and 49 families with one genus (monogeneric). Among the various habits, herbs contributed more than 50 % of plant species (53.59 %).

The plant diversity study of the Gir forest is claim to be the first of its kind. The diversity indices viz Simpson's index (D=0.361), Simpson's index of diversity (1-D= 0.639), Simpson's reciprocal index (1/D = 2.77), Shannon-wiener index (H= 1.194) and Evenness index (E= 0.861) were calculated.

The biological spectrum of the flora of Gir forest falls under the category of therophytic type of phyto-climate, which indicates hot and dry climate in the region, which is congenial for the growth of annuals and perennial herbs.

Key Words: Gir forest, deciduous forest, Asiatic lions, floristic composition, diversity indices.

INTRODUCTION

The loss of natural habitat and species extinctions around the world have served to focus intense international attention on the issue of biodiversity, culminating in the negotiation of the United Nations Conservation on Biological Diversity. It is widely recognized that the path to sustainable development requires the conservation and wise use of the world's biodiversity. Indeed, measuring and monitoring biodiversity must be the first step towards effective conservation and sustainable development. However, because of the overwhelming variety of life on earth and spanning, as it does. The complete range of biological organization form genes to landscapes, measurement of biodiversity poses some serious problems.

The Gir forest is situated in the Saurashtra region of Peninsular Gujarat between 20041'N and 21051'N latitude and 700051'E and 71050'E longitude. The total area of Gir forest occupies approximately 1412.13 km^2, adjoining Junagadh and Amreli districts. The entire part of the tract is hilly and elevation varies form 152 to 513 meters above mean sea level. There are few notable plains in Gir forest especially near Jesadhar. Similar plains also occur at Timberva region. In the river system, Hiren, Machhundri and Singwada are important rivers.

The soil of this forest is formed partly form sedimentary rocks and partly form igneous rocks. The soil layer is very thin, black-cotton soil is generally found in the valleys and low lands and the red soils in the uplands. The red soil is light and porous. Soil is moderately fertile and usually mixed up with good proportions of stones.

A year can be divided into three distinct seasons. The winter (Maximum 19° C, Minimum 10° C) from November to February is followed by the summer (Maximum 40° C, Minimum 34° C) from March to May. The monsoon from June to October contributes to the Southwest monsoon (the average annual rainfall is 700 mm). The rains during monsoon are highly erratic, both in intensity and intervals between two successive rains.

MATERIALS AND METHODS

Several field trips of duration ranging from 2 to 10 days were made at regular intervals to various parts of Gir forest extending over a period of more than three years, and made intensive and extensive collection of plants. For collection and preservation the procedures given by Jain and Rao (1977) and Balgooy (1987) were generally followed. The trips were arranged in such a way so as cover all the six localities and collect most of the plants in flowering or fruiting stages. All the specimens collected were serially numbered. The field notes were taken regularly, included habitat, color of the flowers, association and other pertinent features. Efforts were made to identify the plants from the fresh material; those that could not be satisfactorily identified in the field were brought to the laboratory and identified by checking it with monographs, herbarium specimens and other available literature.

Collected plants were properly processed, numbered and prepared for herbarium. The herbarium specimens were labeled and deposited in Herbarium Section of the Department of Life Sciences, Bhavnagar University, Bhavnagar. For species diversity various diversity indices were calculated

Floristic and vegetational studies

Flowering plants of Gujarat had been studied earlier by Cooke (1908), Saxton and Sedgwick (1918), Sutaria (1958), Gandhi (1958), Chavan (1961), Oza (1962), Jain (1963), Vaidya and Vora (1964), Santapau and Janardhanan (1966), Sabnis (1967), Bedi (1968), Inamdar (1968), Bhatt (1969), Bhatt and Sabnis (1974), Parabia (1974) and Shah (1978); published a 'Flora of Gujarat State' in two volumes.

Cooke (1901-1908) and Talbot (1902) recorded plants of Saurashtra. The floristic studies on different parts of Saurashtra have been made by Borgeson (1929), Rao (1941), Rhizada and Santapau (1955) and Thaker (1966). Santapau (1953) published a preliminary list of 980 plant species from Saurashtra region. The first volume of "The Flora of Saurashtra" was published by Santapau (1962). Bhatt and Imandar (1971) made some more additions to the flora of Saurashtra. The vegetation of Bhavnagar was worked out by Murthy (1957) and reported 378 plant species from the area including cultivated plants. Vora et al. (1981) reported 341 flowering plants of Ghoghamahal of Bhavnagar district. The floristic and ecological study of Bhavnagar and its surroundings has been studied by Oza (1991), and reported 528 flowering plant the area. Desai (1992) has recorded five new taxa from Bansda forest, for the first times, which are new additions to the flora of Gujarat State.

An assessment of floristic studies

The present work is based on the result of three years of intensive and extensive study of the plants of this area. The analysis of the total plant species found are presented in the following way:

The present study enumerated 431 species, which belonged to 294 genera and 94 families of flowering plants (Table-1). Out of which, Dicotyledons contributed 371 plant species belonging to 246 genera and 80 families, which is quite higher than that of Monocotyledons. The Monocotyledons contributed only 60 species belonging to 48 genera and 14 families. Leaving Poaceae and Cyperaceae, Monocotyledons are poorly represented. Proportion of the Monocotyledons to Dicotyledons was recorded 1:5.71 of families, 1:5.12 of genera and 1:6.18 of species (table-2). The ratio of the total number of genera to species was recorded 1:1.46, which is rather low in comparison to a corresponding ratio for whole India (1:7), but it more or less in conformity with the ratio of Delhi state (1:1.63), as reported by Maheshwari (1963) and W. Rajasthan (1:1.9), reported by Bhandari (1978).

A comparison of the present study with other regions of Gujarat State gives very interesting results. The ten dominant families, which were found in the Gir forest, have been compared with other regions of Gujarat viz., Ghoghamahal (Vora *et al.*, 1981); Victoria Park reserved forest (Patel,

Table-1: Number and percentage of families, genera and species of each class

	Families		Genera		Species	
	No.	%	No.	%	No.	%
Dicotyledons	80	85.10	246	83.67	371	86.07
Monocotyledons	14	14.90	48	16.33	60	13.93
Total	94	100.0	294	100.0	431	100.0

Table-2: Ratio between Monocotyledons and Dicotyledons

Family	1 : 5.71
Genera	1 : 5.12
Species	1 : 6.18
Genera to Species	1 : 1.46

Table-3: Number and percentage of genera and species of the dominant families

Sr. No.	Families	No. of Genera	%	No. of Species	%
1.	Fabaceae	24	8.16	45	10.44
2.	Poaceae	26	8.84	31	7.19
3.	Asteraceae	19	6.46	22	5.10
4.	Acanthaceae	15	5.10	21	4.87
5.	Euphorbiaceae	11	3.74	22	5.10
6.	Amaranthaceae	9	3.06	14	3.24
7.	Convolvulaceae	6	2.04	15	3.48
8.	Malvaceae	6	2.04	12	2.78
9.	Rubiaceae	7	2.38	10	2.32
10.	Cucurbitaceae	7	2.38	9	2.08

1982); South Gujarat (Yadav, 1979); Saurashtra region (Santapau and Janardhan, 1966) and Gujarat State (Shah, 1978) (Table-6).

It is evident form this table that most of the dominant families in the present study, have been reported earlier form different regions of Gujarat State, but the order of dominance is different. However, the legumes and grasses occupy the first and second place in all these regions of Gujarat, except South Gujarat, where grasses occupied the first position. Table-3 shows the number and percentage of genera and species of the ten dominant families found in the present study.

The floristic composition analysed critically and it showed that 42 families were monotypic, 49 families were monogeneric, 14 with two genera, ten with three genera, four with four genera, four with five genera, four with six genera, three with seven genera, one with eight genera, one with nine genera, and five with ten or more then ten genera. Monotypic genera contributed only 14.28% in the Gir forest region, which was less when compared with South Gujarat (29.8%), but were more or less in conformity with Bhavnagar (12.63%), as reported by Oza (1991). The other families which showed their remarkable percentage were as follows: families with one genera, two genera, three genera and four genera were 52.68%, 15.05%, 10.75% and 4.30% respectively. Families with five to nine genera are in the range of 1.30 to 4.30% and the families with more than ten genera were little high (5.37%).

Table-4: Number and percentage of species belonging to the different habits

Habit	Total Species	(%) Percentage
Herbs	231	53.59
Trees	82	19.03
Shrubs	62	14.39
Climbers and Twiners	56	12.99
Total	431	100.0

Table-5: Biological spectrum

Life forms	Ph	Ch	H	Cr./G	Th	Phytoclimate
Normal Spectrum (%)	46	9	26	6	13	
Total species	150	23	15	6	237	Therophytic
% species	34.81	5.34	3.48	1.39	54.98	

The habit approach revealed that out of 431 flowering plants, herbs contributed more than 50% of plant species i.e. 53.59, followed by trees, shrubs and climber or twiners which were 19.03%, 14.39% and 12.99% respectively (Table-4). These results were in close agreement with earlier reports of Pandya (1972) and Oza (1991).

The higher percentage of herbs in the area could be attributed to edaphic and peculiar climatic conditions like mager rainfall and high temperature. During summer, when the temperature become severe and the soil become intolerable to plants, as a result only short living plants like annual herbs are only favored. Because, herbs can complete their life cycle before the commencement of the dry season and set seed during the summer. Hence the proper utilization of climate and edaphic factors are often responsible for such type of distribution pattern. While studying the temperate and tropical forest types, Smith (1973) reported that an increased structural integrity of the forest will lead to the proper utilization of climatic and edaphic conditions.

The plant species are presented in the Appendix-1 with reference to their name of the family, genera and species; and also their vernacular name, habit, flowering and fruiting period, Life forms of Raunkiaer (1934), Life forms of Numata (1954), Life forms of Numata and Asano (1969) and Rare or Common.

Table-6: Comparison of ten dominant families of Gir forest with different regions of Gujarat State.

Sr. No.	Gir Forest	Victoria Park	Ghogha Mahal	South Gujarat	Saurashtra Region	Gujarat State
1.	Fabaceae	Fabaceae	Fabaceae	Poaceae	Fabaceae	Fabaceae
2.	Poaceae	Poaceae	Poaceae	Fabaceae	Poaceae	Poaceae
3.	Asteraceae	Euphorbiaceae	Euphorbiaceae	Cyperaceae	Asteraceae	Cyperaceae
4.	Acanthaceae	Asteraceae	Asteraceae	Asteraceae	Malvaceae	Asteraceae
5.	Euphorbiaceae	Malvaceae	Malvaceae	Euphorbiaceae	Acanthaceae	Acanthaceae
6.	Amaranthaceae	Acanthaceae	Convolvulaceae	Acanthaceae	Convolvulaceae	Euphorbiaceae
7.	Convolvulaceae	Convolvulaceae	Acanthaceae	Scrophulariaceae	Euphorbiaceae	Malvaceae
8.	Malvaceae	Amaranthaceae	Amaranthaceae	Malvaceae	Cyperaceae	Convolvulaceae
9.	Rubiaceae	Cyperaceae	Cucurbitaceae	Convolvulaceae	Rubiaceae	Scrophulariaceae
10.	Cucurbitaceae	Tiliaceae	Tiliaceae	Labiatae	Cucurbitaceae	Cucurbitaceae

The biological spectrum presented in Table-5 showed the highest percentage (54.98%) of Therophytes. Other groups of life forms in order of importance are given as follow: Phanerophytes (34.81%), Chamaephytes (5.34%), Hemicryptophytes (3.48%) and Cryptophytes or Geophytes (1.39%). Hence, the area falls under the category of stherophytic type of phytoclimate, which indicates hot and dry climate in the region, which is congenial for the growth of annuals and perennial herbs.

A comperation of the biological spectrum of the flora of Gir forest with different regions of Gujarat (Table-7) reveals that among all life forms, therophytes contributed the highest percentage in all the region of Gujarat including the present study. According to Danserau (1957) the predominance of therophytes indicates warm climate. Bharucha and Dave (1944), Pandeya (1964), Yadav and Singh (1977) and Agrawal (1984) attributed the preponderances of therophytes to the influence of heavy grazing or biotic interferences. However, Meher-Homji (1964) and Daubenmire (1968) attributed the occurrence of therophytes to aridity factor.

Increase in the percentage of therophytes and decrease in that of phanerophytes and hemi cryptophytes in the flora of Gir forest was caused by deforestation, intensive utilization of land for cultivation and grazing by livestocks. The result of the present study is in conformity with the conclusion drawn by Meher-Homji (1981).

Plant Diversity Indices

It is a measured by a number of diversity indices. There has been a lot of development and proliferation of diversity indices over the years. These indices are mathematical expressions that combine three

Table-7: Comparison of biological spectrum of Gir forest with different regions of Gujarat

S. No.	Regions	Ph	Ch	H	Cr./G	Th	Phytoclimate
1.	Present work	34.81	5.34	3.48	1.39	54.98	Therophytic
2.	Victoria Park	25.26	7.51	6.48	9.21	51.54	Therophytic
3.	Ghoghamahal	6.72	6.30	8.82	4.20	49.58	Therophytic
4.	Saurashtra	4.90	20.00	2.00	2.70	39.20	Therophytic
5.	North Gujarat	4.75	16.30	2.00	2.80	47.20	Therophytic
6.	Gujarat State	4.90	17.80	1.40	4.60	43.80	Therophytic

Table-8: Plant Diversity Indices

Category	# Found	Pi	Pi^2	Pi In (Pi)	Measure	Value
Herbs	231	0.536	0.287	- 0.334	S	4
Shrubs	62	0.144	0.021	- 0.279	D	0.361
Climbers	56	0.13	0.017	- 0.265	1 - D	0.639
Trees	82	0.19	0.036	- 0.316	1 / D	0.277
Total	431	1			H	1.194
					E	0.861

components of community structure, i.e. richness (number of species present), evenness (the distribution of individuals among species), and abundance (total number of plants present).

The plant diversity studies of the Gir forest (Table-8) is claimed to be the first of its kind. The diversity indices, Simpson's index (D=0.361): Simpson's index is a dominance measure as refer by Magurran (1988) and a probability measure as referred by Sai and Mishra (1986). The higher value at Gir forest could mean the high dominance of the occurring species. There were herbs with Pi^2 = 0.287 indicating that they were in higher number than all others. Herbs were followed by trees Pi^2= 0.036, trees were followed by shrubs with Pi^2= 0.021 and lastly climbers with Pi^2= 0.017. Index of similarity (0.639) showing high diversity. Reciprocal index was 2.77. Shannon-wiener index was 1.194 indicating low diversity. This is lower than the range 1.5 to 3.5 as mentioned by Margalef (1972). Finally Evenness index was 0.861 indicates that abundance of different species are similar in proportion to all species.

REFERENCES

Agrawal, K.B. (1984). Plant, Climate, Species diversity and concentration of dominance in some grazinglands of Garhwal Himalayas. *Geo-bios,* **11**: 168-173.

Bedi, S. J. (1968). Addition to the flora of Pavagadh pt. II. *J. Bom. Nat. His. Soc.* **62**: 2.

Bedi, S. J. (1968). Vegetational and floristic studies on the Ratanmahal and surrounding hills in Panchmahal District, Gujarat State. *Ph D Thesis, M S University Baroda.*

Bhandari, M. M. (1978). Flora of the Indian Desert. *Scientific Publishers, Jodhpur.*

Bhatt, R. R. (1969). A contribution to the vegetation and flora of Khedbrahma region in North Gujarat. *Bull. Bot. Surv. India* **2(3-4)**: 311-321.

Bhatt, R.R. and Sabnis, S.D. (1974). The forest vegetation and phytogeography of Khedbrahma of North Gujarat. *The Botanique* **6(1)**: 37-48.

Bhatt, D.C. and Imandar, J.A. (1971). Further addition to the flora of Saurashtra. Va*k.* 271-277.

Bharucha, F.R. and Dave, R.N. (1944). The biological spectrum of grassland association. *J. Univ. Bombay* **13**: 15-16.

Borgesen, F. (1929). Notes on the vegetation of Dwarka on the west coast of India. *J. Ind. Bot. Soc.* **8**: 1-18.

Chavan, A.R. (1961). Flora of Devgadh Hill, Devgadh Baria, Gujarat State. *J.M.S. Univ. Baroda.* **10(3)**: 43-57.

Cooke, T. (1901-1908). The flora of the Presidency of Bombay. *Vol.-I & II. London. (B S I reprinted 1958, Vol. I-III, Calcutta).*

Dansereau, P. (1957). Description and recording of vegetation upon a structural basis, *Ecology* **32**: 172-229.

Daubenmire, R. F. (1968). Plant Communities, A textbook of Plant Synecology. *Harper and Row, New York,* 300 pp.

Desai, M. J. (1992). New records for Gujarat flora from Bansda forest. *J Econ. Tax. Bot.* **16(3)**: 551-552.

Imandar, J.A. (1968). A prelimnary survey of the flora of Dharampur forest in Gujarat. *Bull. Bot. Surv. India,* **10**: 126-132.

Jain, S.K. (1963). The vegetation of Dangs district in Gujarat. *Bull. Bot. Surv. India,* **5**: 351-361.

Magurran E Anne (1988). Ecological diversity and its Measurements. *Princton University Press,* 41 Williams street, Princeton, New Jersey.

Maheshwari, T.K. (1963). The flora of Delhi, *C S I R, New Delhi.*

Margalef, R. (1972). Homage to Evelyn Hutchinson or why is there an upper limit to diversity. *Trans, Connect, Acad. Arst Sci* **44**: 211-235.

Meher-Homji, V.M. (1981). Environment implications of life-form spectra from India. *J. Econ. Tax. Bot.* **2**: 23-30.

Murthy, M.H.S. (1957). The vegetation of Bhavnagar and its biological spectrum. *Vidya. J. Guj. University,* **1**: 42-46.

Numata, J. and Asano, S. (1969). Biological flora of Japan, *Tokyo.*

Numata, M. (1954). Some special aspects of the structural analysis of plant communities. *J. Coll. Arts & Sci. Chidoa Univ.* **1**: 194-202.

Oza, G.M. (1962). The flora of Pavagadh, Gujarat State. *Ph D Thesis, M S University, Baroda.*

Oza, R.A. (1991). Taxonomical and Ecological study of the flora of and around Bhavnagar. *Ph D Thesis, Bhavnagar University, Bhavnagar.*

Pandya, S.C. (1964). Ecology of grass land of Sagar, M P. IIA. Composition of the fenced grassland association. *J. Indian Bot. Soc.* **43**: 577-605.

Pandya, S.M, (1972). Flowering Plants of the Gujarat University campus and its surrounding (Ahmedabad). *Vidya. J. Guj. University* **15(2)**: 21-48.

Parabia, M.H. (1974). A contribution to the Cyperaceae of Gujarat. *Ph D Thesis, S P University, Vallabh Vidhyanagar.*

Patel, R.M. (1971). Flora of Balsar and Environs, Gujarat. *Ph D Thesis, S P University, Vallabh Vidhyanagar.*

Patel, B.P. (1982). Ecological survey of the reserved forest near Bhavnagar. *Ph D Thesis, Bhavnagar University, Bhavnagar.*

Raizada, M.B. and Santapau, H. (1955). Contribution to the flora of the Gir forest. *Indian For. Rec.* **4**: 105-170.

Rao, C.C. (1968). Biological spectrum of Karamnasa watershed flora (Var ansi; India) *In: Proc. Sym. Recent Advances in Tropical Ecology, Pt. II. Mishra R and Gopal B (eds.) I S T E, B H U, Varansi.*

Rao, R.A. (1941). List of some more common plants of the desert area. *Misc. Bull. I C A R,* 43.

Raunkiaer, C. (1934). The life forms of plants and statistical plant geography. *Oxford University Press.*

Sabnis, S.D. (1967). A study of flora and vegetation of Baroda and environs. Including an account of the Cyperaceae of Gujarat. *Ph D Thesis, M S University, Baroda.*

Sai, V.S. and Mishra, M. (1986). Comparison of some indices of species diversity in the estimation of the actual diversity in a tropical forest: A case study. *Trop. Ecol.* **27**: 195-201.

Santapau, H. and Janardhanan. K.P. (1966). The flora of Saurashtra. *Bull. Bot. Surv. India,* **8**: 1-58.

Santapau, H. (1953). Plants of Saurashtra, a preliminary list. *Saur. Res. Soc. Rajkot.*

Santapau, H. (1962). The flora of Saurashtra. *Part-I, Rajkot.*

Saxton, W.T. and Sedgwick, L.J. (1918). Plants of Northern Gujarat. *Rec. Bot. Surv. India,* **6**: 207-323.

Shah, G.L. (1978). Flora of Gujarat State, I & II. *S P University, Vallabh Vidhyanagar.*

Smith, A.P. (1973). Stratification in temperate and tropical forests. *Am. Nat.* **107**: 671-683.

Sutaria, R.N. (1958). A text book of Systematic Botany, *Khadayata Book Depot, Ahmedabad.*

Talbot, W.A. (1902). The trees, shrubs and woody climbers of the Bombay Presidency. Govt. Phootozinocagraphic Press, Poona.

Thaker, J.I. (1966). Flora of Barda, *Sastu Sahitya Press, Ahmedabad.*

Vaidya, B.S. and Vora, A.B. (1964). Some new records of angiosperms for Ahmedabad. *Vidya. J. Guj. Univ.* **7**: 193-196.

Vora, A.B. and george, V.C. (1987). The distribution of various life forms in the ground flora under different canopies of Panchmehal forests, Gujarat, *Ind J. For.* **10(3)**: 223-225.

Vora, U.A., Patel, B.P. and Patel, B.K. (1981). The vegetation of Ghogha mahal and its biological spectrum. *Geobios*, **8**: 211-214.

Yadav, S.R. (1979). A contribution to the floristic and phytosociology of some parts of south Gujarat. *Ph D Thesis, S p Univ. Vallabh Vidhyanagar.*

Yadav, P.S. and Singh, J. S. (1977). Grassland vegetation; its structure, function, utilization and management. Today and Tomorrow's. *Printer and Publishers, New delhi,* 132 pp.

SECTION TWO

WETLAND ECOLOGY

Advancing Frontiers of Ecological Researches in India, 247-257, 2006
Edited by A.K. Kandya & Asha Gupta
Published by Bishen Singh Mahendra Pal Singh, Dehra Dun, India

Conservation of Wetlands - Global And Indian Perspectives

R.S. Ambasht & Navin K. Ambasht*
Botany Department, Banaras Hindu University Varanasi
**Botany Deptt., Christ Church College, Kanpur*

INTRODUCTION

Wetlands, as the name suggests are moist, water saturated or water covered lands up to shallow depths. As we became aware about the ecological importance of wetlands, a number of definitions were given. Depending upon the user needs the definitions were modified from time to time. Smith (1996) has quoted Cowardin *et al's* (1979) definition as *"land where the water table is at, or above the land surface for long enough each year to promote formation of hydric soils and to support the growth of hydrophytes"*. Wetlands have also been regarded as the transitional zone between terrestrial and aquatic ecosystems. The U.S. Army Corps have a big wetland unit and their definition is. *"those areas that are inundated or saturated by surface or ground water at a frequency and duration sufficient to support, and under normal circumstances do support, a prevalence of vegetation typically adopted for life in saturated soil conditions."* They include swamps, marshes, bogs and similar areas". Both the natural and man-made depressions, filled with shallow water are included in wetlands. Wetland ecology got a global boost with the emergence of Ramsar convention and Ramsar Bureau (Switzerland). Scientists from around the world mainly interested in water fowls like ducks, flamingos, egrets, lapwings, cranes, etc. held the historic meeting at the Iranian city of Ramsar near Caspian sea coast. There the Ramsar convention was adopted on February 2, 1977 and to mark this historic event, the International Wetland Day is observed every year on 2nd February. There they came up with a comprehensive and compromise definition of wetlands and later published by Ramsar (1990). *"Areas of marsh, fen peatland or water whether natural or artificial, permanent or temporary, with water that is static or flowing, fresh, brackish or salt including areas of marine water, the depth of which at low tides does not exceed six meters"*. This definition has led ecologists to include all shallow lentic and lotic water covered areas including coastal mangroves riparian and river corridors, lakes, ponds, pools as wetlands. They no more represent only transitional zones. Ramsar convention is the strongest international force in the wetland conservation. In all about 118 countries are now signatory or contracting parties to have a global concern and control over important wetland sites. Those sites which are internationally recognised as of global importance from the view point of resources, usefulness, sites of water fowls, migratory birds for overwintering and breeding, fish and prawn resources, etc. are given the status of Ramsar site. India has lakes Chilika (1,14000 ha, Orissa), Keoladeo - Ghana (3000 ha, Bharatpur - Rajasthan), Sambhar (7000 ha, Rajasthan), Loktak (26000 ha, Manipur), Harike (4,100 ha, Punjab) and Wular (18,900 ha, Jammu & Kashmir) as Ramsar sites. Chilika is a unique lake of fresh and brackish water, known for rich production of fishes and prawns, as well as for large flocks of thousands and thousands of migratory birds-often hunted for their delicious meat. At Bharatpur, Keoladeo National park is a protected site known as amongst the richest breeding place for egrets, flamingos and cranes. Loktak is an extensive congregation of wetlands upon which the economy of Manipur state greatly depends. Formation of "Phundis" or floating islands is characteristic of this wetland.

Important wetlands and their Types

Mitsch *et al.* (1994) estimated 7 to 8 million km^2, as global area under wetlands. Even though the total area may not be very high and only about 6% of the land surface, yet their production is much higher exceeding 1000 gC m^2 (Mitsch et al 1994). The present day cover area under Wetlands has greatly reduced due to over exploitation, siltation, filling up for house construction, etc. The grass covered humid sites are known as *wet meadows* and *wet prairies* in which there are water logged pockets. Intermittently flooded shallow meadows of Mediterranean climate is known as *Vernal pools.* In the U.S. wetlands dominated by trees are called *Swamps* while in Europe swamps may also include forested fens and *Phragmite reeds.* In the USA shallow lakes adjacent to watercourse are called *Slough.* Salt laden wet grasslands are called *Salt Marsh,* the halophytic nature may be due to tides or other sources. Deep water enclosed or partially open on deltaic coast in Europe is called *Lagoon* while similar habitat in Australia is called *Billabong.* All such wetlands where there is accumulation of partially decomposed plant material is called *peat.* Such *peats* which have acidophilic mosses like *Sphagnum* and which have poor or no inflow-outflows are called *Bog.* But, when the peat receives inflows or drainage from surrounding areas, it is called *Fen*. Forested wetlands of Alnus and willows are called *Carr* particularly in Europe. *Moor* is synonymous with peatlands with rich accumulation of dead plant material. The Alaskan and Canadian peatlands are known as *Muskeg*. Shallow marsh like ponds are called *Potholes*. Partially submerged coastal halophyte tree stands are called *Mangroves* in the tropics. River bank with high water table is called *riparian* (rip = stream) wetlands. All stream side vegetation falls in this category.

Wetlands are divided into five classes by Mathews and Fung (1987). These are (1) Forested Bogs, (2) Non- forested Bogs, (3) Forested Swamps, (4) Non – forested Swamps and (5) Alluvial formations. Almost 80 – 90% of the world's wetlands are located in boreal region of northern latitudes between 50 – 70 ° N, while the rest are in tropical and subtropical region (Mathews, 1990). Ramsar Bureau and the World Conservation Union (IUCN) have published a number of volumes on wetlands, e.g. (1) Wetlands of Africa, (2) Wetlands of Asia and Oceania, (3) Wetlands of Europe, (4) Wetlands of Neotropics and North America etc. McComb and Lake (1990) have published an inventory of Australian wetlands. Ambasht and Ambasht (1998) have given a comprehensive account of Indian wetlands and Ambasht (1998) on Global water and wetland Resources. Mitsch (1994) has edited Global Wetlands volume based on INTECOL wetland conference held at Columbus, Ohio, USA.

Coastal wetlands and brackish back water lakes are of much global importance. The Mangroves are home of extensive halophytic vegetation and several wild animals including the Royal Bengal Tiger, crocodiles, turtles, fishes and shrimps. Forests produce numerous medicines and the wood is used to produce charcoal.

Fresh water wetlands are broadly classified into (1) free flowing *Lotic* like rivers, streams, canals, rivulets, etc. and (2) *Lentic* i.e. surrounded by land from all sides giving it a non-flowing stationary and calm water characteristics. These are lakes, ponds, puddles, constructed or impounded dams, etc. All lakes are collectively known as *Lacustrine* as against flowing bodies *Riverine* and marshy saturated but no free overlying water as *Palustrine*.

Among the largest wetland sites on global scale are the European and west Siberian low lands spread over 780,000 km^2 plus peats and bogs over 220,000 km^2. The flood-plains of Amazonia in the S. America are distributed over 800,000 km^2. The *Pantanals* or marshes of South America cover 140, 000 km^2. Hudson Bay peatlands extend over 320,000 km^2 area. Flood plains of Mississippi and swamps and sudds of Nile river are also spread over large areas. The magnificent deltas of Ganga, Indus and Brahmputra rivers are greatest deltaic wetlands of the world. Paddy fields, which are essentially a kind of man managed wetland are responsible for largest share of agro-ecosystem in terms of the area of cultivation and food production. Man made wetlands are classified into five types by Dugan (1990) into (1) aquaculture ponds, (2) agriculture farm ponds, tanks, rice fields, ditches and canals, (3) salt pans and salines, (4) mine created pools, urban and industrial excavations, sewage farms, settling tanks and oxidation basins and (5) hydrodams and other water storage for human consumption.

River discharges have converted huge area of land into wetlands. In tropics, data compiled by Serruya and Pollingher (1983) show that in the volume of water discharge, Amazon tops with 6575 km^3, Ganga with Brahmputra and Meghna 1250 km^3. They have recorded water volumes in lakes and reservoirs, 80,000 x 10^6xm^3 for Bhakra and 2000 x $10^6 m^3$ for Chilika in India. African wetlands are much bigger like Tanganyika with 23,100,000 x 10^6 x m^3; Victoria 2,700,000 x 10^6 m^3; Kariba and Nasser each with 157,000 x 10^6 x m^3. Some of the large Indian wetlands beside the six Ramsar sites mentioned earlier are Koleru (90,000 ha), Sarsailum (6,17,000 ha) and Nagarjun Sagar (2,84,000 ha) in Andhra Pradesh. Others are Kaziranga (28,500 ha), Kolar (6,700 ha) Nalsarovar (18,400 ha) and Rann of Kutch (1,00,000 ha).

Ecological Services and Values of Wetlands

Wetlands are not just storage places of water to be exploited by mankind in periods of dryness, but they perform a number of ecological functions in the proper maintenance of ecological balance, quality of water and the hydrological cycle.

(1) Wetlands are hot-spots of biodiversity because they combine the best of terrestrial and aquatic habitats. The number of species in these ecotones is higher than the adjacent ecosystems. Higher diversity and better overall conditions of environment and resources, there are greater interactions amongst populations. Wetlands are, therefore, rich spots of formation of new ecophenes, ecotypes and species. They have greater role as potential centres of origin of species and evolutionary processes.

(2) Wetlands, especially the river-sides and lake margins have been the centre of human settlement and human culture. River courses have played important roles in transports across different places by boats and barges, when roads were not so developed and forests were extensive and dense. Riparian vegetation in their natural state of luxuriant growth acted as effective binder of soil and filter out wastes, excessive nutrients, pollutants, etc. from reaching the main body of water. The ecological integrity of wetlands for water quality and rich crop of fishes, are maintained by embankment vegetation.

(3) Wetlands are the natural homes of wild life, both plants and animals which are currently threatened towards extinction. Large variety of wild relatives of cultivated crops especially rice do exist in India. They are necessary 'gene pool' for present and future use in biotechnological transfer of desired traits in cultivated high yielding varieties. Successful transfer of disease resistant gene (of Virus stunt Disease) from the Indian wild *Oryza nivara* into cultivated *O. sativa* is responsible for higher rice production all over the world.

Many wild rice varieties which grow by themselves in wetlands are harvested by villagers and tribals all over the tropical world and eaten. These rice varieties gregariously grow and attract several migratory birds which feed on their grain as well as on small insects and fishes that thrive in such wetlands.

(4) Wetlands hold much of rain and runoff water and act as important checks against excess water flowing down to rivers and thus control floods. Dumping of wetland depressions by mankind in recent decades has added to floods in two ways: firstly by direct excess water inflow and secondly by carrying silt in the aquatic ecosystem allowing the bottom to rise or upwell. This leads to reduced water storage and carrying capacity causing more floods for the same water input by rain fall.

(5) Embankment vegetation increases water infiltration and therefore ground water recharge is an important function of wetlands. Excessive exploitation of ground water and poorer re-charge due to disappearance or shrinkage of area under wetlands has now posed a serious problem to man for deficiency of domestic and irrigation water. The problem is further compounded by the indiscriminate use of persistent pesticides which have now reached to ground water

storages. Wetlands play importance role in filtering out harmful chemicals and heavy metals from reaching the ground water segment.

(6) Wetlands are also the store-house of many medicinal plants. In the moist humid and saturated condition several very effective medicinal plants like *Bacopa* and *Hydrocotyl* grow. Mangroves - a special kind of coastal halophytic wetlands are also rich source of medicinal plants as also of fuel energy by way of wood and charcoal.

(7) Wetlands of larger size located in hilly terrain are used to release of water to hydroelectric installations, may be in cascades or steps. Such multipurpose hydroprojects are of great economic importance for (a) generation of electricity, (b) fish production, (c) irrigation water through net work of canals, (d) flood control and (e) ensuring water supply in drier seasons. Some countries of northern Europe and Canada generate most of their electric needs from the pollution free hydro-electric generating units. Norway obtains its 99% and Canada 91% of electricity from hydroelectric project.

(8) Wetlands all over the world are of much tourist interest for recreation, natural beauty, aquatic sports, boating, house boats and floating restaurants. Lakes in the valleys of outer ranges of Himalaya like in Kashmir, Himachal, U.P. and the North East India are real big tourist places visited by millions of Indian and foreign tourists. Earnings in summer tourist season is enough to sustain local people round the year. Eco-tourism has become an important industry, where besides the care and comfort of visitors, special care is taken to maintain the quality o lake environment and biotic resources. Anti-pollution steps are effectively taken. Many rivers have much religious importance and pilgrims visit them during bathing festivals. Kumbh mela congregation at certain places like Prayag, Haridwar and Ujjain at 6 and 12 years intervals attract millions of persons from all over the country. These are among the largest human gatherings at river sides in the world.

Wetland Conservation

Conservation term has its root in Latin words *'Con'* meaning together and *'server'* meaning to guard. *Conservator* was the title given to British officers stationed in India to safeguard the natural wealth such as forests, river and wetlands, wildlife and mineral wealth of the country. The term conservation has undergone very significant changes in its meaning over the past few decades. There was a sense of rationed or regulated allocation and use of a limited stock of resources.

For U.S. national policy, John F. Kennedy defined it as *"The wise use of natural environment: it is in the final analysis, the highest form of national thrift- the prevention of waste and despoilment while preserving, improving and renewing the quality of usefulness of all our natural resources"*. The emphasis is now on *wise use.* In fact no conservation strategy can be without the element of 'use' otherwise the resource is bound to be lost. *"Use it or Lose it"* is the maxim. In modern conservation strategy the use has to be based on *public participation* in sharing the benefits and losses, responsibilities and accountabilities. Community benefits and long range sustainability are key factors. In fact the current conservation effort as pleaded by UNEP (1992) is based on *"shift from defensive posture of protecting nature from impacts of development to an offensive effort seeking to meet peoples needs for biological resources while ensuring long-term sustainability of earth's biotic wealth"*.

Keeping in view the long term sustainability of wetlands for their ecological integrity, resource regeneration, pollution free water and aquatic plant and animal products i.e. without biologically magnified heavy metals, pesticides and toxic chemicals, the management plans are to be drawn. Therefore, the stress and degrading forces have to be identified in drawing conservation plans. Wetlands are regarded as *'kidney'* of the ecosystems that filters out wastes and harmful material. Forests are regarded as 'lungs' that oxygenate the system while acting as the CO_2 sink. Wetlands with their macrophytes and phyto-plankton also contribute enough to 'lungs' function. We have already listed important wetlands values and functions. In context of the conservation aspects we briefly discuss here for riverine (lotic) and lacustrine (lentic) wetlands.

The term riparian was rather unknown just a few decades ago, but due to sudden recognition of the value of this controversial parcel of land, riparian habitats have now drawn world attention. It is not just an ecotone or habitat. Lowrance et al (1985) regard it as a complex assemblage of organisms and their environment existing adjacent to flowing water. Rise and fall in seasonal flood and drought phase bring out high degree of connectedness and isolation periodically. They are sites of pulses of floods, exposure, soil erosion, silt deposition, migration and establishment of diaspores of plants and animals brought by river water from great distances. Vegetal cover is the key ecological component that plays critical role in the conservation of riparian as well as lentic wetlands. Ambasht (1958, 1970) provided a new concept of conservation value in terms of percentage of soil, water or different nutrients for embankment plant populations and communities and extended the work to watershed hilly terrains (Ambasht, 1985). He has also provided suitable field and culture experimental methodology to quantify conservation. *Cynodon dactylon,* a common grass is found to conserve 95% to 97% of soil on erosion from sloping topography against natural and simulated rainfalls. Its roots alone bind 77% soil and the rest is due to cushioning effect of shoots and mulch or over lying dead organic matter. *Cyperus rotundus* a sedge weed readily convering moist lake and river embarkment conserves soil between 82 to 93%. *Saccharum benghalense* and *S. spontaneum* are grown on erosion prone riparian lands and through their tall tussock forming nature conserve up to 95% soil. In winter and summer the shoot is harvested for thatching of huts. They survive burning events and with new vigorous growth of fresh shoots after the fire the productivity rises. Graminae and Cyperaceae members are also efficient water conservers as they increase infiltration and retard lateral runoff. Water conservation is above 70%. Some dicotylendon weeds like mat forming *Phyla nodiflora* on river banks conserve soil (94%) and water (73%) efficiently (Ambasht *et al* 1984). *Digitaria adscendens* is found to have soil conservation potential of 87% and water 65% on habitat disturbance in rather polluted industrial region along Rihand River, Ambasht with his scholars established the conservation efficiency for nitrogen and phosphorus as well, for five weeds - *Leonotis nepetifolia, Cassia tora, Ageraturm conyzoides, Parthenium, hysterophorus* and *sida acuta* (Kumar *et al* (1992 a, b). Simulated rainfall with sufficiently high kinetic energy generation was repeatedly done. Most of the effective causative features in favour of conservation such as canopy cover, litter mass, antecedent soil moisture, standing crop biomass, density of fine root (tips) were measured and through M-Stat. programme, their relative values in overall conservation have been established.

Leonotis was most efficient and *Sida acuta* least. Water is found to carry more of phosphorus than the eroded soil does. Details are given by Kumar, Ambasht and others (1996 and 1997).

Lazarus (1990) has estimated that 25,000 million tons of soil are bein washed away each year on world-wide scale. This colossal loss is largel due to wetland losses and over-exploitation of wetland embankme vegetation. For India about Rs.700 million worth of soil loss through erosion was estimated in 1982 (Anonymous, 1982) and it must have increased three to four folds due to actual doubling of soil loss and fall in rupee value. The kidney function of stream side vegetation is reflected in Verry and Timmons (1982) study who found 36 to 60% of nutrient flowing down to streams is filtered out by the vegetation. Peter John and Correll (1984) have found 89% nitrogen being taken by riparian forest plants in Maryland, USA. Kadlec and Kadlec (1979) regard wetland margins as nature's treatment systems against pollution. Ambasht and Srivastava (1994) have found that in heavy metal polluted industrial belt, *Polygonum amphibium* on banks of Rihand River at Obra super thermal power plant accumulate iron up to 1875 µg/g. Use of this species in harvesting out iron and transferring to iron poor agroecosystem is doubly useful. There is always a struggle between adequate plant cover and erosional forces and man unfortunately in his greed or ignorance has sided more with erosional forces. The trend has to be arrested and reversed before the battle of wetland and riverside erosion is lost. There are new models being developed for wetland management. *Phragmites australis* is a very good soil conserver in River Murray embankments in Australia (Frankenberg, 1989).

Besides the problems of siltation, erosion, nutrient enrichment and eutrophication, biological magnification of persistent chemicals, wetlands are experiencing dumping of solid wastes, industrial and municipal effluents, etc. For these Ambasht and Ambasht (2005) have given detailed methodologies with treatment techniques and purification plants.

ACKNOWLEDGEMENTS

RSA is thankful to the Indian National Science Academy for the award of Senior Scientist scheme and to the Banaras Hindu University for conferring Professor Emeritusship.

REFERENCES

Ambasht, R.S. (1958). Root habits in response to erosion, silting and inundation. Proc. Nat. Acad. Sciences. 38:101-105.

Ambasht, R.S. (1970). Conservation of soil through plant cover of certain alluvial slopes in India. *Proceedings of the IUCN XI Technical meeting (pp.44-48). Gland. Switzerland.*

Ambasht, R.S. (1985). Primary productivity and soil and nutrient stability in Indian hilly savanna lands. In J.C. Tothill & J.J. Mott (Eds.), *Ecology and management of world savannas.* Australian Academy of Sciences. Canberra, Pp. 217-219.

Ambasht, R.S. (1998). World water and wetland resources. In R.S. Ambasht (Ed.), *Modern trends in Ecology and environment*. Backhys, The Netherlands , pp. 115-130.

Ambasht, R.S. & Ambasht, N.K. (1998). Ecology of Indian Wetlands. In S.K. Majumdar, E.W. Miller, & F.J. Brenner (Eds.), Ecology of wetlands and associated systems , Pennsylvanis Academy of Sciences. pp. 104 -116.

Ambasht, R.S. & Ambasht, P. K. (2004). Environment and pollution. 4 edn. CBS. Publ., New Delhi.

Ambasht, R.S., Singh, M.P. & Sharma, E. (1984). Soil, water and nutrient conservation of certain riparian berbs. *Journal of Environmental Management,* **18**: 99-104.

Ambasht, R.S. & Shrivastava, N.K. (1994). Restoration strategies for the degrading Rihand River and reservoir ecosystems in India. In W.J. Mitsch (Ed.). *Global wetlands: old world and New, Elsevier, Amsterdam,* pp.483-492.

Anonymous (1982). The state of India's environment – a citizen's report. Centre for science and environment. Ambassador Press, New Delhi.

Cowardin, L. M., Carter, V., Golet, F. C. and La Roc (1979). Classification of wetlands and deep-water habitats of the United States. US Fish and wildlife service, Washington, D.C.

Dugan, P. J. (1990). Wetland conservation - a review of current issues and and required action. IUCN, Gland, Switzerland.

Frankenberg, J. (1989). The role of gegetation in protecting river bank erosion. In – The MDFRC Annual Report (1988 – 89) , Albury, Australia, pp. 10-13.

Kadlec, R.H. and J. A. Kadlec (1979). Wetland and water quality. In P. E. Greeson, J.R. Clark and J.E. Clark (Eds.) Weland functions and values : The state of our understanding. American water resource association, Minneapolis, pp. 436-456.

Kumar, R., R. S. Ambasht and N.K. Shrivastava (1992a). Conservation efficiency of five common riparian weeds in movement of soil, water and phophrus. J. Appl. Ecol. **29** : 737-744.

Kumar, R., R.S. Ambasht and N.K. Shrivastava (1992b). Nitrogen conservation efficiency of five common riparian weeds in a run-off experiment on slopes. J. Environ. Management. **34**: 47-57.

Kumar, R., R.S. Ambasht, A.K. Shrivastava and N. K. Shrivastava (1996). The role of some riparian wetland plants in reducing erosion of organic C and selected cations. Ecological Engineering, **6**: 227-239.

Kumar. R., R.S. Ambasht, A.K. Shrivastava, N.K. Shrivastava and A. Sinha (1997). Reduction of nitroge losses through erosion by *Leonotis nepataefolia and Sida acuta* in simulated rain intensities. *Ecological Engineering,* **8:** 233-239.

Lazarus, D.S. (1990). Save our soil. *Our Planet,* **2**(4),:10-11.

Lowrance, R., Leonard, R. & Sheridan, J. (1985). Managing riparian ecosystem to control nonpoint pollution. *Journal of Soil Water Conservation*, **40** (1) : 87-91.

McComb, A.J. & Lake, P.S. (1990). Australian Wetlands NSW Collins. Australia pp.258.

Mathews, E. (1990). Global data bases for evaluating trace gas sources and sinks. In A. F. Browman (Ed.). Soils and greenhouse effects. John Wiley & Sons. Chichester pp. 311-325.

Mathews, E. & Fung, I. (1987). Methane emission from natural wetlands : global distribution, area and environmental characteristics of sources. Global Biogeochem. Cycles. **5** : 3-24.

Mitsch, W.J. (Ed.) (1994). Global Wetlands. Elsevier : Amsterdam. pp. 967.

Mitsch, W.J., Mitsch, R.H. & Turner, R.E. (1994). Wetlands of the old and new Worlds. In W. J. Mitsch (Ed.). Global Wetlands. Elsevier : Amsterdam pp. 3-56.

Peterjohn, W. T., & Correll, D. L. (1984). Nutrient dynamics in agricultural watershed : observation on the role of a riparian forest. *Ecology*, **65:** 1466-1475.

Ramsar. (1990). *Proceeding of the fourth meeting of contracting parties.* Gland. Switzerland.

Serruya, C. & U. Pollingher (1983). Lakes of Warm Belt. pp. 569. Cambridge Univ. Press, London.

Smith, R.L. (1996). Ecology and Field Biology, (5th Ed.), Harper Collins, New York.

UNEP (1992). The world environment 1972-1992, Chapman and Hall, London.

Verry, E.S. and D.R. Timmons (1982). Water borne nutrient flow through an upland peatland watershed in Minnesota. Ecology **63** : 1465-1467.

Advancing Frontiers of Ecological Researches in India, 259-270, 2006
Edited by A.K. Kandya & Asha Gupta
Published by Bishen Singh Mahendra Pal Singh, Dehra Dun, India

Productive Status of the Macrophytes in Ikop, Lake, Manipur

Ch. Nivanonee and B. Manihar Sharma
Dept. of Life Sciences, Manipur University Canchipur -795003, Manipur, India.

Abstract: The Lake Ikop is a freshwater, saucer-shaped eutrophic natural lake. It is situated in the South-Eastern part of Imphal at a distance of about 40 km. with an area of 13.5 sq. km in the Thoubal District, Manipur. It lies at an altitude of 772 MSL and at an intersection of $93^0 52'$ to $94^0 05'$ E longitude and $24^0 31'$ to $24^0 40'$ N latitude. Net primary productivity was estimated for 19 macrophytes which were found to occur dominantly during Oct. 1998 - Sept. 2000. Maximum monthly net primary productivity were exhibited by *Azolla pinnata, Echinochloa colonum, Jussiaea repens, Potamogeton crispus* etc. in the different study sites. The total (all species combined) monthly net primary productivity varied from 2.18 to 153.62 gm^{-2} $month^{-1}$ during the first year of study and 2.06 to 144.65 gm^{-2} $month^{-1}$ during the second year of study. The Annual net primary productivity varied from 2.28 gm^{-2} yr^{-1} to 133.04 gm^{-2} yr^{-1} (*Eichhornia crassipes*) in site IV during the first year of study and 2.07 gm^{-2} yr^{-1} to 137.13 gm^{-2} yr^{-1} during the second year of study.

INTRODUCTION

Primary productivity is the basis of all the dynamic functions of the biosphere. It involves trapping of radiant energy of the sun and its transformation into high potential biochemical energy through the process of photosynthesis. Net primary production is the most important biological phenomenon in nature on which the entire diverse array of life depends either directly or indirectly. Odum (1971) defined net primary productivity as the role of storage of organic matter in the tissues, in excess of the respiratory utilization by the plants during the period of the measurement. It may also be defined as the net rate of gain of carbon by the vegetation over a given time interval (Long and Hutchinson 1991). Due to the increasing demands on the aquatic resources, ecologists have been attempting to understand the status of primary productivity of the fresh water ecosystems. Macrophytes are the important constituents for the regulation of the cycling of minerals and organic compounds in natural aquatic systems and in establishing trophic level of the food chain.

The determination of the plant biomass (Organic matter content of the plants) has become one of the important measures towards the assessment of the Primary Productivity of the plant communities. Net production has been estimated by adding all the positive values of the differences in biomass values when values for successive months are compared. The values of net primary productivity are assessed at both the individual species level and community level. The values are calculated on the daily basis ($gm^{-2}day^{1}$) at monthly intervals as well as on the annual basis ($gm^{-2}\ yr^{-1}$).

Due to fast depletion of vegetational cover throughout the world, Leith (1975, 1992) computed less productivity of the biosphere (101.8×10^9 t yr^{-1}) in the early 1990's. In spite of all these inconsistencies and rapid destructions of World vegetations, Mackenzie *et. al* (2001) holds that the appropriate current estimate of the net primary productivity of the terrestrial and marine biomes of the world stands at 120×10^9 and $50 - 60 \times 10^9$ t yr^{-1} respectively. Studies on the primary production of the macrophytes of the freshwater bodies have been made only during the past decade and were inspired by the International Biological Programme (IBP) of the UNESCO.

In the North-Eastern region of India which is one of the 25 "Biodiversity Hotspots" of the world (Myers *et al*, 2000), the bioresources are found to be much degraded. The situation is more serious in the North-East especially in Manipur as most of the few existing lakes including Loktak, (a Ramsar site) Waithou, Ikop, Utrapat, Kharungpat, Poiroupat etc. have been seriously degraded due to eutrophication (Sharma, 1999).

The present study has been undertaken with an attempt to share our knowledge of the wetlands of India by assessing the magnitude of biomass and the net primary productivity of the different macrophytes of the Ikop Lake, Manipur which lies at an intersection of $93^0 52'$ to $94^0 05'$ E longitude and $24^0 31'$ to $24^o 40'$N latitude, at an elevation of 722 MSL. The morphometric and bathymetric characters of Ikop lake are presented in Table-1.

MATERIALS AND METHODS

Plants were sampled on monthly basis during the tenure of October 1998 to September 2000. The samples were collected from the four different study sites I -Tarapan, II-Tarahumdoipan, III-Ganjapan, IV-Patchao, in triplicates using 25cm x 25cm square quadrat (Fig. 1). The samples were

Table-1: Morphometry of Ikop lake, Manipur

Maximum length (km)	7.5
Maxium breadth (km)	1.8
Surface area (sq.km)	13.5
Maximum depth (m)	1.58
Minimum depth (m)	0.27
Mean depth (m)	0.93
Mean depth/maximum depth ratio	0.58
Basin shape	Saucer
Basin Slope	Gentle
Shore-line length (Km)	27.5
Shore-line development ratio	2.12
Volume of lake (km^3)	0.013
Surface area/volume ratio	1038.5
Development of Volume	1.75

kept in polythene bags marked with waxpencil and brought to the laboratory where plant materials were washed to remove the adhering silt, soil, marl, other plant and animal debris and were sorted out into different species. Excess of water was drained using blotting papers. Fresh weights of the different plant species were taken after which the labelled samples were packed in paper bags and kept in the oven at 80° C for 48 hrs. Later on dry weights were taken. Biomass were assessed on the dry weight basis and the values have been expressed in gm^{-2}. Standing crop biomass was measured by Harvest method (Odum 1956) using Ekman Dredge (Misra 1968, Ambasht 1970). Biomass values were used for the calculation of the primary production. Net production was estimated by adding all the positive values of differences in biomass when values for successive months were compared. The values of the net primary productivity were assessed both at the individual species level as well as at the community level. The values were calculated on the daily basis (gm^{-2} day^{-1}). The annual net primary productivity were also calculated for the total species (combined).

RESULTS

Biomass and Net Primary productivity was estimated for nineteen (19) different macrophytes. The values varied considerably throughout the study period as expressed in the Table-2, 3, 4, 5, 6.

Table-2: Variation in the values of biomass (gm^2) of the macrophytes in the Ikoplake, Manipur, during first year.

	SPECIES	SITES			
1.	Alternanthera philoxeroides	2.37-6.16	1.09-3.10	2.24-8.39	4.69-8.65
2.	Azolla pinnata R.Br.	1.30-27.20	3.22-55.32	1.64-110.38	1.20-32.783
3.	Ceratophyllum demersum Linn.	1.54-62.55	1.12-26.75	3.64-87.69	1.42-95.13
4.	Echinochloa colonum (Linn.) Link	2.77-6.72	2.80-43.451	1.89-41.21	4.79-9.91
5.	Eichhornia crassipes (Mart.) Solms.	3.66-6.41	4.78-14.02	2.86-14.28	3.98-77.05
6.	Enhydra fluctuans Lour.	-	3.81-33.32	1.73-12.09	3.64-56.59
7.	Hydrilla verticillata (Linn.f.) Royal	1.30-114.68	1.44-235.40	1.40-97/47	1.47-35.58
8.	Hygrorhyza aristata (Retz.) Nees	1.71-9.79	3.83-8.58	7.65-13.44	1.28-41.45
9.	Jussiaea repens Linn.	3.86-38.48	2.35-27.18	3.56-35.78	1.79-6.95
10.	Ipomoea aquatica Forsk.	1.33-4.05	-	-	-
11.	Limnophilla heterophylla Benth.	-	1.85-4.24	-	-
12.	Nymphoides indicum (Linn.) Kuntz	-	3.08-14.15	-	5.71-38.61
13.	Potamogeton crispus Linn.	6.16-51.58	2.24-45.72	3.651-13.12	4.06-38.25
14.	Salvinia cucullata Roxb.	1.29-12.32	2.35-15.27	4.92-32.26	2.42-21.75
15.	Salvinia natans Hoffm.	-	-	1.97-6.44	1.83-5.58
16.	Scirpus supinus Linn.	30.81-33.37	15.34-35.56	18.48-43.54	2.18-17.28
17.	Trapa natans Linn.	1.01-3.01	2.84-20.88	1.08-7.02	1.75-12.90
18.	Ultricularia flexuosa Vahl.	4.20-8.06	1.40-8.48	-	-
19.	Vallisneria spiralis Linn.	2.39-5.96	2.45-13.62	-	-
	All species (Combined)	29.96-205.93	44.11-295.62	43.35-236.36	42.47-194.73

In the present study, during the first year of study, the maximum values of monthly primary productivity was contributed by *Azolla pinnata* in Site -III (99.66 gm^{-2} $month^{-1}$) which was successively followed by 77.05 gm^{-2} in *Eichhornia crassipes and Azolla pinnata* (44.83gm^{-2}). The least peak value was reported in *Potamogeton crispus* (42.05gm^{-2}) in Site-I. The daily net primary productivity values of the different species varied from 0.06 gm^{-2} day^{-1} (May in site-I) to 4.93gm^{-2} day^{-1} (March in Site-III). The values of annual net primary production were also comparatively analysed and the highest value was observed in *Eichhornia crassipes* (133.04 gm^{-2} yr^{-1} Site-IV) successively followed by *Ceratophyllum demersum* (119.48 gm^{-2} yr^{-1} Site-IV), *Azolla pinnata* (188.22 gm^{-2} yr^{-1} Site-III). The least annual net primary production was recorded in *Utricularia flexuosa* (2.28 gm^{-2} yr^{-1}). The percentage contribution varied from 0.42% (*Utricularia flexuosa*) to 19.45% (*Eichhornia crassipes)* in all the study sites during the first year of study.

Table-3: Variation in the values of Biomass (gm^{-2}) of the macrophytes in the Ikop lake, Manipur during second year

SPECIES	SITES			
	I.	II.	III.	IV.
1. *Alternanthera philoxeroides*	1.85-4.91	5.76-10.52	2.41-7.55	2.08-11.41
2. *Azolla pinnata*	1.30-3.86	1.35-2.07	5.88-39.36	0.58-2.545
3. *Ceratophyllum demersum*	1.10-107.99	2.45-41.88	1.34-37.52	2.91-43.83
4. *Echinochloa colonum*	2.66-7.44	2.52-12.19	4.88-12,52	2,25-7.36
5. *Eichhornia crassipes*	2.66-15.28	6.37-22.69	6.98-28.74	3.59-11.67
6. *Enhydra fluctuans*	1.99-4.75	3.80-15.86	9.31-16.79	5.32-16.91
7. *Hydrilla verticillata*	1.69-88.83	1.12-67.74	0.98-40.02	0.79-44.96
8. *Hygrorhyza aristata*	2.10-11.80	1.77-14.08	2.41-15.19	3.59-15.47
9. *Jussiaea repens*	3.37-5.32	-	-	-
10. *Ipomoea aquatica*	1.84-9.38	3.22-19.69	2.39-13.04	3.47-12.34
11. *Limnophila hetrophylla*	-	3.26	-	-
12. *Nymphoides indicum*	-	6.91-11.17	6.23-30.332	2.10-24.81
13. *Potamogeton crispus*	0.88-51.58	8.14-22.79	2.30-29.76	2.10-40.01
14. *Salvinia cucullata*	1,72-6.48	2.80-13.18	0.05-16.59	2.05-25.73
15. *Salvinia natans*	-	1.26-4.24	3.82	-
16. *Scirpus supinus*	3.64	3.22/5.51	4.48-9.13	5.58-7.80
17. *Trapa natans*	1.68-12.11	4.90-35.71	1.76-8.06	3.50-29.07
All Sps. (Combined)	16.81-64	29.03-154.40	64.34-107.84	53.0-159-21

During the second year of study the values of daily net primary productivity values of all the species (combined) varied from 0.06 gm^{-2} day^{-1} to 4.80 gm^{-2} day^{-1}. The values of annual net primary productivity varied from 2.07 gm^{-2} yr^{-1}(*Azolla pinnata*) to 137.13gm^{-2} yr^{-1} (*Hydrilla verticillata*).

On comparative analysis of the two consecutive years of study the values of monthly net primary productivity was found to be higher in the first year than in the second year. Similarly the values of the daily net primary productivity of all the species (combined) were also found to be higher in the first year of study. However the values of annual net primary productivity were found to be higher in the second year of study with the peak value of 139.13gm^{-2} yr^{-1} (*Hydrilla verticillata*). The variations in the daily net productions in all the sites during two consecutive years of study are set in Fig.2 and 3.

Table-4: Variation in the monthly values of net primary productivity (gm^{-2} $month^{-1}$) (Daily NPP Values in gm^{-2} day^{-1} are presented in parenthesis) during first year

SPECIES	SITES			
	I	II	III	IV
1. *Alternanthera philoxeroides*	0.18-3.01 (0.00-0.08)	0.42-2.68 (0.01-0.08)	0.49-4.63 (0.01-0.14)	5.05-8.65 (0.16-0.28)
2. *Azolla pinnata*	1.30-22.49 0.04-0.72	10.49-44.83 0.34-1.44	1.46-99.66 0.04-3.21	0.25-25.22 0.00-0.82
3. *Ceratophyllum demersum*	0.01-7.79 0.00-0.25	3.19-16.02 0.10-0.51	1.04-3.65 0.03-0.12	1.24-74.92 0.04-2.49
4. *Echinochloa colonum*	7.78-26.80 (0.25-0.89)	0.40-38.27 (0.01-1.23)	1.37-36.66 (0.04-1.88)	1.10-7.47 (0.03-0.24)
5. *Eichhornia crassipes*	1.69-5.18 (0.05-0.17)	6.93-9.54 (0.22-0.31)	0.22-13.65 (0.00-0.45)	11.91-77.05 (0.38-2.48)
6. *Enhydra fluctuans*	-	0.03-33.32 (0.00-1.07)	10.01-10.71 (0.32-0.35)	5.13-46.59 (0.17-1.50)
7. *Hydrilla verticillata*	0.50-6.39 (0.01-0.21)	0.33-10.78 (0.01-0.35)	1.00-5.52 (0.30-0.18)	0.81-5.13 (0.20-0.17)
8. *Hygrorhyza aristata*	0.48-5.03 (0.01-0.17)	6.46-8.58 (0.21-0.29)	1.26-9.82 (0.04-0.32)	1.61-27.61 (0.05.0.92)
9. *Jussiaea repens*	1.39-38.48 (0.04-1.24)	2.38-23.26 (0.07-0.75)	0.69-30.12 (0.02-0.97)	1.02-3.68 (0.03-0.12)
10. *Ipomoea aquatica*	1.76-2.29 (0.05-0.07)	- -	- -	- -
11. *Limnophila hterophylla*	- -	1.01-4.24 (0.03-0.14)	- -	- -
12. *Nymphoides indicum*	-	8.58-12.13 (0.28-0.40)	-	12.55-33.51 (0.04-1.11)
13. *Potamogeton crispus*	22.06-42.05 (0.73-1.40)	2.24-41.41 (0.07-1.38)	0.78-9.27 (0.02-0.30)	0.45-20.21 (0.01-0.67)
14. *Salvinia cucullata*	0.15-8.00 (0.00-0.25)	0.66-5.54 (0.02-0.17)	2.41-9.26 (0.08-0.29)	0.20-6.30 (0.00-0.20)
15. *Salvinia natans*	-	-	1.97-5.20 (0.06-0.16)	0.10-3.43 (0.00-0.11)
16. *Scirpus supinus*	2.29-30.81 (0.09-0.99)	2.61-33.56 (0.08-1.08)	1.97-23.15 (0.06-0.74)	1.17-23.92 (0.03-0.77)
17. *Trapa natans*	1.41-1.60 (0.04-0.05)	0.17-17.40 (0.00-0.56)	1.97-5.01 (0.06-0.16)	0.18-12.90 (0.00-0.41)
18. *Utricularia flexuosa*	-	2.28 (0.07)	-	-
19. *Vallisneria spiralis*	2.79-3.57 (0.09-0.11)	-	-	-
All species (combined)	2.18-88.49 (0.06-2.81)	5.42-96.83 (0.16-3.11)	14.03-153.62 (0.44-4.98)	19.10-128.06 (0.61-4.25)

Table-5: Variations in the values of Net Primary Productivity (gm^{-2}month^{-1} during 2nd year) (Daily NPP values in gm^{-2} day^{-1} in parenthesis)

SPECIES	SITES			
	I.	II.	III.	IV.
1. *Alternanthera philoxeroides*	0.42-1.85 (0.01-0.06)	1.75-5.76 (0.05-0.18)	2.41-2.70 90.08-0.08)	3.76-7.65 (0.12-0.24)
2. *Azolla pinnata*	1.30-2.10 (0.04-0.07)	0.72-1.35 (0.02-0.04)	7.64-28.64 (0.24-0.92)	2.45 (0.07)
3. *Ceratophyllum demersum*	1.86-85.57 (0.06-2.85)	0.50-26.16 (0.01-0.87)	0.53-7.00 (0.53-7.00)	1.09-13.77 (0.02-0.45)
4. *Echinochloa colonum*	1.12-4.78 (0.03-0.15)	0.88-11.4 (0.02-0.35)	4.88-12.52 (0.15-0.40)	0.09-5.03 (0.00-0.16)
5. *Eichhornia crassipes*	15.28 (0.50)	0.11-12.04 (0.00-0.40)	5.50-23.24 (0.17-0.77)	1.42-6.78 (0.04-0.21)
6. *Enhydra fluctuans*	0.17-2.25 (0.00-0.07)	1.40-6.53 (0.04-0.21)	4.00-16.74 (0.12-0.55)	0.08-16.91 (0.00-0.54)
7. *Hydrilla verticillata*	2.74-61.68 (0.09-1.72)	0.35-50.94 (0.01-1.69)	5.10.-34.70 (0.17-1.15)	0.09-38.16 (0.00-1.27)
8. *Hygrorhyza aristata*	0.34-11.80 0.01-0.39	1.77-14.08 0.05-0.46	1.58-15.19 (0.05-0.50)	0.08-11.55 (0.00-0.38)
9. *Ipomoea aquatica*	2.00-3.37 (0.06-0.11)			
10. *Jussiaea repens*	0.42-5.58 (0.01-0.18)	0.46-14.96 (0.01-0.49)	0.83-8.79 (0.02-0.28)	1.24-8.82 (0.04-0.29)
11. *Limnophila heterophylla*	-	3.26 (0.10)	-	-
12. *Nymphoides indicum*	-	0.20-11.17 (0.00-0.37)	3.71-20.24 (0.11-0.65)	3.08-8.02 (0.09-0.260
13. *Potamogeton crispus*	37.21-43.61 (1.24-1.45)	0.57-9.85 (0.01-0.31)	3.17-24.29 (0.10-0.88)	3.17-16.72 (0.10-0.53)
14. *Salvinia cucullata*	0.23-4.74 (0.00-0.14)	0.30-13.18 (0.01-0.43)	1.03-7.52 (0.03-0.24)	0.41-19.74 (0.01-0.63)
15. *Salvinia natans*	-	2.38-2.98 (0.07-0.09)	0.38 (0.12)	
16. *Scirpus supinus*	3.64 (0.12)	0.93-4.24 (0.03-0.13)	4.48-4.65 (0.14-0.15)	2.22-5.58 (0.07-0.18)
17. *Trapa natans*	0.33-6.49 (0.01-0.21)	4.90-30.81 (0.16-0.99)	1.18-2.66 (0.03-0.08)	0.68-12.95 (0.12-0.41)
All species (combined)	2.00.144.65 (0.07-4.80)	9.28-93.58 (0.25-3.08)	33.12-58.27 (1.04-1.90)	17.6192.78 (0.56-3.01)

DISCUSSION

Daily Net Primary Productivity

In the present study the maximum net primary productivity during the two years of study was contributed by *Eichhornia crassipes* with values

Table-6: Annual net primary productivity of the Macrophytes of Ikop lake ($gm^{-2}yr^{-1}$)

	First Year				Second Year			
	S.I	S.II	S.III	S.IV	S.I	S.II	S.III	S.IV
1. *Alternanthera philoxeroides*	8.23	3.10	9.92	13.70	6.40	10.73	7.55	16.21
2. *Azolla pinnata*	29.72	55.32	118.22	39.68	5.16	2.07	47.00	2.45
3. *Ceratophyllum demersum*	19.53	39.21	13.47	119.48	102.29	61.04	38.39	22.13
4. *Echinochloa colonum*	43.42	61.41	52.09	27.54	10.32	19.89	22.55	15.52
5. *Eichhornia crasssipes*	9.62	25.71	32.94	133.04	15.28	22.22	68.00	23.83
6. *Enhydra fluctuans*	-	50.68	27.32	77.38	6.07	22.53	30.10	-
7. *Hydrilla verticillata*	15.26	36.26	16.94	14.47	137.13	103.94	60.04	70.04
8. *Hygrorhyza aristata*	22.29	15.04	24.56	56.08	16.34	20.77	22.86	30.14
9. *Jussiaea repens*	18.73	63.16	51.97	14.36	5.37	-	-	-
10. *Ipomoea aquatica*	4.05	-	-	-	16.77	37.70	-	-
11. *Limnophila heterophylla*	-	5.25	-	-	-	3.26	21.47	25.02
12. *Nymphoides indicum*	-	20.71	-	69.16	-	11.37	47.65	42.65
13. *Potamogeton crispus*	64.11	52.50	15.05	38.70	80.82	26.35	35.11	43.18
14. *Salvinia Cucullata*	16.95	20.27	48.42	28.28	14.54	42.85	21.44	-
15. *Salvinia cucullata*	-	-	7.17	3.53	-	7.79	3.82	47.03
16. *Scirpus supinus*	33.80	55.86	43.54	35.36	3.64	6.92	9.13	7.80
17. *Trapa natans*	3.01	25.19	8.99	13.08	12.11	35.71	8.06	43.64
18. *Utricularia flexuosa*	-	2.28	-	-	-	-	-	-
19. *Vallisneria spiralis*	6.36	-	-	-	-	-	-	-
Total	355.07	531.95	470.60	683.84	433.24	435.14	443.17	406.0

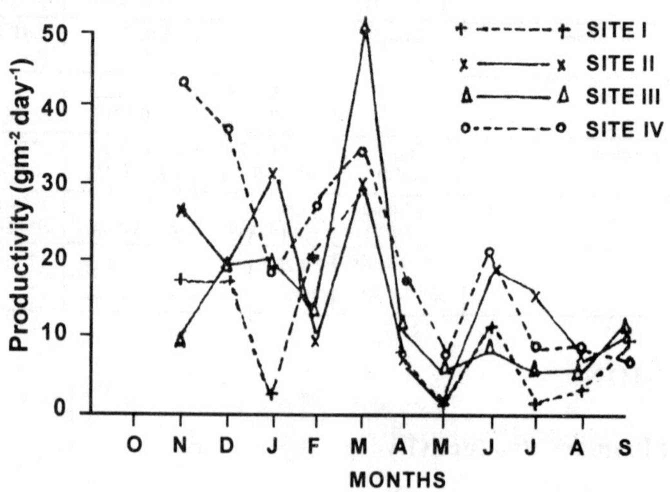

Fig. 2. Variations in Daily Productions (1998-1999)

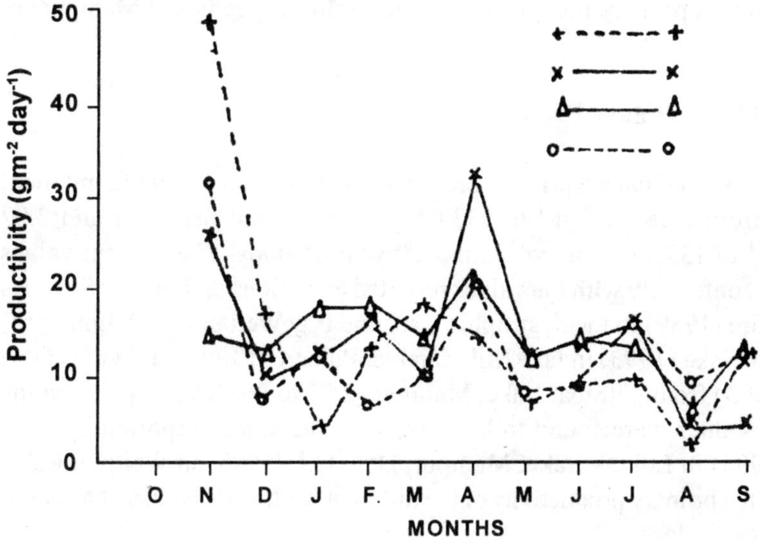

Fig. 3. Variations in Daily Productions (1999-2000)

varying from 0.00 to 2.48 gm^{-2} day^{-1} to 0.06 to 0.77gm^{-2} day^{-1} in the first years of study. The present findings were found to be in conformity with the values reported by Doast and Childers 1999 (0.17 to 1.61 gm^{-2} day^{-1}).

The values of daily net primary productivity in *Hydrilla verticillata* and *Alternanthera philoxeroides* of were found to vary from 0.01 to 0.27 gm^{-2}day^{-1} and 0.00 to 0.28 gm^{-2} day^{-1} respectively. These values were found to be in conformity with the values reported from Utrapat lake, Manipur by Devi O.I. (1998) with the values of 0.08 to 0.73gm^{-2} day^{-1} and 0.11 to 0.79 gm^{-2} day^{-1} respectively. In *Ceratophyllum demersum,* the values varied from 0.00 to 2.85 gm^{-2} day^{-1} in the two consecutive years of study and the value was found higher than the values of 0.01-0.78gm^{-2}day^{-1} reported from the Loktak Lake, Manipur by Devi N.B. (1993). For *Echinochloa colonum* the values of daily net primary productivity varied from 0.01 to 1.23 gm^{-2} day^{-1} and the values of 0.08-0.81 gm^{-2} day^{-1} reported from Waithou Lake, Manipur by Devi O.L. (1993). Likewise for *Limnophila heterophylla* the values of 0.001 to 0.31 gm^{-2}day^{-1} reported by Devi N.B. (1993) from Loktak Lake Manipur were found generally in conformity with the values reported in the present study (0.043 to 0.14 gm^{-2} day^{-1}). For *Potamogeton crispus,* the values of daily net primary productivity varied from 0.01 to 0.80 gm^{-2} day^{-1} during the study period which were found to be in conformity with the values reported by Billore and Vyas (1982, 0.16 to 0.66 gm^{-2} day^{-1}).

Daily net primary productivity of the freshwater bodies of Manipur are set in Table-7.

Annual Net Primary Productivity :

The total annual net primary productivity of the total species (combined) varied from 2.28 gm^{-2} yr^{-1} to 133.04 gm^{-2} yr^{-1} in the first year and 2.07 gm^{-2} yr^{-1} to 137.139 gm^{-2} yr^{-1} in the 2nd year of study. The present values were in conformity with the values reported by various authors viz Schalles and Shure (1989) in the dystrophic Caroline Bay Wetland (140.4gm^{-2} yr^{-1}) and Illkowska (1993) in lake Mikolajakie, Poland 9130.66gm^{-2} yr^{-1}), Devi K.I. (1993) from Ultrapat Lake, Manipur (2.97 to 265.46$gm^{-2}yr^{-1}$) But the present values were found to be lower than the values reported by Devi N.B.(1993) in Loktak Lake, Manipur, Devi O.I. (1993) in Waithou Lake. Annual net primary productivity of the different freshwater bodies of Manipur are set in Table-7.

CONCLUSION

The trophic status of the Ikop Lake in the present investigation may be assessed by analysing the magnitudes of the productivity of the macrophytes both at the individual and community level over a period of two years.

Many ecologists have correlated the lake productivity with the successional status of the lakes. The least productive lakes with the daily

Table-7: Daily (gm^{-2} day^{-1}) and Annual (gm^{-2} $yr.^{-1}$) net primary productivity of the macrophytes in the different wetlands of Manipur

	WETLAND	DAILY N.P. (gm^{-2} day^{-1})	ANNUAL NPP (gm^{-2} yr^{-1})	REFERENCE
1.	Loktak Lake			
	a) Phumdi zone	0.03-9.79	737.64 - 1240.64	
	b) Non Phumdi zone	0.01-5.04	181.37-358.92	Devi N.B. 1993
2.	Waithou Lake	3.70-4.39	1350.51-1601.04	Devi O.I. 1993
3.	Utrapat Lake	0.05-2.88	2.97-265.49	Devi K.I. 1998
4.	Sanapat	0.02-3.22	1.66-182.85	Bebika & Sharma (2002)and Bebika (2001)
5.	IKOP LAKE	0.06-4.93	2.07-137.13	Present Study

net productivity range of 50 to 300 g Cm^{-2} day^{-1} are designated as Oligotrophic while the values varying from 250 to 1000 g Cm^{-2} day^{-1} as Mesotrophic (Wetzel 1975 and Likens 1975). The lakes are termed as Eutrophic when the values of daily net production vary from 600 to 8000 g Cm^{-2} day^{-1} (Likens 1975). The eutrophic lakes are known to be the most productive ones. Likens (1975) stated that a lake can be conservatively characterised to be culturally eutrophic when the annual gross primary productivity exceed 250 g Cm^{-2} or the annual net primary production exceeds 150 g Cm^{-2}. Likens (1975) reported an average value of 1000 g Cm^{-2} day^{-1} (equivalent to 2 g dry matter m^{-2} day^{-1}) for designating the lakes to be eutrophic.

In the light of these considerations it may be inferred that the trophic status of Ikop lake under the present investigation is found to be eutrophic. The eutrophic nature of the lake under investigation may further be substantiated by the findings in the physiographic features.

It is of much relevance to point out that cultural eutrophication is rampant in the lake which is mainly caused by a number of attributing factors such as

(i) raising of massive areas in the lake for pisciculture. (ii) Encroachment of most of peripheral regions for paddy cultivation. (iii) Overuse of inorganic fertilizers (iv) Misuse and overuse of pesticides, insecticides and weedicides in the cultivated areas of the lake. It is noteworthy that toxic chemicals used in the cultivated areas drain down into the lake due to natural leaching especially during the rainy season. This may be most important factor for gross pollution of the lake which calls for the urgent need for remedial action to protect the lake from further deterioration. It is suggested that conservation measures may be formulated and implemented for protecting this age-old, historical eutrophic lake from extinction.

REFERENCES

Ambasht, R.S. (1970). Freshwater Ecosystem. pp.124-160. In: R. Misra (Eds). Ecology, Study of Ecosystem. A.H. Wheeler and co. (P) Limited, Allahabad.

Bebika, Ch (2001) Variations in species distribution and Primary production of the macrophytes in Sanapat Lake, Manipur. Ph.D. Thesis, Manipur University, Manipur.

Bebika, Ch and B.M. Sharma (2002). Variation in the net primary production of the macrophytes in the Sanapat Lake, Manipur. J.Curr. Sci.**2(2)**: 211-218.

Billore, D.K. and L.N. Vyas (1982). Distribution and production of Macrophytes in Pichhola lake, Udaipur (India). pp. 45-54. In : B. Gopal, R.E. Turner, R.G. Wetzel and D.F. Whigham (Eds). Wetlands Ecology and Management. National Institute of Ecology and International Scientific publications, India.

Daost, and D.L. Childers (1999). Controls and emergent macrophyte composition, abundance and productivity in fresh-water. Everglades Wetland Communities. Wetlands **19**(1):262-275.

Devi, Ch. Umabati (2000). Phytosociology and primary production of the macrophytes in the Freshwater ecosystem of Canchipur, Manipur. Ph.D. Thesis, Manipur University, Manipur.

Devi, K. Indira (1998). Ecological studies of freshwater macrophytes in Utrapat lake, Manipur, Ph.D. Thesis, Manipur University, Manipur.

Devi, N. Beenakumari (1993). Phytosociology, primary production and nutrient status of macrophytes of Loktak lake, Manipur. Ph.D.Thesis, Manipur University, Manipur.

Devi, O. Ibeton (1993). Distribution, primary production and nutrient status of the macrophytic communities in Waithou lake, Manipur, Ph.D.Thesis, Manipur University, Manipur.

Illkowska, A (1993). Temperate freshwater Ecotone. pp.17-34. In : B. Gopal, A. Hillbricht-Illkowska and R.G Wetzel (Eds.) Wetlands and Ecotones. Studies on land-water Interactions. National Institute of Ecology and International Scientific Publications, New Delhi.

Leith, H. (1975). Primary productivity in ecosystem. Comparative analysis of global pattern pp. 67-68. In W.H. Van Dobben and R.H. Lowe Mc Connel (Eds). Unifying concepts of Ecology. DrW. Junk Pulbl. The Hague.

Lieth, H. (1992). Geological Research Problems for the coming decade. pp. 3-17. In K.P. Singh and J.S. Singh (Eds.) Tropical Ecosystems : Ecology and Management. Wiley Eastern Limited, New Delhi.

Likens, G.E. (1975). Primary production of inland aquatic ecosystem. In: H. Lieth and R.H. Whittaker (Eds.) Primary productivity of Biosphere. Springer-Verlage. Heidelberg, New York.

Long, S.P. and Hutchison (1991) Primary production in grasslands and coniferous forests with climatic change. An-overview. Ecol. Appl. **1(2)**:139-156.

Mackenzie, A; Ball, A.S. and Virdee, S.R. (2001). Instant Notes in Ecology. BIOS Scientific publishers, Oxford and Viva Books Pvt. Ltd. New Delhi.

Misra, R. (1968). Ecology workbook, Oxford and IBH. Publ. Co., New Delhi.

Myers Norman; Russell A. Mittermeler; Cristina G. Mittermeler Gustavo A.B. da Fonseca and Jenifer Kent (2000). Biodiversity hostspots for conservation priorities. Nature Vol. V03:853-858.

Odum, E.P. (1971). Fundamentals of Ecology. W.B. Saunders Co., Philadelphia, U.S.A.

Odum, H.T. (1956). Primary production in floating Waters. Limnology and Oceanography. **1**:137-144.

Sharma, B.M. (1999). Wetland Ecosystems of Manipur: An Ecological appraisal. In R.K.R. Singh and H.P. Sharma (Eds.) Wetland of Manipur Vol-I P.13-19. Manipur Assn. For Sciences and Society (MASS), Imphal.

Schalles, J.F. and D.F. Shure (1989). Hydrology, Community structure and productive patterns of a dystrophic Carolina Bay Wetland. Ecol. Monogr. **59(4)**: 365-385.

Wetzel, R.G. (1975). Limnology. Saunders, W.B. Philadelphia.

Advancing Frontiers of Ecological Researches in India, 271-282, 2006
Edited by A.K. Kandya & Asha Gupta
Published by Bishen Singh Mahendra Pal Singh, Dehra Dun, India

Dampi watershed studies with GIS & Remote Sensing Techniques

Arun Kumar, R.K. Chingkhei, Th. Dolendro and J. Khongsai
Department of Earth Sciences Manipur University, Imphal 795 003
Email arun117@hotmail.com

Abstract: Present studies are carried out in Dampi watershed, which is a tributary of Manipur river Basin, Manipur. The morphometric analysis, neotectonic setup, digital image processing and Digital Elevation Modelling are included in the watershed studies and reveal interesting results which are very useful to understand many aspects of Dampi watershed. The morphometric analysis reveals that the watershed is in its mature stage and shows the slow disposal of water, which results in high permeability of the subsoil. The agricultural activities produce a high return due to the high soil moisture. The neotectonic studies display the terrain under the influence of the recent tectonism and the river follow the E-W trend in such a way that northern part is up thrusted block with number of structural hills with steep slopes whereas the southern part is down thrusted block with denudational hills. The landuse covers patterns are analysed by digital image processing of the IRS LISS III and the major landuse patterns are dense forest and agricultural activities. The application of digital elevation model helps in understanding the geomorphic setup of the watershed.

INTRODUCTION

Drainage basin is technically a well defined area on the land (the surface of the geo-sphere) by the watershed, which is the imaginary line passing through the highest points that separate the land draining to one river from land draining to adjacent rivers. The drainage basin is, thus, defined as the land area that receives, collects, and concentrates water falling from the atmosphere, together with groundwater emerging from springs in the geo-sphere, to give river discharge. The drainage basin also acts rather like a conveyor belt for material in solution (solutes) and for sediment, which is why it is a fundamental geomorphologic unit. Water moving through the drainage basin can transport the material in solution (as solutes) or mechanical properties in suspension (suspended sediment) or by being rolled along the bed of the river as bed load. The volume of the material moved as solutes, suspended sediments, or bed load can be totalled for each year and dividing this total volume by the area of the drainage basin gives an indication of the rate at which erosion is taking place, represented in terms of the depth of erosion from the drainage basin as a whole. In fact, the sediment is derived from particular parts of the drainage basin rather than from the entire surface as also explained in this paper.

Review of Previous Work

The various types of drainage studies are being conducted so far but the present area seems to have interesting characteristics of its own and offers the scope for applying a number of techniques for understanding the impact of active tectonics in the watershed. In recent years, a large number of watershed studies have undertaken on small watersheds as they represent a distinct identity but they have been ignored at the national level because of their geographic isolation from the main geographical land of the country. Maidment (1993), Chalam *et. al.,* (1996) have established some of the relationship between hydrologic and geomorphic variables using statistical methods. Srinivasan *et. al.*, (1999) studied the morphology of Mamundiyar river basin for groundwater targeting and reported that the triangular areas between the sub-basin which are characterized by low drainage density, low stream frequency and low relief can be demarcated as potential groundwater zones. Diwan and Gupta (2002) reported that high value of average drainage density and high drainage frequency above 3.0 in metamorphic rocks indicates impermeable surfaces, suitable for small surface water impounding structures. Kumar *et al.* (2001) have done morphometric analysis of Ajay river basin using GIS packages ILWIS and observed that this methodology is very efficient for estimation of geomorphologic parameters. Remote sensing has emerged as a powerful tool in recent past for natural resource assessment with the ability of obtaining systematic, synoptic, rapid, and repetitive coverage. GIS has made remote sensing a unique technology and widened the spectrum of remote sensing applications in natural resource management. The watershed studies based on remotely sensed (RS) data and Geographical Information System (GIS) recently carried out in Barak River basin (Chingkhei, 2004), displays very significant results of the monitoring of the forest plantation. The Digital Elevation Model is very useful attempt to drape the various themes on the elevation model to view and analyses the impact in the drainage basin. Thus attempts have been made to study the geomorphology of the watersheds in different parts of the country.

Study Area and its Environment

Dampi watershed is the right bank tributary of Manipur River Basin. It lies between longitude 93 45' 14"E and 93 51'00"E and latitude 24 19' 54"N and 24 23' 65"N. The geographical area is approximately 37.75 sq km covered under the Bishnupur District of Manipur. The watershed constitutes a typical geomorphic setting which has undergone several modifications in shaping its landforms. The number of abnormal setting of the landforms is observed which support for the impact of the neotectonic activities in the watershed. The moderately steep denudational hills constitute the southern part of the watershed whereas the erosional hills exhibit its steep topography. A number of narrow gorges in the watershed display the evidences of active tectonic in the watershed. Startigraphically, the rock formations of the watershed are included in the Disang Group of Tertiary in age. The river valley has the alluvial deposits of Recent Age. The erosional hills constitute a thick vegetative cover and the denudational hills are prone to erosional hazards. The alluvial fans offer a site for permanent terrace cultivation. The river valley has a good yield for the paddy cultivation. The watershed is included under the sub humid agro-climatic region. The average annual rainfall is 1300 mm. The accessibility of the watershed is by road from Bishnupur and Churachandpur district.

The study is taken up with the following aims:

1. The quantitative study of drainage basin
2. Neotectonic studies in the watershed
3. Digital Image processing for landuse patterns
4. Digital Elevation Model in the watershed

MATERIALS AND METHODS

The digital satellite data and Topographic sheets of Survey of India (SOI) on 1:50,000 scales are used for the present studies. The standard GIS software such as MapInfo Version 5.1 and image processing software like Erdas Imagine 8.4 are used to analyze the watershed characteristics. Using the direct and indirect measurements of the eight, third order drainage basins, which constitute the Dampi watershed, the drainage basin morphometry analysis is done. The watershed is covered under 83 H/15 SOI Toposheet and IRS LISS III Scene 112/52. The 20 meters contours of the Dampi terrain are digitized for generating the DEM (Digital Elevation

Model) for the watershed, which is very useful for geomorphic studies. The field studies are also included to find out the accuracy of satellite processed thematic maps of the watershed.

RESULTS AND DISCUSSIONS

Morphometric Studies

The morphometric analysis is carried out for Dampi watershed by measuring eight third order drainage basins, which fall under different lithotectonic unit (Fig. 1). Various important morphometric parameters are used in this study for analysis. The basin morphometry includes the analysis of the characteristics of linear, aerial and relief aspects of fluvially originated drainage basins. These parameters are calculated by using the SOI Toposheet on 1:50,000 scale and given in table 1. The comparative study of these morphometric parameters of eight sub-basins of Dampi watershed indicates an interesting result, which can be used to establish regional indicators for a similar agro-climatic region. The drainage density displays the erosional characteristics of the sub-basin whereas the basin shape is used to understand the tectonic controls on the sub-basin.

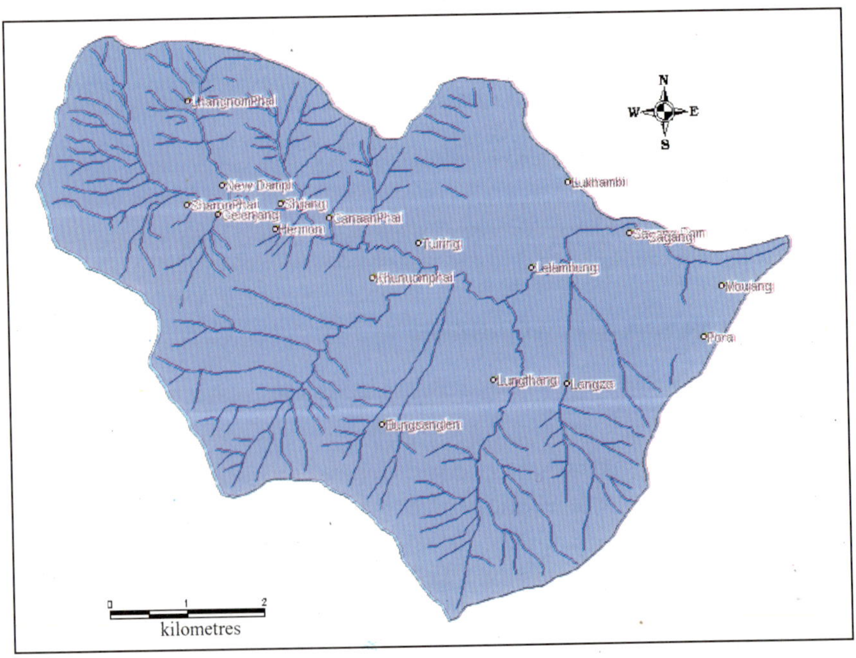

Fig. 1. Dampi watershed

Table-1: Morphometric Parameters

Sl. No.	Geomorphic Parameter	Sub Watershed Division of Dampi Watershed							
		A	B	C	D	E	F	G	H
1.	Stream Number, F1 (in Km)	15	17	10	24	21	16	17	8
2.	Stream Number, F2 (in Km)	5	5	2	7	6	5	4	2
3.	Stream Number, F3 (in Km)	1	1	1	1	2	1	1	1
4.	Stream Number, F1 (in Km)	0	0	0	0	1	0	0	0
5.	Total Stream Number, (in Km)	21	23	13	32	30	22	22	11
6.	Stream Length, L1 (in Km)	7	9	3	10	8	5	5	2
7.	Stream Length, L2 (in Km)	2.5	3	0.5	4	0.75	1.5	1	1
8.	Stream Length, L3 (in Km)	2.5	2	3.5	4	2.5	2	2	0.5
9.	Stream Length, L4 (in Km)	0	0	0	0	1.5	0	0	0
10.	Total Stream Length,(in Km)	12	14	7	18	12.5	8.5	8	3.5
11.	Perimeter Lp (in Km)	8.472	10.24	7.7	10.49	8.736	6.308	5.183	3.865
12.	Area, A (in Km^2)	3.61529	5.153	2.97401	5.68771	3.95944	2.11185	1.8778	0.79979
13.	Maximum Basin Relief, Lb (in m)	3.5	4	3	3.5	2.5	2	1.5	1.25
14.	Maximum Basin Relief, H (in m)	320	520	420	420	520	320	320	420
15.	MainStream Length, Lms (in Km)	3.67	4.369	2.922	4.316	2.894	2.547	1.575	1.429
16.	Bifurcation Ratio, Rb	3:05	3.4:5	0.209722	3.43:7	3.5:3	3.2:5	4.25:4	0.16806
17.	Length Ratio, Rt	2.8:1	4.5:0.67	0.86:7	2.5:1	3.2:3.3 3:0.6	2.5:1.33	2.5:2	04:01.0
18.	Form Factor, Rf	0.295	0.322	0.33	0.464	0.634	0.528	0.835	0.512
19.	Drainage Density,Db	3.319	2.717	2.354	3.165	3.157	4.025	4.025	4.376
20.	Drainage Frequency, Df	5.809	4.464	4.371	5.626	7.577	10.417	11.716	13.754
21.	Constant of channel Maintenance	0.301	0.368	0.425	0.316	0.317	0.248	0.235	0.229
22.	Elongation ratio, Re	0.612	0.64	0.649	0.768	0.898	0.82	1.03	0.807
23.	Circulatory Ratio Rc	0.796	0.785	0.795	0.806	0.808	0.818	0.938	0.823
24.	Relief Ratio, Rh	0.091	0.13	0.14	0.12	0.21	0.16	0.22	0.34
25	Ruggedness Number , Rn	0.96	1.04	0.84	1.26	1.56	1.28	1.28	1.68
26	Relative Relief, Rr	0.038	0.051	0.005	0.04	0.06	0.051	0.062	0.109

From the above analysis/interpretation of the selected sub-watersheds, it is seen that higher drainage frequency in the lower order indicates a well-dissected terrain in their upper reaches. Lower drainage frequency value in the higher order also suggests an advance stage of the drainage or the mature stage. The study of drainage density also supports that high drainage density value is in the well-dissected area and low value at the lower gradient or lower reaches. From the study of shape or form of the sub-watershed drainage it is seen that highest value area have concentration of water within their basin area for a longer duration. The ruggedness number is highest in "H" whereas low in "A" sub-basin represent the rugged and gentle topography respectively. The high circulation ratio correlates to mature and old topography, which allows the slow disposal of water. The high bifurcation ratio indicates the high run off, less infiltration and mature nature of topography. Study indicated that the geologic processes have distorted the basin with bifurcation ratio up to 5. The constant of channel maintenance is the reciprocal of drainage density and its high value suggests the permeable sub soil and concentration of water in the channel for a larger time. It can be concluded that the Dampi watershed basin is composed of various sub-watershed having differential morphometric characteristics and erosional surfaces, which have been developed under variety of conditions. Thus, the studied area shows a polycyclic drainage character. It is found that the drainage characteristics play an important role to distinguish a drainage basin from other. It is a microscopic measurement and individual stream is developed and being modified according to its geotectonic setting which varies quite often.

Neotectonics of Watershed

Neotectonics is the study of crustal movements that occurred both in the Earth's recent past and are continuing at the present day. These movements, which are driven directly or indirectly by global plate motions (tectonics), result in the vertical and horizontal warping, folding or faulting of the Earth's surface. For the most part, plate motions give rise to the slow, progressive deformation of the Earth's outer shell, the ongoing effects of which are detectable only by high-precision space-borne and land-based monitoring over periods of years or even to decades. Neotectonic activities have reshaped the landforms of the terrain. The number of river captures of many streams has been noticed in the watershed. The regional strikes of the ranges are N-S whereas the Dampi river flows in E-W trend. However, its tributaries follow N-S trend, which seems to be captured later by the Dampi river. The Dampi river follows a E-W trending fault which seems to be a normal fault. E-W faulting of Dampi river is comparatively younger than its tributaries which have a NNE-SSW trend. There is a closure of a Dampi synclinorium, which is identified, by its hinge. The axis of the synclinorium seems to have is E-W trend whereas the remaining part is eroded away.

There is a striking change in the southern and northern part of the watershed. The northern part of the watershed constitutes narrow erosional hills with steep scarps (Fig.2). The southern part of the watershed constitutes comparatively large denudational hills, which are gradually terminated by a series of coalescing alluvial fans zones. Streams originated from these denudational hills derive the materials for the alluvial fans. These fans serve as an ideal place for man-made terrace cultivation. The surfaces of the fan, which are less prone to erosion, are occupied by the terrace cultivation. The coalescing fan zone seems to be affected by neo tectonic activities (Fig.2). The trends of these fans are not parallel to these denudational hills that indicate the presence of later disturbances after the fan formation. They are included in the recent tectonics.

Fig. 2. Erosional hills with coalescing alluvial fans

The above examples and facts can be used to formulate the conclusion that the tectonic structure determine the places where river valley are found and the subsequent motion of these structures affects the course of evaluation to the Dampi valley. The river can be referred to indicators of tectonic structures because they distinctly react to the situation of their valley to the meandering and to the structures of any type of tectonic disturbances. They are linked mostly to the axial parts of step fold, or to the most mobile sections with a constant tendency to relative subsidence. The Dampi river almost follows the fault where the northern portion displays a block whereas the southern part indicates the subsidence and denudational control topography (Fig 3). The denudational approach still prevails to a large extent in interpreting the origin of all forms of relief in geomorphology.

Fig. 3. Meandering patterns of the Dampi river affected by neotectonic activity

But in the same way it is also harmful to overestimate the exogenous factor in interpreting the origin and evaluation of the river basin. Since the Imphal valley itself has undergone number of tectonic disturbances and the Dampi watershed is located in its southwestern margin of its periphery, which believed to be witnessing the similar episodes of neotectonic activities. The origin of the Loktak lake is a tectonic one rather than the geomorphologic process. The lake has a tectonic origin and its uplifted block at Thanga hillock has undergone several phases of tectonics in the recent times. Remnants of older synclines are indicative of the later subsidence whereas the sharp relief forms along the uplifted block are indicative of representing scarps and vertical facets. The abnormal behaviour of the river and its tributaries display the typical neotectonic setting in its watershed.

Image Classification in Dampi Watershed:

The Digital image processing has been done using the LISS III satellite data for identifying the land use patterns in the Dampi watershed. The classification is useful to delineate the various land use categories in the watersheds (Fig 4 and 5). For this purpose, the representative signatures of the various landuse patterns have been identified in the satellite imagery using field data (Fig 5). The agricultural area dominates the land use practices in the watershed due to the availability of alluvial plains developed by the river. The agricultural activities are not restricted on these plains but the villagers have adopted the terrace cultivation on the coalescing alluvial fans region. The erosional hills have steep slopes and therefore the dense vegetation is developed on these slopes. The sparse vegetation and degraded vegetation is developed on the denudational hill slopes and alluvial fan zones. The moderate vegetation is very less occupying the minimum area comparing from the all landuse categories. It indicates that the exploitation of the forest cover is not in the scientific manner and thereby the dense vegetation cover is depleted into the degraded vegetation. However, the temporal changes have not been analysed so far. This might be useful for assessing the rate of the forest degradation in the watershed. The statistics is given below (Table-2). The field photographs display the typical existing landuse patterns in the watershed as well as the verification of the accuracy of the digitally processed satellite data into the number of land use categories.

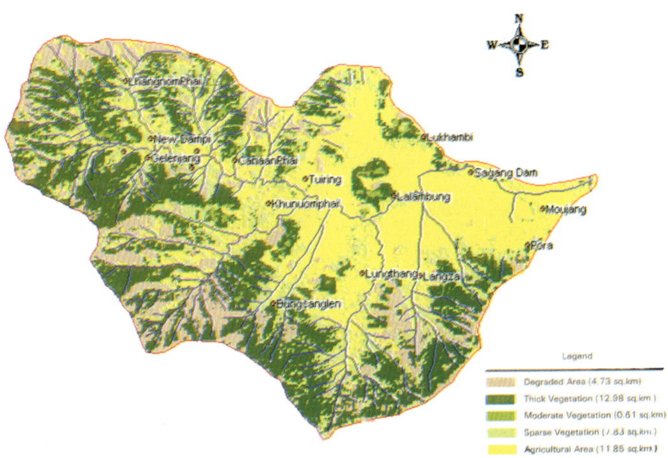

Fig. 4. Landuse map of Dampi Watershed

Table-2: Land use Statistics in the watershed

Sl. No.	Landuse categories	Area in sq km
1.	Agricultural Land	11.85
2.	Thick Vegetation	12.98
3.	Moderate Vegetation	00.61
4.	Sparse Vegetation	07.83
5.	Degraded Vegetation	04.73

Fig. 5. Representative landuse Patterns and their signatures in the IRS-1D data

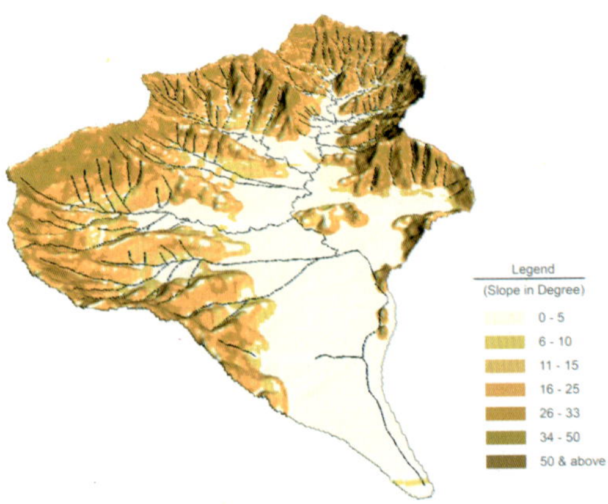

Fig. 6. Digital Elevation Model draped with Slope

Fig. 7. Digital Elevation Model (DEM) draped with landuse pattern (IRS LISS III Data)

Dem in Dampi Watershed

Elevation data forms one of the important sources of ancillary data, which is most widely used in earth related investigation. The most common form for the representation of elevation data in digital format is the grid type of raster form. The elevation matrix consists of the well-known regular grid of elevations, representing a sample of the real topography.

Studies on geomorphology require an aerial look of the watershed in the present studies. Using the topographic maps of SOI the contour on 20 mts intervals are digitized and the DEM was developed. This DEM is further draped on the various surface themes such as slope, satellite images, landuse covers, relief and drainage network.. The Digital Elevation Model is developed in the watershed and the following maps are generated:

The aid from the DEM is very useful to understand much aspect of the watershed studies such as impact of neotectonic activities, landuse patterns and erosion hazards in the watershed. The precise generation of DEM has also helped to reveal the watershed characteristics especially on landforms and the impact of tectonic on them.

CONCLUSIONS

The watershed studies carried out in the Dampi watershed indicated a typical geomorphic setup in a neotectonically active area. The region is seismically active and many large and two great earthquakes have rocked NE India. The present studies are useful to understand the impact of tectonic activities on the small watershed as the watershed characteristics display the impact of such activities. There is also a very important application of morphometric studies in the watershed, which indicates the present state of erosion and availability of surface water. The recent tectonic activities have modified the existing landforms and they become potential sites for agricultural activities for the dwellers of the watershed. Use of digital image processing, in analysing the landuse patterns of the watershed, revealed that the agricultural activities and dense forest cover constitute the major geographical part of the watershed, which indicates good economic returns for the villagers. The watershed characteristics can further be regionalized to generate the thumb rules for watershed development strategies covering ungauged and unexplored watersheds.

ACKNOWLEDGEMENT

Financial assistance provided by SERC, DST, New Delhi for strengthening the GIS and Remote Sensing Lab to carry out these investigations is thankfully acknowledged.

REFERENCES

Chalam, B.N.S., Krishnaveni, M. and Karmegam, M. (1996). Correlation Analysis of Runoff with Geomorphic Parameters. *J. Appl. Hydrol*, IX (3-4), pp24-31.

Chingkhei, R.K. (2002). Application of GIS techniques in evaluation and monitoring of vegetation in Barak Basin unpublished Ph. D thesis pp 121.

Diwan H.D. and Gupta M.P. (2002). Geometric Characteristics of Kurung river basin in Chattisgarh state Curr. Geog. Publ. No. 1 89-100.

Kumar, C.P. and Purandara (2001). Simulation of Soil moisture movement in hard rock watershed using ILWIS model Technical Report I National Institute of Hydrology Roorkee.

Maidment, D.R. (1993). Handbook of Applied Hydrology. McGraw Hill Inc., New York.

Srinivasan, P.R. and Dhakshinamoorthy, N. (1999). Hydrogeochemical studies of groundwater in Mamundiyar river basin Tamilnadu Vol. II p 595-603.

SECTION THREE

WEED ECOLOGY

Advancing Frontiers of Ecological Researches in India, 285-294, 2006
Edited by A.K. Kandya & Asha Gupta
Published by Bishen Singh Mahendra Pal Singh, Dehra Dun, India

Management of Weeds in Dry Seeded Rice

Veenapani Dubey
C.M.D. Post-graduate College, Bilaspur, C.G..

Abstract: Weed management practices have been reviewed in dry seeded rice. Strategy for future research has been suggested.

INTRODUCTION

Dry-seeded rice is referred to as un-germinated rice seeded in unsaturated wetland fields which are prepared dry and will subsequently become flooded. This type of rice has been practiced in rainfed areas of north-east India, South eastern India, Bangla Desh, Myanmar and Thailand for centuries.

Chhattisgarh is a major rice growing state in India Bilaspur is situated almost in the center of Chhattisgarh .It is situated between $21^{0}37'$ and $23^{0}7'$N latitude and $81^{o}12'$E &$81^{O}40'$ E longitude. Its average height is approximately 293.3 meters above sea level .Formations of Bilaspur district are classified into Gondwana, Cuddapah & Dharwars with Archeans.

The type of rice culture used are mainly "direct-seeded" and "transplanted". Direct seeded rice suffers a lot by weed infestations than the transplanted rice. In dry seeded rice, weed seeds and crop seeds germinate and the weeds being more vigorous, almost suppress the crop. A heavy infestation of weeds is characteristic of dry-seeded rice. Grasses make 85-89% of the total weed population and 90-96% of the total dry matter (Table-1). *Echinochloa colonum,* the predominant grass species reduced rice yields by 74- 98%.

The development of a suitable weed control technology is essential to the successful dry seeded rice crop.

With this view the studies were done to evolve a successful weed control technology so that the weed infestation in the direct seeded rice is managed to a sustainable limit at a very low cost inputs.

Table-1: Weeds growing with dry seeded rice (Mukhopadhayay *et al.* 1972)

Weed-class	Sp. No.		Weed Weighs(gms/m²)	
	1968	1969	1968	1969
Grasses	7	6	55.3	62.8
Broad leaved weeds	10	13	2.5	5.5
Sedges	2	2	7.0	2.0
Total	19	21	69.8	70.3

Reduction in Weed problem

It has been observed that weeds in dry-seeded rice should be suppressed during early stages of crop growth because grain yield reduction resulting from weed competition is most severe at this time. Crop should be maintained free from weeds during first one to one and half months of crop growth. This has also been reported adequate by Moody (1981).

Land Preparation

Preparation of land prior to seed sowing is a common practice here in most of the regions namely Bilaspur, Chichirda, Bodri, Amora, Janjgir .The field is left fallow during the dry season after the rice harvest. This results in the increase of weed problem in dry seeded rice crop because of the weeds growing in dry season.

It is advisable to have a weed free fellow during dry season. This reduced many weed problems as it significantly reduced the number of perennial weeds. (Moody & Milan, IRRI, 1979-80). It reduces the weed infestation in areas where upland crops are grown during dry season.

The Stale-seedbed technique

It is possible to reduce weeds in the dry seeded rice crop by stale seed bed technique involving the removal of successive weed flushes before planting .Chemical ,mechanical or manual methods may be used to remove the weeds .Chemicals have the advantage of not bringing more weed seeds to the soil surface where conditions favour germination .The herbicides should be applied or the cultivation should be done when most of the weed seeds in the surface soil have germinated and are in the two to five leaf stage .The dry seeded rice crop is sown with minimum of soil disturbance either before or after the weeds have been killed with herbicides or controlled by cultivation. Theoretically less weeds have been killed with herbicides should grow in association with the rice because most of those in the favorable zone for germination have been killed. It may be possible to kill weed growth before planting in regions where light rains occur for extended period before the onset of the monsoon.

Table-2: Effect of method of land preparation &weeding regimes on fresh wt of *E. colonum* (6 wks after Crop Emergence)

Fresh wt gm / m^2

Weed control method	Conventional land preparation	Stale-seedbed
Butachlor (2 kg ai/ha) f.b hand weeding	80 b	44 b
Butachlor	1172 a	228 a
Unweeded	1120 a	244 a

Mean followed by the same letter are not significantly different at 5% level; fb = followed by; ai = Active ingredient.

Stale -seed bed technique has no advantage if planting is delayed after start of rains. Delay in planting results in more weed control difficulties. The reasons being the field becomes too wet soon after crop establishment.

Zero-tillage

Wijewardane (IRRI) reported excellent crop growth and a few weeds in rice that had been direct-drilled in a field where the weeds have been killed with a non selective ,non residual herbicide before planting. Moody failed to establish a rice crop and to control weeds using the same technique in trials in Nigeria and the Philippines.

The success of a zero tillage seeding technique depends on availability of suitable herbicides to control the weeds. The likelihood of success is greatly decreased if perennial weeds are present.

When the soil surface remains undisturbed and it becomes compacted by rainfall, conditions become less favorable for weed seed germination .Besides when cultivation is eliminated, the greater part of the original weed-seed reserve in the soil can be disregarded. With zero tillage it is usually observed that the weed spectrum is changed with conventional tillage but the number of weeds is not always reduced.

Planting time

It has been observed that delay in the planting of dry seeded rice can result in more weeds. Planting should be therefore done as soon as possible after the rains.

Application of Nitrogen fertilizer

Early Nitrogen application may intensify the weed problem. Moody (1977) also studied that weed growth is stimulated if nitrogen is incorporated into the soil. Ahmed (1979) reported that weed counts and weights in dry seeded rice increased linearly as the rate of basal application of nitrogen increased from 0 to 40 Kg/ha. The first weeding was easier and faster in the plots received no basal nitrogen than in those that received 40kg nitrogen / ha as a basal application because the grassy weeds were smaller and easier to distinguish from the rice.

Table-3: Relative contributions of fertilizer weed & insect control &their combinations on the grain yield of dry seeded rice

Input	Yield (t / ha)*
Fertilizer + insect control + weed control	4.2 a
Fertilizer + weed control	3.9 a
Weed control	2.9 b
Weed control+ insect control	2.8 b
Fertilizer + insect control	2.0 bc
Fertilizer	1.9 c
Insect control	1.2 c
Unweeded check	1.2 c

*means followed by the same letter are not significantly different at 5% level

Water

Weeds in dry seeded rice are controlled better if flooding occurs soon after stand establishment. Harwood (1979) recommended that dry seeded rice be sown only in fields that can be flooded within 30 days after planting. Heavy weed infestation in the early stages of crop growth can not be controlled by submergence later on, Ahmad (1979) reported that the ponding potential of the field had no effect on weeds growth. Civico and Moody, (1979) observed that greater water depths were needed to suppress weed growth when flooding was delayed after weed emergence. However it is not very much important to emphasise on early flooding as a means of weed control.

Cultivar grown

The need for adequate weed control for all types of rice culture, particularly dry seeded rice has assumed more importance with the introduction of the short modern cultivars.

However, for rice there seems to be a strong negative correlation between competitive ability and yield. Hence a better competitor should be chosen if all other attributes are equal.

A cultivar emerging rapidly, having high seedling vigor is expected to compete better against weeds.

The competitive effect of cultivars against weeds should be examined at a low level of weeding or weed competition.

Plant spacing

Increasing seeding rate or decreasing row spacing has very little effect on weed suppression for dry-seeded rice. This is because of the intense weed pressures. However, close plant spacing combined with timely weed control inputs than wider weed spacing is better.

Mechanical weed control

Broadcasting is the common practice of seeding for dry seeded rice. But it makes distinguishing between rice and grass weeds difficult and causes trouble in weeding. Farmers in Chhattisgarh use mechanical weed control methods. It is common practice for farmers to pass a spike toothed harrow through their broadcast seeded fields five -fifteen days after crop emergence.

In Bilaspur, Chhattisgarh the farmers control weeds by ploughing and cross ploughing their fields after flooding when rice is 30-40 days old. It is known as "biasi" in Chhattisgarh.

Sowing in rows is recommended for the practice to be completely successful .However suitable equipment has to be developed before row seeding can gain wide spread acceptance (Moody & Milan, 1979).

It seems very appropriate at this time that further research is needed to determine the successful use of inter-row cultivation in dry seeded rice.

It is required to develop an equipment that can work under a wide range of soil conditions.

Manual weeding

Hand and manual weeding are the most common methods of removing weeds from dry seeded rice of Chhattisgarh. These practices have many drawbacks; such as -

1. They are slow and laborious
2. Complete removal of weeds is difficult.
3. Regeneration of weeds from vegetative parts left in the field.
4 Non availability of labors in time.
5. Costly labourers.
6. Scarcity of labors.
7. Labors are costly because frequent hand weedings are required and the amount of labor required for the purpose is huge.
8. They are difficult to perform under unfavorable weather conditions.
9. Costs increase if weeding is delayed.
10. Crop injury is possible.

It is advisable to use hand weeding in addition to other methods of weed control so that labor required is reduced subsequently.

Herbicide

Herbicides are a logical alternative to hand weeding. Cost of production of dry seeded rice can be reduced through use of herbicides. (Estornios *et al. 1981).* Herbicide combinations are generally superior to single herbicide treatments for weed control in dry seeded rice (Moody, 1981) When a herbicide is incorporated before planting ,it is less effected by weather uncertainties and weed control is more consistent.

Results with herbicides in dry seeded rice have been inconsistent from site to site and from year to year at the same site. The inconsistency could be accounted for varying weed species , their population and soil and climatic conditions. Flexibility in use to suitable conditions and maximize efficiency is critical to the acceptance of herbicides for dry seeded rice culture (Moody, 1981).

However, herbicides are not widely used in dry seeded rice in Chhattisgarh because of the factors like lack of reliability, phytotoxicity, unavailability of suitable herbicide and insufficient data on the economics of chemical weed control. Parker (1968) has stated that there is enormous scope for a relatively inexpensive herbicide in dry seeded rice. Therefore herbicide should be combined with other weed control practices such as hand weeding. Tables 4 & 5.

Table-4: Effect of different rates of seeding on weed weights & yield of different rice cultivars grown under dry seeded rainfed – bunded conditions (Siriwardana)

Cultivar	Seeding rate Kg / ha	Fresh weight gm / m^2	Yield kg/ha.
62-355	50	7248	312
62-355	100	5148	417
62-355	150	5000	225
62-355	200	2500	558
62-355	400	2948	242
Bg-348	50	8748	92
Bg-348	100	7248	83
Bg-348	150	8200	103
Bg-348	200	7600	91
Bg-348	400	8148	833
-	-	8848	-
62-348	100*	0	2967

Propanil + MCPA applied for weed control

Table-5: Effect of various weed control practices on yield of Saket-4 rice in dry seeded irrigated conditions

Treatment	Yield t /ha
Weeded tree times	4.6
Butachlor + hand weeding	3.9
Nitrofen + hand weeding	3.8
Propanil + MCPA	3.1
Pendimethalin + Hand Weeding	2.8
Nitrofen	2.7
Butachlor	2.2
Pendimethalin	0.9
Unweeded	0.2

Dubey (1987) has studied various aspects of ecology of rice weeds in Bilaspur. It was observed that Saturn was proved to be very effective herbicide against all types of weeds (grasses, sedges & broad-leaved weeds). It also reduced percentage germination of *E. colonum* and *P. scrobiculatum* in laboratory conditions as well as in field.

Table - 6. Effect of different weed control treatments on % germination & seedling growth of E. colonum (Dubey 1987).

Treatment	Concentration (ppm)	% Germination	Reduction in length (mm)	
			Radicle	Plumule
2-4-D	5	94.00±3.46	-	-
	10	89.83±2.48	-	-
	25	77.33±2.08	2.50	3.0
	50	67.00±2.00	2.57	3.24
	100	60.33±5.78	3.70	3.74
	250	40.33±1.52	5.75	5.82
	500	22.33±3.05	13.66	14.22
	1000	15.33±1.52	21.51	25.00
MCPA	5	91.00±2.08	-	-
	10	88.88±2.61	-	-
	25	76.33±1.75	2.70	3.27
	50	65.00±3.00	3.25	3.74
	100	57.44±6.34	4.83	5.00
	250	40.00±1.09	17.88	20.12
	500	21.25±2.31	26.53	31.08
	1000	14.00±3.00	32.37	34.52
Saturn (50 % E.C.)	5	84.55±6.36	3.95	6.12
	10	73.66±2.08	13.33	15,02
	25	59.16±0.57	20.0	22.50
	50	34.10±5.79	51.90	53.29
	100	27.40±3.85	67.75	70.13
	250	5.90±0.30	90.08	89.95
	500	0.0	100	100
	1000	0.0	100	100
Butachlor	5	85.66±4.93	3.92	6.00
	10	78.00±1.00	12.50	21.00
	25	60.91±2.51	19.37	21.00
	50	46.00±3.00	47.59	50.08
	100	33.50±1.32	57.72	63.27
	250	16.00±1.00	73.12	81.02
	500	13.33±1.52	89.27	87.95
	1000	00.00	100	100
	Control	99.00±1.0	0.0	0.0

Results are expressed as mean ± SD

Table -7. Effect of different herbicide treatments on growth attributes of *E. colonum.*

Herbicide	Conc. (ppm)	Dry Wt (gm)	Height (cm)	Spread (cm)
2 – 4 D	100	4.70	45.02	4.25
	250	4.70	44.0	4.2
	500	4.65	40.5	4.0
	1000	4.32	39.5	4.0
MCPA	100	4.65	45.0	4.2
	250	4.60	42.0	4.2
	500	4.59	36.6	4.0
	1000	4.25	33.75	4.0
Saturn	100	3.87	39.00	3.15
	250	3.17	21.7	2.29
	500	3.10	11.50	1.95
	1000	Plant did not survive		
Control		4.75	46.0	4.26

Future Research

There is need of research aiming at the prevention of loss of yield caused by weeds in dry -seeded rice and to evolve most economic and feasible methods for management of weeds.

The strategy for future research to achieve this objective should be directed towards following lines :

1. Introduction and development of the levels of agricultural technology under the prevailing economic social and political checks.
2. To evaluate the interrelationships of weed management practices and other advanced technology inputs.
3. Assessment of the overall effect of various weed- management systems on socio- economic status of the small farmer, his family, the associated labor pool and the community specially in Chhattisgarh.

REFERENCES

Ahmed, N.U. (1979). Weeds in cropping system as affected by hydrology and weeding regime with emphasis on dry-seeded rice. MS thesis, University of Philippines at Los Banos, Laguna, Philippines. 241p.

Civico, R.S.A. & Moody, K. (1979). The effect of the time and depth of submergence on growth and development of some weed species. Philipp. J. Weed Sci. **6**: 41-49

Dubey, V. (1986). Ecological studies of weeds of paddy fields of Bilaspur. Ph.D. Thesis, R.S. Univ., Raipur.

Estorninos, I. E.; Jr. D.C. Navaroz and Moody, K. (1981). Farmer's concept about weeds and weeds control practices in rainfed areas of Philippines.

Moody, K. (1979). "Exploiting cultivar differences to improve weed control"-Paper presented at the International Rice Research Conf. 16-20. April 1979.

Moody & Milan (1979). "Weed control in rainfed rice" Pages , 235-245. In IRRI 'Rainfed lowland Rice.'

Moody, K. (1981). 'Weed control in dry seeded rice' IRRI, Los Banos Laguna Philippines. "Report of a workshop on cropping systems research in Asia."

Parker, C. (1968). The development of herbicide use in southern Asia. Proc. Br. Weed control Conf. **9**: 1265-1270.

Rathi, K.S. and Tiwari, A.N. (1979). Weed management in upland direct seeded paddy under irrigated conditions. Indian J. Agric. Res.**13**: 111-112.

Advancing Frontiers of Ecological Researches in India, 295-304, 2006
Edited by A.K. Kandya & Asha Gupta
Published by Bishen Singh Mahendra Pal Singh, Dehra Dun, India

Biology and Adaptive Strategies of Some Weeds of Central India

Archana Verma* and Tara Nigam**

Abstract: Regular field visits were made for the study of biology and adaptive strategies of some common weeds of Kharif and Rabi crops, viz., *Alysicarpus longifolius, Anagalis arvensis, Melilotus indica, Chenopodium album, Cyanotis axillaris, Ageratum conyzoides, Cyperus rotundus* and *Cynodon dactylon.* Except *C. axillaris,* and *C. rotundus,* which reproduce at expotential rate by vegetative means, all the other weed species studied are prolific seed producers and having a good reproductive capacity. Some weed species possess hard seed coat dormancy and required a after ripening period also. All these characters of weed species are adaptive strategies, which prevent them to germinate under unfavourable environmental conditions and hence, ensure their survival.

INTRODUCTION

Limited intercultural operations permit the weeds to grow profusely in different crops during different seasons. Uncontrolled weed growth reduces the grain in wheat upto 20-50% (Phogat *et. el.* 1991, Singh and Ghosh 1992 and Kumar and Singh 1997) and in groundnut, weeds accounts for 45% (Bhan and Singh 1993). Weeds are an example of successful struggle for existence. Weeds are hardy and persistent because they have ability to withstand climatic, edaphic and biotic stresses and on the other hand able to repeatedly invade an environment even when it was apparently removed from the seen by some eradicating agent including man. In each possible microecological situation on the earth, there is a highly adapted group of weeds ready to invade our crop and non-crop situations. Weeds have large ecological amplitude, hence they have unique potentialities for adoption. They survive almost in any environment and adjust themselves to the changed conditions (Sen and Kasera 1998, 96; and Sen *et al.* 1999).

A sound knowledge of the biology of weeds, with particular reference to their survival strategies (i.e. propagation, reproductive capacity

* Department of Botany, Govt. Girls Autonomous College of Excellence, Sagar (MP)
** Department of Botany, Vanijya Mahavidhyalaya, Satna (MP)

(Salisbury, 1942), dispersal mechanism and dormancy) is essential for planning their effective management packages. With disregard to such a study, we can at best temporarily destroy the existing weed growths, without being able to reduce their frequent recurrence. This will mean an unending and haphazard warfare against weeds, which can ill afford (Gupta 2000). An assessment of potential and real aggressiveness and susceptibility of a weed species under varying conditions of a habitat, can be documented by ecological life cycle (Sen 1981). The present investigation is aimed to know the biology and adaptive strategies of some important weeds associated with different Rabi and Kharif crops in semi arid Central part of India viz. *Alysicarpus longifolius, Anagalis arvensis, Melilotus. indica, Chenonodium album, Cyanotis axillaris, Ageratum conyzoides. Cyperus rotundus* and *Cynodon dactylon.* The adaptive strategies of weeds were studies in terms of seed characters, viability, germinability, reproductive capacity and dormancy mechanism (Kasera *et. el.* 1999). Any mechanism, which reduces the mortality rates in the life cycle of weeds, can be considered as one of the important strategies of the same.

MATERIALS AND METHODS

Agricultural fields and wastelands of nearby areas of Sagar Distt. were regularly visited for the study of various phenological parameters of weeds. The agricultural fields are of Tili, Jat Patharia, Babupura, Ratona and Rahatgarh regions. Seeds were collected and stored in polythene bags at room temperature. Viability of seeds was tested by 0.1 % T. T .C. (Porter *et. al.*, 1947). The weight of 100 seeds was measured by single pan electric balance. Germination experiments were performed in sterilized petridishes lined with filter paper. The filter paper was moistened with required amount of distilled water. The experiments were performed in triplicate with each petridish containing 25 seeds. Petridishes with seeds of Rabi and Kharif Crop weeds were kept in seed germinater at $28 \pm 1.5^{0}C$ and $35 \pm 2^{0}C$ respectively. A visual detection of radicle was taken as criteria for germination (Verma 1982). The seeds of all the weed species were treated with conc. Sulphuric acid for 40 minutes for scarification of hard seed coat and to remove dormancy.

RESULT AND DISCUSSION

A record of phenological observations of all the selected eight weeds was maintained from July to June in a yearly cycle. The observations are presented in the form of phenograms (Fig. 1). Data related with seed

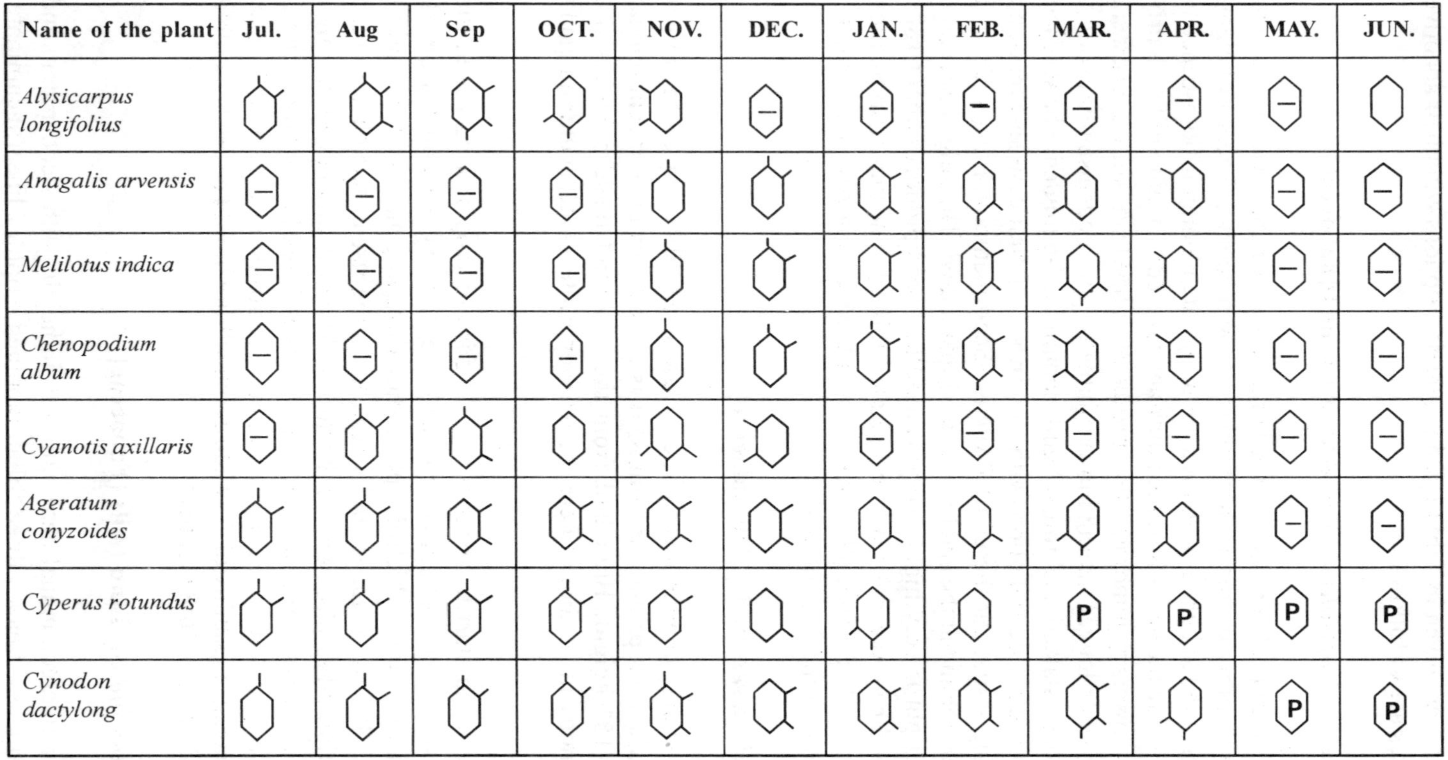

1. Emergence from seed or root stock 2. Vegetative growth 3. Flowering P = Perennation
4. Fruiting 5. Seed maturation 6. Death – = Plant not seen

Fig. 1. Monthly Phenological observations in the form of Phenograms of different weed species.

characters are given in Table-I. Individual descriptions with their adaptive strategies are as follows:

***Alysicarpus longifolius.* Wight & Arn Prodr [Fabaceae]**

It is an annual herbaceous weed of crop and vegetable fields of Kharif season. It rarely occurs in waste places. Stem about 4-5 feet in height, ascending and subglabrous leaves are unifoliate oblong. Inflorescence is dense spicate receme. Flower are pink and pressed to a more or less hairly rachis. Flowers zygomorphic, short pedicellate bracteate. Calyx pubescent as long as or longer than the lower joints of pod. Pods usually exerted, very shortly stalked, glabrous, slightly monilifer, 4-6 jointed. The seeds of the weed start germination with the on set of first few showers and flowering initiate at the end of August. Seed maturation takes place at mid October and at the end of November plant died. The value of average seed out put is very high. Every plant of the weed produces about 1880 seeds and cent percent seeds are viable [Table-1] and start germination after a period of 6-8 months of dormancy. The weed passess unfavourable conditions in the form of seeds.

***Anagalis arvensis.* L. [Primulaceae]**

It is a very common weed of cold season i.e. Rabi crops especially it infest wheat crops and grasslands. It is an erect or procumbent annual herb 5-15" height. Branched from the base and these branches are tetragonous and glabrous. Leaves sessile opposite, ovate, acute and gland dotted.

The weed start to emerge after 30 days of crop sowing i.e. in the last of November. Seedlings usually make vegetative growth upto last week of December after which the plant start flowering and upto the end of January flowering reaches in its peak. With the rise of temperature seed maturation takes place and at the end of March plant start to dry after producing seeds in the soil seed bank. The seeds are black and trigonous having a beak. They are born in capsule. Capsule size of a small pea. Seeds having a viability of 70% and germinability 68%. The seeds required a period of rest of 6-8 months.

***Melilotus indica.* Linn (All) [Fabaceae]**

It is an erect annual herb upto 3 ft. in height with slender stem and pale branches. Leaves three foliate, toothed, stipules linear, acuminate.

Table-1: Average weight, size, shape, colour, percentage germination and reproduction capacity of some weed seeds.

Name of the weed species	Weight* [mg/100 seeds]	Length	Size** (mm)	Breath	Shape	Colour	No. of Seeds per plant (appro.)	Percentage Germinability	Reproductive Capacity
Alysicarpus longifolius	154.55	1.65		1.05	Ellipsoidal	Brown	1880	60	1240.8
Anagalis arvensis	74.5	0.68		0.35	Trigonous	Black	40	68	26.22
Melilotus indica	170.0	1.82		1.05	Ellipsoidal	Brown	240	100	540
Chenopodium album	12.65		0.18#		Orbicular	Dark Brown	72000	35	25200
Cyanotis axillaris	30.22		0.27#		Rounded	Dark Brown	300	52	156
Ageratum conyzoides	12.25	0.34		0.15	Elongated	Black	1500	40	600
Cyperus rotundus#									Propagate by tubers
Cynodon dactylon#									Propagate by runners and rhizomes.

* Mean of 10 observations.

** Mean of 25 observations.

*** Mean of triplicate, each with 25 seeds.

\# Seeds are not important.

Flower in slender spicate close racemes, pedicel short. Pods glabrous, ellipsoidal, compressed tapering at both the ends. One seeded some times two seeded. It is a very common weed of Rabi crop prefers moist places. Seedlings appear in field about a month after crop sowing i.e. last week of November. The seedlings grow vegetatively upto first week of January after which the plant start flowering. By the mid February, flowering reaches in its peak. Fruiting closely follows flowering, by the end of March ripened pods dehisce. The seeds showed 100% viability and 100% germinability, and reproductive capacity value is 540. Seeds having a dormancy of 6-8 months.

***Chenopodium album.* L. [Chenopodiaceae]**

It is one of the most noxious weed of Rabi crops especially of wheat and also infest gardens and waste lands. The weed is a annual erect herbaceous plant having a height of 1-2 ft. Some times coated with white mealy granular pubescence. Stem ribbed, leaves very variable, entire, toothed or lobed, rhombic oblong or deltoid ovate. Petiole slender and generally as long as the blade. Flower forming compact panicle spikes. Seeds black, depressed with an obscure horizontal keel.

The weed starts germination simultaneously with the crop grains and continuously compete with them for nutrient and moisture. The seed shedding of weed commences from mid March, and by the harvest time (Mid April), bulk of the seed is already shed from the plant. The seeds stay dormant until November when temperature conditions become favourable for their germination. The average weight of 100 seeds is 12.65 mg. Seeds are shiny black smooth and orbicular, having a viability 50% and germinability .45% [Table-I]. When a single plant of *C. album* grow in open habit, free from any other competition has a potential to produce about 72,000 seeds. (Standifer 1980 and Maniyappa 1973). It is a C_4 plant hence its light and CO_2 compensation points are low whereby it continue to be productive under very low light and CO_2 levels as found under crop canopy (Paul and Elmore 1984).

***Cyanotis axillaris.* Roem & Seh [Commelinaceae]**

It is a very common weed of Kharif crops. It is glabrous, succulent creeping and rooting herb prefers damp and swampy soils. It has numerous erect and spreading branches. Leaves are sheathing, linear, acuminate and 3-4" long tapering from base to apex.

Weed seedlings appear in cultivated field, 30 days after crop sowing i.e. beginning of August. It become rare at the end of October and almost disappears in the following months. It flowers within few days and disperses the seeds shortly afterwards. Seedling usually makes vegetative growth upto mid September after which the plant start flowering. Flowers are violet blue small, clustered in the inflated sheaths. The species is very sensitive to soil moisture conditions and over come the dry period in the form of dormant seeds within the soil bed. Phenological observations [Fig. 1] suggested that the weed completes its life cycle along with the crop and a considerable proportion of seeds is dispersed in the field before the crop is harvested.

***Ageratum conyzoides.* L. [Asteraceae]**

It is an erect annual, hispid herbaceous plant. It is a troublesome weed of cultivated fields though it prefer moist conditions but invade both Rabi and Kharif crops also occurs abundantly in waste lands.

Leaves are opposite or upper leaves are alternate, margins crenate and ciliate. Heads many, arranged in terminal corymb. All florets are tubular. Achenes black, pappus 5 scales. [Table-I].

It starts germination in the first week of July and grows vegetatively upto September end then starts flowering which is at peak in January. Seed maturation takes place in March and upto the end of April the plant die. Seeds are equipped with special organs that keep them float in air for considerable distances. These special organs are modified persistent calyx, which is modified into parachute like structure pappus. Seeds have a dormancy period of 2-4 month and by this it over come the extreme hot and dry conditions of the environment.

***Cyperus rotundus* L. [Cyperaceae]**

The weed is locally known as Montha or purple nut sedge. It is a glabrous erect and perennial herb with linear leaves, apparently looking like a grass. Stem single, three angled. Umbell simple or compound. Nuts ovoid-ellipsoid.

It is one of the most difficult weeds of the world, particularly on heavy soils. Though it infests all summer and rainy season crops, Kharif crops are worst affected.

It reproduces extensively through its underground bulbs, which form chains and ramify into the soil upto 60 cm. deep, although majority of these tubers are found in the top 12 cm. of soil. The viability of the weed seeds is hardly 2%.

It is a C_4 plant. The C_4 physiology confers many characters that are advantageous to them under varied field conditions.

***Cynodon dactylon* Pers. [Poaceae]**

The weed is popularly known as doob or burmooda grass. It is a common troublesome grass of cultivated fields and its uprooting is very difficult. It is a perennial prostrate herb, stem slender, creeping rooting at nodes. Culms forming matted tufts with slender erect or ascending flowering branches, 3-12" in height. Leaves normally linear or lanceolate. Sheaths tight, glabrous or hairy.

It is a perennial weed and persists because of its ability to reproduce by vegetative means. It possesses both rhizomes and runners. The runners are located on the soil surfaces while rhizomes grow below the soil surface. In adverse environmental conditions the weed go dormant underground and regrow in favourable time with full vigour. Tillage helps the weed to grow in numbers by fragmenting their vegetative propagules. Each such fragment than puts forth new mother plant. The seeds being less important for its propagation.

Thus it may conclude that weeds are prolific seed producers and also having a safe mechanism to over come adverse environmental conditions through dormancy. Production of few thousand of seeds per plant is common in them (Standifer 1980 and Maniyappa 1973). The weed seed bank in soil ensures their continued progenies year after years. Weed seeds have very high dissemination capacity. As it is clear from the Table-1 that almost all the weeds have seeds of very light weight so they can disseminate very easily by wind. All the weed species studied having seeds, fairly resist to adverse external conditions. They can retain their germinability for a considerable period and their hard seed coats play an important role for the same (Sen 1977). The extended viability of weed seeds in soil is a very important tool with weeds to ensure their everlasting existence against all odds of nature. If all weed seeds present in soil were to germinate at one time, a single tillage operation could wipe them out of seen. But it is not so. Only a fraction of weed seeds present in soil germinate at anyone time, leaving others to germinate later (Gupta 2000). One interesting feature of seed production of weeds specially of *A. longifolius, A. arvensis* and *C. album* is that under stress condition they terminate their vegetative" growth early and ensure production of atleast a few seeds before withering away.

Freshly harvested seeds of *A. longifolius* and *M. indica* could not germinate even after scarified by con. H_2SO_4, clearly indicate after ripening period of the seeds and showed that only the hard coat is not responsible for dormancy but it is the after ripening processes of seeds also (Dubey & Mall 1972). A notable feature of weed species studied is their ability to set seeds quickly. The nature of many germination regulating mechanisms in seeds leads itself to an interpretation in terms of survival value of a species.

In this way it is clear that weeds have highly adaptive strategies against environment and in agriculture we are in confrontation with a very strong enemy in weeds.

REFERENCES

Bhan, V.M. and V.P. Singh (1993). Integrated Weed Management (IWM) an approach. Proc. Int. Symp. Indian Soc. Weed Sc. - Hissar, Nov. 18-20, Vol. I, 289-297.

Dube, P.S. and L.P Mall (1972). Ecology of weed seeds. I. Role of temperature and depth of burial in soil. Oekologia. **5** : 105-110.

Gupta, O.P. (2000). Weed Management Principles and Practices. Agrobios Jodhpur. 25-41.

Kasera, P.K.S. Mohhamad, D.D. Chouhan and D.N. Sen (1999). Biology and adaptive strategies of some weeds in Indian arid Agroecosystem. World Weeds (VI) 1999 pp. 89-97.

Kumar, and Govindra Singh (1997). Efficacy and selectivity of tralkoxydim, alone or in combination with isoproturon in wheat. Indian J. of Agron **42(2)**: 306-309.

Maniyappa, T.V. (1973). Seed protection potential of some common weeds and screening of herbicides for their control. Paper read at All Indian Weed Control Seminar, 11-14 June Hissar (India).

Paul, R and Elmore, C.D. (1984). Weeds and The C_4 syndrome. Weeds Today **15(1)**: 3-4.

Phogat, S.B., V.M. Bhan and Balbir Singh, (1991). Efficacy of some pre and post emergence herbicides either alone or in combination for weed control in wheat. India J. Agron, **36(1)**: 103-104.

Porter, R.H., M. Durell and H.J. Romn (1947). The use of 2,3,6-Triphenyl tetrazolium chloride as a measure of seed germinability Pl. Physiol. **22**: 149-159.

Salisbury, E.J. (1942). 'The Reproductive Capacity of Plants. Bell of Sons. Ltd. London.

Sen, D.N. (1977). Environment and Seed Germination of Indian Plants. The Chronica Botanica Co. New Delhi. pp. 116.

Sen, D.N. (1981). Ecological Approaches to Indian Weeds. Geobios. International Publ. Jodhpur. pp. 301.

Sen, D.N. and P.K. Kasera (1988). Biology of some important Kharif and Rabi weeds in India and arid zone. Proc. of 8th Intern. Symp. on Biology, Ecology and Systematics of Weeds. Dijjion France Vol. 2/2 pp. 325-333.

Sen, D.N. and P.K. Kasera (1996). Ecological Studies in Weeds of Irrigated Fields of Rabi Season in Arid Agroecosystem. In Current Researches in Plant Sciences. Vol. II (eds. T.A. Sarma, S.S. Saini, M.L. Trivedi & M. Sharma). Bishen Singh, Mahendra Pal Singh, Dehradun pp. 239-248.

Sen, D.N., P.K. Kasera and S. Mohammad (1999). Perennation of Weeds in Indian Desert. In Biodiversity, Taxonomy and Ecology (eds. P.K. Tandon and P. Singh) Scientific Publishers. Jodhpur, pp. 355-364.

Singh, R.D. and A.K. Ghosh (1992). Evaluation of herbicides for control of wild oats *(Avena ludviciana)* in wheat *(Triticum aestivum)* Indian J. Agron. **37(2)**: 327-331.

Standifer, L.C. (1980). A technique for estimating weed seed population in cultivated soils. Weed Se. **29(2)**: 134-158.

Verma, A. (1982). Ecophysiological studies of some weeds of Central India with special reference to Alysicarpus longifolius. Ph.D. thesis, Dr. H.S. G. V.V., Sagar.

SECTION FOUR

WILDLIFE ECOLOGY

Advancing Frontiers of Ecological Researches in India, 307-327, 2006
Edited by A.K. Kandya & Asha Gupta
Published by Bishen Singh Mahendra Pal Singh, Dehra Dun, India

Population Dynamics of the Central Indian Barasingha (*Cervus duvauceli branderi*) in Kanha National Park

Rajesh Gopal
Project Tiger, MoEF, Govt. of India, Bikaner House, Annexe V, Shajahan Road, New Delhi 110 011

Abstract: Population dynamics of the Central Indian Barasingha, an endemic species of the swampy grasslands of Kanha National Park has been discussed as this species has been rated as endangered by IUCN. Natality, mortality, sex-ratio and frequency of birth at various sites of the park are estimated for different periods.

INTRODUCTION

The Central Indian barasingha **(Cervus duvauceli branderi)** is a denizen of swampy grasslands, which is endemic to Kanha National Park (Madhya Pradesh). Due to its scarce population, the species has been rated as endangered by the IUCN. Apart from the Central Indian sub-species which is adapted for the hard ground conditions, there are two more sub-species, viz. **Cervus duvauceli duvauceli**-occurring in the Dudwa National Park and its adjoining areas in U.P. and Nepal, and **Cervus duvauceli ranjitsinhi**- occurring in the Kaziranga and Manas National Parks of Assam.

The present study was carried out in the Kanha National Park of Madhya Pradesh from October 1991 to December 1994. The park has a typical Central Indian setting, and forms part of the central highlands, stretching across the state of Madhya Pradesh from west to the east. The protected area falls in two districts - viz. the southern part of Mandla and the north-eastern part of Balaghat.

Kanha National Park is one of the first nine Tiger Reserves in the country, where “Project Tiger” was initially launched during the year 1973-74. With the inclusion of the bulk of “Halon Valley” and parts of the “Banjar Valley”, the park was enlarged to the present size of 940 sq.km.

Conservation History

The National Park has a long and glorious conservation history of almost five decades. Some portions of the Banjar and Halon Valleys were declared as sanctuaries in 1935. Later, in 1955, the Banjar Valley sanctuary (253 sq.km.) was elevated to a National Park.

Floral and Faunal attributes

The core zone (National Park) is almost free of biotic disturbances, as most of the settlements (26) have been relocated under Project Tiger. The evacuated village sites, old shifting - cultivation areas and frost-holes have become heterogenous grass meadows, and constitute almost 11.8% of the total park area. These meadows arc main habitats for the wild ungulate population.

The forest vegetation comprises of pure sal (*Shorea robusta*), mixed forest with sal, mixed forest with bamboo (*Dendrocalamus strictus*), and miscellaneous species.

Forest Types :

(1) Moist Peninsular Sal Forests (3C/C2)

A. High level sal (3C/C2 c i)

B. Low level sal (3C/C2 c ii)

C. Valley sal (3C/C2 c iii)

(2) (A). Southern Tropical Moist Mixed Deciduous Forest (3A/C2 a)

(B). Southern Tropical Dry Deciduous Mixed Forest (5A/C - 3)

Primary consumers (Herbivores): Sambar, Chital, Central Indian barasingha, Barking deer, Blackbuck, Chousingha, Blue bull, Gaur, Wild pig, Hanuman Langur, Mouse deer, Hare, Rhesus monkey.

Secondary Consumers (Carnivores) : Tiger, Panther, Wild dog, Jungle Cat, Grey wolf, Mongoose, Porcupine.

Scavengers : Hyena, Jackal.

Omnivores : Sloth bear.

The Central Indian barasingha (*Cervus duvauceli branderi*) is a highly endangered species, which is endemic to Kanha National Park.

Seasons

There are three distinct seasons, viz. winter, summer and the monsoon. The average rainfall is around 1124 mm. The maximum temperature goes upto 43°C in mid-May, while the minimum touches - 2°C in January. Hail storms are not uncommon in the tract. The important rainfed rivers are : Banjar, Semrahi, Kharadi, Minkur, Ghangar, Baghmar, Bhapra, Salghat, Magar nala, Chuhri, Desi, Neela, Surwahi, Sulkum (Surpan), Dudhania, Taraiya, Kashmiri, Rohini and Halon.

METHODS

The direct diurnal observations were made with the help of a binoculars (10 x 50) and spotting-scope. Supporting photographs were recorded using a 35 mm camera with a 70-210 mm zoom lens. Some events were video-recorded also. A night - viewing device was used for studying the nocturnal activity patterns. All the diurnal observations were made either from a vehicle or an elephant back; those done through the night-viewing device were made from a machan. The direct observations, as above, were the basic, primary methods, which covered all the seasons during the period of study. The field observations were compiled in a PC, and graphically projected as the mean - monthly values for elucidation.

The post-natal care and the survival success of the young ones were studied by direct observations and working out the female : fawn ratio. Some of the non-parametric statistical tests used were : the Spearman correlation test, the U-test and the sign-test.

Various formulae and indices were used for studying the population dynamics. The frequency of births per half month and the calculation of the mean date of birth were done as suggested by Caughley (1977). The number of deer seen at a grassland during the observation was the most commonly used density index. The reproductive status of the Central Indian barasingha was calculated and the regression for probit "y" against "x" was computed, wherever required.

A life-table was calculated from a random sample of dead animals, by ageing the mandibles. Using the data, a survivorship curve was plotted. The population structure was worked out from the mean monthly herd data. The seasonal variation in mortality rate (for adults and fawns) was analysed. The age - structure of the female population which was reared in the Kanha enclosure was calculated from the available data, using the method suggested by Eberhardt (1981). The annual park census data was plotted (1953 to 1994) to ascertain the overall trend of the population. Since the Central Indian barasingha is a highly endangered sub-species with no captive population, the observations were restricted, by and large, to those made on free ranging animals.

Population Dynamics

Natality

The Central Indian barasingha is monestrous, producing only one fawn in a year. At a time, the hind gives birth to a single fawn, and twins have not been observed. The gestation period is almost eight to eight and a half months. The hinds usually conceive in the early part or the middle of January, and new fawns appear from the end of August, reaching a maximum in the middle of September. Fawning, to some extent, continues till the second week of November. Yearlings do not participate in the rut, and the hinds conceive after completing two years of age. Lactating hinds with new fawns also actively participate in the rut. Postmortem of a lactating hind (with full udder) in April revealed a foetus almost two months old. Young males (two years old) are seen in the breeding herd, but do not succeed in copulation owing to their lower hierarchical status.

In the Central Indian barasingha, there is a definite season of birth, following an annual rutting cycle. The season of births has been precisely described, and the average birthday has been calculated as the mean or median of one or more seasons, and the dispersion of the season has been measured by the standard deviation of the dates of birth, (Tables 1 to 3). Using this direct method, the median birth dates for 1992 and 1993 have been computed as the 17th of October and the 18th of October respectively. Another indirect method was also used, and the reproductive status of the barasingha on six different dates during the birth season has been computed (Tables 4 to 5), for 1991 and 1993. The mean prevalence of pregnancy and the mean duration of visible pregnancy, apart from the incidence of pregnancy have been worked out (Tables 6 to 7). The reduction in the potential birth rate has also been highlighted (Tables 8 to 9). The doe-fawn ratio, as computed from the mean monthly values, are shown in Table 10.

The post-natal care in the Central Indian barasingha is conspicuous and important. Pregnant hinds usually stay away from the mixed herds, and bed separately in high grass, close to rivulets. Many such pregnant hinds were seen bedding separately in the high grassland at Sonf. The fawns are littered in these grassy beds, and the "after birth" is carefully removed from the vicinity. Predators and scavengers usually avoid such high grass. The grassy beds of a hind are usually connected by grassy tunnels, and the vulnerable fawn remains concealed in such a habitat for almost 75 to 80 days.

Sex-Ratio

In the yearling class, the sex-ratio is close to 1:1 (Tables 11 to 13), if we ignore the slight preponderance of males (1.22 male/female). The mean sex-ratio of all the classes is close 1:2 (i.e.2.35 hinds/stag) or about 46 stags for 100 hinds. The adult sex-ratio, in general, is 1:2 (1 male: 2 females).

Age Distribution

The age structure of the population, at four different habitats (viz. Oarie, Ronda, Sonf and Sondar-Mukki), was determined by classifying the stags and hinds. The stags were categorized by counting the number of tines and the comparative body size. The hinds were grouped on the basis of relative body size. The yearling males were identified by the antler knobs or spikes. The fawn and the "unsexed" young categories were grouped as per 1:2 ratio (i.e. 1 male: 2 females) (Tables 14 to 15). In general, the age distribution is characterized by a preponderance of older age groups, though the proportion of yearling class (especially females) is good at Oarie, Sonf and Sondar-Mukki.

Mortality

The seasonal variation in the rate of mortality has been analyzed for the years 1990-91 and 1991-92, by separately categorizing the data pertaining to fawns, adult males and adult females. The data has been presented in a combined manner (Table 16) by pooling all age groups and sex classes.

Life Table

A life table has been constructed from a random sample of animals that have died, as ascertained by ageing the mandibles collected in the field. Since the first age group is usually under-represented in the samples, an independent estimation of the first year mortality was done, which indicated almost 50% mortality. A *kdx* series was constructed, assuming a zero rate of increase for the population (Table-17). Since, in reality the population has been increasing steadily for some time, a different constant age distribution occurs, which has been computed by working out the lx series and multiplying each age group by e^{rx} (Table 18).

The observed exponential rate of increase, r :

The log plot of census data from 1953 to 1994, the average annual rate of increase and the exponential rate of increase were computed. Likewise, the average annual rate of increase and the exponential rate have been worked out for the enclosure figures also. This presents a situation devoid of predation.

The observed exponential rate of increase "r", was also calculated from the park census figures for various animals. This value has been worked out for the park enclosure figures also. The value of "r" is higher for the enclosure animals (0.1327), when compared to the free ranging ones (0.046774).

The age structure of the female population pertaining to the enclosure has been analysed from the available data as suggested by Eberhardt (1981). Time units of 1 year have been used, and it has been assumed that the population has grown geometrically. This implies that a constant fraction "r" is added to the total every year. It has also been assumed that the gains exceed losses resulting in the annual increase.

DISCUSSION

The reproductive behaviour, in general, is in accordance with the findings of the earlier workers (Schaller, 1967; Martin, 1978; Sinha, 1979). The rutting starts in the first week of December, and lasts till the second week of February. In Dudwa forest, the rutting season of the swamp deer *(Cervus duvauceli duvauceli)* starts in September and continues upto November, with a peak lasting from mid September to October (Singh, 1982). Dominant stags lead the rutting activities, followed by the subordinated bucks. There is no definite "harem" formation, though the dominant buck gives a short chase to other bucks lower in the hierarchy. However, the dominant bucks do not "possess" a number of does, but simply have the first priority for copulating with an oestrous hind. Schaaf (1978) and Singh (1982) have also not reported "harem" formation in the Northern swamp deer (*Cervus duvauceli duvauceli*). Lactating hinds do take part in rutting, and in this respect the observations differ from those of Martin (1978). Twins have not been observed, and the hind is monestrous. The behaviour patterns associated with rutting are : sparring/ combats, rutting calls (bugling), antler thrashes against vegetation, antler display, flehmen, wallowing, urine sniffing, antler rubbing, antler thrashes against mud and mud throw, chives and mounting. Clutton-Brock *et al.* (1982) have also made similar observations in the red deer (*Cervus elaphus*). Geist (1982) has discussed in detail the sexual competition amongst elk deer and in the mule deer.

There is a definite birth season, following an annual rutting cycle. The gestation period ranges from eight to eight and a half months. Zukerman (1953) has indicated a gestation period of 240 to 250 days for the swamp deer. The present study confirms this apart from the findings of other workers (Martin, 1978; Singh, 1982). The Central Indian barasingha is monotocous, and gives birth to a single fawn at a time - as indicated by Schaller (1967) and Martin (1978), and also by Singh (1982) for the Dudwa swamp deer. The "post-natal care" is conspicuous and important, and the "hider strategy" is adopted to conceal the vulnerable fawns, as observed in the elk deer by Geist (1982).

In the young class, the sex-ratio is close to 1:1, but in the adult class it tends towards 1:2 (i.e. 1 male for 2 females). The age distribution shows a preponderance of older age groups, thereby indicating considerable mortality in the recruitment class (average annual mortality rate = 15%). Amongst the new born, the mortality is almost 50% - mainly attributable to the lack of adequate high grass fawning cover leading to depredation by jackals. The log plot of census figures since 1953 indicates a rise and a decline leading to an all time low count in 1971, followed by a rise and subsequent flattening with an onward declining trend. The observed exponential rate of increase "r" is higher for the population which was in captivity (0.1327) when compared to the free ranging ones (0.046774).

Table-1: Central Indian barasingha : Frequency of births per half month (new borns as observed daily) (SONF : 1991)

Half month	Period Code (x)	x^2	Number of births (f)	fx	fx^2
August (b)	0	0	0	0	0
September (a)	1	1	5	5	5
September (b)	2	4	2	4	8
October (a)	3	9	22	66	198
October (b)	4	16	15	60	240
November (a)	5	25	5	25	125
November (b)	6	36	3	18	108
December (a)	7	49	0	0	0

*Ef = 52; *Efx = 178; *Efx^2 = 684

Mean date : $\frac{*Efx}{*Ef} = 178/52 = 3.42$

Variance : $\frac{*Efx^2 - (*Efx^2) / *Ef}{*Ef}$

$$= \frac{684 - (178)/52}{51} = 1.46$$

Sheffard's Correction :

Variance = 1.46 – 1/12 = 1.38
Standard Deviation "s" = v 1.38 = 1.17

Mean = 3.42 periods (1 period = 15.3 days)
Or 3.42 x 15.3 = 52.3 or days after the mid point of period 0.

i.e. 23 August + 52 days = 15th October
Standard Deviation = 1.17 period x 15.3 = 17.9 days
S.E. = v 17.9 = 4.2 or 4 days

Median date of birth:

$$M = L + \frac{i - 1/2}{f}$$

(*E = Sigma)

where L= the true lower limit of the class in which the sample median lies;
i = the serial number of the desired observation within the class;
f = absolute frequency of the class (median)

$$\text{i.e. } 2.5 + \frac{23 - 0.5}{22} = 2.5 + 1 = 3.5 \text{ periods}$$

i.e. 3.5 x 15.3 = 53.55 or 54 days
i.e. 54 days after 23rd August = 17th October.

Due to strict concealment by the hinds, and also on account of lesser mobility, the new fawns are usually spotted after a fortnight subsequent to birth. Therefore, a period of 15 days needs to be subtracted from the mean birthday calculated as above:

Modified mean birthday = 30th of September.
Modified median birthday = 2nd of October.

Table–2: Central Indian barasingha : Frequency of births per half month (new borns as observed daily) (SONF : 1993)

Half month	Period code (x)	x^2	Number of births (f)	fx	fx^2
August (b)	0	0	0	0	0
September (a)	1	1	2	2	2
September (b)	2	4	4	8	16
October (a)	3	9	20	60	180
October (b)	4	16	8	32	128
November (a)	5	25	3	15	75
November (b)	6	36	1	6	36
December (a)	7	49	0	0	0

*Ef=38; *Efx=123; *Efx²=437

$$\textbf{Mean date: } \frac{*Efx}{*Ef} = 123/38 = 3.24$$

$$\textbf{Variance: } \frac{*Efx^2 - (*Ef)^2 / *Ef}{(*Ef) - 1}$$

$$= \frac{437 - (123)^2 / 38}{37} = \frac{437 - 398.13}{37} = 1.05$$

(*E = Sigma)

Sheppard's correction :

Variance = 1.05 – 1/12 = 1.05 – 0.083 = 0.967
Standard Deviation "s" = v0.967 = 0.983
Mean = 3.24 periods (1 period = 15.3 days)

Or 15.3 x 3.24 = 49.6 or 50 days after the mid point of period 0.
i.e. 23 August + 50 days = 13th October
Standard Deviation = 0.983 period x 15.3 = 15.03 or days
S.E. = v15 = 3.88 or 4 days.

Median date of birth

$$M = L + \frac{i - 1/2}{f} \text{ i.e. } 2.5 + \frac{23 - 1/2}{20} = 2.5 + 1.1 = 3.6 \text{ periods.}$$

i.e. 3.6 x 15.3 = 55.08 or 55 days.
i.e. 55 days after 23rd August = 18th October.
Modified mean birthday = 28th of September.
Modified median birthday = 3rd of October.

1991 (SONF) Table – 3-A

Period	Number of births	Percentage
August 16 to 31	0	0.0
September 1 to 15	5	9.62
September 16 to 30	2	3.85
October 1 to 15	22	42.31
October 16 to 31	15	28.85
November 1 to 15	5	9.62
November 16 to 30	3	5.77
December 1 to 15	0	0.0
Total :	**52**	

1993 (SONF) Table–3-B

Period	Number of births	Percentage
August 16 to 31	0	0.0
September 1 to 15	2	5.3
September 16 to 30	4	10.5
October 1 to 15	20	52.6
October 16 to 31	8	21.0
November 1 to 15	3	7.9
November 16 to 30	1	2.6
December 1 to 15	0	0.0
Total :	38	

Table–4: The Reproductive status of the hard ground barasingha on six different dates during the birth season , 1991 (SONF)

Date (x)	Day	Log day (x)	x^2	No. of pregnant hinds P	No. lacta-ting L	Total B	% birth to date 100 L/B	PROBIT % (y)	xy
3.9.91	1	0.00	0.0	28	2	30	6.67	3.52	0.00
11.9.91	9	0.95	0.90	15	4	19	21.05	4.19	3.98
18.10.91	47	1.67	2.79	10	14	24	58.33	5.20	8.68
9.11.91	69	1.83	3.34	6	9	15	60.00	5.25	9.60
17.11.91	77	1.88	5.53	4	21	25	84.00	5.99	11.26
25.11.91	85	1.92	3.70	0	17	17	100.00	8.71	16.72
*Ex=8.25 *Ex^2=14.26							*Ey=32.86 *Exy=50.24		

Calculation of regression for probit (y) against x :

n = 6
*Ex = 8.25
x = 1.38
*Ex^2 = 14.26
(*Ex)^2 / n = (8.25)^2 /6 = 11.34
SSx = (*Ex^2 – (*Ex)^2 /n = 2.92

y = 5.48
*Ey = 32.83
*Exy = 50.24
(*Ex) (*Ey)/n = 8.25 x 32.86/6 = 45.18
SSxy = *Exy – (*Ex) (*Ey)/n
= 50.24–(8.25)(32.86)/6
= 50.24 – 45.18 = 5.06

The slope b of the regression line and the y intercept are calculated from SSx, the sum of squares of x, and SSxy, the sum of products of x and y:-

b = SSxy / SSx

i.e. b = 5.06 / 2.92 = 1.73

a = y - b x = 5.48 -1.73 x 1.38 = 3.1

The median date of birth "M" = (6-a) / b
= (6-3.1) / 1.73
= 1.68

This is the median on the day scale transformed to logs.
The true median = Antilog 1.68 = 47.86 or 48 days.
i.e. 48 days after 3.9.91 is the 21st of October.

(* E = Sigma)

The standard deviation of the season of births = "S" = 1 / b = 1 / 1.73 = 0.58 log days.

An approximation to the true "standard deviation" is :

$$S = \frac{\text{antilog (M-S)} - \text{antilog (M+S)}}{2}$$

$$S = \frac{\text{antilog (1.68-0.58)} - \text{antilog (1.68+0.58)}}{2}$$

= 84.7 or 85 days.

Table-5: The Reproductive status of the Central Indian barasingha on six different dates during the birth season 1993 (SONF)

Date	Day (x)	Log day (x)	x^2	No. of pregnant hinds	No. lactating L	Total B	% birth L/B	Probit %	xy
6.9.93	1	0.00	0.0	14	2	16	12.5 or 13	3.52	0.00
14.9.93	9	0.95	0.90	10	5	15	33.3 or 33	4.19	3.98
10.10.93	35	1.54	2.40	8	9	17	52.9 or 53	5.20	8.68
22.10.93	47	1.67	2.80	7	11	18	61.1 or 61	5.25	9.60
6.11.93	62	1.79	3.20	4	8	12	66.6 or 67	5.99	11.26
15.11.93	71	1.85	3.40	0	19	19	100.00	8.71	16.72

*Ex=7.8 *Ex²=12.7 *Ey=32.91 *Exy=46.77

Calculation of regression for probit (y) against x :

n = 6	y = 5.5
*Ex = 7.8	*Ey = 32.91
x = 1.3	*Exy = 46.77
$*Ex^2 = 12.7$	(*Ex) (*Ey) /n = 7.8 x 32.91/6 = 42.8
$(*Ex)^2 / n = (7.8)^2 /6 = 10.14$	SSxy = *Exy–(*Ex) (*Ey) /n

$SSx = (*Ex^2 – (*Ex)^2 /n = 2.56$

b = SSxy / SSx = 3.97

i.e. b = 3.97 / 2.56 = 1.55

(* E = Sigma)

a = 5.5 – 1.55x1.3 = 3.5

The median date of birth "M" = (6-a) / b
= (6-3.5) / 1.55
= 1.61

The true median = Antilog 1.61 = 40.74 or 41 days.
i.e. 41 days after 6.9.93 is the 18th of October.

The standard deviation
of the season of births = "S" = 1/b = 1/1.55 = 0.65 log days.

An approximation to the true "standard deviation" is :

$$S = \frac{\text{antilog (M-S)} - \text{antilog (M+S)}}{2}$$

$$= \frac{\text{antilog } 0.96 - \text{antilog } 2.26}{2}$$

= 86.44 or 86 days.

Table–6: Central Indian barasingha : Frequency of births , 1991 (SONF)

Pregnant hinds are *spotted* over four months – (viz. from August to early November)–of the year. Samples taken during each of these months are as follows:-

Months	Number of pregnant does	Total number of does	Ratio of pregnant to total females
August	0	11	0.00
September	15	49	0.31
October	6	9	0.60
November	2	6	0.33

The mean prevalence of pregnancy over this period is:-

$$p = \frac{0 + 0.31 + 0.6 + 0.33}{4} = 0.31$$

The mean duration of visible pregnancy "D" = 1 / 4 = 0.25

[Since pregnant hinds were spotted over four months of the year- with the assumption that pregnancy can be detected over the last month of gestation —— D is measured as a fraction of the length of season over which the pregnancies were South – Caughley (1977)]

The incidence of pregnancy "I" = P / D
= 0.31 / 0.25 = 1.24 or 1
(i.e. an average hard ground barasingha hind produces 1 fawn / year).

Table–7: Central Indian barasingha : Frequency of births (1993) SONF

Pregnant hinds are *spotted* over four months – (viz. from August to early November)–of the year. Samples taken during each of these months are as follows:-

Months	Number of pregnant does	Total number of does	Ratio of pregnant to total females
August	2	27	0.07
September	9	18	0.50
October	7	15	0.4
November	4	13	0.31

The mean prevalence of pregnancy over this period is:-

$$p = \frac{0 + 0.31 + 0.6 + 0.33}{4} = 0.31$$

The mean duration of visible pregnancy "D" = 1 / 4 = 0.25

The incidence of pregnancy "I" = P / D
= 0.34 / 0.25 = 1.35 or 1

Table–8: Herd Composition (Mean Monthly Figures)

Sep. 1994	Males (Adult+ Sub-Adult)	Males (year -ling)	Females (Adult+ Sub-Adult	Females (year -ling)	Total fawn (old+ new)	No. of Pregnant females	No. of new fawns	Total No. of animals
Sonf	13	2	42	12	16	15	3	85
Sondar (Kisli)	11	0	3	0	0	0	0	14
Total	24	2	45	12	16	15	3	99

Table–9: Estimation of the Reduction in the potential birth rate. (Ref. Table – 14) September, 1994 (Mean monthly figures pertaining to Sonf & Sondar (Kisli) combined)

Total Number of animals : 99

No. of Adult + Sub-Adult males : 24

No. of yearling males : 2

No. of Adult + Sub-Adult females : 45

No. of yearling females : 12

No. of fawns (old + new) : 16

(1) Count : 99

(2) Proportion of immature groups in the population :

yearlings : 2 + 12 = 14
fawns : 16
14 + 16 = 30 = 30.3%

(3) Proportion of adults in the population :

$$100 - 30.3 = 69.7\%$$

(4) Sex-Ratio of Adults found :

Males :	Females
24 :	45 or 1:1.9 or 1:2

(5) The actual number of females :
Proportion of adults : 69.7%
i.e. 69 adult animals are present in the population.

Considering the sex-ratio of 1 stag : 2 hinds, this amounts to 46 hinds & 23 stags.

(6) Average litter size : 1 per hind.

(7) Potential number of young born to the population if all the adult females became pregnant = 46

(8) The proportion of pregnant females in the year = 15 i.e. 32.6%

(9) Expected number of births : 15

(10) Loss of potential births : 46-15 = 31

Table-10: No. of Fawns per 100 Hinds (Mean Monthly Value)

Months	Females		Fawns 100 hinds
DEC-92	102	25	24.51
JAN-93	1144	201	17.57
FEB-93	228	76	33.33
MAR-93	83	14	16.87
APR-93	32	0	0.00
MAY-93	0	0	0.00
JUN-93	341	69	20.23
JUL-93	256	26	10.16
AUG-93	829	98	11.82
SEP-93	532	77	14.47
OCT-93	478	71	14.85
NOV-93	381	104	27.30
DEC-93	890	345	38.76
JAN-93	470	260	55.32

Table–11: Herd Data–August/November (Average figures for the month)

Year	August			November		
	Fawns	Does	Fawns/Doe (F)	Yearling Bucks	Does	yearling Bucks/ Does (Y)
1992	3	12	0.25	4	32	0.13
1993	5	16	0.31	5	30	0.16
1994	11	23	0.48	8	44	0.18

Table–12: Central Indian barasingha : Calculation of Sex-Ratio among yearlings from herd composition

Year	(August) Fawns/Doe (F)	(November) yearling Bucks/Doe (Y)	Yearling Bucks as proportion of all yearlings (B)	Yearling Sex-Ratio (Males/Female) (B / [1-B])
1992	0.25	0.13	0.58	1.38
1993	0.31	0.16	0.58	1.38
1994	0.48	0.18	0.47	0.89
Averages :	**0.35**	**0.16**	**0.54**	**1.22**

(In the yearling class, the average sex-ratio is 1.22 males per female).

Table–13: Sex-Ratio in central Indian Barasingha

S No.	Place	Year/Month	No. of hinds per stag (Mean Monthly value)	No. of stags per 100 hinds
1.	Sonf	Aug.'92	2.00	50.0
2.	Sonf	Nov.'92	2.40	41.7
3.	Sonf	Dec.'92	1.87	53.5
4.	Sonf	Feb.'93	3.27	30.6
5.	Sonf	Mar.'93	1.70	58.8
6.	Sonf	Jun.'93	3.08	32.5
7.	Sonf	Nov.'93	2.35	42.6
8.	Sonf	Aug.'94	2.76	36.2
9.	Sonf	Sep.'94	2.40	41.7
10.	Oarie	1994	1.25	80.0
11.	Sondar (Mukki)	1994	2.74	36.5
Mean :			**2.35**	**45.8**

Table-14: Population Structure in Sonf (Age Structure) (Unsexed & fawn categories have been grouped as per 1 male : 2 females ratio)

	Males						Females			
(>5)	(3-5) (Age in years)	(2-3)	(1-2)	(0-1)	(>5)	(1-2)	(3-5) (Age in years)	(0-1)		
Large	Medium	Small	Spike	stag	Male	Large Classi- during the month	Yearling Male	Medi- um	F+NF Female	Total No. +UN
23	17	18	22		207	79	154		51	571
54	37	32	52		235	134	455		98	1097
15	36	25	0		152	73	244		67	612
7	29	29	0		193	77	205		67	607
97	19	27	0		202	55	75		92	567
663	92	74	27		770	119	0		343	2088
859	230	205	101	239	1759	537	1133	479	718	5542
15.50%	4.15%	3.70%	1.82%	4.32%	31.74%	9.69%	20.44%	8.64%	12.96%	100.00%

Table-15: Population structure for sondar (m) (age structure) (Unsexed & fawn categories have been grouped as per 1 male : 2 females ratio)

Males					Females					
(>5)	(3-5) (Age in years)	(2-3)	(1-2)	(0-1)	(>5)	(1-2)	(3-5) (Age in years)	(0-1)		
Large	Medium	Small	Spike	stag	Male	Large Classi- during the month	Yearling Male	Medi- um	F+NF Female	Total No. +UN
1	0	0	0		10	3	1		3	18
15	21	4	12		72	57	37		25	243
19	3	0	2		50	49	10		2	135
15	0	0	0		62	25	0		12	114
20	4	2	1		44	9	0		15	95
90	11	0	0	0	164	36	0		52	353
160	39	6	15	36	402	179	48	73	109	958
16.70%	4.07%	0.63%	1.57%	3.79%	41.96%	18.68%	5.01%	7.59%	11.38%	100.00%

Table–16: Analysis of Seasonal Variation in Mortality Rate (1990-92) (Pooled Data) (All Age Groups & Sex Classes) (Average annual mortality rate: 11%)

Months	Mortality Frequency (f'm)	Exposed to risk (Fm)	Mortality rate (f'm/Fm)	Survival rate 1 – (f'm/Fm)
November	1	418.18 or 418	0.002	0.998
December	4	417	0.009	0.991
January	27	413	0.065	0.935
February	1	386	0.0025	0.998
March	3	385	0.0077	0.992
April	3	382	0.0079	0.992
May	2	379	0.005	0.995
June	1	377	0.0026	0.997
July	1	376	0.0027	0.997
August	-	375	0.0	1.0
September	3	375	0.008	0.992
October	-	372	0.0	1.0
Total :	***E = 46**		**ii = 0.891**	
(CHECK : 0.11+0.891 = 1)				

(* E = Sigma)

Table–17: Life Table for the Central Indian barasingha – assuming zero rate of increase

Age (x) Years	No. of mandibles collected	Kdx	klx	kqx	log kqx	log klx
0	(4)	500	1000	500	2.6990	3.000
1	4	125	500	250	2.3979	2.6990
2	2	62.5	375	166.67	2.2217	2.5740
3	5	156.25	312.5	500	2.6990	2.4955
4+	5	156.25	156.25	1000	3.0000	2.1931
Total :	**16 over one 1000 Year of age**					

[Life table for hard ground barasingha population (both sexes combined) – assuming to be stationary in size with a stable age distribution]

Table–18: Life Table for the Central Indian barasingha

Age (x) Years	N	kdx'	kdx'e^{rx}	kdx	klx	kqx	log kgx
0	*(4)	500	500	458.7	1000	541.3 or 541	2.7332
1	4	125	133	122	541.3	225.4 or 225	2.3522
2	2	62.5	70.7	64.86	419.3	154.68 or 155	2.1903
3	5	156.25	188	172.48	354.44	486.62 or 487	2.6875
4+	5	156.25	198.5	182.11	181.96 or 182	1000	3.0000
Total :	**16**	**1000**	**1090.2**	**1000.15**			

(Both sexes combined)

* **Since the first age group in usually under represented in the mandible collection, an independent estimation of the 1st year mortality was done.**

[Life table for hard ground barasingha population both sexes combined – assumed to have a stable age distribution]

N = no. of mandibles collected.

kdx' = deaths/1000, assuming a stationary population.

kdx'e^{rx} = adjustment for the population rate of increase as obtained from the census data.

kdx = adjusted deaths per 1000.

klx = adjusted survival per 1000 born.

kqx = adjusted age specific mortality.

e = base of the natural algorithm.

x = age.

r = infinitesimal rate of increase.

ACKNOWLEDGEMENT

The author is grateful to Prof. G.P. Mishra and Late Prof. B.K. Shrivastava for their valuable guidance and encouragement during the course of the above study, which forms part of his doctoral work in Dr. H.S. Gour Vishwavidyalaya, Sagar during the period from 1991 to 1995.

REFERENCES

Caughley, Graeme (1977). Analysis of Vertebrate Population. John Wiley & Sons – A Wiley Interscience publication.

Eberhardt, L.L. (1981). Population Analysis. In "Wildlife Management Techniques", 3rd Edition, Ed. Robert H. Giles, The Wildlife Society, Washington, D.C.

Schaller, George B. (1967). The deer and the Tiger. The University of Chicago Press.

Martin, C. (1978). Status and ecology of the Barasingha (*Cervus duvauceli branderi*) in Kanha National Park, India and relevance to Management. Biology and Management of the Cervidae. C.M. Wemmer ed. Smithsonian Institution Press.

Sinha, N.K. (1979). A Contribution to the study of Indian Wildlife – A Thesis submitted to the University of Jabalpur for the degree of Doctor of Philosophy.

Singh, V.P. (1982). Bio-ecological studies on *Cervus duvauceli duvauceli*, swamp deer (Barasingha) in Dudwa Forest near Indo-Nepal border. Ph.D. thesis, pp. 182, Kanpur University. Kanpur

Schaaf, D. (1978). Population size and structure and habitat relations of the Barasingha (*Cervus duvauceli duvauceli*) in Sukla Phanta Wildlife Reserve, Nepal. Ph.D thesis, Michigan State University, USA.

Clutton-Brock, F.E. Guinness & Albon, S.D. (1982). Red deer. Behaviour and ecology of two sexes. Edinburgh University Press.

Geist, V. (1982; 1981). Adaptive behavioural strategies. Chapter 2 in Elk of North America, Ed. J.W. Thomas and D.E. Toweill. Harrisburg, Pennsylvania : Stackpole Books.

Zukerman, S. (1953). The breeding season of mammals in captivity, Proc. Zool. Soc. London, **122 (1)**: 827-950.

Advancing Frontiers of Ecological Researches in India, 329-335, 2006
Edited by A.K. Kandya & Asha Gupta
Published by Bishen Singh Mahendra Pal Singh, Dehra Dun, India

Comparative Study of Wildlife Populations in and Around Kanha National Park

*** H.S. Negi & ** S.P. Bajpai**

Abstract : A comparative account of the wildlife populations in three sites, i.e. core and buffer zones of Kanha National Park and surrounding territorial forest divisions, has been described. The forests fall in the categories of Moist Peninsular Sal Forests and Southern Tropical moist/dry deciduous forests.

INTRODUCTION

When the relationship between the wild animal and habitats is considered, a basic datum of the greatest importance is that of population numbers and their distribution in space. The relationship between an area and the population inhabits in it constitutes the population density. An important part of population study refers to the determination of total number of individuals of a wildlife species in an area. The expression becomes more meaningful and advantageous when expressed in numbers per sq. km. Further, population density is an important measurement in making quantitative comparisons between areas or ecosystems. Population density is both one of the most simple and synthetic way in which the nature of relationship between a given area and all the wild animals making use of it can be understood. Expressions of population density and abundance of wild herbivores have been used to relate estimates of carrying capacity for different habitat types, and for developing management and conservation strategies for wildlife species and habitats under threat.

Carnivores depend solely on the ungulate populations, which form their prey base, for survival. An understanding of the structure and organisation of herbivore populations in a wildlife protected area or managed forests is inevitable for its long-term conservation. This provides information not only on the internal dynamics of various components of the populations of a given species but also indicates how various components of a population interact with each other and their environment along the seasonal gradient.

* Deputy Director, Kanha Tiger Reserve, Mandla (MP)
** Professor, Department of Botany, Dr. HS Gour Vishwavidyala, Sagar (MP)

This paper deals with the comparative study of the populations of some ungulate species in 3 different types of Study Sites. The study was conducted between the years 2000 and 2002.

Study Area:

This study was conducted in Kanha Tiger Reserve and the adjoining parts of the East and West Mandla Territorial Forest Divisions. The profile of the study area is as under (Gopal & Shukla, 2001):

Study Site-I (Kanha National Park): This study site of 940 sq. km. is internationally renowned for its rich habitat, floral and faunal attributes. Besides harbouring a viable population of the tiger and the endangered hard ground barasingha, the National Park also holds typical floral and faunal attributes of Central Indian highlands. After Champion and Seth (1968), the following forest types have been identified in the Tiger Reserve.

(1) Moist Peninsular Sal Forests (3 C/C2)

A. High level sal (3 C/C2 ci)

B. Low level sal (3 C/C2 cii)

C. Valley sal (3 C/C2 ciii)

(2) (A) Southern Tropical Moist Mixed Deciduous Forest (3 A/C 2 a)

(B) Southern Tropical Dry Mixed Deciduous Forest (5 A/C-3)

However, a working classification in the field suggests that the forests are mainly of two types: Sal and Miscellaneous.

As the Core Zone (National Park) of a Project Tiger area should have the least disturbance, all forestry operations have been subordinated to wildlife management practices. As such, no silvicultural operations are carried out and cattle grazing and MFP collection are also prohibited.

Study Site-II (Buffer Zone): The Buffer Zone Division of 1009 sq. km. surrounding the Core is treated as a "multiple use area" is the Study Site - II. This zone is characterised by an interspersion of revenue and forest lands, and harbours numerous villages. This zone has been notified by the Govt. as a separate division under administration of Kanha Tiger Reserve.

Study Site-III (Territorial Divisions): Situated contiguously to the Buffer Zone Division along the West, North and North-Eastern boundaries, the parts of the Bamhni, Maharajpur, and the Motinala and Jagmandal Forest Ranges of the West and East Mandla Territorial Divisions respectively constitute the Study Site-III. The total area of this site is 274.575 sq. km. While the geological, physiographical, and hydrological features of this Study Site are almost identical with the other two Study Sites, it is not a wildlife protected area and all the necessary silvicultural operations and forestry operations are carried out as per the respective Working Plans. Being managed forests; no special protection/ managerial measures are taken for the conservation of the wildlife populations of the area. Wildlife conservation practices in this Study Site are limited largely to routine protection against various modes of poaching, and annual population estimation of tiger and panther population.

MATERIAL & METHODS

The vehicle count method was employed for density estimation of wild animal species in the study area. As all the three Study Sites had a reasonably good network of forest roads, traversing the entire area, covering all the vegetative cover types and topographical features, the vehicle count method (Sale & Berkmuller, 1988 and Rodgers, 1991) was adopted for the estimation of population density in the present study.

Road transects were decided on the map and in the field so that they can be run in a reasonably well stratified sampling area. To avoid any confusion during the counts, the time of counts, identification of the starting and termination points and length of roads were planned. Systemic counts were made with the help of two assistant observers from a four-wheeler. The counting was done between 5.30 am and 8.30 am and 4.30 pm and 6.30 pm in the summer, and 6.30 am and 9.30 am, and 4.00 pm and 5.30 pm in the winter. The speed of the vehicle was maintained between 10-20 km./hr. to facilitate accurate animal counting and recording of the perpendicular distance between the vehicle and the animal group sighted. It was desirable to include as much data as possible since the precision of a density estimate is proportional to the square root of the number of animals counted during a survey (Caughley, 1977). Range finders were also used to measure the distance. The habitat type was also identified and noted down all along. The total road lengths of 140, 68 and 37 km. in the three respective Study Sites were run in the summer and winter (Table-1).

The following modified formula was applied for computing ungulate species density per sq. km.:

D = n / L x 2 x fy

Where, n = No. of Animals Seen on the Transect

L = Transect Length

2 = Two Sights of the Transect

f = Function of y, the Mean Perpendicular Sighting Distance

y = 1.2

Table 1. Road Transects for Population Estimation

Study Sites	Road Transect Code	Length (Km.)	No. of Replicates (Summer: 2000)	No. of Replicates (Winter: 2000-01)	No. of Replicates (Summer: 2001)	No. of Replicates (Winter: 2001-02)	Total Distance (Km.)
I	RT1	56	10	8	9	7	1904
	RT2	37	10	8	12	7	136
	RT3	28	10	7	10	8	980
	RT4	19	8	7	10	8	627
II	RT1	33	10	6	8	5	957
	RT2	14	10	8	10	8	504
	RT3	21	7	8	8	10	693
III	RT1	16	10	5	10	8	528
	RT2	8	10	8	12	8	304
	RT3	13	8	10	10	10	494
Total Sampling:							**8360**

Burnham *et al.* (1980) opine that transect lines should be randomly placed with respect to the animal distribution as this placement actually justified extrapolating results to an area larger than just the immediate vicinity of the transect line. As the forest roads at the Study Sites were used as transects, a certain bias was imminent, as forest roads do not traverse evenly through all the habitat types. Therefore density estimates were computed for the entire respective Study Sites.

RESULTS & DISCUSSION

The results of population density per sq. km. of different ungulate species are presented in the Figures 1 to 3. The mean density per sq. km. of the chital was found to be the highest at the Study Site-I in both the seasons, suggesting a large population of the species. The relatively low densities of chital at the other two Study Sites suggest far smaller populations of this species. In general, the mean population densities of all the target ungulate species were lower at the Study Sites-II and III. This can be attributed to the excellent conservation initiatives, and fewer human settlements at the Study Site-I. Though the Study Sites-II and III are not wildlife protected areas, they still supported reasonably good ungulate populations for their future multiplications. The prey densities recorded in the summer was higher than in the winter. It is probable that due to extreme cold in the winter, fewer animals were sighted. As the barasingha population is restricted to only a few habitats at the Study Site-I, it was not observed at the Study Sites-II and III. Similarly, the sightings of Nilgai were very low due to its small population and conformity to a single area of the Study Site-I. The winter density figures for the Study Site-I, however, were found lower than those based on the annual population estimates of the year 2002, arrived at by the Tiger Reserve Management. Similarly, except for the sambar and bison, the winter density figures for the Study Site-II also recorded the same trend when compared to the departmental population estimates. This can be attributed to the fact that the annual population estimation in the Tiger Reserve is based on the block census method, and the total counts are made for each ungulate species. As far as the Study Site-III, a managed forest area, is concerned the population estimates were far lower than those computed for the Study Site-I.

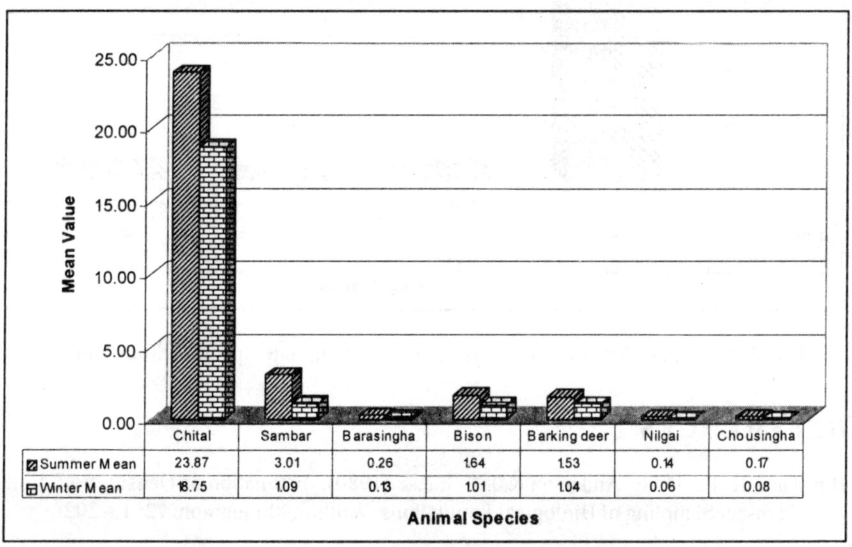

Fig. 1. Seasonal Mean Density (D)/ Sq. Km. of Ungulate Species (Study Site-I)

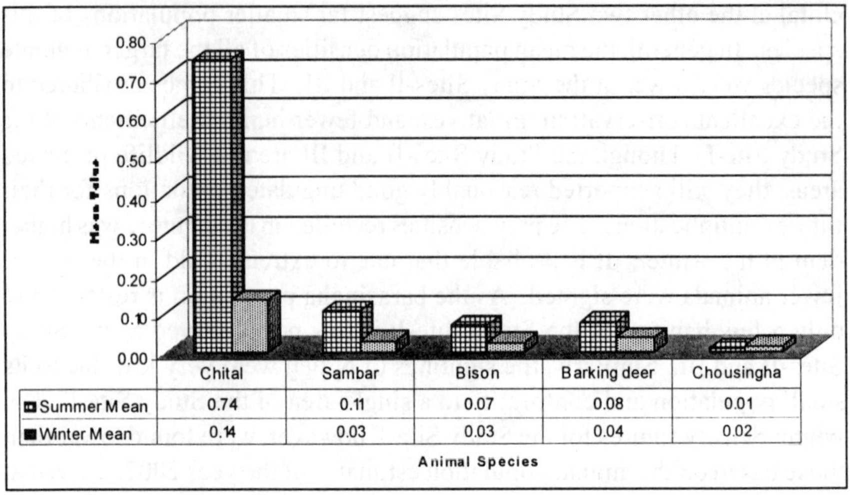

Fig. 2. Seasonal Mean Density (D)/ Sq. Km. of Ungulate Species (Study Site-II)

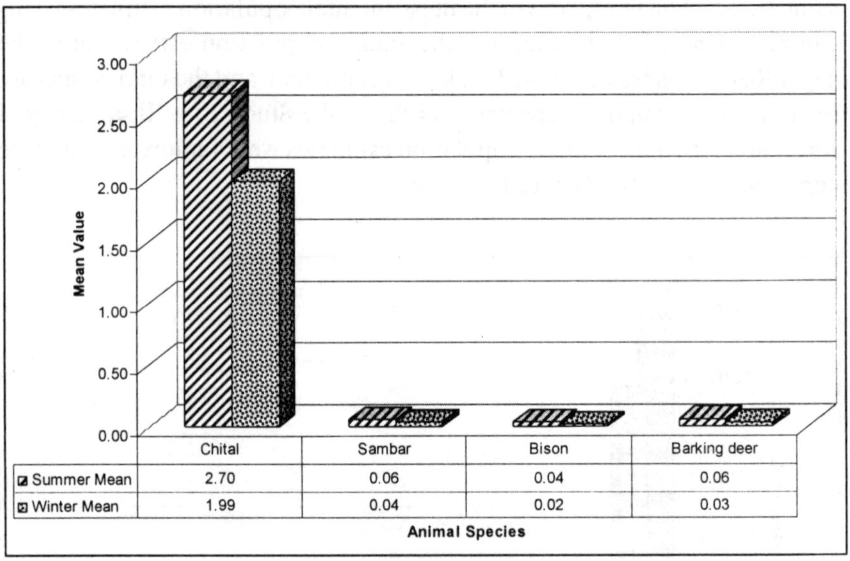

Fig. 3. Seasonal Mean Density (D)/ Sq. Km. of Ungulate Species (Study Site-III)

REFERENCES

Burnham, K. P., D. R. Anderson & J. L. Laake (1980). Estimation of Density from Line Transect Sampling of Biological Populations. Wildlife Monograph. **72**: 1 – 202.

Caughley, G. (1977). Analysis of Vertebrate Population. John Wiley & Sons, New York. pp 234.

Champion, H. G. & S. K. Seth (1968). A Revised Survey of Forest Types of India. Manager of Publication, Govt. of India, Delhi.

Gopal, Rajesh & Rakesh Shukla (2001). Management Plan for Kanha Tiger Reserve (for the period 2001-02 to 2010-11). Project Tiger Directorate, MoEF, Govt. of India, Bikaner House, Annexe V., Shahjahan Road, New Delhi 110.011.

Rodgers, W. A. (1991). Techniques for Wildlife Census in India – A Field Manual, Wildlife Institute of India, Dehra Dun.

Sale, J .B. & K. Berkmuller (1988). Manual of Wildlife Techniques for India. Field Doc. No. 11, Wildlife Instt. of India, Dehradun.

Advancing Frontiers of Ecological Researches in India, 337-345, 2006
Edited by A.K. Kandya & Asha Gupta
Published by Bishen Singh Mahendra Pal Singh, Dehra Dun, India

Research and Monitoring Protocol for the Adaptive Management of the Kanha Wildlife Ecosystem

Rakesh Shukla
Research Officer, Kanha Tiger Reserve, Mandla, M.P.

Abstract: Details have been given in the present paper about various approaches of adaptive management in Kanha National Park, applied from time to time. Monitoring of different parameters indicated the success and failure of these approaches and therefore modifications in working plan and/or methodologies were adopted at times. Increased biotic pressure and reducing number of wild animals, shrinking area of grasslands and number of invading ungulates etc. have been discussed.

INTRODUCTION

Most ecosystems throughout the world are under the threats of various environmental, biotic, economic and demographic factors. The approaches to protect and manage ecosystems, in general, with all their biotic and abiotic components, have been and are being suggested to conserve them, for posterities, for their sustainability and the equitable sharing of the benefits arising out of the utilization of genetic resources.

An ecosystem approach is based on the application of scientific methodologies dealing with the biological processes, structures, functions, organisms and interrelationships among them. The approach also takes into account that humans with their cultural diversities are also an essential component of several ecosystems. Adaptive management is another approach to the management of an ecosystem. This has been defined in various ways by different organizations to suit the mandate assigned to them. One of the definitions is "the rigorous combination of management, research, and monitoring so that credible information is gained and management activities can be modified by experience (Anon, 1995). Smith and Maltby (2003) opine that any adaptive management (or learning-by-doing) system needs to include strong links between managers and the latest scientific expertise on ecosystem functioning and processes. The effectiveness of this kind is dependent in part on the ability of the latter to translate science into practical guidelines and tools. Further, monitoring underpins adaptive management and it depends on adequate feedback mechanisms. Without such mechanism managers cannot respond to change.

Though the Kanha ecosystem is an example of *in-situ* conservation where species are preserved as part of a functioning ecosystem (Shukla, 1998), this is probably for the first time that this paper reviews the research and monitoring protocol under the adaptive management approach to the Kanha ecosystem.

The Kanha Ecosystem and its Significance

Kanha Tiger Reserve, situated in the Mandla and Balaghat districts of Madhya Pradesh, and comprised of the Core Zone (a National Park of 940 sq. km.) and the Buffer Zone (a multiple use area of 1009 sq. km.), is one of the finest wildlife protected areas in south-east Asia and an excellent example of successful in-situ conservation. Though 3 distinct forest types have been identified (Champion & Seth, 1968), the working classification in the Tiger Reserve suggests that the forests are mainly of two types: sal and mixed deciduous.

The Tiger Reserve is managed under a core-buffer strategy, in which the Core Zone is retained free from all exploitations and biotic disturbances, and is insulated from the external influences by a buffer belt. The Buffer Zone harbours ecodevelopment activities through Ecodevelopment Committees for eliciting public cooperation in wildlife conservation (Gopal, 1993).

Kanha Tiger Reserve is a typical geo-physiographical representative of the Central Indian Highlands. The long history of stringent protection against biotic pressures and a range of 10 diverse habitat types within the broad forest cover types along with the positively oriented surrounding villages in the buffer zone ensure the status of Kanha as a promising centre of biodiversity (Shukla, 1998). Besides an endemic small population of the hard ground barasingha (Gopal, 1995) and a viable population of the endangered Indian Tiger rendering this area highly significant, the Reserve also harbours a wide range of faunal species some of which also figure in the IUCN Red List (Gopal & Shukla, 2001). There are 43 species of mammals and several of reptiles. The floral diversity comprises of 609 species and 10 varieties of angiosperms belonging to 386 genera and 104 families, and 17 species of pteridophytes belonging to 11 genera and 9 families (Lal et al., 1986). The above flora of the Reserve also includes around 50 species of aquatic plants and 18 species of rare plants (Parihar & Kotwal, 1989). The Reserve is also rich in lesser fauna with many species of lesser mammals, snakes, insects, molluscs, amphibians, and fishes. Besides, around 300 species of birds have been also been reported in the Tiger Reserve (Panwar, 1982).

Adaptive Management: A Case in Point

In general, the conservation in Kanha wildlife ecosystem has aimed at the maintenance of overall species diversity and the preservation of species extinction. The ecosystem, however, has undergone several ecological setbacks in the past three decades whose partial or full restoration has taken considerable managerial efforts, meticulous planning and timely follow ups. Kanha still faces problems, the chronicity of some of which may also thwart the current conservation efforts to achieve the stated objectives. The Management has also learned lessons from the past expensive ecological fluctuations and uncertainties, and has strived to evolve a research, monitoring and inventorying protocol, under constant refinement, to meet the challenges arising out of the reality of a dynamic and changing ecosystem with non-linear processes. Some of the ecological vacillations faced by the Management are as under:

- A sharp decline in the only world population of the hard ground barasingha (*Cervus duvauceli branderi*) was observed in 1970 when only 66 animals survived (Panwar, 1977 & 1978; Gopal, 1995 &1997).

- Increasing biotic pressure around the wildlife ecosystem with a typical Indian demographic growth and faulty animal husbandry practices (Gopal & Shukla, 2001).

- Most of the anthropogenic grasslands, old abandoned village sites, in the arrested stage of ecological succession, are gradually turning into forest, reducing the total area of grassland habitat, so crucial for the survival of ungulate species.

- General degradation of grasslands due to increasing grazing pressure of thousands of ungulates (Gopal & Shukla, 2001).

- Acute sal heartwood borer (Hoplocerambyx spinicornis) infestation in the ecosystem between 1997 and 1999, affecting thousands of sal trees (Gopal & Shukla, 2001).

Research & Monitoring Protocol

Wise decisions about the use of wildlife resources and their manipulation depend upon the results of quality research and monitoring. There is now growing acceptance that ignorance of science, like ignorance of law is an unjustifiable excuse for environmental abuse. A strong research arm of wildlife wing is of utmost importance for guiding a programme of scientific and rational wildlife resource management in a wildlife protected area. The Reserve Management seriously recognizes the need for research, monitoring and inventorying activities in order to make responsible decisions in its management, planning and operating practices, as well as to gain in-depth scientific understanding of a dynamic ecosystem. These programmes can provide essential information and help accomplish conservation objectives and complete gap analysis for the Tiger Reserve. The Management also ensures cooperation with a wide range of researchers in premier institutions, universities and other govt. and non-govt. agencies. The Management has evolved the protocol on the following considerations and guidelines relevant to the Kanha ecosystem:

Research : Research is one of the major issues in the Plan Outline of the Project Tiger Document, 1972, and is recognized as one of the most important components of wildlife conservation in protected areas. The Document envisaged that the scientific staff of the reserves would undertake research programmes aimed at evaluating systematic factors and influences, for devising pragmatic management practices to cover specific populations and the entire ecosystem. Research constitutes a very important aspect of effective management of wildlife protected area and is crucial for the success of any Tiger Reserve. In Kanha, research projects have been made compatible with the objectives of the protected area. The Management also appreciates the fact that research functions only when its results/ findings are brought together into the decision-making processes.

Monitoring : The Reserve Management also undertakes the monitoring or natural resources designed to perceive changes and quantify trends in resource conditions. This is carried out by collecting and analyzing simple repeated ecological observations in prescribed formats from the same sites (patrolling camps and beat headquarters) throughout the ecosystem to assess and evaluate changes in conditions and progress towards meeting management objectives. This provides valuable information on the relationships between changes in resource conditions and their causes. Monitoring activities also form a feedback between natural resource conditions and management objectives and can serve both to initiate management actions and to evaluate their effectiveness. The main purpose of the ongoing monitoring is to provide a rational basis for taking management actions. In short, the Management believes that the use of monitoring information enhances confidence in wildlife managers' decisions and improves their ability to manage the resources of the ecosystem.

Inventorying : The Management also undertakes inventorying by acquiring, mapping, managing and analyzing information on natural features of the wildlife ecosystem. The process includes the presence, distribution and conditions of plants, specially ground flora, animals, water bodies, and habitat, including human disturbance and illicit grazing etc. In general, Inventories are designed to contribute to a statement of the condition of the resources of the ecosystem. The main difference between monitoring and inventorying is while the former focuses on detecting temporal change, the latter on determining spatial occurrence and distribution.

Initiatives under the Protocol

The Tiger Reserve Management has taken the following initiatives under the research and monitoring protocol for the Kanha ecosystem:

Research:

Doctoral Theses: Kanha Tiger Reserve has been developed as an excellent research centre of the Project Tiger. This has become a study centre for national and international wildlife researchers. The Management encourages problem solving researches relevant to the ecosystem. Besides ongoing researches, there are 9 Ph.D. and 1 D.Sc. theses on various aspects of wildlife in the National Park.

Short Term Projects: There are over 100 research and technical papers relating to the short term projects on wildlife ecology of the ecosystem published in the various national and international journals. Besides, there is also a wide range of status papers, surveys, concept papers, schemes and review articles on Kanha, ensuring the proper dissemination of field knowledge.

Monitoring:

Highly Endangered Species: The highly endangered tiger population is specially monitored for its dynamics and movement patterns, which are duly recorded in the specified proforma-returns, which incorporates various parameters of direct and indirect sightings. The pugmark method for estimating tiger population has also been refined and upgraded and duly documented using computer software. Long-term ecological studies are also under way to document important information on densities, social organization, prey-predator relationship and other ecological parameters.

Endemic Species: The National Park supports the only endemic population of the Central Indian Barasingha (Cervus duvauceli branderi). All the three metapopulations (Gopal, 1995) of this small population have to be monitored very carefully for their fluctuations, trends, mortality and movement patterns. Such qualitative and quantitative information play a very important role in the successful conservation of this species which almost reached the brink of extinction in the early Seventies.

Development of a Spatial Database: Conservation planning and management necessitates detailed inventories of flora and fauna, and appraisal of limiting factors prevailing in the habitat vis-a-vis the spatio-temporal use of the habitat by different animal species. The information which is sought from remote sensing is of spatial nature. As habitat evaluation requires the combining of various variables and spatial analysis, the application of the GIS domain becomes imperative. This represents the latest development in gap analysis technology, using computers to integrate the entire body of data on the natural environment with valuable information on distribution of vegetal cover types. The GIS can also highlight correlations among the abiotic and biotic elements of a protected area. As the GIS contain both spatially and geographically referenced information, it can provide any combination of data at any scale.

The following approaches were employed to achieve the broad objectives of conservation and resource management (Gopal & Shukla, 2001).

- Use of Geo-coded Multi-spectral False Colour Composite (FCC) of IRS-1B/1C/1D for mapping (NRSA)
- Preparation of base map from Survey of India's Topographical Sheets (1: 50,000 scale)
- Updating spatial information using a Global Positioning System (GPS)
- Reconnaissance survey of the study area for correlating the image characteristics and ground features to prepare interpretation key
- Delineation of physiography, current drainage and water bodies using satellite data

- Digitisation of all thematic details viz. compartments, road network, location of villages in the Tiger Reserve, contours, drainage and others using the GIS (ArcView) from 1: 50,000 Survey of India topo-sheets
- Collection of data relating to vegetation from at least 316 sample plots followed by analysis
- Collection of data relating to animal occurrence from transects and plots and their analyses, including the latest version of DISTANCE software, to prepare the animal distribution and abundance maps
- Integration of animal maps with vegetation and compartment maps for obtaining habitat suitability, distribution and abundance maps for major animal species
- Assessment of various habitat parameters and suitability ranking of habitats for producing habitat suitability maps
- Survey of non-vegetal unique habitats (compartment-wise) using GPS and their mapping, apart from data collection on the use of these habitats by different animal species
- Mapping of landuse practices in the Buffer Zone and evaluation of their impact on the Tiger Reserve and connected corridors for producing thematic maps on anthropogenic related factors (grazing, MFP collection, fuel wood collection and the like)
- Linking of spatial and non-spatial data in the GIS domain for preparing habitat suitability and management maps

Since the ecosystem is always dynamic it is imperative to update and repeat the analysis. The GIS readily facilitates this, and thus such a database in the GIS domain becomes valuable for PA management.

Tiger Habitat Evaluation & Monitoring System: The Project Tiger Directorate, GOI, has also undertaken an ambitious pilot project to institutionalize a protocol and monitoring design for evaluating and monitoring tiger habitats in all the wildlife ecosystems and general forest areas (Gopal & Jhala, 2004).

Inventorying:

Status Surveys: Kanha is a typical tropical ecosystem and harbours considerable heterogeneity. The Management has, over the years, encouraged intensive floral surveys by premier institutions. These surveys have provided excellent databases on the general vegetal status of the wildlife protected area, identifying various plant, grass, forb, and sedge species up to their families and genera. Besides, such surveys are also undertaken for preparation of Management Plans. The current Management Plan (Gopal & Shukla, 2001) has undertaken a sampling intensity of 0.026% of the total vegetation area for assessing the vegetal status of the Tiger Reserve. Such documentation also includes the phytosociology and diversity indices of forest and ground flora.

Photographic Album: Drawing significantly on the resource persons of various established research institutions, a photographic album of ground flora has also been prepared to facilitate easy identification by the park staff. Further additions to the above have also been ensured by the Management.

Indices: Wildlife conservation being the main objective of the wildlife protected area, the Tiger Reserve has to employ several methodologies for evaluating the richness of wild animal species. This annual and periodical evaluation includes estimation of many attributes of population of each species found in the Tiger Reserve. Besides, presence / absence status of the species of relatively low importance is also documented. Several avian surveys have also been undertaken in the Reserve and have provided authentic checklists of birds. All the patrolling camps have been provided with monitoring / inventory registers requiring a host of simple information relating to animal and plant species to be filled in daily by the park staff. This body of information forms a very effective database for management policies.

Cover Types: The ecosystem, being rich in floral and faunal attributes, has invited many studies / reconnaissance in the past relating to habitat / cover mapping using remote sensing techniques (Gopal & Shukla, 2001). The current Management Plan has identified 8 vegetal classes and 16 cover types through the Geo-coded Multi-Spectral Satellite False Colour Composite of the IRS-1B/1C/1D taken by the NRSA, Hyderabad. The extent of each cover type including area percentage has also been evaluated.

Grasslands: Grasslands in Kanha constitute the most important part of the habitat mosaic sustaining the prey populations and, in turn the predators and co-predators. These grasslands are being invaded by the forest crop and have gradually shrunk to around 7% of the National Park. This shrinkage has been documented through the digital image analysis by the IRS 1D LISS III Multi-Spectral Data from the NRSA, Hyderabad. This phenomenon has been taken note of for future expansion of grasslands for retaining the typical habitat mosaic and carrying out effective conservation practices.

REFERENCES

Anon. (1995). Sustainable Ecosystem Management in Clayoquot Sound: Planning and practices. Victoria, BC. Canada.

Champion, H.G. & S.K. Seth (1968). A revised survey of forest types of India. Govt. of India Press, FRI, Dehra Dun, Uttranchal

Gopal, Rajesh (1993). Project Tiger & People - Kanha experience, Kurukshetra.

Gopal, Rajesh (1995). The Biology & Ecology of Hard Ground Barasingha in Kanha National Park. Ph.D. Thesis, Dr. H. S. Gour Vishwavidyalaya, Sagar (MP).

Gopal, Rajesh (1997). Geographical Isolation of the Central Indian Barasingha (*Cervus duvauceli branderi*) and Conservation Imperatives for the Future. Proceedings of the 11th World Forestry Congress.

Gopal, Rajesh & Rakesh Shukla (2001). Management Plan for Kanha Tiger Reserve (for the period 2001-02 to 2010-11). Project Tiger Directorate, GOI, New Delhi.

Gopal, Rajesh & Y. V. Jhala (2004). Monitoring Tiger Status and Habitat: A Field Guide. Project Tiger Directorate, GOI, New Delhi.

Lal. J.A. Kumar & P.C. Kotwal (1986). The Botany of Kanha Tiger Reserve, Madhya Pradesh, India. Bio. Mam. **12 (1):** 1- 85.

Panwar, H.S. (1977). Success with the Branderi Barasingha (*Cervus duvauceli branderi*) in Kanha National Park. The Working Committee of the IUCN/ SSC/ Deer Specialist Group held at Longview, Washington.

Panwar, H.S. (1978). Decline and Restoration Success of the Central Indian Barasingha, *Cervus duvauceli branderi*. Proceedings of the Working Committee of the IUCN/ SSC/ Deer Specialist Group. Sept.29 to Oct. 1, 1977. Longview, Washington.

Panwar, H.S. (1991). Kanha National Park: A Handbook. Developed by Centre for Environ. Education, Ahemdabad.

Parihar, A. S. & P. C. Kotwal (1989). Management Plan of Kanha National Park and Project Tiger Kanha for the Period 1989-90 to 1998-99. Project Tiger Directorate, GOI, New Delhi.

Shukla, Rakesh (1998). Biodiversity Conservation with Special Reference to Kanha Tiger Reserve. Indian Forester. **124 (10):** 819-824.

Smith and Maltby (2003). Using the Ecosystem Approach to Implement the Convention on Biological Diverstiy. Key issues and case studies. IUCN, Zurich, Switzerland.

Advancing Frontiers of Ecological Researches in India, 347-359, 2006
Edited by A.K. Kandya & Asha Gupta
Published by Bishen Singh Mahendra Pal Singh, Dehra Dun, India

Insects and their adaptation with reference to diapause

B. Prasad
Department of Life Sciences, Manipur University
Imphal-795003, (Manipur)

Abstract: Diapause is an unique phenomenon of insects for survival during adverse conditions. It is actually an adaptive mechanism of dormancy and is determined genetically. The author has discussed embryonic, larval, pupal and adult diapause and also various types of diapause.

INTRODUCTION

The insects are the unique creatures of diversed groups among animals. They can be easily recognised by the presence of 3 pairs of jointed legs and highly cuticularised body covering. They belong to one of the biggest class - insecta (Hexapoda) under the largest animal phylum - Arthropoda (Joint footed organisms). Having their evolutionary origin during carboniferous period of Palaeozoic era, the insects have conquered almost all the ecological habitats. hence, there is no exageration in referring them as omnipresent. Their highly adaptable features have enabled them to lead an aerial, arboreal, burrowing, parasitic, terrestrial, sub-terranean, aquatic as well as cecidogenic (gall forming) ways of life. According to an estimate, out of 40 millions species known in the world, more than half of the living organisms belong to the phylum Arthropoda alone which has been estimated to be 80% of the total fauna. That is why, it has been regarded as the biggest phylum in the animal kingdom. Insects constitute the biggest class not only under this phylum alone but even the whole animal kingdom. They comprise of approximately 7,50,000 to 8,50,000 described species so far under 30 orders. Such an estimate indicates the unique richness of the insect diversity having more than 6% of the global species. In order to achieve this level of tremendous diversity, the insects and plants both must have entered into a process of symbiosis for their co-existence.

Life grows only in a congenial environment as it evolved when the environment became suitable. As such, life and environment are inseparably interlinked. The author intends to discuss the adaptations in insects which are not connected to their morphology, anatomy or else but something beyond these which is of developmental nature. We are all aware that during certain seasons/periods of the year, many of the insects disappear from our sight. It might be because of heat, cold or certain other invisible factors. All do not die, where do they go? For instance, the most familiar/noctural insect, Cockroach, is not seen even during night in winter. Most of the mosquitoes do not bite during certain parts of the year. There may be many factors/ reasons, but one of these is attributed to a phenomenon of diapause which is an unique adaptation among insects. Metamorphosis is yet another phenomenon during their post-embryonic developmental stages. Both of these are of developmental nature.

Diapause in insects was for the first time, pointed out by Hennegut (1904). Since then it has been well studied and has been regarded as an unique adaptation among these creatures. It has been defined variously as under :

- It is a developmental phenomenon for survival during adverse conditions.
- It is a phenomenon of suppressed development.
- It is an adaptive mechanism of dormancy during the period of unfavourable conditions.
- It is a state of arrested growth and development.
- It is a phenomenon of quiscence which may occur due to sudden rise or fall in temperature.
- It is not a rhythmic event, but it occurs but once in the life of an insect.
- It is a genetically determined state of suppressed development, the expression of which may be controlled by environmental factors.

Simple quiscence such as during heat or cold, differs from diapause in that the quiscence is a direct response to deleterious conditions. Quiscence/ hibernation/estivation is terminated as soon as the environmental conditions return to biologically acceptable range. Diapause, on the other hand, typically begins long before the onset of unfavourable conditions and may not be terminated until long after the disappearance of such conditions.

Types & occurrence of diapause

In temperate climate where winter temperature is too low for development, certain insects enter in a state of diapause at some stage of their life cycle. On the other hand, the insects living in warmer regions, undergo diapause depending on the severity of their environment. However, many insects inhabiting tropical regions get adapted to survive with out diapause. In most of the insects, this phenomenon occurs only in one stage, characteristic of the insect species. Thus, it may be grouped as follows :

A. **State-wise diapause**:- Based on the life history stages, it has been reported to be of the following types:

1. **Embryonic or Egg diapause**:- The insects belonging to the species like *Bombyx mori*, *Melanoplus maxicanus*, *M. differentialis*, *Aedes species*, etc. undergo diapause during their egg stage/embryonic stage when the embryo is fully developed. The significance of entering diapause during this initial stage of developmental cycle is to delay the embryonic development. Example - *Bombyx mori* (Kogure, 1933).

2. **Larval/Nymphal diapause** : Larval stages/Nymphal stages are juvenile/ young stages in insect life cycles. Larvae are the juvenile feeding stages of holometabolous insects like Butterflies and Moths. These stages are markedly different from their adults. The nymphs, on the other hand, are also the younger stages of hemimetabolous insects. They are almost similar to their parents except in size and development of wings. Many insects undergo diapause during these stages of their life cycles and become quiscent and non-feeder during certain parts of the season/ year. For instance the rice stem borer *Chilo suppresalis*, sugarcane stem borer, *Diatraea saccharalis* and mosquito, *Aedes triseriatus*. The larval diapause is induced by the prevailing environmental conditions in order to prolong the larval/nymphal duration.

3. **Pupal diapause** : Pupal period in insects is the resting stage in almost all the holometabolous insects. However, this stage is lacking in Heminetabolons insects. Inspite of being initially a resting stage in the life cycle, many insects enter diapause during their pupal period in order to prolong the pupal period and metamorphosis. For instance, most of the lepidopteran (butterflies & moths) insects undergo diapause in this stage of their life cycle. *Pieris brassicae, P. rapae, P. napae, Antheraea proylei, A. pernyi, A. mylitta, Hyalophora cecropia* etc. enter diapause in their pupal stages.

4. **Adult diapause**: The insects belonging to the orders Heteroptera, Coleoptera & Diptera undergo diapause even during their adult stages in order to overcome unfavourable conditions by delaying their reproductive activities. Adult diapause is most frequently induced by photoperiods experienced during larval stages as well as pupal stages. Ex. *Aedes togoi, Culex pipiens, An. maculipennis* (Mosquitoes), *Drosophila phalerata, Musca auturmnalis* (Dipteran flies).

Diapause therefore, is an unique adaptation in insects life which may occur at any growth stage (Egg, Larva, Pupa and Adult) at least once within the developmental history. However, diapause occurs only at a quite specific stage. Relatively a very few insects have been found to be capable of diapausing at more than one point in their life cycle. The classical example is *Bombyx mori*, the mulberry silkworm. It displays a facultative embryonic diapsuse in response to photoperiod experienced by females during late embryonic and late larval life. The adult diapause is most frequently induced by photoperiods experienced during the larval stages (Kogure, 1933). This type of parental determination of diapause in the progeny has been reported in a number of insect species.

B. Types of diapause depending upon the instinct :

Depending upon the instinct of the insect species there may be two types of diapause:

1. **Obligatory diapause** : Some of the insect species undergo diapause instinctively in one or the other stage of their life cycle without fail in every generation. Such species are generally univoltine or monovoltine and the diapause in such species is an obligatory diapause. Such a diapause is intrinsic in nature and genetically induced. In such a case, every individual of every generation undergoes a period of diapause as part of its life history, regardless of the environmental conditions prevailing during its development. Muller (1957) described it as Parapause. It is a type of diapause which is genetically induced and not subjected to environmental modification. The termination of parapause is controlled by an environmental factor-either temperature or daylength but not both. Such a diapause occurs in certain silkworms belonging to *Antheraea* species and Eggs of *Orgyria* spp. This is a regular phenomenon in many insects. The eggs of *Orgyria* which are laid during late summer and antumn do not hatch until the following May. There is a single generation of adults on the wings from July to October.

2. **Facultative diapause** : This is an occasional phenomenon in some insects in which this is not of a regular occurrence. This is characteristic of bivoltine and multivoltine species. Some generations may be completed without any diapause stage while in other generation, some or all insects may enter diapause. This is entirely an extrinsic phenomenon and is entirely under the control of the environmental factors. An univoltine but potentially multi voltine population is one in which diapause is facultative, but the population is adapted to a geographical area in which the yearly cycle of day length, temperature, and so on, never includes values outside of the diapause inducing range. Thus, there may be genetically similar univoltine and multivoltine populations of the same species. As per the Muller's classification (Muller, 1957) diapause is induced by photoperiod and is terminated only by several weeks or months of exposure to low temperature. While oligopause may be induced in response to seasonal changes of temperature and or photoperiod and is terminated by a reversal of inducing factors. Typical example of such a diapause is *Bombyx mori*.

Initiation and Mechanism of diapause

The diapause is a genetically determined state of suppressed development, the expression of which may be controlled by the environmental factors. It is an important adaptive mechanism for dormancy during the periods of unfavourable environmental conditions, like low winter temperatures, extreme summer heat, periods of drought and non-availability of food. This phenomenon gets signal mainly through the above mentioned unfavourable situations. The onset of diapause sets in even while the prevailing conditions are still suitable for morphogenesis. Therefore, even before the diapause begins, most of the insects lay down the increased fat reserves in their body probably to meet the demands of their body during the diapause period. The onset of diapause in obligatorilly diapausing insects is genetically determined while in facultalively diapausing insects the metabolic changes are initiated by some signal from the environment which indicates the advent of unfavourable conditions. The effects of these internal and external phenomena are integrated in the nervous system which controls the activity of the neuroendocrine system. The neuroendocrine cells of the brain stimulate the neurosecretory cells in the sub-oesophageal ganglia to secrete diapause inducing hormone. The failure of growth in absence of hormones is suggestive of the diapause and infact, it may result from the failure of neuroendocrine activities of brain. The diapause is initiated when the neurosecretory cells become inactive because of the accumulation of fat reserves in the body, slightly before the onset of diapause. The inactivity of neurosecretory cells affect the neuroendocrinological activities which adversely affect and disturb the normal sequence (Fig. 1 & 2) of the activities.

Various hypotheses have been putforth to explain the inactivity of the neurosecretory cells - i) Long photoperiod and high temperature maintains normal neurosecretoty activities where as short photoperiod and low temperature inactivates the neurosecretory cells. ii)According to another view, the neurosecretory cells are inactivated by the accumulation of food reserves in the body. Breakdown of these food reserves stimulates the neurosecretory cells and initiate the termination of diapause. iii) According to yet another view, diapause involves a complex of synthetic reactions with several phases aerobic and anaerobic and reversible and irreversible. Low temperature during diapause slows down the breakdown reactions so that the substances produced by synthetic reactions accumulate. Diapause in eggs which are only just begining to develop must involve a completely different mechanism due to the absence of endocrine system at this time. In *Bombyx mori* eggs, the diapause is caused by diapause inducing hormone produced by neurosecretory cells in the sub-oesophageal ganglia of the female parent.

Processes during diapause development

Diapause is a developmental stage in the insect life. The Physiological processes involved in the completion of developmental stage are collectively known as diapause development. This term was coined by Andrewartha, (1952) who defined it as the physiological development or physiogenesis which goes on during diapause stage in preparation of the active resumption of morphogenesis. It is now well known that the insect in diapause is not physiologically inactive. Although, there are physiological differences in physiological processes between diapausing and non-diapausing stages, especially in neutroendrocrine functions, neurosecretory functions are apparant during diapause. The insect is capable of co-ordinated behavioural and sensory responses specially during larval and pupal diapause. The developmental processes involved during diapause occur in the following 4 (four) distinct phases in a sequence :

i) **Preparatory phase** : During the onset or even just before the onset of diapause, the insect ready to enter into diapause, gets the environmental signal through the daylength or photoperiod especially alongwith other unfavourable conditions. After this, accumulation of metabolic reserves particularly fat occurs. For instance, the Pink bollworm, *Pectinophora*

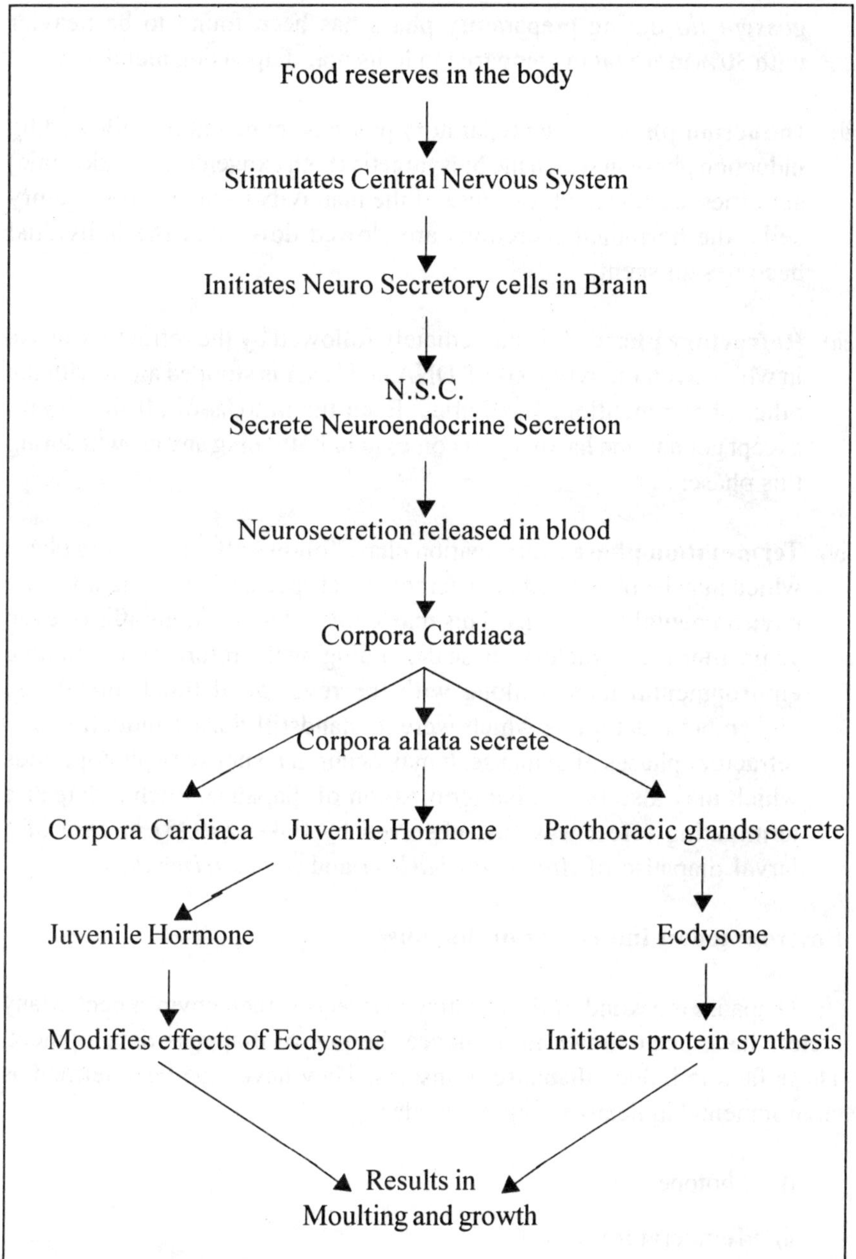

Fig.1. Sequence of events during normal conditions before diapause.

gossypiella during preparatory phase has been found to be heavier with 30% more fat in comparision to its non-diapausing members.

ii) **Induction phase** : The preparatory phase is immediately followed by induction phase in which the biosynthetic (both exogenic and endogenic) activities are lessened. Because of the inactivity of the neurosecretory cells, the hormonal secretions are slowed down and the individual becomes quiscent.

iii) **Refractory phase** : It is immediately followed by the refractory phase in which even the synthesis of DNA and RNA is stopped alongwith the other above mentioned activities. Even the mitosis of all the tissues except gonads and haemocytes comes to halt stopping any growth during this phase.

iv) **Termination phase** : Termination phase follows after refractory phase which may be prolonged in different insect species depending upon the environmental conditions. This may come after week, months or even years after the refractory phase depending on the return of the suitable environmental factors along with the reversal of biochemical and biosynthetic activities which were at standstill during induction and refractory phases of diapause. It may occur in response to photoperiods which may also hasten the termination of diapause. Such a diapause terminating effect was first observed by Baker (1935) in case of a larval diapause of *Anopheles barberi* and *Aedes triseriatus*.

Environmental indicators of diapause

Diapause is a wonderful adaptation in insects in their environment. Many factors have been shown to influence the rate of diapause development. These factors induce diapause in insects. They have also been termed as environmental indicators. These include :

i) Photoperiod

ii) Temperature

iii) Nature of the Nutrition

iv) Physiology of parents

v) Sensitive stages

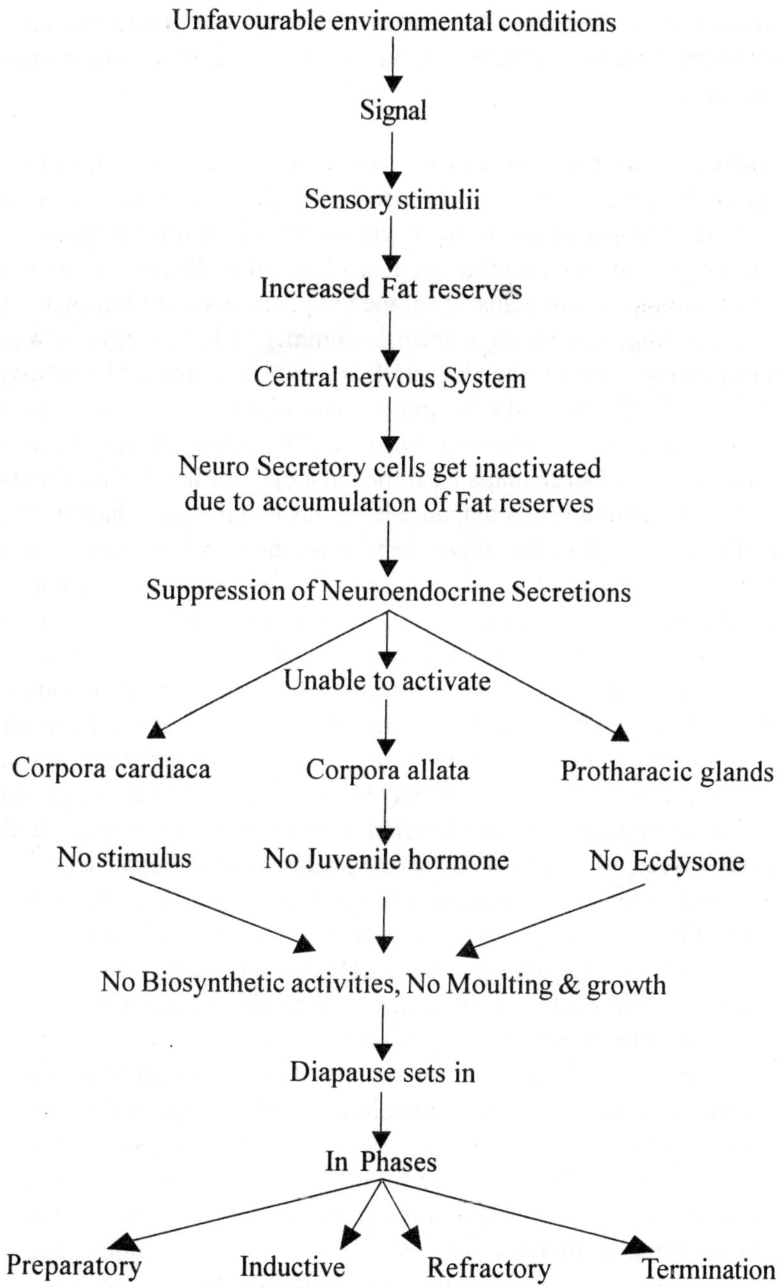

Fig. 2. Sequence of events during onset of Diapause.

In some cases there is either concurrent or sequential interaction among those several factors, making diapause development a very complex phenomenon.

1. **Influence of Photoperiod on diapause** : The synchronisation of growth and reproduction in insect species with favourable seasons within their range of dispersal and of their activity with the daily rhythm of light and darkness, are of extremely high survival value. Diapause and many forms of polymorphism in insects of the temperate zone are season bound. Outside the tropics, long days occur in summer and short days in winter with increasing or decreasing daylength in spring or autumn. The relatively short days of autumn herald the approach of winter and for many species they act as a stimulus initiating diapause. The pupae of *Acronycta*, for instance, do not undergo diapause if the daylength during the larval period exceeds 16 hrs, but do enter diapause if the day length is less than 16 hrs. In case of the cabbage worm, *Pieris brassicae*, the pupal diapause was not induced at any temperature when the insects were reared under constant light (LL). In constant darkness (DD) and low temperature, high incidence of diapause development was observed. Short day photoperiod (12D:12L) caused 100% diapause at 24°C temperature. Whereas 10 hr dark and 14 hr light or 6 hr dark and 18 hr light induced no diapause at all. Essentially similar reactions occur in many other insects in temperate regions in which a small difference in photoperiod may bring about complete change in the type of development. The photoperiod interacts with temperature and in long day insects, the critical day length, below which diapause occurs is often longer at lower temperature. The geographical cycle of daylight and darkness (Photoperiod and scotoperiod) provide a channel of communication between an organism and its environment (Way and Hopkins, 1950). Through this channel, the daily seasonal and latitudinal informations may be communicated to the living system which exploits the sources of temporal informations. The daily photoperiod supplies the time signals needed for the synchronisation of the internal functions. Such biological responses to daylength and the photoperiod have been found to play essential role in distribution, mating behaviour and developmental sequences including diapause. In case of *Acronycta ruminis*, short day (less photoperiod) initiates diapause whereas in *Bombyx mori*, exposure to long days stimulate diapause in the eggs of next generation. In many cases the photoperiod plays vital role in insect life.

2. **Influence of temperature on diapause**: Temperature like photoperiod also plays a vital role during diapause development in insects. In temperate

regions, the, high temperature suppresses and low temperature enhances the tendency to enter diapause. In some insects, like photoperiod, the temperature also acts as a sign stimulus. For instance, in *Bombyx* a high temperature induces diapause while low temperature prevents it. The temperature and photoperiod interact with each other and if a long day insect like *Diataraxia*, is bred under experimental conditions of light and temperature, short day length is dominant and induces diapause irrespective of temperature unless the later is very high. On the other hand, with long photoperiod, temperature is dominant, low temperatue inducing diapause and high tempeature prevents it. In case of *Momoniella*, temperature acts independently of photoperiod and chilling the females causes her to lay eggs which will give rise to diapausing eggs. In continnous darkness (DD), however, low rearing temperatures resulted in a high incidence of diapause, but at 20°C or higher, only non-diapausing pupae were obtained. Short day photoperiod (12D :12L) caused 100% incidence of diapause at all temperatures below about 25°C, but no diapause occured at temperatures above 28°C.

3. **Influence of dietary factors on diapause** : The diet of a phytophagous insect does not remain biologically uniform during the growing season as the plants are developing and changing biological systems. The food plants of spring generation in a multivoltine insect species may be quite different chemically from that on which late summer or early autumn generations feed. Though, the photoperiod and temperature are the major factors controlling the determination of diapause, a modifying influence of the diet has been observed in a number of cases. Andrewartha (1952) suggested that in individuals destined to enter diapause, the food reserves in them accumulate so that the neurosecretory cells are not stimulated. The food reserves can only be broken down after diapause development ends and only then the neurosecretory cells are stimulated and the processes leading to morphogenesis, are resumed. Complete or partial absence of food materials stimulate the appearance of diapause in the mosquito, *Aedes triseriatus*. The different nutritional constituents play their individual role in determining/inducing diapause. For instance, the oil content, in the diet of Pink bollworm, was founed to influence the induction of diapause. The effect of the liquid content of the diet was found to be of secondary importance. The question of the role of the host plants on the diapause determination has been investigated by a number of workers. Jermy & Saringer (1955) could however, detect no nutritional effect. These workers concluded that the temperature and photoperiods were the only inducing factors. But, however, a co-relation between the incidence of diapause and the lipid protein ratio

was detected in certain beetles by some other workers in the field. Similarly, the quality of the food has been found to influence the diapause in many lepidopteran insects.

4. **Influence of Parental Physiology on diapause** : The physiological state and age of the parents have been found to exert a great influence on the process of diapause (Simmonds, 1948). For instance, in adult Locusts, the solitary phase females lay 100% diapausing eggs whereas the females in Gregaria phase lay only 42% diapausing eggs. The rearing conditions affect the physiology of adult and in turn, the incidence of diapause in eggs which these insects lay. Accordingly older gregaria females lay more diapausing eggs than the younger females. In certain insects where diapause is reported in the eggs like those of *B. mori*, it undergoes a completely different mechanism due to the absence of endocrine system at this stage. In them the egg diapause is caused by a diapause inducing hormone produced by the neurosecretory cells in the sub-oesophageal ganglia of the parent. So parental physiology is also an important aspect of diapause induction in any of the life history stages.

5. **Influence of Sensitive stages** : The environmental factors initiating diapause, act before the onset of unfavourable conditions and they may act sometimes before diapause begins. As a result, these stimulii may operate on a different stage from that which undergoes diapause. For instance, *Bombyx mori* in which appropriate stimulation on the eggs of the present generation, causes diapause in the eggs of next generation.

Significance of diapause

Diapause, as has been said earlier, is a wonderful adaptation among insects which without involving any morphological or anatomical changes enables the insects to overcome the odd environmental conditions. It is said that had there been no such phenomenon among insects, there would have been complete loss of certain insect populations. Because of its adaptive nature, it brings about delay in development to facilitate survival through the critical unfavourable conditions. It brings about the synchronised emergence and helps in finding out suitable mates in the population. In temperate regions, the diapause is concerned with survival during cold winters but in tropics, it facilitates survival during dry seasons during which there is acute shortage of food and moisture. It is comparable to migration and enables the insects to live in an area which at times, is unsuitable for them. Therefore, it is a developmental phenomenon which occurs once atleast in the life cycle of most of the insects.

REFERENCES

Andrewartha, H.E. (1952). Diapause in relation to the ecology of insects. *Biol. Rev.*, **27**:50-107.

Baker, F.C. (1933). The effects of photoperiodism on resting tree hole mosquito larvae, *Can. Ent.*, **67**:149.

3. Beck, S.D. and Honey, W. (1960). Diapause in European corn borer, *Pyrausta nubilalis* (Hubn.). *J. Insect Physiol.*, **4**:304-319.

Chapman, R.F. (1973). *The insects : Structure and Functions*. The English Universities Press Ltd., London. pp-717-731.

Corbet, P.S. (1956). Environmental factors influencing the induction and termination of diapause in the Emperial dragonfly, *Anas imperator* Leach (Odonata - Aeshnidae). *J. Exptl. Biol.*, **33**:1-14.

David, W.A.L. and Gardiner, B.O.C. (1952). Laboratory rearing of *Pieris brassicae* L. and *Apantelis glomeratus* L. *Proc. Roy. Ent. Soc., London*, **27**:54-56.

Dickson, R.C. (1949). Factors governing the induction of diapause in the Oriental fruit moth. *Ann. Ent. Soc. Ame.*, **42**:511-537.

Fucuda, S. (1952). Function of the pupal brain and sub-oesophageal ganglion in the production of diapausing and non-diapausing eggs in silkworm. *Ann. Zool. Japan*. **25**:149-155.

Jermy, T. and Saringer, G. (1955). Die Rolle der photoperiode in der Auslosung der diapause des *Lepteriolarsa decimlineata. Acta. Agron. Acad. Sci., Hung.*, **5**:419-440.

Kogure, M. (1933). The influence of light and temperature on certain characters of the silkworm, *B. mori. J. Dept. Agri., Kyushu Imp. Umi.*, **4**:1-93.

Lees, A.D. (1955). *The physiology of diapause in Arthropods*. Cambridge Univ. Press. England. pp-151.

Muller, H.J. (1957): Uber die diapause von *Stenocranus minutus* Fabr. 9Hom-Auchonorrhyncha). *Beitr. Entomol.*, **7**:203-225.

Simmonds, F.J. (1948). The influence of maternal physiology on the incidence of diapause. *Phil. Trans. Roy. Soc. London*. 23, 385.

Williams, C.M. (1956). Physiology of insect diapause. X-An endocrine mechanism in the diapausing pupae of *Cecropea* silkworm. *Silkworm. Biol. Bull*, 110-201-218.

Way, M.J. and Hopkins, B.A. (1950). The influence of photoperiod and temperature on the induction of diapause in *Diataraxia oleracea* L. (Lepidoptera). *J. Exptl. Biol*, **27**:365-366.

REFERENCES

Advancing Frontiers of Ecological Researches in India, 361-365, 2006
Edited by A.K. Kandya & Asha Gupta
Published by Bishen Singh Mahendra Pal Singh, Dehra Dun, India

Wild Life of Manipur - Problems and Prospects

K. Premkumar Singh
Forest Department, Govt. of Manipur, Imphal

Abstract: Manipur is a small state in the north eastern fringe of India bordering Myanmar. It is included in the same zoological zone as that of the forest region of the Himalayas. It has a variety of vegetation ranging from tropical to temperate type of vegetation. The distribution and character of faunal species is also similar to that of the Himalayas, but in addition distinctly. Malayan types of fauna are found all through the state with an intermingling of peculiar forms. This is the reason for diversity and peculiarity of wild life found in Manipur.

MAMMALIAN FAUNA

Many of the species found in Manipur are endemic to the region. Sangai (*Cervus eldi eldi*) is the most endangered and endemic species found in Manipur. It is found in nature only at Keibul Lamjao National Park. Hoolock gibbon, the only ape species found in India occurs in all the hill districts of Manipur. The Slow Loris, Stump tailed macaque, Pig tailed macaque are the rare primates that occur in the forests of Manipur. Among the carnivores clouded Leopard and golden cat are rare nocturnal animals found in Manipur. Tiger and Leopard, still make a quite living in the remote parts of this state. Of the two, leopard are more common whereas the tigers are seldom reported. Barking deer, Sambar, Hog deer and Serow are some of the ruminant species found in Manipur. Serow lives mostly in the steep rocky mountains of all the hill districts of Manipur. The Himalayan Black bear is another animal, which commonly occurs in the hills of Manipur. The elephant makes seasonal migration in to the state in the Indo-Myanmar border areas along the Taret river course and also in the Western part of the state from Assam.

The Himalayan Black Bear is another animal which commonly occurs in the hills of Manipur. The rare & elusive Malayan sun Bear is restricted to the Indo-Myanmar boarder areas particularly the Yangoupokpi Lokchao Wildlife Sanctuary.

Though true badgers are not found in India, the Hog Badger and Ferret Badgers are widely distributed in the hill of Manipur. The elusive lesser carnivore, the ferret Badger are seldom sighted owing to its nocturnal way of life. The Pangolin, Binturong, spotted Linsang are small schedule -I animals found in the Forest of Manipur. The spotted Linsang and Himalayan yellow-throated martin are found in the watershed of Barak River. The Pangolins occur in the Indo-Myanmar border areas and the Binturong though seldom sighted are found in the forest of Senapati district.

AVI-FAUNA

Manipur is rich in Avi-fauna also Mrs. Hume's bar back pheasant (*Syrmaticus humiae*), Blyth's tragopan, Khalij pheasant, Red jungle fowl and Burmese Peafowl are some of the important pheasants found in Manipur.

The Great Indian Horn bill, the Refus Neck hornbill, Wreathed Hornbill, Lesser pied Hornbill, the Brown back horn bill are also found in the Keilam, Hill Range of Churachandpur district.

The spot bill duck, shoveller , Gadwall, Common teal , Wigeon, Pintail, Blue winged teal, Ruddy shelduck Red Crested Pochard white eye pochard, Lesser whistling teal T. Hooded Crane , Indian white breasted water hen, Indian purple moorhen , Coot, Indian pond heron, Chinese pond heron, are some of the Water birds found in Manipur

REPTILIAN FAUNA

Beside mammalian fauna and Avi- fauna, Manipur is also the home of many reptilian fauna. The snake group includes the large sized, fierce looking, though non-poisonous pythons. Other snakes like cobra , Kraits viper, etc. are the important Snakes of Manipur. Among the lacertilians the most common species are Garden Lizard Land monitor Lizard and Water monitor lizard. A variety of tortoise and turtle are the chelonian reptiles found in the lakes ,streams and other water bodies of the State.

PROBLEMS AND PROSPECTS

It is unfortunate that a large number of Avi-fauna and fauna have become extinct from the state of Manipur. The Javan Rhinoceros was seen roaming along the Khuga river valley in Churachandpur District as late as the beginning of 20th Centuary, but is now extinct from Manipur.

The Banteng known as "Santhou" in Manipur was last seen thirty years ago in New Somtal area.. In all probability, these animals have vanished from Manipur soil forever. Many other endangred species like Hoolock gibbon, Slow Loris, Stump tailed macaque, clouded Leopard, golden cat, spotted linsang, Malayan sun bear, Hog badger, Malayan giant, squirrel, Serow etc. are living in a precarious condition in the Forest of Manipur.

Among the birds, the white winged wood duck, Pink headed duck, Grey leg goose, Mallard, Brahmany duck, clucking teal, Plover, Hooded Crane, White ibis, Indian shag, Open bill stork, black necked sports and a host of others have become extremely rare and threatened.

Table-1: Wild Animals found in Manipur

Common Name	Zoological name
Sangai (Brow antlered deer	*Cervus eldi eldi*
Hoolock gibbon	*Hylobates hoolock*
Slow loris	*Nyetecebus coucang*
Stump tailed macaque	*Macaca speciosa*
Pig tailed macaque	*Macaca nevuestrainu*
Clouded leopard	*Neofilis nebulosa*
Goldan cat	*Feties temenincki*
Tiger	*Panthera rigris*
Leopard	*Pantherea paredus*
Barking deer	*Muntiacus muntijak*
Sambhar	*Cervus unicolar*
Hog deer	*Axis procinus*
Himalayan black bear	*Helaretors malayans*
Elephant	*Elephus maximus*
Hog badger	*Aretonys colaris*
Ferrat badger	*Melogale personata*
Pangoline	*Manis cressicaudata*
Binturong	*Artictes binturong*
Spotted linshang	*Prionodon pardicolor*
Himalayan yellow threated matin	*Martes flavigula*
Mrs. Humes bar back pheasant	*Syrinaticus humiae*
Blyth's tragopan	*Tragopan blythii*
Khalij pheasant	*Lophura leucomelana*
Red Jungle fowl	*Gallus gallus spadiccus*
Burmese peafowl	*Pavo muticus*
Great India Hornbill	*Buceros bicornis*

Common Name	Zoological name
Lesser piedhornbill Hornbill	*Anthracocerus malabaricus wreathed*
Spot bill duck	*Anas poecilarhvncha poecilarhvncha*
Shoveller	*Anas chpeatodia*
Gadwal	*A. Strepera strepera*
Wigeon	*A. Penelope*
Blue winged teal	*A. querquedula*
Pintial	*A. acuta*
Ruddy shelduck	*Tadorna tadorna*
Red creasted poachard	*Netta rufina*
White eyed pochard	*Anas nvraca*
Leser whistling teat	*Dendroeygna javanica*
T. Hooded crane	*Grus monacha*
Indian white breasted watuehen	*Amaurarnes phoenicurus*
Coot	*Fulica atra*
Indian pond heron	*Aredeola grayii*
Chinese pond heron	*Aredeola bacchus*
Open bill stock (Tharoichabi)	*Anastomus oscitans*
Black napped blue	*Monarcha zurea*
White winged wood duck	*Cairina scutalata*
Pink headed duck	*Rhodonessa caryophyllacca*
Black necked crane	*Grus nigricollis*
Cobra	*Naja naja kauthura*
Krait	*Bugarus fascratus*
Viper	*Tumeresurus monticola*

The reason for diminishing our wildlife is not far to seek. An explosion of human population with diverse and intense human activities had far reaching consequences on wildlife. Deforestation for commercial and shifting cultivation resulting in habitat destruction coupled with indiscriminate hunting have threatened many species with extinction. Reckless use of insecticides and chemical fertilizers is also another factor for diminishing wildlife.

In order to check the reckless killing of wildlife in the state , as early as 1931, Captain C.L.W. Harvey , the then President of Manipur State Durbar promulgated the Game Rules of Manipur. The Wildlife Protection Act 1972 was introduced in Manipur State from 15th May 1973 thereby offering legal protection to all forms of wildlife in the State, and under section 64 of the said Act, Wildlife (Protection) Manipur Rules, 1974 have come into force w.e.f 21st June, 1974. The Manipur Zoological Garden was established on 2nd October, 1976. for preservation and display of indigenous species.

The Keibul Lamjao area covering an area of 40 sq. kms. was declared as a National Park in 1977 for preserving Sangai , the most threatened and endemic species and its unique habitat.

All out efforts are being made by the Forest Department with the active co-operation of the local people to save Sangai from extinction. As a result the population of Sangai has increased from 14 heads in 1974 to 180 in 2003.

The moist deciduous forest in the Indo-Myanmar border area harbour a variety of rare fauna and flora. An area of 184.4 sq kms. of these forest was declared as a wildlife sanctuary in 1989. With proper Management and legal Protection as Wildlife sanctuary, there is hope that these rich forests with their denizens could be preserved for posterity.

The Keilam hill Range that host-endangered species like clouded leopard. Hoolock gibbon, Hornbills etc. has been notified as wildlife Sanctuary. The Siroy Kasom hill Range, the habitat. of the endangered Lily, Siroy Lily (*Lilum macklineae*) and also the home of Mrs. Humes Bar Backed pheasant and Blyth's tragopan are notified as proposed National Park. Similarly, the Bunning hill range, Zeliad Lake and its surrounding area, and Jri Makru Forest has also been notified as proposed wildlife sanctuaries.

Though all out effort are made by the department, preservation and conservation of fauna and flora in a sustainable way will not be possible without the active co-operation of the local people.

REFERENCES

Anonymous (1994). Forest Deptt., Govt of Manipur Record of wild animals donated to the Manipur Zoological garden, Iroishemba.

Gupta, Asha (2003). Inventorization of Forest Wealth of Yangoupokpi Lokchao Wildlife Sanctuary Project report submitted to Forest Deptt, Govt of Manipur, Imphal

Tombi H. Singh and R.K.S. Singh (1994). Loktak Lake, Manipur. World Wide Fund for Nature, India Mammalian Fauna of Keibul Lamjao National Park, Reptilian fauna of the Keibul Lamjao National Park. Avi fauna of the Keibul Lamjao National Park.

SECTION FIVE

MICROBIAL ECOLOGY

Advancing Frontiers of Ecological Researches in India, 369-395, 2006
Edited by A.K. Kandya & Asha Gupta
Published by Bishen Singh Mahendra Pal Singh, Dehra Dun, India

Microbial Community Dynamics of Compost Ecosystem

Seema Rawat and B.N. Johri*
Department of Microbiology
College of Basic Sciences & Humanities
G.B. Pant University of Agriculture &Technology
Pantnagar-263145, U.S. Nagar, (Uttaranchal)

Abstract : Compost is an important man-made ecosystem, which is of great relevance from microbial ecological point of view as it represents a complete spectrum of microbial diversity. It harbours both mesophilic and thermophilic microflora; former constituting the pioneer community while latter, the climax community. The quickly changing physico-chemical conditions during composting are likely to govern the microbial community composition and succession, which in turn regulate the compost quality. The understanding of *in situ* functionality of each microbial component provides a way to manipulate the compost environment and hasten the process of composting besides improving the quality of compost. The use of cultivation-based and non-cultivation-based approaches provides us a complete insight into the community structure and function.

Key Words: Composting, compost ecosystem, microbial diversity, mesophilic microflora, thermophilic microflora, pioneer community, climax community, microbial community succession.

INTRODUCTION

The large amounts of organic material produced in nature are decomposed on a regular basis by microorganisms and converted into humus, indicating existence of operative cycles of nutrient turnover in the natural ecosystems. The natural process of decomposition, however, takes place at a slow pace which can be accelerated by creating conditions ideal for growth and activity of the decomposers. This accelerated process of decomposition of organic matter by a mixed population of microorganisms in a warm, moist aerobic environment, has been termed as "composting". It can be either 'passive' or 'active'; involving complex physico-chemical interactions between the organic matter and decomposer, resulting in the

* Corresponding Author

end products: compost (soil-like material composed of more resistant residues of the organic matter, breakdown products and dead and living microorganisms along with products formed from further chemical reaction between these materials) which can be utilized as fertilizers, substrates for mushroom production or biogas (methane), carbon dioxide, water and heat. The overall process of composting has been illustrated in Fig. 1. Regardless of the products formed, the active component mediating biodegradation and conversion process during composting is resident microbial community. Therefore, optimization of compost quality is directly linked to composition and succession of microbial communities in the composting process. This means that tools are required to monitor and characterize microbial communities during composting and to relate them to compost quality. While analysis of the presence and distribution of different operational taxonomic units (OTU) within a population provides insights into the ecological functioning of communities, the study of taxonomic structure alone limits insight into the ecological relevance of the community structure. The exclusion or addition of different microflora does not necessarily changes

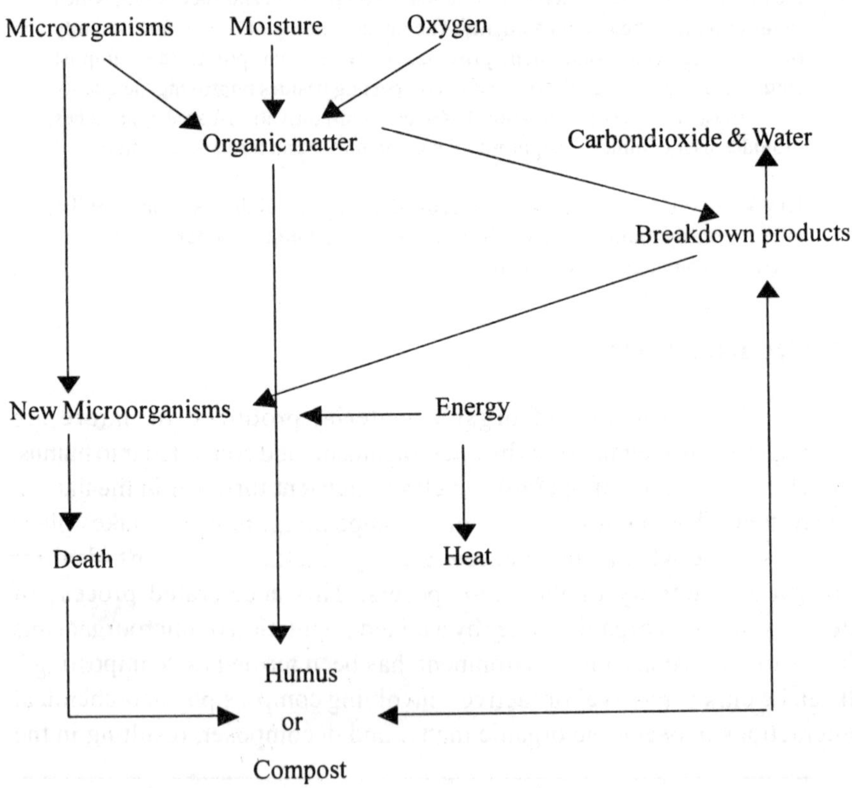

Fig. 1: Process of composting (Biddlestone and Gray, 1985)

the resultant functions of the community (White and Findlay, 1988). The change in abundance of an individual group may not necessarily equate to meaningful shift in the community function. Cultivation approaches employed in the past to study microbial populations inevitably favoured growth of some community members due to selective nature of media and thus it could have excluded majority of the indigenous microbes (Troussellier and Legendre, 1981). Moreover, the time consuming nature of isolate-based methods severely limits the spatial and temporal intensity of sampling and the associated description of the population structure (Garland and Mills, 1991).

Cultivation-independent methods have now been used to characterize the microbial community succession and structure. These include assessment of microbial diversity based on *in situ* extracted phospholipids (Herrmann and Shann, 1997; Boggs *et al.*, 1998; Klamer and Baath, 1998), measurement of carbon source utilization by substrate extracted microbial cell consortia (Inssam *et al.*, 1996; Boggs *et al.*, 1998), and nucleic acid-based techniques (Kowalchuk *et al.*, 1999; Peters *et al.*, 2000).

This chapter deals with the physico-chemical aspects of composting along with microbial community structure and functionality in compost ecosystem. Available information on several aspects of compost ecosystem is still scanty due to horizontal advancements in some areas; also it represents a complex ecological system from the point of microbial distribution and activity. Readers will find a bias towards mushroom compost ecology especially pertaining to thermophilic mycoflora since this ecosystem provides a good material for study of microbial ecology owing to the controlled conditions under which the crop is grown and the relatively short time required to complete the ecological cycle; secondly, this is also on account of our own active engagement during the past many years with the ecology of thermophilic mycoflora in this habitat.

COMPOSTING

Composting represents an astonishing example of solid-state fermentation (SSF) wherein a crude variety of wastes such as sewage sludge, refuse, animal manure, industrial wastes, food wastes, leaves, tree bark, agriculture residues, abattoir residues etc. can be treated through microbial route irrespective of their suitability as feed-stock for compost production (Satyanarayana and Grajek, 1999). Refuse (municipal solid waste) is partly compostable but poses problems due to its extreme heterogeneity as it consists of food scrapes (garbage), paper, glass, plastic, metal, sweepings, yard waste, ash etc. The organic rich fraction after separation can however be composted; the whole municipal solid waste can also be passed through the composting stage (mass composting), possibly with subsequent segregation (Satyanaryana and Grajek, 1999).

The basic aims of composting according to Miller (1994) are: (1) achievement of a suitable bulk density (compost makes a more physically stable landfill and can be easily stored, transported and disposed off than the original material as bulk density of former is higher), (2) modification of complex polysaccharides and plant materials, (3) biological removal of readily available nutrients to avoid overheating, (4) building up of an appropriate biomass and a variety of microbial products, (5) establishment of selectivity, (6) conversion of nitrogen into stable organic form, and (7) sanitation i.e., killing of pathogenic microbes, larvae and weeds.

The composting is carried out *via* either batch or a continuous mode. Batch mode is a long process comprising of four sequential phases, the mesophilic phase, thermophilic phase, cooling phase and the curing phase. The initial stage of decomposition of mass of organic matter, initiated by mixing and wetting the substrates, the mesophilic stage is governed by the mesophilic microflora, which uses up the readily available nutrients. The aerobic fermentation (composting) commences as a result of growth and activity of microorganisms resulting in release of heat, ammonia and CO_2 as by products along with other unpleasant smelling compounds. The metabolic activity of the mesophilic microorganisms results in rise in temperature and paves way for development of thermophilic microflora, which initiates the second phase of composting. This phase of composting starts very rapidly and may last days, or weeks or even months. It is the thermophilic stage that results in maximum decomposition of organic matter besides sanitation. During the cooling stage, mesophilic microflora recolonizes the compost and it is during this phase that more resistant organic matter is degraded. The final phase of composting is called curing, aging or maturing stage, which is long and an important one since it provides a safety net for destruction of the pathogens. Uncured compost can produce phytotoxins, besides depriving soil of oxygen and nitrogen and can contain high levels of organic acids. In the continuous mode of composting, similar phases occur but they are not as apparent as they are in the batch mode and these could occur concurrently rather than sequentially. A well-designed continuous system can eliminate the need for a mesophilic stage and operate continuously at thermophilic temperatures. This mode offers a means of decomposition of putrescible materials quickly under close process control. Municipal compost is prepared by batch mode while garden composting is performed in a continuous mode. Mushroom compost and Vermicompost, though prepared by batch mode, are significantly different from general municipal compost. Mushroom compost is prepared very rapidly (18-24 d) and does not involve curing stage. It is prepared, either by long method (LMC) or short method (SMC), from various agro-residues viz., wheat straw/ paddy straw/ sugarcane bagasse as a base material along with other additives viz., chicken manure, calcium ammonium nitrate, urea, superphosphate, muriate of potash, wheat bran, gypsum as additives. LMC is the primitive, cheap method involving only one phase (without pasteurization) (Mantel *et al.*, 1972). SMC is a quick method constituting of a general advance in controlled composting (Sinden and Hauser, 1950) and involves two sequential phases: phase I (an uncontrolled self-heating process initiated by mixing and wetting the ingredients as they are stacked in windrows which are periodically turned and watered at approximately two days interval) and Phase II which is the indoor process of pasteurization, carried out in tunnels.

Vermicomposting is a method of decomposing organic materials viz., straw, shredded newspapers, saw dust and horse manure using surface feeding worms especially *Eisenia foetida* (redworm). The worms primarily feed on the microbes that are actively involved in the break down of the organic matter.

PHYSICO-CHEMICAL ASPECTS

The quickly changing physico-chemical conditions during composting select for a succession of microbial communities and thus exert a profound effect on the entire process. Amongst the various parameters, composting is affected mainly by temperature, ammonia, carbon dioxide, moisture and C: N ratio of the substrate. The mass of decomposing organic materials is an exception to most ecosystem as it results in not only intensive heat production due to the metabolic activity of microorganisms but also acts as an an effective retention system resulting in a significant rise in temperature. Generally, self-heating occurs when organic materials are assembled provided there is sufficient mass, atleast one ton, for insulation, and that moisture, aeration and nutrition level are adequate (Satyanarayana and Grajek, 1999).

The temperatures in the composting stack usually reach as high as 70^0C within 2 or more days (Finstein and Morris, 1975). A peak temperature of 70^0C can be attained within 4 days in metro-waste composting system meant for processing 150 tons of general municipal refuse daily (Kane and Mullins, 1973). A maximum temperature of 67^0C was observed in wheat straw composting after 8 days, remained above 50^0C for about 3 weeks (Chang and Hudson, 1967); on the other hand, composting of grass cuttings can attain 66^0C within 16 h, reaching a maximum of 76^0C after 48 h before a decline. The difference in temperature recorded is attributed to the type of material and other factors, such as age of the plant material (Satyanarayana and Grajek, 1999).

The size of the compost pile is crucial for not only temperature build up but also for maintenance of appropriate microbial equilibrium and successional pattern. In phase I of mushroom compost, a gradient of 20^0C (outer region of stack) to 70^0C (center) exists while phase II runs at lower temperature of 50 to 60^0C (Johri and Rajni, 1999; Peters *et al.*, 2000). The high temperature (75-80^0C) of compost pile is necessary to induce Millard reactions i.e., fixing of free ammonia through reactions with carbohydrates and lignatious polymers. Millard reaction transforms carbohydrates into chemical form which is accessible to the mushroom crop but not to competitors, and hence, provides for compost selectivity. High temperature is also necessary for the chemical incorporation of nitrogen into stable form within the compost (Johri and Rajni, 1999).

The initial pH of the compost varies between 6.0 and 7.0 and rises by 0.9-1.8 units during the first week due to the release of ammonia but later drops off (Chang and Hudson, 1967). During vermicomposting the pH changes from acidic to neutral while in municipal waste (Stutzenberger *et al.*, 1970), poulty manure-saw dust mixtures (Galler and Davey, 1971) and grass composts (Forsyth and Webley, 1948), initial pH of 4.5-6.0, stabilizes within 7.0–9.0, possibly due to the loss of organic acids through volatilization and microbial decomposition and the release of ammonia through mineralization of organic nitrogen. Peters *et al.* (2000) have reported the decrease in pH from 5.6 to 4.5 within the first two days of mushroom composting followed by an increase to 7.0 while Rawat (2004) reported variation in pH from 6.7 to 8.6 (end of phase I) with phase II composting pH stabilizing at 7.8.

A moisture content of 50-60% is ideal for composting and microbial activity. The ratio between dry matter, water and air is an important factor in composting because raw materials have different moisture contents. In mushroom compost, the optimum moisture is generally 75% at the time of filling, 69% at the time of spawning and 66% after spawn run (Johri and Rajni, 1999). Rawat (2004) reported a variation of 60% (zero day) to 56% (filling) in moisture content during phase I and 62% to 56% during phase II. An increase in electrical conductivity (8.28 to 33.86 mmhos.cm^{-1}) occurs from zero day to end of phase I of mushroom compost with a sharp decline to 23.11 at peak-heat stage, and slight increase (25.60) at the end of phase II (Rawat, 2004).

The changes in oxygen concentration in the compost atmosphere between turnings represent net result of O_2 utilized by microorganism and that replenished by convection and diffusion through compost. The change in O_2 concentration is related with temperature. In the center of compost pile where the temperature is highest, anaerobic conditions exist. When the stack is turned, heat loss occurs and the resulting temperature of about 50^0C stimulates microbial activity and depletes oxygen. Turning of the stack also results in a change in physico-environment whereby the existing temperature gradients are disturbed. The loss and recovery of oxygen in the compost atmosphere is thus largely due to combined influence of temperature, which affects microbial activity and changes the ventilation pattern (Johri and Rajni, 1999). The CO_2 evolution on the surface of compost pile shows good correlation with microbial activity compared to the number of propagules. For example, it has been shown that whereas the total number of propagules is low during peak heating, CO_2 evolution is high due to higher rate of respiration of the abundant thermophilic microflora (Satyanarayana and Johri, 1984). Herman *et al.* (1977) reported that decomposition during composting was dependent upon carbon, nitrogen, lignin and carbohydrate composition of the organic material and expressed this relationship as:

$$CO_2 \text{ evolved} = \frac{\%\ \text{carbohydrate}}{\text{C/N residues} \times \%\ \text{lignin}}$$

The rate of decomposition in compost pile is maximum during the first few days followed by a decline and ceases after sometime although large quantities of cellulose and hemicellulose are still present. In mushroom compost, about 60-70% polysaccharides are consumed by compost microorganisms, however, only 15-25% of total polysaccharides are used during mushroom production. The absolute amount of lignin remains unaltered though changes occur in the degree and condensation of lignins during composting because general microflora is devoid of basidiomycetes (Iiyama *et al.*, 1996).

The rate of decomposition is markedly influenced by size as well as C: N ratio of the composting material. The variation in composition (% by dry weight basis) of municipal compost to garden composts ranges from 25.0-80.0 (organic matter), 8.0-50.0 (carbon), 0.4-3.5 (nitrogen); a C: N ratio of 30: 1 is considered ideal for the activity of most microbes (Biddlestone and Gray, 1985).

MICROBIAL ECOLOGY

Compost is an interesting example of man-made ecosystem that harbours a complete spectrum of microbial diversity. The microbial abundance, composition and activity changes substantially during the composting process and are correlated with high microbial diversity and low activity in mature compost.

Microbial community succession during composting is a classical example of how the growth and activity of one group of organisms can create conditions necessary for the growth of others. Several generations of microorganisms succeed each other during composting wherein each crop of microbial form utilizes the available material in the substrate as also the cellular components of its predecessors for growth, spread and sustenance. The study of community structure and diversity by various workers (Bilai, 1984; Straatsma *et al.*, 1994 a, b; Beffa *et al.*, 1996; Peters *et al.*, 2000) has been instrumental in manipulating the compost environment in order to quicken the composting process and to improve the compost quality.

A. Structural Diversity

The compost is a rich reservoir of microbial types, comprising of mesophilic and thermophilic bacteria, fungi and actinomycetes. The mesophilic microflora forms the pioneer community while thermophiles form the climax community.

Amongst bacteria, *Bacillus* is a predominant component of the compost ecosystem. In composting of table scraps, 87% of isolates, recovered between 49^{0}C to 67^{0}C were *Bacillus* namely, *B. circulans*, *B. coagulans*,

B. licheniformis, *B. stearothermophilus*, and *B. subtilis* (Strom, 1985). The predominance of *Bacillus* in hot composts was also characterized by Blanc *et al.* (1997); four species namely, *B. stearothermophilus*, *B. pallidus* and *B. thermodenitrificans* were identified using amplified ribosomal DNA restriction analysis (ARDRA).

In garden and domestic waste composts, 76.1% of the thermophilic bacterial flora isolated from thermogenic, post-thermogenic and maturation phase of compost, was assigned by their RAPD (Random amplified polymorphic DNA) profiles to just two species, *B. licheniformis* and *B. thermodenitrificans.* The other species assigned by RAPD profiles were *B. licheniformis* (7.2%), *B. sporothermodurans* (4.9%), *B. thermosphaericus* (7.4%) and *Terrabacter tumescens* (5.0%). The viable cell numbers ranged from 1.4 to 150×10^{6} cfu g compost^{-1} with the highest two counts being from two-week and four-week old compost samples. The predominance of *B. thermodenitrificans* (accounting for 68.9% of all isolates overall) strongly supports its significant role in the composting process. Fluorescent direct counts (6.4 (+/- 2.5) $\times10^{10}$ cfu g dry wt^{-1}) for thermophilic heterotrophic aerobes in a freeze-dried 74^{0}C compost sample were higher than the plate counts (2.6(+/- 1.0) x 10^{8} cfu g dry wt^{-1}) (Dees and Ghorse, 2001). The phylogenetic analysis revealed that the two most abundant isolates belonged to the genera *Aneurinibacillus* and *Brevibacillus*, which are not commonly associated with the hot composts. The clone library contained sequences that clustered within the genus *Bacillus*, an exception was a *Lactobacillus*-type sequence but none of the sequences could be assigned to obligate thermophilic *B. stearothermophilus* which is commonly believed to dominate the hot compost (Dees and Ghorse, 2001). *B. licheniformis* is the dominant component of waste feather composts. In a monitored composting bin system, bacilli were present only in fresh wastes and during the peak-heating phase of biowastes (vegetable, fruit and garden waste) composting, the bacterial community structure based on denaturing gradient gel electrophoresis (DGGE) pattern during the steady state of semicontinuous treatment of organic solid waste showed DNA sequence, closest to *B. licheniformis*. Fluorescence *in situ* hybridization (FISH) analysis showed that the bacterium corresponding to the major band accounted for 30% of total eubacterial cell count at the steady state, indicating it to be the key microorganism in the biodegradation process (Haruta *et al.*, 2002). Addition of the compost to soil caused a substantial shift in community composition in soil, mainly within the members of the *Cytophaga-Flavobacterium- Bacteroides* group.

In mushroom compost, bacteria comprises of much lower proportion of total biomass although it is difficult to define the specifics of the domineering species (Johri and Rajni, 1999). Straatsma *et al.* (1994a) have however reported a biomass ratio of 1.0:1.8 of bacteria to fungi after phase II while according to Weigant (1991) the ratio was 1.0:0.9 in conventional compost and 1.0:2.3 in the experimental compost. While fast growing *Arthrobacter* and *Pseudomonas* constitute the pioneer flora (Stanek, 1969, 1972) that rapidly colonizes and degrades the easily available organic matter ('r strategists'), *Bacillus* represents the dominant component (Libmond *et al.*, 1995; Rastogi, 1996; Tewari, 2000; Rawat, 2004). *B. psilocybe* and fluorescent pseudomonads are present in the casing soil and mediate fruiting induction (or sporophore initiation) (Miller *et al.*, 1995).

In a recent study conducted in our group (Rawat, 2004), mesophilic bacterial counts (log_{10} cfu) in mushroom compost were found to vary from 9.34 (zero day) to 6.55 (drench) and thermophilic bacteria from 6.68 to 5.50, respectively; pseudomonads count varied from 8.40 to 6.53. Maximum mesophilic bacterial diversity was observed at first turning of phase I compost (Shannon's index H' = 1.33) and that of thermophilic bacterial diversity at second turning of phase I compost (H' = 0.92). Isolates identical to *Bacillus badius* dominated the entire composting process of mushroom cultivation. The other organisms with high similarities belonged to the ?-*Proteobacteria* and were related to *Azotobacter salinestris* and *Pseudomonas stutzeri*. SSCP (Single stranded conformation polymorphism) profiles and sequencing of subcloned products has revealed the presence of lactobacilli at the early stage of composting whereas during the heating phase, 5 of 12 molecular isolates were closely related to *Bacillus* and one to *Clostridium* (Peters *et al.*, 2000).

Thermophilic spore forming bacilli are very active during the composting temperatures of 50^0C and 65^0C (Nakasaki *et al.*, 1985; Strom, 1985). However, during the thermogenic phase temperature reaches upto 65^0 to 80^0C, temperatures at which no fungi and thermophilic bacilli are known to grow. The presence of the genus *Thermus* in thermogenic (65 to 82^0C) composts was for the first time reported by Beffa *et al.* (1996). The nutritional characteristics, total protein profiles, DNA-DNA hybridization and RFLP (Restriction fragment length polymorphisms) profiles of 16S rDNA showed that *Thermus* strains isolated from hot composts were closely related to *T. thermophilus* HB8, except strain JT4. All isolates except JT4 showed 65 to 71% DNA-DNA homology with *T. thermophilus* and 50 to 62%, with *T. aquaticus*; JT4 showed 72% homology with *T. aquaticus*. A moderately high number of obligately autotrophic *Hydrogenobacter* spp. and facultatively autotrophic *B. schlegelii* growing at temperatures above 70^{0}C have also been reported from hot composts (Beffa *et al.*, 1996). In a more recent study a number of enterobacteriaceae members and that belonging to the genus *Lactobacillus* were observed in the initial stage of composting of organic matter at 41^{0}C. Low G+C gram-positive bacteria constituted a dominant fraction of the bacterial community during the hot composting phase (55-70^{0}C) (Alfreider, 2002).

In contrast, in the municipal solid waste compost, *Escherichia coli* and faecal *Streptococci* population decreases considerably during the thermophilic phase. *Staphylococci* dominate during the mesophilic phase and at the beginning of the thermophilic phase whereas bacilli predominate during the remainder of the compost cycle. *Salmonella* disappears completely from the compost by the 25th day when temperature reached 60^0 C (Hassen *et al.*, 2001). In spent mushroom compost, a major waste by-product of the mushroom industry, a total bacterial count of 7.01 $\log_{10}$ cfu g^{-1} has been reported. Two novel species within the genera *Microbacterium* and *Stenotrophomonas* were reported alongwith *B. licheniformis*, *B. subtilis*, *Enterobacter* sp., *Klebsiella* sp., *Microbacterium* sp., *Paenibacillus lentimorbus, P. mevalonii, Sphingobacterium multivorum* and *Stenotrophomonas* sp. (Watabe *et al.*, 2004). *B. licheniformis* is known to be a common bacterial isolate of the spent mushroom compost (Kleyn and Wetzler, 1981).

Considering the importance of nitrogen balance during composting, the process of nitrification has been studied however the specifics are not clear yet (O'Dwyer *et al.*, 1989; Inbar *et al.*, 1993; Mahimairaja *et al.*, 1994). However, a few ammonia oxidizers and nitrite oxidizers belonging to undetermined genera have been reported from poultry dung-pine sawdust mixture. Although the nitrifiers have been recognized as potentially important organism in composts and composting materials, but their species composition, distribution and activity has not been properly assessed. Kowalchuk *et al.* (1999) have however detected ammonia oxidizer-like sequences belonging to ?-subgroup proteobacteria in mushroom compost, chicken manure and activated sludge samples. Four of the seven-sequence cluster proposed by Stephen *et al.* (1996) were detected viz., *Nitrosospira* cluster 3 and 4 and *Nitrososomonas* cluster 6 and 7. The culture-based method detected proportionately more ammonia oxidizers affiliated to *Nitrosospira* cluster 3 and *Nitrososomonas* cluster 7. The increase in proportion of *Nitrosospira* cluster 4 organisms relative to cluster 3 with a longer incubation time suggested that the members of the former group may require a long period of time to become active in the culture medium. *In situ* hybridization analysis revelaed the presence of *Nitrosospira* clusters 3 and 4 in the mushroom compost; latter was not detected in 6th day sample suggesting its greater susceptibility to high temperatures than the former. However, there are no known examples of thermophilic ammonia oxidizing bacteria; cell activity is probably impaired or destroyed by high temperatures (Focht and Verstraete, 1977). The cell survival at high temperatures may thus be facilitated by the formation of microniches (Derix *et al.*, 1990) and the moisture content could contribute towards the cell survival (Gromov and Pavlenko, 1989). The detection of *Nitrosospira* cluster 4 in the prepared compost both by PCR and MPN suggests growth of ammonia-oxidizers at the ambient temperature during the second week of composting, the time of greatest ammonia consumption, however, reasons for the growth of these cells are not clear (Kawalchuk *et al.*, 1999).

The phospholipids fatty acids (PLFA) markers of bacteria (15:0 and 17:0) were observed to decline alongwith an indicator of aerobic bacteria i.e. 16:1?7C during the composting of dairy manure and pine shave beddings (Boggs *et al.*, 1998). The biomarkers of gram-positive bacteria were predominant along with actinomycetes at thermophilic stage (i.e. first two weeks) and those of gram-negative bacteria at the curing stage of the composting of rice straw (Cahayani *et al.*, 2002).

Actinomycetes constitute an important microbial component of the composting process and contribute extensively by their ability to produce antibiotics and enzymes, and the ability to degrade complex and recalcitrant molecules especially lignocellulosics. Predominant forms in the zero day mushroom compost are represented by the genera, *Saccharopolyspora* and *Streptomyces* (Lacey, 1974). Species of *Streptomyces* make up the pioneer and dominant microflora of mushroom compost and waste feathers compost.

Thermophilic actinomycetes grow extensively during phase II of mushroom compost as evident from the white wefts of mycelium ('fire fang') in the compost. *Nocardia brasiliensis*, *Pseudonocardia thermophila*, *Saccharomonospora viridis*, *S. rectivirgula*, *S. diastaticus*, *S. griseus*, *S. rectus*, *S. thermoviolaceus*, *S. thermovulgaris*, *S. violaceoruber*, *Thermoactinomyces chromogena*, *T. curvata*, *T. fusca*, *T. thalophilus* and *T. vulgaris* have been reported from phase II compost (Fergus, 1964; Kleyn and Wetzler, 1981). Actinomycete population in composts is often dominated by the species of *Thermomonospora* (McCarthy and Broda, 1984). The population count of mesophilic actinomycetes (log_{10} cfu) was found to vary from 5.49 to 5.48 (zero day to drench) while that of thermophiles from, 3.67 to 4.52, respectively (Rawat, 2004).

The spores of actinomycetes are easily dispersed into the air during the spawning process, giving large concentrations in the air of the spawning shed (Crook *et al.*, 1996). The concentration of actinomycetes spores could reach a level of 7.4×10^8 spores m^{-3}, close to spawning lines (Lacey, 1974) while in the cropping sheds it touches 10^4 cfu m^{-3} (Crook *et al.*, 1996). *S. diastaticus* and *T. vulgaris* were the predominant component of spent mushroom compost. Other actinomycete isolates included *S. albus*, *S. griseus*, *T. thalophilus*, *T. chromogena* and *T. fusca* (Kleyn and Wetzler, 1981). The community succession of actinomycetes in mushroom compost was investigated by Peters *et al.* (2000) employing SSCP wherein four of the five molecular isolates were closely related to *S. nodosus*, *S. thermodiastaticus*, *S. vulgare* and *Detolaasinbacter* sp. The domestic waste compost in U.K. is dominated by *T. vulgaris* and *T. thermophilus*. (Crook *et al.*, 1987). *T. vulgaris, T. alba, T. curvata* and *T. lineata* constitute the dominant component of German domestic waste compost (Resz *et al.*, 1977). Lacey (1990, 1995) observed the predominance of *Saccharopolyspora rectivirgula*, *Saccharomonospora*, unidentified white thermophiles, *Thermoactinomyces* spp. and *Streptomyces* spp. in experimental composts indoors while during turning of compost outdoors, *Streptomyces* spp., *Saccharomonospora* spp., *Thermomonospora* spp. and *Thermoactinomyces* spp. were the dominant components; population sizes ranged from 1.8 to 4.9 x 10^5 cfu m^{-3} air.

The range of actinomycetes isolated from domestic waste composts was similar to that found in mushroom spawning sheds but with the maximum concentration of the order of 10^6 cfu m^{-3} air during both air classification indoors and turning outdoors. *Streptomyces* spp. was dominant in composts made from lucerne, oat straw and maize stalks while *T. vulgaris* was dominant in grass composts (Lacey, 1997). During shredding of fresh green waste, counts of airborne actinomycetes were $< 5\times10^4$ cfu m^{-3} while 10^6 cfu m^{-3} counts were observed during turning of the compost piles. *Saccharomonospora* spp. formed the most abundant taxon during shredding with frequent recovery of *S. albus* and *Thermoactinomyces* sp. During urban waste composting, Streptomycetes and a group of unidentified actinomycetes that represented several taxa became most abundant.

Amycolata, Microtetraspora and *T. chromogena* colonies and a new species of *Thermocrispum* were also reported (Korn-Wendisch *et al*., 1995). *Thermomonospora* spp., including *T. chromogena* and *Microtetraspora* were less numerous than is sometimes found in mushroom compost. The populations at shredding were less thermotolerant than those dispersed from compost piles.

In a recent study of mushroom composting Rawat (2004) found that the mesophilic fungal counts (log_{10} cfu) in mushroom compost varied from 4.83 to 5.01 (zero day to drench) and that of thermophilic, from 4.62 to 4.47. Maximum structural divergence and species richness amongst mesophilic fungi was observed in zero day morphotypes (H' = 1.55; R1 = 2.54) while the least in peak-heat morphotypes (H' = 0.68; R1=0.64). Amongst thermophilic fungi, fourth turning of phase I compost exhibited maximal diversity (H'=2.14) and the least in peak heat stage morphotypes (H'= 0.75). *Aspergillus* spp., *Rhizopus oryzae*, *Trichoderma viride*, *Chaetomium* sp., *Penicillium* sp. and *Alternaria* sp. were observed as dominant mycoflora in the initial stages of composting. *Acremonium* sp. and *Paecilomyces variotii* were the additional mesophilic fungi in the later period of composting (Dhar, 1976). The recolonization of mesophilic mycoflora in the compost can occur at the time of spawning due to fall in temperature and if free cellulose and carbohydrates are still present in compost (Eicker, 1980; Vijay and Gupta, 1992). Most of the mesophilic fungi during the course of composting and at spawning stage exert a deleterious effect on the growth of *A. bisporus* mycelium, with ultimate reduction in yield (Vijay and Gupta, 1992).

Composts harbour upto 10^6 propagules of thermophiles (Maheshwari *et al*., 1987); thermophilic fungi are therefore, primarily considered as compost fungi. In the beginning of phase II of mushroom compost, thermophilic fungi and actinomycetes extensively colonize the plant matter until temperature reaches 60^0C, as an outcome of slow peak-heating for about two days (Rawat and Johri, 2002). The high temperature of the first indoor period of phase II kills most of the pathogenic and non-pathogenic microorganisms, except the spores of actinomycetes and thermophilic fungi such as *Scytalidium thermophilum* (Straatsma *et al*., 1991); the latter was most abundant at the end of phase II compost (abundance = 0.68). Thermophilic fungi of the *Torula-Humicola* complex are a necessary and dominant component of the community during Phase II. Dominance of *S. thermophilum* has been reported by several workers (Straatsma *et al*., 1991; Vijay, 1996; Rajni *et al*., 1998) while *H. grisea* var. *thermoidea* and

H. insolens have been described by others (Fergus, 1964). They are inherently close partners in the degradation processes in compost and provide selectivity to compost (Straatsma *et al.*, 1989; Opden Camp *et al.*, 1990).

Thermophilic fungi grow extensively during the last phase of composting from the spores that survive the pasteurization temperature (Straatsma *et al.*, 1989). Thus, thermophilic mycoflora forms the climax community and contributes significantly towards the quality of compost. However, their presence throughout the course of composting is largely responsible for maintenance of biological equilibrium that ultimately leads to unique selectivity wherein *A. bisporus* multiplies without competition. These fungi influence growth of *A. bisporus* at three distinct levels (Weigant, 1992): First, they decrease concentration of ammonia in compost which otherwise would counteract the growth of the mycelium. Second, they immobilize nutrients in a form, which improves apparent availability to the mushroom mycelium. Third, they exert direct growth promotory influence on the mushroom mycelium viz., *S. thermophilum*. The course of fungal succession is partially dependent on the ecophysiological conditions in compost (Satyanarayana *et al.*, 1992). The pioneer thermophilic mycoflora of compost comprises of fast growing and rapidly sporulating fungi such as *A. fumigatus* and *Rhizomucor* spp. with a pH optima below 7.0 and temperature optima of about 40^{0}C. When self-heating and ammonification starts and pH reaches 9.0, the pioneer flora disappears and paves way for *Talaromyces thermophilus* and *Thermomyces lanuginosus* during massive heat production as these fungi exhibit high thermal death point and pH tolerance, possess moderate growth rate but do not degrade cellulose. At the end of the composting process, about 50-70% of the compost biomass is constituted by thermophilic fungi (Sparling *et al.*, 1982; Weigant, 1992). While most of the species are eliminated, *S. thermophilum* appears as near exclusive species after phase II composting and constitutes a climax species in the mushroom compost along with thermophilic actinomycetes (Straatsma *et al.*, 1994b). The number of CFU of *S. thermophilum* in fresh matter of phase II is about 10^6 g^{-1} compost (Bilai, 1984), however, actinomycetes and bacteria appear to play a decisive role in successful colonization by this thermophile. Antagonisms, also plays a significant role in determining the population structure. The volatiles of *Sporotrichum thermophile* and *Chaetomium thermophile* can inhibit conidial germination of *Humicola lanuginosa* by impairing essential metabolic processes whereas *C. thermophile* suppresses mycelial growth of *H. lanuginosa* and *Torula thermophila*. However, effect of the fungistatic volatile factors in compost ecosystem is only marginal in view of high temperature at which they grow (Satyanarayana and Johri, 1981).

Species of *Aspergillus* and *Mucor* were predominant during the composting of biowaste (vegetables, fruit and garden waste) after thermophilic phase (Ryckeboer *et al.*, 2003). *A. fumigatus* and *H. grisea* var. *thermoidea* were the dominant member of the spent mushroom compost. Others were *A. flavus, A. nidulans, A. terreus, A. versicolor, Chrysosporium luteum, Mucor* spp., *Oidiodendron* spp., *Paecilomyces* spp., *Penicillium chromogenum, P. expansum, Trichoderma viride* and *Trichurus* spp. (Kleyn and Wetzler, 1981).

In paddy straw compost, *C. thermophile*, *Humicola* spp. and *S. thermophile,* were abundantly present (Satyanarayana, 1978). Satyanarayana and Johri (1984) observed that the colonizing ability of thermophilic fungi on paddy straw was directly proportional to the inoculum concentration. For example, colonization by *H. lanuginosa*, *S. thermophile* and *T. thermophila* increased with higher inoculum dose. During peak-heating period, only a few thermophilic fungal propagules were present exhibiting high rate of respiration. However, Moubasher *et al.* (1982) reported that thermophilic fungi were not present at peak high temperature in wheat and broadbean straw composts. When it cooled down to 51.5^{0}C, *Myriococcum albomyces, P. dupontii* and *S. thermophile* were found in abundance.

Klamer *et al.* (2001) studied the succession of mycoflora during eight months of composting of *Miscanthus* straw and pig slurry in well insulated containers. Before peak heating, *A. fumigatus* and *R. pusillus* were dominant. Forms developing after peak heating could be divided into two groups: those appearing from day 15 to 27, and others developing from day 50 to 225. The first group was dominated by, *P. variotii*, *S. thermophilum* and *T. lanuginosus* and the second by, *Acremonuim* spp. and *T. lanuginosus.* The Brillouin diversity index changed with temperature; diversity was high before peak heating, low during elevated temperature, and increased again during the third phase of composting. Temperature was the main controlling parameter that changed fungal community during the first month of composting. Based on principal component analysis (PCA), it could be concluded that when the temperature reached an ambient level, only minor change in fungal community was detected although some changes in certain species were still discernible. In our group Rawat (2004) observed 95 distinct bands while studying the fungal community succession in mushroom compost using SSCP; though only 34 fungal isolates were recovered. PCA analysis of SSCP profiles clustered PW I and PW II community together; T_1, T_2, T_3, T_4, fill, peak-heat, con and spawn community in one cluster. Casing soil community was however more distant and different from others; spawn-run, crop, air and drench community were placed in one cluster. A 15-66% similarity was observed amongst the community of different stages of mushroom compost.

Scytalidium thermophilum in Compost Management

S. thermophilum is a natural inhabitant of compost ingredients and drainage from compost, and has been documented to be present throughout composting (Table-1). Among various groups actively engaged in studying the ecology of mushroom compost, Straatsma's group in Holland had laid considerable emphasis on population dynamics of *S. thermophilum* for improved compost management. The disappearance of ammonia and selectivity of compost for the growth of *A. bisporus* mycelium that occurs in Phase II at temperature 45-55^{0}C is linked to the presence of *S. thermophilum* (Ross and Harris, 1983). The density of *S. thermophilum* was found to be positively correlated with mushroom yield (Straatsma *et al.*, 1989); Respiratory CO_2 of *Scytalidium* was documented to be the likely reason for the growth promotory effect of this fungus on mushroom yield (Weigant, 1992).

Table-1: Population of *S. thermophilum* in mushroom compost at various stages of its preparation.

Source of fungi	Log_{10} $CFUg^{-1}$	*S. thermophilum* frequency		*S. thermophilum* recovery (%)[a,b]	
Wheat straw	3.4	Infrequent	6	n.d.	n.d.
Drainage from compost	3.9	Infrequent	2	n.d.	n.d.
Horse droppings	2.3	Infrequent	3	n.d.	n.d.
Ingredient mixtures, start self-heating	3.3	Half	26	28	3
End phase I	2.4	Abundant	12	3	8
End phase II	6.5	Exclusive	5	90	54

[a] Standard deviations of CFU counts and of recovery are 1.00 and 17.3 respectively. Compared with data of ingredient: mixture, CFU counts and recovery were significantly lower at the end of phase I and higher at the end of phase II.

[b] Counts of fungal recovery from washed particles are restricted to *S. thermophilum* n.d.- not determined

Source: Straatsma *et al.* (1994a)

Rajni *et al.* (1998) observed nearly similar microbial distribution pattern that was comparable to Straatsma *et al.* (1991) with predominance of *S. thermophilum* while the inputs in the European and Indian composts are substantially different. They also observed the morphological variability of this fungus in compost with differences in their growth rate and promotion of mycelial growth. In twenty days schedule of compost preparation, *S. thermophilum* was detected from the first turning (i.e., after 5th day of composting) till fifth turning (i.e. 17th day) whereas species of *Paecilomyces* were present only during the last turning i.e., 20th day. Two isolates of *Malbranchea cinnamonea* were recorded between the second and fifth turning stage (8th to 17th day). The presence of *Paecilomyces* sp. and *M. cinnamonea*, which are normally slow growers, provided an opportunity to evaluate their influence on *in situ* mycelial extension of *A. bisporus*. At 24th day of composting the population of *S. thermophilum* was 10^8 propagules g^{-1} of compost.

B. Functional Diversity

Compost is not only structurally diverse but also functionally active. The *in situ* functionality of each microbial component especially the extracellular enzymatic machinery viz., polysaccharases, proteases and lipases, plays an important decisive role in successful colonization and succession in mushroom compost (Rajni *et al.*, 1998; Johri *et al.*, 1999; Rawat and Johri, 2002; Rawat and Johri, 2004). Infect, compost harbours a number of guilds.

The wide exploration of enzymatic machinery of the individual microbial component and their *in situ* enzymatic action in their own niche would help in greater understanding of the relationship between structural and functional diversity. However, the available information appears largely biased towards the functional role of thermophilic mycoflora of compost probably because of their being the dominant component of the functional niche occupied by compost microbiota. The wide enzymatic potential exhibited by these fungi along with enzyme multiplicity with different phsico-chemical characteristics helps in the functioning of each component under different biophysical conditions. The increase in cellulolytic and amylolytic activity during composting is a reflection of change in the population and community structure of the resident microflora. The enzymatic diversity of compost microbiota is known to result in specificity and successional change besides providing a niche to various species to survive in the absence of simple sugars (Rawat and Johri, 2002).

In compost the pioneer microflora in general can utilize simple sugars but such biota disappear soon and only those organisms with wide polysaccharolytic ability persist. The cellulolytic and hemicellulolytic ability of thermotolerant *A. fumigatus* allows it to persist in the wheat straw compost whereas thermophilic *M. pusillus* does not recur even after temperature become suitable for growth due to lack of polysaccharolytic ability. *H. lanuginosa* persists throughout composting due to its ability to lead commensal life with others along with its cellulolytic and hemicellulolytic ability. *C. thermophile*, *H. insolens*, *H. lanuginosa* and *T. dupontii* develop abundantly in the 'plateau' period and rapidly utilize cellulose and hemicellulose. When compost temperature drops, thermophilic *S. thermophile* and mesophilic *Coprinus cinereus* and *Clitopilus pinsitus* appear which can utilize cellulose and hemicellulose in wheat staw at a slower rate (Chang and Hudson, 1967).

The success of microflora in competitive saprophytic colonization (CSC) such as that operative in plant residues and soil depends upon its intrinsic ability to decompose that substrate and the ability to succeed in the competition (Garret, 1963). The strongly cellulolytic and hemicellulolytic, *A. fumigatus*, *S. thermophile* and *T. thermophila* exhibit greater colonization ability than the weakly cellulolytic, *H. lanuginosa* (Satyanarayana and Johri, 1984). The dominance of *S. thermophilum* has been attributed to the presence of complete complement of polysaccharolytic enzyme machinery (Grajek, 1987; Rawat, 1998; Khokar, 1999; Tewari, 2000). The hydrolytic potential of such thermophilic fungi is responsible for solubilization of complex ingredients of compost and making the nutrient available to *A. bisporus*. The stimulation of growth of mushroom mycelium by cellulose decomposing mycoflora has been well-documented (Stanek, 1969).

Considering the fact that culturable microbial populations are limited on account of our poor understanding of their nutritional requirements, detailed, *in situ* enzymatic investigations are likely to provide a better understanding of the relationship between structural and functional diversity of thermophilic fungal community. Iiyama *et al.* (1996) observed that the loss of cellulose and lignocellulose and increase in protein content during the composting period was a result of increased polysaccharolytic activity of the fungal biomass; this resulted in increased level of reducing sugars. In mushroom compost, the level of enzymes are found to increase from zero day to the end of phase I compost thereafter it decreases continuously; results are well corroborated with the population structure (Rawat, 2004).

It has been observed that *in situ* changes in lignocellulose, cellulose and loss in weight during composting is corroborated with the activity of thermophilic fungi. The analysis of plant residues after decomposition by pure thermophilic fungal cultures resulted in biochemical changes that were similar to those observed during composting of organic materials by natural mixed microflora (Satyanarayana, 1978). During the 24 day composting sequence of button mushroom, Rajni (1999) reported that the level of organic carbon decreased from 18.12 to 10.57% while that of cellulose and lignocellulose from 32.3 to 23.0% and 52.4 to 43.1%, respectively. An increase of 77.5% and 86.0% in protein and reducing sugar level was observed. The level of amylase, endo-cellulase and exo-cellulase also changed. Based on the regression analysis between the population of *S. thermophilum* and chemical parameters, a 92.5% change in lignocellulose was found as a result of rise in population of *S. thermophilum* in compost. The levels of dehydrogenase activity decreased by 72%; protease by 32%; xylanase and cellulase by 50% during peak-heat stage of phase II while during the post peak-heat stage of phase II mushroom compost, levels again increased by 71%, 23% and 33%, respectively. These changes were found to be well corroborated with the change in population structure of compost microflora (Tewari, 2000).

CONCLUSIONS

Compost represents an interesting example of thermogenic, solid-state fermentation process that results from a succession of microbial communities. While it is a complex ecosystem, study of structural diversity complemented with functional diversity provides a fair picture of the species spectrum and their associated interactions within the substrate. There is a further need to use both cultivation- dependent and cultivation independent approaches to monitor the microbial ecology of this ecosystem as former gives only limited insights into it. The survival strategies and competitive behaviour of microflora in this interesting niche requires to be understood. The *in situ* functionality of bacteria, actinomycetes and mesophilic fungi is still poorly known whereas tapping the biopotentiality of microflora in this niche would help in hastening the composting process besides improving the quality of compost by pre-inoculation of microflora. This approach while being widely practised in mushroom compost, has currently become popular in other composting systems as Effective Microorganism Technology but is at its nascent stage. New vertical dimensions will add to our existing knowledge, which would throw light into the adaptive behaviour of the microflora in compost ecosystem besides helping unravel new and novel forms. The present movement of organic agriculture augurs well for use of microbial consortia in ecotechnological endeavours of which composting is an essential component. Use of modern molecular tools will permit better and more defined picture of community level changes as also the role of dominant forms in deriving strategies for more effective degradative processes and conservation tools.

ACKNOWLEDGEMENT

Thermophilic fungal research of BNJ cited in this chapter has been supported by INSA, DST, ICAR and CSIR whereas more recent molecular microbial ecology component has been supported by the Deptt. of Biotechnology and Ministry of Environment and Forests. Efforts of a series of doctoral and postdoctoral research students especially Dr. T. Satyanarayana is greatfully acknowledged since these early researches laid the foundation of nearly three decades of "Thermophilic saga".

REFERENCES

Alfreider, A.; Peters, S.; Tebbe, C.; Rangger, A. and Insam, H. (2002). Microbial community dynamics during composting of organic matter as determined by 16S rDNA analysis. ***Science and Utilization.*** **10:** 303-312.

Beffa, T.; Blanc, M. and Aragno, M. (1996). Obligately and facultatively autotrophic, sulfur and hydrogen-oxidizing thermophilic bacteria isolated from hot composts. *Arch. Microbiol.* **165:** 34-40.

Beffa, T.; Blanc, M.; Lyon, P.F.; Vogt, G.; Marchiani, M.; Fishcer, J.L. and Aragno, M. (1996). Isolation of *Thermus* strains from hot composts (60 to 80^{0}C). *Appl. Environ. Microbiol.* **62:** 1723-1727.

Biddlestone, A.J. and Gray, K.R. (1985). Composting. *In*: Comprehensive Biotechnology. (Ed.: M. Moo-Young). Pergamon Press, New York. vol.4. pp. 1059-1070.

Bilai, V.T. (1984). Thermophilic micromycete species from mushroom composts. Mikrobiol. Zh. (Kiev). **46:** 35-38.

Blanc, M.; Marilley, L.; Beffa, T. and Aragno, M. (1997). Rapid identification of heterotrophic thermophilic, spore forming bacteria isolated from hot composts. *Int. J. Bacteriol.* 47: 1246-1248.

Boggs, L.C.; Kennedy, A.C. and Reganold, J.P. (1998). Use of phospholipids fatty acids and carbon source utilization patterns to track microbial community succession in developing compost. *Appl. Environ. Microbiol.* **64:** 4062-4064.

Cahayani, V.R.; Watanabe, A.; Matsuya, K.; Asakawa, S. and Kimura, M. (2002). Succession of microbiota estimated by phospholipids fatty acid analysis and changes in organic constituents during the composting process of rice straw. *Soil Science Plant Nut.* **48:** 735-743.

Chang, Y. (1967). The fungi of wheat straw compost II. Biochemical and physiological studies. *Trans. Br. Mycol. Soc.* **50:** 667-677.

Chang, Y. and Hudson, H.J. (1967). The fungi of wheat straw compost. I. Ecological Studies. *Trans. Br. Mycol. Soc.* **50:** 649-666.

Crook, B., Higgins, S. and Lacey, J. (1987). Airborne microorganisms associated with domestic waste disposal. Unpublished Report to the Health and Safety Executive, Contract No. I/MS/126/643/82.

Crook, B.; Goodman, G.C.; Griffin, P.; Lacey, J. and Topping, M.D. (1996). Mushroom worker's lung, a clinical, serological and environmental study of work conditions in a commercial mushroom farm. *Occup. Enviorn. Med.*

Dees, P.M. and Ghorse, W.C. (2001). Microbial diversity in hot synthetic compost as revealed by PCR-amplified rRNA sequences from cultivated isolates and extracted DNA. *FEMS Microbiol. Ecol.* 35: 207-216.

Derix, P.J.L.; OpDen Camp, H.J.M.; Vander Drift, C.; Van Griensven, L.J.L.D. and Vogels, G.D. (1990). Biomass and biological activity during the production of compost used as a substrate in mushroom cultivation. *Appl. Environ. Microbiol.* **56:** 3029-3034.

Dhar, B.L. (1976). Studies on microflora of mushroom (*Agaricus bisporus*) (Lange) Sing. Compost. M.Sc. Thesis. H.P. Univ., Solan.

Eicker, A. (1980). Mesophilic fungi associated with cultivation of *Agaricus brunnescens*. *Trans. Brit. Mycol. Soc.* **74:** 465-470.

Fergus, C.L. (1964). Thermophilic and thermotolerant molds and actinomycetes of mushroom compost during peak heating. *Mycologia.* **56:** 267-284.

Finstein, M.S. and Morris, M.L. (1975). Microbiology of municipal solid composting. *Adv. Appl. Microbiol.* **19:** 113-151.

Focht, D.D. and Verstraete, W. (1977). Biochemical ecology of nitrification and denitrification. *In*: Advances in Microbial Ecology (Ed.: M. Alexander). Plenium Press, New York. **1:** 135-214.

Forsyth, W.G. and Webley, D.M. (1948). *Proc. Soc. Appl. Bacteriol.* 34p.

Galler, W.S. and Davey, C.B. (1971). *In*: Livestock Waste Management and Pollution Abatement. Publ. Proc. 271, Amer. Soc. Agr. Eng. St. Joseph, Michigan. pp. 159-162.

Garland, J.L. and Mills, A.L. (1991). Classification and characterization of heterotrophic microbial communities on the basis of community-level sole-carbon-source utilization. *Appl. Environ. Microbiol.* **57:** 2351-2354.

Garret, S.D. (1963). Saprophytic competition in the soil. *In*: Soil Fungi and Soil Fertility. Pergamon Press. pp. 115-132.

Grajek, W. (1987). Production of D-xylanases by thermophilic fungi using different methods of culture. *Biotechnol. Lett.* 353-356.

Gray, K.R. and Biddlestone, A.J. (1974). Decomposition of urban waste. *In*: Biology of Litter Decomposition. (Eds.: C.H. Dickinson and G.J.F. Pugh). Academic Press, London vol. 2. pp. 743-775.

Gromov, B.V. and Pavlenko, G.V. (1989). Autoecology of bacteria: water activity. *In*: *Bacterial Ecology*. Leningrad University Press, Leningrad, USSR. 78p.

Haruta, S.; Kondo, M.; Nakamura, K.; Aiber, H.; Ueno, S.; Ishii, M. and Igarshi, Y. (2002). Microbial community changes during organic solid waste treatment analyzed by double gradient-denaturing gradient gel electrophoresis and fluorescence *in situ* hybridization. *Appl. Microbiol. Biotechnol.* **60:** 224-231.

Hassen, A.; Belguith, K.; Jedidi, N.; Cherif, A.; Cherif, M. and Boudabous, A. (2001). Microbial characterization during composting of municipal solid waste. *Bioresource Technol.* **80:** 217-225.

Herman, W.A.; McGill, W.B. and Dormaar, J.F. (1977). Effects of initial chemical composition of roots of three different grass species. *Can. J. Soil. Sci.* **57:** 205-215.

Herrmann, R.F. and Shann, J.F. (1997). Microbial community changes during composting of municipal solid waste. *Microb. Ecol.* **33:** 78-85.

Iiyama, K.; Stone, B.A. and Macauley, B.J. (1996). Changes in the concentration of soluble anions in compost during composting and mushroom (*Agaricus bisporus*) growth. *J. Food Sci. Agric.* **72:** 243-249.

Inbar, Y.; Hadar, Y. and Chen, Y. (1993). Recycling of cattle manure: the composting process and characterization of maturity. *J. Environ. Qual.* **22:** 857-863.

Inssam, H.; Amor, K.; Renner, M. And Crepaz, C. (1996). Changes in functional abilities of the microbial community during composting of manure. *Microb. Ecol.* **37:** 77-87.

Johri, B.N. and Rajni. (1999). Mushroom compost: Microbiology and Applications. *In*: Modern Approaches and Innovations in Soil Management. (Eds: D.J. Bagyaraj, Ajit Verma, K.K. Khanna and H.K. Kheri). Rastogi Publ. Merrut, India. 345-358 p.

Johri, B.N.; Satyanarayana, T. and Olsen, J. (1999). Thermophilic moulds in biotechnology. (Eds: B.N. Johri, T. Satyanarayana and J. Olsen). Kluwer Academic Publishers, Dordrecht, The Netherlands. 354p.

Kane, B.E. and Mullins, J.T. (1973). Thermophilic fungi in a municipal waste compost system. *Mycol.* **65:** 1087-1100.

Khokar, A. (1999). Production and characterization of lipase from *Scytalidium thermophilum*. Ph.D. Thesis, G.B. Pant Univ. Agril. Technol., Pantnagar, India. 116 p.

Klamer, M. and Baath, E. (1998). Microbial community dynamics during composting of straw material studied using phospholipid fatty acid analysis. *FEMS Microbiol. Ecol.* **27:** 9-20.

Klamer, M., Lind, A.M., Games, W., Balis, C., Lasaridi, L., Szmidt, R.A.K., Stentiford, E. and Lopez-Real, J. (2001). Fungal succession during composting of miscanthus straw and pig slurry. Proc. Int. Symp. on composting and use of composted materials for horticulture. Ayr, U.K., 5-11 Apr. 1997, No. 469. 405-412.

Kleyn, J.G. and Wetzler, T.F. (1981). The microbiology of spent mushroom compost and its dust. *Can. J. Microbiol.* **27:** 748-753.

Korn-Wendisch, F.; Rainey, F.; Kroppenstedt, R.M.; Kempf, A.; Majazza, A.; Kutzner, H.J. and Stackebrandt, E. (1995). *Thermocrispum* gen. nov., a new genus of the order Actinomycetales and description of *Thermocrispum municipale* sp. nov. and *Thermocrispum agreste* sp. nov. *Int. J. Syst. Bacteriol.* **45:** 67-77.

Kowalchuk, G.A.; Naoumenko, Z.S.; Derikx, P.J.L.; Felske, A.; Stephen, J.R. and Arkhipchenko, I.A. (1999). Molecular analysis of ammonia-oxidizing bacteria of the beta subdivision of the class *Proteobacteria* in compost and composted materials. *Appl. Environ. Microbiol.* **65:** 396-403.

Lacey, J. (1974). Allergy in mushroom workers. *Larcet.* **1:** 366.

Lacey, J. (1974). The microbiology of hay and straw. *In*: Aspergillosis and Farmer's Lung in Man and Animals. (Eds.: R. DeHuller and F. Suter). Hans Huber, Bern. 16-26.

Lacey, J. (1997). Actinomycetes in composts. *Ann. Agric. Environ. Med.* **4:** 113-121.

Lacey, J. and Williamson, P.A.M. (1995). Airborne microorganisms associated with domestic waste composting at castle bromwich. *CWM Paper* 110/93. London.

Lacey, J., Williamson, P.A.M., King, P. and Bardos, R.P. (1990). Airborne microorganisms associated with domestic waste composting. Warren Spring Laboratory Report No. LR 808 (MR).

Libmond, S.; Savoie, J.M. and Elliott, T.J. (1995). Stimulation of mushroom composting by a polysaccharidase complex induction of cellulase production in *Bacillus subtilis.* Proc. 14th Int. Cong. Sci Cult. Edible fungi. UK. pp. 195-202.

Maheshwari, R., Kamalam, P.T. and Balasubramanyam, P.V. (1987). The biogeography of thermophilic fungi. *Curr. Sci.* 56: 151-155.

Mahimairaja, S.; Bolan, N.S.; Hedley, M.J. and Macgregor, A.N. (1994). Losses and transformation of nitrogen during composting of poultry manure with different amendments: an incubation experiment. *Bioresource Technol.* **47:** 265-273.

Mantel, E.F.K.; Agarwala, R.K. and Seth, P.K. (1972). A guide to mushroom cultivation. Farm Bull, No. 2. Ministry of Agriculture, Farm Information Unit, Directorate of Extension, New Delhi.

McCarthy, A.J. and Broda, P. (1984). Screening for lignin-degrading actinomycetes and characterization of their activity against (^{14}C) lignin labelled wheat lignocellulose. *J. Gen. Microbiol.* **130:** 2905-2913.

Miller, F.C. (1994). Conventional composting system. *In*: *Agaricus* compost. (Ed.: N.G. Nair). Australian Mushroom Growers Association, Windsor, NSW, Australia. pp. 1-18.

Miller, N.; Gillespie, J.B.; Doyle, O.P.E. and Elliott, T.J. (1995). *Mushroom Sci.* **14:** 313-321.

Moubasher, A.H., Aboel-Hafez, S.I.I., Aboelfattah, H.M. and Moharrarh, A.M. (1982). Fungi of wheat and broad bean straw composts II, Thermophilic fungi. *Mycopathol.* **84:** 61-72.

Nakasaki, K.; Shoda, M. and Kubota, H. (1985). Effect of temperature on composting of sewage sludge. *Appl. Environ. Microbiol.* **50:** 1526-1530.

O' Dwyer, B.; Morgan, M.A. and Toole, P.O. (1989). Changes in physical and chemical properties of composts based on cattle slurry and milled peat during incubation. *Acta Hortic.* (The Hague). **238:** 181-182.

Opden Camp, H.J.M.; Stumm, C.K.; Straatsma, G.; Derikx, P.J.L. and Van Griensven, L.J.L.D. (1990). Hyphal and mycelial interactions between *Agaricus bisporus* and *Scytalidium thermophilum* on agar medium. *Microb. Ecol.* **19:** 303-309.

Peters, S.; Koschinsky, S.; Schwieger, F. and Tebbe, C.C. (2000). Succession of microbial communities during hot composting as detected by PCR-Single Strand-Conformation Polymorphism-based genetic profiles of small subunit rRNA gene. *Appl. Environ. Microbiol.* **65:** 930-936.

Rajni, Rastogi, S.; Johri, B.N. and Singh, R.P. (1998). Microbial dynamics and its influence in cultivation cycle of *Agaricus bisporus* (Longe) Imbach. *Mushroom Res.* **7:** 63-70.

Rajni (1999). Molecular ecology of *Scytalidium thermophilum*. Ph.D. Thesis. G.B. Pant University of Agriculture and Technology, Pantnagar, India. 88p.

Rastogi, S. (1996). Population dynamics of microbial flora in compost and casing soil. M.Sc. Thesis. G.B. Pant University of Agriculture and Technology, Pantnagar. 88p.

Rawat, S. (1998). Production and characterization of amylases from *Scytalidium thermophilum*. M.Sc. Thesis. G.B. Pant Univ. of Agric. & Technol. Pantnagar, India. 98p.

Rawat, S. (2004). Microbial diversity of mushroom compost and xylanases of *Scytalidium thermophilum*. Ph.D. Thesis. G.B.Pant Univ. of Agric. & Technol. Pantnagar, India. 256p.

Rawat, S. and Johri, B.N. (2002). Ecology and structural diversity of thermophilic fungi. *In*: Frontiers of Fungal Diversity in India. (Eds.: G.P. Rao, C. Manoharachari, D.J. Bhat, R.C. Rajak, T.N. Lakhanpal). Int. Book Distributing Co. Lucknow, India. pp. 205-232.

Rawat, S. and Johri, B.N. (2004). Xylanases of thermophilic moulds and their application potential. *In:* Handbook of Fungal Biotechnology. (Ed.: D. K. Arora). Marcel Dekker Inc. New York, USA. vol.**20:** 299-314.

Resz, A.; Schwanbeck, J.; Knossel, D. (1977). Thermophile actinomyceten aus miilkompost: Temperaturanspruche und proteolytische Aktivitat. *Forum Stadte-Hyg.* **28:** 71-73.

Ross, R.C. and Harris, P.J. (1983). An investigation into the selective nature of mushroom compost preparation. *Scientia Horticulturae.* **20:** 61-70.

Ryckeboer, J.; Mergaert, J.; Coosemans, J.; Deprins, K. and Swings, J. (2003). Microbiological aspects of biowaste during composting in a monitored compost. *J. Appl. Microbiol.* **94:** 127-137.

Satyanarayana, T. (1978). Thermophilic microorganisms and their role in composting process. Ph.D. Thesis. Sagar University, Sagar. 212 p.

Satyanarayana, T. and Grajek, W. (1999). Composting and solid state fermentation. *In*: Thermophilic moulds in Biotechnology. (Eds.: B.N. Johri, T. Satyanarayana and J. Olsen). Kluwer Academic Publishers, Netherlands. pp. 265-288.

Satyanarayana, T. and Johri, B.N. (1984). Ecology of thermophilic fungi. *In*: Progress in Microbial Ecology. (Eds.: K.G. Mukherji, V.P. Agnihotri and R.P. Singh). Print House, Lucknow. pp.349-361.

Satyanarayana, T.; Johri, B.N. and Klein, H. (1992). Biotechnological potential of thermophilic fungi. *In*: Handbook of Applied Mycology (Eds: D.K. Arora, R.P. Elander and K.G. Mukherjee). Marcel Dekker Inc., New York. pp. 729-761.

Satyanaryana, T. and Johri, B.N. (1981). Volatile sporostatic factors of thermophilic fungal strains of paddy straw compost. *Curr. Sci.* **50:** 763-768.

Sinden, J.W. and Hauser, E. (1950). The short method of composting during fermentation. *Mushroom Sci.* **1:** 52-59.

Sparling, G.P.; Fermor, T.R. and Wood, D.A. (1982). Measurement of the microbial biomass in composted wheat straw and the possible contribution of the biomass to the nutrition of *Agaricus bisporus. Soil Biol. Biochem.* **14:** 609-611.

Stanek, M. (1969). Diewitkung der zelulose setzenden mikroorganismen auf das waehstun des champignons. *Mushroom Sci.* **7:** 161-172.

Stanek, M. (1972). Microorganisms inhabiting mushroom compost during fermentation. *Mushroom Sci.* **8:** 797-811.

Stephen, J.R.; McCaig, A.E.; Smith, Z.; Prosser, J.I. and Embley, TM. (1996). Molecular diversity of soil and marine 16S rRNA gene sequences related to beta-subgroup ammonia-oxidizing bacteria. *Appl. Environ. Microbiol.* **62:** 4147-4154.

Straatsma, G., Gerrits, J.P.G., Gerrits, T.M., Op Dencamp, H.J.M. and van Griensven, L.J.L.D. (1991). Growth kinetics of *Agaricus bisporus* mycelium on solid substrate (mushroom compost). *J. Gen. Microbiol.* **137:** 1471-1477.

Straatsma, G.; Gerrits, J.P.G.; Augustijn, M.P.A.M., OpDenCamp, H.J.M., Vogels, G.D. and Van Griensven, L.J.L.D. (1989). Population dynamics of *Scytalidium thermophilum* in mushroom compost and stimulatory effects on growth and yield of *Agaricus bisporus. J. Gen. Microbiol.* **135:** 751-789.

Straatsma, G.; Olijnsma, T.W.; Gerrits, J.P.G.; Amsing, J.G.M.; OpDen Camp, H.J.M. and Van Griensven, L.J.L.D. (1994b). Inoculation of *Scytalidium thermophilum* in button mushroom compost and its effect on yield. *Appl. Environ. Microbiol.* **60:** 3049-3054.

Straatsma, G.; Samson, R.A.; Olijnsma, T.W.; Opden Camp, H.J.M.; Gerrits, J.P.G.; Griensven, L.J.L.D. van, Van Griensven, L.J.L.D. (1994a). Ecology of thermophilic fungi in mushroom compost, with emphasis on *Scytalidium thermophilum* and growth stimulation of *Agaricus bisporus* mycelium. *Appl. Environ. Microbiol.* **60:** 454-458.

Strom, P.F. (1985). Identification of thermophilic bacteria in solid waste composting. *Appl. Environ. Microbiol.* **50:** 906-913.

Stutzenberger, F.J.; Kanfman, A.J. and Lossin, R.D. (1970). Cellulolytic activity in municipal solid waste composting. *Can. J. Microbiol.* **16:** 553-560.

Tewari, P. (2000). Thermophilic microorganisms from mushroom compost and their polysaccharolytic activites. G.B.Pant Univ. of Agric. & Technol. Pantnagar, India. 99p.

Troussellier, M. and Legendre, P. (1981). A functional evenness index. *Microb. Ecol.* **7:** 283-296.

Vijay, B. (1996). Investigations on compost mycoflora and crop improvement in *Agaricus bisporus* (Lange). Ph.D. Thesis. H.P. Univ. Shimla.

Vijay, B. and Gupta, Y. (1992). Studies on fungal competitors of *Agaricus bisporus. Ind. Phytopathol.* **45:** 228-232.

Watabe, M.; Rao, J.R.; Xu, J.; Millar, B.C.; Ward, R.F. and Moore, J.E. (2004). Identification of novel eubacteria from spent mushroom compost (SMC) waste by DNA sequence typing: ecological considerations of disposal on agricultural land. *Waste Manage.* **24:** 81-86.

Weigant, W.M. (1991). A simple method to estimate the biomass of thermophilic fungi in composts. *TIBTECH.* **5:** 421-426.

Weigant, W.M. (1992). Growth characteristics of thermophilic fungus *Scytalidium thermophilum* in relation to production of mushroom compost. *Appl. Environ. Microbiol.* **58:** 1301-1307.

White, D.C. and Findlay, R.H. (1988). Biochemical markers for measurement of predation effects on the biomass community structure, nutritional status, and metabolic activity of microbial biofilms. *Hydrobiol.* **159:** 119-132.

Advancing Frontiers of Ecological Researches in India, 397-406, 2006
Edited by A.K. Kandya & Asha Gupta
Published by Bishen Singh Mahendra Pal Singh, Dehra Dun, India

Ecology of Terrestrial Blue Green Algae: A Review

Jaishree Dubey
Department of Botany
Dr. H.S. Gour Vishwavidyalaya, Sagar 470003

Abstract: The paper provides information on almost all important aspects of terrestrial Blue Green Algae, i.e. Ecological distribution, historical account, fossil record, biological diversity, nitrogen fixing ability, use as biofertilizers and their other roles in the soil.

INTRODUCTION

In almost every soil, in samples obtained from each continent and from the most remote islands, one finds the presence of algae, yet their position as indigenous soil inhabitants has often been the subject of considerable controversy. For the microbiologist, the soil environment is unique in several ways: it contains a vast population of bacteria, actinomycetes, fungi, algae and protozoa. It may be considered as the most active of biological interactions in nature. Algae are never as numerous as other microorganisms. The lack of sufficient appreciation of this group can be attributed in part to small population.

Ecological Distribution

Ecologically the algae are a world wide group occurring on the land surface, on all types of soil, on permanent ice and snow fields, but their major centre of distribution is in the waters. Under natural conditions, algae grow as mixed communities which may include many species and genera.

Algae are photoautotrophic microorganisms and the occurrence of these has been recorded in soils throughout the world, but no definite geographical localization of families, genera or species has yet been presented. As a group, the algae are moderately adaptable to environmental changes, persisting in unfavourable circumstances such as in alkaline and desert soils.

Soil algae refer to various ecological groups, viz.

1. Soil algae proper- found below the soil surface.
2. Terrestrial algae- growing abundantly on soil surface as crusts or films
3. Aquatic terrestrial algae- growing on the surface of permanently damp soil.

The tropical soils have a greater preponderance of the blue green algae (BGA) than the green ones, presumably because of high temperature and humidity. The algal growth is generally rich at the surface, their abundance decreasing rapidly with depth. The origin of the subterranean algae may be air borne, water borne, biotic or all three. Several workers have recorded the microstratification of algal populations within the upper few mm or cm of soil profiles, rocks or gravel deposits (Friedmann and Galun, 1974; Melting, 1981). The upper 1.5 cm zone of various British soils harbours diverse populations of several green and yellow green algae (Broady, 1979).

In semiarid environments, examination of soil has shown that two upper surface tends to be colonized mainly by the Cyanophyceae (0-2 cm), whereas the greatest cell numbers of Chlorophyceae and Bacillariophyceae occur at 4-6 cm depths (Nordin and Blinn, 1972). Schubert (1984) on the same soils previously studied by Nordin and Blinn (1972) has shown that soil algae fluctuate with the growing season and soil profile.

The Cyanophyceae dominate in deserts, forming surface crusts. The Oscillatoriaceae are common in cultivated and noncultivated desert in Arizona, whereas the other cyanophytes are common in virgin or fallow areas (Cameron, 1963). In the Israeli Negav desert, Friedmann *et al.* (1967) found the Chlorophyceae to be dominant and the blue green, yellow green and diatoms to be less common. In the coastal areas of Texas, the green and blue green algae are the most common and most diverse (Dykstra *et al.*, 1975). In several tropical regions, *Chlamydomonas sp* are ubiquitous, and *Schizothrix calcicola* is one of the most common blue greens.

The terrestrial blue greens and other soil algae show good adaptations to live and grow in those climates and micro environments in which available water is the main limiting factor and some soil algae can survive very long periods without water. Species of *Stichoccus, Microcoleus* and *Schizothrix* have a marked capacity to resist extremely low temperature (even-150°C) in crushed soil. Likewise, many soil algae when dry can tolerate abnormally high temperatures; i.e. *Scytonema* tolerate short exposure to 110°C, whereas *Spongiochloric typica* cells can survive a year in air dried soil that has been dried even at 100°C (Trainor and Mc Lean, 1964).

Neutral and weakly or moderately alkaline soils provide good conditions for the development of BGA; these algae and some diatoms can tolerate fairly high ionic strengths. Nitrogen fixing BGA generally grow better and fix more nitrogen in water- logged soils as compared to oxidizing or aerobic soils (Fogg *et al.*, 1973). These Cyanobacteria are restricted to the photic zone and are usually located within the upper few mm in fine soil or few cm in the coarse soil (Goyal, 1997). Viable propagules of algae have however, been reported upto a depth of one metre (Kamat and Patel, 1973; Friedmann and Galun, 1974). They are carried down by agricultural activity, root growth and soil fauna (Lund, 1967).

Historical account

The History of algae is as old as that of Botany. The first writings on algae appeared in the ancient Chinese classics, where references to the collection, eating and the medicinal uses of algae are found. There are also references in Roman and Greek literature about the use of algae for cosmetic purposes. The Chinese named them as 'Tsao' for having aesthetic significance whereas the Hawaiians called them 'Limu' as they used algae as a food. In the north coast of France, algae were known to be used as early as twelfth century. Although Linnaeus (1753) first introduced the name Algae but the delimitation of this group was made by De Jussieu (1789). Jussieu's characterization of algae was not of much value since it was based entirely on macroscopic features. Actually the great pioneer in the study of algae was Agardh (1824), who first started describing algae considering almost all aspects of their life history.

Blackmann (1990) postulated a flagellate ancestry for the algae, expressed earlier by Klebs (1892), and suggested that there are three distinct tendencies, the volvocine, tetrasporine and chlorococcine which were recognized in the evolution of green algae.

In India Royle, first reported the occurrence of some members of the lemaniaceae in the mountain streams of the Himalayas. Following him were Decaisse and Montagne who described *Actinotrichea* fingillis and *Calothrix indica* in 1842 and 1844. Griffith (1849) reported a number of species of *Chara* from Bengal. Carter (1858) collected *Eudorina elegans* from the pools. From 1919 onwards more Indian workers got interested in algae. During 1919 to 1932. Ghose carried on investigations on the BGA of Punjab. M.O.P. Iyengar was another pioneer investigator who published a good number of papers on algae of south India.

BGA are considered as the most primitive photosynthetic prokaryotes, are supposed to have appeared on this planet during the Pre-Cambrian period (Olson, 1970). Possibly, these are the first photosynthetic microorganisms which persisted over a period of 2-3 billion years performing an important role in evolution of the higher forms. They are believed to have evolved from the gliding photosynthetic bacterium, *Chloroflexis* (Pierson & Castenholz, 1971). Their resemblances with gliding bacteria in gross morphology and with autotrophic bacteria in fixing both CO_2 and N_2 and with higher plants by possessing chlorophyll-a and photolyzing water provocated the scientists to consider them as the central niche in the process of plant evolution.

Fossil Record of the BGA

The great antiquity of the BGA is well documented, through the fossil record is fragmentary and biased towards the formations where preservation of the morphology was possible. The earlier unicellular and filamentous forms attributed to the cyanobacteria were found in sedimentary rocks formed 3500 million years ago. Endolithic forms that reproduce with baeocytes were observed in rocks formed 1500 million year ago. Heterocystous forms and forms with true branching seem to have appeared more recently, after the Precambrian (Schopf and Walter, 1982).

Biology of BGA

The BGA are now commonly referred to as Cyanobacteria (Haselkorn, 1978) comprising about 2,500 species under 150 genera. Individually all the BGA are microscopic. The characteristic feature of BGA are: the cell constituting the thallus are prokaryotic; lack flagella; where locomotion occurs, it is of gliding, jerky type; the main photosynthetic pigments are chlorophyll-a, carotene and unique xanthophylls namely myxoxanthin and mycoxanthophyll, together with phycobiliproteins, c-phycocyanin (blue) and c-phycoerythrin (red); there are no membrane-bound chromatophores; myxophycean starch and cyanophycin are the unique food-storage compounds; many filamentous BGA possess specialized cells known as heterocysts. Sexual reproduction is absent but genetic recombination has been reported. During recent years, genetics of BGA has progressed rapidly, and developmentally regulated gene rearrangements have been established (Haselkon, 1992).

Cyanobacteria: use as biofertilizers

Cyanobacteria are ancient prokaryotic organisms with an oxygen-evolving photosynthetic apparatus which is similar to that of higher plant chloroplasts. Their simple growth requirements together with highly evolved capacity of adaptation to changing environmental conditions enable them to colonize in all kinds of habitats. In tropical waterlogged and deepwater rice-fields, they grow luxuriently and contribute a significant part of free-floating and epiphytic populations (Singh, 1961). Such rice-fields are rich in organic dead matters and exposed to bright light and high temperature which provide conducive conditions for cyanobacterial growth. Most of these cyanobacterial species are capable of nitrogen-fixation because of the possession of nitrogenase, an enzyme restricted to a few selected prokaryotes. Their dual abilities of nitrogen fixation and photosynthesis make them a highly productive and efficient biological system. Therefore, diazotrophic cyanobacteria have a great potential as biofertilizer and their use decreases the demand for chemical fertilizer.

As nitrogen is the most common limiting nutrient for plant productivity, it has led to increased use of fertilizer input. The fertilizer nitrogen consumption in India was only 0.06 million tons in 1952 and reached to 9.5 million tons in 1994-95. It is estimated that for feeding a population of 1.4 billion by 2025, our country will need to produce 3.1 mt food grain and for achieving this, India will need at least 45 mt plant nutrients, out of which at least 35 mt should be chemical fertilizer source and rest 10 mt nutrient should come from other sources like biofertilizers or manure (Kanwar, 1997). Besides the economic and environmental concern for heavy use of chemical fertilizers, sustainability also requires that ecologically sound alternative must be sought. Biological nitrogen fixation which accounts for 65% of the nitrogen currently utilized in agriculture, seems to be the best means of reducing external fertilizer inputs.

The agronomical potential of BGA was recognized by De (1939) and since then lot of work has been done to study the response of BGA inoculations. The practice of inoculation of soils with BGA has been referred to as algalization (Venkataraman, 1961) and it has been reported to have beneficial effects on grain yield in a number of countries including China, Egypt, India, Japan, Philippines and Russia.

Many Indian scientists have systematically surveyed and documented the diversity of cyanobacterial population from paddy field soils with a view to select efficient nitrogen fixers (Pandey, 1965; Kolte and Goyal, 1986; Anand, 1989; Singh and Bisoyi, 1989; etc.). Application of cyanobacterial biofertilizers developed as a technology through the efforts of Venkataraman (1972, 1981) and mass cultivation of selected heterocystous forms were undertaken at agricultural research institutes (Kannaiyan, 1995; Goyal, 2000). Improvements in the biofertilizer production technology include the use of polypacks as low cost bioreactors and immobilisation of selected isolates in artificial and natural matrices (Kannaiyan, 1995; Kannaiyan *et al.*, 1997).

In India, a good deal of work has been done on the algal flora of paddy field (Singh, 1939; Mitra, 1951; Tiwari, 1972; Sardesh Pandey and Goyal 1981; Kolte and Goyal, 1986; Sharma and Komi, 1992; Nayak *et al.*, 1996; Singh *et al.*, 1997; Rout and Dey, 1999). However, the information available on the algal flora of other crops is very meagre (Marathe, 1967; Marathe and Ananthani, 1972; Bongale and Bharati, 1980).

Role of Cyanobacteria in soil

The cyanobacteria form an inexpensive farm grown input which helps in better crop nutrient management while working in perfect harmony with the nature. They provide 15-25 kg biologically fixed N per hectare per season in a linear fashion, add organic matter, synthesize and liberate amino-acids, vitamins and auxins, reduce oxidizable content of the soil, provide oxygen to the submerged rhizosphere of rice plants, remove soil compaction, ameliorate salinity and buffer the pH, solubilize phosphates and increase the fertilizer utilizing efficiency of the crop plants. The waterlogged conditions, high humidity and temperature and shade provided by the crop canopy, afford congenial conditions for rapid multiplication of these highly stress compatible cyanobacteria. Unlike chemical fertilizers, the algae colonize the soil and perennate during dry season, an attribute which leads to bioenrichment of the soil over a period of time. Successful establishment of the promising strains of the algae results in an increase of up to 30 percent in the crop yield (Singh, 1961; Goyal, 1993; Venkataraman, 1972). Cyanobacteria produce humus on soil surface after death and dissolve certain soil minerals, maintaining a reservoir of elements in a semi-available form for higher plants. A 50-70% increase in soil aggregation capacity due to algal inoculation has been observed. This is attributed to the action of polysaccharides released by the algae and presence of interwoven filaments of cyanobacteria growing on soil surface.

There have also been reports that cyanobacteria can benefit crop plants by producing plant growth regulators (PGRs) and their action was likened to those of vitamin B_{12}, gibberellins and amino acids.

Nitrogen fixing ability: an unique character

Several cyanobacterial species are able to grow at the expense of atmospheric nitrogen under aerobic condition, and many more are able to perform N_2 fixation when anaerobic condition is provided artifically. Heterocysts are specialised and differentiated cells whose principal function is the fixation of Di-nitrogen (N_2) to ammonia. This function is catalysed by the enzyme Nitrogenase which is synthesized in Heterocysts. The genes responsible for the production of nitrogenase are expressed only in heterocysts. However, the nitrogen fixing function of heterocysts depends upon the vegetative cells for the supply of electrons. It is not known whether detached heterocysts can fix nitrogen.

Nitrogen fixation by Cyanobacteria in associations

Though cyanobacteria are capable of independent existence, they form symbiotic associations with a wide variety of plants and animals. Such cyanobionts show higher heterocyst frequency and increased nitrogenase activity. Moreover, the fixed nitrogen is transferred to the partner either as nitrogen derived ammonia or as amino acids. One such association that has been extensively used for paddy cultivation in Asia is the *Azolla-Anabaena* system (Kannaiyan, 1995).

One of the recent approaches aims at improving the nitrogen economy of crops by establishing artificial symbiosis between the roots of crop plants and nitrogen fixing cyanobacterial strains. In associations involving whole plants, cyanobacteria have been isolated from stomatal pores, xylem vessels and primary cortical parenchyma. Colonisation of roots of rice plants of *Nostoc* isolated from *Anthoceros* has been attempted (Rai *et al*., 1996a). Cyanobacterial association with the pneumatophores of mangrove plants *Avicennia* sp. has been investigated. Non-heterocystous, filamentous cyanobacteria colonising the pneumatophores contribute considerably to the nitrogen status of mangrove ecosystem (Anand *et al.*, 1996).

REFERENCES

Agardh, C.A. (1824). Systema algarum. Vol. 1, 312 pp.

Anand, N. (1989). Handbook of Blue green Algae. Bishen Singh Mahendra Pal Singh, Dehra Dun, pp. 74.

Anand, N., C. Murugesan, V.S.R. Murugesan, S. Geetha, and N. Mourugasamy (1996). Nitrogen fixation by Cyanobacteria in mangrove ecosystem in South India, Abst., In: Internet. Symp. Cyanobacteial Biotechnol, Bharathidarsan, University, Tiruchirapalli, pp. 59.

Blackmann, F.E. (1900). The primitive algae and the flagellate. An account of modern work bearing on the evolution of algae. Ann. Bot., **14:** 647-688.

Bongale, U.D. and S.G. Bharati (1980). On the algal flora of cultivated soils of Karnataka State, India. Phykos, **19:** 95-109.

Broady, P.A. (1979). Qualitative and quantitative observations on green and yellow green algae in some English soils. Brit . Phycol. J., **14:** 151-160.

Cameron, R.E. (1963). Algae of Southern Arizona. I. Introduction: Blue green algae. Rev. Algol., **6:** 282-318.

Carter, H.J. (1858). *Eudorina*. Ann. and Mag. Nat. Hist.3 Ser., **2:** 237-253.

De Jussieu, A.L. (1789). Genera plantarum. Secumdum ordines natural disposita. Paris, pp. 498.

De, P.K. (1939). The role of blue green algae in nitrogen fixation in rice fields. Proc. Roy. Soc. London, 127 B: 129-139.

Dykstra, R.F, F.J. Mac Entée, and H.C. Bold (1975). Some edaphic algae of the Texas Coast, Texas J. Sci., **26:**171-177.

Friedmann, E.I., Y. Lipkin, and R. Ocampa-Paus (1967). Desert algae of the Negev (Israel). Phycologia, **6:**185-200.

Freidmann, E.I. and M. Galun (1974). Desert algae, lichens and fungi. In Brown, G.W. (Jr.) (ed.) Desert Biology. II, Academic Press, New York.

Fogg, G.E., W.D.P. Stewart, P. Fay and A.E. Walsby (1973). The blue-green algae, Academic Press, London.

Griffith, W. (1849). Notulae ad Plantas Asiaticas (Characeae). Culcutta, Vol. **1:**275-283.

Goyal, S.K. (1993). Algal biofertilizer for vital soil and free N. Proc. Indian Natl. Sci. Acad., B**59:** 290.

Goyal, S.K. (1997). Algae and the soil environment. Phykos, **36:** 1-13.

Goyal, S.K. (2000). Preparation of soil based algal inoculum: Biofertilizers: BGA and *Azolla*. IARI, New Delhi–110012.

Haselkorn, R. (1978). Heterocysts. Annual Review of Plant Physiology, **29:** 319-344

Haselkorn, R. (1992). Developmentally regulated gene rearrangements in prokaryotes. Annual Review of Genetics, **26:**113-130.

Kamat, N.D. and M.Z. Patel (1973). Soil algae of rice field at different depths. Botanique **4:** 101-106.

Kannaiyan, S. (1995). Algal biofertilizer technology for rice crop. In : Rice Management Biotechnology. Ed. S. Kannaiyan, Associated Publ. Co., New Delhi, pp.299-307.

Kannaiyan, S., S.J. Aruna, S. Merriana, Prem Kumari and D.O. Hall (1997). Immobilised Cyanobacteria as a biofertilizer for rice crops. J. Appl. Phycol., **9:**167-174.

Kanwar, J.S. (1997). Algal biofertilizer in rice fields. J. Indian Soc. Soil Sci., **45:** 417-428.

Klebs, G. (1892). Flagellatenstudien II. Ztschr. Wiss. Zool., **55:**353-445.

Kolte, S.O. and S.K. Goyal (1985). Natural variation in the N_2 fixing potential of Cyanobacteria. Phykos, **25:** 166-172.

Linnaeus, C. (1753). Species plantarum. 2 Vols. Stockholm, Ed. C.L. Willdenow.

Lund, J.W.G. (1967). Observation on soil algae. New Phytol., **46:** 35-60.

Marathe, K.U. (1967). Studies on soil algae of India. J. Univ. Bombay, **35:**100-105.

Marathe, K.U. and Y.S. Ananthani (1972). Observations on the algae of some Indian arid soils. Botanique, **3:**13-19.

Melting, B. (1981). The systematics and ecology of soil algae. Bot. Rev., **47:** 195-312.

Mitra, A.K. (1951). The algal flora of certain Indian soils. Indian J. Agric. Sci., **21:** 357-373.

Nayak, H., J.K. Sahu and S.P. Adhikary (1996). Blue green algae of rice fields of Orissa state II. Growth and nitrogen fixing potential. Phykos, **35:** 93-110.

Nordin, R.N. and D.W. Blinn (1972). Analysis of a saline tall grass prairie ecosystem IV. Preliminary investigations on soil algae. N. Dak. Acad. Sci., **25:** 8-17.

Olson, J.M. (1970). The evolution of Photosynthesis. Science, **168:** 438-446.

Pandey, D.C. (1965). A study on the algae from paddy field soils of Ballia and Ghazipur districts of Uttar Pradesh, India. Nowa Hedw, **10:** 299-334.

Pierson, B.K. and R.W. Castenholz (1971). Bacterio-chlorophylls in gliding filamentous procaryotes from hot springs. Nature New Biology, **233:** 25-27.

Rout, J. and A. Dey (1999). A Study of algal flora from rice fields of Irongmara (Barak valley Assam), Phykos, **38:**19-25.

Rai, A.N., M. Borthakur and D. Paul (1996). Symbiotic cyanobacteria: Biotechnological applications. J. Sci. Industrial Res., **55:**742-752.

Sharma, B.M. and P.N. Komi (1993). Nitrogen- fixing blue green algae from rice farms of Jammu and Kashmir. In: Biofertilizer Technology Transfer. Ed. V. Gangawane, pp. 187-190.

Schopf, J.W. and N.T. Walter (1982). Precambrian microorganisms and evolutionary events prior to the origin of vascular plants. Biol. Rev., **45:** 319-352.

Schubert, L.E. (ed.) (1984). Algae as Ecological Indicators. Academic Press, London.

Singh, R.N. (1961). The Role of Blue green algae in nitrogen economy of Indian Agriculture. ICAR Publ. New Delhi, pp.175.

Singh, P.K. and R.N. Bisoyi (1989). Blue green algae in rice fields. Phykos, **28:**181-195.

Singh, N.I., H. Dorycanta, G.A. Devi, N.S. Singh and M.S. Singh (1997). Blue green algae from rice field soils of Nagaland. Phykos, **36:** 115-120.

Sardeshpande, J.S. and S.K. Goyal (1981). Distributional pattern of blue green algae in rice fields soils of Konkan region of Maharashtra State Phykos, **20:**102-106.

Tiwari, G.L. (1972). A study of blue green algae from paddy field soils of India. Hydrobiologia, **39:** 335-350.

Trainor, F.R. and R.J. Mc Lean (1964). A study of a new species of *Spongiochloris* introduced into sterile soil, Am. J. Bot., **51:** 57-60.

Venkataraman, G.S. (1961). The role of blue green algae in agriculture. Sci. & Cult., **27:** 9-13.

Venkataraman, G.S. (1972). Algal biofertilizer and rice cultivation. Today and Tomorrow Printers and Publishers, New Delhi, pp 75.

Venkataraman, G.S. (1981). Blue green algae for rice production, FAO soils bulletin no. 46.

Advancing Frontiers of Ecological Researches in India, 407-419, 2006
Edited by A.K. Kandya & Asha Gupta
Published by Bishen Singh Mahendra Pal Singh, Dehra Dun, India

Tetrazolium Reduction and Chlorophyll Stability of Cyanobacteria in the crusts on the Exposed Soil Surfaces Subjected to Heat Stress

Bagmi Pattanaik and S.P Adhikary
P.G. Department of Botany & Biotechnology, Utkal University, Bhubaneswer-751004

Abstract : Cyanobacteria inhabiting the crust were able to reduce tetrazolium salt forming formazon crystals within 1-24 hours of TTC treatment suggesting that the cells therein remained viable even when subjected to extreme conditions. However, heating the crusts in dry state at 60^0 C was detrimental for their survival. The chlorophyll stability index (CSI) of crusts and the corresponding organisms in culture were adversely affected when subjected to heat treatment at 60^0C for 1 hour, however, the injury due to dry heat was more pronounced than that of the wet heat treatment.

INTRODUCTION

The cyanobacterial species form predominant flora in tropics as crust/ mat in various habitats under extreme environmental conditions. Such crusts/ mats frequently occur on exposed rock surfaces of arid deserts, inselbergs and also on the surfaces of fallow rice field soils (Mazor *et al.*, 1996; Büdel *et al.*, 1999; Adhikary and Sahu, 2000). Survivability of these cyanobacteria in the crust in these habitats possibly depends on the capability to overcome several adverse conditions, e.g. high temperature, intense solar radiation and water stress etc. There are reports that cyanobacteria have the remarkable ability to survive in extremes of dehydration and heat (Brock, 1975; Grilli-Caiola *et al.*, 1996). They can thrive even at temperature exceeding 60-65^0C coupled with high light intensity prevailing on the exposed surfaces during summer months of the tropics and impart a blackish-brown colouration to the corresponding substratum in summer months (Tripathy *et al.*, 1997). However, they change their colour to bluish-green soon after getting monsoon rain due to revival of growth of cyanobacteria on the substratum. This also form a suitable ecological niche for growth of other microorganisms living in the crust as dormant state in association with the dominant cyanobacteria. Upon revival on the onset of a congenial condition, the organisms absorb nutrients as well as moisture and also secrete extracellular polymeric substances leading to their further colonisation on the surfaces (Ernst *et al.*, 1987). At times they can even penetrate into the substratum and survive during dry periods. The cyanobacteria in the crust remain dormant with low physiological activity and can protect against stress injury by preventing or repairing protein aggregation and denaturation by maintaining reduced state (Tripathi and Talpasayi, 1980). These cells resume their active metabolism after their revival in response to suitable conditions. Reducing activity of viable cells can be best studied by tetrazolium reduction test. Tetrazolium salts (Triphenyl tetrazolium chloride) are low potential redox indicators which on reduction change into coloured water-insoluble formazons. These are used routinely for demonstration of dehydrogenase activity in cells and in enzyme preparations (Fay and Kulasooriya, 1972). Hence, TTC reduction test was used as a tool to understand the metabolic status of the crusts from different habitats and also after their treatment under dry heat and wet heat conditions.

Cyanobacteria form characteristic coloured crusts tightly adhering to the brick surfaces of the Lalitgiri excavated sites and also on the dried soils of fallow rice fields and in their natural habitats they are exposed to high solar radiation. These organisms possess the capacity to thrive successfully and retain their metabolic activity even under the adverse environmental conditions, hence resume their growth soon after wetting or after removal of the stress situation. Further, there are several reports in which chlorophyll stability at high temperature has been correlated with resistance of plants to drought, i.e. less the loss in chlorophyll at high temperature the more is the capacity of the plants to resist drought. In all these reports the chlorophyll stability index (CSI) was determined by measuring the chlorophyll extract of heated and unheated leaf samples and correlated with their capacity to resist drought (Murthy and Majumdar, 1962). Later Tripathy (1983) determined the chlorophyll stability value of subaerial cyanobacteria occurring on roof top of buildings and tree barks and attributed the value to drought tolerance.

TTC reduction activity of these organisms under heat treatment as an index of maintaining metabolic activity in dried state and chlorophyll stability of the component organisms in the crusts at high temperature under dry and wet heat conditions was also analysed.

MATERIALS AND METHODS

Cyanobacterial crusts collected from brick substratum of Lalitgiri excavated sites and fallow rice field soils were used in the experiment. For

TTC reduction activity (reaction with 2,3,5,Triphenyl tetrazolium chloride) of the crust and the corresponding dominant cyanobacterial species these were soaked with 1% (w/v) TTC for varied time periods and observed under microscope for detection of reddish formazon crystals. Effect of heat on the TTC reduction activity was examined by exposing the crust as well as the corresponding organism to dry and wet heat at 60^0C for up to 2 hours. For dry heat the organisms was incubated at desired temperature in an oven, and for wet heat treatment, the organisms were soaked with distilled water and then heated in a water bath pre-set to the desired temperature. To determine the chlorophyll stability of the crust and the organisms, a known amount of crust (30 mg) and organism (fresh weight 30 mg) was taken, heated in dry or wet state at 60^0C for 1 hour and then extracted with 90% (v/v) methanol. The amount of chlorophyll-*a* was determined following Mackinney (1941). The chlorophyll stability index was calculated using the formula of Tripathi (1983) as follows:

$$\%\,\text{loss in Chl-}a = \frac{\text{Chl-}\mu a\ (\text{g/g})\ \text{of unheated organisms - Chl-}a\ (\mu\text{g/g})\ \text{of heated organisms}}{\text{Chl-}a\ (\mu\text{g/g})\ \text{of unheated organism}} \times 100$$

RESULTS AND DISCUSSION

TTC reduction activity of the crust and their corresponding cyanobacterial isolates

Usually the crusts with the cyanobacterial components occurred being exposed to adverse conditions in the natural environment. To know their metabolic activity while inhabiting in the crust, the crust and the corresponding dominant cyanobacterial species were treated with 2,3,5, Triphenyl tetrazolium chloride (TTC) for different durations. After treatment with 1% (w/v) TTC, the cells of the cyanobacteria in the crust/mat were observed under microscope for detection of formazon crystals indicative of the reducing activity of the cells. The crust from the brick surfaces of excavated sites of Lalitgiri and from the soils of rice fields, and their corresponding dominant cyanobacterial species were treated with TTC and examined. The crusts from Lalitgiri excavated sites and from the fallow rice fields showed reduction activity to TTC after a prolonged period of treatment. The TTC reduction activity of the cells in the crust from rice field soils was quicker (about 60 min) than those of the crust from the Lalitgiri excavated sites; the filament inside the crust of the later required 24 hours to form pinkish colouration of the cells (Plate 1, Fig. a,d; Table-1). When the cells from the culture were treated with 1% (w/v) of TTC, in both the organisms

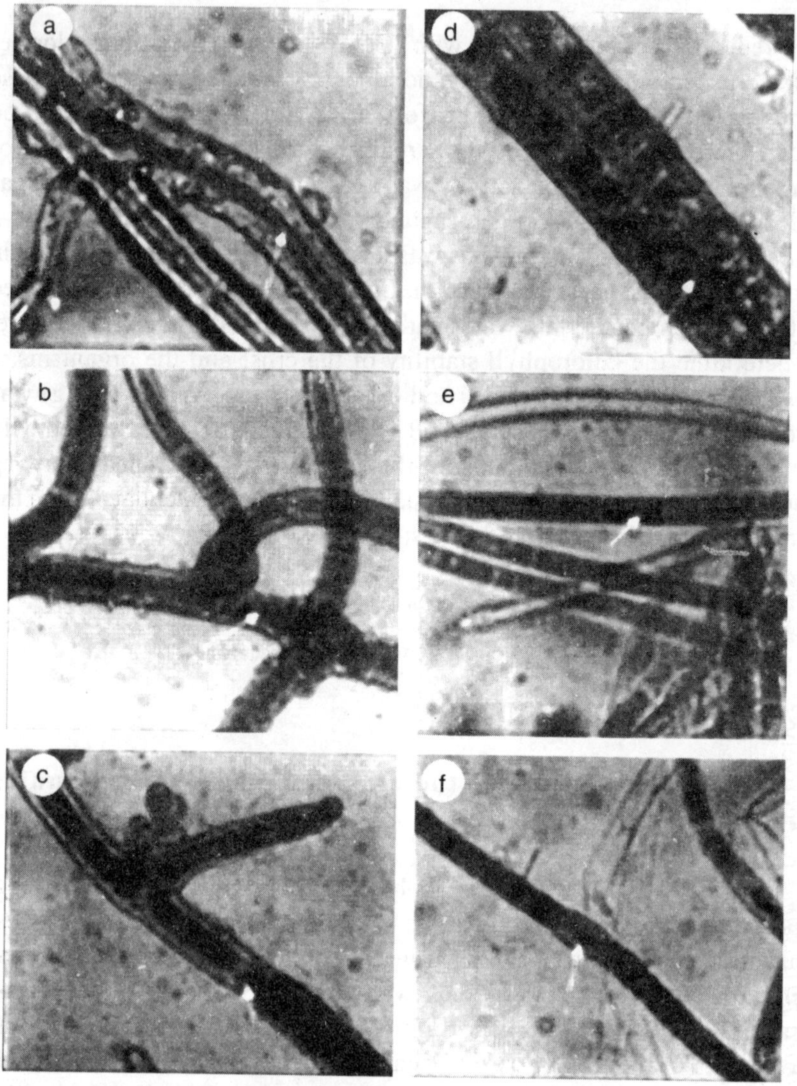

Plate 1, Fig. a-f : TTC reduction activity of the cells of cyanobacteria in the crusts from excavated sites at Latitgiri and rice field soils, and the coerresponding organisms *Tolypothrix arenophila* and *Aulosira fertilissima* after treatment with 1% (w/v) TTC in different time intervals.

a. Cells of *Tolypothrix arenophila* in the crust showing pinkish appearance after treatment with TTC up to 24 hour (×780); b. Pinkish appearance of the heterocysts of *Tolypothrix arenophila* after treatment with TTC up to 5 min (×380); c. Crystal formation in all cells of *Tolypothrix arenophila* after treatment with TTC up to 60 min (×380); d. Cells of *Aulosira fertilissima* in the crust showing crystal formation after treatment with TTC up to 1 hour (× 780); e. Pinkish appearance of the vegetative cells *Aulosira fertilissima* after treatment with TTC up to 10 min (×780); f. Crystal formation in all cells of *Aulosira fertilissima* after treatment with TTC up to 20 min (×780).

Table-1 : TTC reduction activity of the crusts from Lalitgiri excavated sites, rice field soils and the corresponding cyanobacteria *Tolypothrix arenophila* and *Aulosira fertilissima* respectively

Duration of TTC treatment	Heterocyst(H) / Vegetative cell (V)	Lalitgiri excavated site		Rice field soils	
		Crust	Corresponding organism ***T. arenophila***	Crust	Corresponding organism ***A. fertilissima***
5 min	HV	No reaction	Pinkish	No reaction	No reaction
		No reaction	No reaction	No reaction	No reaction
10 min	H	No reaction	Pinkish	No reaction	Pinkish
	V	No reaction	No reaction	No reaction	Pinkish (Pl 1, Fig.e)
20 min	H	No reaction	Pinkish	No reaction	Crystal formation
	V	No reaction	Pinkish	No reaction	Crystal formation (Pl 1, Fig.f)
30 min	H	No reaction	Crystal formation	No reaction	
	V	No reaction	Pinkish (Pl 1, Fig.b)	No reaction	
60 min	H	No reaction	Crystal formation	No reaction	
	V	No reaction	Crystal formation (Pl 1, Fig.c)	Pinkish	
24 h	H	No reaction		Pinkish	
	V	Pinkish (Pl 1, Fig.a)		Crystal formation (Pl 1, Fig.d)	
48 h	H	Crystal formation		Crystal formation	
	V	Crystal formation		Crystal formation	

the heterocysts first became pinkish coloured followed by crystal formation, pinkish appearance of vegetative cells and formation of crystals in all the cells of the filament in sequence with respect to the time of incubation (Plate 1, Fig. b,c,e; Table-1). The cells from the culture showed quicker reduction activity than the crust indicating the cells inside the crust were activated slowly than the filaments grown in the culture.

To understand whether pre-heating of the crust in dry or wet state has any difference in their response to TTC reduction, equal amount of crust and the corresponding cyanobacterial filaments were taken and incubated in a hot air oven (dry heat) at 60^0C for up to 2 hours. On pre-heating the crusts in dry state TTC reduction in the cells was not observed (Plate 2, Fig. a; Table-2). Possibly, the cells did not revive from the crust to became metabolically active in the absence of water. However, when heated in wet state, the cells of the crust from Lalitgiri excavated site showed TTC reduction after 2 hours of treatment (Plate 2, Fig. b). The corresponding organism *T. arenophila* showed metabolic activity up to 1h of dry heat and wet heat treatment (Plate 2, Fig. c,d). To the contrary, the crust from rice field soils when subjected to dry heat did not react to TTC (Plate 3, Fig. a). However, the cells of the organism in the crust showed TTC reduction after 1 h of wet heat treatment till up to 2 hours of treatment (Plate 3, Fig. b). The corresponding organism of rice field crust, *A. fertilissima* showed metabolic activity up to 1 hour of wet heat treatment. Further increase in the duration of heat treatment had an adverse effect (Table-2). But when subjected to dry heat treatment for 10 min the cells did not react to TTC (Plate 3, Fig.c). Formazon crystals were found in its cells heated in dry state for 1 h (Plate 3, Fig. d). However, with increase in the duration of dry heat treatment, metabolic activity of the organism was ceased suggesting heat injury to the cells of the organisms.

Chlorophyll stability of the crust from exposed surfaces and their corresponding cyanobacterial species under wet and dry heat treatment

The dominant cyanobacteria *Tolypothrix arenophila* and *Aulosira fertilissima* of the crust from brick substratum of excavated sites of Lalitgiri and the fallow rice field soils respectively exposed to 60°C in dry and wet state for 1h resulted decrease in chl-*a* content over the unheated cells. In general chlorophyll stability index (CSI) was higher in dry heat treated crust than the corresponding wet heat treatment. Chlorophyll stability index was also higher in the cultured material in the wet heat treated sample than

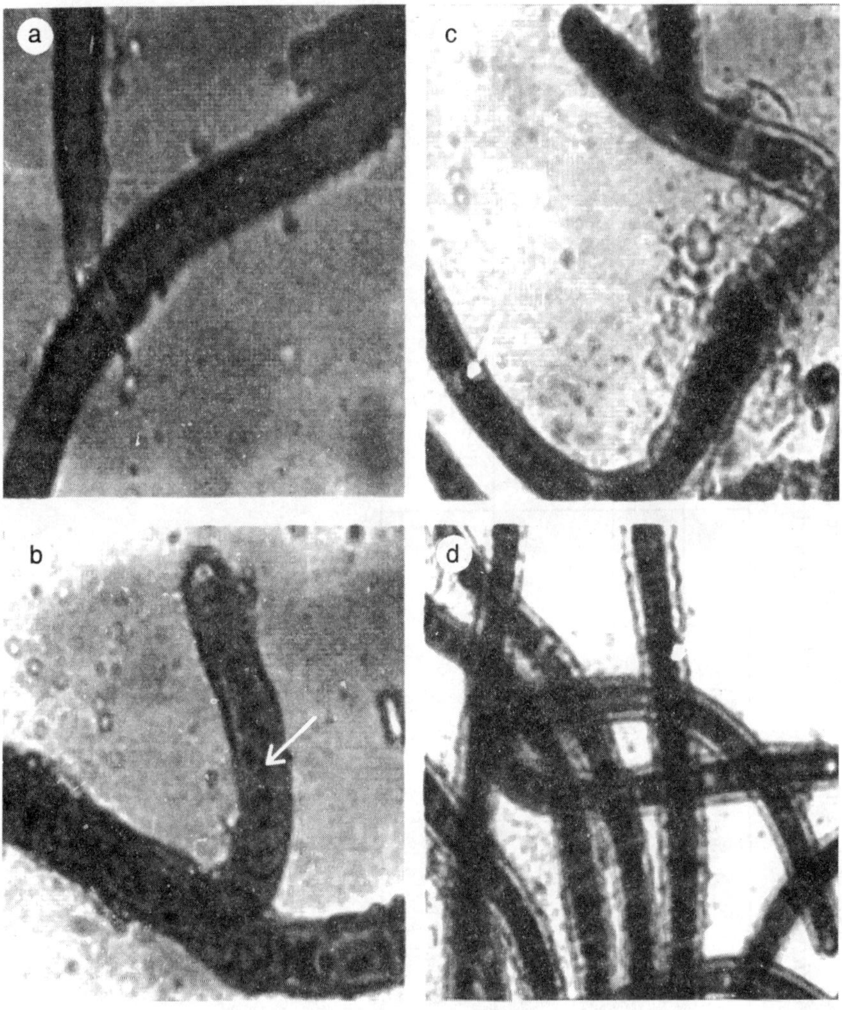

Plate 2, Fig. a-d : Effect of heat on the TTC reduction activity of the cells of cyanobacteria in the crust from excavated sites at Lalitgiri and the corresponding organism *Tolypothrix arenophila* .

a.No crystal formation in the cells of *Tolypothrix arenophila* in the crust after treatment with TTC in dry heat for 10 min (×780); b. Crystal formation in the cells of *Tolypothrix arenophila* in the crust after treatment with TTC in wet heat for 1h (×780); c & d. Crystal formation in the cells of *Tolypothrix arenophila* from culture after treatment with TTC in dry heat for 10 min (c) (×380),and 1h (d) (×380).

that of heat treated samples in dry state except in case of *T. arenophila* isolated from the crust of Lalitgiri excavated sites where the effect of heat was similar under both dry and wet heat conditions (Table-3). The preheated crust and the corresponding organisms were cultured in the BG 11 medium to findout the viability of the cells after heat treatment. The

Table-2 : TTC reduction activity of the crusts from Lalitgiri excavated sites, rice field soils and the corresponding cyanobacteria *Tolypothrix arenophila* and *Aulosira fertilissima* respectively after dry heat and wet heat treatment for different durations

Treatment	Duration of heat pre-treatment	Lalitgiri excavated site		Rice field soils	
		Crust	Corresponding organism ***T. arenophila***	Crust	Corresponding organism ***A. fertilissima***
Dry heat	10 min (60^{0}C)	(Pl 2, Fig.a) 1 h 2 h	+ (Pl 2, Fig.c)	(Pl 3, Fig.a) +	(Pl 3, Fig.c) + (Pl 3, Fig.d)
Wet heat (60^{0}C)	10 min 1 h 2 h	 (Pl 2, Fig.b) +	+ (Pl 2, Fig.d) +	(Pl 3, Fig. b) + +	+ +

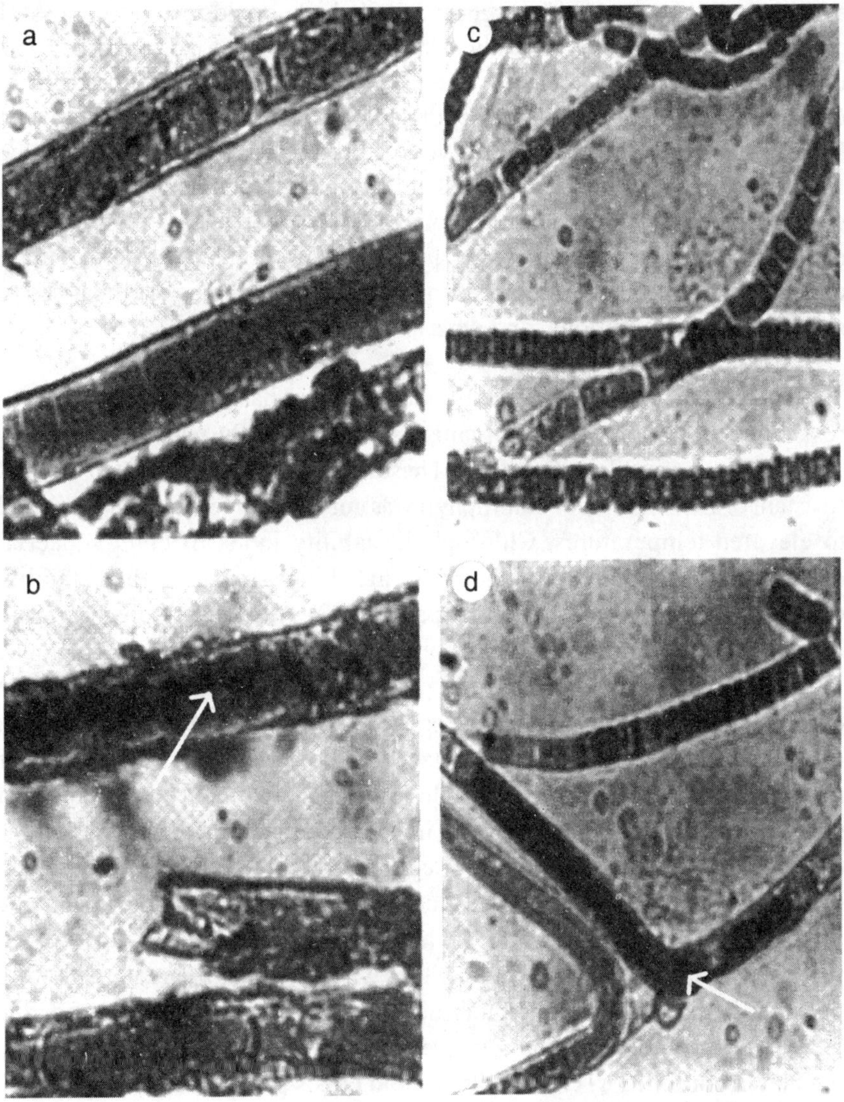

Plate 3, Fig. a-d : Effect of heat on the TTC reduction activity of the cells of cyanobacteria in the crust from rice field soils and the corresponding organism *Aulosira fertilissima*.
a.No crystal formation in the cells of *Aulosira fertilissima* in the crust after treatment with TTC in dry heat for 10 min (×780); b. Crystal formation in the cells of *Aulosira fertilissima* in the crust after treatment with TTC in wet heat for 1h (×780); c.No crystal formation in the cells of *Aulosira fertilissima* from culture after treatment with TTC in dry heat for 10 min (×780); d .Crystal formation in the cells of *Aulosira fertilissima* from culture after treatment with TTC in dry heat for 1h (×780).

organisms in the crust of Lalitgiri excavated site could recover its loss of chlorophyll stability after growing in the culture medium for 5 days at 25 ± 1^0C under 7.5 W/m^2 light intensity. The CSI of the crust was decreased from 68.0 and 64.0 to 28.5 and 20.0 under dry and wet heat treatments respectively. But such a revival was not observed in the pre-heated crust of the rice fields; the status of the heat treated cells remained unchanged even after growing them under normal conditions (Table-3). The loss of chlorophyll-*a* due to dry and wet heat in these organisms was prominently observed upon cultivation of the heat injured cells in culture medium being acclimatised to normal growth conditions for 5 days (Table-3).

The chlorophyll stability at high temperature has been correlated with resistance of photosynthetic organisms to drought (Purushothaman *et al.*, 1974; Talpasayi and Tripathi, 1987). These reports have shown that in drought resistant organisms loss of chlorophyll was considerably less when subjected to elevated temperatures. Chlorophyll stability index of eight subaerial cyanobacteria collected from their natural habitats showed greater stability compared to those from cultures (Tripathi, 1983). Among naturally occurring cyanobacterial forms high chlorophyll stability was observed in those inhabiting adverse habitats. In the present study chlorophyll stability of cyanobacteria occurring in the crust on the exposed brick surfaces of Lalitgiri excavated sites and from the soil surface of fallow rice fields during summer months was studied by exposing them to heat treatment in dry and wet state. It was observed that the cells in the cyanobacterial crusts as well as the corresponding organisms from culture were more susceptible to heat injury when heat was provided under dry conditions in the oven; the immediate injury was comparatively higher to the cells in the crust. However, the effect was alleviated to some extent when the heat treated crusts were cultured in mineral medium under normal conditions. To the contrary, the heat injury to the cells from culture of *T. arenophila* and *A. fertilissima* isolated from Lalitgiri crust and rice field crusts persisted. After removing the heat stress, the heat pre-treated cells of these organisms could not recover in culture under controlled conditions. Recovery of the crusts but not of the corresponding organism from heat injury might be because the cyanobacteria in the crust already exposed to adverse conditions like water deficit, heat, solar radiation etc. under natural conditions, thus were adapted to thrive upon receiving congenial environment by recovering from the stress effects which is not the case on those acclimatised to culture conditions. Comparatively higher injury to chlorophyll-*a* by heat treatment in dry state to the crust as well as to the organisms from culture than to the corresponding

Table - 3 : Chlorophyll stability index (CSI) of the cyanobacterial crust from excavated sites at Lalitgiri, rice field soils and their corresponding organisms

Source of crusts	Cyanobacterial crust					Organism isolated	Corresponding cyanobacteria species from the crust				
	Chl-*a* (µg/g) Unheated	Chl-*a* (µg/g) Dry heat	% loss (CSI)	Chl-*a* (µg/g) Wet	% loss (CSI)		Chl-*a* (µg/g) Unheated	Chl-*a* (µg/g) Dry heat	% loss (CSI)	Chl-*a* (µg/g) Wet heat	% loss (CSI)
	(i) Immediately after heat treatment										
Excavated sites at Lalitgiri	595	190.4	68	214.2	64	*Tolypothrix*	1666 *arenophila*	1071	35.7	1071	35.7
Rice field soils	324.8	163	49.8	298.7	8.0	*Aulosira*	892.5 *fertilissima*	714.0	20.0	833	6.6
	(ii) After 5 days of culture of the heat treated cells in BG 11 medium										
Excavated sites at Lalitgiri	833	595	28.5	666.4	20.0	*Tolypothrix*	2142.0 *arenophila*	595.0	72.2	618.8	71.1
Rice field soils	397.0	203.5	48.7	366.5	7.6	*Aulosira*	1071.0 *fertilissima*	499.8	53.3	571.2	46.6

30 mg air dried crust and 30 mg of fresh weight of the corresponding organisms of the crust were used; heat treatment at 60°C for 1 hour.
CSI - chlorophyll stability index.

wet heat treatment may be because under dry heat, in the absence of moisture, the viable cells were adversely affected due to the stress. Thus heat treatment of the hydrated cells of all the organisms under wet heat treatment showed comparatively higher chlorophyll stability. Similar process might be occurring in natural habitats that during different seasons when the moisture saturated crusts in winter and summer, and water saturated ones in rainy season were subjected to elevated temperature, loss of chlorophyll was never cent percent, resulting in survival of the organisms enduring the adverse climatic conditions on the exposed surfaces.

ACKNOWLEDGEMENTS

The authors thankful to the Head of the Department of Botany and Head of the Department of Biotechnology, Utkal University, Bhubaneswar for providing laboratory facilities. Thanks are also due to the University Grants Commission, New Delhi for financial assistance through a DRS-SAP grant.

REFERENCES

Adhikary, S.P. & Sahu, J.K. (2000). Survival strategies of cyanobacteria occurring as crust in the rice fields under drought conditions. Ind. J. Microbiol. **40:** 53-56.

Brock, T.D. (1975). Effect of water potential on a *Microcoleus* (Cyanophyceae) from a desert crust. J. Phycol. **11:** 316-320.

Büdel, B. (1999). Ecology and diversity of rock-inhabiting cyanobacteria in tropical regions. Eur. J. Phycol. **34:** 361-370.

Ernst, A.; Chen, T.W.& Böger, P. (1987). Carbohydrate formation in rewetted terrestrial cyanobacteria. Oecologia, **72:** 574-576.

Fay, P.& Kulasooriya, S.A. (1972). Tetrazolium reduction and nitrogenase activity in heterocystous blue-green algae. Arch. Mikrobiol. **87:** 341-352.

Grilli-Caiola, M.; Billi, D.& Friedmann, E.I. (1996). Effect of desiccation on envelopes of the cyanobacterium *Chroococcidiopsis* sp. (Chroococcales). Eur. J. Phycol. **31:** 97-105.

Mackinney, G (1941). Absorption of light by chlorophyll solutions. J. biol. Chem. **140:** 315-322.

Mazor, G.; Kidron, G.J.; Vonshak, A.& Abeliovich, A. (1996). The role of cyanobacterial exopolysaccharides in structuring desert microbial crusts. FEMS Microbiol. Ecol. **21:** 121-130.

Murty, K.S. & Majumdar, S.K. (1962). Modifications of the technique for determination of chlorophyll stability index in relation to studies of drought resistance in rice. Curr. Sci. **31:** 470-471.

Purushothaman, D.; Kesavan, T.; Marimuthu, T. & Oblisami,G. (1974). Chlorophyll stability index of some algae. Curr. Sci. **43:** 159-160.

Talpasayi, E.R.S.& Tripathi, S.N. (1987). Chlorophyll stability in some subaerial cyanobacteria. Proc. Ind. Natl. Sci. Acad.: B, **53:** 527-529.

Tripathi, S.N. (1983). Effect of temperature on chlorophyll stability of some subaerial blue-green algae. Z. Allg. Mikrobiol. **23:** 443-446.

Tripathi, S.N. & Talpasayi, E.R.S. (1980). Sulphydryls and survival of subaerial blue\ green algae. Curr. Sci. **49:** 31-32.

Tripathy, P.; Roy, A. & Adhikary, S.P. (1997). Survey of epilithic blue-green algae (cyanobacteria) from temples of India and Nepal. Arch. Hydrobiol. Algological Studs. **87:** 43-57.

Advancing Frontiers of Ecological Researches in India, 421-436, 2006
Edited by A.K. Kandya & Asha Gupta
Published by Bishen Singh Mahendra Pal Singh, Dehra Dun, India

Enterobacter: An emerging nosocomial pathogen

K. K. Patel*, R. Murthy*, M.L. Naik, S. Sinha*** & J. P. Mittra***

* *Dept. of Clinical Microbiology,*
Chhattisgarh Institute of Medical Sciences, Bilaspur (C. G.)
*** *Director General CCOST, Raipur (C. G.)*
*** *SOS in Life Sciences, Pt. R.S.U. Raipur, (C.G.)*

Abstract: The members of genus *Enterobacter* are motile gram negative enteric bacilli belonging to the family Enterobacteriaceae. *Enterobacter* Spp. are not considered to be the primary human pathogens; however, like most Enterobacteriaceae they are capable of causing opportunistic infections in hospitalised or debilitated patients. A National Nosocomial Infection Surveillance System (NNIS) study has found that *Enterobacter* accounts for 5 to 11% of all nosocomially transmitted blood, wound, and respiratory tract and urinary tract infections. Until recently epidemiological studies on genus *Enterobacter* have been based essentially upon the study of phenotypic traits such as biochemical profiles, antibiotic resistance, serological, bactriocin and phage typing. Although often very useful, inability to demonstrate strain to strain variation, unavailability of specific reagents, poor sensitivity to phage, non-specific agglutination are some of the technical pitfalls inherent to the phenotypic studies. With the advent of molecular methods like plasmid-analysis, multilocus enzyme electrophoresis, ribotyping etc., it would be possible to trace the out breaks of disease or monitor the incidence of nosocomial enterobacterial infections. The single most alarming trend regarding the genus *Enterobacter* is the increasing incidence of serious life threatening infections caused by strains resistant to multiple antibiotics. Some of the processes responsible for quick generation of resistance are plasmid -lactamase and chromosomal cephalosporinase, variability of target of the antibiotic and modification of envelope permeability, including alteration of porins (OMPs) and expression of drug efflux. Enterobacter spp. appears to be well adopted for survival and threats to cause immense mortality & morbidity by the proliferation of highly drug resistant strains both in the community and hospital environment.

Key Words: *Enterobacter,* nosocomial infection, opportunistic infection, antibiotic resistance, lactamase, cephalosporinase.

INTRODUCTION

Enterobacter Spp. have gained immense recognition as emerging pathogen in recent years. These organisms seem to have innate resistance to older anti-microbial agents as well as have the propensity to rapidly develop resistance to newer anti-microbial agents. It also appears that *Enterobacter spp*., including multiply resistant strains, have spilled over into the community, occasionally infecting otherwise healthy individuals. More recently, recognition of relatively high rate of co-infection with other pathogens, predominance in liver and lung transplants infection, etiologic role in cotton fever, and increasing incidence in a variety of clinical syndromes.[1]

Because of dramatic changes and expansion of the knowledge of *Enterobacter* spp., it is an initiation to review the recent literature on biology, typing clinico-pathogenicity and anti-biogram of the genus.

Biology & Taxonomy

The genus *Enterobacter* belongs to the family Enterobacteriaceae and can be readily distinguished from the genus *Klebsiella* in that the former is motile, usually ornithine decarboxylase positive, and urease negative.[2] Additionally, most *Enterobacter* spp. are resistant to cephalothin and cefoxitin whereas, *Klebsiella* spp. are often susceptible to these agents.

There are 14 species or biogroups of *Enterobacter* listed in the most recent edition of Manual of Clinical Microbiology. Not all of these have been implicated as causative agents of diseases in humans. Among those that have the most commonly encountered species include *Enterobacter cloacae, E. aerogenes, E. agglomerans* and *E. sakazakii*, *E. taylorae, E. gergoviae, E. asburiae*, and *E. amnigenus* are only rarely isolated from clinical specimens.

E.aerogenes and *E.cloacae* are by far the most frequently encountered human pathogen among the genus *Enterobacter.* These two species can be readily differentiated by tests for lysine decarboxylase and arginine dihydrolase. *E.sakazakii* can be differentiated from *E.cloacae* by its inability to ferment D-sorbitol and its production of a yellow pigmented. *Enterobacter agglomerans* represent a heterogeneous group of diverse organisms that are often yellow pigmented, grow at 4 °C, and are usually negative in decarboxylase/ dihydrolase tests. (Farmer, 1995).

Natural habitat

Enterobacter spp. are widely distributed in the environment and are found in soil, water and sewage. *E. cloacae* and *E. aerogenes* are part of the normal flora of the gastrointestinal tract and are found in a high proportion in sewage sample at concentrations of up to 10^7organism g^{-1}. Environmental sources include rotting vegetable matter and dairy products.

General Epidemiology

By the 1970s, it was established that *Enterobacter* spp. could be nosocomial pathogens, although they were much less commonly encountered than *Escherichia coli* and *Klebsiella* strains. However, the importance of *Enterobacter* spp. as nosocomial pathogens was highlighted in the National Nosocomial Infections Surveillance System (NNIS) data published.

Data form isolates recovered from the ICU revealed that *Enterobacter spp.* were not only the third most common pathogen recovered from the respiratory tract (11.1% of all the isolates) but also the fourth most common pathogen recovered from surgical wound (10.3%), the fifth most common pathogen recovered from the urinary tract (6.1%), and the blood (5.3%).

Although community–acquired infections with *Enterobacter spp.* do occur, the majority of infections with this organism are nosocomial.[7,8] Patients at increased risk of acquiring an *Enterobacter* infection include those with a prolonged hospital stay, especially if a portion of it is spent in an ICU. The presence of a serious underlying illness, especially malignancy, burns, and diabetes, also increases the risk of infection. Immunosuppression from any cause, prematurity and low birth weight in neonates, and the presence of a foreign device are also associated with increased risk of acquisition of an *Enterobacter* infection. The single most frequently cited risk factor for acquisition of an *Enterobacter* infection is the prior use of anti-microbial agent in the patient involved.

Enterobacter infection can be acquired from either endogenous or exogenous sources. Given the ubiquitous nature of the organism, it is found in the feces of humans and animals and in water, plants and plant materials, insects, and dairy products. Single-sources out-breaks have been traced to contamination of intravenous solution, blood products. distilled water, endoscopes, cotton swabs, hands of personnel, hydrotherapy water, stethoscope, cryopreserved pancreatic islet infusion, lipoidal solutions, and devices used for monitoring intra-artial pressure.

Colonization of the gastrointestinal tract and other body sites with *Enterobacter* spp. occurs frequently in the seriously ill patients may be colonized with more than one strain at any given time. Thus, it appears that severe debility, coupled with the suppressive effects of antibiotics on the normal flora, provides an excellent opportunity for the colonization by *Enterobacter* spp.

Clinical Manifestation

Enterobacter spp. have been implicated in a broad range of clinical syndromes. The literature is replete with descriptions of bacteremia and infections of skin and soft tissues, respiratory tract, urinary tract, bone and joints, central nervous system, gastro-intestinal tract, and other organs. In general, the characteristics of infection due to *Enterobacter* spp. resemble those due to other facultative gram-negative bacilli. However, there are some distinctive features that may serve as guideposts to selection of therapy or to planning of institutional control measures.

a. Bacteremia : The rates of bacteremia tend to cluster around 1 per 1,000 admissions for university hospitals or tertiary-care centers. These rates tend to be two-to three fold higher in specialized units, such as cancer centers, and two - to three fold lower in community hospitals. Although rate are lower in the community, the problem is significant with a growing trend.

Enterobacter bacteremia tends to occur more commonly in males in a ratio of 1.3 to 2.5:1.0. Males predominate among both infected adults and children. Bacteremia is more commonly encountered at the extremes of age, i.e., in neonates and the elderly. Majority of bacteremia are acquired institutionally *E.cloacae* predominates in most of the series followed in order by *E. aerogenes E. agglomerans*, *E. sakazakii* and others 14 to 53% of bacteremias appear to that involve *Enterobacter spp.* are ploymicrobial. From the companion organisms in the ploymicrobial bactereremias appear to be randomly distributed between gram-positive and gram-negative isolates.

Signs, symptoms, and laboratory findings: The incubation period of *Enterobacter* bacteremia estimated from common source outbreaks in which the organism was infused directly into the bloodstream the time for appearance of signs and symptoms varies from as short as 2h to as long as 20 days, with most of them appearing occurring in a few hours to 2 days.

Fever is the hallmark of bacteremia in both adults and children. Reported rates have ranged from 83 to 87 % in children and from 92to 98% in adults or mixed populations.

(a) Hypotension or shock has been reported in 9 to 34% of the adults or mixed populations. The frequency is similar in children with 8 to 28 %.

(b) Leukocytosis occurs in approximately two-third of patients with bacteremia. Leukopinia has been reported in 9 to 17% of individuals of a broad range of ages.

(c) Thrombocytopenia.

(d) Jaundice and Hemorrhage have each been noted in a few series.

The syndrome of disseminated intravascular coagulopathy has been recognized in 0 to 6% of bacteremic episodes most of the usual cutaneous manifestation associated with bacteremia have been noted occasionally. These include purpura fulminans, hemorrhagic bullae, and ecthyma gangrenosum. Cyanosis and mottling has been encountered in two thirds of bacteremia children.

b. Lower Respiratory Tract Infection :

Most of the species of *Enterobacter* have been implicated in a wide spectrum of lower respiratory infection including asymptomatic colonization of respiratory secretions, purulent bronchitis, lung abscess, pneumonia, and empyema. The incidence of lower respiratory tract infection due to *Enterobacter* spp. appears to have increased steadily over the last four decades. Estimates of the incidence of *Enterobacter* spp. in nosocomial respiratory infection in the 1970s ranged from less than 2 to 9%. The rates increased from 9.5% in the early 1980s to 11% in 1986 to 1990. *Enterobacter* spp. have recently surpassed *Klebsiella* spp. to become the third most common cause of nosocomial respiratory tract infections in the United States. The recent recognition of an important role of *Enterobacter* spp. in community- acquired pneumonia in Spain is disquieting. *Enterobacter* spp. have recently been recognized as major pathogens in lung transplant recipients. Approximately 40% of recipients will develop acute bacterial pneumonia in 2 weeks immediately following transplantation. The clinical and laboratory manifestations of *Enterobacter* pneumonia differ little from those observed for pneumonia due to gram-negative bacilli.

The widely accepted criteria for the diagnosis of pneumonia are new infiltrates and positive cultures of transtracheal aspirates or sputum plus blood, demonstration of tachypnea and tachycardia. Hemoptysis appears rarely. Leukocytosis with a shift to the left in differential cell count is usual (82 to 100%).

C. Infection of Skin and Soft Tissues :

Enterobacter spp. have been implicated as causes of an array of clinical syndromes involving the skin and soft tissues: cellulitis, fasciitis, abscesses, emphysema, myositis and wound infection.

Institutionally acquired infections of surgical wounds and burns

The proportion of nosocomial wound infections due to *Enterobacter* spp. has increased throughout recent decades. Result of the NNIS program for 1986 to 1990 indicated that *Enterobacter spp* were the fourth most common cause of surgical wound infection in ICU and the most common gram negative organism implicated.

d. Endocarditis :

Endocarditis, due to gram negative bacilli, appears to be increasing in incidence. The risk appears highest in intra-venous drug abusers and individuals with prosthetics valves. Prior to 1980, there were only anecdotal reports about *Enterobacter* endocarditis associated with penetrating foreign bodies, mechanical and porcine prosthetic valves, intravenous drug abuse, and cardiac surgery.

e. Intra-abdominal Infections :

Enterobacter spp. have been implicated in Intra- abdominal Infections. This is consistent with their residence the colonic flora of many humans.

f. Urinary tract Infection :

The clinical manifestations of urinary tract infection due to *Enterobacter* spp. differ little from those of infections due to other gram negative bacilli. The spectrum of illness ranges from asymptomatic bacteriuria to pyelonephritis and urosepsis. Prior to 1980, *Enterobacter spp* accounted for 0 to 14% of infections.

More recently *Enterobacter* spp. have accounted for 2.4% of childhood urinary tract infection in Saudi Arabia and 6 to 7% of nosocomial infections in the U.S. incidence of *Enterobacter* spp. among nosocomial urinary pathogens is slowly increasing over the years.

Multiple drug resistance has been observed in nosocomial *Enterobacter* urinary isolates. The role of antecedent anti-microbial administration in selecting for resistance has been emphasized.

g. Central Nervous System Infection :

Enterobacter spp. have been implicated as etiologic agent in a variety of central nervous system infections. Meningitis, ventriculitis, brain abscess, and infections proximate to foreign bodies have been reported episodically over the years.[7]

More recent reports have focused upon an overall increase in the incidence of meningitis due to enteric bacilli, emergence of resistance among strains of *Enterobacter* use of novel regimens for treatment, and the special problems posed by *E. sakazakii*.

h. Ophthalmic Infection :

Enterobacter spp. have been implicated in a variety of infectious processes involving the eyes and periorbital tissues.

Although *Enterobacter spp*. account for only a small fraction of the cause of endophthalmitis however, like *Pseudomonas* spp. they are among the most aggressive pathogens, they may be resistant de novo or become resistant during therapy, and may cause outbreaks arising from environmental contamination.

i. Septic-Arthritis and Osteomyelitis- *Enterobacter* spp. have been implicated in a variety of syndromes that involve the bones and joints, although relatively infrequent, severe septic arthritis, osteomyelitis, infection of multiple bones and joints in infants and children vertebral osteomyelitic hip infection have been reported over the past three decades.

j. Pathogenicity - Few studies have investigated potential virulence associated with the genus *Enterobacter* in human infections. The lack of significant research in this area is probably a reflection of the predominant role of these organisms as nosocomial pathogens.

Many *Enterobacter* species cause the *in vitro* hemagglutination of erythrocytes obtained from a number of vertebrate species. These hemagglutinins are either mannose–sensitive (MSHA) or mannose–resistant (MRHA), and their phenotypic expression strongly correlates with production of type 1 or type 3 fimbriae, respectively. Only *E. gergoviae* consistently expresses MRHA. Most *E. cloacae* isolates produce a hydroxyamate-type siderophore, aerobactin, which is commonly associated with microbial species causing invasive disease.

Several other pathogenic properties have been identified in random isolates of *E. cloacae* in the feces of an 11- month-old boy. *BamHI* – digested DNA from this strain reacted with an *E. coli* a-hemolysin-specific probe. Paton and Paton described the isolation of a Shiga-like toxin (SLT) producing strain of *E. cloacae* from an infant with hemolytic–uremic syndrome.

Probably the most intriguing virulence factor identified so far in *E. cloacae* is the outer membrane protein, OmpX 155. This 17-kDa protein, encoded by a chromosomal gene, causes decreased porin production (OmpF, OmpC) in *E. coli* transformants, leading to increased antimicrobial resistance to several b-lactam compounds.

LABORATORY DIAGNOSIS & TYPING METHODS

Laboratory Identification :

Isolation -Recovery of *Enterobacter* spp. from environmental sources and the gastrointestinal contents of humans had no apparent significance. Development of enrichment techniques and selective media designed for the specific isolation of *Enterobacter* from such sample has not been attempted, only recovery.

The majority of *Enterobacter* isolates are both sucrose-positive (with the exception of *E. cancerogenus*) and lactose-positive (with the exception of *E. cancerogenus* and *E. hormaechei*).

Therefore most species, such as *E. cloacae* and *E. aerogenes* appear as pink (lactose-positive) or yellow (sucrose-positive) colonies on standard selective media containing these sugars (e.g., MacConkey agar, Hektoen enteric agar).

On non selective media such as trypticase soy or nutrient agars, *E. sakazakii* and many isolates of *E.agglomerans* produce a nondiffusible yellow pigment.

IDENTIFICATION

Commercial Systems :

Enterobacter is one of the most difficult groups to correctly identify to species level using commercial automated, semiautomated, or manual identification systems.

Commercial systems in general do not identify species accurately residing in the genus *Enterobacter* to a satisfactory degree (i.e., to >90% accuracy).

Most manual systems (API 20E, API Bio Merieux, Hazelwood, MO; Crystal E/NF, Becton Dickinson, Cockeys ville, MD; RapID OnE) appear to be better equipped to identify *Enterobacter* species accurately than fully automated panels such as the MicroScan WallkAway and Vitek units.

Conventional Media

Because of the taxonomic uncertainties that exist within this group, it is currently impossible to construct accurate defining criteria for inclusion in the genus *Enterobacter*. Table-1 gives some of the major biochemical features that have traditionally been associated with inclusion in the genus *Enterobacter*. Most *Enterobacter* isolates are indol-negative and *o*-nitrophenyl-β-D-galactopyranoside (ONPG)-positive, reduces, trehalose, and cellobiose.

A majority of *Enterobacter* isolates can be identified to species with the use of a small number of additional biochemical characteristics including amino acid decaboxylases and sugar fermentation tests.

Typing :

Typing method is important tool for establishing the source and modes of transmission for epidemic strains.

Table-1: Biochemical reactions for the genus *Enterobacter*

Test *Enterobacter sakazakii*	*Enterobacter aerogenes* *Enterobacter cloacae*		*Enterobacter gergoviae*	
Mannitol motility	±	d	±	±
Simmons' citrate	+	+	+	+
Urease	-	d	±	-
H_2S from TSI	-	-	-	-
ONPG	+	+	+	+
Phenylalanine	-	-	-	-
Arginine dihydrolase (ADH)	-	+	-	+
Lysine decarboxylase (LDC)	+	-	d	-
Ornithine decarboxylase (ODC)	+	+	+	+
Gas from glucose	+	+	+	+
Acid form : arabinose	+	+	+	+
: rhamnose	+	+	+	+
: raffinose	+	d	+	+
: salicin	+	+	+	+
: sorbitol	+	+	-	-
: trehalose	+	+	+	+
: xylose	+	+	+	+
Methyl red	-	-	-	-
Voges-proskarur	+	+	+	+
Indole	-	-	-	-

Symol: d : 16-84% strains positive ,
"+" 85-100% strains are positive
"-" 0-15 % strains positive

Typing of genus *Enterobacter* can be done by a number of methods as given below.

❖ **Biotyping :** Schemes have been developed as a first approximation of strain relatedness in outbreak situations. The most common system used to asses strains identity has been the API 20E strip (Vitek Bio-Merieux, Hazelwoos, MO) Gaston and associates analyzed API 20E profile for 790 isolates of *E.cloacae* not related to outbreaks. Thirty-six profiles were found, with one bio-type (API profile 3305573).

De Champs and colleagues used both the API 20E and API 50 CH Galleries to biotype 23 imipenem-resistant and imipenem-susceptible *E. arcgenes* strains.

Old proposed a biotyping system for *E.cloacae* using seven differential tests on conventional media. Off 110 isolated typed, 86 (78%) resided in biotype 4, again emphasizing the lack of discriminatory value of most of the phenotype typing schemes.

- **Serotyping :** A serogrouping scheme based on the presence of somatic antigens was devised for *E. cloacae* in 1983, Twenty-eight serogroups were originally established after antigenic analysis of more than 300 *E.cloacae* isolates, with serogroups O: 3(21%) and O: 8(13%) predominating. [26] This scheme has since been expanded to include two additional serogroups, for a total of 30. Serogrouping has a greater discriminatory value than bio-typing in determining strain relatedness and at least 85% of all *E. cloacae* are typable.[27] Therefore, serogrouping has distinct advantage over biotyping but may not be able to stand alone as an epidemiologic typing system in outbreaks due to common serogroups (O: 3,O: 8,O: 13) or when isolates are not typable or spontaneously auto-agglutinate.

- **Phage typing :** A Phage typing scheme has been constructed for *E.cloacae*. Gaston developed a set of 25 bacteriophages through which he had typed 76 (83%) of 92-field isolate. 28 Although this system is able to type serologically rough strains and those that fail to react in specific somatic antisera, it is less reproducible than biotyping.

- **Bacteriocin typing :** A 1982 German publication found that only 79% of more than 300 *E. cloacae* isolates were susceptible to 1 or more of 16 different bacteriocins; 52 provisional bacteriocin types were defined in this study.

Bacteriocin typing has also been applied to *E. aerogenes;* 70(90%) of 76 isolates could be assigned to 1 of 15 sensitivity patterns based on bacteriocins secreted by one of five producer strains.

- **Plasmid Profile :** A number of molecular methods have been used as aids in determining strain relatedness among clusters of *Enterobacter* isolates associated with human infections. Of this group, plasmid analysis is by far the most widely employed technique. Most strains of *E.cloacae* harbor extrachromosomal elements ranging in molecular mass form 3 to 150 MDa. Because of the frequent plasmid carriage by *Enterobacter* species, plasmid analysis is an extremely useful fingerprinting tool. However, Clark and associates are of the opinion that because of plasmid instability, alternative method may be necessary to confirm primary result.

- **Multilocus Enzyme Electrophoretic Testing :** Gaston and Warner used multilocus enzyme electrophoresis (MEE) mobility of 12 different cellular enzymes for 62 *E. cloacae* strains. Each enzyme generated between two and five different electromorphs. A subset of four enzymes (lactate dehydrogenase, 6-phosphogluconate dehydrogenase, glutamate dehydrogenase, and an unidentified enzyme band) were found useful in typing most *E. cloacae* strains that belonged to serogroups other than O: 3/O: 8.[32] Although MEE is a valuable technique, it is relatively difficult to develop and labor intensive.

- **Analysis by Pulse Field Gel Electrophoresis (PFGE) :** These studies have indicated considerable DNA polymorphism in the clinically important *Enterobacter* genus and good correlation between isolates from single patients, so they seem to provide extremely discriminatory result and highly useful epidemiological information.

- **Ribotyping :** Ribotyping is a common epidemiologic technique used with other gram negative bacteria. It has also been found to be highly discriminatory in distinguishing strain relatedness in *Enterobacter*-associated outbreaks or incidents.[33] It can offer advantage over other molecular strategic (e.g. Restriction enzyme analysis) in both specificity and ease of gel interpretation.

- **PCR Based methods :** PCR and *Rep* PCR are rapid and useful methods for tracing epidemic strains during outbreaks on a day-to-day basis. A comparative study of biotyping, anti-biogram susceptibility, whole cell protein gel electrophoresis of chromosomal DNA and PCR with arbitrary primer of *Enterobacter* spp. isolates can be performed to determine the best markers for epidemiology purposes. Biotyping whole cell protein and plasmid analysis was the least discriminatory method whereas anti-microbial susceptibility and PCR with arbitrary primer showed moderate discriminatory power. Typing based on pulse field gel electrophoresis of chromosomal DNA appeared to be best discriminatory method.

Anti-microbial Susceptibility

The single most alarming trend regarding the genus *Enterobacter* is the increasing incidence of serious, life- threatening infections caused by strains resistant to multiple antibiotics. In the latest NNIS report, *Enterobacter* spp. were the third most common cause of gram-negative nosocomial infections. This distressing fact is exacerbated by the increasing tendency of *E.cloacae* and *E. aerogenes* to gain resistance to many commonly administered antibiotics, including third-generation cephalosporins. Such resistance often arises *de novo* in-patients receiving empiric therapy for systemic infections. Later, these drug- resistant enterobacters, primarily residing in the gut, may emerge to produce severe or fulminate illnesses in debilitated persons.

Cephalosporin resistance in *Enterobacter* spp. is caused by constitutive expression of a chromosomal β - lactamase or similar activity carried on a plasmid.[1] Plasmid- mediated β - lactamase activity most commonly involves extra-chromosomal and Plasmid- mediated extended-spectrum β- lactamase (ESBL) activity imparts general resistance to most β- lactam antibiotics except imipenem.

While rare, resistance to imipenem in *Enterobacter* has been documented in most instances, imipenem resistance to *Enterobacter cloacae* has been associated with elevated expression of chromosomal β- lactamase and permeability changes in the outer membrane.

Introduction of a plasmid carrying a wild-type *amp* D allele into this imipenem-resistant isolate restored full susceptibility to β- lactamase antibiotics.

Conclusion & Perspective on the future

Enterobacter spp. appears well adapted for survival and proliferation in new millennium. A meticulous attention to principles of antisepsis may reduce the occurrence of the relatively infrequent outbreaks that are traceable to human vectors or environmental contamination.

Selective decontamination of the gastrointestinal tract and avoidance of the use of agents that lower the gastric pH to reduce oropharyngeal colonization are rational, but unproven, approaches that should be subject to controlled clinical trials.

The evidence that *Enterobacter spp.*, including multiplication of resistant strains, are increasingly important etiologic agents in community-acquired pneumonia in Spain and are responsible for soft tissue and urinary tract infections in otherwise healthy individuals in North America is indeed disquieting.

Many other fundamental questions remain unanswered. What pathogenic mechanism(s) sets *Enterobacter spp.* apart clinically from other gram negative enteric bacilli? What favors the survival and transmissibility of the organism in solutions and on surfaces of caterers or medical devices? What are the mechanism and factors favoring the emergence of resistance to "fourth generation" cephalosporins, carbapenems, and fluoroquninolones? Can further emergence of resistance be minimized? What controls the expression of the inducible β - lactamase in *Enterobacter* spp. Only with additional basic research and innovative approaches be designed for therapy and ultimately for prevention of therapy may be designed for the prevention of nosocomial infection by the *Enterobacter* spp.

REFERENCES

Adegbola, R.A. and Old, D.C. (1983). Fimbrial haemagglutining in *Enterobacter species. Jgen Microbial,* 129:2175-2180.17.

Anderson E.L., B.I. Asmar, and A.S. Dajani (1994). Increasing *Enterobacter* bacteremia in pediatr. *Infect. Dis. J.* **13:**787-792.

Bdey, G.P., I.S. Elting and Rodriguez. (1991). Bacteremia caused by *Enterobacter*: 15 years of experience in a cancer hospital. *Rev. Infect. Dis.* **13:**550-558.

Bouza, E., M.G. de la Torre, E.L.A. Erice J.M., Diaz-Borrego, and L. Buzon (1985). *Enterobacter* bacteremia- an analysis of 50 episodes. *Arch. Intern. Med.* **145:**1024-1027.

Burchard, K.W., D.T. Barrall, M, Reed, and G.J. Slotman (1986). *Enterobacter* bacteremia in surgical patients, *Surgery* **100:**857-861.

Chow, J.W., M.J. Fine, D.M. Shlaes, J.P. Quinn, D.C. Hooper, M.P. Johnson and R. Ramphal (1991). *Enterobacter* bacteremia: clinical features and emergence of antibiotic resistance during therapy. *Ann Intern Med.* **115:**585-590.

Chow, J.W., V.L. Yu and D.M. Shlaes (1994). Epidemiologic perspectives of *Enterobacter* for the infection control professional. *Am J. Infect. Control.* **22:**195-201.

Clark, N.C., Hill, B.C., O'Hara, C.M., Steingrimsson, O, Cooksey, R.C. (1989). Epidemiologic typing of *Enterobacter sakazakii* in two neonatal nosocomial outbreaks. *Diagn. Microbiol. Infect.* **13:** 467-472.

Falkiner, F.R. (1992). *Enterobacter* in hospital *J. Hosp. Infect.* **20:**137-140.

Farmer, J. J.II. (1995). *Enterobacteriaceae*: Introduction and identification In *Manual of Clinical Microbiology* (6th edn.) pp 438-449 eds P R Murray. E J Baron, M A Pfaller, F C Tenover and R H Yolken (Washington D C: American Society for Microbiology).

Framer JJ III, Davis B.R., Hickman-Brenner FW, *et al.* (1985). Biochemical identification of new species and biogroups of *Enterobacteriaceae* isolated from clinical specimens. *J Clin. Microbiol.* **21:**46-76.

Fung, C.P., L.S. Wang, V.C. Juang, G.Y. Liu and D.L. Cheng. (1998). *Enterobacter cloacae* bacteremia: clinical analysis of 41 cases. Clin. *Med. J. (Taipei)* **42:**297-304.

Gallagher, P.C. (1990). *Enterobacter* bacteremia in pediatric patients. *Rev. Infect. Dis.* **12:**808-812.

Garaizar, J, Kaufmann, M.E., Pitt, Tl (1991). Comparison of ribotyping with conventional methods for the type identification of *Enterobacter cloacae*. *J. Clin. Microbiol.* **29:**1303-1307.

Gaston, M. A. and Warner, M. (1989). Electrophoretic typing of *Enterobacter cloacae* with a limited set of enzyme stins. *Epidemiol. Infect.* **103:**255-264.

Gaston, M.A. (1987). Isolation and selection and selection of bacteriophage-typing set for *Enterobacter cloacae. J. Med. Micro***biol.** 285-290.

Gaston, M.A. (1988). *Enterobacter*: An emerging nosocomial patogen. *J. Hosp. Infect.* **11:** 197-208.

Gaston, M.A. (1998). *Enterobacter*: an emerging nosocomial pathogen. *J. Hosp. Infect.* **11:**197-208.

Gaston, M.A., Bucher, C. Pitt T.L. O. (1991). Serotyping scheme for of Enterobacter cloacae. *J. Clin. Microbiol.* **18:**1079-1083.

Haddy, R.I.M.L. Cecii, L.L. Norris and R.J. Markert. (1991). *Enterobacter* bacteremia in the community hospital. *J. Fam. Pract.* **32:**601-606.

Haertly, R., Bandlow, G. (1993). Epidemiological fingerprinting of *E. Cloacae* by small fingerprinting restriction endonuclease analysis and pulsed-field gel electrophoresis of genomic restriction fragments *J. Clin. Microbiol.* **31**:128-133.

Hawkins, R. E. C, R. Lissner, and J.P. Sanford (1991). *Enterobacter sakazakii* bacteremia in an adult, *South Med. J.* **84:**793-795.

Jarvis, W.R. and W.J. Martone (1992). Predominant pathogens in hospital infections. J. Antimicrob, Chemother, **29 (Suppl.A):**19-24.

John, J.F.J., R.J. Sharbough and E.R. Bannister (1982). *Enterobacter cloacae*: bacteremia, epidemiology, and antibiotic resistance. *Rev. Infect. Dis.* **4:**13-282.

Karnad, A., S. Alvarez, and S.L. Berk (1987). *Enterobacter* pneumonia. South. *Med, J.* **80:** 601-604.

Lindh, E., K. Dorndusch, K, Jalakas, and A. Forsgren (1990). Antibiotic susceptibility and β - lactamase production in clinical isolates of *Enterobacter spp.*- APMIS **98:**462-470.

Miller, J.M. and Rhoden, D.L. (1991). Preliminary evaluation of Biology, a carbon source utilization method for bacterial identification. *J. Clin. Microbiol.* **29:**1143-1147.

Modi, N., Damjanovic, V, Cooke, R.W.I. (1987). Outbreak of cephalosporin resistant *Enterobacter cloacae* infection in a neonatal intensive care unit. *Arch. Dis. Chid.* **62:**148-151.

O'Hara, C.M., D.L. Rhoden and J.M. Miller (1992). Revaluation of the API 20E identification system versus conventional biochemical for identification of members of family *Enterobacteriaceae* :A new look at an old product. *J. Clin. Microbiol.* **30:**123-125.

Old D. (1982). Biotyping of *Enterobacter cloacae J. Clin. Pathol.* **35:** 875-878.

Pareja, A, C. Berna, A. Leyva, G. Picdrola, and M.C. Maroto. (1991). Etiologic study of patients with community-acquired pneumonia. *Chest. 101:* 1207-1210.

Pfaller, M.A. Sahn D, Ciaglia, C. *et al.* (1991). Comparison of the Auto SCAN-W/A rapid bacterial identification system for identification of gram negative bacilli *J. Clin. Microbiol.* **29:**1422-1428.

Sanders Jr. W E and Sanders C. C. (1979). *Enterobacter spp.*: Pathogens poised to flourish at the turn of the century. *Clin. Microbiol. Rev.* 10 220-241.

Stoorvogel, J, van Bussel MJAWM, Tommassen J, and van de Klundert JAM (1991). Cloning of a beta-lactam resistance determinant of *Enterobacter cloacae* outer membrane protein (OmpX) *J. Bacterial.* **173:**156-160.

Traub, W.H., Sphor, M., Blech, R. (1982). Bacteriocin typing of clinical isolates of Enterobacter cloacae. *J. Clin. Microbiol.* **16:**885-889.

Advancing Frontiers of Ecological Researches in India, 437-444, 2006
Edited by A.K. Kandya & Asha Gupta
Published by Bishen Singh Mahendra Pal Singh, Dehra Dun, India

Studies on the microbial flora of two weed fishes with reference to their environment

Ch. Sarojnalini, T. Suchitra and N. Jaya
Department o Life Sciences,
Manipur University, Imphal -795003
E-mail-sarojnalini-ch.@yahoo.com.

Abstract : The microbial qualities of two weed fishes found in Manipur viz, *Puntius sophore* (Hamilton Buchanan) belonging to the family Cyprinidae, and *Colisa fasciatus* (Schneider) belonging to the family Belontidae were analysed. Total plate count of bacteria were high (10^7/g) in *P. sophore* and *C. fasciatus*. The fishes were caught from polluted aquatic environment. Fungal count was higher in *P. sophore* (10^4/g). *Staphylococcus aureus* and faecal *Streptococci* count reached upto 10^5/g whereas in fresh *C. fasciatus* the faecal *Streptococci* count reached upto 10^6/g. However in the sundried products, all the microbial load decreased due to the reduction of moisture content of the fishes. Typical *E. coli* was not detected in sundried *P. sophore* and pathogenic *Salmonella* were also not detected in all the fish samples. *Aspergillus, Penicillium* species were dominant among the fungi isolated in both the fresh and sundried samples.

INTRODUCTION

The recognition of environment and its deterioration in the form of pollution to land, water and atmosphere was taken into account in the western countries during the industrial revolution and with the development of science and technology in the twentieth century. The ecosystem is a life support system consisting of biotic and abiotic components-man, animals, plants, micro-organisms, soils, water and atmosphere all interacting together having biotic diversity and material cycles so that the system itself is kept viable.

At present, the human population consumes a considerable amount of protein derived from the aquatic ecosystem and planners are thinking of harvesting more and more food materials from the aquatic life to meet the growing demand. The major community in water-mud-silt-detritus constitute bacteria and fungi. As any life has to depend for the characteristics and constituance on the physical environment in which it exists so also the interaction between environment and biota; it is pertinent to study the microbial flora and elaborate it, which form a definite infrastructure in aquatic ecosystem. It will not only help for assessing the quality of aquatic environment but also such aquatic life as fish which meets the food requirement of human being.

The non-specific response to invading micro-organisms is also somewhat dependent on the nutritional status of the host. In general, nutritional defeciencies result in antagonism to infection when the infectious agent is dependant on specific host-enzyme system of metabolites or has a greater requirement for a particular dietary nutrient than the host. Quality control of fresh fishes aimed at minimizing contamination and proliferation of micro-organisms, which is directly related to the environment of the fishes. Mainly the weed fishes such as *Puntius sophore*, *Colisa fasciatus* and other small fishes of local origin get spoiled soon, after caught. The environment of a fish can influence the microflora and thus develops an important step towards spoilage. The micro-organisms present in the fishes must be related with the environment of the fish. Fish from polluted water, on the other hand may contain additional varieties of spoilage bacteria, and some of the indicator species which may be pathogen (Thatcher, 1971).

The health hazards associated with the consumption of polluted waters and contaminated food are well known. Microorganisms are found in all environmental conditions on the body surface and its environment. They may cause serious epidemic diseases (Collins *et al.*, 1989). Therefore, this paper reports about the status of micro-organisms associated with fresh and sundried *Puntius sophore* and *Colisa fasciatus*. Use of sundried *Puntius* in the preparation of a fermented fish paste "Ngari" for human consumption was reported by Sarojnalini and Vishwanath (1988). Composition and nutritive value of this species was discussed by the same authors in the year 1994.

MATERIALS AND METHODS

The study was conducted at Imphal, the capital of Manipur (latitude : 23.80^0N and longitude : 93.03^0E to 94.78^0 E). The state enjoys subtropical climate. *Puntius sophore* and *Colisa fasciatus* which were found in the Imphal market were collected randomly and brought to the laboratory for analysis.

Total colonies of fungi and bacteria were determined by the dilution plate method of APHA (1976) using PDA (Potato Dextrose Agar) and

Trypton Glucose Agar medium respectively. Fungal colonies were identified based on Gilman (1957) and Ellis (1971, 1976) by staining with lactophenol Cotton blue. The isolated bacterial colonies were identified at IMTECH, Chandigarh,

Pathogenic bacteria like *E.Coli, Staphylococcus aureus,* faecal *Streptococci* and *Salmonella* sps. were tested as per the method of APHA (1976). Most probable number count of Coliform were determined by incubating diluted sample in Brilliant Green lactose using Durham's tube. *E. coli* were determined by incubating in Eosine Methylene blue agar medium, *Salmonella* was determined by incubating in salenite cystein broth for 24 hrs for enrichment at 37^0C and then inoculated to Brilliant Green Agar medium. For *Staphylococcus aureus* sample were inoculated in egg yolk tellurite enriched Baired Parker Agar to observe development of elevated black colonies.

RESULTS AND DISCUSSION

Table I shows the Microbial qualities of fresh and sundried *Puntius sophore* and *Colisa fasciatus*. The total plate count of bacteria in both the samples i.e, *Colisa* and *Puntius* reached upto 10^7cfu/g. The count was slightly higher in *Puntius sophore* in both the fresh and sundried products. The differences in the bacterial count might be due to the difference in aquatic environment. Due to contamination some bacteria are normally present in intestine and gills. These bacteria helps to increase in microbial load. But soon after death, bacteria and enzymes invade the tissue and other organs and finally enter the flesh through any open wounds. *Staphylococcus aureus* and faecal *Streptococci* counts were high in Fresh *Colisa* (10^5-10^6cfu/g) but reduced to 10^5 cfu/g in sundried product. *Staphylococcus* in fishery product is not a serious problem but food poisoning may occur if the products are handled carelessly during processing. In *P. sophore* also the value reached upto 10^5 cfu/g. Faecal *Streptococcus* count was highest in fresh *Colisa* (10^5- 10^6 cfu/g) but slightly lower in sundried products (10^5cfu/g). In fresh and sundried *Puntius sophore* also the value reached upto 10^5 cfu/g. *Coliforms* were detected in all the samples analysed. They were considered undesirable because they may indicate faecal contamination. Typical *E. coli* were detected in all the samples except sundried *P. sophore.*

E. coli and faecal *Coliforms* are abundant in faeces and sewage. *E. coli* population were high in the area where sewage contamination occurs.

Table-1 : Microbiological qualities of fresh and sundried *Puntius sophore* and *Collisa fasciatus* in c fu/g

Samples	Fresh *P. sophore*	Sundried *P.sophore*	Fresh *C. fasciatus*	Sundried *C. fasciatus*
TPC	Range 3.6×10^6 - 4.4×10^7	2.0×10^6 - 5.6×10^7	7.6×10^6 - 1.30×10^7	4.0×10^5 - 5.4×10^5
(Bacteria)	Mean 2.9×10^7	4.1×10^6	1.0×10^7	4.8×10^5
TPC	Range 2.0×10^3 - 8.0×10^4	6.0×10^4 - 9.0×10^4	1.6×10^3 - 3.0×10^3	1.0×10^2 - 2.2×10^2
(Fungi)	Mean 2.7×10^4	7.3×10^4	2.2×10^3	1.6×10^2
Staphylococcus aureus	Range 2.5×10^4 - 1.2×10^5	1.8×10^5 - 2.5×10^5	1.3×10^6 - 3.1×10^6	1.5×10^5 - 3.5×10^5
	Mean 1.1×10^5	2.1×10^5	1.9×10^5	2.5×10^5
Faecal *Streptococci*	Range 2.7×10^5 - 4.3×10^5	1.0×10^5-5.0×10^5	5.0×10^5-1.4×10^6	1.3×10^4-1.6×10^5
	Mean 3.4×10^5	3.1×10^5	7.8×10^5	7.0×10^4
Coliforms (MPN)	Range 7.5×10 - 2.4×10^2	4.3×10 - 2.4×10^2	2.4×10^2 - 1.1×10^3	2.1×10 - 1.1×10^2
	Mean 1.30×10^2	1.2×10^2	6.0×10^2	5.0×10
E. coli	Detected	Not detected	detected	detected
Salmonella	Not detected	Not detected	Not detected	Not detected
Values are mean for 3 samples				

TPC = Total Plate Count

Table-2: Fungal colonies isolated from fresh and sundried *P. sophore* and *C. fasciatus*.

Fresh *P. sophore*	*Penicillium diversium, P. helicum, Aspergillus sydowi, A niger, Fusarium oxysporium, Cladosporium cladosporioids.*
Sundried *P. sophore*	*Rhizopus nigricans, R. oryzae, Aspergillus sydowi, A. fumigatus, Penicillium fellutanum, P. citrinum,P. regulosum, Cladosporium sps, Helminthosporium nodulosum and Cunninghimela sps.*
Fresh *C. fasciatus*	*Aspergillus sps, Penicillium sps, Cladosporium* sps and white sterile mycelium.
Sundried *C. fasciatus*	*Aspergillus niger, A sydowi, A. repens, Penicillium citrinum, Fusarium dimerum, Gliocladium penicilloides* and white sterile mycelium.

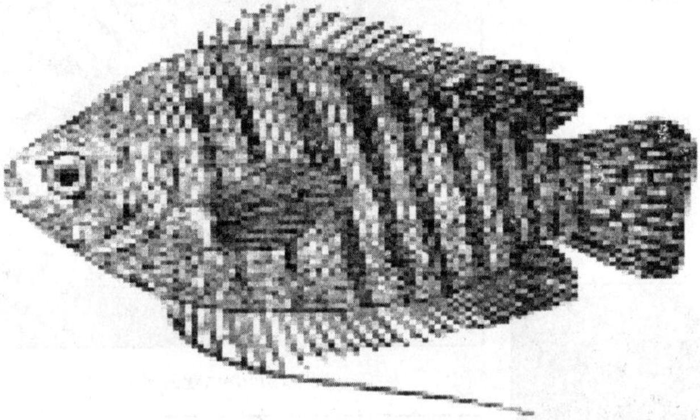

Fig 1. *Colisa fasiatus* (Schneider)

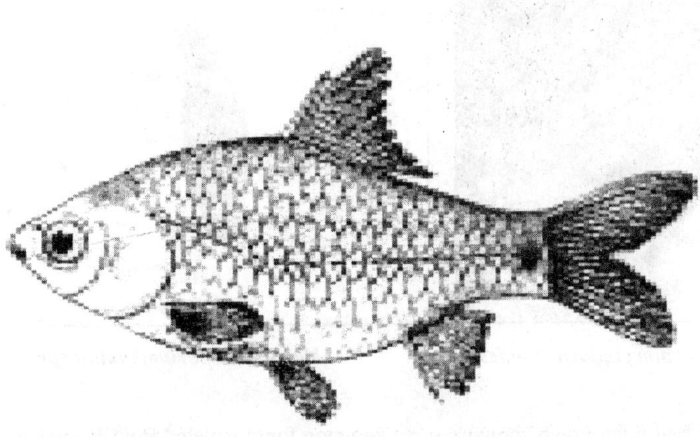

Fig 2. *Puntius sophore* (Hamilton Buchanan)

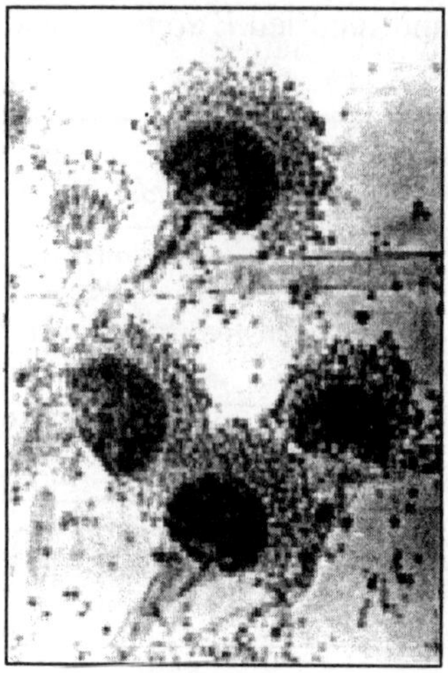

Aspergillus sydowi

Aspergillus niger

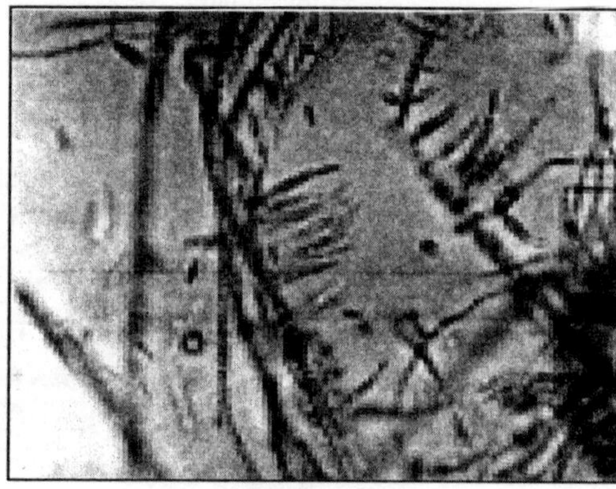

Fusarium oxysporum

Cladosporium cladosporioides

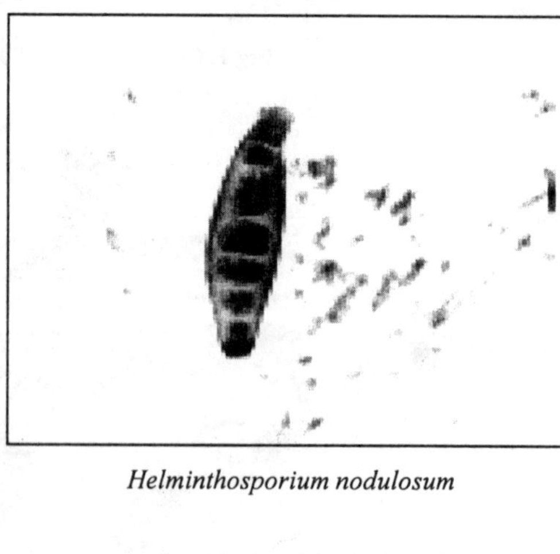

Helminthosporium nodulosum

Plate I . Photomicrograph showing some common fungi isolated from *P. sophore and C. fasciatus*.

A polluted water has total coliform count of >1000/100ml and a faecal *Coliform* count >200/100ml (Cabelli, 1978). In the present finding coliforms were found to be high but within the standard range (*Coliform* >10^2) as suggested by Shewan (1971). Pathogenic *Salmonella* were not recovered from all the sample analysed. The dominant bacteria includes *Bacillus, Micrococcus, Flavobacterium* and *Pseudomonas* in all the fresh samples whereas *Bacillus* sps. and three sps of *Micrococcus* were dominant in the sundried products.

The fishes caught in different environments have widely varying flora. Different fungal colonies belonging to different genera were isolated. *Aspergillus* and *Penicillium* species were the dominant fungi in all the fish samples. It might also be related to season, handling and processing etc. (Shewan, 1961) as well as the container and environment of the market. We also propose to study the incidence of aromonas in relation to organism of faecal origin and presence of organic matter in several aquatic habitat with different loads of pollution.

The actual initiation of diseases in fishes is a complex process which involves more than mere contact between the host and the parasite. Host susceptibility, parasite virulence, and environmental factors must all interact. Thus, saprophytic bacteria such as *Pseudomonas, Aeromonas, Myxobacteria sp. Vibrio anguillasum* are normally present in natural water at all times but do not usually cause disease problems unless environmental stress of some type intervenes.

ACKNOWLEDGEMENT

Authors are grateful to UGC-Special Assistance Programme for financial assistance and Dr. M. Shyamkesho Singh, Associate Professor, Department of Life Sciences, Manipur University for the identification of fungus.

REFERENCES

APHA (1976). *Compendium of Methods for the Microbiological Examination of Foods. (Speck, M.L., Ed.).* American Public Health Association, Washington, DC, USA.

Bramstead, F and Averbach, M. (1962). The spoilage of freshwater fish. *In Fish as Food.* Vol. I (Brongstrom, G. Ed). Academic Press, New York, 725 pp.

Cabelli, V. (1978). New standards for enteric bacteria *In water Polluion Microbiology* Vol. 2, ed Chell, R. pp 233-271, John Waley, New York.

Collins, C.H. Lyne, P.M. and Grange, J.M. (1989). *Microbiological methods* 6th Edn. Butlernorth and Co. (Publishers) Ltd. London, 409 pp.

Ellis, M.B. (1971). Dermatiacous Hypomycetes. p. 608, CMI, Kew, Surrey, U.K.

Ellis M.B. (1976). More Dermatiacous Hypomycetes. p. 540. CMI, KE, Surrey, U.K.

Gilman, J.C. (1957). *A Manual of Soil Fungi.* The Iowa State University Press, Iowa, USA.

Nair, K.K.S and Nair, R.B. (1988). Bacteriological quality of freshwater fish from Krishnarajendra Sagar Reservoir. Fish Technol; 25(I).

Sarojnalini Ch. and Vishwanath W. (1988). Composition and digestibility of fermented fish foods of Manipur. J. Food. Sci. Technol. **25** : 349-351.

Sarojnalini Ch. and Vishwanath W. (1994). Composition and nutritive value of sundried *P. sophore*. J. Food. Sci. Technol. **31(6):** 480-483.

Shewan, J.M. (1961). The microbiology of sea water fish, In *Fish as food.* Vol. I (Borgtron G. Ed). Academic Press New York, pp. 487-560.

Shewan, J.M. (1962). Food poisoning caused by fish and fishery product. *In fish as food.* Vol 2. (Borgstron G. Ed). Academic Press, New York, pp. 443-466.

Shewan, J.M. (1971). Microbiology of fish and fishery products. J. Apple. Bacteriol, **34 (2):** 299.

Thatcher, F. (1971). Hygienic and Safety Aspects of Quality Control *In fish inspection and quality control*. FAO pp, 222-225.

Advancing Frontiers of Ecological Researches in India, 445-448, 2006
Edited by A.K. Kandya & Asha Gupta
Published by Bishen Singh Mahendra Pal Singh, Dehra Dun, India

Observations on the Morphology and Taxonomy of an Interesting Blue-green Algae

K.V.S. Reddy, (Mrs.) K.M. Reddy
L.A.D. College for Women, Nagpur
Dr. Ambedkar College Nagpur.

Abstract: An interesting blue-green alga was observed in the liquid cultures of samples collected from Yeotmal and Mangrurpir, situated in the Vidarbha region of the Maharashtra State, India. These two places are at a distance of about 100km from each other. The presence of this alga in the cultivated soils of such widely separated soils is quite interesting.

Thallus

The alga is found mostly attached to the sides of the culture flasks. In the initial stages of growth, the thallus is ball like, later on expanding into a soft mucilaginous, yellowish stratum.

Filaments

Filaments are 18.3 - 22.0 μ broad, invested with a thick sheath. The sheath is 7-10 μ thick and is lamellated and constricted at regular intervals; thus having a beaded appearance. The lamellations are of mixed type, converging near the point of constriction in the lower part of the filament. In the middle and apical parts of the filaments, the lamellations appear to be divergent. In some filaments, the sheath extends well beyond the apices of the trichomes. Trichomes are constricted at the cross walls and are 3.32 - 7.5 μ broad. Vegetative cells are barrel shaped or elliptical. At times, cells have slightly pointed ends. Vegetative cells are 3.32 - 12.0 μ in length.

Heterocysts

Herterocysts are both intercalary and terminal. Terminal heterocysts are sometimes paired; spherical to barrel shaped and measure 5.0 - 7.4 μ in diameter, 5.8 - 6.6 μ broad and 6.6 - 7.4 μ long. Terminal heterocysts are formed quite early in the developmental stage. Sometimes, intercalary heterocysts are not ensheathed.

Akinetes

Akinetes are formed in chains away from the heterocysts and are centripetal in development. All the cells of the trichomes are potentially capable of getting transformed into akinetes. Akinetes are 6.64 - 8.3 μ broad and 8.3 - 13.0 μ long, mostly ellipsoidal to sub spherical, with granular contents.

Discussion

Literature dealing with the genus *Anabaena* was reviewed and no alga similar to the present from was found. Laloraya (1971) reported some new species and varieties of *Anabaena*. None of the forms described by him has a sheath as described in the present form. They also differ in size and shape of cells and akinetes. Hibbert (1976 a, b & c) recorded three new species from Minnesota Lake. In all the three species there is a thick mucilaginous sheath surrounding the trichomes, but lamellations are lacking in the sheath. Furthermore, the present alga differs from them in the shape as well as general dimensions of the akinetes.

The alga comes nearer *Anabaena doliolum* Bharadwaja (Desikachary 1959) in measurements and the shape of the akinetes; but differs from it in the presence of a thick beaded sheath with divergent lamellations, and in having slightly longer cells and bigger spores. It resembles the genus *Petalonema* Berk due to the presence of a thick divergently lamellated sheath. Since *Petalonema* Berk belongs to Scytonemataceae, false branching is present in it and akinetes are absent. Thus, though apparently, the present alga appears like *Petalonema* Berk, it differs from it due to the absence of false branching and presence of akinetes. Due to these features, this alga is placed in the family Nostocaceae; and because of somewhat loosely arranged, straight trichomes and soft thallus, the alga has been attributed to the genus *Anabaena*. Since no *Anabaena* species has been reported with such a broad and beaded sheath with divergent lamellation, as found in *Petalonema*, this alga has been named *Anabaena petalonemoides* sp. nov.

Diagnosis

Thallus yellowish, in the form of mucilaginous balls or forming a stratum. Filaments straight or slightly irregularly curved, 18.3-22.0 μ broad; sheath beaded, 7-10 μ thick with divergent lamellation. Cells elliptical to

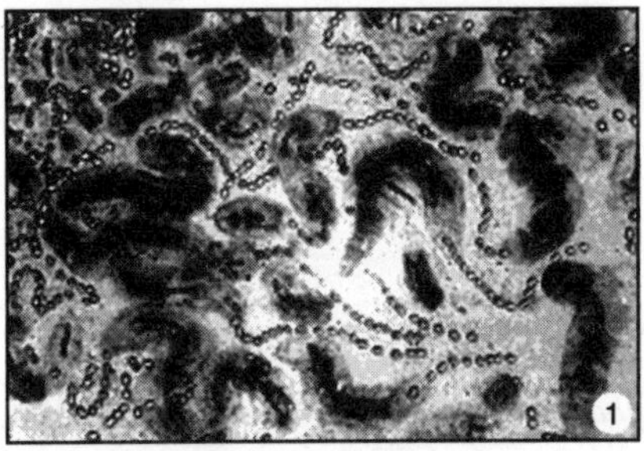

Fig. 1. Fimaments in a Colony x BC

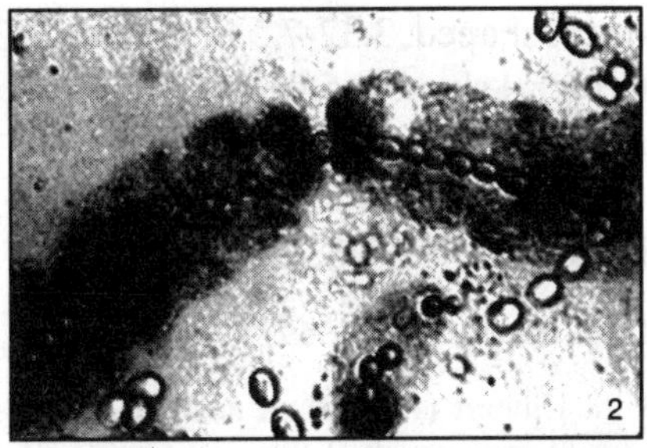

Fig. 2. Trichome with inlencalary heterocys 1x320

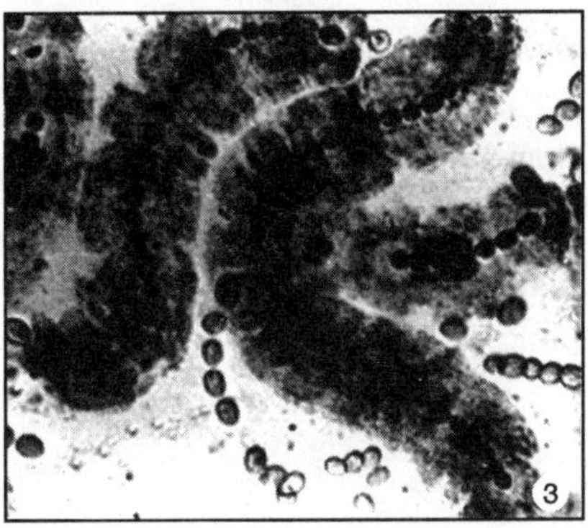

Fig. 3. Trichomes with basal heterocyst and beaded sheath x 320

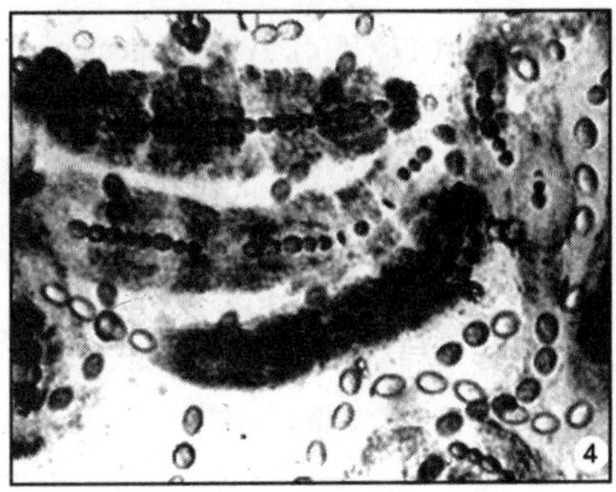

Fig. 4. Sheath showing divergent lamelation x 320

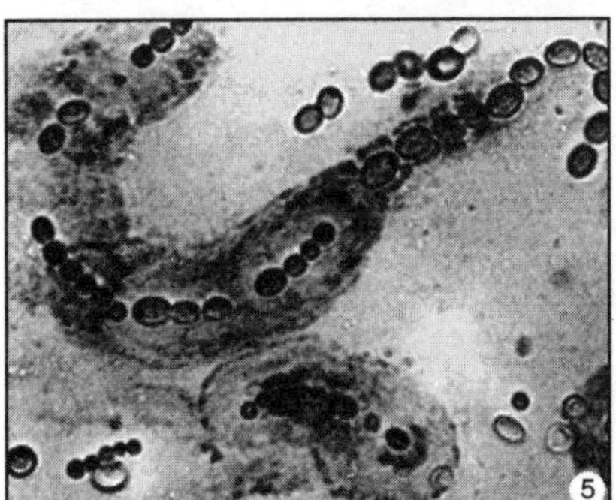

Fig. 5. Spores in chains x 320

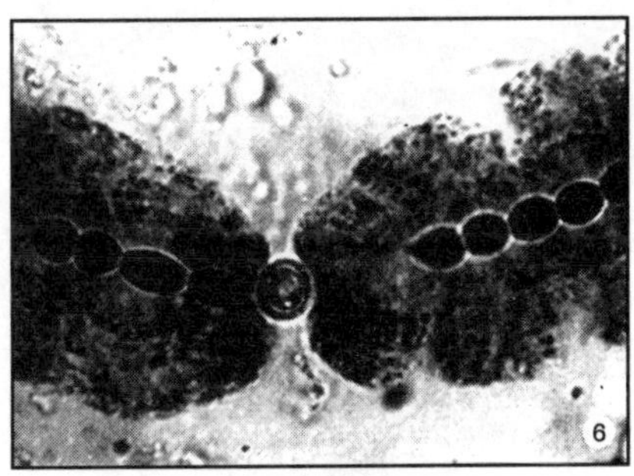

Fig. 6. Fig. 3 onlarged to show heteracyst and sheath x 800

Fig. 1-6. Photomicrographs of *Anabaena petalonemoides* sp. nov

barrel shaped, 3.32-7.5 μ broad and 3.32-12.0 μ long. Heterocysts spherical to barrel shaped, 5.0-7.4 μ in diameter, 5.8-6.6 μ broad and 6.6-7.4 μ long. Akinetes elliptical to barrel shaped, in chains away from the heterocysts.

REFERENCES

Desikachary, T. V. (1959). Cyanophyta I.C.A.R., New Delhi

Hill, Hibbert (1976 a). A new species of *Anabaena* (Cyanophyta, Nostocaceae) from Minnesota lake 1. Phycologia **15** : 61-64.

Hill, Hibbert (1976 b). A new species of *Anabaena* (Cyanophyta, Nostocaceae) from Minnesota lake, II. Ibid **15** : 65 - 68.

Hill, Hibbert (1976 c). A new species of *Anabaena* (Cyanophyta, Nostocaceae) from Minnesota lake III. Ibid **15** : 69-71.

Laloraya, V. K. (1971). Some new *Anabaena* from paddy fields of India. Phykos **10** : 118 - 126.

Advancing Frontiers of Ecological Researches in India, 449-456, 2006
Edited by A.K. Kandya & Asha Gupta
Published by Bishen Singh Mahendra Pal Singh, Dehra Dun, India

Degradation of Cyanide by algae and their nutrients

S. D. Shankhadarwar, P. B. Nandkar
Dept. of Botany, Nagpur University, Nagpur

Abstract: Nitrogen source is essential for the growth and multiplication of Prokaryotes and Eukaryotes. In its absence (*Oscillatoria curviceps. Phormidium foveolarum* and *Chlorococcum humicolum* are capable of utilizing nitrogen from the supply of sodium cyanide in the range of 5 to 30mg/Lt in BG-11 ^{0} medium. The study deals with cyanide degradation in presence of nutrients and algae. Microbial degradation was faster and quicker than chemical degradation,-reported in growth period in BG-11° medium at pH 7.6.

INTRODUCTION

In industrial wastewater operation transports, many chemicals disturb the ecological balance of the environment, some of chemicals leach heavy metals which would otherwise remain immobile. A few chemicals and heavy metals may be toxic by causing aquatic and land biota to sicken or to die. Sodium cyanide is found predominantly in industrial effluent by metallurgical operation, electroplating and international mining community to extract gold. Although sodium cyanide is highly toxic, it is commonly used in electroplating industries. Cyanide is singly charged anions containing unimolecular amounts of carbon and nitrogen triply bonded together CN^-. It is strong ligand. capable of complexation at low concentration with any heavy metal. The methods of cyanide remediation have been used physical and chemical separation, oxidation and bioremediation. Bioremediation is increasingly being considered as less expensive alternative to physical and chemical process. Biological degradation of cyanide methods may be both cheaper and more environmentally acceptable than chemical methods. Bioxidation methods are process of breaking and transforming hazardous material into simple non-toxic substances by a biological treatment.

A various species of bacteria, fungi, algae, yeast, and other plants along with their enzymes and amino acids, are known to oxidize cyanide naturally, extensive literature on subject has been generalized in several survey

(1.3.5.6.7.9. 10. 12. 13. 14. 19). However, the impact of algae on this is difficulty to evaluate due to lack of knowledge of the action of chemical degradation factors (15) However relatively little work appears to show that characterization and development of bacterial cell free extracts or enzymes able to degrade cyanide. The role of micro organisms and microbial enzymes in the degradation of cyanide has been studied for long time micro organism are known to posses various enzymes capable of converting cyanide into carbon and nitrogen which serve as substrate (10.11) such enzymes include formamide/ hydrolase (E.C.4.5.1.6) L-3 cyanoalanine synthese E.C.(4.4.1.9.) rhodanase E.C.(2.8.1.1) thiosulphate sulfotransferase E.C.(3.5.5.3) 12.13.14.18.21. The purpose of this work was to study the role played by algal community and the impact of chemical and physical factors in degradation of cyanide.

MATERIALS & METHODS

Algal samples from industrial wastewater were collected and isolated in the algal laboratory. Isolates of *Oscillatoria curviceps*, *Phormidium foveolarum* and *Chlorococcum humicolum* can easily be grown in the BG11° medium. Carbon and nitrogen sources were omitted and replaced by sodium cyanide in BG11° medium.

Cyanide concentrations in the experiment were selected in the range of 5 to 30 with algal biomass in the range of 20 to 30 mgl^{-1}. All experiments were conducted in triplicates for chemical and algal degradation of cyanide The concentrations of cyanide and algal growth in the test experiments were recorded on 3rd, 5th , 7th and 10th day. All experiments were conducted in the range of 25° to 30° C.

Analytical method: -Determination of cyanide concentration by silver nitrate method (APHA. 1989)

Results

The influences of nutrient of BG11° on growth of *Oscillatoria curviceps*, *Phormidium foveolarum*, and *Chlorococcum humicolum* are reported in this study. All experiments on degradation and biodegradation were conducted in the range of 5 to 30 mgl^{-1} cyanide at 25-30° C and 20-30μ mol photon m^{-2} s^{-1} flux light intensity

Degradation of cyanide by Chemicals:

The cyanide concentration in the range of 5 to 30 mgl^{-1} was chosen to find the degradation by chemicals. In this degradation, the cyanide concentration on 3rd, 7th and 10th day, decreased to 4.4 mgl^{-1}, 4.1 mgl^{-1} and mgl^{-1}, 4.0 mgl^{-1} from original 5 mgl^{-1} concentration: 9.0 mgl^{-1}. 7.8 mgl^{-1} and 7.0 mgl^{-1} from 10. mgl^{-1}, 13.8mg^{-1}, 11.4 $mg^{-1}$and 11.0 mg^{-1} from 15 mg^{-1}, 18.4 mg^{-1}, 17.4 mg^{-1}, 17.00 mg^{-1}. from 20 mg^{-1}, 24 mg^{-1}. 9.5 mg^{-1} and 19.0 mg^{-1} from 25 $mg^{-1}$and 29 mg^{-1}. 26 $mg^{-1}$and 24.0 mg^{-1} from 30 mg-1 respectively. Overall chemical degradation was 3 to 12% on 3rd day. 12 to 24% on 7th day and 15 to 30% on 10th 'day of experiment.

Biodegradation of cyanide by *Oscillatoria curviceps*:

Cyanide concentrations on 3rd, 7th and 10th day of culture decreased to 3.7 mg^{-1} '. 0.3 mgl^{-1} and zero from original 5 mg^{-1} CN: 8.1mgl^{-1} 0.8 mg^{-1}. zero from 10 mg^{-1}, 12.7 mg^{-1}. 0.9 mg^{-1}.zero from 15 mg^{-1}. 10. mg^{-1}, 5 mg^{-1}: 1.8 mg^{-1}. zero from 20 mg^{-1}.18.0 mgl^{-1} : 2.3 mgl^{-1}. zero from 25 mgl^{-1} and 24.5 mgl^{-1}. 2.4 mgl^{-1}. zero from 30mgl^{-1} respectively.

Biodegradation of cyanide by *P. foveolarum*. :

Cyanide concentrations on, 3rd 7th arid-l0th day of culture decrease to 4.0 mgl^{-1}. 0.8 mgl^{-1} . zero from original 5 mgl^{-1} 8.5 mgl^{-1} . 1.0 mgl^{-1} . zero from 10 mgl^{-1} 13 mgl^{-1} 3.0 mgl^{-1} zero from 15 mgl^{-1} 15.6 mgl^{-1}. 2.2 mgl^{-1}. zero from 20 mgl^{-1} 18.0 mgl^{-1} 2.0 mgl^{-1}. zero from 25 mgl-1 and 28 mgl^{-1}. 2.4 mgl^{-1} zero from 30 mgl^{-1} respectively.

Biodegradation of cyanide by *Chlorococcum humicolum*

Cyanide concentration on 3rd . 7th and 10th day of culture decrease to 4.2 mgl^{-1} 1.8 mgl^{-1}.zero from original 5mgl^{-1} 9.0 mgl^{-1}. 4.0mgl^{-1} . zero from lOmgl^{-1} 13mgl^{-1} 4.0 mgl^{-1} zero from 15mgl^{-1} 18mgl^{-1}. 6.0. zero from 20 mgl^{-1}: 23.0mgl^{-1} 6.7mgl^{-1}.zero from 25 mgl-1. 28mgl^{-1}. 6 mgl^{-1} zero from 30 mgl^{-1} respectively. Overall cyanide degradation by algae showed 6 to 30% on 3rd day. 60 to 94% on 7th day and complete degradation 100%' on 10th day. (Observation are graphically represented in fig. l.1 to l.6).

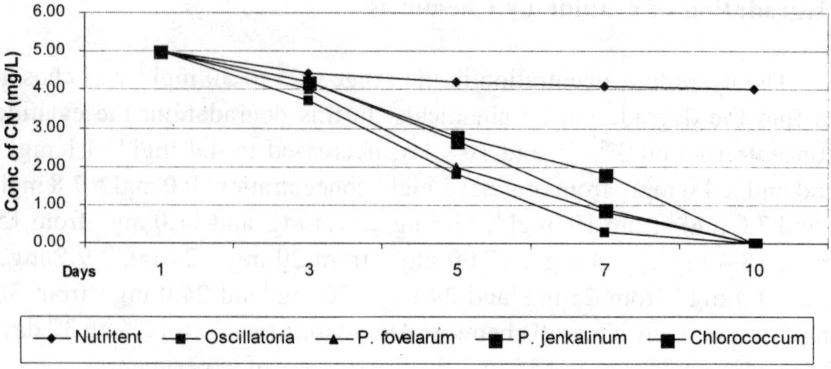

Fig. 1.1. Degradation of CN (5 mg/L) by different algae.

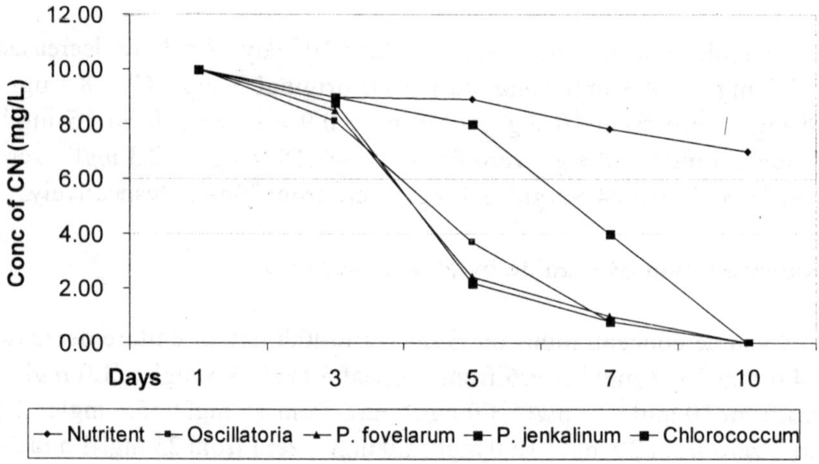

Fig. 1.2. Degradation of CN (10 mg/L) by different algae.

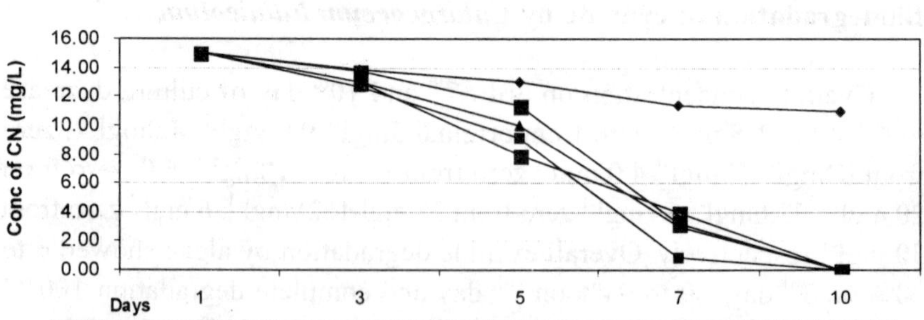

Fig. 1.3. Degradation of CN (15 mg/L) by different algae

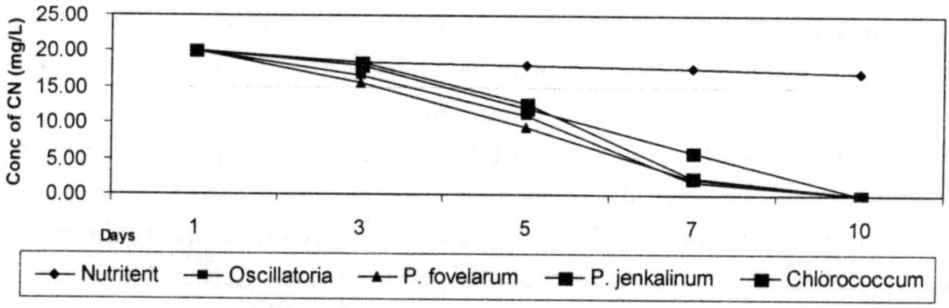

Fig. 1.4. Degradation of CN (20 mg/L) by different algae

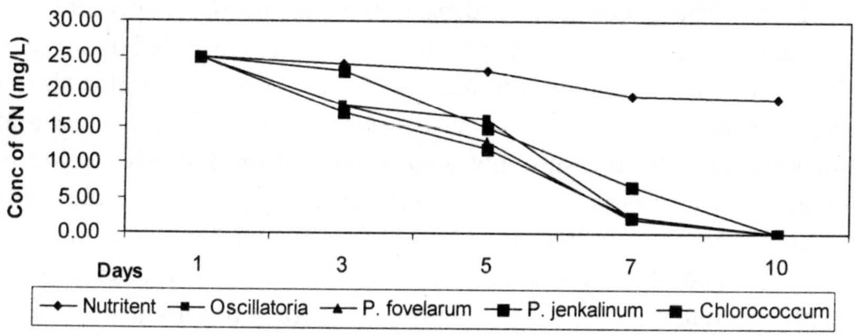

Fig. 1.5. Degradation of CN (25 mg/L) by different algae

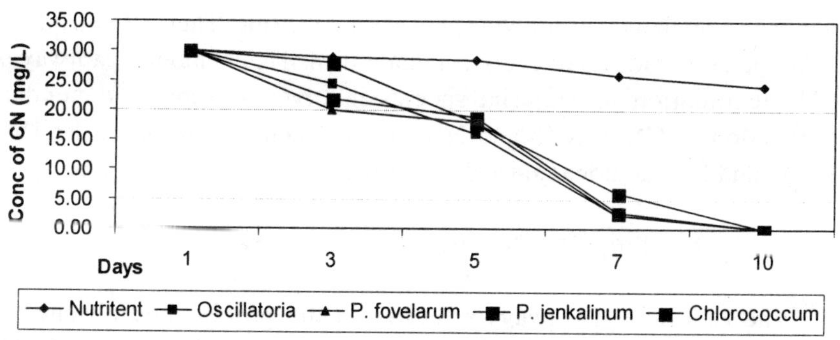

Fig. 1.6. Degradation of CN (30 mg/L) by different algae

RESULT & DISCUSSION

In cyanide degradation various procedures are in practices. However, current and past remediation methods for waste and process wastes containing cyanide are enlighted with a view to destroy the cyanide. Cyanide biodegradation by algae is faster and quicker than chemical degradation. The chemical degradation shows partial degradation from 5 to 30%. which might be due to chemical-reaction occurring in the BG 11°

medium with added cyanide. This partial degradation reaches to a stage of final degradation if *Oscillatoria curviceps* or *Phormidium foveolarum* or *Chlorococcum humicolum* added individually in its algal medium alongwith cyanide. The degradation increases rapidly at all the concentrations of cyanide from 5 to 30 mgl^{--1} of 5th to 7th day and is reached to maximum level (100%) on the 10th day *Oscillatoria* degrades, more cyanide than *Phormidium* and *Chlorococcum* whereas *Phormidium* degrades more cyanide than the degradation by *Chlorococcum* Overall *Cyanobacteria* degradation of cyanide faster than the green algal degradation. The major role of algae is the cyanide (CN) splitting into carbon and nitrogen elements may be due to .enzymes recorded in nitrogen deficient medium. Algal growth measured on biomass basis also shows increase in the growth substantiates in *Oscillatoria, Phormidium* and *Chlorococcum*. These results are also similar to the recorded value in case of *Pseudomonas putida. Pseudomonas sutzeri Pseudomonas flurorescens* NCIMB-11764. (10.1 1.12.'3.14. 1518). Isolated the enzymes formamide hydrolase (E.C. 4.2.1.66) L-3-cyanoalanine synthase (E.C. 4.4.1.9). rhodanase (2.8.1.1). cyanase (cyanate_amidohydrolase: E.C 3.5.5.3). These enzymes are responsible to convert cyanide into carbon and nitrogen which serves as substrate (10) and cyanide converts into end products, which induce ammonia and carbon dioxide. They are considered to be the end products of cyanide metabolism. The metabolic pathway of CN' degradation in bacterial species involves an oxygen dependent conversion of CN^- to CO_2 and NH_3. *Pseudomonas flurosecens* NCIMB 11764 and Pseudomonas *sutzeri* AK31 (11.12.)

$$HCN+0_2+2H+ NADH—— 002 + NH_4 + NADP$$

The two other pathways effective in the process and absence of oxygen have their own reaction producing formamide of formate and NH_3. The utilization of CN as source of nitrogen is believed to proceed through $NH_{3.}$ It follows from our study that algal degradation of CN^- is possible only under condition that prevents its chemical decomposition. Therefore a feasible approach to tackling the problem of the high content of cyanides in liquid waste would be to convert them to carbon and nitrogen compound in an area of interest having high contents of cyanide with aid of algal community.

REFERENCES

A. P. H. A. (1989). Standard methods for the Examination of wastewater. 17th edition. Washington. USA.

Akcil. A. (2002). Cyanide versus environment of Turkey. Final discussion mining Environmental management united Kingdom_**9(6):** 22-23.

Andrade. M.C., Figerra. M.M. and Linardi, V. R. (1995) Short communication: Utilization of Ammonia. Generated from Abiotic cyanide Degradation, by *Rhodotorula rubra*. wid. *J. Microbiol Lett.* **(156):** 61-67.

Allen. J. and Strobel. (1966). The assimilation of HCN by Variety of fungi. *Can. J. Microbiol.* **12:**414-416.

Blumenthal. S. Goldschmidt *et al.* (1968). Cyanide metabolism in higher order plants III. The biosynthesis of B-cyanoalanine. ,/. *Bio. Chem.* .**243:**5302-5307.

Castric., P.A. (1981). The Metabolism of hydrogen cyanide by bacteria p333-261. In B Vennnellend, et al cyanide in Biology Academic press ,NEW. York

Chan., J.L. and kunz. D.A (1997). cyanide utilization in *Pseudomonas fluorescens* NCIMB11764 involves a papative sidephore F.EM.S. Microbiol. Lett. 156: 6 -67.

Grigov'eva. U.V. Avakyan. Z.A. Turova. T.P. K.ondrat'eva. T.P. and Karavaiko, G.1. (1999).' The search for study of Microorganism, that degrade cyanide and Thiocyanates. *Mikroblogiya*. **68(4):** 453- 460.

Fallen. R. D., Copper. D. A., Speece. R. and Henson. M. (1991). Anaerobic Biodegradation of Cyanide, under methanogenic conditions. *Appl.* and *Environmental Microbiology*. pp. 1656-1662.

Harris. R.E. Konwles C.J. (1983). Isolation and growth of *Pseudomonas* species that utilizes cyanide as source of Nitrogen. *J. Gen, Microbiol* 129. 1005-101 I.

Karavaika. G.1. Kondrat'eva. Savari E. E. Grigoo'eva. and Avakyan (2000). Degradation of Cynide and thiocyanate. *Microbiology*. **69(2) :** 167 -173.

Ingvorsen. K.B. Hojer. Pedrson. and Godifredson (1991). Novel cyanide - hydrolyzing enzyme from AlcaligensXylosoxidans sub sp. Denitrification. *Appl. Environmental Microbiol* **57:** 1783-1789.

Konwles C.J. (1976). Microorganism and cyanide. *Bacterial Rev*. **40:** 652-680.

Kunz. D. A. Nagappan. O. Silva, Avolas and Delong. G. T. (1992). Utilization of cyanide as nitrogenous substrate by *Psedomonas fluoroescens* NCIMB 11764. evidence for multiple pathway of metabolic conversion. *Appl. Environ.* microbial **58:** 2022 - 2029.

Mudder, T.I. Whitlock. (1984). Biological Treatment of cyanidation waste water. SME preprint No. Littletom. co.

Rollinson. G.R. Jones. M.P. Medous. R.E. and Harris and Konwles. C.J. (1987). The Growth of a cyanide - utilizing strain of *Pseudomonas flurorescens* in liquid culture on nickel- cyanide as source of nitrogen. *F.E.M.S. Microbiol .Lett* . **40:** 199 - 205.

Skowronsksi. B. and Strobel. G.A. (1969) Cyanide. Resitance and cyanide utilization by a strain of *Bacillius Pumilus can J Microbiol* **15:** 93-98.

Ware, G.C. and Painter (1995). Bacterial Utilization of cyanide *Nature* 175-900.

SECTION SIX

POLLUTION ECOLOGY

Advancing Frontiers of Ecological Researches in India, 459-471, 2006
Edited by A.K. Kandya & Asha Gupta
Published by Bishen Singh Mahendra Pal Singh, Dehra Dun, India

Management of Wastes : An Ecosystem Approach

Sanat K. Jain

International Instt. For Environmental Mgmt. & Research And Consultant, Envirotech Associates, A-7, Inside Toran Gate, A.B. Road, Indore - 452008, M.P.

Abstract: Details of the indigenous and exogenous, efficient and eco-friendly micro-organisms have been discussed for treatment and disposal of various kinds of wastes, i.e. gaseous, liquids and solids so that they can be converted into useful materials for agriculture, poultry, animal husbandry, public health and aqua-culture etc. Such a system of waste management is successfully working in 116 countries of the world under the view that waste is actually a resource out of place.

INTRODUCTION

To begin with the waste, we can say that it is that material which is not in the use of human beings or else, use of such materials is not known to us. In this context, the term waste becomes a relative word depending upon its "use or no use".

In general, there are three types of wastes, viz.- gaseous, liquid and solids. But due to technological developments, a fourth type of waste, viz. Radio-active or Nuclear Waste (a Techno-Waste), is also existing in our world. All these are posing serious problems for disposal.

In fact, the mad race for industrialization and economic developments have resulted in over exploitation of natural resources, thus leading to a situation where two worlds of man - the biosphere, lithosphere & hydrosphere of his inheritance and the techno-sphere of his creations, are out of balance.

Before dealing with the disposal and management of wastes, let us know some facts about our *environment*. As we know that the environment has two components – biotic: consisting of living beings and abiotic : consisting of non-living matters (organic and inorganic). The biotic and abiotic components function together as an ecological system –

ecosystem, displaying a complex interaction between themselves and existing at an *equilibrium* under Natural conditions. Any change brought about by man's activities or other circumstances places a "*Stress*" upon the system, disturbing the equilibrium, evoking a suitable "*Response*" from it and in turn leading to a *New Equilibrium* condition which may or may not be beneficial to the original components of *System* including man.

From what has been discussed about, it is apparent that if mankind is to survive, developmental process is to be continued and life on earth is to be *sustained*, there is an urgent necessity to reorient our thinking in respect of *Use and Processing of Our Resources* that would satisfy human needs with minimum resource waste while sustaining Economic growth and development within the carrying capacity of the ecosystems (Odum, 1971).

Therefore, today's need for an integrated outlook on environmental or waste management is better met by the "community" approach of ecology than by the "organism" approach of traditional biology. Let us examine few aspects of ecological significance and ecosystem (Fig.1).

Biotic Community

It is an assemblage of populations living in the Prescribed area or the Physical habitat. Various new factors come into play when an organism has to live in a community since communities display a definite functional unity with characteristic trophics (nutritional), substances and patterns of energy flow.

Abiotic

The abiotic component of ecosystem consists of water, gases, minerals, etc. on which various physical factors can have a direct bearing and thus effect the biotic component of that system, depending upon the Limits of Tolerances of organisms (rather than Liebig's Law of Minimum).

Several examples from natural environment can also be given to illustrate the effect of tolerance on the relative occurrence of different species. For example – salinity differences, nitrifying & de-nitrifying organisms etc.

Energy Flow & Nutrient Cycling

Now let us see how energy flow & nutrient cycling are achieved in the ecosystem and how man's activities in general (and the discharge of Wastes in particular) can be fitted in. (Fig. 1 and 2).

It is clear that the biotic component of any ecosystem consists of –

Primary Producer

which are photosynthetic, autotrophic organism mainly green plants and algae which can grow on simple inorganic substances (CO_2, NH_3 etc.) and synthesize/ prepare food in the presence of sunlight.

Consumers

which are heterotrophic organisms which ingest other organisms or the particulate matters, e.g. zooplnakton, worms, fishes, hares, cows etc.

Decomposers

which are again heterotrophic organisms –mainly bacteria and fungi which breakdown the complex compounds into the simpler ones and release Energy & Nutrients. These are in fact, micro-consumers, and have high rates of metabolism and Turnover. In the process, they form simple inorganic substances assembled by the producers. One of the important characteristics of decomposers –as a group is their ability to Function under the variety of environmental conditions. Organic matter essentially contains carbonaceous (Starch), nitrogenous (Protein) and sulfurous (Oil & fats) material together with phosphorus, iron and other elements. Microbial decomposition gives various end-products depending on whether the decomposition is aerobic or anaerobic.

In aerobic decomposition carbohydrates are broken down by aerobes to yield simple end-products such as CO_2 along with energy for their growth. Nitrogenous materials are broken down into:

Ammonia —Nitrites ——Nitrates.

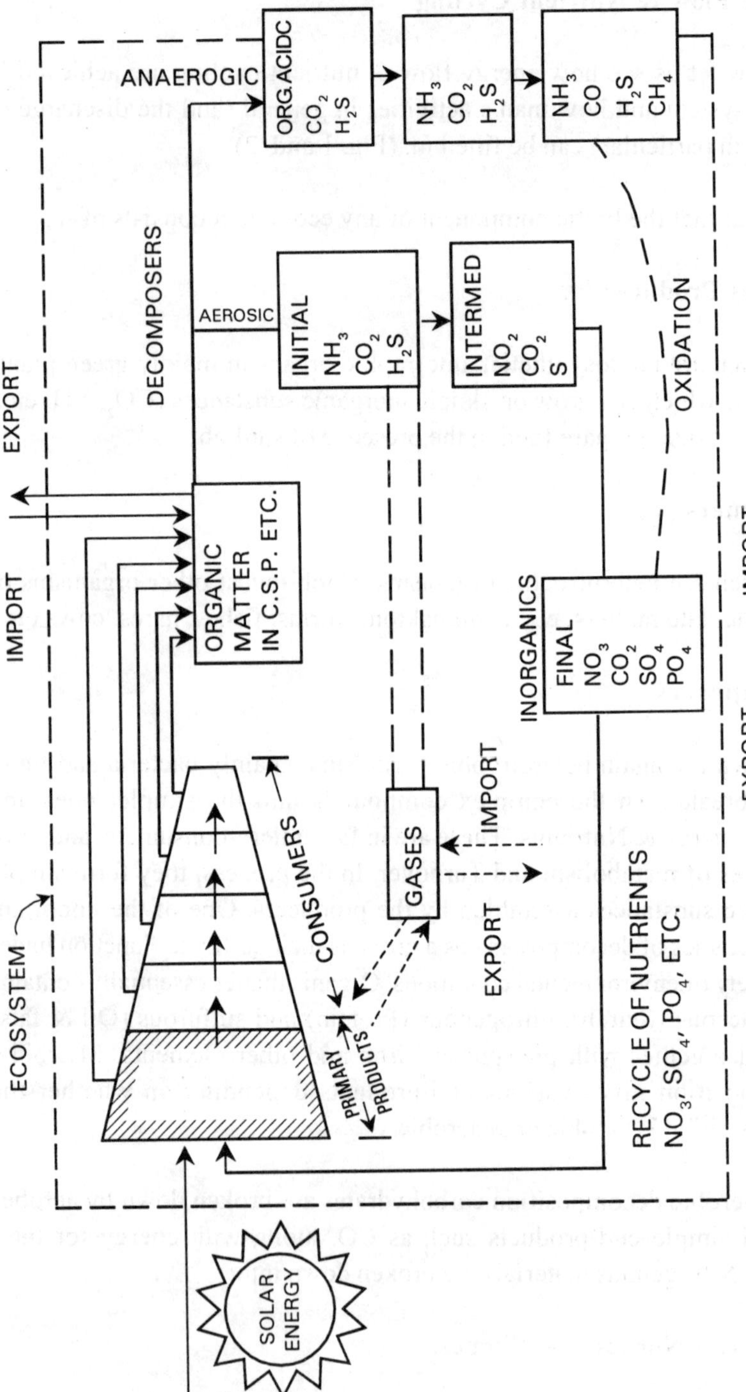

Fig.1 Energy flow and Recycling of nutrients in an Ecosystem

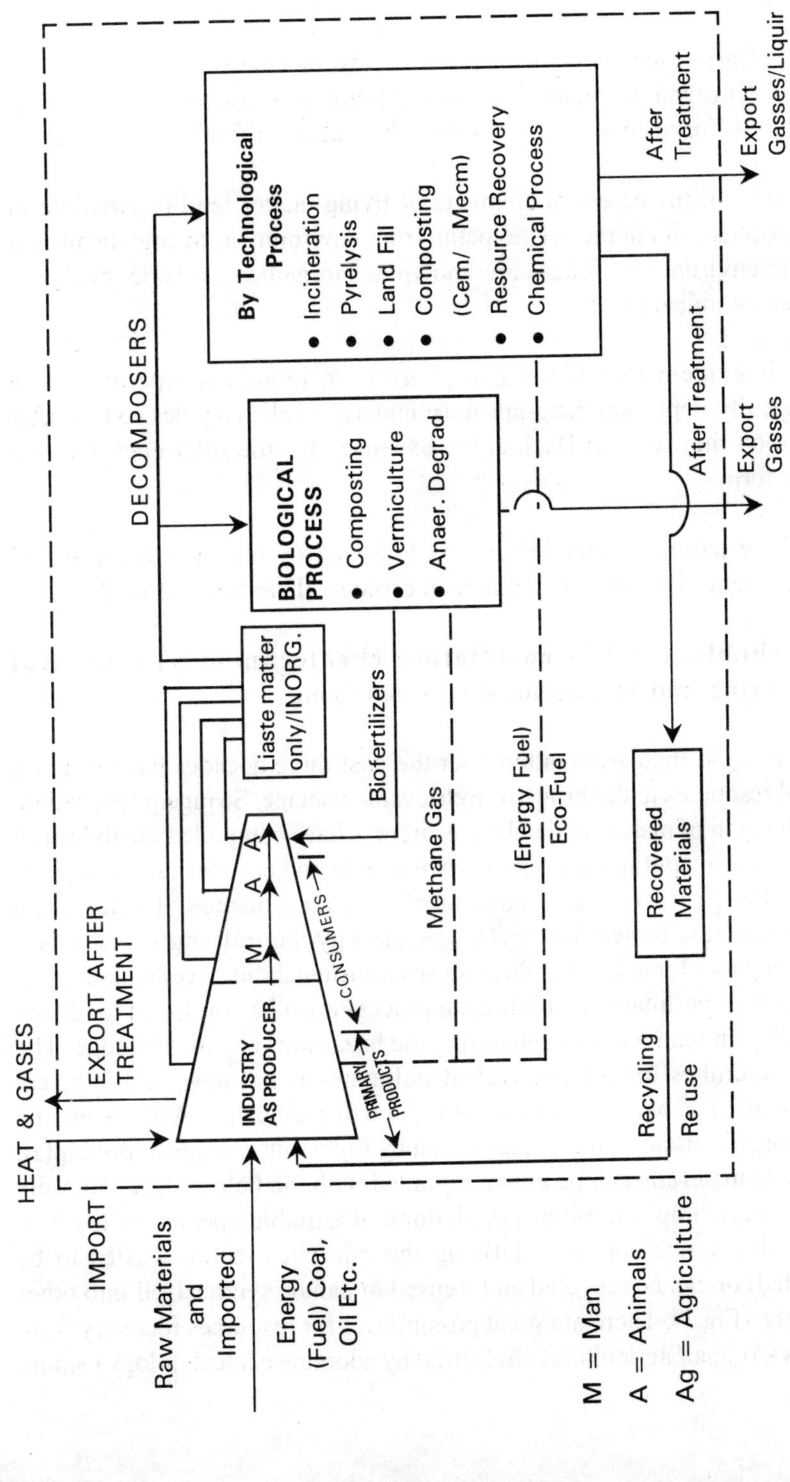

Fig.2 Ecomodel for the Waste management (Based on principles of Ecosystem)

In **Anaerobic decomposition**

Carbonaceous matter ——— acids, methane
Nitrogenous matter ——— ammonia, nitrogen
Sulfurous matter ——— hydrogen sulfide

Thus, all the essential elements of living matter tend to circulate in the biosphere in characteristic paths from environment to organisms and back to environment. These are known as inorganic – organic cycles or bio-geo-chemical cycles.

It is evident that if the Biosphere is to remain in operation, the biologically important materials must undergo cyclic changes so that after utilization they are Put Back at the expense of some solar energy into a usable form.

These natural principles can be followed by man for disposal and management of wastes to maintain an ecological harmony. (Fig. 2)

Biotechnology / Bioremadiation (Ecotechnology) / natural engineering and management of ecosystem.

The rapid industrialization over the past three decades has left pure, natural resources to the brink of irretrievable damage. Stringent legislation, globalization of market and public awareness have compelled the industrial community to continuously improve, innovate and use cleaner technology to reduce pollution and consumption of resources. Under such circumstances, biotechnology/ecotechnology/natural engineering is a promising tool for achieving this industrial sustainability. Ecotechnology or treatment of pollutants by biological process must be employed to reduce the pollution load without generating the hazardous chemical sludge. The use of microbes for the removal of pollutants is not new, as biological treatment of sewage has been practiced for decades. Also from the foregoing discussions, it is apparent that by following ecological principles, the waste materials can be either degraded with the help of Decomposers (i.e. by enriching microbial populations of suitable species in the bio-degradable wastes or acclimatizing the microbes in the wastes to be degraded) or can be recycled and Reused or can be synthesized into other products. (Fig.3). There are great possibilities for resources recovery from Wastes -Animal/ agricultural / industrial by adopting ecotechnology (Smith, 1996).

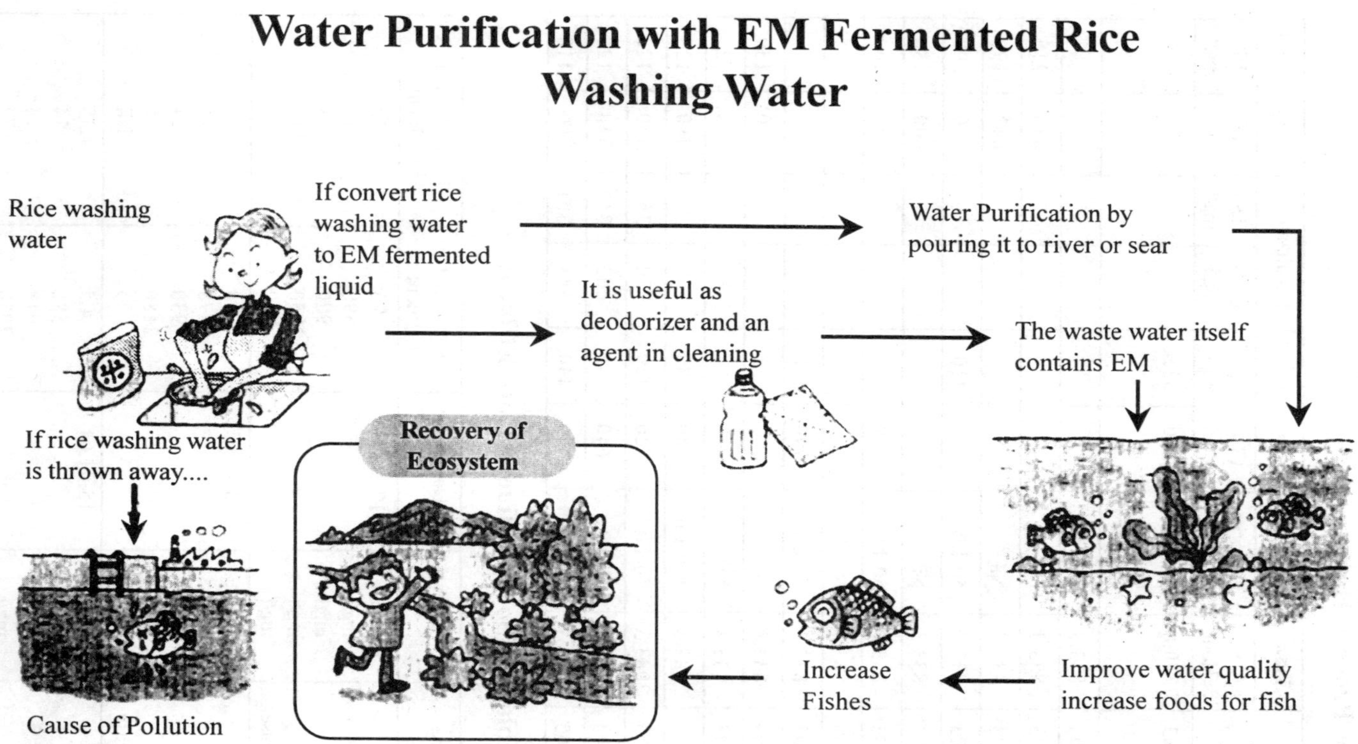

Fig.3 EM can be propagated using rice washings. This will involve the participation of households.

Table-1: Results of various Effluent Treatment Plants with EM culture

A. **Textile Process House**

Date	Flow	Make-Up			Equalization		Treated Water	
	Ml/ Hour	Quantity	COD mg/l	TSS mg/l	COD mg/l	TSS mg/l	COD mg/l	TSS mg/l
18.10.2003	770.8	190 Lit. Equuali-zation	680	325				
19.10.2003	833.3	Nil						
20.10.2003	833.3	Nil					640	210
21.10.2003	833.3	Nil					840	190
22.10.2003	833.3	50 Lit.	910	410			720	200
23.10.2003	833.3	Nil					600	180
24.10.2003	833.3	50 Lit.	800	375			520	160
25.10.2003	833.3	Nil						
26.10.2003	833.3	50 Lit.	790	350				
27.10.2003	833.3	Nil					490	150
28.10.2003	833.3	Nil					410	120
29.10.2003	2183.3	100 Lit.	780	380			480	135
30.10.2003	2083.3	50 Lit.	720	320	770	363	360	120
31.10.2003	2083.3	50 Lit.	640	305	663	312	340	120
01.11.2003	2083.3	50 Lit.	940	440	584	301	260	110

B. **Vegetable Oil (Soya) Refinery Plant Effluent**

Parameters	Unit	Date	Untreated Effluent	Final Treated Effluent
pH	pH scale	19.11.03	4.5	7.50
TSS	mg/l		590	84
BOD	mg/l		900	28
COD	mg/l		2050	242
Oil & Grease	mg/l		20	Nil
pH	pH scale	18.12.03	5.0	7.44
TSS	mg/l		578	73
BOD	mg/l		930	24
COD	mg/l		2180	198
Oil & Grease	mg/l		25	Nil
pH	pH scale	15.01.04	4.5	7.35
TSS	mg/l		580	54
BOD	mg/l		918	14
COD	mg/l		2080	172
Oil & Grease	mg/l		18	Nil

C. **Sugar Mill Effluent**

Parameters	Unit	Date	Untreated Effuent	Final Treated
pH	pH scale	08.12.03	6.8	7.35
TSS	mg/l		325	68
BOD	mg/l		635	26
COD	mg/l		1340	185
Oil & Grease pH	mg/l pH scale	12.01.04	10 6.4	Nil 7.42
TSS	mg/l		280	48
BOD	mg/l		618	16
COD	mg/l		1275	156
Oil & Grease	mg/l		18	Nil

D. **Domestic Sewage Effuent**

Parameters	Dosing Date	Inlet to Anae Pond	Inlet to Facult Pond	Inlet to Maturation P.	Final Outlet
BOD mg/l TSS mg/l Faecal Coli per 100 ml	26.11.02	235 056 $13x10^4$	105 160 $23x10^3$	07 20 $70x10^2$	05 14 $13x10^2$
BOD mg/l TSS mg/l Faecal Coli per 100 ml	12.12.02	120 172 $50x10^4$	68 48 $30x10^4$	13 18 $17x10^4$	02 18 $22x10^2$
BOD mg/l TSS mg/l Faecal Coli per 100 ml	07.01.03	90 78 $8x10^4$	40 68 $6x10^3$	11 60 $3x10^2$	07 48 11
BOD mg/l TSS mg/l Faecal Coli	11.02.03	115 164 $94x10^3$	64 42 $60x10^2$	16 34 $20x10^2$	11 26 03

Ecotechnology or Bioremediation is a treatment process that uses naturally occurring microbes (yeast, fungi or bacteria) to break-down or degrade, hazardous substances into nontoxic substances. In a polluted environment certain bacteria, fungi, protists and other microbes are active breaking down organic pollutants. These technologies assist microorganisms growth and increase microbial populations by creating optimum

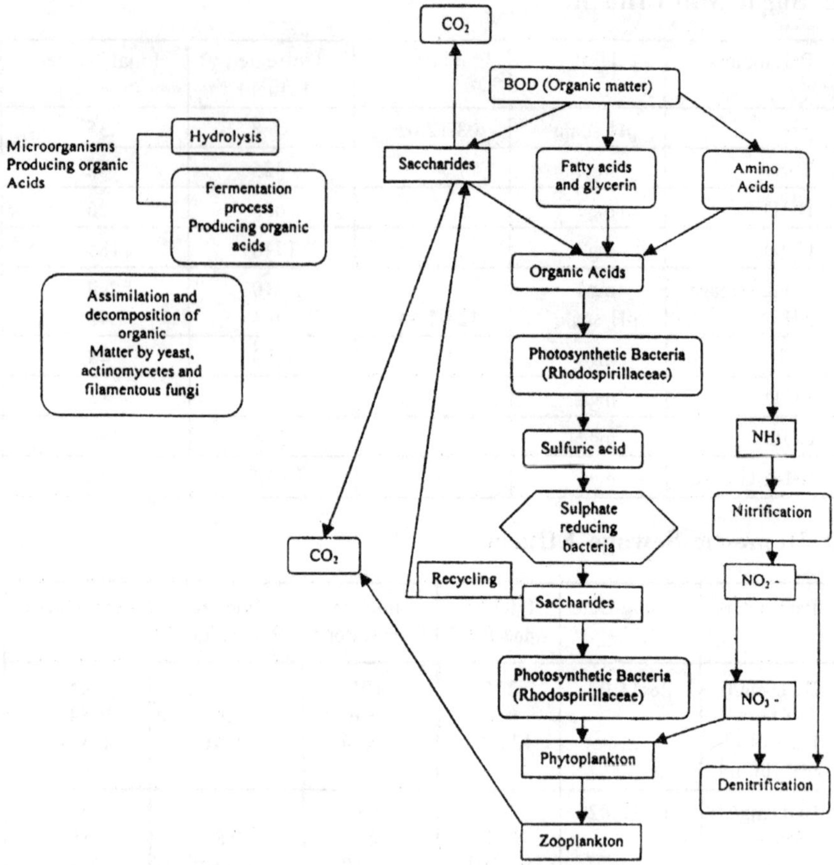

Fig.4 Mechanism in the EM Waste water treatment (EM treatment of odor, waste water and Environmental problms - Dr. Teruo Higa and Mr. Nobumasa Chinen, College of Agriculture, University of the Ryukyus, Japan)

environmental conditions to detoxify the maximum amount of contaminants (Scragg, 1999).

There are two types of microorganisms :

a) **Indigenous microorganisms :** Those microorganisms that are found already living at a given site. To stimulate the growth of these, a proper environment, temperature, oxygen and nutrient content are to be supplemented.

b) **Exogenous microorganisms :** If the biological activity needed to degrade a particular contaminant is not present at site, microorganisms

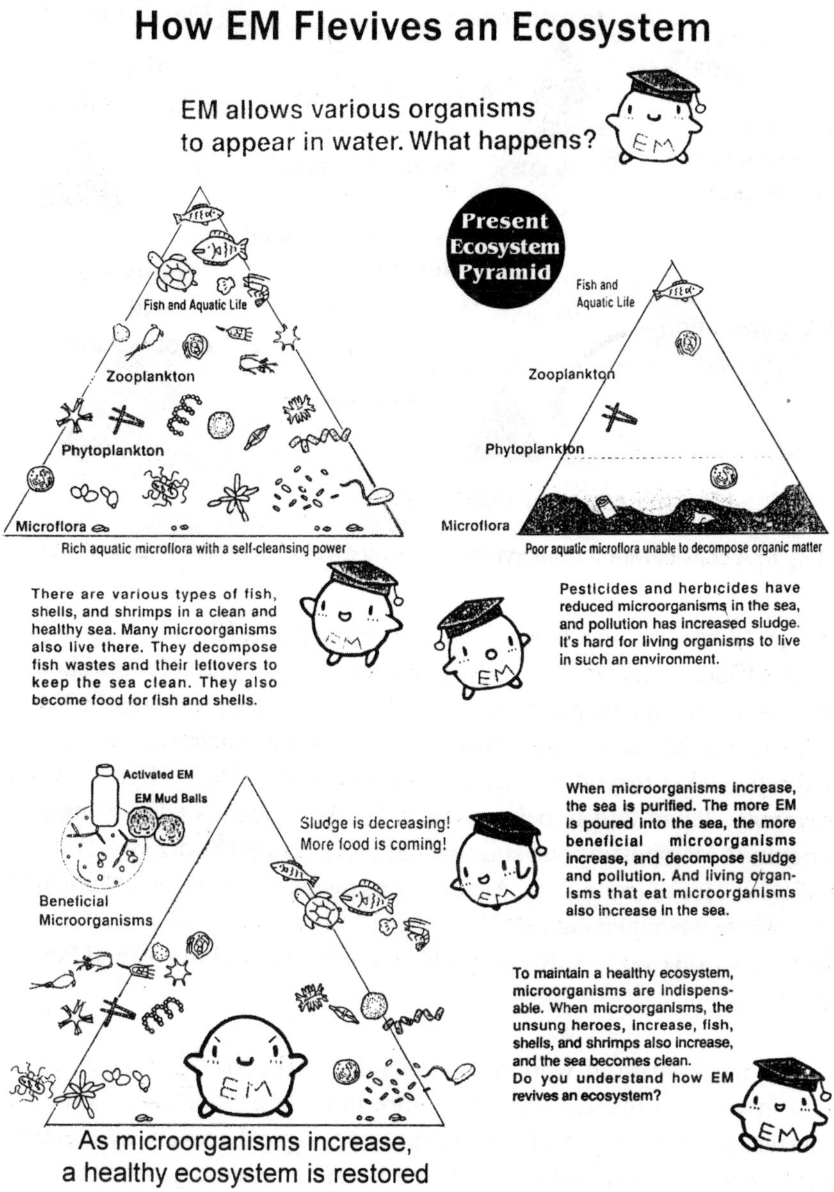

Fig. 5

from other locations whose effectiveness has been tested, can be added to the contaminated site (or wastewater), these are called Exogenous microorganisms. (Kalavathy, 2003)

The latter type of microbial consortium of efficient & ecofriendly microbes called EM has been used by the author in the treatment of

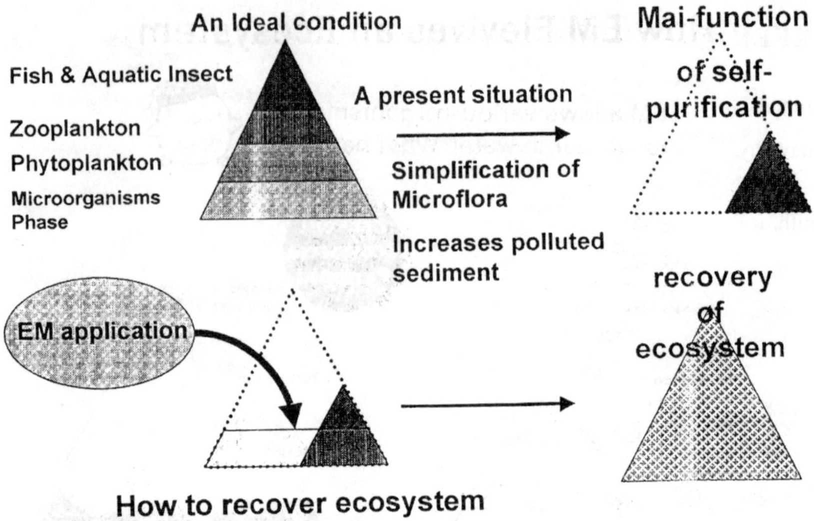

Fig. 6: A consideration from pyramid of ecological system (drainage texture)

Sewage (domestic and hotels), various industrial effluents, viz.-sugar factory effluent, vegetable oil (Soya) Refinery Plant, Textile Process house and empty gel capsule plant. The results of these are presented in Table 1. This microbial consortium comprising of Strains of Lactobacillus, yeasts, ray fungi and Photosynthetic bacteria is developed by Dr. T. Higa at Rykus University, Japan and is being used in the betterment of Environment, Agriculture (composting and soil amendments), Animal husbandry, Poultry, Medical & Health (for AIDS), Aquaculture, swimming pools and Municipal Solid Waste Management (MSW). This EM is already being used in 116 countries world over to solve the problems of Pollution, Poverty and Public Health.

The mechanism of working of this microbial consortium is given in fig. 4 in which it is clear that these work in succession as it is a mixture of anaerobes and aerobes, both and boost ecological activity by creating an optimum environmental conditions for beneficial microorganisms to grow keeping the ecosystem well balanced as shown in fig. 5. EM simply enhances the natural cleaning process.

In the last, it can be said that waste is a resource out of place.

REFERENCES

Odum, E. P. (1971). Fundamental of Ecology, Saunders Publ. Philadelphia, USA

Kalavathy, S. (2003). A cleaner method for cleaning the Environment .In - Environment Science & Engg. 1 (2) : 41-49.

Smith, J. E. (1996). Biotechnology, 3rd Ed. Cambridge Univ. Press, London.

Scragg, A. (1999). Environmental Biotechnology. Longman Publ.

Ecopure, March , 2003.

Advancing Frontiers of Ecological Researches in India, 473-488, 2006
Edited by A.K. Kandya & Asha Gupta
Published by Bishen Singh Mahendra Pal Singh, Dehra Dun, India

Triazoles : A tool for Ecological Research

V. P. Singh
Institute of Environment management and Plant Sciences
Vikram University, Ujjain-456010

Abstract: Several triazole derivatives were developed and recommended for use worldwide as either fungicide, herbicide, or plant growth regulators. Apart from these three applications, they are very much useful for ecological and environmental studies of plants. The same aspects have been discussed in the present paper in detail.

INTRODUCTION

In 1954, the first compound of triazoles- Amitrol was produced and it was introduced in market as herbicide and growth regulator by some American company, but in 1959 this compound was withdrawn from the market because it induced thyroid tumor in rat. After 32 years, in 1986 again American company released another triazoles derivative - Paclobutrazol (PBZ) and the first paper appeared in California Agriculture in 1986 by Gary W. Hickman- " A new growth regulator for green house plants; produces more compact, shorter plants without any side effects. Although before this publication, a group of Canadian and Japanese scientists was working on derivatives of triazoles as plant protectants

In 1981, this compound was introduced basically as fungicide, diverse with respect to chemicals structure and spectrum of biological activity. Almost all of these chemicals exhibit varying degree of systemicity and they have controlled wide range of fungal diseases in plants and animals. Commercial products have both fungitoxic and plant growth-regulating properties, irrespective of whether they were released for one use or the other.

Triazoles belong to class of compounds known as the ergosterol biosynthesis inhibitors which constitute the largest and most important group of systemic compound which contain a compound which inhibit GA_3 biosynthesis such as Paclobutrazol (PBZ) and S 3307. They have been released as plant growth retardant and their chemical structure is similar to those of the fungicides-triadmefon (TRIADF), triadimenol (TRIADN) and S-3308. From a comparative study of these compound (Fletcher and Hofstra, 1985), found that triazoles have been characterized as either herbicides or plant growth regulators (PGR) in varying degree they exhibit both biological properties.

Table-1: Some Important triazole derivatives

S.N.	Name	Chemical Name
01	BAS III. W	1-phenoxy-3(1H-1,2,4-triazol-1-yl) 4-hydroxy-5, 5-dimcthyl hexane
02	Bitertanol	β-[(1,1-Biphenyl)-4-yloxy]-α - (1, 1-dimethyl)-1H-1,2,4-triazole 1-ethanol
03	Fluotrimazole	1-(3-Trifluro methyl triphenyl) 1,2,4-triazole
04	LAB 150978	1-(4trifluromethyl)-2-(1,2,4-trizolyl-[1]-3-5-methyl-1,3-dioxyan-5-yl) propen-3-ol
05	Paclobutrazol pp 333	(2RS, 3RS)-1-(4-chlorophenyl)-4, 4-dimethyl-2-(1,2, 4-triazol-1-yl) Pentan 3-ol
06	Propiconazole	(±)-1-[2-(2,4-dichlorophenyl)-4-prophy-1,3-dioxolan-2-ylmethyl]-1, 2,4-triazole
07	Tetraconazole	[M-14360, (±)-2-(2,4-dichlorophenyl)-3(1H-1,2, 4-triazol-1-yl) prophy 1,1,2,2-tetrafluoroethylether]
08	Triapenthanol	(E)-(±)-8-(cyclohexyl-methylene)-α - (1,1-di-methyl-ethyl)-1H-1,2,4-triazole-1-ethanol
09	Triadimefon	1-(4-chlorophenoxy)-3,3-dimethyl-1(1H,2,4-triazol-1-yl)-2-butanone
10	Triadimenol	β-(4-chlorophenoxy-)-α- (1,1-dimethyl ethyle)-1H-1, 2,4-triazole-1-ethanol
11	Uniconazole	(E)-1-(p-chlorophenyl)-4, 4-dimethyl-2-(1,2,4- triazol-1-yl) pentan-3-ol

The plant growth regulating properties of sterol inhibiting fungicides has been recently review by Fletcher *et al.* (2000). TRIAD in addition to its established fungicidal properties also protects plant from injury due to :

a. Drought

b. Chilling

Table -2. Some common names of triazoles and countries of origin.

Trade Name	Common name	Country of Origin
Bayleton	Triadimefon	Germany
Baytan	Triadimenol	Germany
Sumagic	XE-1019	Japan
Spotless	XE-779	Japan
Clipper	Paclobutrazol	England
Vigil	Diclobutrazol	England
Tilt	Propiconazole	Switzerland
Sonax	Etaconazole	Switzerland
Nuster	Flusilazole	USA
BAS III		Germany
LAB 150978		Germany

c. Air pollutants- SO2

d. Ozone

e. UV-radiation

f. Heavy metals (Singh, 1993; Thomas and Singh, 1997 ; Purohit and Singh, 2000)

g. Salinity (Fletcher, 1986 ; Arya, *et al.* 1995).

Mode of action:

Fletcher and Hofstra (1988) and Fletcher et al (2000) have reviewed the specific mechanisms of Inhibition of sterol biosynthesis by the group of compounds under discussion. The primary site of interference by these inhibitors occurs in the sterol biosynthesis pathways and the accumulation of sterol intemedials. This is general agreement that the triazoles are inhibitors of C-14 demethylation reactions (Fletcher, 1985). The demethylation of the ergosterol precursor at the C-14 position of 24-methylenedihydrolanosterol proceeds via several oxidation steps catalyzed by a cytochrome-P450 mixed function oxygenase (Koller, 1987). The relative activity of enantiometric forms of TRIAD fungicides has been correlated to their affinity with this target enzyme. In plants, triazole compounds interfere with the biosynthesis of gibberellin by inhibiting the oxidation of ent-Kauerne to ent-Kaurenoic acid. The isoprenoid pathways contain polymers derived from a basic 5-carbon isoprene unit. The pathway generates animal, fungal insect and plant hormones as well as other

(1) Triadimefon

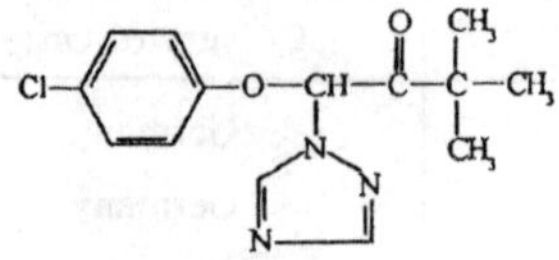

(2) Triadimenol

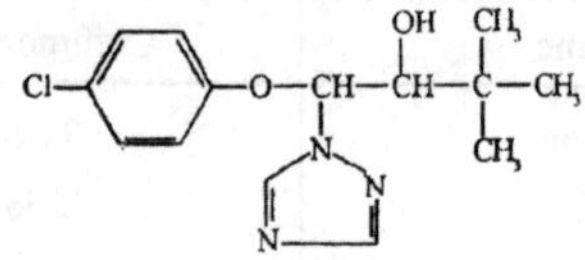

(3) Bitertanol

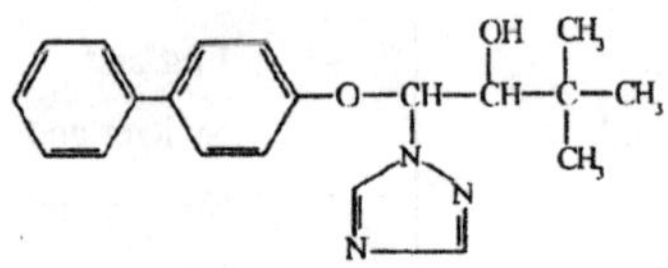

(4) Diclobutrazol

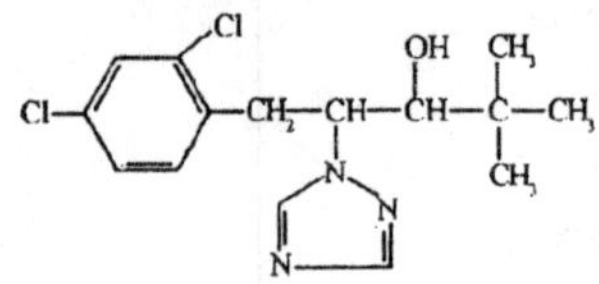

(5) Propiconazol

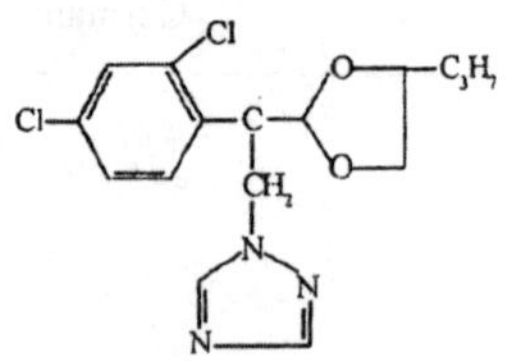

(6) Etaconazol

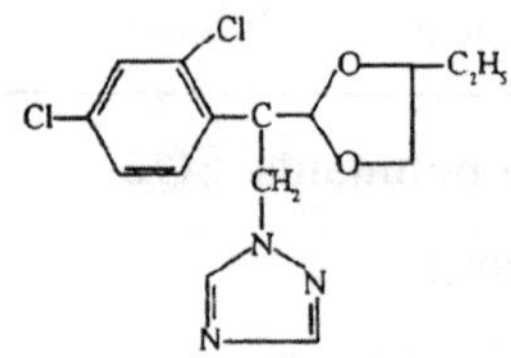

(7) Fluotrimazole

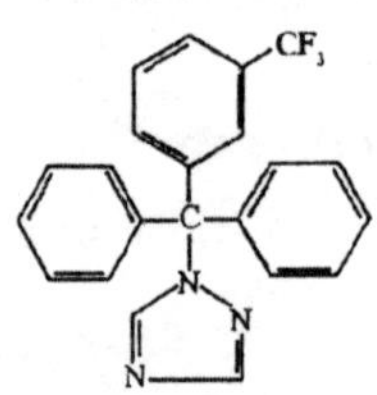

(8) Paclobutrazol

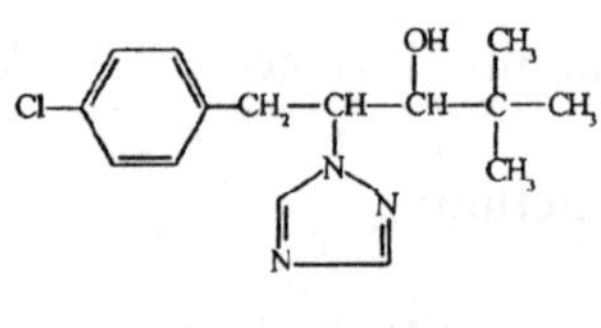

(9) S-3307

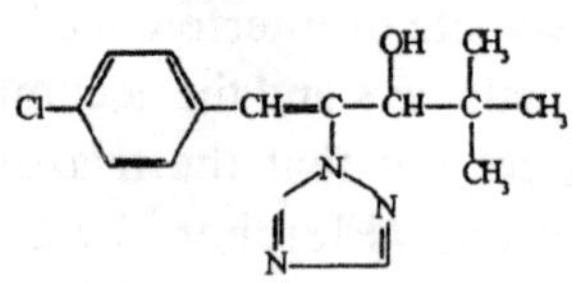

(10) S-3308

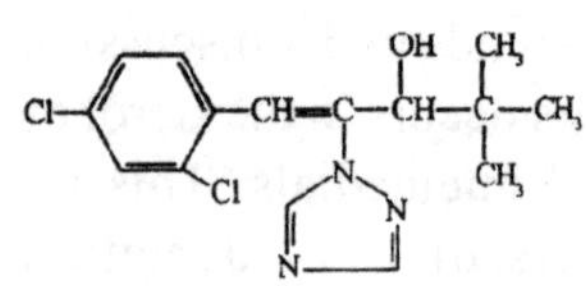

Source: Fletcher (1985) and Fletcher and Singh (1989)

Fig. 1. Chemical structures of some triazole growth regulators

metabolites including Vit. A and E, phytoalexins, steroids, allelopathic compounds and insectanti-feedants (Fletcher, 1985; Goodwin and Mercer, 1990). Specific products of the isoprenoid pathway that are inhibited by the triazoles are ergosterol in fungi and gibberellins in plants.

It is not surprising therefore; the triazoles have a broad spectrum of activity. In plants, ABA, which has been implicated in the hardening process, and cytokinins, which delay senescence, is increased. By contrast, levels of ethylene, a stress hormone, are decreased. Triazoles increase antioxidants such as cartooned, Vit. C and E and reduce damage caused by several stresses including parquet, an active oxygen generating herbicide. Increased in Cytokinins (Fletcher and Arnold, 1986) tocopherol (Mackay et al, 1987) and a transient rise in ABA (Asare-Boamah et al, 1986) have been observed after triazoles treatment.

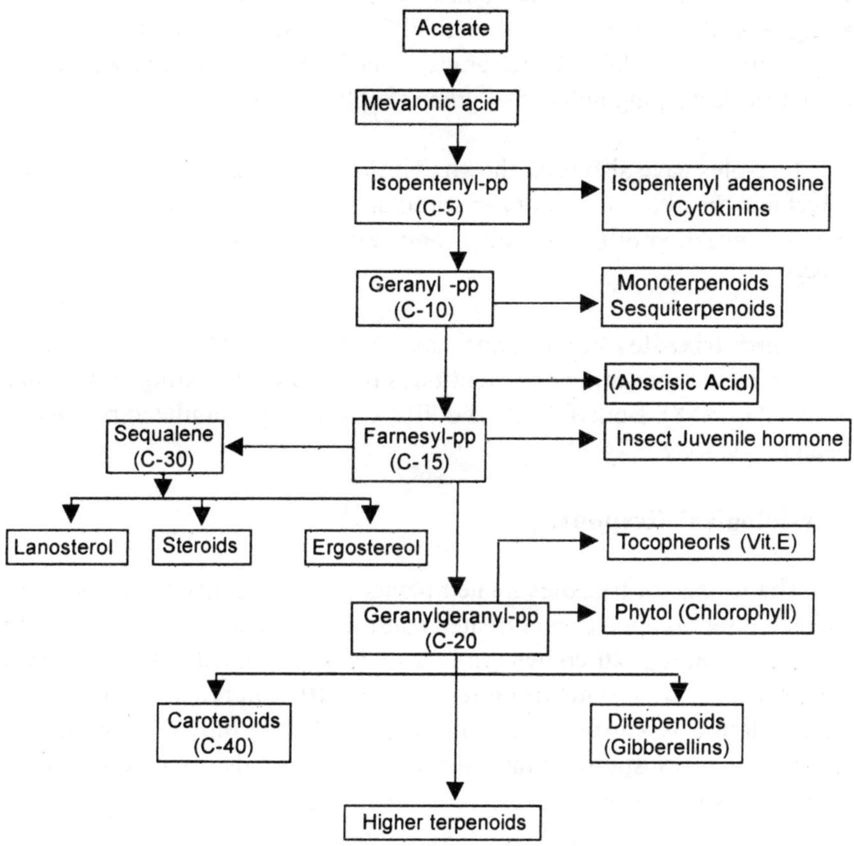

Fig. 2. The isoprenoid pathways

General Plant Response to triazoles:

Triazoles are more potent than most other growth retardant and relatively low rates are required to inhibit shoot growth. However, even at high application rates they generally are not phytotoxic. The most pronounced effect of triazoles on plants is a reduction in height, with the treated plants being greener and more compact. Reduction in shoot growth by triazoles occurs primarily as consequences of reduced internodes.

Triazoles have several morphological effects on leaves. They reduce leaf area, but increase epicuticular wax, width and thickness (Gao et al, 1987). Treated leaves also tend to be darker green, possibly because of an increase in chlorophyll content per unit leaf area (Fletcher and Arnold 1986). Both TRIAD and Paclobutrazol (PBZ) increased chloroplast size along both the long and short axes. PBZ increased the stomatal lamella and number of thylakoids per granum stack, thereby increasing the CS area. But decreasing number of grana (Sopher *et al.* 1999).

Triazoles have also been shown to influence root growth, although the effect may be either inhibitory or stimulatory, depending on the plant and the concentration of the triazoles compound used (Fletcher and Hofstra, 1988).

Some triazoles compounds have been found to be effective in promoting the formation of adventitious roots in plant cutting (Davis and Sankhala, 1988). Both TRIAD and PBZ effectively stimulated rooting in bean hypocotyle.

Physiological Response:

The foliage of triazoles treated plants typically exhibits intense dark green colour compared to untreated controls. In most cases this is due to enhanced chlorophyll content and more densely packed chloroplast in a smaller leaf area. The triazoles also greatly increase anthocyanin accumulation in carrot tissue cultures. Depending on the species, leaf conductance, transpiration rate, and water use may also be modulated the treatment with triazoles.

Triazoles as plant multiprotectant:

A seed treatment that incorporates triazoles by soaking was developed and found to be more effective than conventional treatment. Compared to controls, treated plants were short compact with thicker, greener leaves, more epicuticular wax, larger chloroplast stimulated root growth. Some of these morphological changes may partly account for the stress protective properties of triazoles. Soon after introduction of TRIADF and TRAIDN as fungicides, it was recognised that they had PGR properties.

1. Inhibition of shoot growth without the change of number of nodes.
2. Maintained or slightly promoted root growth, main root often thicker.

Apart from morphological effects, a whole series of physiological alternations have been reported.

1. Retardation of senescence with enhanced concentration of chlorophyll, protein and mineral elements in the plant tissue (Sankhala *et al.* 1985 and Kraus et al, 1991).
2. Improved resistance to environmental stress conditions like, Chilling, drought, heat stress, water logging, air pollutants, UV-radiation heavy metals and salinity.
3. Reduction in water consumption.
4. Reduced the transpiration rate.
5. Altered nutrient uptake from soil.
6. Promoted flowering modified sex expressions.
7. Stimulated translocation of assimilates to seeds

Because of these properties this group of PGR's have found wide spread use in agriculture and horticulture in:

1. Improving lodging resistance and canopy structure of crop plants.
2. Reducing the vegetative growth in the favour of the generative (Pea nuts, cotton, Orchard trees. Ornamental plants).
3. Controlling the growth of trees, bushes and hedges to save trimming cost and,
4. Improving the quality of seedlings for mechanical transpiration.

Triazoles *Vis a Vis* antioxidant

A variety of toxic oxygen species (e.g. Superoxide, hydrogen peroxide hydroxyl radicals, etc.) are produced in plants exposed to environmental stress. The toxicity of active oxygen lies in fact that it can react with numerous cell components thereby causing a cascade of oxidative reactions resulting in the inactivation of enzyme, lipid peroxidation, DNA strand breaks and pigment bleaching. Plant has evolved mechanisms to scavenge these active oxygen species and thereby protect various cellular constituents.

These antioxidant defenses fall into three classes:

1. Lipid soluble, membrane associated antioxidants (Alpha tocopherol and carotenoid).
2. Water soluble reductants (Ascorbate and Glutathione)
3. Enzymatic antioxidants (Peroxidase and Catalase, Glutathione reductase, Superoxide dismutase).

From the preceding discussion it seems compelling to conclude that at biochemical level, the protection offered by triazoles depends heavily on the modulation of the activity and levels of antioxidants.

To sum up. Triazoles seem to induce increased stress resistance by:

1. Increasing photosynthetic efficiency, thereby reducing the generation of free radicals.
2. Increasing the levels of antioxidant involved in both the prevention of free radical generation.
3. Dissipating excess light energy *via* Xanthophyll cycle.
4. Stabilizing the membrane structure even under stress conditions and
5. Increasing the root/Shoot ratio and more effective transpirational cooling.

1. Triazoles as an Environmental stress Protectants:

From previous discussion it has been seen that the triazoles have been used extensively in protecting plants from various stresses. Various research finding in India and abroad strongly support the effectiveness of triazoles in protecting plants from various and diverse environmental stresses including drought, heat, chilling, UV-radiation, air-pollutants, Ozone, water logging, heavy metals and salinity.

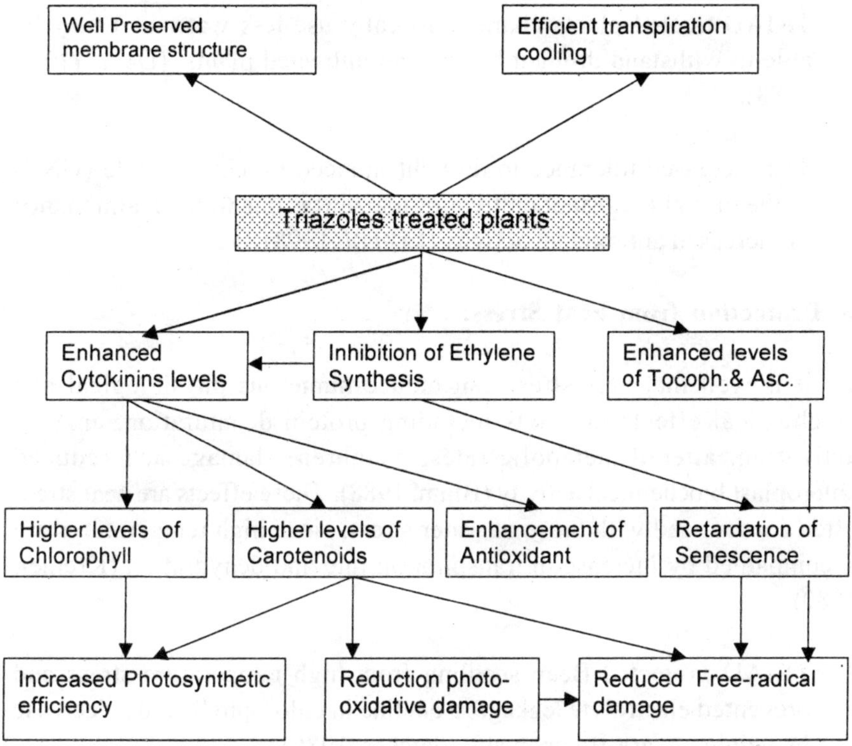

Fig. 3. Mode and mechanism of triazole mediated stress

A. Protection against drought stress: When plants are subjected to drought stress, the treated plants shows, there is an inhibition of growth, increase in ABA levels, stomatal closure and senescence of the lower leaves.

1. TRIAD treatment is known to improve the survival of plants during periods of drought (Fletcher et al 1988). Previous studies with TRIAD showed that it reduced transpiration in soybean, radish and Pea and increased yield under water stress.

2. Reduction of transpiration and protection from drought in bean by TRIAD was associated with reduction in shoot weight, length, leaf area and increased diffusion resistance, indicating partial closure of stomates, and transient rise in ABA levels (Asare-Boamah *et al.*, 1986).

3. TRIAD treated plants characteristically use less water and may be able to withstand drought batter than untreated plants (Davis *et al.*, 1988).

4. The increased tolerance to drought induced by Uniconazole (UNZ) in the drought resistant cultivar of maize was due to the maintenance of increased antioxidant activity (Li *et al.* 1998).

B. Protection from heat Stress:

It is well that heat stress can induce numerous physiological and biochemical effects in plants including protein denaturation, enzyme activation, altered metabolic rates, membrane damage and reduced chloroplast biochemical activity (Blum, 1988). These effects are heat stress often confounded with those of water stress, since high temperatures are accompanied by increase in transpiration rate and dehydration (Kremer, 1980).

1. TRIAD protected Bean seedling from high temperature stress and prevented electrolyte leakage, a decline in chlorophyll, and a decrease in weight (Asare-Boamah and Fletcher, 1986).
2. UNZ induced thermo-tolerance in wheat seedling in part; to increase transpiration rate of treated pants, thereby resulting in lower leaf temperature (Booker *et al.* 1991).
3. One of the major sites of heat stress damage is the chloroplast and therefore, reduced leaf temperature may be important in maintains a functional photosynthetic system.
4. Wheat seedling treated with PBZ was found to have higher levels of antioxidant enzymes.
5. Evidence also indicate that protection of wheat seedling from high temperature damage by triazoles compounds may be related to increase in the activities of important antioxidant enzymes such as SOD, Ascorbate peroxidase, glutathione reductase, peroxidase and catalase.
6. Thus triazole-induced protection from heat stress may be related to a more efficient free radical scavenging system.

C. Protection from chilling:

There are some common features observed due to chilling injury such as alternations in membrane properties, malfunction of mitochondrial respiration and inhibition of hill reaction in the chloroplast. Chilling induced membrane injury has been linked to oxidation events like generation of SOD radical's singlet oxygen, H_2O_2 and hydroxyl radicals.

1. Many research findings support the delaying of chilling injury by triazoles. (Whitaker and Wang, 1987; Senaratna *et al.* 1988 ;Pinhero and Fletcher, 1994).
2. Protection against chilling injury by triazoles is attributed to an inhibition of chilling induced degradation of membrane lipid.
3. Another suggestion is given which is based on the increment of the level of several free radicals scavenging system like – tocopherol and ascorbic acid in triazoles treated plants.

D. Protection from UV-radiation:

Some of the deleterious effects of UV-B radiation on plants include DNA damage, lipid peroxidation, and inhibition of photosynthesis of pigments.

1. Several mechanisms for tolerance to UV-B have been proposed, including enhanced synthesis of UV-B absorbing pigments, leaf thickening, increased synthesis of cuticular wax and reduced leaf expansion. All of these strategies reduce the UV-B quanta intercepted by leaf mesophyll tissues.
2. Enhanced free radicals scavenging capacity by triazoles to control oxidative damage incurred from UV-B radiation was also investigated by Kraus and Fletcher (1994) and Abbas and Zaidi (1997).

E. Protection from Ozone stress:

The common symptoms injury are the leaf chlorosis, necrotic lesions, solute leakage, reduced hill activity, a decrease in phospholipids and a concomitant increase in free fatty acids indicative of free radicals damage.

1. Some studies have been shown that the triazoles like PBZ UNZ TRIAD protect plants from injury due to ozone toxicity.

2. In 1987, Mackay *et al.* reported in their experiment that no chlorosis or lesions were seen in the S-3307 treated leaves after the exposure to the same dosage as given to non-treated plants.
3. In this finding, solute was also found to increase with ozone treatment in non-treated plants.

F. Protecting from water logging :

The systems of water logging injury includes stomatal closure and clorosis of leaves, decrease stem elongation with an increase in stem thickness, wilting epinasty and leaf abscission, decreased root elongation with increased ethylene production of adventitious roots and increased susceptibility to pathogens. It was reported that PBZ protected wheat seedlings from injury due to water logging. All these symptoms of damage were reduced by the PBZ treatment at three different growth stages (Webb and Fletcher, 1996).

G. Protection from Salinity stress:

Salinity effects many aspects of plants metabolism and accumulation of various organic solutes. The normal biosynthetic machinery of the plant is disturbed under saline conditions. It has been reported that adverse effects of salinity on plants could be to some extent, overcome to treating plants with triazoles derivatives (Trivedi *et al.* 1988, Purohit and Singh, 2000).

Purohit and Singh (1999) reported that sodium chloride stress decreased protein content, peroxidases and SOD activity in Bhindi in contrast to TRIAD treatment increased protein synthesis, antioxidant enzyme activities and lowered ATP breakdown and thereby ameliorated the NaCl stress in Bhindi seedlings.

H. Protection from heavy metals:

This work was conducted in my laboratory under many research projects sponsored by ICAR and UGC. As it has been widely reported by various workers that heavy metal availability induces a series of biochemical and physiological alternations in plants. These alternations present some common characteristics like- membrane damage, alternations of enzyme activities, inhibition of root growth, disturbances of hormonal balance, deficiency of essential nutrients, inhibition of rate of photosynthesis and respiration, changes in photo-assimilates, translocation and alternations in water relations. Our finding support that triazoles treated crop plants are able to tolerate heavy metal toxicity better than their non-treated counterparts. (Singh, 1993; Thomas and Singh, 1995,1996; Purohit and Singh, 1998, 1999, 2000).

1. In 1999, Purohit and Singh have reported that UNZ treatment resulted in significantly higher chlorophyll contents and hill activity in all cadmium + UNZ treated plants.
2. In other experiment, where plants were treated with Cd, Zn and Pb, in this study UNZ considerably reduced the inhibition of chlorophyll synthesis induced by heavy metals.
3. UNZ also increased the free proline level, enhanced the cytokinin levels and accumulates the ABA.
4. In our laboratory work, the most important effect of triazole treatment was observed that plant treated with triazoles has less uptake and accumulation of heavy metals than non-treated.
5. Though the reason for reduced accumulation could not be known, but some speculation can be given to decline the metal accumulation in triazoles treated plants. A surplus of free heavy metal in the cytosol induces the synthesis of phytochelatins by activating the enzyme phytochelatin syntheses.

There are three possibilities to protect the plants from heavy metals:

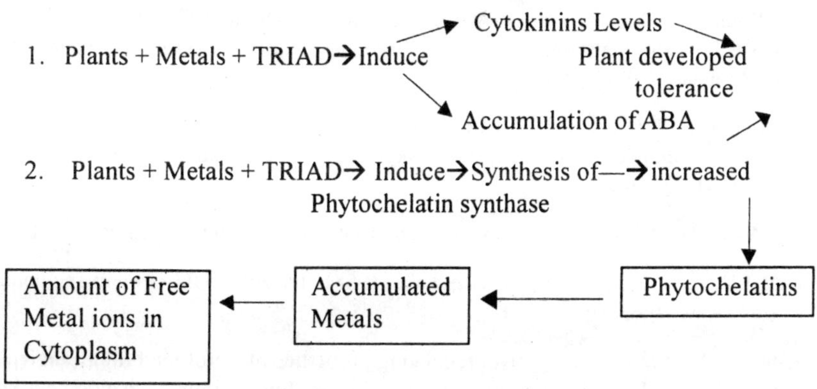

(Y-Glu-Cys) units are produced by plants on exposure to heavy metals.

3. The 3rd possibility: Many anatomical changes like thickness of mesophyll tissue changes in number of vascular bundles stomatal aperture, etc. have been described in the triazoles treated plants.

It may be the reason that treated plants may have more effective internal. System to restrict metal translocation from root system to shoot, for example the endodermis may be more effective in preventing this support. This speculation could be suggested due to our some significant findings in which percentage translocation of heavy metals was evaluated to be comparatively lower than non-treated.

ACKNOWLEDGEMENT

Author is thankful to ICAR, New Delhi for granting projects and to Prof. RA Fletcher, Department of Environmental Biology, University of Guelph, Ontario, Canada to give opportunity to work under CIDA/NSERC research associateship.

REFERENCES

Abbas, S. and P.H. Zaidi (1997). Chemical manipulation through triadimefon against UV-B induced injury to membrane components of *Cicer arientinum* L. Plant Physiol. Biochem. 24:64-66.

Arya, N., R. Thomas and V.P. Singh (1995). Protection of *Fagopyrum esculentum* Moench. From different salinity gradient by triadimefon. VI International sysmposium on Buckwheat, Shinshu University, Ina Nagana Japan. 345-347.

Asare-Boamah, N.K. and R.A. Fletcher (1986). Protection of bean seedling against heat and chilling injury by triadimefon. Physiol. Plant. **67**:353-358.

Asare-Boamah, N.K., Ghofstra, R.A. Fletcher and E.B. Dumbroff (1986). Triadimefon protects bean plants from water stress through its effects on Abscisic Acid. Plant Cell Physiol.**27:** 383-390.

Booker, H.M., J.J. Gillespie, G. Hofstra and R.A. Fletcher (1991). Uniconazole induced thermo-tolerance in wheat seedlings mediated by transpirational cooling. Physiol. Plant. **81**:335-342.

Blum, A. (1988). Plant Breeding for stress environment. CPC Press Boca. Raton FL.

Davis, T.D., G.L. Steffens and N. Sankhala (1988). Triazole plant Growth regulators. Hort. Rev. **10:63-103**.

Fletcher, R.A. (1985). Plant growth regulating properties of sterol inhibiting fungicides pp. 103-113. In SS Purohit hormonal regulation of plant growth development Vol-II Agrobotanical Publishers. Bikaner, India.

Fletcher, R.A. and V. Arnold (1986). Stimulation of Cytokinins and chlorophyll synthesis in Cucumber cotyledons by triadimefon. Physiol. Plant. **66:**197-200.

Fletcher, R.A. and G, Hofstra (1985). Triadimefon - a plant multiprotectants. Plant Cell Physiol. **26:**775-780.

Fletcher, R.A. and G, Hofstra (1988). Triazole as potential Plant protects, pp. 321-331. In D. Berg and M. Plempel (Eds) Sterol synthesis inhibitors in plant protection. Ellis Howard Ltd. Cambridge. UK.

Fletcher, R.A., A. Gilley, N. Sankhala and T.D. Davis (2000). Horticulture Review. **24:**56-137.

Fletcher, R.A., N.K. Asare- Boamah, L.C. Kreig, G. Hofstra, E.B. Dumbroff (1988). Triadimefon stimulates rooting in bean hypocotyle. Physiol. Plant. **73:**401-405.

Gao J, G. Hofstra and R.A. Fletcher (1987). Anatomical changes induced by triazole to wheat seedlings. Can. J. Bot. **66:**1178-1185.

Goodwin, T.W. and E.I. Mercer (1990). Introduction to plant biochemistry. 2nd Pergaman Press. Oxford UK.

Hickman, G.W. (1986). A New growth regulators for green house plants. California Agriculture. **73:**46-48.

Koller, W. (1987). Isomers of sterol synthesis inhibitors fungicidal effects and plant growth regulators activities. Pesticides Sci. **18:**120-147.

Kraus, T.E. and R.A. Fletcher (1994). Paclobutrazol protects wheat seedling from heat and paraquat injury: is de-toxification of active oxygen involved. Plant Cell Physiol. **35:**45-52.

Kraus, T.E. and D.P. Murr and R.A. Fletcher (1991). Uniconazole inhibits stress-induced ethylene in wheat and soybean seedlings. J. Plant Growth Regulators **10:**229-234.

Kremer, W.P.J. (1980). Drought stress and the origin of adaptations pp.7-20. In NC turner and PJ Kremer (Eds) Adaptation of plant to water and high temperature stress. Willey Tornato,

Li I, J. Van Staden and A.K. Jager (1998). Effects of plant growth regulatior on antioxidant system in seedling of two maize cultivars subjected to water stress. Plant growth regulators. **25:**81-87.

Mackay, C.E., T. Senaratna, B.D. McKersie and R.A. Fletcher (1987). Ozone induced injury to cellular membrane in *Triticum aestivum* L. and protection by triazoles S-3307. Plant Cell Physiol. **28**:1271-1278.

Pinhero, R.G and R.A. Fletcher (1994). Paclobutrazol and ancymidol protect corn seedlings from high and low temperature stress. J. Plant Growth regulators. **15:**47-53.

Purohit, S. and V.P. Singh (1998). Induced tolerance by uniconazole against cadmium phytotoxicity in *Abelmoschus esculentum* L. International J. Pollution Research. **4:**45-57.

Purohit, S. and V.P. Singh (1999). Uniconazole induced protection of *Abelmoschus esculentum* L. against Cadmium stress. Photosynthetica. **34:**116-119.

Purohit, S. and V.P. Singh (2000). Uniconazole induced changes in chlorophyll and free proline contents of metal stressed in *Abelmoschus esculentum* L. Asian J. Micro. Biotech. & Env. Sci. 179-181.

Sankhala, N, T.D. Davis, A. Updhyaya, D. Sankhala, P.H. Weser and B.N. Smith (1985). Growth and Metabolism of Soybean as affected by Paclobutrazol. Plant Cell Physiol. **26:**913-921.

Senaratna, T.R., C.E. Mackay, B.D. McKersie and R.A. Fletcher (1988). Uniconazole induced chilling tolerance in tomato and its relationship to antioxidant content. L. Plant Phsiol. **133:** 56-61.

Singh, V.P. (1993). Uniconazole (S-3307) induced cadmium tolerance in wheat. J. Plant Growth Regul. **12:**1-3.

Singh, V.P. and R. Thomas (1997). Effect of Cd, Mn and Zn on the vivo NRA in triazole treated *Cucumis sativus* L. seedling. Indian J. Plant Physiol. **40:**124-128.

Sopher, R.C., M. Krol and R.A. Fletcher (1999). Chloroplastic changes associated with Paclobutrazol induced stress protection in maize seedlings. Can. J. Bot. **77:**1-12.

Trivedi, S., D. Sankhala, N. Sankhala, U. Upadhya, D.J. Weber and B.N. Smith (1988). Effect of Paclobutrazol and salt on growth and lipid constituents of moth bean seedlings. Proc. Plant Growth Reg. Soc. Am. **15:**39-46.

Webb and R.A. Fletcher (1996). Paclobutrazol protects wheat seedlings from injury due to water-logging. Plant Growth Regul. **18:**201-206.

Whitaker BD and CY Wang (1997). Effect of paclobutrazol and chilling on leaf membrane lipids in *Cucumber* seedlings. Physiol. Plant. **70:**404-411.

Thomas, R. and V.P. Singh (1995). Effects of the triazole derivatives on mercury-induced inhibition of chlorophyll and carotenoid accumulation in *Cucumber* cotyledons. Indian J. Plant Physiol. **38:**313-316.

Thomas, R. and V.P. Singh (1996). Reduction of Cadmium induced inhibition of chlorophyll and carotenoid accumulation in *Cucumis sativus* L by Uniconazole (S-3307). .Phtosynthetica. **32:**145-148.

Advancing Frontiers of Ecological Researches in India, 489-493, 2006
Edited by A.K. Kandya & Asha Gupta
Published by Bishen Singh Mahendra Pal Singh, Dehra Dun, India

Environmental Conservation: Important Consideration

R.K. Jain and R.K. Srivastava
P.G. Deptt. of Botany and Environmental Science,
Govt. Model Science College (Autonomous) Jabalpur (M.P.)
(Center of Excellence, NAAC Accredited 'A' Grade)

Abstract : In the present paper, several important aspects relating to the sustainability of environmental protection have been discussed in perspective of Agenda-21 of United Nation's Environmental Development Program.

INTRODUCTION

Among all the nine planets in the universe, only the earth is bestowed with benevolent environment. The earth environment with biotic and abiotic components, provides resources for development and survival of microorganism and assimilates the residue produced during the course of its activities. However, in the context of development, it should be understood that the earth environment has got the limited capacity for supporting resources and assimilating wastes (Emlen 1972).

Nicholas Polunin, the noted environmentalist has said that **"enlightened understanding by a human being of his or her environment is prerequisite to saving it."** In India, we do have definite Environment Conservation ingrained in our history, culture, religion and philosophy. It is a separate matter that we forgot it during the colonial rule, and, like them, became exploiters rather than being user of environment.

In the initial years of economic development, the focus was to improve the standard of living by increasing production, employment and more resource allocations. Less attention was paid to environmental pollution, as environmental impact were not apparent and resource utilization was within the repleneshible and assimilative capacity of the earth. However, with exorbitant increase of population, the production of goods and service has increased manifold to meet the minimum needs of people, thus, the resource utilization has increased and assimilative capacity of earth has come down.

The present planning process allocate the monetary resources to difficult sectors to increase the production levels, as it is behind to enhance the quality of life. However, the ecological resources and supportive capacity of the environment does not find consideration in the process. Hence, there is a need to review the process of developmental planning. While the concept is not new, reported efforts in its application to developmental planning in India are few (Khanna, *et al.* 1992).

Some peoples may find it ironic, though others will see it as merely logical, that man is the earth's only problem and yet offers the only possible solution. There is, of course, no doubt that man's capacity for altering the biosphere is greater than his ability to control the changes he has brought about. There is hardly a spot on earth where man has not left any mark, and the landscape in many areas has changed beyond recognition. The rapidly increasing human population is the biggest threat to biodiversity, that the earth has even faced. International conventions, world commissions and UN agencies are created assuming that a rational, top down process will bring us closer in achieving the objectives derived from the wisdom of the planners and analysis. But the actual development and process nearly always seen to follow different rules.

"Sustainable Development", according to the World Commission On Environment and Development (1981-88) is **"development that meets the needs of the present without compromising the ability of future generation to meet their own needs".** The need of the present means the need of the world's poor to which over riding priority should be accorded. Basically when we talk of sustainable development, the easiest definition is that, we the present generation have inherited a certain amount of ecology and environment in terms of land, water and air, when we leave it to next generation, we should leave it at least in same condition, if not in a better condition than that we inherited.

Agenda 21 of the **United Nation Conference on Environment and Development** held at the **Rio de Janeiro, Brazil in Jan. 1992,** trained the focus on population impact on sustainable development. Growth of world population and production combined with industrial consumption pattern placed increasingly severe stresses on the life supportive capacity of our planet. The world population in expected to scale 8 billion by 2020.

It goes without saying that the unchecked demographic growth means pressure on all natural and material resources with inescapable impact on sustainable development. Besides, there will be sharp dilution of all our economic and social gain. The results are too apparent –

- Over 300 million constituting forty percent of our population are below the poverty line.
- About 1.5 lakh villager out of 5.5 lakhs are without safe drinking water.
- Our literacy is only 50% while that of women is only 39%.
- One out of every five household has no roof over it's head.

Ecology and economy are inextricably intertwined and the pressure on the economy exerted by an exploding population in a developing country like India, cannot but take a heavy toll on the environment. The poor depend on natural resources for their day to day existence. They invade forests for fuel, resort to cultivation in erosion prone areas, overuse some of the exhaustible and non-renewable resources. As noted by the world commission, "but poverty itself pollute the environment creating environmental stress in a different way. Those who are poor and hungry will often destroy their immediate environment in order to survive. They cut down forests, their livestock will overgraze grasslands, they will over use marginal land, and in growing numbers, they will crowd in to congested cities. The cumulative effect of these changes so far reaching as to make the poverty itself a major global problem (Narayanan 1999).

The poor and faminised are understandably more concerned about today than tomorrow, they are too much locked in contact with the demand of the present to care for the posterity of tomorrow. It is paradoxical that while the poor are the experts for environmental damages, they themselves are also the victims of environmental degradation perpetuated partly by them and mostly by others.

There is a growing realization that the development processes in India in last several years have not been ecofriendly. Substance is said to be built on the continued capacity of nature to renew its forests, fields and rivers. These resource systems are intrinsically linked in life producing and life conserving culture.

Environmental damage can also be construed as a resource economics problem in which a combination of availability, accessibility, and renewability of available natural resource determine the degree of environmental protection that is needed. Emphasis here is on the instrumental value of the environment as a source of natural resources for human consumption.

Externalities, unsustainable consumption and a failure to distinguish man made capital from natural capital are fundamental to the definition of damage employed in the perspective. Those who stress this definition are not blind to the contribution of the contamination and eco-simplification perspective of environmental change, but their preference is to relate to nature as consumers and to reduce harm to the environment by regulating consumptions of scarce natural resource.

The destructive impact of consumption involve pollution or loss of complexity only indirectly, the key factor is the instrumental value placed on non-renewable or slowly regenerating natural resources that can be mixed, milked, pumped, sowed or uprooted, hunted fished, collected or otherwise consumed for human benefit. Whereas environmental pollution and losses in ecological diversity are typically, unintended consequences or side effects of growth and development, in the consumption or exploitation. They are intended objects of destructive behavior forest cleared for timber, oceans over fished for food and a host of other environmental sacrifice that are typically justified or ground of human wants and needs.

Environmental protection, according to the consumption perspective, requires the calculation of maximum sustainable yields for renewable resources and long term conservation strategies for the management of non-renewable resources.

The dawn of 21st century is clouded by unheard of perils to the performance of the natural world. In a globally interconnected economy, fast deforestation, falling water tables, erratic climatic change and environmental pollution could harm economics around the world in the decades to come (Hempel 1998).

It has been known from the vedic times that nature and human kind form an inseparable part of the life support system. This system has five elements as air, water, land, flora and fauna which are interconnected, interrelated and interdependent and have coevolved and are co-adapted. Deterioration in one invariably affect the others. If the deterioration is for a short term and the life support system has enough resilience, it repairs itself and revert to original state. However, if deterioration continue, the whole system including all life is thrown out of gear. The crux of the matter is, after all environment for whom!. The answer is for the living system, of which humankind is an integral part.

One must recognize that the situation has not reached the point of no return and by resorting to appropriate environment management by integrating the economic and environmental system and by creating environmental awareness among the mass of people and by changes in the mindset everywhere, we can choose ecologically compatible path of development.

REFERENCES

Emlen J. M. (1972). Ecology: An evolutionary approach. Wesley Publisher. New York.

Khanna, P. *et al.* (1992). An approach to carrying capacity based developmental planning process, technical paper. NEERI; Nagpur

Agrawal, S.K. (2003). Environmental consumerism in 21st century, in Environmental Scenario for 21st century. APH Publisher New Delhi.

Hempel, L.C. (1998). causes of Environmental destruction, in Environmental Governance the global challenge. Affiliated East West. New Delhi.

Naraynan, A.K. (1999). Sustainable development and population explosion in Environment Education for Sustainable future, ed. D. Bandhu and R. Dayal. Indian Environmental . Society. New Delhi.

Advancing Frontiers of Ecological Researches in India, 495-506, 2006
Edited by A.K. Kandya & Asha Gupta
Published by Bishen Singh Mahendra Pal Singh, Dehra Dun, India

Environmental Auditing

G. P. Prasain & Kh. Dhiren Meetei
Department of Commerce, Manipur University, Imphal.

Abstract: Recently, environmental auditing has come to be seen as a tool for generating such information and for assessing enterprises for the potential environmental that may cause, their environmental liabilities and the degree of compliance with environmental standards and legislation. Users of the environmental audits are the companies, customers, banks and other lending institutions, local and state governments and the general public. Environmental audits help to decrease environmental and public health risks and assist in developing environmental management at the company level. The World Bank increasingly requests that environmental audits be done for certain types of projects. The audits are a part of the overall environmental assessment process and are complement for normal environmental assessment studies, depending upon the type of project.

Till 1980s, companies concerns for environmental protection were primarily limited to avoiding major disasters and regulatory enforcement. This defensive reaction was aimed at fending off environmental costs that were perceived as limiting their competitive advantage. During the last decade, successful companies have taken a more proactive, creative approach to looking at their environmental performance. Many companies are managing their environmental effects in a structured way by incorporating them into their standard business practices.

Conducting an environmental audit is no longer an option but a sound precaution and proactive measure in today's heavily regulated environment. Indeed, evidence suggests that environmental audit has a valuable role to play, encouraging systematic incorporation of environmental perspectives into various aspects of an organization's overall operation, helping to trigger new awareness and new priorities in policies and practices.

INTRODUCTION

The qualities of the environment and structure environmental legislation have led to a demand for reliable environmental information on various industries and other types of business. Recently, environmental auditing has come to be seen as a tool for generating such information and for assessing enterprises for the potential environment that may cause, their environmental liabilities and the degree of compliance with environmental standards and legislation. Users of the environmental audits are the companies, customers, banks and other lending institutions, local and state governments and the general public. Environmental audits help to decrease environmental and public health risks and assist in developing environmental management at the company level. The World Bank increasingly requests that environmental audits be done for certain types of projects. The audits are a part of the overall environmental assessment process and are complement for normal environmental assessment studies, depending upon the type of project.

Till 1980s, companies concerns for environmental protection were primarily limited to avoiding major disasters and regulatory enforcement. This defensive reaction was aimed at fending off environmental costs that were perceived as limiting their competitive advantage. During the last decade, successful companies have taken a more proactive, creative approach to looking at their environmental performance. Many companies are managing their environmental effects in a structured way by incorporating them into their standard business practices.

A key part of any Environmental Management Systems is the environmental audit-the tool which companies use to evaluate their environmental performance.

Conducting an environmental audit is no longer an option but a sound precaution and proactive measure in today's heavily regulated environment. Indeed, evidence suggests that environmental audit has a valuable role to play, encouraging systematic incorporation of environmental perspectives into various aspects of an organization's overall operation, helping to trigger new awareness and new priorities in policies and practices.

What is an environmental audit?

According to ISO 14001 an environmental audit is a systematic and documented verification process of objectively obtaining and evaluating evidence to determine whether an organization's Environmental Management Systems (EMS) conforms with audit criteria set by the organization, and for communicating the results of this process to management. European Eco-audit and Management Scheme (EMAS) define an environmental audit as "a systematic, documented, periodic and objective evaluation of performance of the organization, management system and processes designed to protect the environment".

An environmental audit is a methodical examination of environmental information about an organization, a facility, or a site, to verify whether, or to what extent, they conform to specified audit criteria. The criteria may be based on local, regional, national or international environmental standards, national laws and regulations, permits and concessions, internal management system specifications, corporate standards, or guidelines of organizations such as the World Bank. The reasons for undertaking an audit and the aims to be achieved will determine the audit criteria. An environmental audit is a management tool for taking inventory of a company's environmental assets and liabilities. It provides information on a company's compliance status at a given point of time, analysis of the perceived implications of the information gathered and, if management so chooses, options that arise from this information and analysis. In short, an audit supplies a snapshot of one segment of management's responsibility.

Environmental audits are being used as a tool and an aid to test the effectiveness of environmental efforts at local level. An environmental audit is a systematic, independent internal review to check whether the results of environmental work tally with the targets. It also focuses on whether the methods used to achieve goals are effective. To be more precise the work of an environmental audit is a study of documents and reports to see whether there are any deviations between targets and achievements. Interviewing key people in the organisation does this. An environmental audit will confirm whether or not the environmental targets have been attained.

Thus, an environmental audit is a management tool to systematically, periodically and objectively, evaluate and record the environmental performance of a company or enterprise.

PRINCIPLES OF ENVIRONMENTAL AUDITING

A prerequisite to successful implementation of any aspect or type of environmental audit is the commitment of management to maintain or move toward sound environmental practices. This shows itself through the operations of the company, facility or site, management attitudes to environmental matters, and the level of commitment shown by staff. As a systematic process of getting and evaluating information in connection with the environmental aspects of an operation, an organization or a site, the environmental audit will generally require:

(i) Sufficient and appropriate information about the operation, organization or site;

(ii) Adequate resources available to support the audit process;

(iii) Adequate cooperation from the company and other entity that is being audited; and

(iv) An audit protocol (e.g., a checklist or questionnaire).

An environmental audit is undertaken by an auditor and is based on goals defined by the client, who might also be the auditee. The audit criteria should be agreed upon between the auditors and the client and communicated to the auditee along with the aims and scope of the audit. The auditors should be objective and independent of the site or activity being audited, although they may sometimes be part of the same company. Information collected during the audit should always be treated as confidential. It is assumed that the auditors who audits the activities of the organization will follow systematic procedures or steps, so that a similar audit performed by others auditors would produce consistent results. Various types of audits will use different techniques and different way to get and evaluate information about the subject matter of the audit.

On the basis of audit criteria, an auditor should collect information and documents evidence to determine whether audit criteria have been met. These findings are the basis of the audit report to the client. An audit takes place within a short period of time and with limited resources. It is thus very important to assess the reliability of audit conclusions, and keep the inherent uncertainties in mind when using the audit results.

REASONS TO UNDERTAKE AN ENVIRONMENTAL AUDIT

Environmental audits can be carried out for various reasons. These could include any of the following reasons:

(i) To verify compliance

(ii) To review implementation of various policies and programme.

(iii) To identify the liabilities of the company in connection with the environmental protection.

(iv) To review the management systems of the company.

(v) To assess the environmental performance of the company in connection with the rules of law at the national level or international level.

(vi) To promote environmental awareness among the various machines of the company and the society.

Any company may want to carry out an environmental audit for one or all of the above reasons. A company's motivation for carrying out an environmental audit will determine the type of audit it chooses to implement.

Why auditing became important?

Environmental auditing is not a new one. It started in the 1970s in the USA in response to national liability laws, and was developed further to response to new rules on major hazards. Environmental auditing started in the big chemical companies, but has increasingly been adapted by a range of companies from manufacturing, retail and service industries. As US multinational corporations started auditing abroad as a way to ensuring standards, the practice spread to Europe. A significant number of UK companies have been started audits for many years.

One of the important events in the history of environmental auditing was the Bhopal incident in 1984. Following the accidental emissions of toxic gases at Bhopal, and the subsequent widespread impacts on people and the environment, the implementation of operational controls to prevent a reoccurrence became essential. This brings to the start of the formal management systems within major industrial units as a way of controlling health and safety and environmental problems. The audit process started as a means of monitoring whether the control systems were being adhered to.

In the European context, disasters such as seveso, chernobyl and comlford are further examples of corporate lack of care and the need for management systems to control industrial pollution. This has been reinforced by events at the 1992 Rio Earth Summit.

The value and importance of environmental auditing is reflected in the development of international and European standards. The international standard for Environmental Management Systems ISO 14001: 1996 consists an audit process both as the means to gain certification through an external assessment and as an internal monitoring function within the corporation. Environmental auditing is also integral part to the European Eco-audit and Management Scheme (EMAS).

TYPES OF ENVIRONMENTAL AUDIT

The various types of environmental audit can be discussed as under: -

(i) **Compliance Audits :** Under the audits the legal obligations of an organization are compared with its environmental performance. Compliance audits are intended to identify whether an organization is aware of all relevant legislation, regulations and legal agreements. Auditors should review the procedural documents, authorisations and consents, records and monitoring equipment. Auditors will interview appropriate staff and make observations from site works to verify legal compliance. Occasionally an auditor will compare findings with the regulator.

(ii) **Environmental Management Audits:** Environmental management audits test whether the environmental policy is "fit for purpose" and appropriate to the activities, products and services of the organization. They can confirm whether the company is implementing the commitments given in the environmental policy and programme and is meeting its environmental objectives and targets.

(iii) **Pre-acquisition Audits:** These audits are typically carried out by, or on behalf of, a prospective buyer or seller of industrial land (particularly mergers, acquisitions and divestitures). The liabilities that can be found out by audit include examples of bad practices, toxic tips, stores of hazardous substances and contaminated land or groundwater. This can lead to reduced property values, exclusion from or cancellation of the transaction, high insurance premiums, problems with bank loans, etc.

(iv) **Supplier Audits :** This audits review the environmental practices and permits of an organization's own purchasing requirements and policy commitments. This ensures that suppliers maintain environmental records and adopt controls to match the organization's own performance standards.

(v) **Environmental Performance Audits :** These audits are systematic review of progress and achievements in improving environmental performance. For example, an audit of a waste minimisation programme would examine waste generated by a site, identity viable actions for reductions, reuse and recycle, and make recommendations for the environmental management programme.

(vi) **Single Issue Audits :** These involve a detailed examination of a particular issue such as a noise emission, an unpleasant odour, use of paper, transport, etc. For example a noise audit would identify all noise sources, establish statutory policy standards and breaches, and verify corrective action and prevention programmes.

(vii) **Hazard Identification Audits :** This focuses on operations that handle hazardous materials and substances. This type of audit is often carried out by insurance business or loss adjusters and covers the environmental risks and implications of accidental releases.

(viii) **Environmental Statement Audits :** This audit involved the verification that the environmental statement (the publicly available report) is a true reflection of current site conditions. The audits cover all Eco-management and Audit Scheme requirements and must be carried out by an external verifier.

WAYS OF CONDUCTING AN AUDIT

All audits are designed to determine actions against intent. The way the audit is conducted depends on the needs of the instructing party. There are broadly two ways of conducting an audit. They are:

(i) **Verification Audits:** It is a strict assessment of actual conditions against requirements, with minimal discussion about remedial actions, typically carried out for :

* ISO 14001 (final assessment)
* EMAS (Verification audit)
* Due diligence prior to the sale of a property or company.

(ii) **Assessment Audits:** It is carried out to measure continuous progress or for benchmarking, using a constructively critical dialogue. They are typically carried out by:

* Corporate auditors for divisional sites.
* Internal auditors to measure own programmes
* Pre-verification auditors before a formal audit.

ENVIRONMENTAL MANAGEMENT SYSTEM

The principles of quality management and the widely used ISO 9000 standards for quality systems have led to the development of specifications for environmental management systems (EMS). An EMS is the organizational structure, responsibilities, practices, and procedures of an organization to fulfil its environmental goals and control its environmental impacts.

A documented management system, with feedback mechanisms like an internal audit function, will provide assurance to company management and all stakeholders that the company shall meet now and in the future, legislative and policy requirements. Such a system enables an organization to establish; environmental policy, objectives at various levels of the organization, controls commensurate with risks and external requirements, records for documenting, performance and feedback mechanisms at appropriate levels of the organisational.

An audit as part of an EMS is a systematic examination to determine the extent to which the EMS corresponds to planned arrangements, (or documented EMS criteria), and if it is implemented effectively to achieve stated policies and objectives. The EMS audit requires that the company has documented its environmental policies and goals, the significant environmental effects of its operation, and the control established to deal with these.

Examples of issues that an auditor could consider when evaluating the performance of a management system are:

(i) Are the environmental policy and corresponding objectives documented, communicated, known and adhered to?

(ii) Are all applicable legislative and policy requirements rerecorded, and is the responsibility for keeping the company up-to-date on such requirements allocated?

(iii) Are training needs systematically assessed, and is necessary training given to all employees?

(iv) Are internal audit programs implemented, containing auditor qualification criteria, audit plans, systems for corrective actions and regular follow-up?

(v) Is an emergency response plan based on a comprehensive hazard assessment developed and regularly reviewed, and are emergency drills conducted?

STAGES OF AN ENVIRONMENTAL AUDIT

An audit may be divided into three stages: pre-audit, site and post-audit activities. As a rule of thumb, a quarter of the time is used before the site visit, approximately half the time on the site, and the rest, for reporting and follow up. The roles and responsibilities during these stages will be different for the auditor, the auditee and the client.

(i) **Pre-audit stage :** The Terms of Reference (TOR) should describe the project, the scope, the objectives and the criteria of the audit, and provide relevant background information. It is important that the procedures for the audit are clearly specified; how investigations should be performed, their depth, and the amount of verification required. Time tables for all audit works and reporting lines should be included along with responsibilities and communication line. The TOR normally also includes a draft budget for the audit.

A lead auditor will be responsible for all stages of the audit. He or she will normally select one or more team members with relevant competence for the audit type as soon as resource needs have been identified. The audit team must possess the necessary auditing skills as well as knowledge of the environmental aspects of the organization, facility, or site under consideration. The lead auditor will communicate with the auditee to arrange all practical arrangements, including data and time for the visit, travel and lodging arrangements, contact person of the auditee, the agenda for the site visit, and any special rules to be obeyed, like safety precautions, or whether photographing is allowed. The auditor should normally receive background information from the client or auditee in order to become acquainted with processes, facilities, past environmental problems and other information deemed relevant for proper preparation of the audit team.

The audit protocol is normally the auditor's primary vehicle for obtaining information covering all relevant aspects of the specified audit. During the pre-audit stage the protocol is prepared, or a standard "blue-print" protocol is modified to fit the specific site. The protocol provides a valuable framework within which he work, but it should not restrict the auditor from identifying and assessing aspects not covered in the protocol.

(ii) Site Audit Stage : The investigations on site for an environmental audit may take from one to ten days, unless drilling and sampling are necessary, which may take one to several weeks to complete. The investigations should commence with and opening meeting attended by site management and all those directly involved in the audit. The purpose of the meeting is to introduce the audit team, describe the objectives, scope, criteria, and working procedures and agree on a work scheduled. A tour of the site or areas subject to the audit is usually made to acquaint the auditors with the location and activities.

Care should be taken not to disturb the activities on site. Information will be adhered in a short time, documents will be requested and personnel will be asked to answer the auditor's questions. The audit situation often places interviewee under stress, even in cases when the environmental performance is acceptable. The auditor should plan his activities well in order to create a minimum of disturbance and ensure a friendly atmosphere during interviews sessions. If the auditor comes across potentially dangerous situation, he or she should inform the company representative acting as liaison to site management and make a note of the finding for the report.

The investigations will depend on the type of audit. Most audits will rely heavily on interview sessions. Document reviews and physical examination of storage areas, treatment plants, etc. are used both to gather information about environmental performance and to verify information received or findings noted. Findings should always be substantiated by concrete evidence. The working papers of the auditors are the basis for the conclusions in the audit report, and should be diligently completed to enable auditors to review results and findings, and to recall how conclusions were reached.

The activities at site conclude with a closing meeting attended by the same people as the opening meeting. The meeting should be brief and should present the results of the audit, as they will appear in the audit report. The auditor should stick to the conclusions of the team unless something is misunderstood.

Observations not supported by evidence may be presented as opportunities for improvements. If suggestions for improvements or remedial action are included in the TOR these may also be outlined in the closing meeting. The lead auditor will notify management as to when a draft audit report will be submitted.

(iii) **Post-audit Stage: -** In the post-audit stage, the audit report is completed on the basis of the conclusions of the closing meeting .A draft report is usually submitted to the auditee and the client for comments. The audit report will state the audit objectives, scope and criteria, identity persons involved from both the auditor's and the auditee's side, the methodologies and procedures applied, and the main findings and conclusions of the audit, with a list of bodies consulted during the audit process. For Bank projects, the report should normally include prioritised recommendations for mitigatory and other actions, and their cost. The report should be on unbiased and objective evaluation, and neither the client nor the auditee should allow changing the main conclusions of the audit team.

REFERENCES

Anonymous (1983). The Times of India Directory and Year Book: 29-31.

Anonymous (1993). Allowing Voluntary Participation by Companies in the Industrial Sector in a Community Eco-management and Audit-Scheme. Published by Council of the European Communities Regulation No. 1836/93.

Anonymous (1994). Specification for Environmental Management Systems. Published by British Standards Institute BS 7750.

Anonymous (1994). Environmental Auditing. Published by Canadian Standards Association CAN/CSA-Z 751-94

Anonymous (1995). Handbook on Industrial Pollution Prevention and Abatement, Environmental Department Published by World Bank.

Buckey, R. (1995). Environmental Auditing-Environmental and Social Impact Assessment. Published by John Wily & Sons. New York.

Clark, J.W.W. Viessman and M.J. Hammar (1971). Water Supply and Pollution Control, 2nd edition, International Textbook Co., Scranton.

Escritt, L.B. (1972). Sewerage and Sewage Disposal, I, Facdonald & Evancs Ltd., London. : 324-364.

Gulsan, S.S. & Kapoor, G. K. (1994). Economic & Other Legislations, Published by Sultan Chand & Sons, New Delhi: 15.10.

Nemerow, N.L. (1979). Liquid Waste of Industry, Theories, Practices and Treatment, Published by Addison-Wesley Publishing Co: 225-276.

Rao, M.L. and A.K. Dutta (1979). Waste Water Treatment, Published by Oxford and IBH Publishing Co.

Sundaresan, B.B. (1977). Seminar on Industrial Wastes, Calcutta, Organised by N.E.E.R.I. and Calcutta Metropolitan Development Authority, Published by A.K. Basu.

Advancing Frontiers of Ecological Researches in India, 507-515, 2006
Edited by A.K. Kandya & Asha Gupta
Published by Bishen Singh Mahendra Pal Singh, Dehra Dun, India

A Preliminary Study on Source Apportionment of RSPM–A Case Study

Geet Govind Sinam
Central Pollution Control Board,
4, Dholpur House, M.G. Road, Agra

Abstract: The aim of the present investigation was to identify a reliable method to apportion the potential source of Respirable Suspended Particulate Matter (RSPM) in ambient air. Adopting a range of approaches (combination of models) apportionment of RSPM was done. It was revealed from the study that RSPM constitute of 28% ultra-fine particle, 34% fine particle, 30% respirable particle, and 8% particle less than 20μ. The study shows that the particle is of local origin and vehicular is not the main contributor.

INTRODUCTION

Ambient air quality monitoring in India started since 1967 (World Bank Report) and there were 328 monitoring stations in operation in 1999 (Annual Report, CPCB, 2000-01). It may be noted that Sulphur dioxide (SO_2), Nitrogen dioxide (NO_2), and Total suspended particles (TSPM) are the parameters monitored on regular basis. Respiratory suspended particulate matter (RSPM), with the particle diameter cut-off somewhere near 10 micron (μm), has been added recently at a number of monitoring stations, totaling 77 by the end of 2000.

Particulate matter is the most serious pollutant in most cities of India. There are many sources of particulate pollution: large industries plant, medium-and small scale industries, refuse burning, household burning of biomass for cooking and heating, vehicular exhaust, re-suspended road dust, construction, particles migrating from other regions, and naturally occurring dust. TSPM is less relevant to the health impact of air pollution than RSPM and smaller particles. Small particles suspended in the atmosphere cause adverse health effects, reduce visibility, and are a nuisance when they deposit onto surfaces. For this reason the trend worldwide is to focus increasingly on measuring the concentration of smaller particles, which can penetrate deeper into human respiratory systems. The study of smaller particles is not only important to human health but is also critical to plant physiology as fine particle clogs stomata opening in leaves thereby affecting transpiration. It is important to know the relative contribution of these different sources.

There are many models used in the apportionment of particles in the ambient air. Some of these models are – conceptual model of chemical mass balance (CMB), enrichment factors (EF), temporal and spatial correlation eigenvectors, particle size etc. Some of these models are very effective and reliable. As the models' were developed by using input from specific area, they were not found to be very effective when applied at different locations. Thus, the effectiveness of these models are apparently geographically limited However they can be used by incorporating some minor modification according to the conditions prevailing in the specified study area, thus enhancing their applicability.

So far few studies have been done on source apportionment on RSPM in India. About dozen or so studies done on source apportionment has been done on TSPM. Sharma and Patil (1992) used chemical mass balance to examine source contributions to TSPM in Mumbai. This method is based on the US-EPA chemical mass balance receptor model to attribute its source. Meenakshy *et al.* (1996) measured 20 elements in Mumbai for over a year and factor analysis was carried out identifying four pollution sources: re-suspended dust, refuse-vegetation burning, sea-salt and road traffic. Soil derived (re-suspended dust) dust was found to be a major contributing source of larger particles (more than 10 μm), whereas refuse burning and road traffic were identified as major sources of smaller size particles (less than 10 μm). Jitendra *et al* (1996) used a multi source Gaussian model to estimate TSPM concentration distribution in Mumbai. Khare et al. (1994) using high volume sampler apportioned certain cations and anions around industrial site in Agra by the principle component analysis method.

Javitz (1988) developed a general methodology for performing simulations of the Chemical Mass Balance (CMB) model and applied to simple and complex local scale scenarios. The simple scenarios consists of Coastal, Coal fired power plant, motor vehicle and vegetation burning sources; the complex scenarios adds Oil fired power plant, Ocean, steel mill, Lead smelters, municipal incinerator and background aerosol sources. Daily receptor filters concentrations of the most commonly measured elements in the primary emission are simulated. The simulations incorporate daily fluctuation in source strength, daily fluctuations in source profiles and measurement error at the receptor. These moderately realistic simulation provide an encouraging picture of CMB accuracy and precision even when the coefficient of variation of temporal source profile is 25% and the coefficient of variation of measuration error is 10%.

Chow (1992) monitored PM_{10} at Rubidoux, Riverside-Mognolia, and at the University of California at Riverside to quantify the variability of both PM_{10} and its chemical composition on a neighbourhood scale (<5 Km) using CMB receptor model.

Chow *et al* (1992) studied on PM_{10} aerosol which was carried out at six sites in California's San Joaquin Valley from 14 June 1988 to 9 June 1989, as a part of the 1988-1989 valley Air Quality Study. Concentration of PM_{10} and $PM_{2.5}$ mass, organic and elemental carbon, nitrate, sulfate, ammonium and elements were determined in 24 hr. aerosols samples collected at three urban and three non-urban locations during this period. The sources, which contributed to ambient concentrations of PM_{10} were determined by CMB receptor model using the source profiles determined specifically for that study area.

McDonald *et al* (2001) applied the CMB technique to an underground gold mine to assess the contribution of diesel exhaust, rock dust, oil mists, and cigarette smoke to airborne fine particulate matter.

Source apportionment is very important in Ambient Air quality studies. This information would be important for devising effective mitigation measures and setting realistic targets, because it matters a great deal whether high ambient concentrations are a result of naturally occurring particles, or even particles migrating from other regions.

MATERIALS AND METHOD

2.1 Study Site

Study was carried out at a location lying upwind to Taj Mahal in the Agra [Lat. 27° 10'N, Longitude 78° 05'E] of Uttar Pradesh, India (Figure-1). The study area has residential cum industrial mixed type of landscape. Taj Mahal, a heritage monument is flanked by green belt on the eastern and western side. River Yamuna runs on the northern side from west to east direction. The southern side has residential area, which is densely populated. The entire study area comes in the sensitive area (reference:

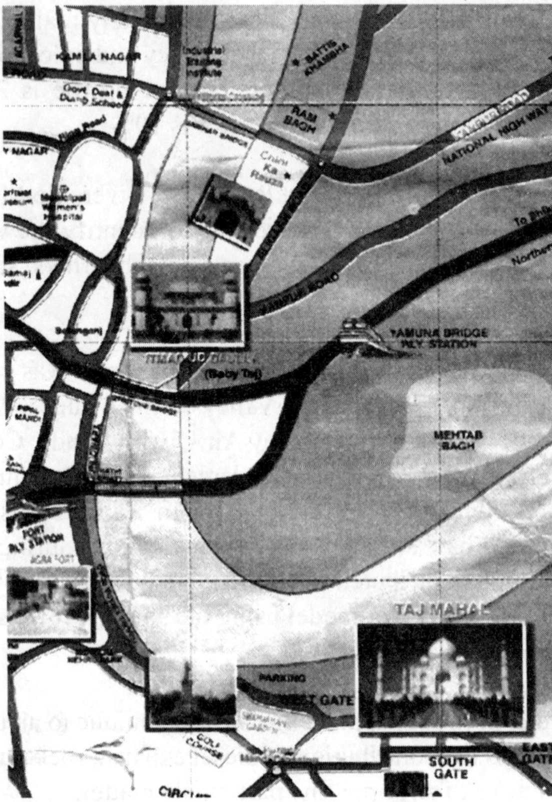

Fig. 1. Location of the study site

TTZ). Climate is humid-continental with 25-30°C mean annual average, and yearly precipitation averages 685mm. The study area is highly polluted.

2.1 RSPM Sampling

RSPM was collected in Respirable Dust Sampler (RDS APM-460DX, Envirotech) using GFA and EPM 2000 glass fiber filter paper. The EPM 2000 filter was used for the characterization of RSPM. The flow rate of the sampler was set at 1 LPM. Sample was taken for 24hr/day and each day divided into three shifts of 8-hr duration. Sampling was conducted continuously through the alternate use of sampler from June 1, to June 30, 2002.

The RSPM concentration is found out by the difference in weight for Pre and Post sampled filter paper. Prior to weighing, filter papers were equilibrated for 24 hr in a desiccator with silica gel.

2.2 Particulate Size Distribution Analysis

Analysis was done by Fritch Particle Sizer Analysette 22 instrument (at JNARDDC- Nagpur). Processed liquid samples were kept in ultrasonic chamber and pneumatically pumped into the instrument. Discrete particles were passed through laser beam and the diffraction of laser energy was recorded to assess its sizes.

2.4 RSPM Apportionment

Since single methodology has not been able to answer all or even most questions, a range of approaches is being adopted to try and identify the source of RSPM.

The first method is by particle size distribution. Based on particle size, the particles are classified under four categories: Ultra-fine particle (0-1μ), fine particle (1-2.5μ), respirable particle (2.5-10μ), and particle less than 20μ (10-20μ). The source for each category is than identified: Ultra-fine - Diesel exhaust, combustion of coal, oil, atmospheric transformation product; Fine – Smelters, steel mill, coal fired thermal power plant; Respirable particle – wind blown dust and grinding operations, pollen and bacteria; Particles less than 20μ - Resuspended dust, street dust, soil dust, and fly dust (CSE). The total fraction of particle for each category is found out from the particle size distribution curve.

Second method is by calculating the correlation coefficient between RSPM and NO2. As vehicular exhaust is the main source of NO2 (U.S. EPA), it can suggest the contribution from road traffic to particulate matter. This method cannot quantify the result but it can highlight the influence of RSPM value by road traffic.

Another method is to compare the value of study sites with the value of background value. Similarly, this method is also not quantifiable but is useful in providing us an indicator. This can give an indication whether the source is from anthropogenic activities or of natural origin.

RESULTS AND DISCUSSION

The mean average of RSPM of the study area and the background is given in Table-1.

Table-1: Physical characterization of RSPM

Sample	Location	RSPM ($\mu g/m_3$)	PM 2.5 (Cum %)	PM 10 (Cum %)	Ratio PM 2.5: PM10	Elemental Carbon avg. ($\mu g/m_3$)
A	Site A	160	64.06	96.68	0.66	0.35
B	Site B	124	64.29	88.31	0.73	0.68
C	Site C	68	59.83	84.62	0.71	0.21
D	Site D	153	59.92	94.79	0.63	0.66
E	Back-ground	93	52.99	94.31	0.56	0.26

24 hourly average concentrations (Source: Prerna)

The Ultra-fine particles whose potential sources are diesel exhaust, combustion of oil, coal, and atmospheric transformation products constitute 28% of the particle (refer Table-2). These may be contributed from the genset which are extensively used in times of load-shedding. The fine particle whose chief source are smelters, steel mill, coal fired thermal power plant constitute 34% of the particles. The presence of foundries in the vicinity of study area may be the reason for this particular particulate matter in RSPM. The respirable portion of the particulate matter whose chief sources are wind blown dust, grinding operations, pollen and bacteria constitute 30% of the RSPM. As the time of the study was during the

Table-2: Particle Percentage at Study Site.

Category/ Source	Size Range	Site A Particle %	Site B Particle %	Site C Particle %	Site D Particle %	Mean %
Ultra-fine: Diesel exhaust, combustion of coal, oil, atm. transformation products.	0-1 μm	29.5	30	27.5	26	28
Fine: Smelters, steel mill, coal fired power plant.	1-2.5 μm	34.5	34.5	32.5	34	34
Respirable: Wind blown dust & grinding operations, pollen & bacteria	2.5-10 μm	33	25	26.5	35.5	30
Particle < 20μ: Resuspended dust, street dust, soil dust, and fly dust.	10-20 μm	3	10.5	13.5	4.5	8

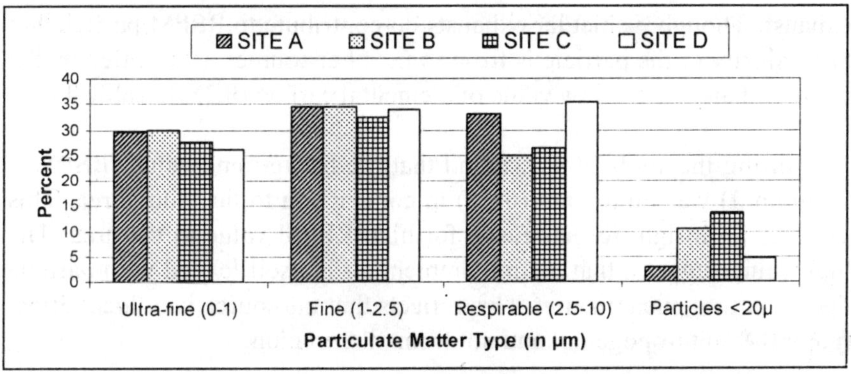

Table-3. Particulate Matter Distribution at Various Sites

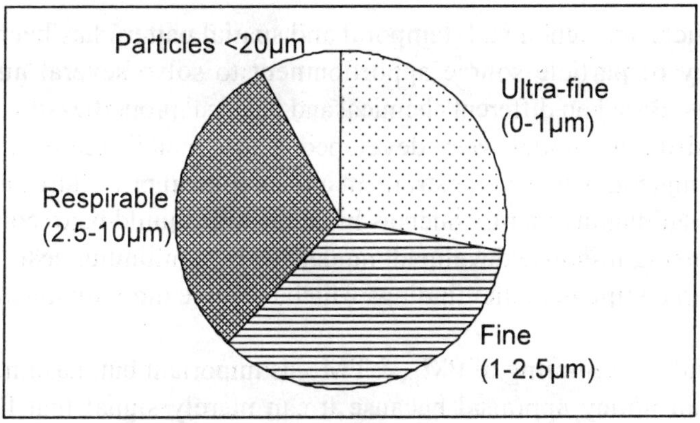

Chart-1. Mean Distribution of Particulate Matter

peak summer, top soil remain dry and are easily blown away by the wind. This may be reason for high proportion of respirable particle in the RSPM. And finally, the particle whose aerodynamic-diameter is less than 20μ, potential sources being resuspended dust, street dust, and fly dust constitute 8% of the RSPM. The low value of Lager particle size may be attributed to the height of the monitoring station from the ground.

The Correlation Coefficient for Site C ($r = 0.519$) shows a strong positive correlation. But the correlation coefficient for Site B ($r = 0.244$). Site D ($r = -0.0475$) and Site A ($r= 0.01$) show very poor relation. Overall, the relationship obtained is poor. From this result, we may presume that other factor is more significant than the vehicular source. As the overall correlation between RSPM and NO_2 is poor, it can be concluded that the pollution due to RSPM in the study area is not primarily due to vehicular exhaust. Though, vehicular exhausts do contribute to RSPM particle but the majority of the particle is from some other sources as revealed by the study and also by the low value of elemental carbon (0.38%, Table-1).

During the study, it was found that the background value (RSPM = 93 μg/m 3) was quite high (60%) in comparison to the study area. This gives us an insight to the reason for high RSPM value in the area. The high value suggests that natural ambient air is itself loaded with particle due to natural phenomenon. This reflects that the source is of local effect (natural & anthropogenic) and not from other region.

SUMMARY AND CONCLUSION

Particle size, chemical, temporal and spatial pattern has been used in the study of particle source apportionment to solve several air quality problems. Based on different chemical and physical properties of particulate matter, different models were developed. These models require extensive monitoring and expensive instrument and analytical procedures. Hence, it is proposed that in country such as India the aim should be to conduct the studies using in-house developed models or by combining best (reliable) models to complement the findings which can give more reliable results.

Regular monitoring of PM_{10} & $PM_{2.5}$ is important but, has limited role to play in policy appraisal because it can merely signal that there is a problem. Monitoring needs to be supplemented by studies to identify sources and assess effects on public health.

REFERENCES

Anonymous. Central Pollution Control Board. Annual Report, 1999-2000.

Anonymous. Central Pollution Control Board (2002). Environmental Management Plan-Agra.

Anonymous. Down To Earth (1998). Ed. 15 Dec. "concern over particles". 38

Anonymous. U.S. Environmental Protection Agency. Air Trends (1995). Summary. Available at site< http://www.epa.gov/air/aqtrnds95/NO2.html >

Anonymous. Tata Energy Research Institute (2001). "Review of past and on-going work on urban air quality in India," report submitted to the World Bank, December. Available at <http://www.worldbank.org/sarurbanair>

Chow., J. C.; Chung Shing Liu, Joe. Cassmassi, John G. Watson, Zhiqiang Lu and C. Pritchett (1992). "A neighborhood scale study of PM10 source in Rubidoux, California". Atm. Env. Vol. 26A. No. 18. 693-706

Chow., J. C.; John G. Watson, Douglas H. Lowenthal, Paul A. Solomon, Karen L. Magliano, Steven D. Ziman and L. Willard Richards (1992). "PM10 source Apportionment in California San Joaquin Valley". Atm. Env. Vol. 26A. No. 18. 3335-3354

D. Saha *et al.* (2002). "A preliminary investigation on the characterization of RSPM in Agra." Prerna. 16

Javitz. H.S, Watson J.G. and N. Robinson (1988). "Performance of the CMB model with simulated local scale aerosols". Atm. Env. Vol. 22. No. 10. 2309-2322

Jitendra Shah and Nagpal Tanvi, ed. (1996). "Urban Air Quality Management Strategy in Asia URBAIR: Greater Mumbai Report," October, Washington D.C.

John G. Watson and Judith C. Chow (2002). "Particulate pattern Recognition". Introduction to Environmental Forensics; Ed. by Brian L. Murphy and Robert D. Morrison. Academic press. NY. 429-448

Meenakshy, V. B.S. Negi and S. Sadasivan (1996). "PM-10 aerosols at Chembur, Bombay: their elemental concentration and sources," Proceedings of the 5th National Symposium on Environment, March, 227-230.

McDonald. J. D, Barbara Zielinska, John, C. Sagebiel and Mark R. McDaniel (2003). "source Apportionment of Airborne Fine Particulate Matter ". J. Atr. & Waste Management. Assoc. 53. 386-395.

Sharma, V. K. and R.S. Patil (1992). "Size distribution of atmospheric aerosols and their source identification using factor analysis in Bombay, India," Atmospheric Environment 26B(1). 135-140.

Sharma, V.K. and R.S. Patil (1994). "Chemical mass balance model for source apportionment of aerosols in Bombay," Environmental Monitoring and Assessment 29(1):75-88.

Advancing Frontiers of Ecological Researches in India, 517-536, 2006
Edited by A.K. Kandya & Asha Gupta
Published by Bishen Singh Mahendra Pal Singh, Dehra Dun, India

Impact of Irrigation of Sugarcane Fields with Pulp and Paper Mill Effluent on Groundwater Quality

P.R. Chaudhari, J.P. Kotangale, Shanta Satyanarayan, and S.B. Lakhe
National Environmental Engineering Research Institute
Nehru Marg, Nagpur 440020

Abstract: The biological and physico-chemical observations on well water samples from agricultural fields irrigated with pulp and paper mill effluent and their comparison with village wells away from the irrigated fields indicate the pollution of ground water due to seepage of effluent. The pollution is visible by the light brown colour of the ground water and high density of phytoplankton population and presence of indicator species, among which the species such as *Chroomonas, Chlamydomonas,* and *Cyclotella* are of special mention. The indicator algal groups of paper mill effluent pollution are chlorophyceae and bacillariophyceae.

Study of Shannon Wiener Index of Diversity and Palmer's Pollution Index shows that the water pollution is mostly due to inorganic pollutants and organic pollution is moderate. The zooplankton indicator groups were copepoda, rotifera and cladocera while indicator zooplankton species were *Cyclops, Ceriodaphnia*, and *Brachionus calyciflorus*. The well waters were also observed to be polluted on the basis of bacteriological quality.

The observations indicate that sandy loam soil which permits easy percolation of surface water to ground water, excess use of effluent for irrigation and shallow water table in the area are responsible for ground water contamination. Therefore, while using industrial wastewater effluent for irrigation, the soil quality, effluent quality and application rate needs to be studied carefully to avoid contamination of groundwater.

INTRODUCTION

Pulp and paper mill industry has been included by Ministry of Environment and Forests (MoEF) in the list of major industries in India, which require environmental clearance because of its potential impact on environment. The wastewater from this industry consists of black liquor from pulp mill unit containing mostly lignin and wastewater from other sections of the plant with waste organic matter and suspended solids. The later wastewater is treated and treated effluent is recycled for washing or reused for irrigation purpose.

Recycling and reuse of industrial wastewater effluent for irrigation purpose is a popular practice in India. Municipal waste has been in use for irrigation purpose for several centuries. Industrial effluents have been tried by several researchers for irrigation. Rohlich (1995) and Mckee (1957) have reported that wastewater from cheese factory and dairy farm favoured the growth of several crops. Anon (1954) and Mc Cornich (1959) observed that wastewater irrigation from pulp and paper mills had no adverse effect on the yield of crops like cotton, corn and grasses etc. Bishop and Wilson (1954) observed that the use of paper mill waste for irrigation of crops like potatoes, corm bean, cabbage, carrots and tomatoes etc. gave results comparable with that of irrigation with plain water. Encouraging results were also found in studies on irrigation of crops like paddy, wheat and groundnut with the wastewater from pulp and paper mills (Khambatta and Ketkar, 1969). Irrigation by paper mill wastewater has shown encouraging results on oats and orchard grasses (Hashimoto and Yokota, 1965), on germination of rice, black gram and tomatoes (Rajannan and Obliswamy, 1979), on populus (Hansen *et al.*, 1980). Reddy *et al.* (1981) observed that irrigation with wastewater is superior to that with fresh water in respect of yield of sugarcane and plant characteristics.

The wastewater also contains toxic materials that reduce crop growth and have inhibiting effect on soil characteristics (Somashekhar *et al.*, 1984a). Jorgensen (1965) reported that a depression on growth of plants with pure wastewater irrigation. Srivastava (1991) reported that high concentration of different constituents of paper mill and chlor alkali plant waste water in dissolved form with high BOD and COD contents seemed to affect adversely the germination and subsequent growth of radish and onion seedling. Hansen *et al.* (1980) stated that the chemical properties of the forest soil changed significantly in relation to sodium, chloride and nitrogen when flooded with the wastewater. In addition to this the sodium chloride content and pH increased through the soil profile and P, Ca, Mg, and organic carbon content increased on the surface. Thus the suitability of wastewater for irrigation depends upon the quality of wastewater and in the purpose of use.

If the quality of wastewater effluent is not suitable for irrigation or it does not meet the standards of irrigation water, the irrigation may result in pollution of ground water and adverse effects on soil and the crop. The present work is undertaken to study the characteristics of pulp and paper mill industry wastewater and its effluent and impact of the use of effluent for irrigating sugarcane fields on groundwater quality.

MATERIALS AND METHODS

A large Pulp and Paper Mill Industry from southern India was selected for study. The mill was designed to produce 50000 TPA of newsprint, 40000 TPA of writing and printing paper with flexibility to meet market demand and operational exigencies. Bagasse pulp account for over 70% in the newsprint furnish and 75% in writing and printing furnish while hardwood (Eucalyptus) and other tropical imported wood / pulp account for the balance. The required hardwood pulp is also produced in the mill.

Pulp and paper mill industry produces black liquor from pulp mill unit having high BOD. This black liquor is a dark coloured liquid containing inorganic compounds and dissolved organic matter. The inorganic compounds are regenerated in chemical recovery plant while organic compounds are burnt in recovery boiler to produce steam. The effluent coming from bagasse handling and fibre preparation has very high BOD and suspended solids. The suspended solids are segregated and treated separately through a system consisting of bar/mechanical screen and clarifier to reduce suspended solids and then subjected to anaerobic treatment to reduce BOD load prior to mixing with other effluents.

Effluent with low BOD and suspended solids from other sections of the plant viz. paper machine, pulp mill and soda recovery plant is passed through bar screen, mechanical screen and detritor and two primary clarifiers to remove suspended solids. The effluent is further treated in an aeration system based on activated sludge process, equipped with 32 numbers of 75 HP fixed type aerators followed by two secondary clarifiers to achieve the required COD/BOD reduction and treated effluent conforms to the limits prescribed by Pollution Control Board. This wastewater effluent is used for irrigation of sugarcane fields around the industry and for green belt and avenue plantation.

The sample of wastewater effluent was collected from the discharge point for irrigation purpose. Well water samples were collected from seven dug wells present in sugarcane fields, which are irrigated with wastewater effluent, and from three dug wells, in villages around the industry. The water samples were analyzed for physico-chemical and biological parameters to estimate plankton (phytoplankton and zooplankton) population, which is the well known indicator or marker group depicting the quality of water and the soil. The dug wells in the vast sugarcane fields were used to collect ground water samples for physico-chemical and biological analysis. The dug wells also contained dark blue-green patches on its walls. These blue-green patches were also analyzed for their composition of algae. The Cauvery River is 3 km away from the industry. This river receives wastewater from textile industries and agricultural run off or seepage from irrigated agricultural fields. The water supply to TNPL is through an intake well located in the bed of river Cauvery. Three Water samples were collected from the river, one each at upstream and downstream of the factory and one sample near intake well over Cauvery River.

Phytoplankton flora was analyzed by Lackay Drop count method. Zooplankton flora was analyzed by Sedgewick-Rafter Cell. Bacteriological analysis was carried using standard membrane filter technique. All the analyses including physico-chemical parameters were carried out as per Standard Methods (APHA, AWWA, WPCF, 1998).

OBSERVATIONS

Waste water Generation and its Treatment

The mill has a modern effluent treatment plant (ETP) designed to treat 20 MGD of effluent generated from various sections of the mills. The effluent with high BOD and suspended solids (from bagasse handling and fibre preparation) is treated separately through bar/mechanical screen and then through anaerobic treatment to reduce BOD prior to mixing with other effluents. The effluent with low BOD and suspended solids from other sections of the plant viz. paper machine, pulp mill and soda recovery plant is passed through screens and primary clarifiers and then mixed with former effluent and treated in an aeration system based on activated sludge process, followed by two secondary clarifiers to achieve BOD/COD reduction to conform to the limits prescribed by Pollution Control Board. This effluent is utilized for irrigating sugar cane fields around the industry.

Physico-chemical Quality of Treated Waste Water

The plankton and bacteriological groups are good indicators of eutrophication and water quality (Chaudhari, *et al.* 1998, 1994; Somashekhar, 1984b) and are used to evaluate the effects of paper mill effluent irrigation on groundwater quality.

The physico-chemical characteristics of final treated effluent are shown in Table-1. The effluent is alkaline with pH ranging from 7.4 to 7.6. The colour is faint brown. The BOD is very less ranging from 4-5 ppm, however, COD/BOD ratio vary from 33.4 to 37 indicating the presence of lignin in the effluent which is degraded biologically at a very slow rate and is responsible for the colour of the effluent. The source of lignin is mostly the raw material like bagasse and hardwood material used for pulping. Total dissolved solids, phosphates, TDS, chlorides and sulphates are higher in concentration in the effluent. All the heavy metals are found to be below stipulated limits prescribed by Tamil Nadu Pollution Control Board. All other parameters are also found to be below the stipulated limits for irrigation water.

Table-1: Physico-chemical Quality of Treated Effluent from Pulp and Paper Mill Industry

Sr. No.	Parameters	Range of Values Throughout the Year	TNPCB Standards
1.	Colour Pt. Co. Units	196-230	—
2.	PH	7.4-7.6	5.5 – 9.0
3.	Total Alkalinity	156-218	—
4.	Sulphide	BDL	2.0
5.	BOD	4.5-5.0	100
6.	Phosphates	4.2-4.3	—
7.	TDS	1594-1648	2100
8.	SS	42-46	200
9.	Chlorides	394-418	600
10.	Sulphate	284-298	1000
11.	Sodium	251-254	—
12.	Potassium	10-12	—
13.	COD	149-157	—
14.	Total ammonia nitrogen	4.0-4.9	—
14.	Heavy Metals		
	Cr	0.012-0.018	2.0
	Zn	0.013-0.109	1.5
	Pb	BDL	1.0
	Cd	BDL	1.0
	Mn	0.006-0.008	—
	Fe	BDL-0.27	—
	Cu	BDL	3.0
	Ni	BDL-0.013	3.0

Physico-chemical Quality of Ground Water

The physico-chemical quality of ground water is shown in Table-2. The well water samples were found to be light brown in colour and some samples were observed to be turbid. The water was alkaline with pH ranging from 7.8 to 8.2. The COD was observed to range from 29.4 to 63 ppm. Conductivity (3.82-4.53 ppm), total hardness (1010-1320 ppm), calcium hardness (560-1250 ppm), chlorides (359-802 ppm), sulphates (210-374 ppm), sodium (255-400 ppm), calcium (224-500 ppm), silica (80 ppm) were found to be higher in well water samples. It was also interesting to observe

Table-2: Physico-chemical Characteristics of Dug Well Water Samples Collected from Areas Contiguous to Pulp and Paper Mill

S.N.	Parameters	Village Wells	Dug wells in agricultural fields
1	Temperature, ^{0}C	29-32	28-29
2	Colour (visible)	Clear toTurbid	Light brown
3	PH	7.9-8.8	7.8-8.2
4	Conductivity, ms/cm	1.40-2.68	2.29-4.53
5	Total alkalinity as CaCO3	223-423	242-308
6	Total hardness	370-560	1010-1320
7	Calcium hardness	160-260	560-1250
8	Magnesium hardness	190-320	70-680
9	BOD	1.14-3.5	<5.0
10	TDS	899-1710	2451-2901
11	SS	<2-5	<10
12	NH_3-N	BDL-9.4	BDL
13	Chlorides	64-310	359-802
14	Sulphate	18.2-219	210-374
15	Sodium	36-255	255-415
16	Potassium	4.5-77	2-12
17	Calcium	64-104	224-500
18	Magnessium	43.7-73.6	16.1-156.4
19	Silica	60-100	80
20	COD	4.08-28.56	29.4-63.0
21	NO_2-N	BDL	BDL
22	Pb	BDL-0.177	0.115-0.487
23	Ni	BDL-0.02	BDL-0.058
24	Fe	BDL	

BDL : Below detectable limit; All values are expressed in mg/L except temperature and pH;

that many well/hand pump water samples collected from adjoining villages were also found to be polluted with higher levels of COD (upto 48.96 ppm), total alkalinity (upto 938 ppm), chlorides (upto 700 ppm), sulphates (upto 493 ppm), and sodium (500 ppm).

Biological & Bacteriological Quality of Ground Water from Villages

A total of 7 algal species were recorded from three well water samples situated in villages (Table-3). Out of these, two species belonged to green algae (chlorophyceae), three to diatoms (bacillariophyceae), one species each to green flagellates (euglenophyceae) and pyrrhophyceae. No blue-green algal species was recorded from the water samples. Water sample nos. V2 and V3 showed the dominance of diatoms while water sample no. V1 showed the dominance of chlorophyceae. *Euglena* sp. (Euglenophyceae) and *Peridinium* sp. (Pyrrhophyceae) were observed in water sample no. V1 only. The dominant phytoplankton species were *Chroomonas* and *Chlorococcum* (Chlorophyceae) in water sample no. V1; *Cyclotella* (bacillariophyceae) in water sample no V2; and *Cymbella, Cyclotella* (bacillariophyceae), *Chroomonas* (chlorophyceae) and *Navicula* (bacillariophyceae) in water sample no. V3. The phytoplankton count varied as 210608 algae/ml in water sample no. V1, 100 algae/ml in water sample no. V2, and 2507 algae/ml in water sample no. V3.

Table-3: Phytoplankton Flora (Number of Individuals/ml) in Water Samples from Dug Well Samples in Villages

Phytoplankton Groups/species	Village Well 1 (V1)	Village Well 2 (V2)	Village Well 3 (V3)
Chlorophyceae			
1. Chroomonas	172168	—	540
2. Chlorococcum	34850	—	—
Bacillariophyceae			
3. Cocconeis	—	—	656
4. Cyclotella	—	100	984
5. Navicula	—	—	327
Euglenophyceae			
6. Euglena	2050	—	—
Pyrrhophyceae			
7. Peridinium	1540	—	—
Total count	210608	100	2507

Zooplankton was recorded in only one village well water sample i.e. water sample no. V1 (Table-4 & 9). The Copepoda was found to be the sole dominant group in this water sample.

These Well water samples were found to be contaminated with coliforms and faecal coliforms (Table-5). The count of total coliforms and faecal coliforms were in the range of 150 to 1400 and 80 to 900 CFU/100 ml respectively. Hand pump water samples were not polluted or very less polluted with coliforms and faecal coliforms except only one hand pump water sample, which was found to be faecally contaminated having total coliform and faecal coliform count as 1400 and 900 CFU/100 ml respectively.

Table-4: Zooplankton Fauna (Number of Individuals / m^3) in the Water Samples from Dug Wells in Villages

Zooplankton groups/Species	Village Well 1 (V1)	Village Well 2 (V2)	Village Well 3 (V3)
Copepoda			
Cyclops sp.	750	—	—
Nauplius larva	1600	—	—
Total number	2350	—	—

Table-5: Bacteriological Quality of Well Water Samples

Description	Total Coliforms CFU/100 ml	Faecal Coliforms CFU/100ml
Village Wells		
V1	150	80
V2	1400	900
V3	1400	500
Agricultural Field Wells		
F1	1800	1210
F2	1220	820
F3	950	400
F4	190	100
F5	2070	775
F6	200	100
F7	2800	600

CFU : Colony forming unit

Biological & Bacteriological Quality of Ground Water from Agricultural Fields Irrigated with Paper Mill Effluent

A total of 18 phytoplankton species were recorded from well water samples from agricultural fields (Table-6). Out of these algal species, 2 species belonged to cyanophyceae, 8 species to chlorophyceae, 5 species to bacillariophyceae, 2 species to euglenophyceae and one species to pyrrhophyceae. Density as well as diversity of cyanophyceae was very low in the water samples. Chlorophyceae was observed to be the dominant group in the water samples along with bacillariophyceae as subdominant group except in water samples nos. F2 and F6 (Table-7). Euglenophyceae group is found as next abundant group in the water samples. Pyrrhophyceae was also observed in considerable number in most of the water samples. *Chroomonas* and *Chlamydomonas* were found to be dominant in most of the samples. *Cyclotella, Sirocladium, Peridinium, Euglena*, and *Chlorococcum* were commonly present in many water samples. The total phytoplankton count varied from 4481-164535 algae/ml.

A total of 9 zooplankton groups were recorded in water samples, out of which copepoda group was dominant in majority of the ground waters. The samples also showed the presence of rotifera and nematoda, which are good pollution indicators (Tables-8 & 9).

All the well water samples were found to contaminated with both total coliforms and faecal coliforms in the range of 190 to 2800 and 100 to 1210 CFU/100 ml respectively (Table-5).

RESULTS AND DISCUSSION

Treated Effluent

The final treated effluent used for irrigation is slightly alkaline, light brown in colour, with very less BOD but marginally high COD and minerals and heavy metal content. However, all the parameters conform to pollution Control Board's standards for irrigation standards (Table-2).

The well water samples from agricultural fields were observed to be light brown in colour. This clearly showed the impact of the seepage of effluent to the ground water polluting it. The soils are mostly sandy loam with rapid infiltration rate, which allows good percolation of effluent to the ground water. The rocky stratum below prevents water to percolate down

Table-6: Phytoplankton Flora (no. of individuals per ml) in the Water Samples from Dug Wells in Agricultural Fields

Algal Groups / Species	F1	F2	F3	F4	F5	F6	F7	Cumulative Count	% Composition of Algae
Cyanophyceae									
1. Arthrospira	1170	—	—	—	—	639	—	1809	0.38
2. Oscillatoria	—	—	—	—	3280	—	—	3280	0.69
Chlorophyceae									
3. Ankistroedesmus	—	—	—	—	1380	—	—	1380	0.29
4. Chroomonas	27669	—	15577	154951	3260	—	9223	210680	44.60
5. Chlamydomonas	—	68867	—	—	—	—	—	68867	14.58
6. Chlorella	5242	—	7379	—	4920	—	—	17541	3.70
7. Chlorococcum	4342	8199	—	—	1575	3074	1680	18870	4.00
8. Oocystis	—	—	—	977	—	—	—	977	0.21
9. Scenedesmus	—	—	—	2460	—	—	—	2460	0.52
10. Sirocladium	—	—	—	31974	—	—	—	31974	6.77
Bacillariophyceae									
11. Actinastrum	1250	—	—	—	—	—	—	1250	0.26
12. Cyclotella	—	—	—	24595	8200	—	4612	37407	7.9
13. Navicula	6149	—	—	—	1750	—	1630	9529	2.02
14. Fragillaria	—	—	2460	—	—	—	—	2460	0.52
15. Stephanodiscus	1620	—	2399	3689	3166	—	—	10874	2.30
Euglenophyceae									
16. Euglena	1537	6559	1640	—	6560	768	4099	21163	4.48
17. Phacus	—	—	—	1230	—	—	—	1230	0.26
Pyrrhophyceae									
18. Peridinium	4612	16397	820	7124	—	—	—	30490	6.46
Total	53591	100022	30275	227000	34091	4481	22781	472241	
Count of Pollution Indicator Species	50721	100022	25416	164535	22725	4481	18169	386069	
Percentage of Pollution Indicator Species	94.6	100	84.0	72.5	66.7	100	79.8	81.75	

which is retained in the plough layer and leads to water logging condition harmful for crop growth. The inner earthen walls of the wells were found to be encrusted with blue-green algal patches and algal mats. These patches were dominated by green and blue-green algae at the expense of nutrients present in seepage effluent in the ground. This made the water aesthetically and physico-chemically unsuitable for drinking purpose. This also indicates excess utilization of effluent for irrigation purpose.

Table-7: Phytoplankton Community Structure in Well Water Samples

Sampling Stations	Total Count (No/ml)	Cyano-phycee (%)	Chloro-phyceae (%)	Bacillario-phyceae (%)	Eugleno-phyceae (%)	Pyrrho-phyceae (%)
			Village wells			
V1	210608	—	98.3	—	0.97	0.73
V2	100	—	—	100	—	—
V3	2507	—	21.5	78.5	—	—
			Agricultural field wells			
F1	53591	2.2	69.5	16.8	2.9	8.6
F2	100022	—	77.0	—	6.6	16.4
F3	30275	—	75.8	16.1	5.4	2.7
F4	227000	—	83.9	12.5	0.5	3.1
F5	37371	8.8	38.6	35.1	17.6	—
F6	4481	14.3	68.6	—	17.1	—
F7	22781	—	47.9	27.4	18.0	6.8

Above observation is supported by phytoplankton and zooplankton community characters and other biological observations indicating eutrophication of ground water in the agricultural fields irrigated by pulp and paper mill effluent. Phytoplankton and zooplankton communities are well known indicators of water pollution and their use for indicating water quality has been made by Chaudhari *et al.* (1994) for evaluating ground water quality.

Phytoplankton Population

The phytoplankton density is directly dependent on the physico-chemical quality of water. Increased level of eutrophication provides more nutrients to algal growth, which absorb them directly from water. Therefore, phytoplankton density increases with the increase in eutrophication of water.

Table-8: Zooplankton Fauna (number of individuals per m^3) in the Water Samples from Dug Wells in Agricultural Fields

Zooplankton group/species	F1	F2	F3	F4	F5	F6	F7	Cumulative Count	% Composition
Nematoda									
1. Diplogaster sp.	—	—	50	—	—	—	150	200	0.03
Rotifera									
2. Brachionus calyciflorus	300	—	100	80000	23400	600	100	104500	16.84
3. Lecane sp.	—	—	—	—	—	50	—	50	0.008
Cladocera									
4. Ceriodaphnia sp.	250	—	—	112000	100800	1850	50	214700	34.60
Copepoda									
5. Cyclops sp.	850	100	50	168000	122400	5200	1400	298850	48.16
6. Nauplius larva	100	—	50	—	—	—	250	400	0.07
Ostracoda									
7. Cypris sp	—	—	—	—	—	—	50	50	0.008
Diptera									
8. Diptera larva	—	—	—	—	—	100	50	150	0.024
9. Plecoptera larva	—	—	—	2000	—	—	—	2000	0.322
Plectoptera									
10. Water mite	—	—	—	—	—	100	100	200	0.0322
Acarina									
Total Number	1500	100	250	362000	246600	7900	2150	620500	100.0

Table-9: Zooplankton Community Structure in Well Water Samples

Sampling points	Total zooplankton per m3	Protozoa (%)	Nematoda (%)	Rotifera (%)	Cladocera (%)	Ostracoda (%)	Copepoda (%)	Ephemeroptera (%)	Diptera (%)	Acarina (%)
Village Wells										
V1	2350	—	—	—	—	—	100.00	—	—	—
V2	Nil	—	—	—	—	—	—	—	—	—
V3	Nil	—	—	—	—	—	—	—	—	—
Agri. Field Wells										
F1	100	—	—	—	—	—	100.00	—	—	—
F2	250	—	20.00	40.00	—	—	40.00	—	—	—
F3	362000	—	—	22.10	30.94	—	46.41	0.55	—	—
F4	246600	—	—	9.49	40.88	—	49.63	—	—	—
F5	7900	—	—	8.22	23.42	—	65.82	—	1.27	1.27
F6	1500	—	—	20.00	16.67	—	63.66	—	—	—
F7	2150	—	6.97	4.65	2.33	2.33	76.74	—	2.33	4.65

Phytoplankton population is commonly used as indicators of pollution in surface and ground water. The ranges of phytoplankton density in water indicating eutrophication levels are given in Table-10. The phytoplankton density in agricultural field well water samples ranged from 4481 to 227000 algae per ml of water sample. Three agricultural field well water samples showed phytoplankton density above 1,00,000 algae/ml water sample showing highly eutrophic water. Four well water samples showed phytoplankton density ranging from 22781 to 53591 algae/ml indicating α - mesotrophic water quality, while two water samples showed phytoplankton density ranging from 2507 to 4481 algae/ml indicating β - mesotrophic water. The quality of all the well water samples thus is either mesotrophic or eutrophic. These values can very well be compared with algal density of 100 and 2507 algae/ML in the oligotrophic village well water sample nos. V1, V2 & V3, which is also substantiated by their physico-chemical characteristics (Table-3). These village well water samples are observed to contain dominance of clean water indicator algal species of bacillariophyceae. Only one well water sample no. V1 was found to be highly contaminated as indicated by its very high phytoplankton count and presence of organic pollution indicator green algae viz. *Chroomonas* and *Chlorococcum* in water samples.

Table-10 : Saprobien Quality of Water

Sr. No.	Saprobien Level	Number of Organisms	BOD (mg/l)	COD (mg/l)
1.	Oligosaprobic (not polluted)	100	3	1 - 2
2.	Intermediate			4
3.	β - mesosaprobic (slightly polluted)	10,000	3 – 5.5	8 – 9
4.	Intermediate zone			10 – 15
5.	α - mesosaprobic (moderately polluted)	1,00,000	5.5 – 14	15 – 20
6.	Intermediate zone			20 – 30
7.	Polysaprobic zone (grossly polluted)	> 1,00,000	14	>30

* Leithe (1975)

Many water bodies in India polluted either by domestic sewage or industrial waste showed the dominance of green algae and blue-green algae. Dal lake is infested by algal blooms of chlorophyceae, cyanophyceae, bacillariophyceae and macrophytes (WWF India 1987; Kundangar *et al.*, 1991; Chatrath (ed), 1992). Khabartal Lake near Begusarai, Bihar is facing heavy biotic pressure. This wetland is extremely useful for wildlife. It harbours many species of algae of chlorophyceae and cyanophyceae (Daniel, 1985; Mishra and Negi, 1985). A large number of beels of Assam which are important for domestic consumption and bathing are affected by pollution from domestic and industrial wastes and agricultural run-off. These beels harbour species of chlorophyceae, cyanophyceae and macrophytes (Yadav and Chaudhari 1984, 1986). Some of the lakes have shown the dominance of blue-green blooms as a result of eutrophication. In Kodaikanal Lake more or less year round blooms of blue-green algae were a peculiar feature. *Microcystis aeruginosa, Spirulina* sp. and *Lyngbya* sp. occurred in this lake, the former in abundance, and later in smaller numbers, less frequently. Ooty lake in Nilgiri Region, which receives the storm water of the tourist town, showed a progressive evolution from a chlorophycean flora through Dinophycean *Ceratium hirundinella* blooms (Ganapati, 1957) to a blue-green *(Microsystis aeruginosa)* blooms. Virtually blankets of this cyanophyceae became a permanent feature from the sixties.

However, in the present case the well water polluted by the seepage of pulp and paper mill effluent, green algal blooms dominate and cyanophyceae is negligible. The bloom forming green algal species in these well waters may be recognized as bioindicators of water pollution due to paper mill effluent. Green algae are followed by bacillariophyceae, euglenophyceae and pyrrhophyceae in their percentage composition in phytoplankton community. Similar observations are also recorded by Sudhakar and Venkateswarlu (1989) in case of studying the impact of discharge of paper mill effluent on the rivers Tungabhadra and Godavari. Indicator rating for phytoplankton groups is given below :

Indicator Rating

Chlorophyceae > Bacillariophyceae > Euglenophyceae > Pyrrhophyceae

The blue-green algae are also indicators of polluted water. However, as the paper mill effluent contains higher concentration of ammonia-N, the well water samples from agricultural fields contains blue-green algae in small amount except in well water sample no. F5 and F6 which may have less amount of ammonia-N concentration. The author in studying stabilization ponds records similar observations where blue-green algae became dominant only in third or fourth pond having stabilized effluent with low ammonia-N. Sreenivasan (2002) reported the dominance blue-green algae *Microcystis aeruginosa*, *Anabaena* etc. in seemingly oligotrophic Berijam Lake in Kodaikanal hills, which was poor in dissolved salts and organic contents.

The algal indicators of eutrophication (Palmer, 1969) recorded in significant amount in the present investigation are *Ankistrodesmus, Chlamydomonas, Chroomonas, Chlorella, Chlorococcum, Scenedesmus* (Chlorophyceae); *Euglena* (Euglenophyceae); *Navicula, Nitzschia* (Bacillariophyceae); and *Peridinium* (Pyrrhophyceae). Percentage composition of pollution/eutrophication indicator algal species in phytoplankton community in polluted well waters varies from 66.7% to 100% (Table-6). This shows that the ground water is significantly affected by the seepage of pulp and paper mill effluent. Based on cumulative percentage composition of algal species, the best algal indicators of water pollution due to pulp and paper are *Chroomonas* (44.6%), *Chlamydomonas* (14.58%), *and Cyclotella* (7.92%).

Phytoplankton : Shannon Wiener Index: (SWI)

The values of Shannon Wiener Index based on phytoplankton flora in agricultural field well waters are given in Table-11. The values range from 1.21 to 2.93. Three well water samples were observed to have SWI values ranging from 2.35 to 2.93, showing that these well waters are on the lower most limit of mesotrophic (b - mesosaprobic) water quality, while four well water samples have Shannon Wiener Index varying from 1.21 to 1.80 are a - mesosaprobic in nature.

Phytoplankton : Palmer's Pollution Index (PPI)

Palmer's Pollution Index in well water samples varied from 5 to 19 (Table-11). Only one well water sample (F5) showed higher values of PPI showing presence of organic pollution. While in other water samples the values are below 15 showing low organic pollution in well water samples. This clearly shows that the pollution of well water is due to nutrient enrichment and low organic pollution.

The foregoing discussion shows that even though the effluent quality of pulp and paper mill conforms to the stipulated quality of irrigation water, it resulted in pollution of ground water although it was observed to be beneficial for the sugar cane crop growing there. Kolte and Arkatkar (2002) reported that proper utilization and application through suitably scientifically designed irrigation system (like high rate transpiration system, sprinkler/ spray irrigation etc.) taking into consideration the important characteristics like effluent characteristics, soil types, slope of land, application rate, choice of crop, scheduled timings of distribution of the effluent and the timely attention to it etc. would result in zero discharge and will prevent natural water pollution at source. Their experiments on the growth of *Glycine max* and *Phaseolus mungo* with varying concentrations of dairy wastewater showed favourable results such as early germination, increased growth of roots, stimulated growth of seedlings and improved radical and plumule length. A case study in evaluation of soil quality at Govt. Milk Scheme, Akola (India), discharging 750 m^3 of effluent daily on land after secondary treatment, has indicated an increased concentration of organic carbon, available P and available K to an extent which grades the soil as 'super high fertile' when compared to the soil fertility status rating.

Table-11: Shannon Wiener Diversity Index and Palmer's Pollution Index of Plankton

Sampling Stations	Phytoplankton			Zooplankton	
	Number of algal species	Shannon Wiener Index	Palmer's Pollution Index	Number of zooplankton species	Shannon Wiener Index
F1	9	2.35	11	4	1.62
F2	4	1.35	6	1	0.0
F3	5	1.80	8	4	1.92
F4	8	1.52	10	4	1.56
F5	9	2.93	19	3	1.35
F6	3	1.21	5	6	1.37
F7	5	2.55	9	8	1.82

Shannon Wiever Diversity Index :

< 1 : Eutrophic water quality
1 – 3 : Mesotrophic water quality
> 3 : Oligotrophic water quality

Palmer's Pollution Index :

1 –15 : Organic pollution in water is negligible or absent
15-20 : Probable presence of organic pollution
> 20 : Presence of high organic pollution in water

Zooplankton Population

Zooplankton population is also good indicator of water quality being next to phytoplankton. The density of zooplankton is generally less than phytoplankton population as it grazes on phytoplankton as food material. Therefore, apart from physico-chemical conditions of aquatic environment, the type of phytoplankton species in the water body also influences zooplankton density and diversity. The zooplankton density varied from nil to 2350 individuals/m^3 in village well water samples (Table-4), while it varied from 100 to 362000 individuals/m^3 in agricultural field well water samples (Table-8). One of the village well water samples (V1) was found to be polluted while other village wells showed no pollution in them. Agricultural field wells nos. F3 to F7 were observed to be polluted on the basis of zooplankton density. It was interesting to note that in agricultural field well water samples F4, F5 & F6, the zooplankton count was exceptionally high than any polluted water body, indicating high level of pollution in the well water samples.

From Table-8, it is observed that copepoda, cladocera and rotifera groups of zooplankton are the indicator group of water pollution and they are dominant in agricultural field wells and polluted village well. Thus, the indicator value of these groups as an indicator of pollution due to paper mill effluent may be given as follows:

Indicator rating : Copepoda > Rotifera > Cladocera.

It is conventionally known that rotifera is the indicator of organic pollution due to domestic sewage. However, in this case, copepoda is observed to be the good indicator of pollution due to paper mill effluent.

Table-8 shows the percentage composition of different zooplankton species in well water samples from agricultural fields. *Cyclops* sp. (copepoda) is observed to have highest percentage composition i.e. 48.16% in zooplankton community followed by *Ceriodaphnia* sp. having percentage composition of 34.60%, and *Brachionus calyciflorus* having percentage composition of 16.84%. These three zooplankton species are the best indicators of water pollution due to pulp and paper mill effluent.

Zooplankton : Shannon Wiener Index: (SWI)

Shannon Wiener Index based on zooplankton community (Table-11) varied from 0.00 to 1.92 showing eutrophic (well no. F2) to mesotrophic (all other agricultural wells and one village well, V1) quality of ground water.

CONCLUSION

The results of the observations show that use of industrial wastewater for irrigation may pollute the groundwater if prior care is not taken to follow scientific irrigation methods. While planning irrigation with industrial wastewater, consideration should be given to proper treatment of wastewater, fertility status of wastewater, soil quality, depth of water table, tolerance and water requirement of the crop.

REFERENCES

Anon (1954). Mill waste grows low and high corn. *Pulp and Paper*, **28** :36.

APHA, AWWA, WPCF (1995). *Standard Methods for Examination of Water and Wastewater*, 18th Edition, USA.

Bishop, F.W. and Wilson, J.W. (1954). Southland paper mill waste treatment and disposal. *Sewage and Industrial Wastes*, **26**: 1485.

Chatrath, K.J.S. (ed.) (1992). *Wetlands of India*. Ashish Publishing House, New Delhi.

Chaudhari, U.S., Johri, Seema and Chaudhari, P.R. (1998). Trophic status of Chatri Lake in the vicinity of Amravati City. *Proceedings of Eighty Fifth Session of the Indian Science Congress*, at Hyderabad, 12998. Part IV, pp. 196.

Chaudhari, P.R., Sarkar, Rekha, Vyas Smita and Sitre, S.R. (1994). Plankton indicators of sanitary status of groundwater sources. *Proc. All India Seminar on Sanitary and Plumbing Services*. 12-13 February 1994, The Institute of Engineers (I), Nagpur Local Center, 23-28.

Daniel, J.C. (1985). India's wetland resources. In: Thorsell J.W. (ed.), *Conserving Asia's National Heritage: The Planning and Management of Protected Areas in the Indo-Malayan Realm*, 39-42, Gland : I U C N.

Ganapati, S.V. (1957). Limnological studies of the upland waters in Madras State. *Arch. Hydrobio*. **53**:3-61.

Hansen, E.A., Sawason, D.H. and Tolsted, D.N. (1980). Irrigation of intensively cultured plantations with paper mill effluent. *Tappi*, Nov., **63(11)** : 139-143.

Hashimoto, T. and Yokota, H. (1965). Edaphological studies on the utilization of waste pulp liquor. *J. Sci. Soil Manure*, Tokyo, **36** : 231-234.

Jhingaran, V. G. (1991). *Fish and Fisheries of India*. Hindustan Publishing Corporation, New Delhi.

Jorgensen, J.R. (1965). Irrigation of slash pine with paper mill effluent. *Bull La St University Div Engg. Res.* (80), 92-99 C.A. 63 (18979).

Khambatta, S.J. and Ketkar, C.M. (1969). Present scheme for treatment and utilization of paper mill effluent on land for irrigation. Paper presented in the *Symposium on Problems Connected with Disposal of Waste from Various Industries Round About Poona*. Nov., pp. 29-30

Kolte, S.S. and Arkatkar, S.S. (2002). Prevention of natural water pollution by utilization and reuse of industrial effluents – a case study. *Proc. of National Conference on Pollution Prevention and Control in India : IAEM*, organized by Visvesvaraya Regional College of Engineering, Nagpur and Indian Association for Environmental Management, Nagpur during march 2-3, 2002:127-132

Kundangar, M.R., Sarwar, S.G. and Shah, M.A. (1991). Ecological Monitoring of Dal Lake, *Quarterly report* submitted to Government of Jammu & Kashmir.

Mckee, F.J. (1957). Dairy waste disposal by spray irrigation. *Sew. Indust. Wastes*, **29**:1-157

Mc Cornich, L.L. (1959). *Effect of paper mills wastewater on cattle, crops and soil*. M.S. Thesis, Lousiana State University.

Mishra, J.K. and Negi, M.S. (1985). *A Report on Kabar Lake Ecosystem*. Wildlife Division, Ranchi.

Palmer, C.M. (1969). A composite rating of algae tolerating organic pollution. *J. Phycol.* **5**:78-82.

Rajannan, G. and Obliswamy, G. (1979). Effect of paper factory effluent of soil and crop plants. *Indian J. Environ. Health*, **2**:120-130.

Reddy, M.R. Jivendra and Jain, S.C. (1981). Paper mills effluent for sugarcane irrigation. *IAWPC Tech. Annual*, **8**:129-146.

Rohlich, G. (1995). Objectives and procedures for a study of spray irrigation of Dairy wastes. *Proceeding 12th Indust. Waste Conf. Purdue University Ext Series* **94** :636.

Somashekhar, R.K., Gowda, M.T.G., Shettigar, S.I.N. and Srinath, K.P. (1984a) Effect of industrial effluents on crop plants. Ind. *J. Environ Health* **26(2)**:126-146

Somashekhar, R.K. (1984b). Phytoplankton constituents as indicator of water quality. A study of River Cauvery. *Int. J. Environ. Studies*, **23**:209-215.

Sreenivasan, A. (2002). Limnology of upland lakes and reservoirs or Tamil Nadu, India. *Indian Hydrobiology*, **5(2):**143-154

Srivastava, R.K. (1991). Effect of paper mill effluent on seed germination and early performance of radish and onion. *Journ. Ecotoxicology and Environmental Monitoring*, **1**:13-19.

WWF India (1987). *Wetlands-Status and Management in India : An Overview*. Environmental Services Groups, Delhi.

Yadava, Y. S. and Choudhary, M. (1984). The Problem and potentialities of beel fisheries of Assam. In : *Seminar on Guidelines for Science and Technology Policies for Assam*, June 4-6, Gauhati (Abst.).

Advancing Frontiers of Ecological Researches in India, 537-544, 2006
Edited by A.K. Kandya & Asha Gupta
Published by Bishen Singh Mahendra Pal Singh, Dehra Dun, India

Aspergillus - An Important Aeroallergen

A.A. Saoji
Department of Botany.
Institute of Science, Nagpur

Abstract: Fungal spores form an important constituent of the bioaerosol. Fungi are often well adapted to airborne dispersal of their spores having either tall conidiophores that penetrate into or through the laminar boundary layer or specialized liberation mechanisms that forcibly project spores through this layer. In the course of evolution the fungi have probably exploited the wind for their dispersal more thoroughly than any other group of organisms and consequently dominate the air spora. Fungal spores in air and sediments dust are of considerable significance due to the fact that numerous species are pathogenic and can cause diseases in human, animals and plants.

Fungal spores are found to be clinically very significant allergens for both type I and type II hypersensitivity. Strom (1924), presented a work in which he claimed that asthma was particularly prevalent in the lowlands and he attributed this to the higher concentration of fungal spores in damp and low-lying places. He found relatively pronounced hypersensitivity to *Aspergillus fumigatus*. Hansen (1928), published reports on several cases of asthma which could be brought about by fungi of *Aspergillus* and *Penicillium* species. Since then there have been many reports on hypersensitivity to fungal spores. It is also reported that human activity accounted for the most long distance travel of fungal pathogens and that in the absence of host plants fungi were dispersed with an extreme slowness.

Prince and Mayer (1976) published reports on several cases of asthma which were brought about by fungi of the *Aspergillus* and *Penicillium* species. *A. niger* and *A. nidulans* are often associated with *A. fumigatus* in Aspergillosis of bird. *A. terreus* has been isolated from cellulitis and subcutaneous granulomas in man and animals, but *A. terreus* and A. clavatus are rare etiologic agents. *A. restrictus* (a member of the ubiquitous, A. glaucus group) was the etiologic agent in a widely disseminated fatal infection.

Aspergillus species have been recognized as important allergen since 1873 (Blackly 1873), *A. fumigatus* Fres., may cause allergy in man (Austwick 1966) and even produce toxins Agrawal and Shivpuri (1974) found that *A. flavus* and *A. tamarii* gave severe reactions, while *A. niger* and *A. versicolor* were only moderately antigenic and concluded that these *Aspergillus* species possess "Species Specific" antigens.

Hesseltine (1974) cited research suggesting that it may cause liver cancer in humans and animals. Williams (1974) reported mycocardial aspergillosis may be a major or fatal complication in a patient with another primary and predisposing disease, or in a patient who is debilitated by cytotoxic or immunosuppressive drugs.

Clinical studies have also been conducted at some centers in India, (Banglore, Delhi, Calcutta, Gwalior and Jabalpur) to identify the potential fungal allergens. Shivpuri (1980), reported *Aspergillus* species, Alternaria, Candida, Helminthosporium, Mucor, Phoma, Rhizopus are identified as most offending fungal allergens causing type I hypersensitivity.

Aspergillus and *Penicillium* spores represent some of the most abundant aeroallergens in different parts of the world. The qualitative and quantitative reports have pointed out that *Penicillium* predominates in most regions, but it is replaced by *Aspergillus* in humid tropics (Lezana – Fernandes *et al.,* 1990; Ghani *et al.*, 1993).

The genus *Aspergillus* is one of the basic components of the atmospheric mycoflora in most species of the world. Previous reports of Nagpur region (Giri, 1995; Kakde 1996) suggest the presence of *Aspergillus* topping the list of aeromycoflora. The species of *Aspergillus* genus form a very large proportion of all the moulds. They are to be found almost everywhere on every conceivable type of substratum. A good working knowledge of the genus is essential to any one under taking serious work on problems of mouldy deterioration. The activities of aspergillii however are not entirely destructive in nature. Numbers of *Aspergillus* species are also useful in the various ways to mankind.

Several workers have studied airborne *Aspergillii*. (Kramer 1979; Hudson, 1969) Rati and Ramalingam (1973) isolated thirty-four species of *Aspergillus* from the outdoor air of Mysore. Rati *et al.,* (1980) recorded twenty-seven species of *Aspergillus* in poultry shed at Mysore.

Giri (1995) and Kakde (1996) reported thirteen species of airborne *Aspergillus* from indoor as well as outdoor environment of Nagpur.

For Nagpur City Chati (1998) has worked out a detailed quantitative and qualitative study of *Aspergillus* in intramural and extramural environment.

Aspergillii are employed in several industrial processes because of their great enzymatic activity. In the East, strains of *Aspergillus oryzae* having long been used for the saccharification of rice starch in the production of sake and similar potable liquors. Strains belonging in the same group are used in the manufacture of soy sauce. More recently the addition of fungal α – amylase to flours at the mill or bakery has been practiced. The enzyme is derived from *Aspergillus* oryzae, *A. niger*, *A. awamori* (Pyler, 1988; Kulp 1993). Recently the xylanolytic system of the fungus *Aspergillus nidulans* was described. Among the endoxylanses produced by this fungus, the endoxylanase X_{24} has appropriate biochemical properties that make it suitable for its application in bread making (Fernandez et al., 1996). The gene encoding this enzyme has been characterized (Perez – Gonzalez et al., 1996). Strains belonging to *A. oryzae* are used in the manufacture of soy sauce and of the enzymic mixture sold under the names "Takadiastase" and "Polyzime". In Japan, A. wentii is employed in processing soybeans because of its ability to loosen the hard tissue of the bean.

The black *Aspergillii* or *A. niger* group, are the fungi, commonly called as black mould. The genus is widely distributed. The air everywhere seems to contain the conidia of this organism. The soil, too, contain the spores of the Aspergillii. The black Aspergillii, long known to be capable of producing considerable amounts of oxalic acid from sugar, are now used for the successful commercial production of citric acid. Citric acid is one of the most important bulk – produced organic acids. Some 400,000 tons are produced per year, largely by processes involving *Aspergillus niger* (Mattey, 1992). Citric acid is primary product of *A .niger* metabolism. In order to force a microorganism to excrete a primary metabolite, the microorganism must be grown under suboptimal conditions for biomass production, for citric acid accumulation, the fungus is grown in a medium lacking Manganese ions but with a high initial sucrose concentration and high dissolved oxygen tension. (Roehr *et al.*, 1992). *A. niger* can detect copper even in traces and this technique is known as bio-assay. (Mulder, 1938).

A number of antibiotics have been isolated from *Aspergillus* Mich cultures (Bilai, 1963). *Aspergillus fumigatus* Fres. Produce fumigatine, helvolic acid (fumigacine), Fumagilline and Gliotoxin. Fumigatin possesses antibacterial activity against several Gram-positive and Gram-negative bacteria. Fumigatin and its derivatives have found no practical application to their toxicity. Helvolic acid is active chiefly against Gram-positive bacteria, pathogenic and phytopathogenic. It is inactive against fungi.

Fumagillin possess comparatively low activity against bacteria and fungi. But it is interesting and important as an antibiotic active against protozoa and viruses (Bilai, 1963). Fumagillin was introduced into medical practice as a therapeutic preparation for amoebic dysentery. It has also proved to be highly effective in treating nosematosis an invasive disease of bees caused by parasitism of the microsporidian *Nosema apis* in the epithelium of their mid-gut.

Aspergillus nidulans (Eidam) Wint, produced nidulin and nor-nidulin (antibiotic) possesses weak antibacterial activity and is inactive against many species of bacteria. Nor-nidulin possesses higher antibacterial activity.

Aspergillus ustus Bain, produce ustin which possesses higher antibacterial activity. *Aspergillus flavus* and *A. humicola* produce humicolin. Humicolin possesses antifungal action. *Aspergillus terreus* Thom, produce geodin, terrecin and terreic acid. Geodin exhibits antibacterial activity chiefly against Gram-positive bacteria. Gram-negative bacteria are sensitive only to much higher concentrations of geodin. Terreic acid exhibits antibacterial activity, but only against certain species of bacteria. *Aspergillus candidus* Link, produce candidulin. Candidulin is active against several Gram-positive and Gram-negative bacteria and fungi. (Yojima, 1955). A*spergillus* ochraceus Wilhelm and *A. melleus,* produces mellein (Ochracin). Mellein (Ochracin) possess weak antibiotic action.

Aspergillus flavus Link, produce Kojic acid, aspergilli (granegillin), hydroxaspergillic, neooxyaspergillic acids and orizazin.

The antibacterial spectrum of aspergillic acid has been studied with respect to many species of bacteria. The hydoxyaspergillic acid is not antibiotically active against bacteria sensitive to aspergillic acid. Neooxy-aspergillic acid possesses a broad spectrum of antibacterial action. It is active against certain types of cholera, streptococcal and dysenteric phages.

Kojic acid develops on various substrates especially in warm regions. It is used in some branches of the fermentation industry to obtain proteolytic and amylolytic enzymes.

Aspergillus velutinus produce velutinin. Velutinin possesses antifungal properties.

Aspergillus niger and several other species are often found on exposed foodstuffs and cause decay. In doing so, some species produce toxic substances called Mycotoxins. The most important of these is aflatoxin.

Aflatoxin was discovered in 1960 when 100,000 or more turkey pullets died in England from a new disease, called on that time the Turkey X-disease (Sargeant *et al.*, 1961). Many ducklings and calves died soon after having consumed Brazilian peanut meal that was discovered to be mouldy. The mould producing the violent toxin was *Aspergillus flavus* and the toxin was aptly named aflatoxin. The peanut fruits and seeds are invaded only after the vines have been pulled and when fruits are still on the plants. Africa, India and South East Asia are regarded as the "high-aflatoxin-risk areas" (Mirocha and Christensen, 1974).

Aflatoxins are highly oxygenated heterocyclic compounds. The eight characterized aflatoxins are B1, B2, G1, G2 (named on the blue or green fluorescence in UV light) M1, M2 (first detected in milk of cows fed on groundnut meal) and B2a and G2a (derivatives of B1 and G2). Among the eight forms, aflatoxins B1, B2, G1 and G2 are more commonly encountered. These may occur together or independently. Hesseltine (1974) cited research suggesting that it may cause liver cancer in humans and animals.

A. fumigatus, A. flavus, *A. niger*, and other species are animal and human pathogens that cause a group of diseases collectively known as aspergillosis (Greenberger, 1993; Shah 1994). *A. fumigatus* is the usual etiologic agent of pulmonary aspergillosis of the typical, severe invasive organs. *A. fumigatus* and *A. niger* may colonize an ectatic bronchus, without invasion of lung parenchyma, forming a compact "fungus ball" (Shah 1996; Panchal *et al.*, 1997). Strains of A. flavus oryzae group, A. phialiseptus, A. clavatus and *A. terreus* also cause pulmonary aspergillosis. *A. niger* is a common agent of otomycosis.

A. niger and *A. nidulans* often are associated with *A. fumigatus* in aspergillosis of birds. *A. terreus* has been isolated from cellulitis and subcuteneous granulomas in man and animals, but *A. terreus* and *A. clavatus* are rare etiologic agents. *A. restrictus* (a member of the ubiquitous A glaucus group) was the etiologic agent in a widely disseminated fatal infection. In a high percentage of cases, aspergillosis is secondary to a primary disease, which has impaired the patient's immunologic defences.

Mycocardial aspergillosis may be a major or fatal complication in a patient with another primary and predisposing disease, or in a patient whom do cytotoxic or immunosuppressive drugs debilitate. (Williams, 1974). Pulmonary aspergillosis is often superimposed upon tuberculosis, silicosis, an anatomical abnormality or deficient immunologic response related to corticosteroid or antibiotic therapy. *A. flavus* and *A. fumigatus* (occasionally other species) may invade and colonize nasal sinuses. In patients with diabetes mellitus or other conditions such as corticosteroids therapy which may reduce resistance, the fungus may produce cellucitis of the face and orbit which may spread upto the central nervous system. Traumatic corneal ulcers, especially if treated with corticosteroids, may become infected with *Aspergillus*. Thickened, disintegrating nails may be invaded by hyphae of *Aspergillus*. But it is difficult to evaluate the etiologic role of fungi (except dermatophytes) which grow in this type of substrate.

Aspergillus was found to be cellulolytic and are biodeteriogens. Mould deterioration of library books, leather and textile is a matter of great concern (Singh *et al.*, 1990). Several species of *Aspergillus* grow on leather and cloth fabrics, reducing their commercial value and imparting a musty odor to shoes and clothing. In humid tropical planets, where *aspergillii* are especially troublesome, people have to keep their wardrobes as dry as possible to prevent shoes and clothing from being covered with a heavy growth of *Aspergillus*. It is common custom in such regions to have all wardrobes specially wired so that one or more electric light bulbs can be kept burning continuously to keep the atmosphere dry.

The *Aspergillus* is proved to be very important hazardous mould, having economic importance and is one of the basic components of atmospheric mycoflora in most parts of the world. The role of this mould in allergy, mycotoxicoses, biodeterioration and pulmonary infection of man and animals is well known. Since this mould possess a very potent enzymes, acids and mycotoxins it can destroy organic materials very effectively.

REFERENCES

Agarwal, M.K. and Shivpuri, D.N. (1974). *Adv. Pollen Spore Res.*, 1, 78-128.

Austwick, P.K.C. (1966). The role of spores in the allergies and mycoses of man and animals – Colston papers 18, 321-338.

Bilai, V. I. (1963). Antibiotic - producing microscopic fungi. Elsevier publishing company, London.

Blackley, C.H. (1873). Hay fever: its causes, treatment and effective prevention: experimental researches (London).

Chati, S.S. (1998). *Aspergillus* studies in intramural and extramural environments of Nagpur (M.S). Ph.D. Thesis, Nagpur university, Nagpur.

Fernandez -Espinar, M.T., Valles, F. Pinaga, J. A. perez Gonzalez and D. Ramon (1996). Construction of an *Aspergillus nidulans* multicopy transforming for the xlaB gene and its use to purify the minor X24 xylanase. Appl. Microbiol. Biotechnol. **44:**338 -341.

Ghani, T., Kumar S, and Kale, S.R. (1993). Aeromycoflora inside and outside a college library. Indian Int. *J. Environ. Hlth. Res.* **3:** 49 -58.

Giri, S.K. (1995). Aeromycological studies in indoor environments of Nagpur (Mah). Ph.D. Thesis, Nagpur university, Nagpur.

Greenberger, P.A. (1993). Allergic bronchopulmonary Aspergillosis. In : allergy - Principles and Practice, 4th edn. Ed. Middlenton E. Jr. Reed C. E., Ellis Franklin Adkiuson N. Jr., Yunginger J. W., Busse, W. W. St. Louis Mosby : 1395-1414.

Hansena, K. (1928). Uber Schimmelpilz asthama verhandal. Deutsch Gesellsch. Inn. Med. Kongr.

Hesseltine, C.W. (1974), Natural occurrence of mycotoxins in cereals. Mycopath, et Mycol. Appl., **53:** 141-153.

Hudson, H.J. (1969). Trans, Br. Mycol. Soc., 52, 153 -159.

Kakde, U.B. (1996). Aeromycological studies of vegetable and fruit market environment, Ph.D. Thesis, Nagpur university, Nagpur.

Kramer, C.L. (1979). Fungus : Moss and fern spores. In Aerobiology : The Ecological Systems Approach. (ed. R. L. Edmonds), PP. 24 -41, 73 -84. Dowden, Hutchinson, and Ross, Inc., Stroudsburg. Pennsylvania.

Kulp, K. (1993). Enzymes as dough improvers, P-152-178. In B.S. Kamal and C.E. Stauffer (ed.) Advances in baking technology VCH. Publisher Inc., New York.

Lezana- Fernandez, M.A. Vetazquez - Monrroy J.O., Alvarez-Lucas, H.C., Fernandez de Hoyos , K and Rodriguez- Alvarex, T.E.F. (1990). Informacion epidemicologia para la elaboracion- del diagnostico de salud en Mexico la Mortalided. Algunos factors condicionantesy.

Mattey, M. (1992). The production of organic acid. Crit. Rev. Biotechol. **12**: 87-132.

Mirocha, C.J. and Christensen, C.M. (1974). Fungus metabollites toxic to animals. *Ann. Rev. phytopathol* **12**: 303-330.

Mulder, E.O. (1938). Sur l'influence de cuivre sur la croissance des microorganisms. Ann. Ferm., n. s. 4: 513-533 (Abs. In Rev. Appl. Mycol. **18:** 241. (1939)).

Panchal, N., Bhagat, R., Pant, C., Shah, A. (1997). Allergic bronchopulmonary aspergillosis. The spectrum of computed tomography appearances Respir Med. **91:** 213-219.

Perez-Gonzalez, J.A., L. de. Graaff, J. Visser and D. Ramon (1996). Molecular cloning and expression in saccheromyces cerevisiae of two *Aspergillus nidulans* xylanase genes. Appl. Environ. Microbiol. **62:** 2179-2182.

Prince, H.F. and Meyer, G.H. (1976). An up-to-date look at mold allergy. *Ann. Allergy* **37:** 18.

Pyler, E.J. (1988). Enzymes in baking, P- 132 -183. In E. J. Pyler (ed.), baking, science and technology, Vol. I. Sosland Publishing Co., Merriam, Kans.

Rati, E. and Ramalingam, A. (1973). Selective methods for the enumeration of airborne *Aspergillus flavus* and Aspergillii J. Mysore Univ. **26 B:** 168- 171.

Rati, E., Jayaprakash, K.B. and Ramalingam, A. (1980). Airspora of Poultry shed at Mysore. Indian J. Microbiol **20:** 6-12.

Roehr, M., C.P. Kubiuk, and J. Kominek (1992). Industrial acids and other small molecules, P-91-131. In J. W. Bennett and M. A. Klich (er.), *Aspergillus*-biology and industrial applications. Butterworth- Hinemann, Stove ham, Mass.

Sargeant, K., Sheridan, A. and O'kelly J. (1961). Discussion Nature, **192:** 1096-1097.

Shah, A. (1994). Editorial – Allergic bronchopulmonary Aspergillosis: An emerging disease in India. *Indian J. Chest Dis. Allied Sci.* **36:** 169 -172.

Shah, A. (1996). Allergic bronchopulmonary Aspergillosis : Diagnosis and Management. Cardio-Thoracic J. **1(11):** 32-43.

Shivpuri, D.N. (1980). Clinically important pollen, fungal and insect allergen for naso-bronchial allergy. Aspects Allergy. *Appl. Immunol.* **13:** 19 -23.

Singh, A.B., Chattergi, M., Singh, B.P. and Gangal, S.V. (1990). Airborne viable fungi in library: Before and after agitation of books. *Ind. J. Aerobiol.* **3(1&2):** 32-38.

Storm Van Leeuwen, W. (1924). Bronchial Asthma in relation to climate. Proc. Royal Soc. Med.

Williams, A. H. (1974). Aspergillus myocarditis. Am. J. Clin., Path 61, 247 -256.

Yojima, T. (1955). Candidulin *J. Antibiot.*, Tokyo. 2, 225.

Advancing Frontiers of Ecological Researches in India, 545-554, 2006
Edited by A.K. Kandya & Asha Gupta
Published by Bishen Singh Mahendra Pal Singh, Dehra Dun, India

Eco-Management of Trees in a Poor-Rich Urban Ecosystem

S.B. Chaphekar
14, Dhus Wadi, Thakurdwar, Mumbai.

Abstract: Hazards of growing trees in urban environment are focused, with case study of Mumbai city. It is emphasized that each of these hazards can be mitigated with determined effort of citizens and authorities, for which citizens have to accept trees as citizens with an equal right to inhabit urban areas.

INTRODUCTION

It is an accepted fact today that the world in general and this country in particular, are urbanizing rapidly, both in terms of area and human population. To cite an example, the number of urban settlements in the state of Maharashtra with population above 10,000, was 163 in 1961; it has increased to 241 in 1991. Of these, big towns with population exceeding one lakh, increased from 12 in 1961 to 27 during the same period. Percentage of urban population of big towns also increased from 60.49 % to 77.83 % during the same period. This rate of growth is higher than the national average. (Diddee, 2002).

As a result, urban problems are also more severe in the State. Whereas craving for proximity to Nature replete with greenery that improves microclimate (Chaphekar and Kulkarni, 1973), is continuously on an increase, thanks to mainly to cultural attachment of man to nature and strong awareness of the same, spaces available for growing and maintaining greenery are continuously on the decline. It is on this background that we should look at the problems of trees in cities and the ecological role that they can and should play.

On the other hand, whereas the Indian Forest Policy states that 33% of land in the country should be covered by forests, the present cover is somewhere near 19% and is still giving way to non-forest land uses, expanding urbanization being a major cause. Under this situation, it becomes imperative that urban authorities make all efforts to compensate for the deficit of national forest area by greening cities nationwide.

Problems Faced by Trees in Urban Areas

There are problems however, that impede planting and maintaining green areas in urban settlements. Here, I will cite some of those encountered in Mumbai, the city that presents all types of environmental problems. With its size, population and commerce increasing over the decades (Table-1), acutely deficient greenery is on the mind of every citizen. It is also a paradox that this is a city which contributes substantially

Table 1: Area and Population of Mumbai over Years

S.No.	Year	Area	Population
1	1931	24.19 sq.miles	11,61,383
2	1951	67 sq.km	29,94,444
3	1971	437.7 sq.km	59,70.575
4	1991	437.7 sq.km	99,57,000
5	2001	437.7 sq.km	1,20,00,000

to national wealth (e.g. > 30% of Income Tax collected in the country is from this city alone). At the same time, almost 50% of the population lives without proper housing. Since several Indian cities are replicating the situation in Mumbai, it is necessary to point out and attend to immediately, the environmental short-comings of this city, so that they are not repeated elsewhere.

A city is defined as an assemblage of non-agricultural elite (Bose, 1965), still it is realized that green spaces and trees have definite role to play in cities (Table-2). The city authorities and NGOs strive to plant and nurture trees (sometimes with big fanfare), but they face several hurdles (Table-3), often leading to a defeat of the purpose. These are discussed in some detail here.

Table-2: Role of Trees in Urban Areas

1.	Beautification of urban settings
2.	Improvise pattern in landscapes
3.	Complement to architectural designs
4.	Habitat for avifauna
5.	Shelter to human population, especially poor, nomads and small entrepreneurs
6.	Softening of micro-climate, through absorption of radiations and increasing humidity
7.	Sorption of air pollutants leading to cleaner environment.

Table 3: Urban Stresses that Affect Plants

1.	Lack of space
2.	Soil characters – Compaction
3.	Soil characters – Concrete structures
4.	Soil characters – Municipal waste
5.	Soil characters – Industrial waste
6.	Quality of available water
7.	Quantity of available water
8.	Chopping, Trimming and Misuse of trees
9.	Road widening, Excavation, Erosion of structures
10.	Salt spray (in coastal areas)
11.	Traffic hazards
12.	Air pollution.

1. **Lack of Space** – A city where human population is continuously on an increase, even pavements are occupied by people in shifts (Table-4) and existing footpaths are shrinking in size; shortage of space for plantation of trees need not be over-emphasized. Also, the chances of trees planted in available space have thin chance to survive.

Table 4: Uses in Shifts of a Piece of Footpath in 'C' Ward in Mumbai

S.No.	Time of day	User / Uses
1	0500 – 0700	Milk vendors
2	0530 – 0630	Sorting of newspapers
3	0630 – 1100	Vegetable vendors, Newspaper vendors
4	0600 – 0830	Tea stall
5	0800 – 0830	School children with escorts waiting for bus,
6	0830 – 1200	Fruit vendors,
7	0830 – 1200	'Q' for bill payment
8	1100 – 1600	Field workers for footpath, telephone repairs
9	1500 – 1900	Vada – Pao stalls
10	1600 – 1630	Mothers waiting for children's return
11	1600 - 2000	Vegetable vendors
12	1600 – 2100	Fruit vendors
13	1930 – 2200	Cooking and catering of fast food
	2200 – 2230	Cleaning for bed-making of the houseless

2. **Soil Characters – Physical -** Whatever be the original type of soil in a city, it is greatly altered during urbanization. Hence each tree is planted in a mix of red earth, sand, manure, etc. as per horticultural practice. Unfortunately, the extreme use of space, trampling, storage of material, etc., put physical pressure on ground near a sapling or the base of a tree trunk resulting in soil compaction. This affects adversely movement of water and nutrients in soil and breathing of roots, causing stunted and uneven growth.

3. **Soil Characters–Concrete Structures** – Concretization / Asphalting of footpaths right up to the base of a tree trunk prevents water from reaching roots. Underground concrete structures like foundations of buildings, pillars and roads also prevent proper growth of roots of trees, affecting growth of shoots.

4. **Soil Characters – Municipal Wastes -** Urban municipal waste is a mix of paper, rubber, leather, kitchen waste, plastic, glass, packaging material, tins and so many things. To the already long list, recently there is addition of computer waste. As a group, this litter affects growth of root systems. Whereas kitchen and garden waste create problems of eutrophication, toxic nature of paints, tins and computers etc. inhibit root growth. Non-degradables like plastic, glass, metal waste etc. obliterate soil aeration, with as yet not-well-understood impacts, though plastics were once blamed for collapsing of trees.

5. **Soil Characters – Industrial Wastes** – Solid waste dumps from industries affect plants in the same way as some of the wastes from municipal dumps. In case of hazardous waste however, death of tree roots takes place.

6. **Quality of available water** – Since municipal authorities are not in a position to supply enough water, citizens are advised to use ground water or recycled water. Organized facility for recycled water is practically non-existent, and where it exists, as in some 5-star hotels, water is mostly recycled in toilet flushes. Ground water drawing has reached such a level, that in many places salinity levels do not permit its use for irrigation (Chaphekar *et al.* 2001). Hence, most of the plants in the city have to do with inadequate irrigation.

7. **Quantity of Water** – Unlike in forests, trees are unable to get enough water from ground in cities. As a result, they are the first to suffer thirst whenever there is any event affecting water supply position.

8. **Chopping, Trimming and Misuse** – These happen to trees anywhere, but more so in cities, as acts of sheer vandalism at worst or for safety of people at best. Trimming of roadside trees is often undertaken for safe movement of vehicles, especially double-decker buses. It also happens sometimes when branches of trees touch windows and verandahs of roadside houses. Summary cutting down of trees is done for new construction projects like buildings, roads and flyovers, railway tracks, bridges, etc. Mutilation of trees is done for personal benefit by unaware / unscrupulous people – vendors for using cut trunks as tables and platforms, branches for hanging their ware and beddings, nailing of trunks and branches for hanging clothes, ware and even deities! Cooking fast-food under trees in commercial areas costs lower branches and leafy twigs, to give pitiable shapes and appearance to trees.

9. **Road widening, Excavations and Erosion of structures** - Construction of infra-structural facilities like widening of roads leads to harvesting of roadside trees. Similarly, excavation for laying of pipes for water supply or drainage, telephone cables, gas pipelines, etc. is a common phenomenon in cities, which expose root systems of trees, thus weakening them. A peculiar phenomenon of collapse of soil under parapets along sea front due to wave action, leaving roots dangling and washed by salty tidal water was noticed in some parts of Marine Drive boulevard.

10. **Salt Sprays** – The salt spray rubbing on laminas cause foliar damage in trees along sea front. This is more evident during late summer and early monsoons, when strong south-west monsoon winds dominate coastal atmosphere.

11. **Traffic Hazards** –Trees found hazardous to vehicular traffic are trimmed frequently. Sometimes they are even removed, the problem getting more serious with increasing number of vehicles in cities, particularly since the last five years. Presently about eleven lakh vehicles ply on city roads on a working day.

12. **Air Pollution** – Most of our cities also double as industrial cities. Urban pollution is hence a mix of industrial effluents and auto-exhaust The more phyto-toxic gases in cities are sulphur dioxide, Ozone and suspended particulate matter (SPM). SO_2 causes interveinal leaf injury

and tip burns. O_3 leads to spotting of mature leaves while SPM gets deposited on laminar surfaces clogging the stomata. All the pollutants weaken trees (Treshow, 1981) and spoil tree aesthetics. Reduction in lead content of petrol in recent years has reduced substantially Pb content of dust in urban atmosphere and on leaf surfaces (Chaphekar *et al.* 2001), compared to the earlier observations (Joshi, 1989). The menace of dust pollution is more severe at road junctions as per findings of recent air monitoring by the Corporation.

13. **Management of Trees in Cities** - From the above stresses that affect growth of trees, few can be readily resolved. From the census of trees carried out in Mumbai in the year 1998-99, it was revealed that there are hardly 5,00,024 trees in the city of 12 million people, i.e. more than 19 persons per tree (Shashikumar, 1998-99). The ratio should be of a reverse nature for the city to be green, with all the advantages to be available to an urban settlement.

1. Since it is not easy to increase space for trees in the island city, we should strive to keep the existing trees in good health, so that they can carry out their ecological role more effectively. For this purpose several suggestions are made to relevant authorities (Chaphekar *et al.* 2001). These are of feasible nature, requiring marginal financial but more human resources. Some are stated below.

2. Compacted soil at the base of the tree should be upturned at least once a month.

3. Asphalting or concretization of footpath should not reach up to the base of tree trunk. A different design of a continuous, one meter wide belt of land rather than a square or round pit should be provided for tree plantation on footpath along a road.

4. Municipal waste should be separated into perishable and non-perishable. The former should be composted while the latter alone should go for landfill. Since all available land in the city is already filled to capacity, it is suggested that the long-lasting municipal solid waste should be sorted for recyclable and non-recyclable material, the latter should be ground to powder form and then used for landfill at a place that should be covered by greenery.

5. Industries should be compelled to deal with their solid waste strictly as per the existing regulations.

6. Available water should be augmented with nitrogen-rich water from sewage treatment plants. Waste water from domestic use should also be recycled after treatment and testing for suitability for irrigation. Suitable DEWAT (decentralized water treatment) designs are available.

7. Deficit in water supply should be made good with rain water harvesting, a huge potential that has not been harnessed so far.

8. Each tree should be under joint observation of authorities and volunteers from reputed NGOs, so as to avoid chopping and misuse by members of the public. The NGOs should be empowered for the purpose. If it is absolutely necessary that a tree has to be removed, feasibility of its transplantation should be explored. For unavoidable trimming, a branch should be sawed (not hacked), as near as the trunk as possible, the cut surface smoothened with sand paper and coated with non-metallic paint.

9. Digging and excavations should be undertaken with diligent care to avoid injury to roots and completed at high speed. The exposed roots should be kept damp with a soaked cloth during the time of exposure.

10. A natural phenomenon of salt spray cannot be avoided on sea front, hence salt spray tolerant species like Bhend (*Thespesia populneoides), Samudrafal (Barringtonia sp.),* Jangli Badam *(Terminalia catappa),* Ber (*Zizyphus jujuba),* Suru (*Casuarina), Parkinsonia,* etc. should be planted.

11. In addition to the recommended strategy for trimming of trees, a strategy should be followed during planting of new trees along roads, i.e. planting of trees that will not affect smooth flow of traffic. Old trees should be accommodated in suitable traffic pattern design when road-widening is being done. For plantation, selection of dust and pollution tolerant tree species, e.g. Satvin *(Alstonia scholaris),* Nerium, Ixora, Kumbha (*Careya arborea),* Yellow Oleander (*Thevetia neriifolia),* etc. should be done. (For an exhaustive list, readers are requested to refer to a C.P.C.B. publication, 'Guidelines for Development of Green Belts,' 2000). A methodical mix of tall and small trees and shrubs in rows forms an effective screen for roadside

residents for protecting them from vehicular pollution. Periodic checks of vehicles for their roadworthiness in terms of exhaust quality and quantity should be done meticulously, Measures for smooth flow of traffic need to be taken, since it is well known that stop-and start traffic pattern generates more pollution .

12. Industries presently contribute about 35% of the total air pollution in the city while rest of it is mostly from automobiles. To reduce pollution in the city, which is sufficiently high to cause foliar injury in many plant species in some places, roadside plantation should be undertaken as given above. Industries should be compelled to take necessary steps to control pollution at source. Whatever low level of pollution still escapes to the environment, should be sorbed by tolerant species (as given in the CPCB publication cited above) in the green belt around big industries and around industrial estates consisting of small-scale industries. Some pollution sensitive species like mango, Indian Cork tree, Drumstick should be interspersed in the green belt, to serve as early indicators of pollution so as to make timely corrections.

DISCUSSION

In the essentially man-created habitat, numerous problems have been unwittingly created by man, on which there are no easy solutions to regain Nature. At the same time, most citizens need trees in their vicinity, for cultural, economic and ecological reasons. It is high time then, that the trees are accepted as essential members of human settlement and hence of community. Once this acceptance is imbibed, viable solutions will be found to provide trees their rightful place in a human settlement – spaces will be provided (not only on paper) – for trees in human habitats, water requirement of trees will be budgeted by municipal authorities and care for plantation and nurturing will go beyond the concern of horticulturists alone.

Once space is earmarked for growing trees, it is possible to expect more trees. In a survey of Thane city next to Mumbai in the year 2000, it was recorded that tree canopies covered about 38% in some residential areas, 37% in some industrial premises, 37% in an educational establishment, but only 10% along new roads where extensive plantation has just been undertaken (State of Environment, Thane, 2000).

Similarly, water can be provided through decentralized rain water harvesting and through recycling of domestic waste water. If these two things are incorporated in designing residential colonies, all the need of irrigation water will be more than fulfilled, as per present calculations.

Care for trees and their roots during construction work is a matter of suitable work ethic and discipline, which develops with a basic philosophy of not harming plants, just as we seek not to hurt workers and other humans during construction work. In the same way, pollution should be controlled at source – industries or automobiles – so that very little pollution takes place that does not prove toxic to plants. An average plant tolerates pollution to a much higher level than what is permissible as per legislated norms (Table-5). Very low level of SO_2 pollution actually helps plants as a source of nutrient sulphur. Hence, trace level SO_2 pollution will actually lead to improvement of the ecological efficiency of plants.

Table 5 : Tolerance of plants to SO2, as compared to CPCB standards

Permissible in Residential areas	–	80 ug/cum for 24 hours	60 ug/cum Annual average
Tolerated by Sensitive plants	1820 ug / cum for 1 hour	468 ug / cum for 8 hours	21-44 ug / cum for growing season

CONCLUSIONS

In view of shrinking forest cover and increasing urbanization, it is necessary to increase urban area under tree cover. There are several aspects of urban environment that impede tree growth. None of them however, are insurmountable, with a determined and dedicated contribution from the citizens. Healthy urban trees greatly improve urban environment, in terms of aesthetics, bird habitat, pollution abatement and moderation of micro-climate. It is possible to implement existing scientific knowledge for the purpose.

REFERENCES

Bose, N.K. (1965). Calcutta: A premature metropolis. Scientific American, 213(3): 91-102.

Chaphekar, S.B. and Kulkarni, S.S. (1973). Light, heat and roadside plants. Indian Forester, 99(9): 579-581.

Chaphekar, S.B. and others (2000). Guidelines for Developing Green Belts. Central Pollution Control Board, New Delhi.

Chaphekar, S.B., Lattoo, C.S. and Khalap. K.P. (2001). Health of Greenery in Brihan Mumbai Area of Mumbai Metropolitan Region. Report to MMRDA, Mumbai.

Didde, J. and Wakhare, A. (2002). Urban Trends in Maharashtra. In: Geography of Maharashtra. p. .231.

Joshi, N.C. (1991). Experiments in Phytomonitoring of Urban Atmosphere. Ph.D. thesis to Mumbai University.

Sasikumar, B. (1998-99). Tree Census of Brihan Mumbai. Report to Tree Authority, Mumbai.

State of Environment of Thane. (2000). Report of Thane Municipal Corporation.

Treshow, M. (1981). Environment and Plant Response. McGraw Hill Publ, N.Y.

Report – Air Pollution in Brihan Mumbai (2002).

Air Pollution - Report on Website of Maharashtra Pollution Control Board, Mumbai.

Photorama of Bombay (1981).

Advancing Frontiers of Ecological Researches in India, 555-559, 2006
Edited by A.K. Kandya & Asha Gupta
Published by Bishen Singh Mahendra Pal Singh, Dehra Dun, India

Zinc accumulation in strains of two Blue-Green Algae: *Chlorococcum humicolum* and *Phormidium foveolarum*

S.D. Shankhadarwar & P.B. Nandkar
Deptt. of Botany, Nagpur University, Nagpur

Abstract: Metal tolerating species are continuously being released into the environmental through industrial effluent discharge. Microbial biomass is well known in the bioabsorption of appreciable amounts of heavy metals. *Chlorococcum humicolum* and *Phormidium foveolarum* tolerant to high level of zinc were obtained in repeated 50 subcultures. The level of Zn just permitting detectable growth was raised from 3 to 20 mgl-1 for *Chlorococcum humicolum* and 2 to 10 mgl^{-1} for *Phormidium foveolarum*

INTRODUCTION

Many algae can uptake metal from the aquatic bodies. The microbial biomass passively bind large amount of metals. The binding of metal referred as bio-absorption (Gadd and White 1993; Horikoshi *et al.*, 1981). Bioabsorption has important implication in wastewater treatment. Shumate *et al. (1978)* have pointed out that micro organism could be used to remove heavy metals from wastewater. Cu, Zn, and Co are the most important metals often found in effluent discharge from industries involved in manufacturing of alloys and in galvanization. Zinc is common heavy metal, environmental contaminant and also as essential trace element for organism. It is mostly used in electroplating operation. It plays a vital role in the physiological process of many organisms. However, high concentration of Zn can be toxic to many organisms. Bioabsorption phenomenon of heavy metals by fungi and bacteria is well documented (Gadd 1988; Leuf *et al.,* 1991; Voleskcy and Holon 1995) but little work has been done using the cyanobacteria. Algae are often dominant organisms at moist sites combining high level of Zn with high pH. The present paper reports isolation of *Chlorococcum humicolum* and *P. foveolarum* resistant to high level of Zn from industrial sewage treatment plant.

MATERIALS AND METHODS

Chlorococcum humicolum and *P. foveolarum* collected and brought into the laboratory from industrial sites. Inhibitory level of Zn was determined by selecting the range of 0.01 to l0 mg zinc to algae in culture flask and growth for 20 days at 26- 30°C with photon flux density 20 to 30 μmol photon m-2s-^{1} Tolerance of Zn was increased from 3 to 20 mgl^{-1} *Chlorococcum* 2 to 10 mg Zn for *Phormidum foveolarum* grown in BG-11 deficient of zinc.

RESULTS

Resistance of *Chlorococcum humicolum* and *P. foveolarum* to zinc increased by repeated sub culturing at inhibitory levels of Zn to the alga. The process was repeated 50 times, leading to gradual increasing strong inhibitory level. Metal resistant algae were analysed for metal on atomic absorption spectrophotometer. All strains were washed three to four times with deionised distilled water. The heavy metals released in the water during washing were treated as adsorbed metals, while the metals that were not released in water during washing were treated as absorbed metals. Total accumulated metals were calculated by adding adsorbed and absorbed amount of heavy metal. *Chlorococcum humicolum,* has accumulated 18.07 mgl^{-1} per dry weight of alga. It includes 11.76 mgl-1 absorbed and 6.31 mgl^{-1} adsorbed zinc per gm dry matter of alga. Similarly *P. foveolarum* has accumulated 7.2 mgl^{-1} per dry weight of alga. It includes 5.7 mgl^{-1}absorbed and 1.5 mgl^{-1} adsorbed zinc per gm dry matter of alga.

Algae	Metal	Strain	Absorbed metal	Adsorbed metal	Total accumulation
C. humicolum	Zn	20mg	11.76	6.31	18.07
P. foveolarum	Zn	10mg	5.70	1.50	7.20

DISCUSSION

Tolerance of algae to zinc raised from 3 mgl^{-1} to 20 mgl^{-1} for *Chlorococcum humicolum, and* 2 mgl^{-1} to 10 mgl^{-1} for *P. foveolarum* in 40 to 50 subcultures to attend maximum tolerance. In this study, morphological changes in *Chlorcoccum humicolum* and *P. foveolarum* were recorded at higher level of metallic concentration. The strains of *P. foveolarum* showed constriction at the septa in trichome and beaded

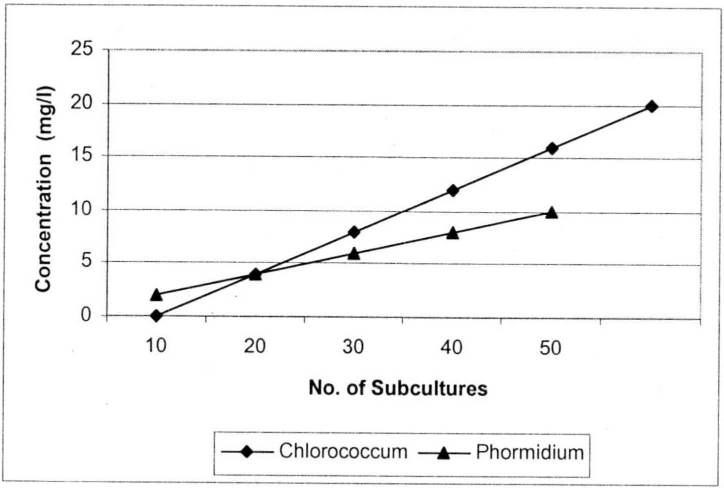

Fig. 1. Isolation of zinc accumulating mutagens of Cyanobacteria

appearance in l0 mgl-1 of zinc whereas in *Chlorcoccum humicolum* no prominent change has been recorded but it produces many autospores in the cells, which may be due to higher concentration of zinc in the growth media. Harding and Whitton (1976) reported increased development of prostrate branches in *Stigeoclonium tenue* as compared to upright growth in streams containing 20 mgl^{-1} Zn. Whitton *et al.* (1981) obtained that Zn tolerant mutants are likely to be filled at many field sites, such as surrounding of mine wastes; typically a wide range of micro-habitats with different metal levels. Mutants of *Anacystis nidulans* tolerant to high levels of zinc were obtained by repeated sub-culturing at concentrations of the metal sufficient to cause a marked reduction in growth rate. The level of Zn just permitting detectable growth was raised from 1.45 to 16.5 mgl^{-1} Zn after 75 subcultures (Shehata & Whitton 1982). Kumar (1964) selected certain strains of *Anacystis nidulans* by successive sub-culturing in liquid media containing increasing concentration of antibiotics. Singh (1978) described that spontaneous mutants of *Nostoc muscorum* grown in presence of these elements not only tolerated them but required them for growth with N_2 or NO_3 as the nitrogen source. Wikbores and Ukeles (1982) investigated the capacities of four algal species i.e. *Monochrysis lutheri Isochrvsis galbana, Dunaliella cuchlora* and *Phaeodutvium triconutum* for their adaptation to grow in an inhibitory concentration of Cu, Cd and Zn in laboratory cultures. All the species showed some capability to adapt to one or more metals in several subcultures. Wild (1968) obtained copper rich soils in Zimbabwe with *Anabanea, Nostoc, Oscillatoria*, *Phormidium* and *Scytonema* for metal uptake by micro

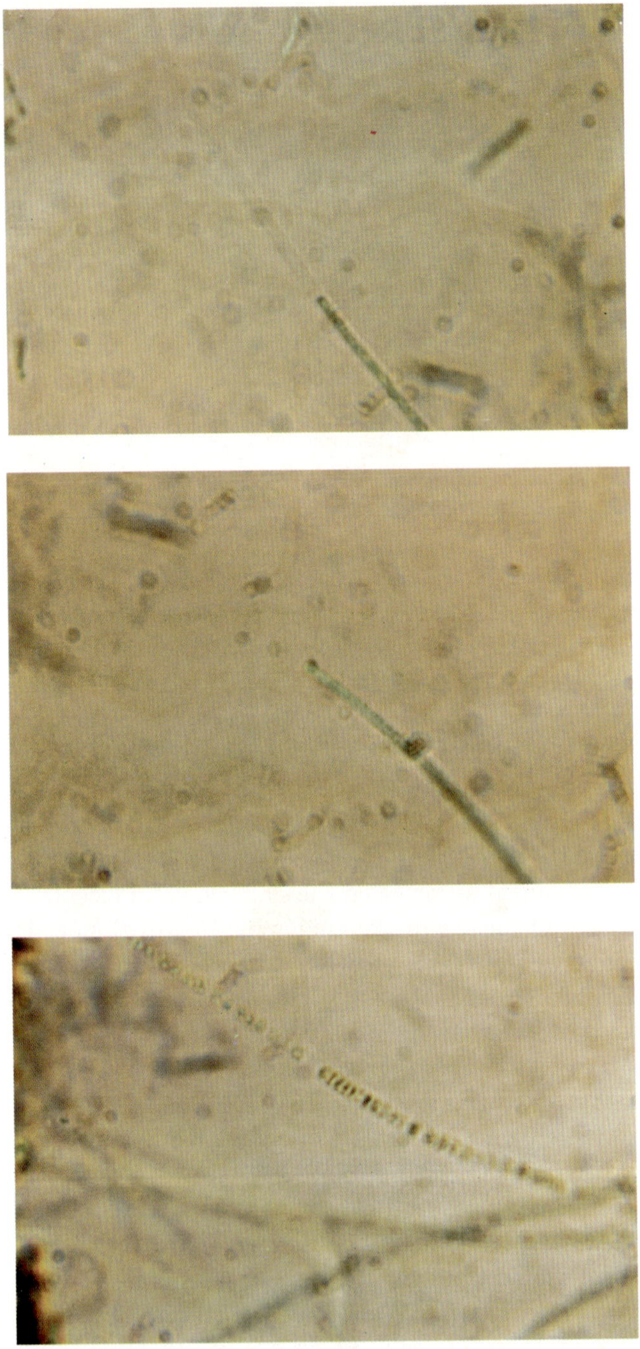

Mutants of *Phormidium foveolarum*; Mutant Strain with Zn 10 mg^{-1}
1. Control 2. Mutant 3. Mutant

organisms based on surface adsorption. Uptake of cadmium by *Chlorella vulgaris* has been reported by Khummongkal *et al.* (1982). The influence of cobalt, nickel, copper and cadmium on *Anacystis nidulans* was tested in the laboratory in order to help the behaviour of blue -green algae at mining sites (Whitton and Shehata 1982).

REFERENCES

Gadd, G.M. (1988). Accumulation of metals by, microorganism and algae. In Biotechnology. A comprehensive treatise Vol. 6b. special microbial processes. eds. H.J. Rehm and G. Reed. VCH verlagsg esellschaft. Weinham. Germany. pp. 401-433.

Gadd. G.M. and White. C. (1993). Microbial treatment metal pollution. Microbiol. Ecol. 4: 303-317.

Harding. J.P.C & Whitton, B.A. (1977). Environmental factors reducing the toxicity of' zinc to *Stigeoclonium tenue* . *Br. Phycol .J.* **12**:17-21.

Horikoshi, T. Nakajima, A. and Sakaguchi. T. (1981). Studies on the accumulation of uranium by microorganism. *Elir. J. Appl. Microbiol. Biotecnol.* **12**: 90-96.

Khummongkl, D. Conterford. G. S. and Fryer, C. (1982). Accumulation or heavy 'metals in unicellular algae. *Biotechnology, Bioengineering* **24**:2640-2660.

Kumar, H.D. (1964). Effects of certain toxic chemical metal and mutagens on the growth' of blue green, alga *Anacystis nidulans. Can J. Bot.* **43**:1523-1532.

Leuf. E. Prey. T. and K. Ubick. C.P. (1991). Biosorption of zinc 'by fungal mycelial wastes. *Appl. Microbiol. Biotecnol.* **34**: 688-692.

Shehata, F.H.A. Whitton. B.A. (1982). Zinc tolerance in strains of blue green alga *Anacystis nidulans. Br. Phycol. J.* **17**:5-12.

Shumate, S.E., Strandberge, G.W. and Parrot, J.R. (1978). *Biotecnol. Bioeng. Symp.* **8**:13

Singh, N.H. Vaishanpayan, A. and Singh, R.K. (1978). Evidences for the involvement of a genetic determinant controlling Fictional specificity of group VI elements in the metabolism of N_2 and $N0_3$ in the blue green alga *Nostoc muscorum. Biochem, Biophys. Res. Comm.* **81**:67-74.

Volesky, B. and Holan, Z.R. (1995). Biosoi'ption of heavy metals. *Biotecnol. Porg.* ll: 235-250.

Weikberes. G.H. and Ukeles. R. (1982). Growth and adaptation in esturine unicellular alga in media with excess copper, cadmium, zinc and effects of metal contaminant algal food on Grassostreae Virginica Larvae. *Ecol. Prog. Ser.* 7: 101.206.

Whitton, B.A., Gale, N.L. & Wixson, B.C. (1981). Chemistry and plant ecology of zinc rich wastes dominated by blue -green algae .*Hydrobiologia.* **83**: 331-341.

Whitton, B.A. and Shehata, F.H.A. (1982). Influence of cobalt, nickel. copper und cadmium on the blue green algae *Anacystis nidulans*. Envorn. Pollution **27**: 275-281.

Wild. H. (1968). Geobotanical anomalies in Rhiodesia. 1. The vegetation of copper Bearing soil. Kirkia. **7**:17.

SECTION SEVEN

GRASSLAND ECOLOGY

Advancing Frontiers of Ecological Researches in India, 563-573, 2006
Edited by A.K. Kandya & Asha Gupta
Published by Bishen Singh Mahendra Pal Singh, Dehra Dun, India

Primary Productivity and Dry Matter Dynamics of Floating Grasslands at Keibul Lamjao National Park (KLNP), Manipur, N.E. India.

Daisy Angom and Asha Gupta.
Ecology Research laboratory
Deptt. of Life Sciences, Manipur University
Canchipur, Imphal-795003, India

Abstract: The changes in hydrological regime of Loktak lake (A Ramsar Site) has affected the vast floating mat at Keibul Lamjao National Park - KLNP (Latitude 23^0 83' N to 25^0 86' N; Longitude 93° 03' E - 93^o78' E) at Manipur, N.E. India which has ceased to experience its annual draw-down cycle since the construction of Ithai barrage in 1984. KLNP, the only natural abode of famous and endangered brow-antlered-deer, *Cervus eldi eldi*, supports unique floating grasslands over a heterogenous mass of organic matter locally called as phumdis. The present study is undertaken to evaluate the primary productivity and dry matter dynamics of grasslands of KLNP. The seasonal and annual dry matter production and its transfer through different producer compartments viz. Live shoot, dead shoot, litter and root were analysed. Net accumulation rates, disappearance rates and turnover were calculated. It was found that 51% to 60% of TNP was shared by root components. The seasonal dry matter accumulation (TNP) ranged from 558.2 gm^{-2} to 4085.5 gm^{-2} with daily production rates varying from 4.53 to 13.45 gm^{-2} day^{-1}. Total dry matter disappearance was about 69% of total input in the system, thus revealing the accumulation of organic matter in the floating phumdis of KLNP.

INTRODUCTION

Total dry matter production is considered to be a measure of its systems efficiency to fix energy in all its compartments (Rochow, 1974). Thus, the net primary production is the rate of storage of organic matter in plant tissue in excess of respiratory utilization during the experimental period (Odum, 1971) and is the main source of energy for other trophic levels in an ecosystem. Hence it is an essential step in understanding the functioning of an ecosystem (Laurenroth and Whitman, 1977).

In production analysis four major compartments viz. above ground (live shoot), standing dead, litter and below ground (roots) are distinguished.

The inter compartmental interactions reflect the functioning of the system in time and space.

Very little information is available on the productivity studies of grasslands in north-eastern India. Some of the earlier works on productivity in the north-eastern regions are those of Ramakrishnan and Ram (1988), Yadava and Kakati (1990) and Devi *et al.* (2001). However till now no attempt has been made to study the dry matter dynamics and productivity in the unique floating grasslands of Manipur which are existing on partially decomposed organic matter locally called as phumdis. Hence, the present paper deals with the dry matter dynamics and productivity studies in floating grassland of Keibul Lamjao National Park (KLNP), Manipur, N.E. India.

Study Area

Manipur, a state in the north-eastern part of India lies in the north of the Tropic of cancer and extends between 23°8′ N to 25°68′ N latitude and between 93°2′E to 94°78′ longitude. The sub-tropical monsoonic climate with moderate temperature, rainfall and relative humidity prevails in the state. Fig. 2. shows the ombrothermic diagram for the study site.

Keibul Lamjao National Park is located under Bishnupur district of Manipur and the only natural habitat of Brow antlered deer, Sangai (plate I, II and III). It is situated at about 47 km from Imphal city and lies between 24°27′ to 24°31′ N latitude and 93°53′ to 93°55′ E longitude. It is

Plate-I: Brow-antlered deer (*Cervus eldi eldi*)

Plate-II: National Park (KLNP) with Pabot hillock, a shelter for the deer

Plate-III: *Zizania latifolia* the favoured food plant of the deer growing luxuriantly on the floating park.

a part of Loktak lake, the only fresh water lake of the state and is one of the Ramsar sites. The park is situated in the south-eastern side of Loktak lake and covers an area of about 40 sq. km (Fig.1).

About 2/3rd of the park is covered by Phumdi (a heterogeneous mass of soil, vegetation and organic matter in different stages of decay) which floats on water. Earlier the park experienced annual draw-down cycle. Thus the phumdi floated when the level of the lake rose and settled at the bottom when the level of the water dropped during the dry periods. However, after the construction of Ithai barrage in 1984 the annual cycle ceased.

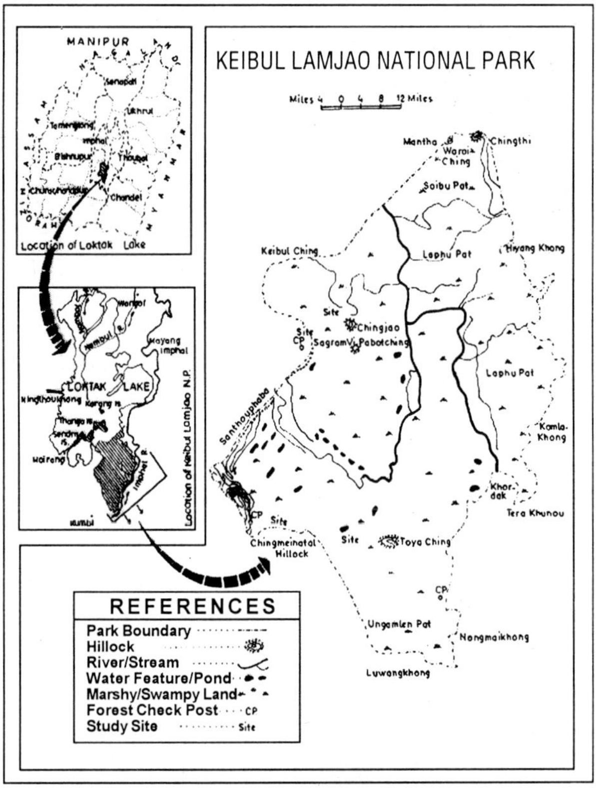

Fig. 1. General Feature of Keibul Lomjao National Park

Phumdi varies in thickness from a few centimeters to two meters. The humus of Phumdi is of black colour and very spongy with large number of pores. About 20% of phumdi remains above water.

Phumdis are nutrient efficient and contain 1.2-1.5% of Nitrogen, 0.045-0.129% of Phosphorous and 1.09-1.31% of Potassium. Under phumdis exist a clear water zone whose depth varies with the fluctuation of water level. The water below phumdi is acidic, pH ranging from 3.8-7.2. Nutrient concentration is low in nutrient except for Sodium. The average annual values of some of the nutrients are : N = 0.085 mgl-1, P = 0.0005 mgl^{-1}, K = 2.35 mgl^{-1}, Ca = 7.77 mgl^{-1}, Na = 12.43 mgl^{-1} (Angom unpublished). Under water there is a nutrient rich sediment.

The phumdis besides providing food and shelter to Sangai, also are the chief sources of vegetables, fodder, fuel and even medicine for the people (Daisy and Gupta, 1999 a, b). However, it has been found that most of the plant collected by man are also the preferred food plants of Sangai. The overlapping of such utilization pattern of the park resources has resulted in declining population of these plants (Daisy and Gupta, 1999a).

While the phumdi sinks with the weight of a man or a cattle, it can readily bear the weight of the Sangai. The deer has developed adaptation to the phumdi. The animal moves on the phumdi in a characteristic style. While the hind legs are bearing the body weight, the forelimbs are raised high and the animal takes a small hop, maintaining the balance not sinking in the phumdi and because of this style of locomotion that Gee (1960) called Sangai, the dancing deer.

MATERIAL AND METHODS

For the present investigation, Harvest method was employed. The Net Primary productivity was calculated using the method described by Milner and Hughes (1968) and the net accumulation and disappearance rates were estimated using the formula given by Sims and Singh (1971, 1978 a and b). Turnover rates of shoot and root components were also found out.

RESULTS

Shoot Net Production (SNP)

The maximum net shoot production value was observed during Rainy season (701.25 gm^{-2}) and the daily production rate recorded was 5.74 gm^{-2} day^{-1}. The minimum value was recorded in winter season (238.95 gm^{-2}) with a daily production rate of 1.94 gm^{-2} day^{-1}. The intermediate value shown in summer season was 582.25 gm^{-2} with a daily production rate of 4.85 gm^{-2} day^{-1}.

The annual net shoot production was 1798.75 gm^{-2} yr^{-1} and the rate of production was 4.92 $gm^{-2}day^{-1}$. Table-1 shows the values of seasonal and annual SNP.

Root Net Production (RNP)

Of the three seasons, maximum root net production was observed during summer season (1032.5 gm-2 yr-1) and the daily production rate was 8.60 gm-2 day-1. It is followed by rainy season (918.25 gm-2) with

Table-1: Seasonal variations in Shoot Net production(SNP), Root Net Production(RNP) turnover rates and Total Net Primary Production(TNP) of KLNP, Manipur.

Season	SNP		Shoot Turnover	RNP		Root Turnover	TNP	
	Net Production gm^{-2}	Rate of Production $gm^{-2}day^{-1}$		Net Production gm^{-2}	Rate of Production $gm^{-2}\ day^{-1}$		Net Production gm^{-2}	Rate of Production $gm^{-2}\ day^{-1}$
Winter	238.95	1.94	0.355	319.25	2.59	0.190	558.2	4.53
Summer	582.25	4.85	0.290	1032.5	8.60	0.493	1614.75	13.45
Rainy	701.25	5.74	0.310	918.25	7.52	0.29	1619.5	13.27
Annual	1798.75	4.92	0.955	2286.75	6.26	0.973	4085.5	11.19

Table-2: System Transfer function of dry matter for KLNP grassland.

Compartment	KLNP grassland
TNP-SNP	0.440
TNP-RNP	0.559
SNP-SD	0.462
SD-L	0.332
SNP-L	0.154
L-LD	1.456
RNP-RD	0.97
TNP-TD	0.687

TNP = Total Net Primary Production
SNP = Shoot Net Primary Production
RNP = Root Net Primary Production
SD = Standing Dead Shoot Production
L = Litter Production
LD = Litter Disappearance
RD = Root Disappearance
TD = Total Disappearance

a daily production rate of 7.52 gm-2 day-1.

The least production was found in winter season (238.95 gm-2) with and the daily production rate was obtained as 2.59 gm^{-2} day^{-1}

The annual root net production was found to be 2286.75 $gm^{-2}yr^{-1}$ with a daily production rate of 6.26 gm^{-2} y^{-1} (Table-1).

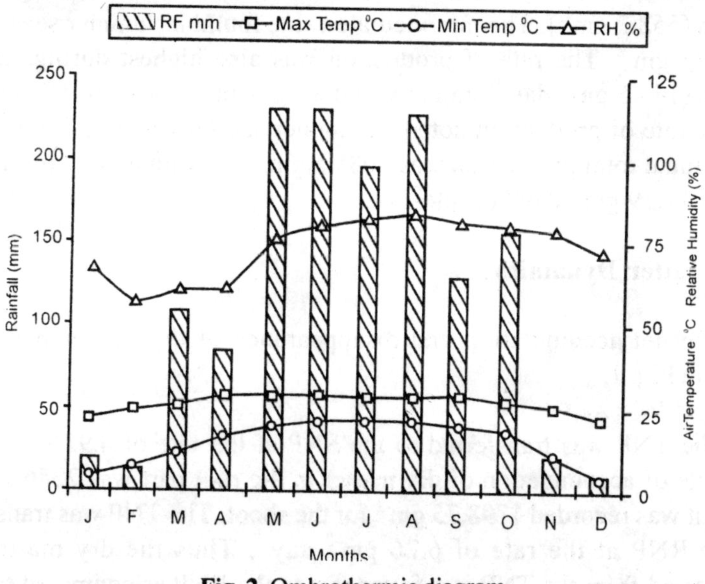

Fig. 2. Ombrothermic diagram.

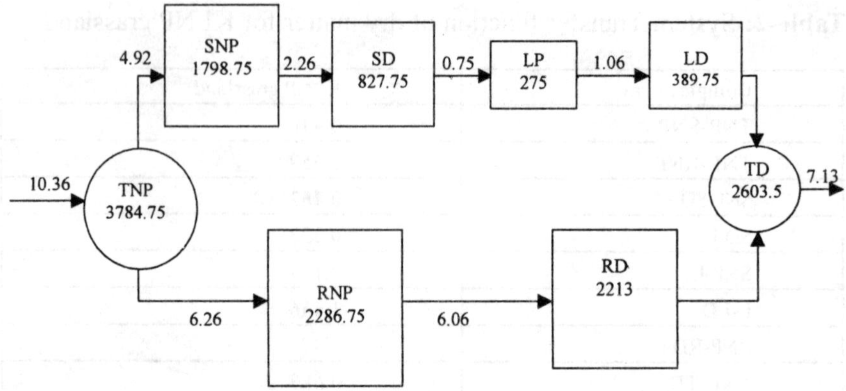

Fig. 3. Dry matter dynamics in floating grasslands of KLNP, Manipur. Values along arrows indicate accumulation and disappearance rates (gm^{-2} day^{-1}) during the study period through the different compartments viz.

TNP = Total Net Production. LD = Litter Disappearance.
SNP = Shoot Net Production. RNP = Root Net Production.
SD = Shoot Disappearance. RD = Root Disappearance.
LP = Litter Production. TD = Total Disappearance.

Total net Primary Production (TNP)

The maximum total net primary production was recorded during rainy season (1619.5 gm^{-2}) while the minimum was observed during winter season (558.2 gm^{-2}). The intermediate value found in summer season was 1614.75 gm^{-2}. The rate of production was also highest during summer season (13.45 gm^{-2} day^{-1}) and lowest during winter season (4.53 gm^{-2} day^{-1}). The rate of production noted in summer season was 13.27 gm^{-2} day^{-1}. The annual total production was 4085.5 gm^{-2} yr^{-1} with a daily production rate of 11.19 gm^{-2} day^{-1} (Table-1).

Dry Matter Dynamics

The net accumulation and disappearance of dry matter have been given in Fig. 1.

The TNP was transferred to the SNP at the rate of 4.92 gm^{-2} day^{-1}. The rate of accumulation of dry matter in the root part was 2286.75 gm^{-2} while it was recorded 1798.75 gm^{-2} for the shoot. The TNP was transferred to the RNP at the rate of 6.26 gm^{-2} day^{-1}. Thus the dry matter was transferred from the TNP at a faster rate to the RNP as compared to SNP.

DISCUSSION

In KLNP total net production (TNP) was found to be 4085.5 gm^{-2}. Out of these 44% was transferred to SNP and 56% to RNP (Table-2). The rate of root disappearance was found to be higher than litter disappearance which indicates the stability of the system and it is in accordance with Sims and Singh, (1971). The total disappearance was slower (7.13 gm^{-2} day^{-1}) than the rate of accumulation (10.36 gm^{-2} day^{-1}). Thus there is more accumulation of organic matter in the grasslands of KLNP.

The recording of highest TNP during the rainy season could be attributed to the suitable ambient temperature and sufficient moisture in the floating grassland of KLNP. Consequently; the plants grow vigorously attaining maximum total biomass in rainy months after which the biomass begins to decline and attains minimum value in winter months. It was also found out that precipitation has favourable impact on Net Primary Production (NPP) as Shoot Net Production (SNP) and Total Net Production (TNP) were found to be higher in rainy season. Some other workers like Pandey and Bhatt (1988), Chaturvedi *et al.* (1988) have also found that NPP is strongly related with precipitation than with other climatic variables. Singh and Singh (1980) opined that with the onset of rainy season with favourable moisture and temperature conditions, the annuals come up and new shoots are formed from dormant perennials resulting in a rapid biomass built up.

Root Net Production (RNP) values were found to be highest in summer and lowest in winter. The decline in cold dry months in root biomass may be due to translocation of reserve food materials of root for growth and development of new tillers of grasses. The maximum biomass in rainy months may be due to massive translocation of assimilates to root biomass prior to the senescence of live shoot biomass as suggested by Bray *et al.* (1959), Billore and Mall, (1976).

Rate of accumulation of dry matter (10.36 gm^{-2} day^{-1}) found in floating grasslands of KLNP (Fig. 3) was found to be higher than other workers like Yadav and Kakati (1984) (6.7 gm^{-2} day^{-1}) in a grassland community at Imphal, Dhaulakhandi *et al.* (2000) (4.5 gm^{-2} day^{-1} and 5.3 gm^{-2} day^{-1} for upper and lower sites) in alpine grassland of Western Himalaya, Devi *et al.*, (2001) (2.04 gm^{-2} day^{-1}) in a grassland community of Canchipur (Manipur), Bawa, (1995) (1.19 gm^{-2} day^{-1}) in a Himalayan grassland at

Shimla and Chaturvedi *et al.* *(1978)* (0.66 gm-2 day-1) in a grazing land under pine forest in Central Himalaya.

However the rate of transfer of dry matter from TNP to RNP (6.26 gm^{-2} day^{-1}) was faster than from TNP to SNP (4.92 gm^{-2} day^{-1}) which indicates that maximum dry matter is not available to ungulates as fodder showing stressed conditions in the floating grasslands of KLNP. Of the total life forms 28.17% of Hemicryptophytes (HC) and 28.46% of Chamaephytes (C) are represented in KLNP which through their adaptations like runners, stolons and rhizomes can survive in the adverse conditions (Daisy and Gupta, 1999b, and Bawa, 1995). Their dominance indicates the changed hydrological regime in KLNP. Disappearance rates of litter (1. 06 gm^{-2} day^{-1}) was found to be slower than root disappearance (6.06 gm^{-2}). This may be due to overall slower turnover of shoot biomass (Table-1). The higher net production rate 10.36 gm^{-2} day^{-1} may be the reason for higher accumulation of organic matter in floating Phumdi of KLNP pointing out lesser microbial activity in grasslands of KLNP.

ACKNOWLEDGEMENTS

One of use, Daisy Angom expresses her thanks to WWF - India for granting her financial assistance. The authors also acknowledge the help bought from park authorities for permitting to work in the Park.

REFERENCES

Bawa, R. (1995). Biomass dynamics of Himalayan grasslands. Int. J. Ecol. Environ. Sci. **21**:25-36.

Billore, S.K. and L.P. Mall (1976). Nutrient composition and inventory in a tropical grassland. Plant and Soil **45**:509-520.

Bray, J.R., D.B. Lawrence and L.C. Pearson (1959). Primary productions in some Minnesota terrestrial communities Oikas, **10**:38-49.

Chaturvedi, O.P., A.K. Saxena and J.S. Singh (1988). Structural and functional analysis of grazingland under pine forest in Central Himalaya. Acta Oekologia **9(2)**:167-178.

Daisy Angom and Asha Gupta (1999a). Utilization of plant Resources by peripheral communities of Keibul Lamjao National Park (KLNP) area. International Conference on Wetland and Aquatic Resources, Nainital, India.

Daisy Angom and Asha Gupta (1999b). Vegetational analysis of grasslands of Keibul Lamjao National Park (KLNP), Manipur N.E. India. International Conference on Wetland and Aquatic Resources, Nainital, India.

Devi, A.R., B.M. Sharma and H.M. Sharma (2001). Net primary production of a grassland community of Canchipur (Manipour), India. Indian Journal of Environmental and Ecoplanning **5**:321-328.

Dhaulakhandi M., Govind S. Rajwar and Pramod Kumar (2000). Primary productivity and system transfer functions in an alpine grassland of western Garhwal Himalaya. Tropical Ecology **41(1)**:99-101.

Gee E.P. (1960). Report on the status of the Brow-antlered deer of Manipur (India). Oct.- Nov. 1959 and March, 1960. Communicated by the survival Service Commission of the International Union for the Conservation of Nature and Natural Resources Journal, Bombay Natural History Society. **57(3)**:597-617.

Lauenroth, W.K. and W.C. Whitman (1977). Dynamics of dry matter production in a mixed grass prairie in West Northern Dakota. Oecologia (Berlin) **27**:337-351.

Milner, C. and R.E. Hughes (1968). Methods for the measurement of the primary production of grassland. IBP Handbook No. 6 Blackwell Scientific Publications, Oxford.

Odum, E.P. (1971). Fundamentals of Ecology. W.B. Saunders, Philadelphia.

Pandey, N. and A.B. Bhatt (1998). Climatic and topographic influences on net primary production under grazed and ungrazed conditions in an alpine grazing land of Garhwal Himalaya. Indian Journal of Forestry Vol. **21(4)**:387-391.

Ram, J. (1988). Phytosociology and primary productivity of an alpine grassland in Garhwal Himalaya. Ph.D. Thesis, Kumaon University, Nainital, India.

Ramakrishnan, P.S. and S.C. Ram (1988). Vegetation, biomass and productivity of seral grassland of Cherrapunji in N.E. India. Vegetatio **74**:47-53.

Rochow, J.J. (1974). Estimation of above ground biomass and primary productivity in a Missouri forest J. Ecol. **62**:567-577.

Sims, P.L., and J.S. Singh (1971). Herbage dynamics and net primary production in certain ungrazed and grazed grassland in North America, p. 59-124. In N.R. French (ed.). Preliminary analysis of structure and function in grasslands. Range Sci. Dep. Sci. Ser. No. 10. Colorado State University, Fort Collins.

Sims, P.L. and J.S. Singh (1978a). The structure and function of ten western North American grasslands III. Net primary production, turnover and efficiencies of energy capture and water use. J. Ecol. **66**:573-597.

Sims, P.L. and J.S. Singh (1978b). The structure and function of ten western North American grasslands IV. Compartmental transfer and energy flow within the ecosystem, J. Ecol. **66**: 983-1009.

Singh, K.P. and R.P. Singh (1980). Biomass and net production of herbaceous layer in tropical dry deciduous forest at Varanasi. Trop. Ecol. **21(1)**:47-58.

Yadav, P.S. and L.N. Kakati (1984). Biomass dynamics and net primary productivity in grassland community at Imphal. Int. J. Biometerol. **28**:123-145 (Supplement).

Advancing Frontiers of Ecological Researches in India, 575-584, 2006
Edited by A.K. Kandya & Asha Gupta
Published by Bishen Singh Mahendra Pal Singh, Dehra Dun, India

Standing State and Cycling of Nitrogen in Soil-Vegetation Compartments of a Grassland Ecosystem at Canchipur, Manipur

A. Radhapyari Devi, B. Manihar Sharma and H. Manoranjan Sharma
Ecology Research Laboratory, Department of Life Science
Manipur University, Canchipur, Imphal, Manipur – 795 003

Abstract: The present study was conducted in a grassland ecosystem located at Canchipur (Manipur) with an objective to estimate the plant tissue of nitrogen (%), its standing state ($g.m^{-2}$), seasonal and annual uptake, transfer and release to soil at two different sites. The concentration of nitrogen in above ground live shoots varied between 0.552% and 1.128% at Site I and 0.675% and 0.964% at Site II. In dead shoots, the concentration varied between 0.491% to 0.864% at Site I and 0.571% to 0.886% at Site II while the concentration in litter varied between 0.467% and 0.722% at Site I and between 0.409% and 0.701% at Site II. In the below ground compartment, the concentration varied between 0.606% and 0.932% at Site I and between 0.579% and 0.711% at Site II. The accumulation of nitrogen in the soil ranged from 607.65 $g.m^{-2}$ (October) to 865.10 $g.m^{-2}$ (July) at Site I and varied between 383.30 $g.m^{-2}$ (October) and 636.25 $g.m^{-2}$ (August) at Site II. Annually live shoots (0.106% to 0.160%), dead shoots (0.069% to 0.124%), litter (0.046% to 0.073%) and belowground parts (0.233% to 0.365%) retained very little amount of total nitrogen available in the plant and soil systems. The soil system retained 99.278% to 99.546% of the total nitrogen. The small fraction of nitrogen present in the biological systems in the present study indicated that the soil serves as major nitrogen reserve and annually some nitrogen is stored in the system for utilization during the next growing season.

INTRODUCTION

Nutrient cycling is one of the most important functional aspects of the ecosystem. The mineral components of an ecosystem operate in a dynamic state through a series of inputs and outputs of the essential elements. As ecosystems progress towards the climax conditions, they tend to become more conservative of nutrients, finally reaching the input-output ratio of one (Odum 1969, Jordan and Kline 1972). Primary productivity of any ecosystem basically depends on the supply of nutrients by the soil system (Whitkamp 1971) and the supply of essential elements and energy input (Rigler 1975).

Studies on the nutrient cycling in Indian grasslands are many (Chaturvedi *et al* 1988, Pandey and Singh 1990, Bhandari *et al* 2000). However, in the north-eastern India very little information is available till date pertaining to the nutrient cycling of grassland ecosystems (Pandey *et al* 1993, Ram and Ramakrishnan 1992, Kakati 1985, Devi, A.R. 1997). Therefore, the present study was undertaken to examine the pattern of storage and cycling of nitrogen in the soil-vegetation compartments of a grassland ecosystem located at Canchipur, Manipur (India).

Study Site

The present study site is located at Canchipur (24°44′ 53.88″ N latitude and 93° 56′ 1.104″ E longitude) in Imphal East district, Manipur at an altitude of 784 m above the mean sea-level. The climate is sub-tropical monsoonic. Mean maximum temperature varied between 21.8°C (January) and 30°C (September) while the mean minimum temperature ranged from 4.5°C (January) to 22.1°C (July). The average annual rainfall is 1330.4 mm. The mean relative humidity varied from 59.65% to 84% during the study period.

The soil is sandy loam in texture with high porosity (73.07%-87.97%). The soil is slightly acidic in nature (5.6–6.9) with an organic carbon content of 1.04 % to 2 77%.

Two grassland sites were selected for the present study. Site I (grazed) is subjected to all types of biotic disturbances whereas Site II (semi protected) is comparatively free from grazing and other biotic disturbances. The two study sites though located adjacent to each other showed marked variations in species composition due to local changes in physiography and soil characteristics.

MATERIAL AND METHODS

Plants biomass was harvested at ground level from nine 50 x 50 cm^2 quadrats in each study site at monthly intervals. Litter was collected separately. The herbage was sorted out into live and dead shoots. For the estimation of below ground plant biomass, three monoliths of 50 x 50 x 30 cm^3 were excavated at each sampling date and roots were washed in water. The collected samples were oven-dried at 80°C and weighed. Observations were taken in the first week of each month (June 1999 to June 2000).

Above ground net production (ANP) was calculated as the sum of increments in the live shoot biomass and concurrent increments, if any in dead shoot biomass (Singh and Yadava 1974). Belowground net production was calculated as the sum of significant increments in total root biomass. The net accumulation and disappearance rates were calculated following Singh and Yadava (1974) and Sims and Singh (1978a & 1978b).

Total nitrogen content of soil (0-30 cm) and plant samples were determined by macro and micro-Kjeldahl digestion method following Jackson (1958) and Misra (1968). Standing state of nitrogen for live shoots, dead shoots, roots and litter were separately determined by multiplying the dry weight of the components with its nitrogen content. Similarly, uptake transfer, return and release of nitrogen were determined by multiplying the nitrogen concentration values with net primary production, litter and litter disappearance values (Singh *et al* 1979).

RESULTS

Nitrogen Concentration

Values of nitrogen concentration (%) in the above ground and below ground parts showed marked seasonal variations (Table-1). Nitrogen concentration decreased as the live shoots turned to litter *via* standing dead. In the present study live shoots contain more amount of nitrogen than roots. The maximum concentration of nitrogen varied between 0.964% in Site II to 1.128% in Site I. The maximum concentration in roots varied from 0.711% in Site II to 0.932% in Site I. Generally Site I (grazed) have higher concentration of nitrogen than Site II (semi-protected) in all the producer compartments.

Accumulation in Soil

The accumulation of nitrogen in the soil showed marked seasonal variation in both study sites (Table-2). In Site I, nitrogen accumulation varied between 607.65 $g.m^{-2}$ (October) and 865.10 $g.m^{-2}$ (July). In Site II, the accumulation varied between 383.30 $g.m^{-2}$ (October) and 636.25 $g.m^{-2}$ (August). Generally Site I has more amount of nitrogen accumulation in the soil than Site II.

Table- 1: Concentration of Nitrogen (%) in the four primary producer compartments in the two study sites

Site	Producer *Compartment*	1999							2000					
		Jun.	Jul.	Aug.	Sep.	Oct.	Nov.	Dec.	Jan.	Feb.	Mar.	Apl.	May	Jun.
I	Live-Shoot	0.972	0.809	0.552	0.935	1.122	0.662	0.967	0.833	0.811	0.728	0.948	0.851	1.128
	Dead-Shoot	0.491	0.628	0.671	0.754	0.864	0.709	0.808	0.756	0.757	0.819	0.824	0.713	0.738
	Litter	0.500	0.573	0.550	0.688	0.722	0.567	0.570	0.697	0.531	0.467	0.687	0.686	0.656
	Belowground	0.606	0.619	0.687	0.744	0.745	0.831	0.932	0.717	0.802	0.871	0.653	0.882	0.843
II	Live-Shoot	0.964	0.866	0.914	0.728	0.678	0.703	0.705	0.686	0.675	0.766	0.696	0.690	0.892
	Dead-Shoot	0.886	0.828	0.668	0.591	0.571	0.602	0.621	0.613	0.588	0.686	0.789	0.788	0.827
	Litter	0.656	0.573	0.550	0.516	0.516	0.500	0.409	0.629	0.639	0.467	0.450	0.637	0.701
	Belowground	0.692	0.671	0.676	0.680	0.652	0.684	0.664	0.579	0.589	0.700	0.686	0.679	0.711

Table-2: Accumulation of Nitrogen ($g.m^{-2}$) in the soil (upto 30 cm depth) in the two study sites

Study Site	1999							2000					
	Jun.	Jul.	Aug.	Sep.	Oct.	Nov.	Dec.	Jan.	Feb.	Mar.	Apl.	May	Jun.
I	765.30	865.10	655.50	778.50	607.65	670.80	847.80	822.60	666.15	754.50	771.50	740.25	779.20
II	625.05	482.50	636.25	502.20	383.30	577.80	587.90	549.75	607.30	506.70	505.05	532.05	478.15

Seasonal Distribution in Plant/Soil System

Of the total nitrogen in the plant/soil system, 0.092% to 0.116% occurred in live shoots in Site I and in Site II it varied from 0.115% to 0.173% in live shoots (Table-3). The percentage in roots varied from 0.177% to 0.285% in Site I and between 0.251% and 0.488% in Site II. The percentage of nitrogen retained by soil varied from 99.481% (winter) to 99.623% (rainy) in Site I and from 99.171% (summer) to 99.474% (winter) in Site II. Thus only 0.377% to 0.829% of total nitrogen was found circulated among the various primary producer compartments at any time.

Uptake, Transfer And Release Of Nitrogen

In Site I, maximum uptake of nitrogen by plants was observed during winter season (2.852 g.m^{-2}) whereas in Site II maximum uptake was observed during rainy season (3.111 g.m^{-2}). Of these 29.50% in Site I and

Table-3: Distribution pattern of nitrogen in plant/soil system at mean standing state

(% of total)

Site	Season	Live-shoot	Dead-shoot	Litter	Roots	Soil
I	Summer	0.092	0.091	0.051	0.248	99.518
	Rainy	0.116	0.052	0.032	0.177	99.623
	Winter	0.103	0.073	0.058	0.285	99.481
	Annual	0.106	0.069	0.046	0.233	99.54
II	Summer	0.123	0.132	0.086	0.488	99.171
	Rainy	0.173	0.123	0.065	0.300	99.339
	Winter	0.115	0.089	0.071	0.251	99.474
	Annual	0.160	0.124	0.073	0.365	99.278

26.46% in Site II were retained in the roots and the rest was translocated to shoots (Table-4). Transfer of nitrogen from live shoots to dead shoot was maximum during winter season (2.043 g.m^{-2}) in Site I and during rainy season (3.576 g.m^{-2}) in Site II. Similarly, transfer from standing dead to litter was also highest during the same seasons in both the study sites. The rate of release of nitrogen was maximum in rainy season and minimum during summer season in both the study sites.

Table-4: Seasonal and annual uptake, transfer and release of nitrogen ($g.m^{-2}.yr^{-1}$) in the two study sites

Compartment	Site	Season			Annual
		Rainy	Winter	Summer	
Uptake in aboveground net production	I II	1.539 2.288	2.011 1.637	1.338 1.131	4.901 5.030
Uptake in belowground net production	I II	0.373 0.823	0.841 1.327	0.200 1.716	1.396 3.855
Total net uptake	I II	1.912 3.111	2.852 2.964	1.538 2.847	6.297 8.885
Transfer from live-shoot to dead-shoot	I II	1.030 3.576	2.043 1.798	1.049 1.448	4.236 4.930
Transfer from dead-shoot to litter	I II	1.232 2.585	1.686 1.983	0.834 1.345	3.747 3.214
Release through litter decomposition	I II	1.232 2.827	1.374 1.676	1.075 1.730	3.672 3.280
Release through root decomposition	I II	0.942 1.203	0.386 1.674	0.311 0.771	1.961 1.941
Total Release	I II	2.174 4.030	1.760 3.350	1.386 2.501	5.633 5.221

On an annual basis, the vegetation retracted a total of 6.297 $g.m^{-2}$ (Site I) and 8.885 $g.m^{-2}$ (Site II). Out of these 4.901 $g.m^{-2}$ in Site I and 5.030 $g.m^{-2}$ in Site II were reflected in the aboveground net production. Thus only 22.17% in Site I and 43.38% in Site II were retained by the belowground parts and rest was translocated to the shoots. In both the study sites, the rates of transfer of nitrogen from live shoots to dead shoots was more rapid than the corresponding transfer from dead shoots to litter. Annually, 65.19% of the total release was through litter decomposition and only 34.81% was through root decomposition in Site I. In Site II, 62.82% of the total release was through litter decomposition and the remaining 37.18% was through root decomposition. Thus, 89.45% (Site I) and 58.76% (Site II) respectively of the total nitrogen retracted by the vegetation was returned to the soil annually.

DISCUSSION

Seasonal variations in nitrogen storage in aboveground and belowground parts were a common feature in both the study sites. Higher nitrogen concentration in the producer compartments of Site I may be due to its relatively higher accumulation in the soil and partly due to the presence of more leguminous forbs in Site I than in Site II (4 in Site I and 1 in Site II). The decline in the concentration of nitrogen as plant material moves from live shoots to litter storage is a common phenomenon in temperate (Bokhari and Singh 1975, Bhandari *et al.* 2000) as well as in tropical grasslands (Pandey and Singh 1990). This decline has been attributed to the withdrawal of nitrogen from the shoots during senescence (Clark 1977) and weathering and leaching (Turkey 1970). Lower concentration of nitrogen in roots than in the live shoots have also been reported both in temperate (Ohlson and Malmer 1990, Chaturvedi *et al.* 1988, Bhandari *et al.* 2000) and tropical grasslands (Billore and Mall 1985, Yadava 1980, Pandey and Singh 1990).

In the present grassland, only 0.377% to 0.829% of the nitrogen resides in the plant system at a given time. These have indicated larger availability of nitrogen in the soil (Gupta *et al.* 1990). Similar higher percentages of nitrogen in the soil have also been reported from different Indian grasslands (Yadava 1980, Pandey and Singh 1990, Karunaichamy and Paliwal 1995, Bhandari *et al.* 2000). Low nitrogen storage observed during the present study could be attributed to the poor growth of legumes (Jones and Woodmansee 1979).

The seasonal and annual uptake, transfer and release of nitrogen in the present study were very much comparable with the values recorded from different eco-climatic zones of India. Uptake of nitrogen was higher in Site II (semi protected) than in Site I (grazed). Higher transfer rates at various compartmental levels could be attributed to channeling of nutrients from previous years (Bhandari *et al.* 2000). The total uptakes recorded in the present study (6.297 – 8.885 gm^{-2}) were well within the range of total nitrogen uptake of 2.90 gm^{-2} (semi-arid) to 27.58 gm^{-2} (dry sub-humid) grasslands of India (Singh *et al.* 1979). The return of nitrogen through litter decomposition (3.280 – 3.672 gm^{-2}) in the present grassland is much lower than the uptake by the aboveground net production (4.901–5.030 gm^{-2}), which thereby suggests the possibility of redistribution of the nutrients within the plant components for efficient use in the subsequent growth season (Pandey *et al.* 1993). The uptake of nitrogen in the aboveground net production (4.901–5.030 gm^{-2}) is always higher than its corresponding uptake by the belowground net production (1.396 – 3.855 gm^{-2}). Similar higher uptake by aboveground net production was also reported by Yadava (1980), Billore and Mall (1985), Gupta *et al.* (1990), Karunaichamy and Paliwal (1995), Bhandari *et al.* (2000) from different Indian grasslands.

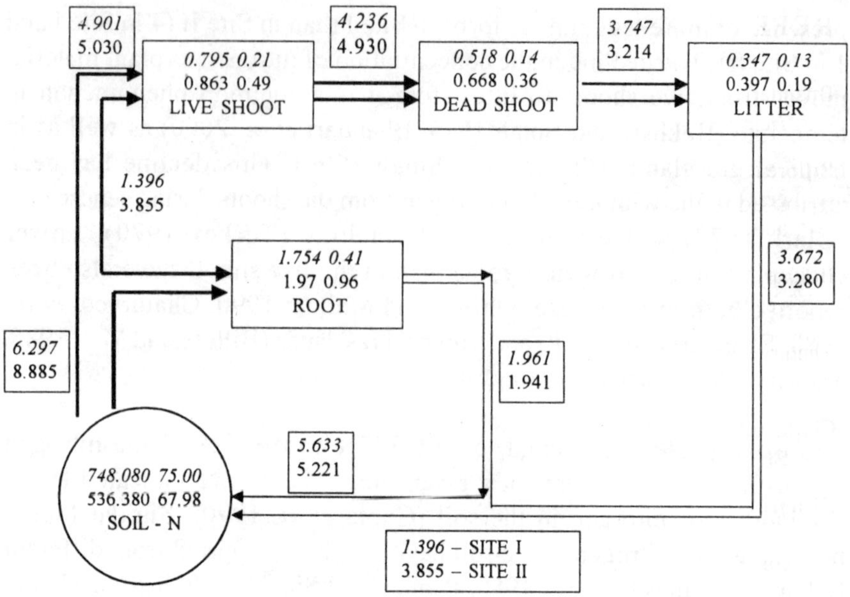

Fig. 1. Uptake, transfer and release of nitrogen. Values in compartments are mean standing crop (gm^{-2}) and those on arrows are net flux rates ($gm^{-2}.yr^{-1}$)

The total release of nitrogen through litter and root decomposition (5.221–5.633 gm^{-2}) was comparatively lower than the total corresponding uptake (6.297–8.885 gm^{-2}). This indicates the possibility of its retention in the vegetation for reuse in the subsequent season (Clark 1977). Thus, 58.76 % to 89.45 % of the total nitrogen retracted by the vegetation was returned to the soil annually and only 10.55% to 41.24% remained in the biological system. Small fraction of nitrogen present in the biological system indicated that the transfer rates, resilience, time and the inter-relationship amongst the various compartments have contributed greatly to the economy of the ecosystem (Bokhari and Singh 1975). Similar observations were also reported for Prairie grasslands (Marrs *et al.* 1981) and sodic grasslands (Gupta *et al.* 1990). Thus, the values of annual uptake, transfer and release of nitrogen in various compartments and its release through litter and root decomposition in the present grassland were found to be significantly higher than those in temperate grasslands (Dahlman *et al.* 1969, Bokhari and Singh 1975, Bhandari *et al.* 2000) and almost comparable with those in the tropical and sub-tropical grasslands (Yadava 1980, Pandey *et al.* 1993, Karunaichamy and Paliwal 1995).

REFERENCES

Bhandari, B.S., J.P. Mehta and S.C. Tiwari (2000). Fire and nutrient dynamics in a *Heteropogon contortus* grazingland of Garhwal Himalaya. Trop. Ecol. **41(1)**:33-39.

Billore, S.K. and L.P. Mall (1985). Nutrient composition and inventory in a tropical grassland under varying grazing stress. pp. 112-125. In : K.C. Misra (ed.) Ecology and Resource Management in Tropics. Vol. I. Bhargava Book Depot, Varanasi.

Bokhari, U.G. and J.S. Singh (1975). Standing state and cycling of nitrogen in soil-vegetation components of prairie ecosystems. Annals of Botany **39**:273-285.

Chaturvedi, O.P., A.K. Saxena and J.S. Singh (1988). Structural and functional analysis of grazingland under pine forest in central Himalaya. Acta Oecologica **9(2)**:167-178.

Clark, F.E. (1977). Internal cycling of ^{15}Nitrogen in short grass prairie. Ecology **58**:1322-1333.

Dahlman, R.C., J.S. Olson and K. Doxtader (1969). The nitrogen economy of grassland and dune sands pp. 54-82. In: Biology and Ecology of Nitrogen. Proceeding of a conference, National Academy of Sciences, Washington DC, USA.

Devi, A.R. (1997). Comparative study of net primary production and nutrient status in the grasslands of Canchipur, Manipur. Ph.D. Thesis, Manipur University, Canchipur, Manipur.

Gupta, S.R., A. Sinha and R.S. Rana (1990). Biomass dynamics and nutrient cycling in a sodic grassland. Int. J. Ecol. Env. Sc. **16(1)**:57-70.

Jackson, M.L. (1958). Soil Chemical Analysis. Prentice-Hall Inc. New Jersey, USA.

Jones, M.B. and R.G. Woodmansee (1979). Biogeochemical cycling in annual grassland ecosystems. The Botanical Review **45**:111-114.

Jordan, C.G. and J.F. Kline (1972). Mineral cycling: some basic concepts and their application in a tropical rain forest. Annual Review of Ecology and Systematics **3**:33-50.

Kakati, L.N. (1985). Structure and functioning of the grassland ecosystem of Manipur. Ph.D Thesis, Manipur University Canchipur, Manipur.

Karunaichamy, K.S.T.K. and K. Paliwal (1995). Nutrient dynamics of Tropical grazingland ecosystem in southern India. Trop. Ecol. **36(2)**:227-235.

Marrs, R.H., R.D. Roberts, R.A. Skeffington and A.D. Bradshaw (1981). Ecosystem development on naturally colonized China clay wastes II. Nutrient compartmentation. Ecology **69**:163-169.

Misra, R. (1968). Ecology Work Book. Oxford and IBH Pub. Co., New Delhi.

Odum, E.P. (1969). The strategy of ecosystem development. Science **164**:262-270.

Ohlson, M. and N. Malmer (1990). Total nutrient accumulation and seasonal variation in resource allocation in the bog plant *Rhyncospora alba*. Oikos **58**:100-108.

Pandey, H.N. and R.N. Singh (1990). Nitrogen storage and cycling in early successional stages of a grassland community at Varanasi, India. Trop. Ecol. **31(1)**:75-81.

Pandey, H.N., R.S. Tripathi and Uma Shankar (1993). Nutrient cycling in an excessively rainfed subtropical grassland at Cherrapunji. J. of Bioscience **18(3)**:395-406.

Ram, S.C. and P.S. Ramakrishnan (1992). Fire and nutrient cycling in seral grassland of Cherrapunji in north-eastern India. Int. J. Wildland Fire **2**:131-138.

Rigler, F.H. (1975). The concept of energy flow and nutrient flow between trophic levels pp. 15-26. In: W.H. Van Dobben and R.H. Lowe-Mc Connel (eds.) Unifying concepts in Ecology. Reports of the 1st Int. congress (Ecology). The Hague.

Sims, P.L. and J.S. Singh (1978a). The structure and function of ten western North American grasslands. III. Net primary production, turnover and efficiencies of energy capture and water use. J. Ecol. **66**:573-597.

Sims, P.L. and J.S. Singh (1978b). The structure and function of ten western North American grasslands. IV. Compartmental transfer and energy flow within the ecosystem. J. Ecol. **66**:983-1009.

Singh, J.S. and P.S. Yadava (1974). Seasonal variation in composition, plant biomass and net primary productivity of a tropical grassland at Kurukshetra, India. Ecol. Monogr. **44**:351-376.

Singh, J.S., K.P. Singh and P.S. Yadava (1979). Ecosystem synthesis. pp. 231-239. In: R.T. Coupland (ed.) Grassland Ecosystems of the World. Analysis of grasslands and their uses. Cambridge University Press, London.

Turkey, H.B. Jr. (1970). The leaching of substances from plants. Annual Review of Physiology **21**:305-332.

Whitkamp, M. (1971). Soil as component of ecosystems. Annual Review of Ecology and Systematics. **2**:85-105.

Yadava, P.S. (1980). Distribution and cycling of nitrogen in soil-vegetation components of a tropical grassland. Trop. Ecol. **21(1)**:24-31.

Advancing Frontiers of Ecological Researches in India, 585-590, 2006
Edited by A.K. Kandya & Asha Gupta
Published by Bishen Singh Mahendra Pal Singh, Dehra Dun, India

Studies on Dry Forage Production and Quality of *Iseilema* Grassland Community as Influenced by Legume Introduction in Bundelkhand Region, U.P.

Neel Ratan and U.N. Singh
Department of Botany,
D.V. Postgraduate College, Orai - 285001 (U.P.) India

Abstract : The influence of legume intercropping of four species of *Stylosanthes* viz. *S. hamata, S. scabra, S. guianensis* and *S. humilis* on the dry forage production and quality of *Iseilema* grassland community was studied during 2000-2001 in medium textured soil under rainfed conditions at Bahadpura Sheep Farm, Orai (U.P.).

The results revealed that introduction of legumes proved advantageous from dry forage production and quality point of view. The dry forage yield increased by 44.2% due to intercropping with *S. hamata* followed by *S. scabra* (30.2%), *S. guianesis* (29.6%) and *S. humilis* (3.02%). The crude protein yield increased by 75.4% in *Iseilema* grass + *S. hamata* followed by *Iseilema* grass + *S. scabra* (53.3%).

Introduction of legumes in natural grassland not only increase the production but also the quality of the forage. It is quickest method to their improvement (Hutton, 1970; shaw and Bryn, 1976). Performance of important forage legumes under cultivable condition in mid-hills of Himanchal Pradesh has been reported by Singh *et al.* (1977) and Katoch *et al.* (1978). Information on the effect of introduction of legumes on dry forage production and quality of natural grassland is, however, lacking. The present investigation is carried out with this objective and the findings of the experiments conducted in

MATERIALS AND METHODS

The study was undertaken during July, 2000 to June, 2001 at Bohadpura Sheep Farm, Orai. One year old protected grassland dominated by *Iseilema laxum* was located at an attitude of 125 m above mean sea level (25^0 59' N lat. and 79^0 37' E long.).

The soil of the study area was of medium textured with 37.18, 6.45, 27.15 and 31.15% coarse sand, fine sand, silt and clay respectively. The pH value was 7.60 and available phosphorus 0.019%. The total nitrogen and organic carbon were 0.05 and 0.65% respectively. The climate of the area is typically monsoonic with a year divisible into three seasons namely rainy (July-October), winter (November-February) and summer (March-June). The average annual temperature is uniformly high (over 25°C) but the mean monthly values vary considerably (13.9°C mean minimum to 34°C mean maximum). The mean total annual precipitation is 1169.2 mm of which 84% falls between July and October. The climatic data are represented in Fig.1-a. The Ombrothermic diagram (Fig.1-b) indicates four wet months (i.e. from July to October) and eight dry months of the year.

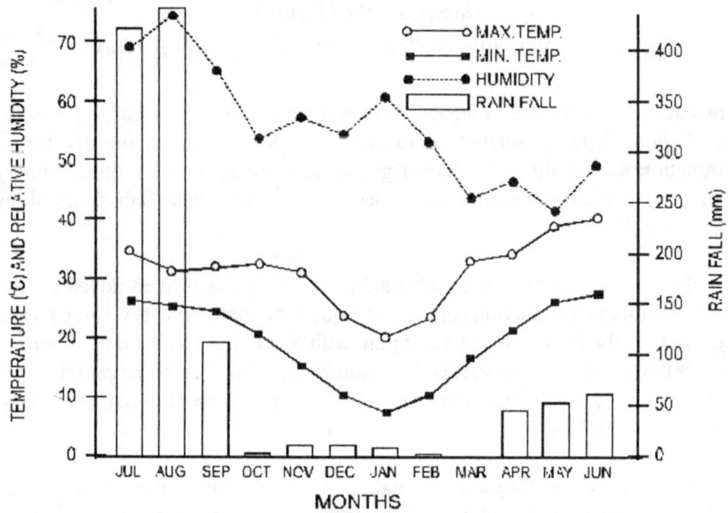

Fig. 1a. Clmetic condition at Orai Station (2000-2001)

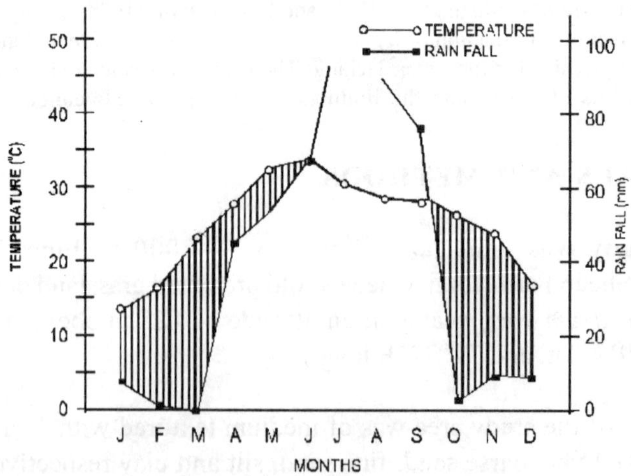

Fig. 1b. Ombrothermic diagram of Orai

Experiments were carried out in a randomised block design with three replications. Four different species of *Stylosanthes* Viz. *S. hamata, S. scabra, S. guianensis* and *S. humilis* were introduced in *Iseilema* grassland community at the rate of 5 kg/ha seeds in rows 50 cm apart after the first soaking premonsoon rains in the beginning of the study year (July).

To obtain dry forage production (grass+legumes) 4 cuttings i.e. on August 9 and 29, September 20 and October 16 were taken as per growth. In each treatment there was one control i.e. without association of introduced legume species indicated as zero (0) treatment. Plant samples of each cutting were pooled treatment wise and oven dried at 70°C for 24 hrs and weighed for dry forage determination. Plant samples were analysed for nitrogen by micro-kjeldahl method and crude protein was calculated by multiplying with 6.25. Soil samples were analysed for organic carbon by Walkley and Black's method and total nitrogen by macro-kjeldahl method.

RESULTS AND DISCUSSION

Dry forage production

With respect to their effect on forage production the performance of all the legumes (*Stylosanthes hamata, S. scabra, S. guianensis* and *S. humilis*) introduced in *Iseilema* grassland community is presented in Table-1. It is evident from the Table that different species of *Stylosanthes* germinated on seeding and showed good growth except *S. humilis* which showed very poor growth and gave negligible forage production i.e. 3.38 t/ha (grass + legume) over the control (3.28 t/ha). Introduction of *S. hamata* has proved to be advantageous giving a considerable total forage production of 4.73 t/ha followed by *S. scabra* (4.27 t/ha) and *S. guianensis* (4.25 t/ha).

Data presented in Table-1 revealed that the forage production by introduced legumes alone was recorded maximum by *S. hamata* (0.63 t/ha) and minimum by *S. humilis* (0.04 t/ha). Forage production by *S. hamata* and *S. scabra* did not differ significantly.

The increase in forage production with introduction of forage legumes have also been reported by several workers (Patil, *et al.* 1977; Rai *et al.* 1980; Singh *et al.* 1981; Rai and Kanodia, 1982; Singh *et al.* 1983; Kanodia, 1984; Rai, 1985 and 86) in different pasture grasses.

Table-1: Dry forage Production (t/ha) of *Iseilema* grassland community with various species of *Stylosanthes* legume

Treatments	Grass	Legume	Total
Grass alone	3.28	-	3.28
Grass + *S. hamata*	4.10	0.63	4.73
Grass + *S. scabra*	3.67	0.60	4.27
Grass + *S. guianensis*	4.10	0.15	4.25
Grass + *S. humilis*	4.54	0.04	3.58
Mean	3.74	0.35	4.02

It is clear from Table-1 that the production of legumes is low except *S. hamata* and *S. scabra*. This is due to competition with grass, since grases are known to be quite aggressive in their growth. However, beneficial effect of grass was observed to intercropping of these legumes. Similar result was also reported by Rai *et al.* (1980) in case of *Cenchrus ciliaris* and *Seteria spacelata* under rainfed conditions.

Crude protein content

Values of crude protein content given in Table-2 revealed that the maximum C.P. content of grass alone was recorded to be 5.66 per cent when intercropped with *S. hamata* followed by *S. scabra* (5.51%), *S. guianensis* (5.4%) and *S. humilis* (5.30%) intercropping.

Table-2: Crude protein content of *Iseilema* grassland community with various species of *Stylosanthes* legume

Treatments	Crude protein content (%)		Crude protein yield (kg/ha)		
	Grass	Legume	Grass	Legume	Grass+Legume
Grass alone	5.50	-	180.40	-	180.42
Grass+*S. hamata*	5.66	13.40	232.06	84.42	316.48
Grass+*S. scabra*	5.51	12.40	202.21	74.40	276.61
Grass+*S. guianensis*	5.40	8.30	221.40	12.45	233.85
Grass+*S. humilis*	5.30	7.30	187.60	2.92	190.52
Mean	5.47	10.35	204.73	43.55	239.58

Increase in crude protein in mixed herbage (grass + legume) on introduction of legumes has been reported by Magoon and Dabadghao (1973) and Katoch *et al.* (1977)

Amongst legumes the maximum crude protein content of 13.40 per cent was recorded in *S. hamata* and minimum in *S. humilis* (7.30%).

On the basis of these findings it may be concluded that introcution of legumes increase the quality of the forage, apart from enhancing the dry forage production of the pasture. The beneficial effect of grass-legume mixture over pure stand has also been reported by several workers (Chauhan and Faroda, 1979; Rai *et al.* 1980; Rai and Kanodia, 1982; Singh, *et al.* 1983 and Rai, 1985).

Crude protein yield

The crude protein yield of grass + legume mixtures differed significantly and the maximum value of 316.48 kg/ha (232.06 kg/ha from grass + 84.42 kg/ha from legume) was obtained with intercropping of *S. hamata* followed by *S. scabra* (202.21 kg/ha from grass + 74.40 kg/ha from legume = 276.61 kg/ha) (Table 2). However, the crude protein yield of grass + legume mixture increased in all the cases and the increase was 75.4, 53.3, 29.6 and 5.6 per cent with intercropping of *S. hamata*, *S. scabra*, *S. guianensis* and *S. humilis* respectively over the grass alone.

ACKNOWLEDGEMENTS

Authors are thankful to the Incharge of Bohadpura Sheep Farm, Orai for kind assistance in field study. The Head of the Institution, D.V. Postgraduate College, Orai is also duly acknowledged for providing laboratory facilities.

REFERENCES

Chauhan, D.S. and R.S. Faroda (1979). 'Studies on intercropping of pasture legume with *Cenchrus* ciliaris grass', *Forage Res.* **5**: 79-80.

Hutton, E.M. (1970). 'Tropical pastures', *Adv. Agron.* **22**: 1-37.

Kanodia, K.C. (1984). 'Research on tropical legumes in India', A review, *Plant and Nature*, **2(1)** : 23-35.

Katoch, D.C., K.K. Dogra and L.N. Singh (1978). 'Forage yield and chemical composition of some tropical pasture legumes in mid-hills of Himanchal Pradesh', *Forage Res.* **4**: 173-175.

Katoch, D.C., L.N. Singh, K.K. Dogra, R.N. Rai and S.S. Negi (1977). 'A note on the effect of introduction of some legumes on the yield and quality of forage from natural grasslands in mid-hills of Himanchal Pradesh', *Ind. J. Anim. Sci.* **47**: 834-836.

Magoon, M.L. and P.M. Dabadghao (1973). 'How to achieve potential grassland production', *Ind. Fmg.* **23**: 15-19.

Patil, B.D., K.D. Kanodia and P. Rai (1977). 'Grasses intercropped with annuals under dryland conditions', *Indian Datruman*, **28** : 509-572.

Rai, P. (1985). 'Forage production of buffel grass as influenced by intercropping with legumes', *Annals Arid Zone*, **24(4)** : 344-345.

Rai, P. (1986). 'Effect of intercropping of annual grain legumes on the productivity of Anjan grass', *Ind. J. of Agron.* **31** : 147-153.

Rai, P., B.D. Patil, K.C. Velayudhan and R. Agrawal (1960). 'Nitrogen equivalence of range legumes introduced in natural grasslands', *Ind. J. Range Mgmt.* **1(2)**: 97-101.

Rai, P. and K.C. Kanodia (1982). 'Search for an ideal legume for *Setaria* mixed pastures', *Forage Res.* **8**: 145-149.

Singh, K.A., K.C. Kanodia and P. Rai (1983). 'Stand geometry of *Cenchrus ciliaris* L. intercropped with *Stylosanthes hamata* (L.) Taub. Effect on forage yield and quality,' *Forage Res.* **9**: 49-56.

Singh, L.N., D.C. Katoch and K.K. Dogra (1981). 'Effect of legume introduction on forage yield and quality of natural grasslands', *Forage Res.* **7**: 71-75.

Singh, L.N., S.C. Bhandari and K.K. Dogra (1977). 'Preliminary evaluation of some temperature pasture legumes and grasses in mid-hills of Himanchal Pradesh', *Forage Res.* **3**: 27-34.

Shaw, N.H. and W.W. Bryn (1976). *Tropical Pasture Research*, C.S.I.R.O., Brisbane, pp. 454.

SECTION EIGHT

CROPLAND ECOLOGY

Advancing Frontiers of Ecological Researches in India, 593-614, 2006
Edited by A.K. Kandya & Asha Gupta
Published by Bishen Singh Mahendra Pal Singh, Dehra Dun, India

Some Environmentally Safe Approaches to Crop Disease Control towards Sustainable Agriculture

Ashok Aggarwal, Vipin Parkash, Dipti Sharma, Seema Sharma and R. S. Mehrotra*
Botany Department, Kurukshetra University, Kurukshetra, Haryana, 136119.

Abstract: Although biological control of plant pathogens has not dramatically solved agricultural problems so far, it is true that we are at a turning point in technology to make significant advances at the end of this century. It may be several years before microbial biocontrol product be accepted by general farmers and made use of. There are several difficulties one of the foremost is the shelf life of the biocontrol product. Secondly the amount of biocontrol produce needed is quite substantial and then its production on a mass scale is full of difficulties. The Environmental and Governmental agencies will clear the product only when it is completely free of any danger involved. Biological control is not a substitute for chemical control but can be definitely a supplement on a large scale. Together with other eco-friendly methods such cultural practices, including soil solarization and others pointed out in this article have a great promise in making our agriculture more sustainable.

INTRODUCTION

Plant diseases take their toll in losses of crop production. Losses of food, fiber and ornamentals are held to about 14% in world production. In order to sustain the living population, which is increasing at an alarming rate, fertilizers and pesticides have been used on a large scale. The pesticide industry emerged between 1918 and 1940. Pesticide use in agriculture was not popular. According to Agnihothrudu (1982) "Labour was cheap, food was abundant at that time. The Second World War stimulated pesticide research. The usefulness of DDT, its wide spectrum of activity, its residual action, all appeared miraculous". In 1962 was published "Silent Spring". Soon pesticides became more suspect. Since the first international symposium on soil borne plant Pathogens in Berkeley in 1963, the increasing use of potentially hazardous fungicides in agriculture has been the cause of growing, worldwide concern. Moreover it is roughly

* Retd. Prof., 263/7 Urban Estate, Kurukshetra, Haryana, 136119.

estimated that the cost involved in the development of each new pesticide and its patenting involves an estimated cost of about US$ 20,000,000. For these reasons there has been an increased effort to develop and use non-chemical methods to plant disease control. We shall discuss some of these methods of plant disease control in this paper.

According to Rechcigl and Rechcigl (1997), plant diseases are one of the major causes of crop loss worldwide and billions of dollars worth of crops are lost each year due to plant diseases. Plant disease management is fundamental not just to sustainable agriculture but also to sustainable growth of agriculture. According to Cook (1997) plant disease management not only be most effective, must be affordable, renewable and socially acceptable. Chemicals have been used for preventing and controlling diseases. But recently the use of Chemicals in agriculture has come under sharp criticism because of their injudicious and improper use and adverse effect on the environment. This concern, coupled with the pressure from increased regulation of the pesticides industry, has prompted scientists to see alternative approaches for managing crop diseases. It will be worthwhile to quote Dr. M.S. Swaminathan which Dr. Nagarajan (1983) has quoted in the preface of his book entitled "Plant Disease Epidemiology", we live in a world of sad scientific ironies and economic enigmas. The region where man first settled down to cultivate plants and thereby initiated what we call agriculture are also the regions which contain the greatest number of hungry people today". It is in this connection plant disease management becomes all the more important and the great relevance of ecological approaches to plant disease control. Rechcigl and Rechcigl (1997) have recently edited a volume "Environmentally safe approaches to crop disease control".

Environmental aspects

The effect of environment on the disease development and control of disease

Disease can be affected by the growth of the crop and its physiological status and both are influenced by environmental factors. The effectiveness of crop spraying depends largely on the weather conditions. So a proper understanding of the interaction between microclimate and spray can help to improve the effectiveness of chemical control measures by better targeting and reduction of spray (Mathews and Hislop, 1994). Knowledge of interaction between host, pathogen and environment can help in the development of control strategies in a number of areas like avoidance of the pathogen, exclusion of the pathogen, eradication of the pathogen and therapy.

Plant disease forecasting is an important component to safer and better use of chemicals for disease control. Work in our country has been done on several important diseases such as late blight of potato, rice blast, rusts of wheat, apple scab and several others. In most of the agriculturally developed countries of the world forecasting and warning simulation models have been developed (McCartneef, 1997). These models help considerably in proper controlling of the plant diseases with lesser usage of chemicals.

Can we really do away with the use of chemical measures?

Fungicides are chemical products active in controlling pathogens, which damage or destroy plant. Today's fungicides after a diverse mode of action and are grouped into few classes based on how they interact with host and pathogen (Siegel and Sisler, 1997). Grover (1986) has in his Presidential Address to the members of the Indian Phytopathological Society has given an overview of the chemicals for the control of plant diseases in India. The fungicides when used at their recommended rate and timing are frequently the most cost-effective and reliable tools to protect crop against their pathogens (Ragsdale *et al.*, 1991). Modern fungicides are used either as soil or seed treatment and as foliar or fruit sprays. All these fungicides must be capable of safe use by applicators, leave no residues in harvested crop, have no significant impact on the environment and be economical to farmers.

When used judiciously, fungicides provide manifold benefits with negligible risks. Without fungicide use food production would decline sharply, resulting in severe cost increases to the consumer. Fungicide application is frequently essential to protect foods from spoilage and contamination by harmful fungal toxins (Graham and Gray, 1995). At least 45 mycotoxins are known carcinogens and have been shown to be present in contaminated foods/grains.

It is inconceivable to believe that we can do away with chemical fungicides for plant disease control. But we have to make the minimum of use of these hazardous chemicals when other methods of controlling plant diseases are not available.

As already stated there is a worldwide swing to the use of environmental friendly methods of protecting over crops from pests and pathogens. The indiscriminate use of pesticides has led to emergence of more resistant pests or pathogens that have developed a built in resistance to some of the frequently used chemicals.

According to Kumar and Mukerji (1996) the third world uses one-sixth of the total pesticides used globally, at least 37,000 people are effected annually, 10,000 of them fatally. Hamlen *et al.* (1997) have presented a balanced view on the importance and complexity of fungicide use. Pesticides are necessary but are not the long-term solution to pest management programmes. The non-target effect, hazardous nature, petroleum-based pesticides will become more expensive. These points very clearly indicate that ecofriendly methods are must if we have to survive safely on this planet. Biological control of plant pathogens is one of the possible alternatives to chemical control and will form an important component of the integrated disease management programme.

Potential Biocontrol Agents

The potential biocontrol agents identified for control of plant pathogens belong mainly to the following genera of microorganisms (Hawksworth *et al.*, 1996).

Fungi – *Trichoderma, Trichothecium, Gliocladium, Aspergillus, Penicillium, Coniothyrium, Corticium, Fusarium, Myrothecium, Phialospora, Pythium, Sphaerellopsis, Sporidesmum, Tuberculina, Chaetomum, Cladosporium, Verticillium.*

Bacteria – *Agrobacterium, Pseudomonas, Bacillus.*

Actinomycetes – *Streptomyces.*

Viruses – *Mycoviruses* ds RNA.

Nematophagoes fungi – include the nematode trapping fungi *Monacrosporium* such as *Arthrobotrys, Dactylella, Genicularia,* endoparasites – *Hirsutella, Catenaria, Meria, Nematoctonus, Nematophthora* and highly specific egg parasites such as *Dactylella oviparasitica, Paecilomyces lilacinus,* and *Verticillium chlamydosporium.*

Fungal pathogens for biological control of weeds have been extensively investigated. According to Jacobsen and Backman (1993) only five biocontrol products were registered for use by the U.S. Environmental Agency (EPA) for control of plant diseases till 1993.

Several books and reviews have appeared on this topic of biological control. The latest one is the edited book entitled "Biological control of soil bone plant pathogens" edited by Hornby (1990), which highlights invited papers on biological control of soil borne pathogens presented in section V and related sections of the 5th International Congress of Plant Pathology held at Kyoto in 1988 in Japan.

Biological control of plant diseases by Bacteria

Backman *et al.* (1997) have recently reviewed the work done till then in an article entitled, "Bacteria for biological control of plant diseases". It is in 1950s, the Soviet Union introduced what were called "bacterial fertilizers", which not only promoted better growth of plants but also paved the way for further research for biocontrol of plant pathogens by bacteria. Bacteria have been developed as biological control for various plant diseases and for non-chemical pest control. Bacteria can suppress some seed and soil borne diseases e.g. *Bacillus subtilis* reduces damping off disease of sugar beet (Dunleavy, 1981). Competition for iron and antibiosis to pathogens are main factors for success of a biocontrol agent. Commercially bacteria are being used to control various plant diseases. Where there are no chemical products working effectively or as an adjunct to chemical control. Research is being carried out to use combination of bacteria to improve their disease controlling power. *Bacillus* is being used as one of the key genera for the control of various foliage diseases e.g. *B. subtilis* for bean rust, *B. cereus* for onion rust and for *Alternaria* on tomato. Appropriately 5 million ha of crops were treated with different strains of *Bacillus* (sold under the trade names Quantum, Kodrak and Epic by Gustafson, Inc.) targeting diseases of roots caused by *Rhizoctonia, Pythium* and *Fusarium,* plus promoting increased root mass and plant vigor through hormone like responses and disease control (Backman *et al.*, 1994). For the biological control of foliage diseases emphasis is being given to increase the colonization of the beneficial bacteria on leaf surface by providing them with food sources that will increase the population of beneficial bacteria. According to Lindow (1992), *Pseudomonas fluorescens* A506 (sold as Blight Ban A 506) is being used for the control of *Erwnia amylovora* cause of fire of apple and pear. This strain (A506) utilizes nutrients and other resources in the blossom environment that are required for the development of pathogenic *E. amylovora* population (Wilson and Lindow, 1993). Several strains of the members of the genus *Psendomonas* are being used as biocontrol agent (Leisinger and Margraff, 1979). Some of the root pathogens controlled are of coriander (Mall, 1968) cotton (Howell and Stepanovis, 1979) wheat (Bagyaraj, 1981) and chickpea (Kaisor *et al.*, 1989). The post harvesting diseases are main areas of concern in the crops of vegetables and fruits and responsible for major loss in their yield. Major infection sites for post harvesting diseases are wounds. From past 15 years various bacteria are being used to control post-harvest diseases e.g. *B. subtilis* B-3 can suppress brown rot of peaches caused by *Monilinia fructicola.* The antagonist population of bacteria exclude pathogen populations through antibiosis and sequestration of resources. It has been reported by various scientists that endophytes provide resistance to plants for various diseases. If the pathogen of other host will be introduced in the plant through natural pores i.e. stomata and hydathodes then the plant will prepare defence proteins which will not only kill the inducing agent but also provides resistance towards true pathogens (Burkett *et al.*, 1996). Some bacteria act as inducers of natural plant defense mechanism, and thus produce systemic resistance against bacterial and fungual pathogens. For example, two PGPR stains, 89 B-27 (*P. fluorescens*) and 90-166 (*Serratia marcescens*) induce resistance against bacterial and fungal pathogens (Raupach *et al.*, 1996). Earlier salicylic acid has been found to be produced after an infection by a pathogen and thereafter this chemical causes induced systemic resistance (ISR) products (phytoalexins etc.) to accumulate but salicylic acid itself was too rapidly degraded for use in agriculture. Now Ciba-Geigy Ltd., Basel has developed a non-toxic chemical activation (Benzothidiazole derivate, trade name Actigard or (GA-245704) that functions to control both fungal and bacterial diseases by using the host plants natural resistance mechanism.

Another classical example in biological control by bacteria is by using *Agrobacterium radiobacter* for suppression of crown gall caused by *Agrobacterium tumefaciens.* In Australia, most agrobacteria around gall-free trees are nonpathogenic, but most around galled trees are pathogenic. When seedling roots of peach were wounded, dipped in a suspension of non- pathogenic strains and planted in soil infested with pv. *tumefaciens*, no galls of the disease developed. This finding led to this effective-biocontrol of the disease. This non-pathogenic bacterium (*A. rodiobacter*) was found to produce bacteriocins called 'agrocins' that are encoded by a plasmid and are responsible for antagonism to the pathogenic *A.*

tumefaciens that appear to compete for the same niche of the rhizosphere. Clare (1993) has reviewed the work done on *Agrobacterium* as biocontrol agent.

Fungal control agents

Mehrotra *et al.* (1997) have reviewed the work on this aspect. Fungi are the important biocontrol agents of various plant diseases and pests. There are three mechanism for biocontrol *i.e.* antibiosis, competition and parasitism or mycoparasitism. In antibiosis, the antagonist produces antibiotic compound or some extracellular enzyme which has adverse effect on pathogen. The fungus *Trichoderma* has been extensively used for biocontrol of several plant pathogens. According to Bliss (1951) sublethal dose of CS_2 gave *Trichoderma* a competitive advantage over *Armillaria mellea, which* is responsible for effective disease control. Parasitism among fungi is called mycoparasitism. *Pythium nunn* is an interesting mycoparasite of various soil borne pathogens. *P. nunn* form coil around the hyphae of attacking *Pythium ultimum* and *P. vexans* (Lifschitz *et al.*, 1984). Pulitz and Baker (1987) reported biological control of *Pythium* damping off of cucumber with *P. nunn.*

Suppressive soils

Suppressive soils are known for various pathogens for example *Gaeumannomyces graminis* var. *tritici* (Shipton, 1975), pathotypes of *Fusarium oxysporum* causing vascular wilt (Toussoun, 1975). *F. solani* f. sp. *phaseoli* (Burke, 1965) root-knot nematode (Stirling and Mankaur, 1978), *P. cinnamomi* (Broadbent and Baker, 1974). Some important antagonists are being used to control seed borne, and aerial pathogens. *Peniophora gigentea* a biocontrol agent of soil borne pathogens has been registered by U.S. Environmental Plant Protection Agency (USEPPA) for commercial use for control of *Heterobasidion annosus* (*Fomes annosus*) of pine stumps (Rishbeth, 1963, 1979). This is widely used in U.K. and U.S. for protection of pines. Soil application of *Trichoderma* and *Gliocladium* have been done by USDA. Beltsville, U.S. for control of damping off and blight of snap beans, damping off of cotton and *Rhizoctonia* disease of potatoes (Lewis and Papavizas, 1987, 1991). Several *Trichoderma* based formulations are now available in our country. Antagonists which are generally applied to seeds are *Chaetomium* sp., *Penicillium* sp. and *Trichoderma* spp. Wiley and Kommedahl (1981) reported that seed treatment of sweet corn and wheat with *T. harzianum* was very effective in crop management. Biological organisms can be coated onto seeds in a number of ways to provide an alternative to chemical pesticide treatment.

Mycorrhizal fungi in control of diseases

Biocontrol is defined as reduction of the amount of inoculum or disease producing activity of a pathogen accomplished by or through one or more organisms other than man. The use of microorganisms to control plant pathogens, fungi and bacteria is an active and promising areas of research (Mukerji and Garg, 1988) The VAM fungi play an important role in enhancing the activity of biocontrol agents, either directly or through an indirect effect on the microflora of mycorrhizoshere (Azcon-Aguilar and Barea, 1995). Mycorrhizal association altered root exudation induced changes in the composition of microbial communities in the mycorrhizosphere (Bansal and Mukerji, 1994, 1996; Mukerji *et al.,* 1998). The role of mycorrhizal fungi in biological control of plant pathogens is well documented. The role of vesicular arbuscular mycorrhizal fungi in controlling soil borne diseases has been recently received by Shukla *et al.* (1996).

Krishna *et al.* (1982) reported for the first time the interaction of VAM fungi and S*treptomyces cinnamomeous* that inhibited the growth of plant pathogenic fungi reduced the sporulation and colonization of *Glomus fasciculatum* on finger millet. Kumar (1997) reported that root rot of sunflower can be controlled by using dual inoculum of *Trichoderma* sp. and *Glomus fasciculatum.* Host plant previously inoculated with VAM fungi exhibited increased resistance to several soil borne root pathogens (Barea and Jefferies, 1996; Chandra, 1992; Davis and Menge, 1980; Jalali, 1991; Linderman, 1994; Morandi, 1996; Schonbeck, 1979 and Aggarwal *et al*., 1999). Uniform reduction of disease symptoms have been described for fungal pathogens viz., *Phytophthora, Gaeumannomyces, Fusarium, Chalara (Thielaviopsis), Pythium, Rhizoctonia, Sclerotium, Verticillium* and *Aphanocmyces.* Besides these VAM fungi are also known to reduce bacterial diseases. The improved nutrition in mycorrhizal plants led to reduction in disease incidence also (Davis *et al.,* 1979 and Ross, 1971).

Viral control agents

The plant diseases caused by virus are responsible for reduction in crop yield by considerable amount. Rush and Sherwood (1997) had received this subject exhaustively. The plant viral diseases can be controlled by cross protection. In this the plant suffering from infection by one virus is infected by another virus which interferes in various ways with infection by first one. McKinney (1929) was the first to report the phenomenon of cross protection. He described the prevention of expression of a strain of Tobacco mosaic virus causing yellow mosaic by light-green mosaic strain of tobacco mosaic virus (TMV). Many hypotheses were put forward concerning the nature of cross protection. Hamilton (1985) indicated that the condition of insusceptibility of the protected plant to super infection by the challenge virus results from a limitation on the completion of the life cycle of the challenger.

In addition to this there is a mechanism which accounts for specificity in cross protection. Ross (1974) reported that available ribosomes are utilized by a rapid increase in RNA of the initial virus and that leaves the challenge virus RNA unbounded and therefore, susceptible to degradation. Palukaitis and Zaitlin (1984) said that for (+) RNA viruses, infection by the challenging virus could be reduced by inhibiting its ability to synthesize (+) RNA.

De Zoeten and Fulton (1975) suggested that the coat protein present in the cell will encapsidate the viral RNA of the challenge strain before it can express itself. One supporting evidence for the involvement of coat protein in cross protection was found by Dodds *et al.* (1985) in Tomato protected with CMV, super infection occurred when the leaves were challenged with viral RNA but not with intact virions.

Systemic spread of the challenge virus can be prevented. The challenge virus can be limited to the site of initial inoculation by saturation or change in the putative receptors involved in the cell to cell membrane of virus. Blum *et al.* (1989) suggested that a viral induced, host specific inhibitor of viral transport. They said that the challenge virus cannot synthesize its movement proteins because the protecting strain induces the host to synthesize an inhibitor.

Cacao swollen shoot in Africa and passion fruit woodiness in Queensland. Australia (Simmonds, 1959) were first viral diseases in which wild strains protected plants and showed good development and yield in field conditions. The use of cross protection for the control of viral diseases of plants have some drawbacks also. During cross protection we must concern the degree of protection the mild strain gives, secondly the strain which is mild in crop of interest may not be mild in other crops in that area.

In addition to virus other agents which are used in cross protection are viroids - satellites and unrelated viruses. Viroids are low molecular weight RNAs which are capable of causing viroid plant diseases. But the use of viroids for biocontrol of plant diseases is not successful because protection is highly dependent on host species, symptoms are delayed but viroid replicates, severe symptoms eventually expressed. For example, Garnsey *et al.* (1988) reported that mild isolates of CEV (Citrus exocortis viroid) was not able to counter cross protection against severe challenge strains.

An unrelated virus can be used to control various viral diseases of plants. Prillwitz and Schlosser (1993) demonstrated that beet soil-borne virus (BSBV) reduces titer and attenuate symptoms development in sugar beet challenged with beet necrotic yellow vein virus (BNYVV).

Satellites are small RNAs ranging in size from 200 nucleotides (nt) to 1500 nt. Satellites require helper virus for replication and lacks significant sequence homology with helper.

There are other methods such as cultural practices including sanitation, eradication, exclusion, quarantine procedures, agronomic practices such as type and length of crop rotation, tillage and residue management, application of fertilizers, planting date, depth of, cultivars selection, frequency of irrigation can be suitably managed for better plant growth and yield and minimum of pests and diseases. Another area which has gained momentum is the use of soil solarization to raise the soil temperature a technique optimized by Katan and his co-workers in Israel (Katan *et al.*, 1976). Another frontline area of research is the genetic engineering of plants so that they become resistant to the pathogens. Perhaps space will not permit us to discuss all these in this review, which has been confined mainly to fungal, bacterial and viral biocontrol agents. However, we shall discuss some of these in the following pages.

Siderophores

The role of siderophores produced by fluorescent Pseudomonads was first reported by Kloepper *et al.* (1980). Recent work has clearly established that siderophores (Microbial Iron-III transport agent) produced by fluorescent Pseudomonads as partly responsible for enhanced plant growth and biocontrol. These siderophores exhibit considerable diversity in chemical structure which depend upon the nature of the organism concerned. Siderophores, their biochemistry and possible role in the biocontrol of plant pathogens had been reviewed by Leong (1986). Beneficial fluorescent *Pseudomonas* and other microorganism can promote plant growth and induce disease suppressiveness by mechanisms other than siderophore production namely through competition for carbon or through production of hormones, antibiotics or bacteriocins.

Most aerobic and facultative anaerobic microorganisms respond to low-iron stress by producing extracellular low molecular weight (500-1000 dalton) Iron-(III)-transport agents which are designated as siderophores. This selectively complex Iron-III is with very great affinity. The function of siderophores is to supply iron to the deficient cell. Siderophore is antibiotic in the sense that the compound produced inhibits growth of another organism nearby.

Ferric siderophores are usually too large to diffuse through the small, water filled porin channels in the cell envelope of gram negative bacteria and hence active iron transport is initiated by outer membrane receptor proteins for ferric siderophores. The synthesis of these receptor proteins are usually also induced by Iron limiting conduction. The inner membrane and periplasmic proteins are also required for ferric siderophores transport and release of iron from the ferric siderophores proceeds via reduction since siderophores have little affinity for Iron (II). The released iron is then incorporated into Iron containing cellular constituents, where it may interact as Iron (II) with repressor protein to regulate the biosynthesis of siderophores and the receptor protein for the ferric siderophores.

The involvement of siderophores produced by certain fluorescent Pseudomonads in biocontrol of *Gaeumnnomyces graminis* var. *tritici*, the cause of take all disease of wheat and barley or *Fusarium oxysporum* f. sp. *lini* cause of flax wilt is very clear. Kloepper *et al.* (1980) have suggested that the siderophores (powerful natural iron chelators) produced by root-colonizing fluorescent *Pseudomonas* spp. may remove iron needed by the take-all fungus. Suppression of diseases caused by other soil borne organisms have also been related to the presence of siderophores. Misaghi *et al.* (1982) reported that water soluble fluorescent ,pigments of 156 *Pseudomonas* isolates was found to be fungistatic to several pathogens including *Phytophthora megasperma, Pythium aphnidermatum* and few others. There is evidence that suppressiveness of soil to take all disease of *Fusarium* wilt could be eliminated by supplementing appropriate amount of Fe, EDTA, to the pathogens suppressive soil. It suggests that microorganisms present in the iron limiting suppressive soil are capable of producing siderophores but that Iron-III can not be utilized by pathogens. A strain B-10 of *Pseudomonas* was able to convert pathogens conducive soil into pathogen suppressive soil which was partially attributed to the production of Iron transport agent which usually serve two purposes: Firstly they enhance growth and yield by supplying iron to the plant and secondly they convert pathogen conducive soils to pathogen suppressive soils by making microbial iron unavailable to the pathogen and also act as potent antibiotics. It is worthwhile to mention that susceptible (to iron deprivation) microorganisms apparently lack the iron assimilation system for these ferric siderophores from beneficial fluorescent Pseudomonads. These deprived microorganisms probably lack specific outer membrane receptor proteins for ferric siderophores. Hubbard *et al.*, (1983) found that *Trichoderma hamatum* failed to control damping off of pea in New York state when applied as a seed treatment. This was ascribed to starvation of the antagonist for iron, caused by certain strains of Pseudomonads thought to produce siderophores in the soil which was already deficient in soil. When iron content of the soil was increased, *T. hamatum* was not inhibited and damping off was controlled.

Soil solarization

Soil solarization involves the covering of soils with clear, thin polyethylene sheets in mid summer, generally for 4-6 weeks, before sowing of the crop. According to Porter (1992) the benefits of plastic mulches for increased production of horticultural crops has been known for many years (Waggoner *et al.*, 1960). In the past, plastic mulches have been used generally as a preplant treatment to increase soil temperatures for early season growth of crops and as a post plant treatment for horticultural crops as well as for ornamental crops in warmer regions of the globe. Disease control by the mulching or trapping the soil with transparent plastic traps during summer is relatively a new innovation. The method known as soil solarization was first used in the field in Israel and in glasshouses in Japan in the mid 1970s and soon afterwards in California (Katan *et al.*, 1976, Kodama. and Fukui, 1979, Pullman *et al.*, 1981). This method has been followed to control several diseases in our country.

Through pollen management

Thakur and Williams (1988) and Thakur *et al.* (1983) have reported excellent control of ergot in pearl millet through pollen management. Rapid pollination reduces ergot infection. Experiments were conducted at ICRISAT center ergot in F_1 hybrids of pearl millet by pollen management. In several experiments at ICRISAT centre, when an ergot loss susceptible early flowering line was strategically planted as pollen donor for susceptible hybrids, ergot severity was significantly reduced by 13% to 51% and grain yields were increased in different hybrids. Thakur (1983) showed similar findings at Hisar in India. This method to control of ergot in pearl millet hybrids is based on pollination induced resistance is quite promising but needs wider testing before it can be recommended for use by farmers. Successful exploitation of pollen management for ergot control will depend largely on whether seed companies and extension agencies can teach farmers to understand the principles involved and to accept the possible non-uniformity in hybrid crops (Mehrotra, 1993).

Genetic Engineering to control plant diseases

Genetic engineering allows the genetic modification of plant by adding new traits through recombinant DNA technology. Genetic engineering makes possible the incorporation of a single gene or a combination of genes into the plant genome in order to confer traits such as disease resistance, herbicide tolerance, tolerance to salt, drought and cold, increased productivity and increased notational value. This technology combines molecular biology, microbiology, tissue culture, cell transformation, plant selection and breeding techniques. Genetic Engineering allows the development of genetically altered plants with defined traits and valuable properties without altering other desirable characters of the recipients. Genetic engineering is of special interest for the straight insertion of disease resistance genus into the genome of susceptible commercial cultivars as a method to protect such plants from phytopathogenic pests.

Control of fungi

Early reports in genetically engineering plants for fungal resistance was mostly on tobacco; later on oilseed rape, potato, rice, cucumber and tomato plants have also been engineered for fungal resistance for example oilseed rape plants and rice plants were engineered using multiple pathogenesis-related proteins against *Rhizoctonia solani* (Broglie *et al.*,

1991; Lin *et al.*, 1995 and Xu *et al.*, 1996). In tomato transgenic plants were resistant to the root fungal pathogen *Fusarium oxysporum* sp. *lycopersici* (Van den Elgen *et al.,* 1993). In case of potato plants containing a gene encoding glucose oxidae from *Aspergillus niger* were resistant to *Phytophthora infestans* (Wu *et al.,* 1995). Numerous genetic studies have been performed on tobacco plants for several fungal pathogens including. *Rhizoctonia solani, Peronospora tabacina, Phytophthora parasitica* var. *nicotianae, Cercospora nicotianae, Botrytis cinerea, Alternaria longipes* and *Ustilago maydis.* Alexander *et al.*, 1993; Zhu *et al.*, 1994; Hain *et al.*, 1993; Kinal *et al.*, 1995; Park *et al.*, 1996; Terras *et al.*, 1995; Jach *et al.*, 1995 and Longemann *et al.*, 1992). The level of protection against fungal pathogens in transgeniç plants is highly variable. Variable differences in degree of protection are influenced by the specific activity of the enzymes or proteins, their localization concentration within the plant cell, characteristics of fungal pathogens and type of interaction between the pathogen and the plant.

Control of bacteria

Crop plants genetically modified for resistance to bacterial pathogen are less in number as compared to transgenic plants against fungal pathogens. Important are in the list includes modified potato against *Erwinia carotovora* spp. *atroseptica* (Diiring *et al.*, 1993), Apple rootstock transgenically developed increased resistance to *Erwinia amylovora* (Novelli *et al.*, 1994). In case pf tobacco transgenic plants developed showed resistance against *Pseudomonas syringae* pv. *tobaci* and *P. syringae* pv. *phaleosicola* (Beffa *et al.*, 1995; Carmona *et al.*, 1993). Transgenic tomato plants were found to be resistant against *P. syringae* pv. tomato (Martin *et al.*, 1993). The levels of resistance could be correlated to levels of protein expression in the different transgenic lines analyzed (Hightower *et al.,* 1994).

Control of nematodes

The potential of genetic engineering to control nematodes is gaining importance. Genetically engineered plants exhibit a reduction in gelling and egg production upon infection, disrupt the feeding behaviour of the nematodes, reducing the fecundity of nematode. Potato plants were engineered against *Meloidogyne* spp. (Atkinson *et al.*, 1995), Transgenic rice plants showed inhibited growth of *Globodera pallida* (Urwin *et al.*, 1995) Tobacco trangenies were affecting against *Meloidogyne incognita*

(Baum *et al.*, 1996). These examples show the possibility of providing nematode resistance to genetically engineered plants.

ACKNOWLEDGEMENTS

Authors, Vipin Parkash and Dipti Sharma are thankful to CSIR New Delhi for financial assistance.

REFERENCES

Aggarwal, A., Kumar, J., Sharma, D. and Mehrotra, R.S. (1999). Biological control at root rot of sunflower using *Trichoderma harzianum* and a mycorrhizal fungus. J. India Bot. Soc., **78:**87-97.

Agnihothrudu, V. (1982). Pesticides, pest control and their future. Curr. Sci., **51(11):**539-546.

Alexander, D., Goodman, R.M., Gut-Rella, M., Glascock, C., Weymam, K., Friedrich, L., Maddox, D., Ahl-Goy, P., Luntz, T., Ward, E., and Ryals, J. (1993). Increased tolerance to two oomycete pathogens in transgenic tobacco expressing pathogenesis – related proteins. In: Proc. Nat. Acad. Sci. USA., **90:**7327.

Atkinson, H.J., Urwin, P.E., Hansen, E., and McPherson, M.J. (1995). Designs for engineered resistance to root-parasitic nematodes, Trends Biotechnol., **13:**369.

Azcon-Aguilar, C. and Barea, J.M. (1995). Saprophytic growth of arbuscular mycorrhizal fungi. In : Mycorrhiza. Eds. Varma, A. and Hock, B., Springer-Verlag, Berlin.

Backman, P.A., Brannen, P.M., Mahaffee, W.F. (1994). Plant response and disease control following seed inoculation with *Bacillus subtilis* in improving plant productivity with rhizosphere bacteria. (Ryder, M.H., Slephers, P.M., and Bowen, G.D., Eds.), CSIRO, Addcide, 3.

Backman, P.A., Wilson, M., Morphy, J.F. (1997). Bacteria for Biological Control of Plant Diseases, In "Environmentally Safe Approaches to Crop Diseases Control (Eds. Rechcigl, N.A. and Rechcigl, J.E.) CRC Press, Boca Raton: 95-109.

Bagyaraj, D.J. (1981). Studies on the biological control of take-all of wheat : 21-24. In: Progress in Soil Biology and Ecology in India (G.K. Veeresh, Ed.) University of Agricultural Sciences, Hebbal, Bangalore.

Bansal, M. and Mukerji, K.G. (1994). Positive correlation between VAM induced changes in root exudation and mycorrhizosphere mycoflora. Mycorrhiza, **5(1):**39-44.

Bansal, M. and Mukerji, K.G. (1996). Root exudates in rhizosphere biology. In: Concepts in Applied Microbiology and Biotechnology, (Eds.) Mukerji, K.G., Singh, V.P. and Dwivedi, S., Aditya Books Pvt. Ltd., New Delhi, pp. 97-119.

Barea, J.M. and Jeffries, P. (1996). Arbuscular mycorrhizas in sustainable soil-plant systems. In: Mycorrhiza-Structure, Function and Physiology, (Eds.) Varma, A. and Hock, B. Springer-Verlag, Berlin. Bethlenfalvay, G.J. and Linderman, R.G. 1992. Mycorrhizae in sustainable agriculture. ASA Special Publication Number **54**: Madison, WI, USA.

Baum, T.J., Hialt, A., Parrott, W. A., Pratt, L.K., and Hussey, R.S. (1996). Expression in tobacco of a functional monoclonal antibody specific to styley secretions of the root-knot nematode. Mol. Plant Microb. Interact., **9**:382.

Beffa, R., Szell, M., Mevwly, P., Pay, A., Vogeli-Lange, R., Metraux, J.P., Neuhans, G., Meins, F., and Nagy, F. (1995). Cholera toxin elevates pathogen resistance and induces pathogenetic – related gene expression in tobacco, EMBO, J., (**14**), 5753.

Bliss, D.E. (1951). The destruction of *Armillaria mellea* in *Citrus* soils. Phylopathology, **41**:665-683.

Blum, H., Gross, H.J., Beier, H. (1989). The expression of the TMV - specific 30 K Da protein in tobacco protoplasts is strongly and selectively enhanced by actinomycetes. Virology, **169**:51.

Burke, D.W. (1965). *Fusarium* root-rot of beans and behaviour of the pathogen in different soils. Phylopathology, **55**:1122-1126.

Broadbent, P. and Baker, K.F. (1974). Behaviour of *Phytophthora cinnamoni* in soil suppression and conducive to root rot. Aust. Jour. Agric. Res., **25**:139-145.

Broglie, K., Chet, I., Holliday, M., Cressman, R., Biddle, P., Knowtton, S., Mauvais, C.J., and Brogie, R. (1991). Transgenic plants with enhanced resistance to the fungal pathogen *Rhizoctonia solani*. Science, **254**:1191-1194.

Burkelt, J., Jetiyarwn, K., Backmann, P.A. and Tuzun, S. (1996). Field evaluation and physiological characterization of induced resistance to control Black-rot of Crucifers. Proc. Am. Phytopathol. Soc., 435 A-2, (Abstr.).

Carmona, M.J., Molina, A., Fernandez, J.A., Lopez-Fando, J.J., and Garcia-Olmedo, F., (1993). Expression of the X-thionin gene from barley in tobacco confers enhanced resistance to bacterial pathogens. Plant J., **3**:457.

Chandra, S. (1992). VAM:Dimensions of application. Indian Phytopath., **45(4)**:391-406.

Clare, B.G. (1993). *Aagrobacterium* : biological plant disease control, in Advanced Engineered Pesticides, (Kim, L., Ed.), Morcel Dekker, New York, 129.

Cook, R.J. (1997). Ecologically sustainable plant disease management. In: Integrated Plant Disease Management for sustainable agriculture (abstracts of symposium and poster sessions) Indian Phytopathological Society, New Delhi, 1.

Davis, R.M. and Menge, J.A. (1980). Influence of *Glomus fasciculatum* and soil phosphorus on *Phytophthora* rot of citrus. Phytopathology, **70**:447-452.

Davis, R.M., Menge, J.A. and Erwin, D.C. (1979). Influence of *Glomus fasciculatum* and phosphorus on *Verticillium* wilt of cotton. Phytopathology, **69**:453-456.

DeZoeten, G.A. and Fulton, R.W. (1975). Understanding generates possibilities. Phytopathology, **65**:221-228.

Diiring, K., Porash, P., Fladung, M., and Lorz, H. (1993). Transgenic potato plants resistant to the phytopathogenic bacterium *Erwinia carotovora*, Plant J., **34**:587.

Dodds, J.A., Lee S.Q. and Tifffany, M. (1985). Cross protection between strains of Cucumber mosaic virus: effect of host and inoculum type on accumulation of virions and DS-RNA of challenge strain. Virology, **144**:301-306.

Dunleavy, J. (1981). Control of damping-off of Sugar beat by *Bacillus subtilis*. Phytopathology, **45**:252.

Garnsey, S.M., Lee, R.F. and Semancik, J.S. (1988). Lack of cross protection between *citrus exocortis* viroid and *citrus viroides* associated with mild symptoms Etrog citron, 5th Int. Cong. Plant Pathology, Kyoto, Japan, 69, (Abstr.).

Graham, J.D. and Gray, G.M. (1995). How should pesticides be regulated? Risks in perspective. Harvard Center for Risk Analysis, 3(1).

Grover, R.K. (1986). Chemicals for the control of plant diseases in India. Indian Phytopath., **39(1)**:1-13.

Hain, R., Reif, H.J., Krause, E., Langerbartels, R., Kindl, H., Vorman, B., Wiese, W., Schmelzer, E., Schrier, P.H., Stocter, R.H., and Stenzel, K. (1993). Disease resistance results from foreign phytoalexin expression in a novel plant. Nature, **361**:153.

Hamilton, R.I. (1985). Using plant viruses for diseases control. Hort Sci., **20**:848-852.

Hamlen, R.A., Labit, B., John, A.B., Holliday, M.J., Smith, C.M., Shillingfora, C.A. (1977). Use of Chemical Measures, In: Environmentally safe approaches to crop Disease Control (Eds. Rechcigl, N.A. and Rechcigl, J.E.) : 65-92, CRC Press Boca Raton. Hancock, J.G. Hilgardia, **45**:107-122.

Hawksworth, D.L., Kirk, P.M., Sutton, B.C., Peglir, D.N. (1996). Ainsworth & Bisbijs Dictionary of the Fungi, 8th edition, CAB International England, pp. 676.

Hightower, R., Baden, C., Penzes, E., and Dunsmuir, P. (1994). The expression of cecropin peptide in transgenic tobacco does not confer resistance to *Pseudomonas syringae* pv. *tobaci*, Plant Cell Rep. **13**:295.

Hornby, D. (1990). Biological Control of Soil-Borne Plant Pathogens, C.A.B. International, Wallingford, U.K., 465.

Howell, C.R. and Stipanovic, R.D. (1979). Control of *Rhizoctonia solani* on cotton seedlings with *Pseudomonas fluorescens* and with an antibiotic produced by the bacterium. Phylopathology, **69**:480-482.

Hubbard, J.P., Harman, G.E. and Harden, Y. (1983). Effect of soil borne *Pseudomonas* spp. on the biological control agent, *Trichoderma hamatum* on pea seeds. Can. J. Bot., **36**:726-836.

Jach, G., Gornhardt, B., Mundy, J., Longemann, J., Pinsdof, E., Leah, R., Schell, J., and Mass, C. (1995). Enhanced quantitative resistance against fungal disease by combinatorial expression of different barley anti-fungal proteins in transgenic tobacco, Plant J., **8**:97.

Jacobsen, B.N. and Backman, A. (1993). Biological and cultural plant disease control: Alternatives and supplements to chemicals in 1 PM systems, Plant Dis., **77(3)**:311-315.

Jalali, B.L. (1991). Side effect of pesticides on mycorrhizal system an overview. In: Current trends in mycorrhizal research. Proc. National Conference on mycorrhiza, H.A.U. 1990, Eds. Jalali, B. L. and Chand, H., TERI, New Delhi, pp. 172-174.

Kaiser, W.J., Hunnan, R.M. and Weller, D.M. (1989). Biological control of seed-rot and pre-emergence damping-off of Chickpea with fluorescent *Pseudomonas.* Soil Biol. Biochem. **21**:269-273.

Katan, J., Greenberger, A., Alon., H. and Grinstein, A. (1976). Solar heating by polyethylene mulching for the control of diseases caused by soil borne pathogens. Phytopathology, **76:**683-36

Kinal, H., Park, C. N., Berry, J.O., Kottin, Y., and Bruenn, J.A. (1995). Processing and secretion of a virally encoded anti-fungal toxin in transgenic tobacco plants: evidence for a K_e x 2_p pathway in plants. Plant Cell, **7:**677.

Kleopper, J. W., Leong. J. T. M. and Schroth, M. N. (1980). Enhanced plant growth by siderophores produced by plant growth promoting rhizobacteria. Nature: **286:**885-886.

Kodama, T. and Fukui, T. (1979). Solar heating sterilization in closed vinyl house against soil borne disease. 1. The movements of soil temperature and determination of thermal lethal conditions for some soil borne pathogens. Bull. Nara. Pref. Agric. Exp. Sta., **10:** 71.

Krishna, K.R., Balakrishna, A.N. and Bagyaraj, D.J. (1982). Interaction between a vesicular-arbuscular mycorrhizal fungus and *Streptomyces cinnamomeous* and their effects on finger millet. New Phytol., **92:** 401-405.

Kumar, J. (1997). Studies on root rot of sunflower with special reference to rhizosphere including mycorrhizae. Ph.D. Thesis, Kurukshetra University, Kurukshetra.

Kumar, R.N. and Mukerji, H.G. (1996). Integrated disease management – Future perspective, In: "Advances in Botany" (Ed. K.G. Mukerji),: 335-347, APS Publishing Corporation, 5 Ansari Road, New Delhi.

Leisinger, T. and Maraff, R. (1979). Secondary metabolites of the fluorescent *Pseudomonas*. Microbial. Rev. **43:**442-442.

Leong, J. (1986). Sidrophores their biochemistry and possible role in the biocontrol of plant pathogens. Ann. Rev. Phytopath., **24:**187-209.

Lewis, J.A. and G.C. Papavizas (1987). Reduction of inoculum of *Rhizoctonia solani* in soil by germlings of *Trichoderma hamantum.* Soil Biol. Biochem., **19:**195-201.

Lewis, J.A., Papavizas, G.C. and Lumsden, R.D. (1991). A new formulation system for the application of biocontrol of fungi to soil. Biocontrol Sci. Tech., **1:**59-69.

Lifshitz, R., Dapler, M., Elad, Y., Baker, R. (1984). Can. J. Microbiol., **30:**1482-1485.

Lin, W., Anuratha, C.S., Datta, K., Potrykus, I., Muthukrish. Nan, S., and Datta, S.K. (1995). Genetic engineering of rice for resistance to sheath blight, Biotechnology, **13**:686-689.

Linderman, R.G. (1994). Role of VAM fungi in biocontrol. In: Mycorrhiza and Plant Health. Eds. Fleger, P. and Linderman, R.G., The American Phytopathological Society Press, St. Paul, USA, pp.1-27.

Lindow, S.E. (1992). Integrated control of frost injury, fire blight and fruit russet of pear with a blossom of application of an antagonistic bacterium. Phytopathology **82:**1129-1132.

Longemann, J., Jach, G., Tommerup, H., Murdy, J., and Schell, J. (1992). Expression of a barley ribosome inactivating protein leads to increased fungal protection in transgenic tobacco plants, Biotechnology, **10:** 305-310.

Studies on wilt diseases of coriander with special reference to variability of the pathogen.

Mall, O.P. 1968. Studies on wilt diseases of Coriander with special reference to variability of the pathogen. Indian Phytopath., **21:**377-387.

Martin, G.B., Brommenschenkel, S.H., Chunwongse, J., Frary, A., Ganal, M.W., Spivey, R., Wu, T., Earle, E.D., and Tanksky, S.D. (1993). Map based cloning of a protein kinase gene conferring disease resistance in tomato, Science, **262:**1432-1436.

Mathews, G.A. and Hislop, E.C. (1994). Application technology for crop protection. CAB International, Wallingford, U.K.

McCartneef, H.A. (1997). The influence of environment on the development and control of disease. In: Environmentally safe Approaches to Crop Disease Control (Eds. Rechcigl, N.A., J.E., Rechcigl, (RC Press, Boca Raton), pp. 3–31.

Mckinney, H.H. (1929). Mosaic diseases in the Canary Island, West Africa and Gibraltor. J. Agric. Res., **39:**557-579.

Mehrotra, R.S. (1993). Innovative Methods for Plant Disease Control – Presidential Address. Plant Dis. Res., **8(2):**1-17.

Mehrotra, R.S., Aneja, K.R., Aggarwal, A. (1997). Fungal control agents. In: Environmentally safe approaches to crop disease control, 111-137, (Rechcigl, N.A., Rechcigl, J.E., Eds.), CRC Press, Boca Raton, 386, pp.,

Mishagi, I. J., Stowell, L. G., Grogan, R. G. and Spearman, L.C. (1982). Fungistatic activity of water soluble fluorescent pigments of fluorescent Pseudomonads. Phytopathology, **33:**36-39.

Morandi, D. (1996). Occurrence of phytoalexins and phenolic compounds in endomycorrhizal interactions and their potential role in biological control. Plant and Soil, **185:**241-251.

Mukerji, K.G. and Chamola, B.P. and Sharma, M. (1998). Mycorrhiza in control of plant pathogens. In: Management of threatening plant deceases of National importance, (Eds.) Agnihotri, V.P., Sarbhog, A.K. and Singh, D.V., Malhotra Pub. House, New Delhi, pp. 297-314.

Mukerji, K.G. and Garg, K.L. (1988). Biocontrol of plant diseases. Vol. I, CRC Press, Boca Raton, Florida, USA.

Nagarajan, S. (1983). Plant disease epidemiology. Oxford and IBH Publishing Co., New Delhi, pp. 267

Novelli, J.L., Aldwinckle, H.S., Destefano-Beltran, L., and Jayness, J.M. (1994). Fransgenic 'Malling 26' apple expressing the attacin E gene has increased resistance to *Erwinia amylovora,* Euphytica, **77:**123.

Palukaitis, P. and Zaitlin, M.A. (1984). A model to explain the "crop protection" phenomenon shown by plant viruses and viroids in Plant Microbe Interactions: Molecular and Genetic Perspectives (Kosuge, T., Nester, E.W. Eds. Macmillan, New York.

Park, C.M., Berry, J.O. and Bruenn, J.A. (1996). High level section of a virally enclosed anti-fungal toxin in transgenic tobacco plants. Plant Mol. Biol., **20:**359-362.

Paulitz, T.C. and Baker, R. (1987). Biological control of *Pythium* damping-off of cucumbers with *Pythium nunn*: Population dynamics and disease suppression. Phytopathology, **77:**335.

Porter, J.J. (1992). Soil solarization and biological control in recent developments in Biocontrol of plant Diseases, Mukerji, K. G., Tewari, J. P. Arora, P. K. and Saxena, G., Eds., Aditya Books, New Delhi, India, 164-173.

Prillwitz, H. and Schlosser, E. (1993). Interactions between beet soil borne viruses (BSBV-2) and beet necrotic yellow vein virus (BNYVV), in Proc. Second. Symp. Int. Working Group Plant Viruses with fungal, Vectors (Hiruki, C. Ed.) American Society of Sugar Beet Technologists, Denver, Co., p. 71.

Pullman, G.S. De Vay, J.E., Garber, R.H. and Weinhold, A.R. (1981). Soil solarization: effect of Verticillium wilt of cotton and soil borne populations of *Verticillium dahliae, Pythium* spp., *Rhizoctonia solani* and *Thielaviopsis basicola.* Phytopathology, **71:** 954-958.

Ragsdale, N.N., Hylin, J.W., Sisler, H.D., Witt, J.M. (1991). Health and Environmental Factors Associated with Agricultural use of Fungicides. Portion USDA/States Natl. Agric. (NADIP), Fungicide Assessment Project, pp. 117.

Raupach, G.S., Liu, L., Murphy, J.F., Tuzun, S. and Kloepper, J.W. (1996). Induced resistance against cucumber mosaic cucumovirus using plant growth rhizobacteria (PGPR). Plant Dis., **80:** 891-896.

Rechcigl, N.A. and Rechcigl, J.E. (1997). Environmentally safe approaches to crop disease control, CRC Lewis Publishers, Boca Raton, pp. 386.

Rishbeth, J. (1963). Stem protection against *Fomes annosus.* III. Inoculation with *Peniophora gigantea.* Ann. Appl. Biol, **52:**63-77.

Rishbeth, J. (1979). Modern aspects of biological control of *Fomes* and *Armillaria.* Eur. Jour. For. Pathol. **9:**331-340.

Ross, A.F. (1974). Interaction of viruses in the host. Acta Hortic., **36:**247-250.

Ross, J.P. (1971). Effect of phosphate fertilization on yield of mycorrhizal and non-mycorrhizal soybeans. Phytopathology, **61:**1400-1403.

Rush, C.M., Sherwood, J.L. (1997). Viral control agents, 139-159, In: Environmentally Safe Approaches to Crop Disease Control (Rechcigl, N.A. and Rechcigl, J.E., Eds.), CRC Press, Boca Raton, Fl.

Schonbeck, F. (1979). Endomycorrhiza in relation to plant diseases. In: Soil Borne Plant Pathogens, Eds. Schippers, B. and Grams, W., Academic Press, New York: pp. 271.

Shipton, P.J. (1975). Take-all decline during cereal monoculture, in Biology and Control of Soil Borne Plant Pathoguss, 137-144, (Bruehl, G.W. Ed.), American Phytopathological Society, St. Paul, M.N.

Shukla, B.N., Khare, M.N. and Vyas, S.C. (1996). Role of vesicular arbusuclar mycorrhizal fungi in controlling soil borne diseases, 223-232. In Advances in Botany (Mukherji, K.G. Ed.), APH Pub. Corporation, New Delhi.

Siegel, H.D. and Sisler, H. (1977). Antifungal Compounds, Vol. 2, interaction in biological and ecological systems, Marcel Dekker, New York.

Simmonds, J.H. Q. (1959). Mild strain protections as a means of reducing losses from the Queensland woodiness virus in the passion vine. Queensland. J. Agric. Sci., **16**:371-375.

Stirling, G.R. and Mankaur, R. (1978). *Dactylella oviparasitica*, a new fungal parasite of *Meloidogyne* eggs. Mycologia, **70**:744-783.

Terras, F.R., Eggermont, K., Kovaleva, V., Raikhel, N.V., Osbron, R.W., Kester, A., Rees, S.B., Torrekens, S., Van Leuven, F., Vanderleyden, J., Cammue, B.P.A., and Brae Karet, W.F. (1995). Small Cysteis–rich anti-fungal proteins from radish: their role in host defense, Plant Cell, **1**:573.

Thakur, D.P. (1983). Epideiology and control of Ergot disease of Pearl Millet, Seed Sci. Tec., **11**:797.

Thakur, R.P. and Williams R.J. (1988). Pollination effects of pearl millet ergot. Phytopathology, **70:** 80-84.

Thakur, R.P., Williams, R.J. and Rao V.P. (1983). Control of ergot in pearl millet through pollen management. Ann. Appl. Biol., **103**:31-36.

Toussoun, R.A. (1975). *Fusarium*- suppressive soils in biology and control of soil-borne plant pathogens: 145-151, (Bruchl, G.W., Ed.), American Phylopathological Society, St. Paul, M.N.

Urwin, P.E., Atkinson, H.J., Waller, D.A., and McPherson, M.J. (1995). Engineered oryzacystatin–J expressed in transgenic having hairy roots confers resistance to *Globodera pallida*, Plant, J., **8**:121,.

Van den Elzen, P. J.M., Jongedijk, E., Melchers, L.S., and Cornelissen, B.J.C. (1993). Virus and fungal resistance: From laboratory to field, Philos Trans. R. Soc. London B, **342**:271,

Waggoner, P.E. , Miller, P.M. and DeRoo, H.C. (1960). Plastic mulchinh principles and benefits. Connecticut Agri. Exp. Station Bulletin, 634, 44-47.

Wiley, H.B. and Kommedahl, T. (1981). Biological seed treatment in sweet corn and wheat as a component of crop management. Phylopathology, **71**:265-269.

Wilson, M. and Londow, S.E. (1993). Interactions between the biological control agent *Pseudomonas fluorescens* A 506 and *Erwinia amylovora*. Phytopathology, **83**:117-120.

Wu, G., Schortt, B.J., Lawrence, F.B., Levine, E.B., Firzsimmons, K.C., and Shah, D.M. (1995). Disease resistance conferred by expression of a gene encoding H_2O_2 – generating glucose oxides in transgenic potato plants. Plant Cell, **7**:1357-1361.

Xu, H. and AbouHaider, M.G. (1996). Synergistic relationship between potato aucuba mosaic virus and potato virus Y. strain O, Can. J. Plant Pathol., **18**:102-105.

Zhu, G., Maher, E.A., Masoud, S., Dixon, R.A., and Lamb, C.J. (1994). Enhanced protection against fungal attack by constitutive co-expression of chitinase and glucanase genes in transgenic tobacco, Biotechnology, **12:** 807-811.

Advancing Frontiers of Ecological Researches in India, 615-620, 2006
Edited by A.K. Kandya & Asha Gupta
Published by Bishen Singh Mahendra Pal Singh, Dehra Dun, India

Response of Pre-treatments on Germination of *Delonix regia* (Boj.) Rafin Seeds

Rajeev Baweley* and U.N. Singh
Deptt. of Botany, D.V. (P.G.) College
Orai (Jalaun) U.P. 285 001

Abstract: In the present work, many methods of scarification viz. dry heat, chemicals, acid treatment, hot water, boiling water and shock treatments; and mechanical scarification were tried to break the dormancy of *D. regia* seeds. Mechanical scarification methods (filing, nicking and clipping) were found to be the best (78 to 85% germination Vs. 8 to 15% in control), producing earlier and higher germination of seeds.

INTRODUCTION

Delonix regia (Boj.) Rafin (Fabaceae) commonly known as gulmohar is a fairly large deciduous ornamental tree, often planted in gardens and as an avenue tree along highways. The leaves are much valued as manure for paddy cultivation as they rot quickly (Troup, 1975).

Seed dormancy of *D. regia* is imposed by the hard impermeable seed coat and hence the seeds record poor germination. Softening of the hard seed coat by various methods may improve the germination percentage of seeds (Pandya, 1989).

MATERIALS & METHODS

The experiments were carried out at Deptt. of Botany, D.V. (P.G.) College, Orai, during the year 2001-2003.

Ripen pods were collected from standing trees of *D. regia* in the month of May 2001. Seeds were separated by breaking the pods with heavy hammer strokes and each seed was separated from the locule and stored in sealed glass bottles at room temperature in the laboratory.

* Mailing address - 89, pathnau, RATH (Hamirpur) U.P. Pin - 210431 INDIA

Various pretreatments were performed with the seeds of *Delonix regia* in three successive rounds. Each round took about 6-8 month period to be completed. Pretreatments were -

Round 1 - Water soaking (40^0C) for various durations.
Dry heat in the oven (40,60 and 80^0C) for various durations.
Chemical treatment (Various chemicals).
Mechanical scarification (Filing).
Control.

Round 2 - Concentrated acid (H_2SO_4, HCl and HNO_3).
Diluted (50% and 10%) H_2SO_4.
Mechanical scarification (Nicking).
Control.

Round 3 - Hot water treatment (60^0C and 80^0C) for various durations.
Boiling water treatment for various durations.
Shock treatment (boiling/ice water) for various durations.
Mechanical Scarification (Clipping).
Control.

Fifty seeds were treated with each of the treatment and replicated four times. Imbibition was recorded after 20 hrs in normal water soaking the seeds. Number of germinated and dead seeds on each day was recorded and the experiment was terminated on 15th day.

RESULT AND DISCUSSION

As shown in various Tables (1 to 3) mechanical scarification was found to be the best among all the treatments of each round. Filing gave 85% (8% in control), nicking 81% (13% in control) and clipping gave 78% (15% in control) germination. Many workers have successfully applied this method to break the dormancy of hard coated seeds of different species e.g.. Bohra *et al.* (1994) in *Prosopis juliflora* and Dharmendra kumar and Pyare Lal (1999) in *Sesbania rostrata.*

Being extremely hard, seed coat of *D. regia* require heavy filing, otherwise many seeds remain unaffected. Nicking by scissors is also require heavy cuts showing the toughness of seed coat.

All the treatments given in Table-1 were not suitable to improve imbibition. Dry heat treatment (40°C, 60°C and 80°C for 5, 10 and 15 days) was very little effective in softening the seeds as most of the seeds remained hard at the end of experiment (77% to 85%).

Table-1: Percentage of germinated, hard and dead seeds of *Delonix regia* as affected by various pretreatments.

Sl. No.	Pretreatments Details	After 15 days		
		Ger.	Dead	Hard
1.	Seeds soaked in water (40°C) for 12 hrs.	5	5	90
2.	Seeds soaked in water (40°C) for 24 hrs	8	9	83
3.	Seeds kept in oven at 40°C for 5 days	7	8	85
4.	Seeds kept in oven at 40°C for 10 days	8	7	85
5.	Seeds kept in oven at 40°C for 15 days	10	9	81
6.	Seeds kept in oven at 60°C for 5 days	9	8	83
7.	Seeds kept in oven at 60°C for 10 days	10	11	79
8.	Seeds kept in oven at 60°C for 15 days	8	12	80
9.	Seeds kept in oven at 80°C for 5 days	9	12	79
10.	Seeds kept in oven at 80°C for 10 days	10	13	77
11.	Potassium nitrate (10%) for 10 hrs	5	10	85
12.	Potassium nitrate (10%) for 20 hrs	7	11	82
13.	Potassium dichromate (5%) for 10 hrs	6	9	85
14.	Potassium dichromate (5%) for 20 hrs	7	11	82
15.	Sodium nitrite (5%) for 10 hrs	8	13	79
16.	Sodium nitrite (5%) for 20 hrs	11	19	70
17.	Thiourea (5%) for 10 hrs	6	8	86
18.	Thiourea (5%) for 20 hrs	8	11	81
19.	Cattle urine for 10 hrs	3	10	87
20.	Cattle urine for 20 hrs	5	12	83
21.	Mechanical Scarification (filing)	85	15	—
22.	Control	8	7	85

No chemical was found to improve the permeability of hard coated seeds of this species (70% to 87%, hard seeds). Only Sodium nitrite (5%) for 20 hrs gave a little improvement (11% germination Vs 8% in control). Negative effect of KNO_3 has been reported in *Acacia* species (Padma *et al.*, 1993) and cattle urine in *Buchanania lanzen* seeds (Choubey *et al. 1997).*

As evident from Table-2, Conc. H2SO4 was found to remove the coat imposed dormancy of *D. regia* seeds when soaked for 5 hrs (80% ger. Vs 13% in control) leaving no hard seed at the end of 15 days. When

Table-2 : Percentage of germinated, hard and dead seeds of *Delonix regia* as affected by various pretreatments.

Sl. No.	Pretreatments Details	After 15 days		
		Ger.	Dead	Hard
1.	Concentrated (98%) H_2SO_4 for 5 min	13	5	82
2.	Concentrated (98%) H_2SO_4 for 10 min	15	8	77
3.	Concentrated (98%) H_2SO_4 for 30 min	18	10	72
4.	Concentrated (98%) H_2SO_4 for 45 min	25	7	68
5.	Concentrated (98%) H_2SO_4 for 60 min	29	4	67
6.	Concentrated (98%) H_2SO_4 for 90 min	35	8	57
7.	Concentrated (98%) H_2SO_4 for 3 hrs	40	15	45
8.	Concentrated (98%) H_2SO_4 for 4 hrs	72	23	5
9.	Concentrated (98%) H_2SO_4 for 5 hrs	80	20	—
10.	Concentrated (98%) H_2SO_4 for 6 hrs	21	79	—
11.	Concentrated (98%) H_2SO_4 for 7 hrs	8	92	—
12.	Concentrated HCl for 60 min	12	3	85
13.	Concentrated HCl for 120 min	15	5	80
14.	Concentrated HCl for 180 min	20	8	72
15.	Concentrated HNO_3 for 60 min	12	5	83
16.	Concentrated HNO_3 for 120 min	14	6	80
17.	Concentrated HNO_3 for 180 min	17	9	74
18.	Diluted (50%) H_2SO_4 for 60 min	7	5	88
19.	Diluted (50%) H_2SO_4 for 90 min	10	5	85
20.	Diluted (50%) H_2SO_4 for 180 min	7	11	82
21.	Diluted (10%) H_2SO_4 for 10 hrs	5	7	88
22.	Diluted (10%) H_2SO_4 for 20 hrs	8	2	90
23.	Diluted (10%) H_2SO_4 for 30 hrs	12	8	80
24.	Mechanical scarification (Nicking)	81	19	—
25.	Control	13	8	79

duration was increased to 6 hours, more seeds became dead showing its injurious effect. Patel and Kukadia (1996) reported a shorter duration (5-10 min) to obtain 46-54% germination in the same species with Conc. H2SO4 Conc. HCl and HNO_3 were not found suitable in this species.

As given in Table-3, hot water soaking (60°C and 80°C) exhibited a negative effect on seeds as number of dead seeds increased after treatment. Higher temperatures and longer durations were found to be more harmful as the most of the seeds became dead after treatment.

Table-3: Percentage of germinated, hard and dead seeds of *Delonix regia* as affected by various pretreatments

Sl. No.	Pretreatments Details	After 15 days		
		Ger.	Dead	Hard
1.	Hot water soaking (60°C) for 5 hrs	5	8	87
2.	Hot water soaking (60°C) for 7 hrs	9	16	75
3.	Hot water soaking (60°C) for 20 hrs	-	60	40
4.	Hot water soaking (80°C) for 4 hrs	17	51	32
5.	Hot water soaking (80°C) for 11 hrs	7	81	12
6.	Hot water soaking (80°C) for 20 hrs	-	96	4
7.	Boiling water dipping	12	6	82
8.	Boiling water soaking for 5 sec	14	10	76
9.	Boiling water soaking for 10 sec	24	21	55
10.	Boiling water soaking for 20 sec	32	23	45
11.	Boiling water soaking for 30 sec	35	25	40
12.	Boiling water soaking for 1 min	39	36	25
13.	Boiling water soaking for 2 min	42	47	11
14.	Boiling water soaking for 5 min	23	70	7
15.	Boiling water/ice water 5 sec - 1 time	15	10	75
16.	Boiling water/ice water 5 sec - 2 times	16	13	71
17.	Boiling water/ice water 5 sec - 5 times	19	16	65
18.	Boiling water/ice water 10 sec - 1 time	26	12	62
20.	Boiling water/ice water 10 sec - 2 times	31	15	54
21.	Boiling water/ice water 5 sec - 5 times	37	20	43
22.	Mechanical Scarification (Clipping)	78	22	-
23.	Control	15	10	75

Boiling water treatment was found moderately effective to soften the hard seed coat. Forty two percent germination (15% in control) was observed after 2 min. soaking the seeds in boiling water. Duration longer than 2 minute soaking in boiling water increased the number of dead seeds (70% in 5 min). This treatment was found to be most effective in *Acacia* species (Singh *et al.*, 1990) and in *Cleistanthus collinus* seeds (Sharma *et al.*, 1999).

Shock treatment (boiling water/ice water) was also moderately effective. This treatment improved the germination upto some extent (37%) but some seeds became dead when the duration of treatment increased (5 time 20% dead seeds). A good number of seeds remained hard (43-75%) after various durations of shock treatments.

REFERENCES

Bohra, M.D.; J.C. Tiwari; U. Burman; N.K. Sharma and L.N. Harsh (1994). Response of pretreatment on germination of *Prosopis juliflora.* Journal of Tropical Forestry, **10(4):** 305-309.

Choubey, Archana; Ram Prasad; O.P. Choubey and N.C. Pant (1997). Some aspects of germination studies in *Buchanania lanzan* Spreng. Seeds. Journal of Tropical Forestry, **13(2):** 65-73.

Dharmendra Kumar and Pyare Lal (1999). Improving germination of *Sesbania rostrata* green manure crop. Seed Research, **27(1):** 20-24.

Padma, V.; B. Murlimohan Reddy and G. Satyanarayana (1993). Breaking dormancy in certain *Acacia* spp. by pre-sowing seed treatments. Seed Research, **21(1):** 26-30.

Pandya, S. (1989). Improving forest production through quality seeds. Journal of Tropical Forestry. **5(2):** 251-255.

Patel, V.S. and M.U. Kukadia (1996). Effect of presowing treatments for *Delonix regia.* Journal of Tropical Forestry, **12(2):** 101-103.

Sharma, Archana; O.P. Choubey and K.P. Tiwari (1999). Pretreatment for hastening seed germination and seedling growth in *Cleistanthus collinus* (Benth). Journal of Tropical Forestry, **15(4):** 247-252.

Singh, Chhavi; H.V. Khajuria and P.K. Ralhan (1990). Germination behaviour of some exotic *Acacia* species. Acta Botanica Indica, **18(1):** 38-40.

Troup, R.S. (1975). The Silviculture of Indian Trees, FRI publication, Dehra Dun, Vol-III.

SECTION NINTH

ECOLOGY MODELING

Advancing Frontiers of Ecological Researches in India, 621-630, 2006
Edited by A.K. Kandya & Asha Gupta
Published by Bishen Singh Mahendra Pal Singh, Dehra Dun, India

Role of Mathematics in Ecology

M. Ranjit Singh
Department of Mathematics
Manipur University, Imphal – 795 003
E-mail : mranjitmu@rediffmail.com

Abstract: In this note some works of compilation on mathematical models of population growth, cell division and predator-prey are discussed. Some related elementary concepts of differential equations are covered in this note and other mathematical terms are mentioned

INTRODUCTION

In recent years, mathematics has made a considerable impact as a tool with which to model and understand Biological phenomena. The subject is being applied more and more to ecology and other biological sciences. Mathematical Ecology deals with competition among species for limited resources, predator-prey & host-parasitic interactions, diversity & stability of ecosystem, etc. In other words, it requires the study of population that interact, thereby affecting each other's growth rates.

From the mathematical point of view one needs to recall some mathematical/statistical terms/formulae as follows : Mathematical symbols; Greek letters; Statistical symbols; Logarithm & Antilogarithm; Reading of logarithm/and antilogarithm table; A.M; G . M; H.M.; Median ; Mode; Measures of Dispersion (or Variation) – mean deviation, standard deviation, variance, tests of significance, student's t-tests, Chi-Square Test, Set & Set operations; Relation & Function; Differentiation & Integration etc.

Further, the basic ideas of differential equations are restated below:

Differential Equation

In a differential equation the variable with respect to which the derivatives are taken is called the independent variable and a variable is called dependent of a derivative of that variable occurs in the equation. In the differential equation.

$$dN/dt = aN - bN^2 \quad \text{(I)}$$

where $N(t)$ represents the population concentration at time t and a & b are constants, N and t are the dependent and independent variables respectively.

If in a differential equation there is only one independent variable then it is called an ordinary differential equation. If there are more than one independent variable, in which case partial derivatives are required, then the equation is known as partial differential equation. The equation (I) is an ordinary differential equation.

Consider, $x\dfrac{\partial f}{\partial x} + y\dfrac{\partial f}{\partial y} = n.f$

where f = f(x, y) is a function of two independent variables, $\dfrac{\partial f}{\partial x}$ & $\dfrac{\partial f}{\partial y}$ are partial derivatives of f with respect to x & y respectively.

The order of a differential equation is the highest order derivative appearing in the equation. In (I), order of $\dfrac{dN}{dt}$ is one so the order of (I) is one. The order of the differential equation $\left(\dfrac{dy}{dx}\right)^2 + y = 0$, where y is a function of x is also one since highest order derivative is $\dfrac{dy}{dx}$ which is of first order.

The degree of a differential equation is the highest exponent of the derivative of the highest order appearing in it after the equation has been expressed in the form free from radicals and fractions as far as derivatives are concerned. For an example, $\left(\dfrac{dy}{dx}\right)^2 + y = 0$ is of second degree equation whereas $x^2\dfrac{d^2y}{dx^2} + 2x\dfrac{dy}{dx} + y = x^2 + 2$ is of order two and first degree equation.

Population Growth Model

In this section we have to show how the study of first order ordinary differential equations may be used to model the growth of population. The size of a population varies in time. It is a matter of interest in several contexts. Let the number of individuals in a given area at time t be P(t). At time t + T the number of individuals P(t + T) so that P(t + T) – P(t) must be the number of individuals that have been added to the population during the time interval T. The longer the interval the more individuals can be expected to arrive and the shorter the time the fewer additions can be expected. Write the change in the time interval T as NT, then.

$$P(t + T) - P(t) = NT \text{ or, } \quad \frac{P(t + T) - P(t)}{T} = N$$

Letting $T \to 0$ we have

$$\frac{dP(t)}{dt} = N$$

(read as derivative of P with respect to t equals to N) (1)

Suppose that the change in size of the population is due entirely to individuals being born. As time progress the fertility of parents may alter so that more or less offspring are born. Then the number born in the interval T may vary as time proceeds. In other words N may alter as t does. Another effect is introduced here that at time t the more births are likely to occur, then N will depend on P(t) also. So, both possibilities can be allowed for by rewriting our differential equation as $\frac{d}{dt}P(t) = N\{t, P(t)\}$.

[This enables to show that N depends on two quantities t & P(t)]
It can be rewritten as

$$\frac{dP(t)}{dt} = N(t, P) \qquad (2)$$

Sometimes, $\frac{1}{P}\frac{dP}{dt}$ is known as the specific growth rate. So, by using this, the another way of expressing (2) is to say that the specific growth rate is N(t, P)/ P. It is reasonable to assume that, in a short time interval, there will be about twice as many births as in a time interval of half its length. Thus one could expect that the number of births would be proportional to P(t)T when T is small. If the fertility of the parents does not change the actual number of births in the time interval, T can be expressed as $N_0P(t)T$.

where N_0 is a constant. Then (2) becomes,

$$\frac{dP(t)}{dt} = N_0 P(t) \tag{3}$$

which states that the specific growth rate is N_0, the same for all times and all sizes of population.

The solution of (3) is

$$P(t) = P_0 e^{N_0 t} \tag{4}$$

where P_0 is a constant. For confirmation, one may check as follows:

Taking the derivative of (4) with respect to t, we have $\frac{d}{dt}P(t) = P_0 e^{N_0 t} N_0$.

At t = 0, P(t) = P_0, the size of the population at t = 0. So from (4), $e^{N_0 t} = 1$. Hence, $\frac{d}{dt}P(t) = N_0 P(t)$ which is (3).

In another way, one can verify as follows:

Integrating (3) with respect to.t between the limits t = 0 to t

$$N_0 \int_0^t dt = \int_0^t \frac{1}{P}\frac{dP}{dt}dt = \int_{P(0)}^{P(t)} \frac{dP}{P} \quad \text{(changing the limits from t to P)}$$

$$N_0 t = \log\left(P(t)/P(0)\right) \tag{5}$$

which is same as (4). The behaviour of the population as time increases according to (4) is shown graphically below:

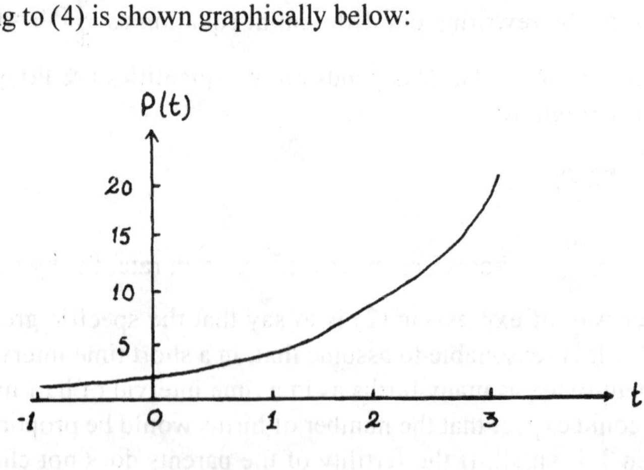

Fig. 1: Graph of exponential growth

As seen in the graph, the size grows steadily and the increase becomes dramatic as time goes on. Of course, in any real situation, there will be a limit to the growth because of a shortage of essential supplies, insufficient food and self-pollution of the environment. Nevertheless, many organisms exhibit exponential growth in their initial stages. It is always easy to check whether a population is growing exponentially by plotting log p against time; a straight line should be obtained. The slope of the line is the specific growth rate N_0 as is clear from (5).

Equation (3) has been derived on the assumption that only births occur. In the case that there are deaths but no births the same equation can be reached. However, N_0 is now a –ve number since the population decreases in the time interval T. It follows from (4) that the population decays exponentially with time from its size at $t = 0$.

Further, we may see other population problems. For instance, we may postulate that the number of deaths in the short-time interval T is $D_0P(t)T$. Similarly, individuals may enter the given area from outside, say I(t)T immigrants in the interval T. Likewise, some may depart from the area giving rise to E(t)T emigrants. Then,

$P(t + T) - P(t) = N_0P(t)T - D_0P(t)T + I(t)T - E(t)T$ which leads to when $T \to 0$

$$\frac{d}{dt}P(t) = (N_0 - D_0)P(t) + I(t) - E(t) \tag{6}$$

More generally, I and E could be made to depend on P so that (6) can be difficult to solve. Equation (6) is called Verhulst's Differential Equation. This equation can be regarded as a particular case of (1).

Cell Division Model

When cells divide, their numbers grow by a process as like as in population growth. A new feature is that the multiplication in numbers is restricted by crowding effects. Biochemically, they may be due to lack of nutrient, shortage of oxygen, change in pH or the production of inhibitors. Whatever the cause, the cells are interacting with one another. Since each cell can interact with p others there are p^2 possibilities. This suggest that in (1) we should put

$$N\{t, p(t)\} = N_0p(t) - ap(t)^2 \tag{7}$$

where N_0 and a are +ve constants. The term involving N_0 is the same as before and accounts for the increase due to division. The term containing a represents the growth caused by crowding. With this substitution (7), (1) gives

$$\frac{dp}{dt} = N_0 p - ap^2 \tag{8}$$

This equation (8) is called the differential equation of logistics.

Integrating (8) w.r.t t between the limits t = 0 to t, we have

$$\int_0^t dt = \int_0^t \frac{1}{N_0 p - ap^2}\frac{dp}{dt}dt = \int_{p(0)}^{p(t)} \frac{dp}{N_0 p - ap^2} \quad \text{(changing the limits from t to p.)}$$

Here, $\int \frac{dp}{N_0 p - ap^2} = \frac{1}{N_0}\int\left(\frac{1}{p} - \frac{a}{ap - N_0}\right) = \frac{1}{N_0}\log\frac{p}{ap - N_0}$ [Integration by partial fractions]

Hence

$$t = \frac{1}{N_0}\log\left(\frac{p(t)\{ap_0 - N_0\}}{\{ap(t) - N_0\}p(0)}\right)$$

whence, $p(t)\{ap_0 - N_0\} = \{ap(t) - N_0\}p(0)e^{N_0 t}$. Consequently,

$$p(t) = \frac{N_0 p(0)}{ap(0) + \{N_0 - ap(0)\}e^{-N_0 t}} \tag{9}$$

This is called the logistic law of growth. The curve of logistic growth is shown in below (fig 2)

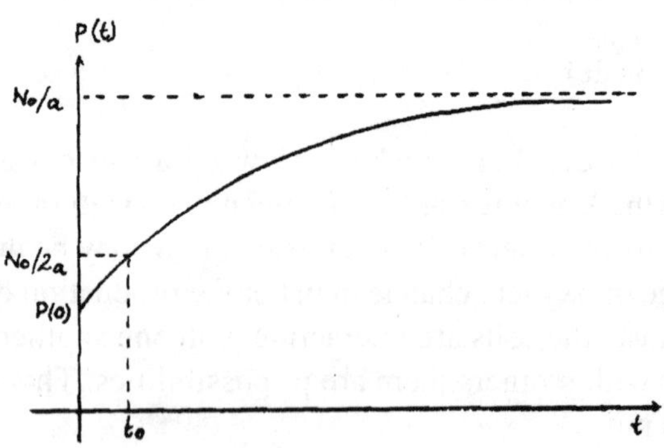

Fig. 2: The curve of logistic growth

Here, we assume $N_0 > a\,p(0)$. The curve rises steadily from the value p(0) at t = 0 to eventual value of N_0 / a, there being no maxima or minima in between. There is, however, a point of inflexion where the curve crosses its tangent at $t = t_0$ where

$$t_0 = \frac{1}{N_0}\log\left(\frac{N_0}{ap(0)} - 1\right) \quad \text{and} \quad p(t_0) = \frac{N_0}{2a}$$

Observe that the final value N_0 / a of p does not involve p(0), so that, no matter what the initial size of the population.

The logistic law assures that all the cells divide at the same rate. This is not always true. There are types in which cells divide faster than the others. Whether the logistic law can be applied still depends upon the difference between the various rates of division present. If the rates are not too far apart, it is probably feasible to take N_0 as their average. For greater deviations it may be necessary to adopt a model in which the statistics of the numbers of cells of a given age and type at a given age and type at a given time play apart.

Predator-Prey Model

The study of the interactions between living organisms and their environment are of fundamental important in ecology. Let us consider two organisms or species characterized, for example by their respective population densities, say X and Y.

Let X be the predator density and Y the prey density. In the absence of predators we expect no inhibition in the growth of the prey. A simple growth relation Y could be

$$\frac{dY}{dt} = kY \tag{10}$$

where t is time and k a +ve rate constant.

The solution of (10) is given by

$$Y = Y_0 \exp(kt) \tag{11}$$

where Y_0 is the initial population density.

If (10) is used to describe the population of Y, then Y will increase exponentially in time. Such growth behaviour is reasonable for a limited time. We expect in practice that Y will well either settle down to some steady state value, fluctuated between various levels or decline. If the first possibility arises, the equation (10) can be replaced by the logistic equation

$$\frac{dY}{dt} = aY - bY^2 \tag{12}$$

which has general solution

$$Y = \frac{aY_0}{\{bY_0 + (a-b)Y_0 \exp(-at)\}} \tag{13}$$

The justification for this type of growth is

(a) when Y is small, (12) reduces to (10) and the growth is exponential.

(b) as t increases Y approaches the value $\frac{a}{b}$ steadily and without oscillation.

In (12), we call 'a' the intrinsic rate of increase and 'k= a / b' the carrying capacity.

Consider the behaviour of the predator species X in the absence of prey. Without prey, the predators X are expected to decrease so the equation is

$$\frac{dX}{dt} = -eX \tag{14}$$

which has solution

$$X = X_0 \exp(-et) \tag{15}$$

where e is a positive constant.

Hence, the decline of population of X are represented in terms of the exponentially decaying solution (15). Such decline is reasonable for some limited period of time. So, we expect X to become extinct in a finite rather than an infinite time of time when starved of prey Y.

Assume the **Voltera-Lotka** Model. In the absence of predation, Y follows a logistic growth curve and the rate at which prey are eaten is proportional to the product of the densities of predator and prey. Then we have

$$\frac{dX}{dt} = -eX + fXY \tag{16}$$

$$\frac{dY}{dt} = aY - bY^2 - cXY \tag{17}$$

where a, b, c, e, f are constants.

The assumption of Volterra and Lotka are valid under the following conditions :

(a) one or both species more at random;

(b) when they meet, there is a constant probability that the predator will kill the prey;

(c) the time taken by the predator in consuming the prey is negligible.

There are a number of variants of the **Volterra-Lotka** model that have been proposed and that attempt to take into account other effects that may influence the predator-prey interaction. For example, when a constant number $\overline{Y}$ of the prey can find some cover or refuge, which makes them in accessible to the predator, then the system (16), (17) is modified to

$$\frac{dX}{dt} = -eX + fX(Y - \overline{Y}) \tag{18}$$

$$\frac{dY}{dt} = aY - bY^2 - cX(Y - \overline{Y}) \tag{19}$$

REFERENCES

Jones, D.S.; Sleeman. B.D. (2003). Differential Equations and Mathematical Biology, Chapman & Hall / CRC.

Kapur, J.N. (1989). Interfaces Between Mathematical and Biological Sciences, Fascinating World of Mathematics Vol. III, Mathematical Sciences Trust Society.

Katz, B. (1966). Nerve, Muscle and Synapse, McGraw Hill, New York.

Lambert, J.D. (1991). Numerical Methods for Ordinary Differential Systems, John Wiley & Sons, New York.

Lambert, J.D. (1973). Computational Methods in Ordinary Differential Equations, John Wiley & Sons, New York.

Melvin, H. (1973). Mathematical Physiology, Elsevier, Amsterdam.

Murray, J.D. (1999). Mathematical Biology, Springer-Verlag, Heidelberg.

Rai, B; Choudhury, D.P.; Freedman, H.I. (2002). A Course in Ordinary Differential Equations, Narosa, N Delhi.

Smith, J.M. (1974). Models in Ecology, CUP.

Zeeman, E.C. (1972). Differential Equations for the heart beat and nerve impulse, Towards a Theoretical Biology, Vol. 4, C.H. Wardington (ed.), University Press, Edingburg.

Advancing Frontiers of Ecological Researches in India, 631-644, 2006
Edited by A.K. Kandya & Asha Gupta
Published by Bishen Singh Mahendra Pal Singh, Dehra Dun, India

Matrix Models in Analysing Population Dynamics

Asha Gupta
Department of Life Sciences, Manipur University
Imphal – 795003, Manipur
E-mail : anjalika_22000@yahoo.co.in

Abstract: Matrix models that integrate population dynamics and population structure are powerful tools for investigating population dynamics. The models play an important role in furthering and understanding population ecology, help in predicting the 'population level of cropping, project future population growth, differentiate age specific characteristics, guide conservation programme and help in better management of species, population and ecosystems. Developed during 1990's, matrix models now find many applications. The paper reviews how matrix models have been used in analyzing population dynamics, mentions some case studies, discusses applications and disadvantages of this family of models and suggests how to enhance their application.

INTRODUCTION

In arable farming, the proper timing in weed control is of paramount importance and of major concern to the farmer looking into the immense use of weedicides and their costs. These concern originate from considerations of a) the yield and quality and thus economic value of crop b) the magnitude and extent of operations required for weed control and hence cost of cultivation and husbandry (Huffaker, 1958; Sagar and Mortimer 1976).

Here, it may be emphasised that mathematical models have an essential role to play in furthering and understanding population ecology. Levin (1968) has suggested that our perfect model would be maximally general, maximally realistic, maximally precise and maximally simple. However, when we examine models on population dynamics, it may be noted that they sacrifice precision and certain amount of realism so as to retain a high degree of generality and simplicity. Begon and Mortimer (1986) suggest that models on population dynamics should be -

a) a satisfying description of the diverse system of the natural and experimental world.

b) an aid to enlightenment on aspects of population dynamics which had previously been unclear and

c) a system which can be easily incorporated into more complex models of intraspecific interactions.

Demographic analysis - Matrix population model

Life-cycle is the fundamental unit of description of organism. The analysis of demographic consequences viz. the vital rates of birth, growth, development and mortality rates help in describing the development of individual through the life cycle. Thus, demography assesses both the dynamics and structure of population. The potential for exponential growth is the most fundamental fact of population dynamics (Cole, 1958). However, population do not continue unlimited exponential growth for long. This was explained, by Verlhaust who derived the logistic equation in 1838 (Hutchinson, 1978). Inter-specific competition and predator-prey interaction were described by Lotka 1924, Volterra 1926 and Gause 1934. Lotka (1924) developed stable age theory in which structure and dynamics of population classified by age groups and living in constant environment was emphasized. Pearl (1928) constructed life table for Drosophila population investigating the impact of many factors like density, starvation, genetics on the survivorship curve. Deevey (1947) in his classic paper included the life table for rotifer, mammals and birds. Further discussion on life table were done by Allee *et al.*, (1949) and Andrewartha and Birch (1954).

Matrix population model were developed in mid of 1940s. The age classified models and density dependent matrix model leading to exponential growth matrix model were derived and attempted directly from graphical description of life style.

Model construction requires not only the knowledge of analytical tools but also concern about data and manipulations needed to transform the data to the required format.

The Matrix

The matrix is a term which describes a rectangular array of number for e.g. A is a matrix of three rows and 4 columns or a 3×4 matrix.

There are 12 numbers or elements in the matrix where aij element is the element in the ith row and j^{th} column of matrix A. Bold letter is used

$$A = \begin{bmatrix} a_{11} & a_{12} & a_{13} & a_{14} \\ a_{21} & a_{22} & a_{23} & a_{24} \\ a_{31} & a_{32} & a_{33} & a_{34} \end{bmatrix}$$

to represent a matrix and its subscript is used to represent its element. A matrix with the same numbers of rows and column is a square matrix. The principal diagonal of the matrix dominate the matrix and comprise of elements of the 1st row and 1st column, 2nd row and 2nd column etc. There may be matrices of only 1 row or 1 column or even there may a matrix with a single element known as scalar. For square matrices, there are Eigen values and Eigen vector and many methods exist for calculation of Eigen values and Eigen vectors.

The equation,

$Av = \lambda v$
indicates that A = square matrix
v = column vector and
λ = scalar.

For every Eigen value λ, there is an associated Eigen vector v. The dominant Eigen value gives the rate at which the population size increases.

Leslie (1948) gave a deterministic model predicting the future age structure of a population of female animal from the present known age structure and assumed risk of survival and fecundity. The model is represented by the equation

$$A_{at} = a_{t+1}$$

Where a_t is the column vector representing population age structure at time 't' and a_{t+1} is a column vector representing the age structure at time t+1. The numbers of animals in the age classes after periods of time can be predicted by repeating the multiplication.

$$a_{t+k} = A^{k}{}_{at}$$

If logarithm of population size is plotted against time, the slope of the line after the stable population structure is equivalent to ln ë, the intrinsic rate of natural increase to dominant Eigen value can be used to estimate the numbers of animals to be removed so as to bring population to its original size.

$$H = 100\left(\frac{\lambda - 1}{\lambda}\right)$$

Where H is expressed as percentage of total population.

Leslie Matrix Model :

One of the earliest forms of matrix was developed by Leslie in 1945, 1948 and popularly termed as Leslie Matrix model. It is given as below.

The population is divided into n+1 (i.e. 0,1,2,3.......n). equal age group so that the oldest group possible or the age group in which all animal

$$\begin{pmatrix} f_o & f_1 & f_2 \ldots\ldots\ldots\ldots\ldots f_{n-1} & f_n \\ p_o & o & o \ldots\ldots\ldots\ldots\ldots o & o \\ o & p_1 & o \ldots\ldots\ldots\ldots\ldots o & o \\ o & o & p_2 \ldots\ldots\ldots\ldots\ldots o & o \\ . & . & . \qquad\qquad . & . \\ . & . & . \qquad\qquad . & . \\ . & . & . \qquad\qquad . & . \\ o & o & o \ldots\ldots\ldots\ldots\ldots p_{n-1} & o \end{pmatrix} \begin{pmatrix} n_{t,}o \\ n_{t,}1 \\ n_{t,}2 \\ n_{t,}3 \\ . \\ . \\ . \\ n_{t,}n \end{pmatrix} = \begin{pmatrix} n_{t+1,}o \\ n_{t+1,}1 \\ n_{t+1,}2 \\ n_{t+1,}3 \\ . \\ . \\ . \\ n_{t+1,}n \end{pmatrix}$$

surveying die, is n. The numbers of animal in the various age classes at time t+1 are obtained by multiplying the number of animals in various age classes at time 't' by a matrix expressing the appropriate fecundity and survival rate for each of class. The f_i (i = 0,1,2,....n) represent the reproduction by the female in the ith age group, and the p_i (i = 0,1,2.......n-1) represent the probabilities that a female in i^{th} age group will be alive in the (i+1) age group.

The Matrix Model

The pictorial depiction of life cycle graph can be shown in the matrix model.

$$n(t+1) = An\,(t)$$

The projection matrix A in the above equation multiplied by n (t) i.e. a vector of stage abundance gives the value for future population abundance at t+1.

The matrix can be generated simply from the life cycle graph by putting co-efficient a_{ij} which is the co-efficient on the arc from n_j to n_i in the life cycle graph.

The projection matrices corresponding to life cycle depiction can be given as below :

Case Study 1

Williamson (1959) provided an application of Matrix model showing the significance of Eigen value and Eigen vector. This model start with an initial population of one old female represented by a column vector. Each old animal produces an average of 12 young before dying. Each middle aged animal produces an average of 9 young before either dying or becoming one age period older with equal probability. Young animal produces no young and probability of 1/3 of reaching the middle group. The model described as below –

$$A = \begin{bmatrix} O & F_2 & F_3 & F_4 \\ P_1 & O & O & O \\ O & P_2 & O & O \\ O & O & P_3 & O \end{bmatrix}$$ for figure 1

$$A = \begin{bmatrix} P_1 & F_2 & F_3 & F_4 \\ G_1 & P_2 & O & O \\ O & G_2 & P_3 & O \\ O & O & G_3 & P_4 \end{bmatrix}$$ for figure 2

$$A = \begin{bmatrix} O & O & O & O & F_1 \\ P_1 & O & O & O & O \\ P_3 & O & O & F_2 & O \\ O & P_2 & P_4 & O & O \\ O & O & O & P_5 & O \end{bmatrix}$$ for figure 3

After one period of time there will be 12 young females i.e.

$$\begin{pmatrix} 0 & 9 & 12 \\ 1/3 & 0 & 0 \\ 0 & ½ & 0 \end{pmatrix} \begin{pmatrix} 0 \\ 0 \\ 1 \end{pmatrix} = \begin{pmatrix} 12 \\ 0 \\ 0 \end{pmatrix}$$

$$\begin{pmatrix} 0 & 9 & 12 \\ 1/3 & 0 & 0 \\ 0 & ½ & 0 \end{pmatrix} \begin{pmatrix} 12 \\ 0 \\ 0 \end{pmatrix} = \begin{pmatrix} 0 \\ 4 \\ 0 \end{pmatrix}$$

Repeating application of the model gives the following prediction in which the population distribution vector is pre-multiplied by the fecundity and survival rates.

$$\begin{pmatrix} 0 & 9 & 12 \\ 1/3 & 0 & 0 \\ 0 & ½ & 0 \end{pmatrix} \begin{pmatrix} 0 \\ 4 \\ 0 \end{pmatrix} = \begin{pmatrix} 36 \\ 0 \\ 2 \end{pmatrix}$$

$$\begin{pmatrix} 0 & 9 & 12 \\ ? & 0 & 0 \\ 0 & ½ & 0 \end{pmatrix} \begin{pmatrix} 36 \\ 0 \\ 2 \end{pmatrix} = \begin{pmatrix} 24 \\ 12 \\ 0 \end{pmatrix} \quad \text{etc.}$$

If for 20 times the iteration is done the number of young, middle aged and old aged maintained a constant ratio to each other.

The dominant Eigen value was found to be 2 which indicated that during each time, the population was doubled. Similarly the dominant Eigen vector indicated the stable structure of population i.e. the proportion of young, middle aged and old animals in the stable population can be shown by the

vector $\begin{bmatrix} 24 \\ 4 \\ 1 \end{bmatrix}$.

Case Study 2

Usher (1966) developed the basic Leslie matrix model for Scot's pine forest which contained an uneven aged and size structure of trees and in which there is no definite regeneration phase. As time period was sufficiently small, it was ensured that tree either remained in the same size class or moved to the next larger size class but it did not move up by more than 1 size class.

It was assumed that forest regenerated naturally and any gap created was used by natural regeneration or by enlargement of crown of the surrounding trees. Thus, the fecundity in the matrix depend on the number of trees that are harvested. If these are ni trees in size class i at time t, then at t+1, when we have a stable population structure, there should be λni trees in that class, and (λ-1)ni of these trees will be harvested to reduce the population to ni. The factor (λ-1) is therefore required to modify the regeneration tenure in the forest row of the matrix model.

Usher's model is :

$$
Q = \begin{pmatrix}
a_0c_0(\lambda-1) & c_1(\lambda-1) & c_2(\lambda-1) & C_{n-1}(\lambda-1) & C_n(\lambda-a_n) \\
b_0 & a_1 & 0 & 0 & 0 \\
0 & . & . & . & . \\
. & . & . & . & . \\
0 & 0 & 0 & a_{n-1} & 0 \\
0 & 0 & 0 & b_{n-1} & a_n
\end{pmatrix}
$$

where, a_i(i=0, 1,...., n-1) is the probability that a tree will remain in the ith size class, b_i (i=0, 1,......n-1) is the probability that a tree will move up one size class; c_i (i=0, 1,......,n) is the number of trees of size class 0 regenerating in the gap caused by the harvesting of a tree. A management decision depends on the number of trees to be left in the largest class and will often be 0. This in turn affects the regeneration terms which becomes C_n ((λ-1+bn) if b_n=1-a_n, and thus this simplifies to $C_n(\lambda-a_n)$.

Usher (1966) found that 17% of tree can be harvested in every 6 year period.

$$H = 100\left(\frac{1.204266-1}{1.204266}\right) \text{per cent} = 16.96\%$$

where $1.204266 = \lambda$.

Case Study 3

Lowe, (1969) studied the population structure of red deer. (*Cervus elaphus*). It is an animal that arrived in Britain during the Atlantic period with greater control exerted by man. The predators were either killed or controlled with progressive destruction of forest. The increasing interest in agriculture and upland forestry and decreasing interest in deer stocking placed considerable emphasis on management of deer population. Lowe (1969) estimated the population structure of deer and constructed the life table from which he estimated the natural mortality increased to 8 and 9 year old animals for both stag and hinds. There was marked difference between mortality rate for stag was about 10% whereas hinds was between 10-20%.

Usher (1973) estimated age specific birth rate using Lowe's data. It was assumed that red deer produced only 1 fawn each year. After matrix computation of 32×32, the dominant Eigen value was obtained as (1.1636). It was found that

$$H = 100\left(\frac{\lambda - 1}{\lambda}\right) \text{per cent}$$

$$= 100\left(\frac{1.1636-1}{1.1636}\right) \text{per cent}$$

$$= 14.1\%.$$

i.e. a harvest of 14.1% can be taken each year. Thus the model can be applied for deer management to determine alternative strategies of harvesting in which exploitation is concentrated on particular age classes. In absence of exploitation, there is an equilibrium population density N* given by the solution to :

$$| N_{(Nt)} - I | = 0.$$

Subscript to the matrix M indicate that its element function of N_t defined as

$$N_t = \sum_i n_i$$

In absence of exploitation the population would be expected to approach an equilibrium.

Case Study 4

Gupta (1990, 1993, 1995, 1997,1998, 1999, 2003 and 2004) made attempts to combine age and structure stage into one and analysed the demography of herbaceous annuals by matrix model.

The Leslie matrix model take the general form of N(t+1) = A.N (t).

Where, N(t+1) and N(t) are vectors representing age or stage class distribution of individuals at time t+1 and t, A is the matrix defining the age or stage specific survival and fecundity values. The rate of emigration and immigration of propagules were assumed to be equal. The model was thus simplified. The asymptotic population growth rate λ is given by the dominant Eigen values of A_1 and the stable age/stage distribution by corresponding Eigen vector. Rate of decay or death rate was designated as $-\mu$ and calculated at each time interval.

The age distribution vector were derived from log graphs where the abscissae demarcated the length of span for each age group scaled in such a way that the length of life span for seedling stage equals one unit and on the ordinates the density of different age groups were plotted in all the herbaceous annuals populations.

The contribution to the population at time t+1 can be obtained by multiplying Nt by the following matrix comprised of a×a as submatrices.

Where, A_A to A_D represent the survivorship of 4 stages during the time interval.

$$\begin{pmatrix} A_A & & & O \\ & A_B & & \\ & & A_C & \\ O & & & A_D \end{pmatrix}$$

For simulations the monthly emergent seedling population for the initial year were analysed in case of all plants and multiplied with survivorship co-efficient.

The matrix model provided a path elucidating the dynamics of population in which reproduction and survivor co-efficient for flowering stages suggested it a critical stage in life cycle of *Parthenium hysterophorus* Linn., a notorious weed that approved that the effective control of the weed may be brought about by regulating it at flowering stage (Gupta 1990, 1993).

Similarly for *Tridax procumbens*, Linn., flowering is a sensitive stage. Thus it appears a division of resources between vegetative and reproductive organ is important for it (Gupta, 1995, 1997), whereas in Bidens pilosa survival co-efficient of juvenile stage was found to be low indicating that competition for nutrient and space (Dense turf) is an important factor in regulating it (Gupta, 1995, 1997). In Amaranthus spinosus (Gupta, 2003) and Portulaca oleracea (Gupta, 2003, 2004), the mortality risk was highest at seedling stage suggesting to have the control measures at seedling stage so as to eliminate the probability of reaching the adult stage thereby setting seeds for further infestation and spread. Burkey *et al*., (1994) demonstrated with a seasonal model how the value of the resource to each individual may be reduced due to its patchy distribution.

It was concluded that the difference among population of various weed species depend on the difference of their transition probabilities of matrices in the same year and on the difference in stochastic processes. As also observed by Aberg (1992), the difference were reflected in value of λ and -μ rates.

The simulation indicated a striking resemblance between observed and projected populations. It indicated that the projection matrix was satisfactory in differentiating the age specific characteristics of cohorts and sub-cohorts. (Gupta 1993, 1995, 1997, 1998, 1999, 2003, 2004).

Gupta (1999) observed that Matrix population models that integrate population dynamics and population structure are a powerful tool for investigating population dynamics as also suggested by (Caswell, 1989). Hence, construction and role of weed population models in designing weed management programmes become significant.

Symposium 'Matrix model in Ecology' organized on occasion of VIII International Congress of Ecology at Florence, Italy on 21st July 1998 acquainted with the examples of application of this important tool on ecological researches (Gupta, 1999). Though in the process, the realism of the model is sacrificed to some extent, it has many advantages like summarizing the expressions and handling of procedures for the solution of problems. Matrix models have found applications in analyzing population dynamics and successional changes, impact of biotic stress on community prey-predator relationships, study of homeostasis and resilience, etc. They have also been used in ecological simulations and have found applications in management, conservation of populations and biota, etc.

Some application include by Aberg (1998) who emphasized the use of matrix population model for analysing the spatio-temporal variation in the perennial plant Ascophyllum nodusum. He analysed demography of size structured population using stochastic model simulation. Boucher (1998) emphasized in his study about matrix modelling of forest tree species that transient dynamic and pattern of environment dependence of matrix entries are a vital guide for conservation than stable stage concentration and elasticities. Cushing (1998), used matrix model to predict asymptotic dynamics in laboratory experiment, from equilibrium to cycle to chaos. Using time series data obtained from controlled laboratory experiment, an appropriate stochastic version of the deterministic model was devised in order to parameterize the model. Schardi (1998) in his work on 'Empirical models in ecology some application of artificial neural network' emphasized that complex processes can be modelled when casual relationships are unknown or not fully understood form large heterogeneous data set and summarize through a simple matrix model by use of artificial neural network.

Schroeder (1998) attempted to couple habitat suitability model with Leslie matrix models and developed a model complex for assessing the effect of management practices in fens with reference to target species- *Conocephalus dorsalis* (Orthoptera) Takada (1998) observed that several theoretical results concerning evolution in density dependent matrix model with stage structure can be obtained. Once the phenology and mechanics of invasion are understood, any generality that emerged may be useful in predicting, managing and possibly preventing the changes that accompany an invasion. Gupta (1999) observed that more application of matrix model may be evolved in future involving scientific management of species, population and ecosystems.

Matrix model is a device which not only precisely describe the population statistic besides stating age specific data, it can also help in finding R by calculating 'dominant latent root' of the transition matrix which is equivalent to finite rate of increase R. (Usher, 1972). The modelling can help in analysis of population behaviour,/help in predicting the level of 'cropping' allowing a particular level of harvest and without driving a population to extinction.

The model works on the assumption that elements of the transition matrix are constant over time; reality of natural population statistics like fecundity and survival may be induced in the matrix model. Similarly the statistics can be made density dependent in case the fecundity and survival element vary in the transition matrix in relation to population size.

Many application of matrix model exist, however disadvantage continues as the determinate exponential model of population growth assumes that population will continuously increase in nature. However by making all elements of the matrix function of some property of population size, a more realistic model can be achieved. Predator –prey system, seasonal and random environmental changes, cropping schedule can all be represented by some development in basic matrix model, dominant eigen value and eigen vector being the expression of the basic matrix.

However, ecologist in India have ignored and neglected this family of models. The greater demand on the mathematical ability of the modeler can be compensated by making available computer programmes to calculate at least the dominant eigen value and eigen vector and may enhance its application.

ACKNOWLEDGEMENT

The author expresses her gratitude to Ministry of Human Resource Development for offering her post-doctoral Fellowship to study at USSR. She expresses her thankfulness to Prof. Maximov and Dr. Tagir Gilmanov of Moscow University for initiating her to matrix model applications.

REFERENCES

Aberg, P. (1992). A demographic study of two population of the seaweed As-cophyllum nodosum. Ecology **73(4)**: 1473-1487.

Aberg, P. (1998). Spatial and Temporal variation in Matrix population Models of Demography. Proc. VII – International Congress of Ecology, Florence, Italy.

Allee, W.C. Emerson, AE, Park O and T. Park (1994). Principles of Animal Theology. W.B. Saunders Philadelphia.

Andrewartha, H.G. and L.C. Birch (1954). The Distribution and Abundance of Animals. University of Chicago Press.

Begon M and Mortimer M. (1986). Population Ecology. Blackwell Scientific Publication U.K.

Boucher, D. (1998). Environment – Dependence of fast growing primary forest trees: Matrix modeling of *Vochysia ferruginea* in Tropical rain forest and *Castanea dentata* in Temperate deciduous forest. Proc. VII International Congress of Ecology, Florence, Italy.

Burkey, T.V. and N.C. Stenseth (1994). Population dynamics of territorial species in seasonal and patchy environments Oikos **69**:47-53.

Caswell (1989). Matrix Population Model-1 Sinauer Asso. Inc. Publ. Sunderland, Massachusetts.

Cole, L.C. (1958). Sketches of general and comparative demography. Cold Spring Harbor Symposia on Quantitative Biology **22:**1-15.

Cushing, J.M. (1998). Using matrix model to predict asymtotic dynamics in laboratory experiments, from equlibria to cycles to chaos. Proc. VII International Congress of Ecology, Florence, Italy.

Deevey, E.S., Jr. (1947). Life tables for natural populations of animals. Quarterly Review of Biology **22:** 283-314.

Gause, G.F. (1934). The struggle for existence Williams and Wilkins, Baltimore.

Gupta, A. (1990). Population Ecology of *Parthenium hysterophorus* Linn. in Herbaceous Communities of Central India (State – Maharashtra) Evolutionary Morphology and Ecology, Acad. of Sc. USSR: 1-36.

Gupta, A. (1993). Population Dynamics studies in Congress Grass (*Parthenium hysterophorus* Linn.). Proc. XV International. Botany Congress, Yokohama, Japan.

Gupta, A. (1995). Population Dynamics of few weeds in Manipur. Project report: CSIR, Delhi.

Gupta, A. (1997). Matrix Model population for two herbaceous species Int. J. Ecology and Environmental Sc. **23:**197-207.

Gupta, A. (1998). Leslie Matrix Model for *Euphorbia hirta* L. Population. Proc. VII International Congress of Ecology. Florence, Italy.

Gupta, A. (1999). Matrix model in Ecology. In Perspective in Ecology Ed. A. Farina, Backhuys Publ. Leiden, NL: 197-203.

Gupta, A. (2003). Matrix Model for Few weed populations from Manipur, North East India. Proc.-International Conference on Ecorestoration, Dehra Dun.

Gupta, A. (2004). A Matrix population model for *Portulaca oleracea* Linn. Proceedings Indian Science Congress - Botany Section.

Hartshorn Gary S. (1975). A matrix Model of Tree population Dynamics in Ecological Studies Ed by: Jacobs *et al.*, Springer Verlag.

Huffaker C. B. (1958). Experimental studies on predation : dispersion factors and predator-prey oscillations. Hilgardia, **27**:345-383.

Hutchinson, G.E. (1978). An Introduction to population Ecology. Yale University Press, New Haven.

Lefkovitch, L.P. (1965). The study of population growth in organisms grouped by stages. Biometrics **21**:1-18.

Leslie, P. H. (1945). On the use of matrices in certain population mathematics. Biometrika, **33**:183-212.

Leslie, P.H. (1948). Some further notes in the use of matrices in population mathematics. Biometrika **35**:213-245.

Levin, S.R. (1968). Evolution in changing Environments. Princeton University Press, Princeton.

Lotka, A. S. (1924). Elements of Physical Biology. Williams and Wilkins, Baltimore. (Printed in 1956 by Dover Publications, New York).

Lowe, V.P.W. (1969). Population dynamics of the red deer *Cervus elaphus* L. Rhum. J. Animal Ecology **38**:425-457.

Pearl, R. (1928). The Rate of Living. Alfred A Knopf, New York.

Sagar G.R. and A.M. Mortimer (1976). An approach to the study of the population dynamics of plants with special reference to weeds. Applied Biology, **1**:1-43.

Scardi, M. (1998). Empirical models in Ecology : an application of artificial neural network. Proc. VII International Congress of Ecology, Florence, Italy.

Schroeder, B. and D. Soendgerath (1998). A multihabitat model coupling matrix models with habitat- suitability. Proc. VII International Congress of Ecology, Florence, Italy.

Takada, T. and H. Nakajima (1998). Theorems on the invasion process in density-dependent matrix models. Proc. VII International Congress of Ecology, Florence, Italy.

Usher, M.B. (1966). A matrix approach to the management resources with special reference to selection forests. J. Applied Ecology **3**:355-367.

Usher, M.B. (1972). Developments in the Leslie matrix models. In : Mathematical Models in Ecology (Jeffers J.N.R. ed). Symposium of the British Ecological Society. No. 12 pp. 29-60. Blackwell Scientific Publications, Oxford.

Usher M.B. (1973). Biological Conservation and Management. Chapman & Hall, London.

Volterra, V. (1926). Varrazione e fluttuazini del numero d' in dividui in specie animali conviventi. A H F Accad. naz. Lincei Memorie (Ser. 6), **2**:31-113.

Williamson, M.H. (1959). Some extension in the use of matrices in population theory. Bull. Math. Biophys., **21**:13-17.

Advancing Frontiers of Ecological Researches in India, 645-650, 2006
Edited by A.K. Kandya & Asha Gupta
Published by Bishen Singh Mahendra Pal Singh, Dehra Dun, India

GPS applications for ecological studies in Manipur

Uttam C. Laisram
Dept. of Science & Technology
Government of Manipur

Abstract: Easy availability and common use of a small, light-weighted and handy equipment (GPS) to find out the global position of any object is a new facility in the hands of geographers, engineers, explorers, field foresters, hunters, surveyors, scientists and commoners. This equipment was very large in size in the earlier days and was used only in navigation (long ship journeys in the ocean) and aviation (long air traveling by aero-planes) where nothing is known exactly about the direction of movement of aircraft or steamer and distance already traveled etc. To reach the targeted location was always very challenging and possessed high level of risk. But, the progress in science and technology in the new millennium has resulted in the development of this useful electronic equipment which is a boon for the man-kind and several tough works have become very easy. The present paper gives detailed information about this most modern equipment and the facilities endowed in it.

INTRODUCTION

GPS, which stands for Global Positioning System, is the only system today available commercially that is able to show your-exact position on the Earth at all times. The GPS system works under a constellation of 24 nos. GPS satellites which cover all areas of the globe, in orbit at about 11,000 nautical miles above the Earth. The satellites transmit signals that can be detected by a portable handheld device known as a GPS receiver, which can determine any location with great precision in terms of longitude, latitude and altitude, as well as time.

GPS is one of history's most exciting and revolutionary developments in location and mapping technology. GPS technology has matured into a resource that goes far beyond its original design goals and nowadays scientists, sportsmen, farmers, soldiers, pilots, surveyors, hikers, delivery' drivers, sailors, dispatchers, lumberjacks, fire-fighters, and people from many other walks of life are using GPS in ways that make their work more productive, safer, and significantly easier. The cost of entry-level GPS receiver is less than Rs. 20,000/-, making it extremely affordable.

The real-world applications of GPS fall into five broad categories.

- Location - determining a basic position
- Navigation - getting from one location to another
- Tracking - monitoring the movement of people and things
- Mapping - creating maps of the world
- Timing - bringing precise timing to the world

Importance of GPS technology in real-world applications

Location : To establish the geographical position of an unknown point in the middle of a desert, a vast forest or in the middle of the ocean, is indeed a daunting task if one is to use conventional methods. The use of GPS makes it as simple as just taking a reading on the GPS receiver in a matter of seconds. This method enables scientists to precisely record the location of geological landmarks, flora and fauna, streams, buildings etc. in terms of longitude, latitude and altitude.

Navigation : In a reverse application, if the target location is known in terms of longitude and latitude, and there is a need to go to that point, say a scientist returning to a forest site after one year with a need to revisit a particular location, the GPS receiver becomes a handy navigator.

Tracking: To keep track of a moving item, the GPS receiver, which furnishes the exact location at any point of time is used along with a transmitting system to give regular feedback on the position of the moving item. This method is used for animal tracking to study their movement behaviour. Tracking of goods trucks, buses, boats etc. are done with this method. For tracking applications, a communication system for returning the location data to the base station is essential.

Mapping: Recording a number of sequential locations along the periphery of a lake, water body, roads, forest area with a specific species of plants, a plantation area etc. forms the basis for mapping by using a GPS receiver. In the case of a periphery along which a vehicle can be driven, the GPS receiver is fitted on the vehicle which is driven while taking recording at fixed time intervals. This data makes mapping a very easy task.

Timing: The GPS satellite system also transmits precise Universal Time for use in various applications which require time records.

Agencies in Manipur using GPS for various projects:

GPS technology has been used in different applications in Manipur. The Agencies which have used GPS in Manipur are:

- World Bank Consultant for Highway Road network in Manipur in association with PWD.
- French Government Consultant for Imphal to Ukhrul Highway in Manipur in association with PWD.
- Manipur Remote Sensing Applications Centre (MARSAC)
- Loktak Development Authority
- Department of Earth Sciences, Manipur University.

GPS applications in Manipur

a) Study for development of Highway Road network of Manipur under World Bank Project

Under this study, an extensive survey of the various road highways in the State was undertaken by the World Bank Consultant. Even though the Survey of India topography maps containing road details are available, the data pertains to the period of the 1970s, making them outdated in respect of roads, which have undergone many changes since then. The study made extensive use of GPS to get the road data. Apart from getting and verifying point based data, it is possible to take continuous readings with the help of the GPS receiver at close intervals as per requirement. The GPS receiver itself or an external antenna connected to it is mounted on the moving vehicle as it drives along the road for which geo-coded data is to be recorded. As the GPS receivers usually do not have large storage capacity, the GPS receiver is connected to a computer notebook, and the data stream is recorded in the hard disk of the computer. With the associated software, the map of the road is automatically drawn. Various landmarks along the route are marked and associated with the relevant descriptions. Using the data recorded from the GPS, the road mapping could be updated in many areas where new roads had been constructed and where diversions had been made over time. The entire data is integrated into a Geographical Information System (GIS) and the digitized road network was prepared with GPS as one of the important tools.

b) Study and design of Imphal to Ukhrul Highway in Manipur under French Government programme

Under a proposal for making a world class road highway between Imphal and Ukhrul, a study of the existing road as well as design was carried out by the Project Consultant under French Government funding. In this study also, there was collection of data inputs to determine the accurate location of the road points. For this purpose, the GPS was used extensively. Being a road highway in hilly terrain, the GPS data pertained not only to longitude and latitude, but also to altitude. The use of GPS enabled the French Engineers to make a 3-D model (wire -mesh model) of the entire road form Imphal to Ukhrul, and a "fly-through" view could be seen on the computer as if one was actually traveling on that highway.

c) Ground truth recording for satellite imageries under different projects

The Manipur Remote Sensing Applications Centre (MARSAC) has carried out a wide range of remote sensing application projects in Manipur. In all cases, the identified items in the satellite imagery is required to be verified in field at several locations, so that inference for existence of similar ground conditions like land cover, forest cover, crops, plantations, water bodies, wasteland etc. For this purpose, the GPS is taken during the field trips and the location of the subject item is recorded by taking the GPS readings. Also the location of specific landmark items can be recorded easily by using the handheld GPS receiver. The GPS receiver is also used to map a particular area like a wetland or the boundary of a plantation.

d) Use of GPS for Bathymetric mapping of Loktak Lake

The Loktak lake covers an area of about 26,600 hectares, with a catchment area of 98,000 hectares, and is the largest natural wetland in north-east India The wetland has been identified as one of the sites for conservation under the Indian National Wetland programme. The Loktak lake was designated as a Ramsar site on 23 March 1990. The Loktak Development Authority is entrusted with management of the lake on a sound ecological basis. The Keibul Lamjao National Park, habitat of the threatened Brow antlered deer (locally known as 'Sangai'), is located in the south-east area of the lake.

Due to rapid deforestation in the catchment area and denudation of vegetation, there has been several changes in ecological character of the lake. The rate of soil erosion has increased considerably during recent years and has resulted in high siltation of the lake with detrimental effect on the lake ecosystem.

To have a reliable information system on the rate of siltation of the Loktak lake, it is necessary to take a recording of the depth of the bed of the lake at regular intervals. The Loktak Development Authority undertook this project some years back, using GPS as one of the important tools. As the Lake is not very deep and the maximum depth is about 5 meters only, it is possible to measure the depth by using a measurement pole from a boat. The problem faced was location of the spot where the measurement was being taken in the various parts of the lake. For this purpose, the handheld GPS receiver was used to record the exact geographical location of every spot where the depth reading was taken. With this system, the readings could be taken in a grid for generation of a lake bed depth map. The GPS enabled the boat to be taken close to each of the pre-determined locations from where the depth readings were to be taken.

e) Using GPS for accurate time measure recording in Seismograph equipment

The Department of Earth Sciences, Manipur University, has a leading role in seismographic recording of earthquake occurrences in Manipur and the North East. Several Seismographs have been installed in different parts of Manipur. One of the vital parameters to recording seismic activities is the accurate time. The difference in time between the seismic signals recorded in different recording locations enable the scientists to calculate and find out the epicenter and other characteristics of the earthquake. Previously, time reference was taken from radio broadcast. The time radio broadcast is still available from Physical Research Laboratory, Ahmedabad, India, among others, but this method required a person to be engaged for manual recording/correction of time. With the availability of time signals over the GPS system, the recording is done directly without manual intervention and has made the time stamping of seismic recording an automatic process.

f) Other potential GPS applications in Manipur

Another interesting application of GPS is in wildlife animal tracking for prevention of straying from the protected areas, as also to study their movement behaviour. The brow-antlered deer "Sangai" which is found in the floating Keibul Lamjao Wildlife Sanctuary in Loktak Lake, is a rare species which is threatened with extinction. For study of the animal movement and its eating habits etc., a GPS-fitted collar can be fixed on some of the animals and the signals can be sent back through a wireless system or through the Cellular phone system. A preliminary proposal had already been discussed. The animals are to be manually caught by netting etc. as use of tranquilizer from a gun is considered dangerous to the animal which may become unconscious and fall into the water of the lake, causing death. This use of GPS for animal movement monitoring has already been done in many wildlife projects around the world and can be effectively used in Manipur.

g) Accuracy and other restrictions

The use of GPS technology in Manipur as mentioned above encompasses almost all aspects of GPS uses. The details of accuracy of readings and methods to increase the accuracy have not been covered in this paper, as it is of a technical nature and is more relevant to those who are actually working on the projects. However, it may be worthwhile to note that the GPS satellite system is under the control of the United States Government and the US Government reserves the right to reintroduce Selective Availability (SA) to make the GPS system less accurate in any selected part of the world at any time, particularly during war, when the GPS can also be easily adapted for use in missiles. Otherwise, the world has been blessed with a smart technology that is already in use extensively for developmental works.

Advancing Frontiers of Ecological Researches in India, 651-670, 2006
Edited by A.K. Kandya & Asha Gupta
Published by Bishen Singh Mahendra Pal Singh, Dehra Dun, India

Impact of The Himalaya on Modification of Ecosystems of Indian Subcontinent and Asia: A Review

O.P. Goel
Visiting Professor
Department of Earth Sciences,
Manipur University, Imphal-795 003

Abstract: The Himalayan mountain chain stretches over 2500 Km in length and is 250 to 300 Km wide, sprawling across the northern edge of the Indian subcontinent. It rises from 500m above mean sea level to a stupendous height of > 8000m and has virtually affected the atmospheric circulation over the Asian Continent including the Indian landmass.

Even though the Himalayas had emerged as a highland by middle Miocene (20 Ma), it was not until the late Miocene (11 to 7.5 M.y.ago-Ma) that it became a mountain barrier high enough to disrupt west-to-east flow of winds and push low-pressure area over northern India, which attracted most summer winds from the Indian Ocean. The Himalayas suffered accelerated erosion in the late Miocene. The change of Palaeoflora in the Siwalik domain from Evergreen-Tall tropical to Tall-grasses. 10 to 7.5 Ma imply sudden change to dry climate characterized by seasonal heavy rains. The sudden appearance of endemic upwelling species at 8.5 to 7.4 Ma in the Indian Ocean indicates activation of upwelling currents, which were set in motion at about 8.0 Ma by southwesterly monsoon winds.

Strong tectonic movements at the beginning of the Quaternary and at 0.8 Ma, as manifested in influence of enormous volumes of sediments in the Siwalik and in the Bay of Bengal, must have lifted up the Himalayas to still greater elevation, and caused diversion of the even flow of moist winds and created large cool areas conducive to precipitation of snow. Cold climatic conditions, which had started 2.5 to 3.0 Ma in the northern hemisphere, culminated in the *Glaciation* of the higher mountains in the Pleistocene. It remains to be seen whether the oscillation of humid and dry phases in the Holocene is in any way related to spurts of tectonic movements on the terrain-bounding active faults of the Himalayas.

This was the last glaciation on the globe and there is absence of glaciation anywhere in the globe during the Mesozoic era.

INTRODUCTION

A variety of proxy data have been presented to demonstrate the connection between Tibetan uplift and climate change on regional and global scales. The ratio of oxygen isotopes ($^{18}O/^{16}O$) in the calcareous shells of foraminifers preserved in marine sediments essentially records past ice-sheet volumes and is a proxy for ocean temperatures. The oxygen isotope record for the past 55 Ma shows a long-term increase in $\delta^{18}O$ values ($\delta^{18}O$ expresses the changes in the $^{18}O/^{16}O$ ratio), indicating progressive cooling, punctuated by times of rapid cooling (Fig.1a). The first major increase in $\delta^{18}O$ values (indicating cooling) occurred about 35 Ma ago and probably represents the initiation of the growth of the Antarctic ice sheet. 20 Ma of cooling followed this. Further increases in $\delta^{18}O$ values occurred during mid-Miocene times (approximately 15-13 Ma ago) and late Pliocene times (approximately 4-2 Ma ago); these increases probably represent further growth of Antarctic ice and the build-up of the northern hemisphere ice sheets, respectively. Tills, moraines, ice-rafted sediments and paleontological

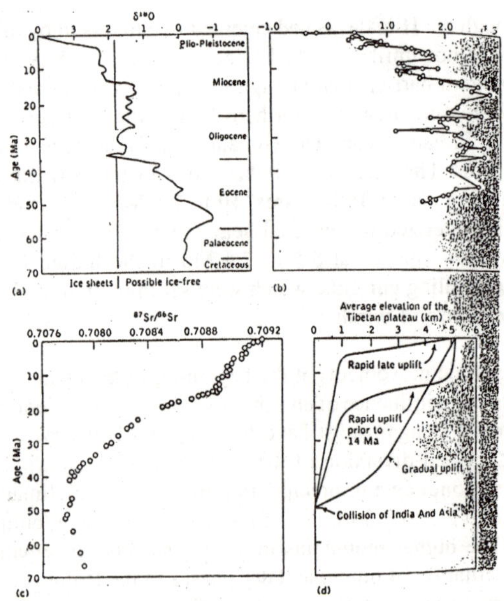

Fig.1. A comparison between Isotope curves and Models for Tibetan Uplift. (a) Simplified compilation of $\delta^{18}O$ measurement from deep-sea cores in the Atlantic Ocean. (b) $\delta\ ^{13}C$ curve for marine carbonates over the past 70 Ma.(c) $^{87}Sr/^{86}Sr$ curve for marine carbonates for the past 70 Ma (after Ray and Ruddiman, 1993). (d) The contrasting models for Tibetan uplift before 14 Ma. Note the correlation between the isotope curves.

evidence provide additional support for this cooling trend and the onset of glaciation in the Antarctic at about 35 Ma ago and in the Northern Hemisphere at approximately 4.5 and 2.5 Ma ago.

It is difficult to assess the relative importance of all these complex processes in driving climate change. Attempts have been made, however, to model the evolution of Cenozoic climate using computer models of the Earth's atmosphere. Using General Circulation Models (GCM), American Earth Scientists have investigated the atmospheric changes that would be associated with the uplift of the Himalayas and Tibetan plateau. Their initial results have shown that many of the global changes in precipitation and temperature that were calculated by the models are consistent with those observed in the geological record over the past 40 Ma. Additional experiments have extended the GCMs to examine the sensitivity of the Indian monsoon. The results showed that strong solar forcing when the elevation was only half its present height could have induced strong monsoons, similar to those of today. The question then arises as to when was Tibet high enough to initiate the monsoons, and was this coincident with the proxy evidence for global cooling and increased monsoon activity at around 7 to 8 Ma ago.

Alternatively, reduced denudation could be related to major vegetational changes, which occurred at about 8 Ma ago. These uplift dates are

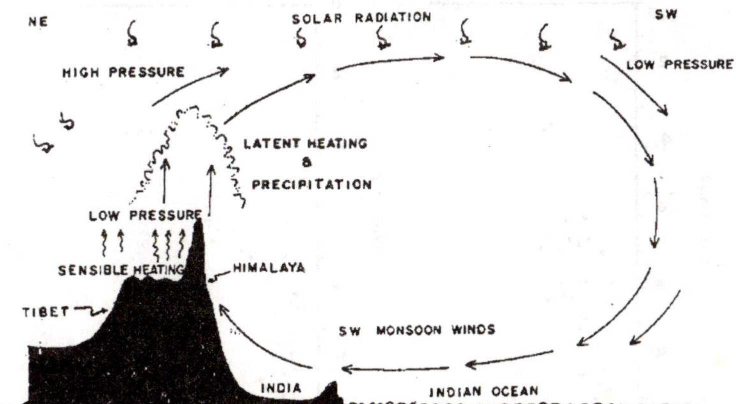

Fig. 2. The >5000 m high mountain barrier made up of the Himalaya and Tibetan Plateau causes not only the change in the wind direction, but also the precipitation of rains and snows. The shaded part in the map (after Molnar *et al.* 1993) shows 2000m and 4000m high areas.

essential for modeling and understanding the role of uplift on climate change.

An alternative view on uplift and climate change proposes that the Quaternary glaciations in the Himalayas could have enhanced uplift. This theory suggests that when the Earth's surface is evenly eroded, and is therefore unloaded, a corresponding readjustment will occur raising the crust to approximately 5/6 th of its original height. If, however, the crust was not evenly eroded, for example by glaciers which deeply incised into

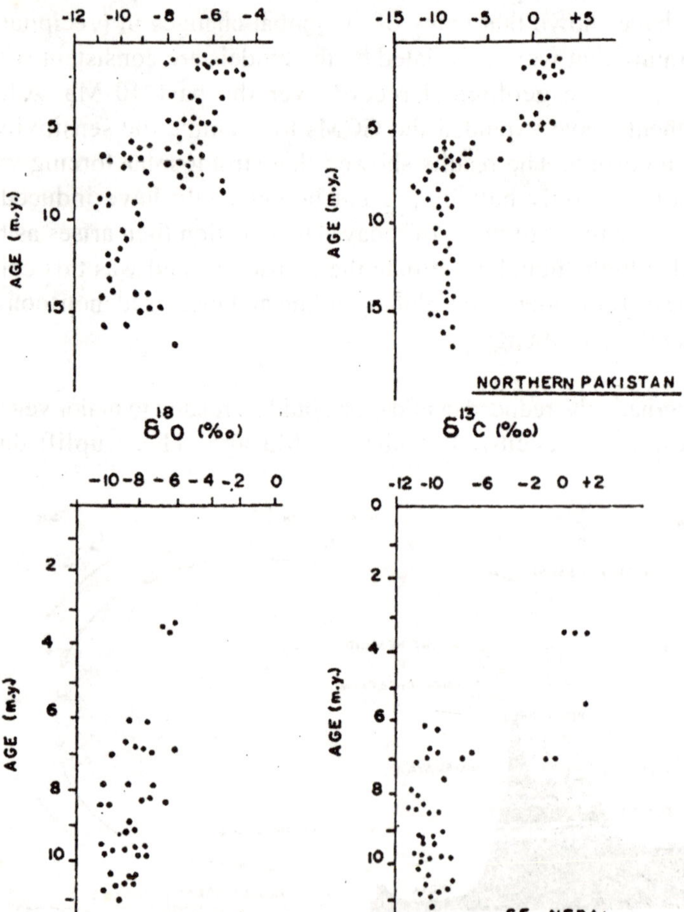

Fig. 3: Emergence of lofty mountain barrier disrupted the west- to- east flow of winds and caused northward movement of the high - pressure area so that the low pressure centre moved over northern India in summer. The low-pressure centre attracts the moisture-laden southwest monsoon air from the Indian Ocean.

the landscape, but the same amount of material was removed, as was the case with even erosion, then the rebound would help produce higher mountain peaks. This process may have helped produce the high peaks in the Himalayas. While this process may provide an alternative mechanism for uplift, the geological data suggest, however, that much of the Himalayas obtained their height before the onset of major glaciation. Unfortunately, however, there are no reliable dates for the oldest glaciation in the Himalayas and Tibet, and the extent of glaciation is poorly constrained. It is therefore difficult to evaluate the connection between glaciation and uplift.

Emergence of the High Mountain Barrier and Tibetan Plateau

The emergence of the Himalayan barrier, welded to the lofty Tibetan Plateau, must have caused profound perturbations in the circulation of winds including moisture- laden air. The 3500 by 1500 km large plateau with its remarkably uniform flatness and elevation =5000 m (Fielding 1996) is beaten increasingly by solar radiation in large amounts. It provides a heat source at mid latitudes (Fig.4) thus opposing Hadley circulation between the equator and the temperature latitudes (Molnar *et al.* 1993).

The formation of the mountain barrier must have disrupted west -to east-flow of winds and caused northward displacement of high-pressure centre that commonly lies near the 30° –35° N latitudes, so that the low-pressure centre moved over northern India in summer (Fig.4). This low-pressure centre attracted moist summer winds from the Indian Ocean to the Indian Subcontinent. Thus developed the unique Monsoon System characterized by the cycle of six seasons, - Vasant, Grishma, Varsha, Sharad, Shisir and Hemant. The climatic change took place over the period 9 to 6 m.y. - roughly at 8 m.y.

Himalayan-Tibetan Uplift and Global Climate Change

During the late 1980s and early 1990s much attention was focused on the potential role of late Cenozoic uplift of the Himalayas and Tibetan plateau in driving global cooling and the growth of large continental ice sheets during Late Tertiary and Quaternary times. The formation of the Himalayas and the Tibetan plateau is thought to have altered global atmospheric circulations and to have changed the concentrations of gases in the atmosphere as a result of changes in the biogeochemical cycles associated with the increased weathering of newly exposed rock surfaces, which in turn led to global cooling.

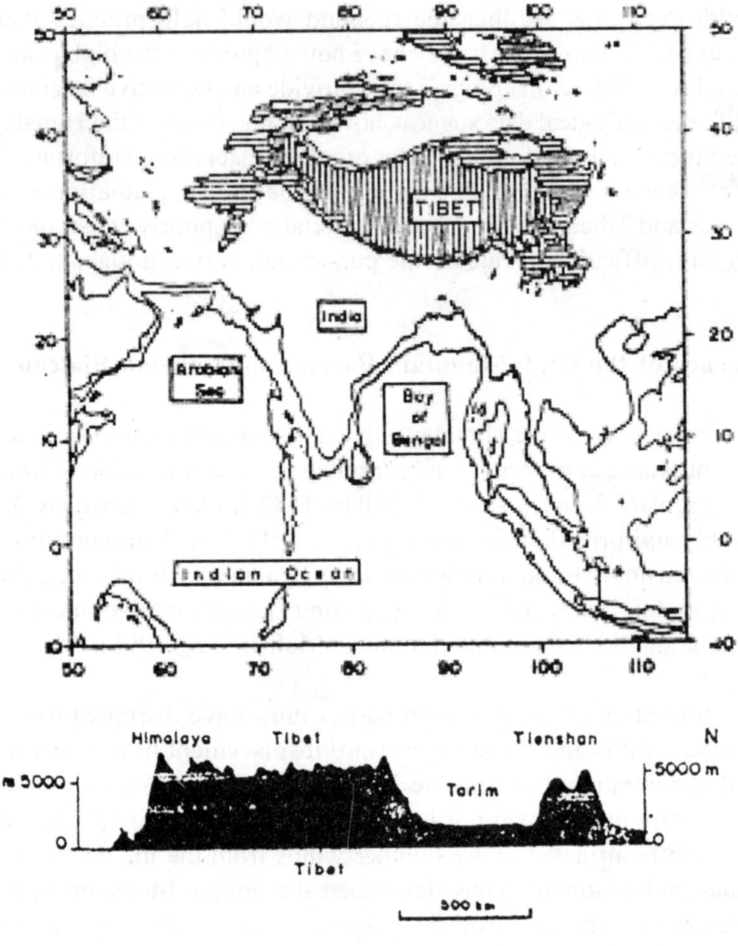

Fig. 4. Isotope ratios of oxygen and carbon of carbonates of the palaeosols in the Siwalik of Potwar un northern Pakistan and of the Bakhiya Khola in south-eastern Nepal indicate that the climate had become warmer and rainfall seasoned by 8.5 to 6 m.y. time. (Based on Quade *et al.* 1989 and Harrison *et al.*, 1993).

There is unequivocal evidence to show that the Indian and Asian continental plates collided between 54 and 49 Ma (Million years ago). Since then, India has continued to move northwards into Asia at a rate of between 40 and 50 mm a year (Raymo & Ruddiman 1992). This has resulted in folding and thrusting of the Indian crust in the Himalayas and southern Tibet. The homogeneous deformation of Tibet entailed shortening and doubling the thickness of the crust, elevating the Himalayas and Tibetan plateau to an average height of between 5000 to 5500 m above sea level. The tectonic forces resulting from the northward movement of India were

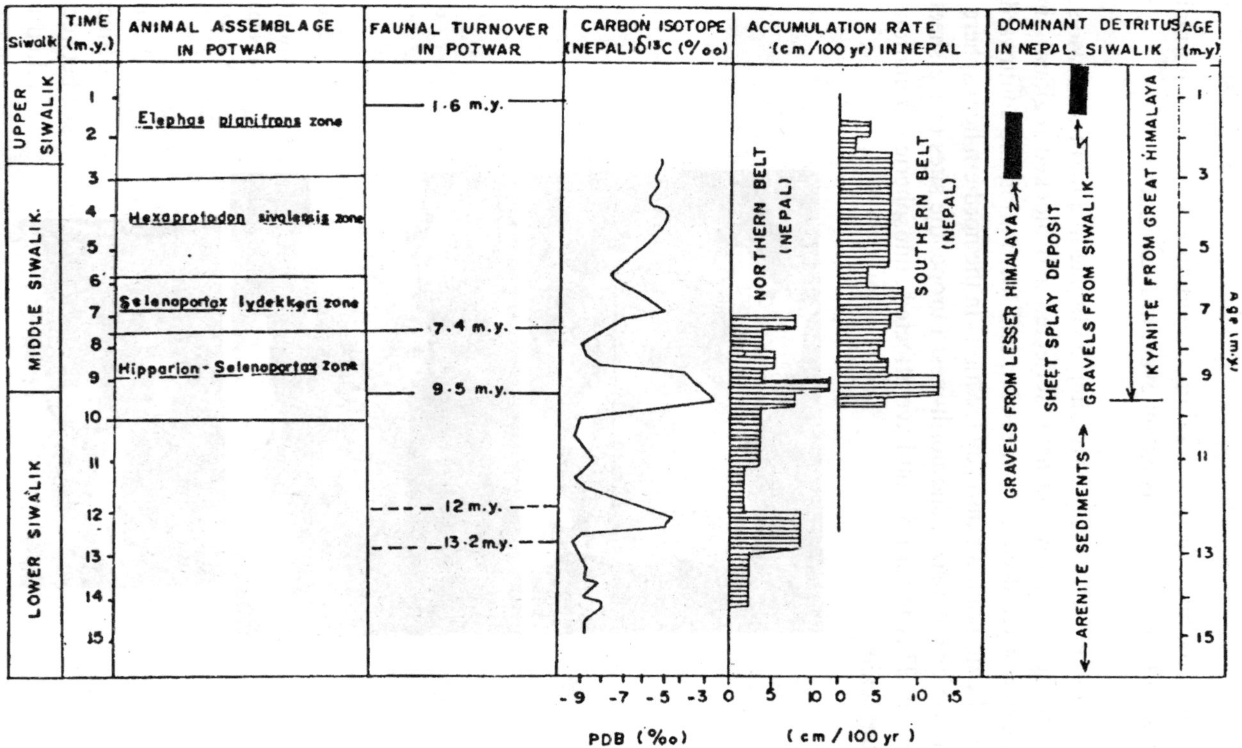

Fig. 5. The Siwalik in south-central Nepal shows that detritals like the diagnostic mineral Kyanite (along with grains of gneiss) make their first appearance in the sediments at 9.2 m.y., indicating exhumation and uplift of the provenance- the Great Himalaya. Concomitantly, the rate of sediment accumulation increased 2.5 folds in the southern belt of the Siwalik at 9.5 m.y. and 7.4 m.y. (modified after Tanaka, 1997).

not, however, in the opinion of some geologists great enough to support the thickened crust, and the plateau began to spread under its own weight.

Currently, the plateau is spreading, extending east-west on normal faults, by about 1 per cent of its extent every million years.

This rise in elevation may have altered climate in a number of ways. First, the rising plateau would have deflected and blocked regional air systems, causing jet streams to meander, and facilitating the southward migration of cold polar air into key locations in the northern hemisphere, particularly North America and north-west Europe, the sites of ice sheet development. This in turn would have affected atmospheric circulations on a global scale.

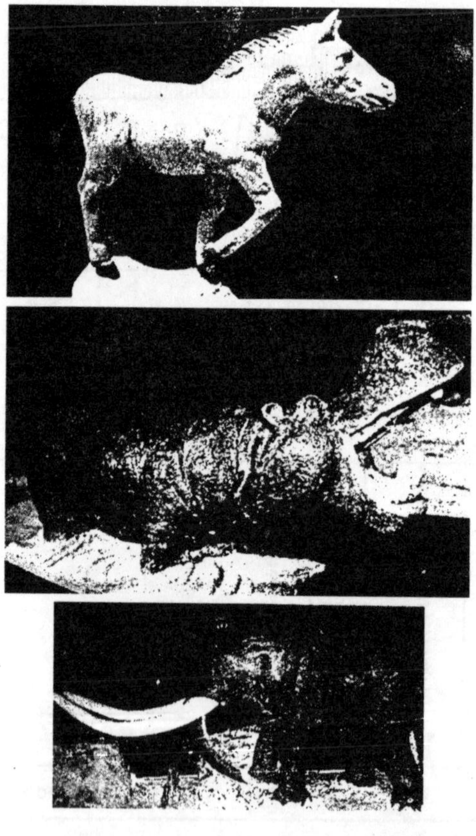

Fig.6. Attracted by extensive grasslands in the Siwalik terrain- as the climate grew warmer with shorter spells of rain- the grazing animals came from far away places. The *Hipparian* came at 9.5 m.y. from Europe and Stegodon, *Hexaprotodon* and *Elephas Planifrons* from Africa appeared 7.4 to 6.0 m.y. ago.

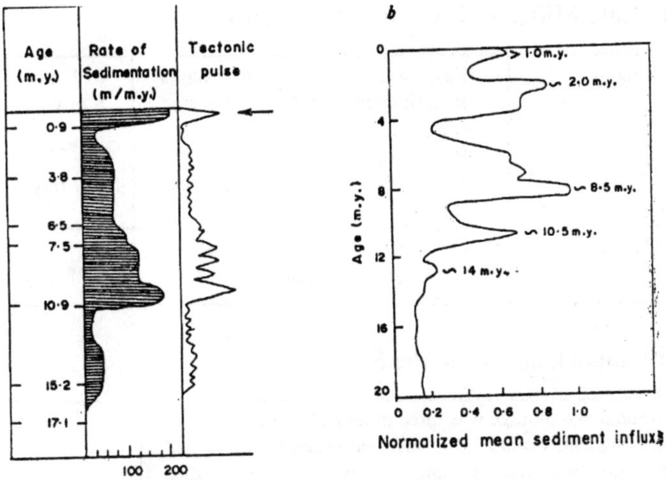

Fig.7. Influx into the Bay of Bengal of terrestrial sediments derived from the Himalayan terrains suddenly increased manifold in the period 11 to 7 m.y., and subsequently at about 0.9 to 0.8 m.y., implying increased erosion due to uplift of the Himalaya, coupled with stronger monsoon rains (a-after Amano and Taira, 1992; and b-after Rea, 1992).

Secondly, the elevated region would have enhanced temperature – driven atmospheric flows as higher and lower atmospheric pressure systems developed over the plateau during the winter and summer, respectively. This would have intensified the Indian Monsoon, leading to heavier rainfall along the frontal ranges of the Himalayas, while the aridity across the Tibet plateau and central Asia would have increased as northward-moving monsoon winds lost their moisture as they were forced over the Himalayas. In Central Asia, the increased aridity would have led to higher temperatures during the summers and extremely low temperatures during the winters. In addition, the plateau would have cooled as it rose higher into atmosphere, and glaciers would have been able to develop. These glaciers would have provided positive feedback, reflecting incoming solar radiation and leading to further cooling.

Thirdly, the greater availability of new exposed rock surfaces and detritus produced by denudation processes would have allowed more chemical weathering to take place. During the chemical weathering atmospheric CO_2 reacts with rock-forming silicate minerals to produce bicarbonates, which are washed into the ocean where they eventually form carbonate rocks. Because CO_2 is an important greenhouse gas, helping to warm the atmosphere, a decrease in the amount of atmosphere CO_2

Table-1: Late Miocene Tectonic Movements

Tibet Himalaya	East-west extension on N-S faults Reactivation of MBT in Pakistan (Fission Track dates) Reactivation of MCT Zone in NW Himalaya and Nepal (Muscovite-cooling ages)	8±1 m.y 8 m.y. 8-6 m.y.
Indian Ocean	Deformation of oceanic crust and covering Sediments	8.75 m.y.

Table-2: Faunal Changes in the Siwalik

Siwalik faunal assemblage was three times richer than that of the present. Compared to just one species of elephant today, there were 30, and there were 15 genera of anthropoid apes. Marginalization or even extermination of indigenous animals, including rhinos, buffaloes and cows.	
Faunal turnover in Potwar Basin	9.5 –7.4 m.y.
Appearance of three-toed Hipparion with pigs	9.5 m.y.
Immigration of *Stegodon* (Proboscid), *Hexaprotodon* (hippo), *Selenoportax* (bovid)	7.4 m.y.
Appearance of *Vishnutherium* (giraffe), *Presbytes* (langur), *Elephas planifrons* (elephant)	6-5.3 m.y.

would lead to global cooling. It has been argued that this process resulted in decreased CO_2 over the past 40 Ma leading to an icehouse effect and the onset of the Quaternary ice age.

Throughout the Cenozoic, other major tectonic events also took place, which could have forced climate change. Continental displacements led to changes in the configurations of the oceans, and seaways opened and closed. These would have given rise to major changes in global and regional oceanic circulations, altering the exchange of heat throughout the globe. The rate of sea-floor spreading also decreased during early Tertiary times, reducing the supply of CO_2 from volcanic eruptions to the atmosphere. As sea-floor spreading was reduced, the oceanic spreading ridges cooled and subsided, lowering sea levels. This produced more land available to be weathered and allowed more plants to grow, storing CO_2 as organic carbon, further reducing the amounts of atmospheric CO_2. The importance of each of these processes, and of mountain uplift, in causing global cooling throughout the Tertiary, and ultimately the onset of the Quaternary ice age, is difficult to prove and assess.

Table-3: Accelerated erosion in the Himalaya and rapid sedimentation

Siwalik Domain: Sedimentation rate increased 2½ folds: From 0.12 to 0.3 mm/y in Potwar Basin	11 m.y.
From 0.2 to 0.5 mm/y in Arungkhola	10 m.y.
First appearance of kyanite and gneiss grains in Nepal Basin	9.2 m.y.
Great abundance of sheet splay deposits in Nepal	10 m.y.
Ganga Basin: Intensity of chemical weathering increased	8-7 m.y.
Bengal fan: Sediment influx increased five folds	11-7 m.y.
Predominance of smectite-kaolinite clays in fine sediments	7 m.y.
Dissolved material, reflected in strontium isotope ratio, abruptly increased	8 m.y.
Accumulation of phosphorus peaked	8 m.y.
Increase in land-derived organic matter (0.5% to 2%)	9-6 m.y.

Table-4: Beginning of Upwelling currents in the Indian Ocean

Globigerina bulloides increased (<5% to >50% of forams)	8.5-7.4 m.y.
Actinomma increased suddenly (amongst radiolarian)	8 m.y.
Collosphaera sp aff.huxleyi increased	8 m.y.
Extinction of *Fohsella* group (deep-dwelling planktonic forams)	12-11 m.y.
Abundance of dominantly biogenic deposition of endemic upwelling species in place of silicicclastic sedimentation that occurred in warm water of low surface productivity	

Such an increase in weathering would deplete the atmosphere of all its CO_2 within a few million years. Negative feedback must therefore operate to return CO_2 to the atmosphere. One possible mechanism is related to imbalances in the organic carbon sub-cycle. $^{13}C/^{12}C$ ratios in marine carbonates reflect changes in the size of the organic carbon reservoir. If, for example, the burial of organic matter decreases, then $\delta^{13}C$ would decrease (a decrease in the $^{13}C/^{12}C$ ratio). $\delta^{13}C$ values in marine carbonates showed a marked decrease during the Cenozoic (Fig.1c), indicating that carbon burial also decreased. This in turn could have caused additional CO_2 to be released into the oceans and atmosphere. Because oxygen is necessary to oxidize organic matter and release CO_2, the decrease in buried organic carbon may reflect an increase in dissolved oxygen concentrations in the oceans. Oxygen solubility increases with decreasing temperature. The dissolution of oxygen into the oceans would therefore have increased as the oceans began to cool. This would lead to the oxidation of organic matter and a greater release of CO_2, into the atmosphere. The residual times of CO_2 (less than 1 Ma) and O_2 (c.10 ma) in the atmosphere would not lead to a perfect balance of these processes. Rather, the balance of the organic sub cycle would become less effective with time. This may have led to more intense forcing of cooling in late Cenozoic times as CO_2 became progressively depleted. Other possible negative feedback mechanisms that could release CO_2 into the atmosphere include the precipitation of silicate minerals in the Deep Ocean, and sea-floor weathering of basalt, but these processes are poorly understood.

$^{87}Sr/^{86}Sr$ isotope ratios recorded in marine carbonates provide an important proxy for estimating the rates of chemical weathering. A marked increase in $^{87}Sr/^{86}Sr$ isotope ratios after Late Eocene times (~ 38 Ma ago: Fig. 1c) may be attributed to increased weathering or a change in the type of rock being weathered (erosion of rock supplies radiogenic strontium from the land) or a decrease in sea-floor hydrothermal supply. The latter

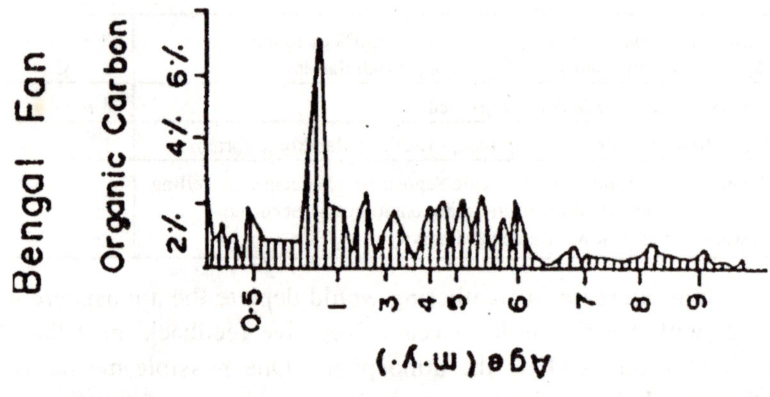

Fig. 8. Orogenic carbon derived from the soils of the Himalayan terrain that were undergoing strong chemical weathering suddenly increased at 9 m.y., substantially at about 6.5 m.y. and very greatly at 0.85 m.y. as the data from Site 117 of the Ocean Drilling Programme (Meyers and Dickens, 1992).

is controlled by changes in the rates of sea-floor spreading. Because sea-floor spreading rates changed little during the past 30-40 Ma, the changes in the $^{87}Sr/^{86}Sr$ isotope ratios may be attributed to increased weathering. Today, the Ganges-Brahmaputra river systems have unusually high $^{87}Sr/^{86}Sr$ isotope ratios, and it is therefore argued that the increased ratios may reflect enhanced supply by glacial erosion associated with the growth of the Antarctic ice sheet.

The timing of uplift is one of the major uncertainties in attributing uplift to climate change. With regard to the timing of uplift, there are three contrasting schools of thought.

Primarily on the basis of mammalian fauna, palaeokarst, and geomorphology could equally, however, be attributed to climate change.

The second school suggests that uplift has occurred rather gradually since Early Eocene times, with substantial elevations being reached by late Eocene times. This is based on lithological variations in lacustrine rocks from eastern China, but there is a large degree of uncertainty with this hypothesis.

The third school, and the one whose view is most generally accepted, considers that rapid uplift occurred after about 25 Ma ago, and that the plateau attained its present elevation by about 14 to 15 ma ago. This is based on the assumption that the present elevation and extensional deformation of the plateau probably resulted from uplift owing to convective thinning of the underlying lithospheric mantle. Post mid-Miocene lavas from the northern part of the Tibetan plateau, derived from lithospheric mantle, owe their origins to the thinning of the lithospheric mantle. These lavas have been dated at about 13 Ma. Furthermore, extensional faulting in the Thakkola graben in Nepal, which is the result of the extensional collapse of the plateau, has been dated at 14 Ma. These examples support the view that maximum elevation must have been achieved prior to 14 Ma. A reassessment of the sediments deposited on the Indo-Gangetic Plain and the Bengal fan has shown that there was a reduction in sediment supply from the Himalayas 8 ma ago. This could be attributed to reduced tectonic activity after 8 Ma, further supporting the view that plateau uplift occurred more than about 8 Ma ago.

Effect of Palaeoenvironment due to rise of the Himalaya

Without the Himalayan ranges, which cause precipitation of moisture from the passing clouds, much of the subcontinent would be a dry desolate land. The high ranges that prevent the SW-monsoon clouds to go north are also responsible for the drier conditions in Ladakh and Tibet in the north and western Rajasthan in the south.

The snow-covered ranges have exercised a moderating influence on the temperature and humidity in the Indian subcontinent. The Himalayas stops the Siberian wintry cold winds flowing into India, so that the winters

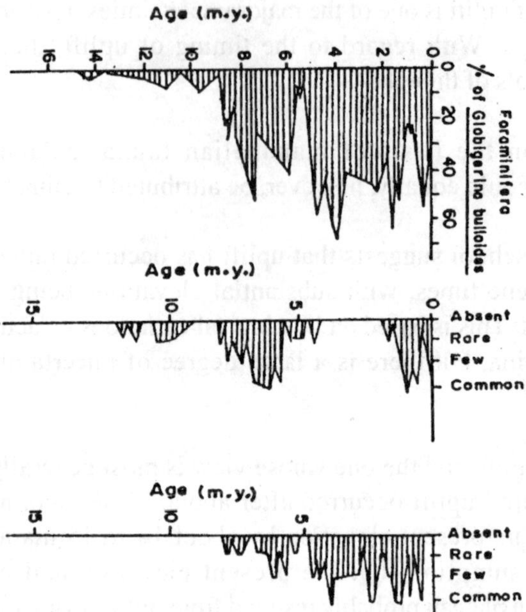

Fig.9. Sudden increase in the short period 8.5 to 7.4 m.y. of endemic upwelling species amongst the planktonic assemblages of foraminifera and radiolaria (at site 722 on the Owen Ridge) in the Indian Ocean imply availability of nutrients, which only the upwelling currents could have brought to the surface 8.5 to 7.4 m.y. ago. Resurgence of upwelling currents means strong atmospheric currents of the monsoon (modified after Prell et al. 1992-and Nigrini, 1991).

in North India are warmer than they would have been if the Himalayas had not been there.

Even though the Himalayas had become a mountain of prominence by Mid-Miocene, it could not have been an orographic barrier high enough then to prevent the movements of heavy-bodied bulky quadrupeds like the Hippopotamus, Rhinoceros, and Elephants in and across the lake basins of Potwar, Jammu, Kashmir, and Kathmandu (Valdiya 1993, 1998).

Revival of movements on faults that delimit the boundaries of four terrains of the Himalayan province was responsible for spectacular uplift of the Himalayas approximately 8.0 m.y. ago (Valdiya 1998, Cochran 1990, Krishnan *et al.* 1998 and Molnar *et al.* 1993).

Simultaneously, the Tibetan landmass (welded to the Himalaya) was split up by N-S tension faults 8±1 m.y. ago, leading to east-west extension of the Tibetan plateau. To the south in the Indian Ocean, the oceanic crust

Table 5: Late Quaternary Tectonic Movements: Consequences

Deformation of the Indian Ocean Crust	4 m.y.
Revival of movement on MCT, MBT, etc. in the Himalaya	4-2 m.y.
Exceptionally severe movements on MBT and in Siwalik	1.6 m.y.
Deformation of the Indian Ocean sediments	0.8 m.y.
Extremely fast erosion in the Himalaya and shooting up of sediment deposition in Bengal Fan from 20-70m/m.y. to >200 m/m.y.	0.9-0.8 m.y.
Peak influx and concentration of land –derived organic matter in the Arabian Sea	0.85 m.y.

along with its superincumbent sedimentary pile was deformed nearly 8 to 7.5 m.y. ago (Table-1).

The implication of this development is that since then long dry periods were followed by extremely wet spells, when heavy rains prompted a variety of natural processes such as strong weathering, erosion, and generation of detritus, reshaping of landscape and modification of ecosystems, both in the rising mountains and in the plains and the ocean.

Expansion of Grassland

As a consequence of changed climatic conditions, there was dramatic transition in the period 10 to 7.5 m.y. of the flora in the foothill domains of the Siwalik where tropical trees gave way to tall grasses with scattered trees and shrubs. The climate became warmer and the rainfalls seasonal. This is evident from the composition of the palaeoflora in the Nepal Siwaliks. The assemblage dominated by tropical Evergreen large trees such as *Dipterocarpus, Shorea, Hopea, Calophyllum* etc. before 10 m.y. gave way to the assemblage in which there was great abundance of moist or dry deciduous vegetation including *Bauhinia, Terminalia* and tall grasses. Increasing abundance of brown mollic soil and marked shift towards heavy carbon isotopes of pedogenic carbonates (Retallack 1995) testify the dramatic floral changes in the period 9.5 to 7.4 m.y. in front of the Himalayan Mountains.

Epochs of High Rainfall and Aridity in The Himalaya

The prolonged period of Late Quaternary dryness, including the time of last Glacial Maximum (20,000 to 16,000 yr. Before Present-B.P.)

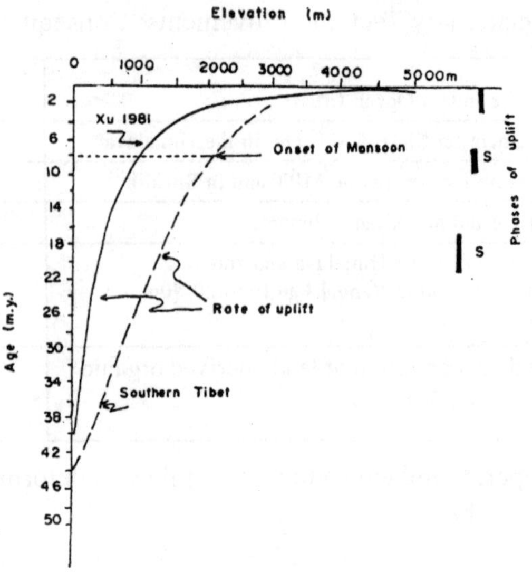

Fig. 10. Tibet-Himalaya landmass must have been lifted up to its tremendous elevation varies rapidly in the last 2 m.y. (after Sakai, 1997). Even though there is slight difference on the perception of the rate of uplift, it will be evident that the Himalaya along with Tibet rose very rapidly in the last 11 to 10 m.y.

persisted until nearly 11,000 yr. BP. Then the climate became progressively wetter and warmer as the SW Monsoon intensified, reaching its peak in the early to middle Holocene. The wet and warm epoch was followed by a time of severe aridity around 3500 yr. BP and lasting until about 2000 yr. BP (Table-6).

Warm-wet conditions started about 10,500 to 10,000 yr. BP when the SW Monsoon intensified, as evidenced by detritus brought by rivers to the Arabian Sea.

The $\delta^{13}C$ values of peats in the palaeolakes (e.g. at Colgrain) of the Nilgiri Hills confirm a very wet spell from 9,000 to 8,000 yr. BP (Rajgopalan *et al.* 1997). Higher rainfall during the period 10,800 to 3,500 yr. BP is testified by the pollen profiles of Lakes- Didwana and Lunkarnsar in Western Rajasthan (Table-6).

Pollens from the Dokrani peat in the Gangotri glacier area in the Great Himalaya (in Garhwal) indicate a warm-wet spell of climate between 6,400 and 4,000 yr. BP. Evidently, the larger part of the Indian subcontinent and adjoining Tibet experienced warm and wet climate 10,000 to 4,000 yr. BP.

Table-6: Holocene Climate Changes

Area	Warm-wet (yr BP)	Hot-dry (yr BP)
Arabian Sea (planktonic forams)	10,500- 5,000	3500
Western coast (terrestrial detritus)	10,500-10,000	
Karwar coast (land-derived pollens)	10,500-5,000	3500
Nilgiri Hills ($\delta^{13}C$ values of peats)	9,000-8,000	5000-2000
Didwana-Lunkaransar lakes in Rajasthan (pollenprofiles)	10,800-4,500	4000-Present
Gangotri glacier (pollens in sediments)	6,500-4,000	3500-3000
Tsokar lake, Ladakh (*Juniperus* flora)	10,000	
BangongCo lake, Aksai Chin (pollens)	6,000 (max.)	4000-3000
South-eastern Tibet (pollens in lakes)	7,500-3,000	3000-1500
Bhimtal basin, South Kumaun (pollens)		3550±120
Rara lake, West Nepal (pollens)		4500
Karewa lake, Kashmir (pollens)		5000

By 3,500 yr. after BP the SW Monsoon had weakened considerably, the upwelling currents became slack and the land witnessed onset of aridity. Reduction of humid conditions and increase in grasslands in the continent at this time is borne out by the pollens washed down to the shelf by rivers (Caratini *et al.* 1991).

The water in the west Rajasthan lakes became increasingly saline and the level dropped around 4,000 yr. BP (Wasson 1995). The pollen testimony demonstrates that the arid conditions prevailed around 3,500 yr. BP (3550±120) in the Bhimtal Basin in south central Kumaun (Kotlia *et al.* 1997), about 4500 yr. BP in the Rara Lake Basin in Western Nepal (Yasuda & Tabata 1988), approximately 5,000 yr. BP in the Karewa Basin in Kashmir (Dodia *et al.* 1985).

The 4,000-2,000 yr. BP period seems to be a time of high aridity all over the Indian subcontinent.

Immigration of Mammals

The tremendous expansion of grasslands, with its rich resources of forage and food in the foothills, attracted grazing animals from the neighbouring lands- from as far as Africa, Europe, Central and Eastern

Asia. The immigration of quadrupeds brought about major faunal ***"turnovers"*** about 9.5 m.y. to 7.4 m.y. ago (Table-2) as discernible in the Potwar basin, Kashmir (Barry *et al.* 1985).

The Siwalik faunal assemblage then was three times richer than that of the present (30 species of elephant compared to just one today) and 15 genera of anthropoid apes (Wadia 1975). Among the marginalized indigenous animals were rhinos, buffaloes and cows.

In northern Pakistan the development of arid conditions caused rapid changes in the rodent species between 9 and 7 m.y. The three-toed horse *Hipparian* and Pigs (Table-2, Fig.7) appeared at 9.5 m.y., having came all the way from Europe. The proboscid elephant-*Stegodon* along with the hippopotamus- *Hexaprotodon* and the elephant *Elephas planifrons,* among many others (Fig.7) came to the Siwalik domain about 7.4 m.y. ago.

The immigration of heavy-footed large-bodied quadrupeds across the Himalayan domain in the late Miocene implies that some segments were simply not high and rugged enough to prevent the movements of the bulky animals as late as 9.5 to 7.5 m.y. ago (Valdiya 1993, 1998). Admittedly, a few corridors could have been enough for migration, however considering the bulky size of the immigrants and their wide distribution in Tibet, in the outer Lesser Himalaya and Siwalik domains, more than corridors are needed to permit this animal influx.

Upwelling Currents in the Indian Ocean

There was a sudden change in the nature of pelagic sedimentation in certain parts of the Arabian Sea between 8.0 and 10.5 m.y.- from the accumulation of opal-rich silicicclastics, which are characteristic of warm water having low surface productivity, to dominantly biogenic deposition of endemic upwelling species (Table-4, Fig. 10). For example, *Globigerina bulloides* increased from <5% to >50% of the foraminiferal population between 8.5 m.y. and 7.4 m.y. (Prell *et al.* 1992, Prell *et al.* 1992).

Spurts of upwelling currents took place in the intervals 8.5 to 7.3 M.y and 6.3 to 5.0 M.y. during which period the Rivers Ganga, Brahamaputra and Irrawaddy dumped large quantities of sediments in the Bay of Bengal. The blowing of the strong southwesterly winds 10.5 to 8.0 m.y. ago indicates that the Himalayan mountain had suddenly risen up and formed a barrier to the then prevailing smooth west-to-east flow of the air currents. However, Srinivasan and Sinha (1998) attribute this change to severance of connection of the tropical Pacific with the Indian Ocean due to northward movement of Australia and resultant evolution of the Indonesian Archipelago, the latter development causing drastic change in the equatorial circulation.

Neogene Paleaobathymetry and **Sea-floor Tectonism** of the Andaman-Nicobar domain also suggest that the Himalaya had started rising up around 9.8 m.y (Srinivasan 1986).

Revival of movement on the many regional faults of the Himalayas made it a lofty barrier. Apatite fission-track dates of nearly 8 m.y. in the Kohat area of Pakistan indicate movement on the Main Boundary Thrust (MBT) at that time and the muscovite-cooling ages from the zone of Main Central Thrust (MCT) in NW Himalayas and Nepal show major peak of activity at about 8 to 6 m.y. (Sorkhabi *et al.* 1998). There is little doubt that strong tectonic movements occurred on the boundary thrusts of the Himalayas all over its extent in the late Miocene.

Note: All Tables from Valdiya, 1999.

ACKNOWLEDGEMENTS

The present paper is an outcome of information available from Internet and Encyclopaedia and mainly based on the work of Prof. K.S.Valdiya. Present author does not claim it to be its original contribution, it is merely a review article for the general reading of the ecologists interested in having information about the Impact of the Himalaya that had affected the climate of the Indian subcontinent including Asia. Author expresses his gratitude to Prof. A.K. Kandya, Professor of Botany, Sir H.S. Gaur University, Sagar for entrusting this job to write a Review article. The help received from Dr. Arun Kumar, Head Department of Earth Sciences, Manipur University, Imphal is thankfully acknowledged for extending all the facilities for completing the present work.

REFERENCES

Armijo, R.P.; P. Tapponier and H. Tongun (1985). J. Geophys. **92:**2787-2838.

Barry, J.C., Johnson, N.M., Raza, S.M. and Jacobs, L.L. (1985). *Geology,* **13:**637-640.

Burbank, D.W.; L.A. Derry and C. France - Lanord 1993. Nature **346:**48-50.

Caratini, C., Fontugne, M., Pascal, J.P., Tissot, C., and Bentalob, I. (1991). *Curr. Sci.*, **61**:669-672.

Cochran, J.R. (1990). *Proc. Ocean. Drill. Program. Sci. Results,* **116**:397-414.

Dodia, R., Agarwal, D.P., and Vora, A.B. (1985). In *Current Trends in Geology*, Today & Tomorrow's, **VI**:101-108.

Fielding, E.J. (1996). *Tectonophysics,* **260**:55-84.

Kotlia, B.S. *et al.* (1997). Paleogeogr. Palaeoclimatol. Palaeoecol. **130**:307-322.

Krishna, K.S. *et al.* (1998). *J. Geophysics. Res.*, **103**:17859-17875.

Meyers, P.A., and Dickens, G. R. (1992). In *Synthesis of Results from Scientific Drilling in Indian Ocean* (eds Duncan, R.A. *et al.*), Am. Geophys. Union, Washington, 295-308.

Molnar, P., England, P. and Martinod, J. (1993). *Rev. Geophysics.,* **31**:357-396.

Nigrini, C. (1991). *Proc. Ocean Drill. Program. Sci. Results,* **117**:89-126.

Prell, W.L. and Kutzbach, J.E. (1992). *Nature,* **360**:647-652.

Prell, W.L., Murray, D.W. and Clemens, S.C. (1992). In *Synthesis of Results from Scientific Drilling in Indian Ocean* (eds Duncan, R.A. *et al.*), Am. Geophys. Union, Washington, 447-469.

Quade, J., Cerling, T.E., and Bowman, J.R. (1989). *Nature, 342:163-166.*

Rajagopalan, Geeta, Sukumar, R., Ramesh, R., Pant, R.K. and Rajagopalan, G. (1997). *Curr. Sci., 73:60-63.*

Raymo, M.E. and Ruddiman, W.F. (1992). *Nature,* **259**:117-122.

Rea, D.K. (1992). In. *Synthesis of Results from Scientific Drilling in Indian Ocean* (eds Duncan, R.A. *et al.*), Am. Geophys. Union, Washington, 387- 402.

Retallack, G.J. (1995). In *Quaternary Environments and Geoarchaeology of India* (eds Wadia, S., *et al.*), Geol. Soc. India, Bangalore, 36-51.

Sakai, H.J. (1997). *J. Geogr.*, **106**:131-144.

Sorkhabi, R.B., Arita, K. and valdiya, K.S. (1998). In *Geodynamics of Northwest Himalaya* (eds Jain, A.K. and Manichavasagam, R. M.), Gondwana Research Group, Trrivandrum.

Srinivasan, M.S. (1986). In *Ophiolites And Indian Plate Margin* (eds Ghose, N.C. and Varadarajan, S.), Oxford-IBH New Delhi, 295-308.

Tanaka, S. (1997). *J.* ***Geol. Soc. Jpn.*** **103**:253-264.

Valdiya, K.S. (1993). *Curr. Sci.,* **64**:873-885.

Valdiya, K.S. (1998). *Dynamic Himalaya,* Universities Press, Hyderabad, p. 178.

Valdiya, K.S. (1999). Rising Himalaya: Advent and Intensification of Monsoon *Curr. Sci.,* **76 (4)**:514-524.

Wadia, D.N. (1975). *Geology of India,* Tata-McGraw Hill, New Delhi, p. 508.

Wasson, R.J. (1995). In - Quaternary Environments and Geo-archeology of India. Eds. S. Wadia *et al.,* Geol. Soc. India, Bangalore, pp. 22-35.

Yasuda, Y. and Tabata, H. (1988). *Proc. Indian Natl. Sci. Acad.,* **A54**:538-548.

SECTION TEN

APPLIED ECOLOGY

Advancing Frontiers of Ecological Researches in India, 673-694, 2006
Edited by A.K. Kandya & Asha Gupta
Published by Bishen Singh Mahendra Pal Singh, Dehra Dun, India

Plant Genetic Resources of North East India and their Sustainable Management

Supriyo Chakraborty
Department of Biotechnology,
Assam University,
Silchar - 788 011, Assam

Abstract : The North Eastern Region (NER) of India comprising of eight states is a vast natural repository of Plant Genetic Resources (PGRs). Genetically diverse cereals, pulses, forest trees, fruits, vegetables, plantation crops, oilseed crops, orchids, bamboos and canes abound in this region. PGRs play a significant role in human welfare through various means *e.g.* sustainable food and nutrition security, crop variety development, sustainable agriculture and as a source of medicinal plants. The NE Region is not only a hotspot of biodiversity but also a hotspot of PGRs. Up till now 3165 plant species have been identified as endemic to the Eastern Himalayas including the North Eastern Region of India and are found nowhere in the World. Genetic resources of a plant species arose in nature as a result of the effect of one or more biological phenomena namely mutation, hybridization and polyploidy. The prime concern of biodiversity around the world today is the loss of genetic resources. It has been estimated that about 140 biological species including plants, animals and microbes are being lost everyday from the earth and this implies the loss or extinction of one species every 12 minutes. This paper discusses the reasons for the loss of genetic resources in the NE Region of India and the means to conserve the genetic resources. It also highlights the sustainable management of plant genetic resources so that the needs of the present generation can be met from the genetic resources without reducing the ability of the genetic resources to meet the diverse needs of our future generations. Our life support systems are under stress and the loss of the biological resources including the genetic resources is increasing day by day. Fertile land forms for crop husbandry are gradually shrinking while the biotic and abiotic threats to food security are expanding. The answer to overcome these threats lies in the sustainable management of plant genetic resources because genetic resources are natural store house of valuable genes that have potential to combat the effects of various stresses in plants in the coming years. Sustainable management of plant genetic resources can only ensure the maintenance of the valuable genes for anticipated use by our posterity. Sustainable management and plant genetic resources are just inter-linked as the two sides of a coin and one can not be separated from the other.

INTRODUCTION

Apart from the scenic beauty of mountains and streams, the North Eastern Region (NER) of India is endowed with numerous plant genetic resources. The NER comprises of eight states namely Assam, Arunachal, Manipur, Meghalaya, Mizoram, Nagaland, Tripura and Sikkim. Genetic resources of a species refer to the sum total of genes present in a species. In simple words, genetic resources refer to the lines, varieties and populations of a species. Genetic resources are also known as gene pool or germplasm or genetic stock. Greater the amount of genetic resources, higher the opportunity of genetic improvement of a species. Genetic resources in a region have three components – plant, animal and microbial genetic resources. The NER with its diverse climatic conditions and geomorphologic features harbours a great variety of plants, animals and microbes. Of these species, some are highly endemic to this region and observed nowhere in the world. This makes North Eastern Region of India a center of attraction to the biologists all over the globe.

We, the human beings, partly meet our requirements of food, fibre and medicine from plant genetic resources. It is indeed difficult to determine the economic value of the goods and services, which the plant resources offer us now or hold potential to offer in future. To estimate the economic value, we have to calculate at first the number of living plant species and the number of lines within each species. Biologists have up till now described a total of about 1.5 to 1.8 million species including plants, animals and microbes. But the true number of living species has been estimated to range from 3.6 million to 100 million. Most biologists prefer to a total estimate of about 30 to 50 million species. There are thousands of species yet to be discovered and those species, which have already been discovered, are yet to be known in respect of their genetic resources. Least known are the microbes. In the case of insects, nematodes and fungi, about 10 percent is known (Wilson, 2001). But less than 1 per cent species is known in bacteria and viruses. This clearly reveals how little we know about the living species and their genetic resources inhabiting the earth.

Life might have originated as a naked mass of protoplasm or more likely as a fragment of self replicating DNA. Creation of a species from the primitive life form took hundreds of millions of years through the operation of evolutionary forces. Over the geological ages many flora came into existence, several early forms were eliminated by nature and were then replaced by more adaptive forms. Several other plant species have still been continuing to exist in almost unaltered forms.

Any biological species never exists in isolation. Different biological species live together in a habitat forming a biotic community. A biotic community and its abiotic environment constitute an ecosystem such as mountain ecosystem, riverine ecosystem and desert ecosystem etc. Each ecosystem is characterized by its own energy flow and food chain. A natural ecosystem is well balanced and maintains equilibrium. The ecosystem is a complex network of different species, which depend either directly or indirectly upon each other. All of them interact with the surrounding environment. Any disturbance in the delicate balance of the ecosystem makes it unstable and affects the community structure leading to extinction of one or more species.

Indian Plant Genetic Resources – A Brief Review

India is a vast repository of biological species and their genetic resources due to its varied physical features, tropical location and diverse climate. About 3,00,000 plant species grow on earth. Of these, nearly 45,000 plant species occur in India alone, which accounts for about 15% of global wealth. Approximately 33% of the plant species found in India are endemic. Moreover, out of 15,000 flowering plant species (belonging to 315 families and 2250 genera), nearly 5000 flowering plant species of India are endemic and they belong to 47 families and 140 genera. Besides these, India provides shelter to almost half of the world's aquatic plant species. In addition, India has nearly 25000 algae, 23000 fungi, 2664 bryophytes, 1600 lichens, 1022 pteridophytes, 850 bacteria and 64 gymnosperms. The exact number of genetic resources or germplasm in each species is not known. This reveals the enormity of India's plant genetic resources. But despite the vast genetic resources of the country, about 450 plant species are threatened with extinction. These species are not getting full liberty of time and space for their survival. Efforts are needed to conserve not only the species but also the genetic resources within each species.

Plant Genetic Resources of North East India – An Outline

The North Eastern Region, the Eastern Ghats and the Western Ghats in India are very rich in biological species and genetic resources. The richness of plant species in North Eastern Region is largely due to its subtropical location, varied physical features like mountains, valleys and rivers and also due to ecological diversity. Rainforests are considered to be the large repositories of biodiversity. In North East India there are quite a few rainforests. The "Jaidihing" rainforest located in Assam - Arunachal border is a reservoir of enormous biodiversity.

The temperature range in this region is very wide. In winter the temperature is as low as zero degree Celsius in the place called "Jero" in Arunachal Pradesh whereas the temperature rises as high as 38 degree Celsius in Nagaon and Kamrup districts of Assam. Winter is very long in some parts of Nagaland. Dibrugarh in Assam is characterized by maximum number of rainy days per annum. Flood caused by the Brahmaputra river every year is a great havoc and submerges large areas in Assam and Arunachal Pradesh resulting in extensive damage to crop and property. Soil erosion due to the Brahmaputra River has posed a great threat to the Jakhalabandha Township in Assam during winter of 2003. On the other hand, the Karbianglong and North Cachar Hills of Assam receive scanty rainfall throughout the year and often remain drought–hit in summer months. Entire North Eastern Region is characterized by high humidity in July-August. In some places, the relative humidity reaches up to 90% alongwith high temperature and bright sun-shine which together create an unfavourable pinching effect on skin in July and August. The warm humid climate of North East India has a great impact on the kind and distribution of plant species.

a) Forest Tree Genetic Resources

The forests of North East India are very rich in the genetic resources of forest trees. Some timber tree species commonly found in the forests of this region are Sal (*Shorea robusta*), Teak or Segoon (*Tectona grandis*), Agar (*Lagerstroemia speciosa*), Nahar (*Mesua ferrea*), Sissoo (*Dalbergia sissoo*), Gamhar (*Gmelina arborea*), Bonsum (*Phoebe attenuata*), Pine (*Pinus khasya*), Lali (*Dysoxylum procerum*), Gonsoroi (*Cinnamomum glanduliferum*), Khokon (*Duabanga sonneratioides*), Titasopa (*Michelia champaca*), Poma (*Cedrela toona*), Boga Poma (*Chickrassia tabularis*), and Amari (*Amoora wallichii*) etc. (Dutta, 1979). Besides timber species, bamboos and conifers are also found in the forests of North East India.

b) Cereal Genetic Resources

Among the cereal crops, rice is the staple food of North Eastern Region. The entire region, particularly Assam, is exceptionally rich in rice genetic resources. At present nearly 10,000 to 15,000 rice germplasms are available in the North Eastern Region.

The physiology and agro ecological situations vary widely in Assam. Under such diverse agro ecological conditions the cultivation of rice over the years has resulted in the development of a multitude of strains with specific adaptation through natural selection and farmers' discretion (Pathak, 2001). Besides these, variations in the traditional cultivation practices of different ethnic groups and their quality preferences are the factors adding to the diversity of the rice not only in Assam but also in the entire region. The ethnic migration and immigration to the North Eastern Region in general and Assam in particular have also led to the introduction of different types of rice cultivars leading to further diversity. Indigenous rice germplasm of Assam is beset with genetic diversity and represents a treasure of valuable gene systems (Sharma *et al.*, 1971; IRRI, 1974; Das *et al.*, 1981). Rice germplasm collections which are maintained at the Central Rice Research Institute, Cuttack (Orissa) include 2054 collections from Assam alone out of a total of 12256 collections from all over India (Paroda and Sharma, 1986). This is probably sufficient to explain the extent of genetic diversity of rice in Assam and adjoining states. At present, about 4000 rice accessions including local varieties, improved varieties and breeding stocks are being maintained at Assam Agricultural University in Jorhat (Pathak *et al.*, 1995). The actual number of germplasm may, however, be smaller, if one takes into account the possibility of duplication of the conserved materials (Ganguli, 1950; Ahmed, 1986). In spite of the substantial number of rice germplasms being maintained at Assam Agricultural University, it is still felt that the stock is far from being exhaustive and sufficient enough.

The extent of genetic diversity among the primitive cultivars and land races of rice is so vast in the North Eastern Region that this region could well be considered as one of the centers of origin of cultivated rice. Rice grains dating back to 900 B.C. were excavated at Ambari in Guwahati. This reveals the cultivation of rice in Assam for quite a long period of time. It has been estimated that the genetic diversity of rice in Assam evolved about 3000 years ago. Assam rice collection is proved to be an important source of valuable genes for resistance to insect pests, diseases, drought and flood. Two wild species of rice namely *Oryza officinalis* and *Oryza granulata* are commonly found in marshy land of Assam. The gene for resistance to brown plant hopper (BPH) has been identified in *Oryza officinalis* whereas *Oryza granulata* has been found to contain the gene for resistance to white backed plant hopper (WBPH). Both BPH and WBPH are important insect pests of rice.

Besides rice, significant genetic diversity of maize is also observed in North Eastern Region. N.I. Vavilov, a Russian plant breeder and geneticist, considered North Eastern Himalayan region comprising of Arunachal Pradesh, Sikkim and adjoining areas of China as one of the secondary centers of origin of maize. Several genotypes of millets also grow in Arunachal, Nagaland and Meghalaya.

c) Grain Legume Genetic Resources:

The North Eastern Region encompassing diverse agro ecological situations, ethnic and cultural diversity is beset with the genetic diversity of several grain legumes namely urdbean or blackgram (*Vigna mungo*), mungbean or greengram (*Vigna radiata*), cowpea (*Vigna unguiculata*), pigeonpea (*Cajanus cajan*), pea (*Pisum sativum*), frenchbean (*Phaseolus vulgaris*), ricebean (*Vigna umbellata*) and fababean.

Sarma *et al*. (1991) reported that 728 germplasms of different pulse crops had been collected from the North Eastern Region. The highest number of genetic resources was collected in rice bean (230), followed by black gram (121), french bean (111), cowpea (67), pigeon pea (64), pea (56), mungbean (42) sembean (29; *Dolichos lablab*) and chickpea (8). Ricebean grows abundantly in North Eastern Hills. High genetic diversity of ricebean and frenchbean are observed in the hilly states of North Eastern Region whereas high diversity of urdbean is found in Assam. Since time immemorial, both greengram and blackgram have been cultivated in North Eastern India. The primary center of origin of greengram

Table-1: Area of diversity of grain legumes in North Eastern India.

Crop	Area of Diversity
Ricebean	All the eight North Eastern States.
Urdbean Frenchbean	Assam,Manipur, Meghalaya, Nagaland, Sikkim, and Tripura. Arunachal Pradesh and Meghalaya
Cowpea	Assam,Manipur, Meghalaya, Nagaland and Sikkim.
Pigeonpea	Assam and Tripura
Pea	Meghalaya
Mungbean	Assam, Meghalaya, Nagaland and Tripura.
Fababean	Arunachal Pradesh

Courtesy: Sarma *et al* (1991) and Gupta (1999)

and blackgram is the Hindustan centre, which covers all the eight North Eastern States. In Karbianglong district of Assam, perennial genotypes of pigeonpea are found.

d) Fruit Genetic Resources

The North Eastern Region of India belonging to the Hindustan centre of origin as stated by N.I. Vavilov is the primary centre of origin of several horticultural crops like banana, orange, sour lime and other citrus species, brinjal, cucumber, Indian lettuce (*Lactuca indica*), yam (*Dioscorea* spp.), Indian radish (*Raphanus indicus*), black pepper (*Piper nigrum*), coconut (*Cocus nucifera*) and turmeric (*Curcuma domestica*) (Singh, 1986).

Among the fruits, citrus is very common in the North Eastern Region. There are as many as 17 citrus species, 53 varieties and 7 probable hybrids found in this region (Bhattacharyya and Dutta, 1957). Assam lemon and "Gol Nimbu" (Citrus *jambhiri*) are commercially grown in Assam. Khasi mandarin is widely grown in Meghalaya. In Assam, Tinsukia and some parts of Jorhat district are rich in the diversity of mandarin and other citrus species. In banana, despite numerous cultivated seedless triploids, *Musa acuminata* and *Musa balbisiana* diploids are found in semi-wild state in this region (Bhattacharyya, 1987). Two genomes of banana are designated as A and B. In Assam, commercially important banana cultivars are Jahaji (AAA), Bar Jahaji (AAA), Malbhog (AAB), Cheni Champa (AAB), Manohar (ABB) and Kachkal (ABB). Semi–wild banana genotypes also grow abundantly in Arunachal Pradesh, Nagaland and Meghalaya. Besides banana, semi-wild species of mango like *Mangifera sylvatica* are found in the evergreen forests of Assam. In addition to these major fruits, some minor fruits are also found in Assam. These minor fruits are leteku (*Baccaurea sapida*), poniol (*Flacorita indica*), Kardoi (*Averrhoa carambola*), thereju (*Prunus jenkinsii*), nagatenga (*Rhus semialata*), outenga (*Dillenia indica*), amara (*Spondias mangifera*), silikha (*Terminalia chebula*), bhomora (*Terminalia belerica*), nagalichu (*Euphoria longana*), etc. (Neog and Mohan, 1994). In the case of pineapple, two commercial cultivars namely "Kew" and "Queen" are widely grown in Assam, Meghalaya, Nagaland and Tripura. Although pineapple is vegetatively propagated to maintain true-to-type character, genetic variation due to mutation could be detected in pineapple if an extensive search is made in the North Eastern Region.

e) Vegetable Genetic Resources

The entire North Eastern Region exhibits a rich diversity in the genetic resources of various vegetables. Some vegetables which exhibit wide diversity are cucurbits, tuber crops and nontuberiferous solanums. Quite a good number of wild and semi-wild relatives of vegetable crops are found in the North Eastern Region.

Among the tuber crops, a wide variability exists in colocasia (Ahinakachu, Dahikachu, Nalkachu, Nagakachu, Kanikachu, Tekelikachu, Kakakachu, Gheekachu etc.). In *Dioscorea alata* and *D. esculenta* the locally available genotypes are Kathaloo, Kunwarialoo and Moaaloo. Bhattacharya (2001) reported about an interesting plant species of the region. Its botanical name is *Vigna vexilata*, which produces above ground edible pods as well as underground tubers.

Wide genetic diversity exists for pumpkin and other cucurbits in the entire North Eastern Region and more particularly in Assam, Nagaland, Meghalaya and Tripura. In bitter gourd, three species namely *Momordica charantia*, *M. dioica* and *M. cochinchinensis*, having spiny to partially spiny appearance and small to large fruits, are available. The nonedible wild relatives of Cucumis (*C. setosus, C. trigonus* and *C. hystrix*) and ridgegourd (*Luffa graveolans* and *L. echinata*) grow abundantly in this region. A good number of wild species of Trichosanthes like *T. lobata, T. cucumerina, T. palmata* and *T. lapiniana* are found in this region in addition to two cultivated species namely *T. dioica* and *T. anguina.*

The North Eastern Region has a rich diversity in the genus *Solanum* that has several wild and semi-wild relatives in the region. The predominant *Solanum* species are *Solanum khasianum, S. ferox, S. incanum, S. gilo, S. torvum, S. verbasifolium* and *S. xanthocarpum*. Within the brinjal (*Solanum melongana*) crop, a wide variability exists in this region and we find different varieties of brinjal like Borbengana, Nalbengana, Thalbengana, Kharuabengana, Kuchiabengana ,Sagalisingia bengana. In okra, wild relative *Abelmoschous ficulneus* is also available in this region.

In the case of spices, the North Eastern Region can be considered as a large reservoir of the genetic resources particularly for ginger and chilli. Commonly available genotypes of ginger of this region are Maran, Tura, Karkai, Bourich, Thingpui, Thinglaidum, Singmakhir etc. and of chillies are Balijuri, Barbih, Latabih, Kanjalakia, Memjalakia, Tiger chilli, Bird's eye chilli, Krishna, Suryamukhi and Balijuri. The North Eastern Region also has a rich diversity of turmeric.

The natural vegetation of the North Eastern Region exhibits a great variety of herbs, shrubs and trees with edible leaves. These plants appear to possess distinct medicinal properties and high nutritional value. Some of the important species of these leafy vegetables commonly used by the local people are Matikanduri (*Alternanthera sessilis*), Pui (*Basella rubra*), Pahari Paleng (*Atriplex hortensis*), Khutura (*Amaranthus spp.*), Nefafu (*Clerodendrum colebrookianum*), Helosi sak (*Enhydra fluctuans*), Mandhania (*Eryngium foetidum*), Chutia lofa (*Fagopyrum esculentum*), Masundari (*Houttuynia cordata*), Brahmi (*Herpestis monnieria*), Kalmow (*Ipomoea aquatica*), Bhedai lata (*Paederia foetida*), Piralee Paleng (*Talinum triangularee*), Jilmil (*Chenopodium album*), Narasingha (*Murraya koenigii*), Bokful for flower (Sesbania *grandifolia*), Modhusoleng (*Polygonum chinensis*) Horu Manimuni (*Hydrocotyl roduntifolia*), Bor Manimuni (*Centella asiatica*), Suklati (*Perilla ocimoides*), Maha Bringharaj (*Wedelia calendulacea*) etc.

f) Plantation Crop Genetic Resources

The North Eastern Region particularly Assam is rich in the genetic resources of a few plantation crops like tea, coconut and arecanut. In tea, besides the two cultivated species namely *Camellia sinensis* (China tea) and *Camellia assamica* (Assam tea), other wild relatives of cultivated tea like *C. kissi, C. caudata, C. sasanqua* and *C. oleifera* are also found. Substantial genetic diversity of coconut (*cocus nucifera*) and arecanut (*Areca catechu*) are observed in the North Eastern Region, particularly in Assam, Meghalaya and Tripura.

g) Oilseed Crop Genetic Resources

In North Eastern Region, several oilseed crops have been grown for a long time. These include edible oilseeds namely rapeseed and mustard, sesame, niger and groundnut and non-edible oilseeds like castor and linseed. This region is said to be the centre of origin of yellow sarson (*Brassica campestris* var. yellow sarson) due to its high genetic diversity found in this region (Maiti *et al*., 1988). Indian mustard (*Brassica juncea*), also known as raya, was introduced from China to India through the North Eastern Region (Maiti *et al*., 1988). Various forms of Indian mustard also occur in this region.

h) Orchid Genetic Resources

The North Eastern Region of India is richly endowed with the genetic resources of orchids. This region inhabits nearly 800 species of orchids out of 1300 species found in India (Singh, 1986). In the world about 25000 to 35000 orchid species belonging to 600-800 genera are available (Garay, 1960). The North Eastern Region owing to its peculiar topographical gradient and varied climatic conditions contain the largest group of tropical, sub-tropical and temperate orchids. The orchids of this region are mostly epiphytic in nature unlike the terrestrial orchids of North Western India. The orchids of North Eastern India differ in size, shape and colour of the flowers. Orchid species belonging to the largest epiphytic genus *Dendrobium* (ca. 70 species) and species of the genus *Paphiopedilum* (Lady's slipper) are commonly found in the North Eastern Himalayas. Some orchids are endemic to the North Eastern Region of India and are not found anywhere else in the world. Some orchid species of this region belonging to four important genera namely *Cymbidium, Dendrobium, Paphiopedilum* and *Vanda* are almost extinct. In order to reduce the loss of orchids from the North Eastern Region of India, steps have been taken to conserve orchid species in orchidaria, sanctuaries and germplasm conservation centers. The ICAR complex for NEH region at Shillong (Meghalaya) maintains a good collection of orchids. Besides this, the government of Arunachal Pradesh has established an orchid sanctuary at Tipi whereas Sikkim Government has established orchid sanctuaries at Deorali and Singtom for orchid conservation. Orchidaria have also been established in Assam Agricultural University and Dibrugarh University.

i) Bamboo and Cane Genetic Resources

Bamboo and cane (*Ratan* species) genetic resources are abundantly found in entire North Eastern Region of India, more particularly in Tripura, Assam, Arunachal and Nagaland. Different kinds of bamboo genera namely *Bambusa, Melocanna, Dendrocalamus* etc. are found in this region. Among the various kinds of cane, the "Sonamukhi" type is highly priced for its good mat-making quality. Different types of bamboo namely Bholuka, Jati, Muli etc. are very common in this region.

j) Sericulture Linked Plant Genetic Resources

Since pre-historic time, the North Eastern Region, more particularly Assam has been known for sericulture. Muga culture and eri culture are very common in Assam. Assam has a rich diversity of the food plants of muga silkworm (*Antheraea assama*). Muga culture is confined mainly to the Brahmaputra valley of Assam (India). Besides the two primary food plants, i.e, Som (*Machilus bombycina*) and Soalu (*Litsaea polyantha*), more than seven secondary food plants of muga silkworm are found in Assam. In the case of eri culture, the eri silkworm (*Philosamia ricini*) feeds mostly on castor (*Ricinus communis*). Besides castor, the eri silkworm can feed on more than 10 different alternative host plant species, which are found in Assam. A few alternative host plants of eri silkworm are *Heteropanax fragrans, Manihot utilissima, Carica papaya, Plumeria acutifolia* and *Ailanthus spp.* etc.

Importance of Plant Genetic Resources

Genetic resources of a species refer to the sum total of genes in the species. The genes present in all the land races, obsolete cultivars, modern cultivars, breeding stocks and wild forms of the species, represent genetic resources of a plant species. Thus genetic resources, in fact, imply the huge collection of different genes in various combinations within the species. Plant genetic resources are very important to mankind in various ways, which are briefly outlined below:

a) Sustainable Food and Nutrition Security:

Food is the basic need to survival. Food security implies physical and economic access to food by all people at all times. Modern technology has made physical access to food possible but economic access to food by the poor people of several third world countries is yet to be achieved. Nutrition security means access to balanced diet.

To order to ensure sustainable food and nutrition security to all people, we concentrate on different plant species and their numerous genetic resources .We can consume nearly 7000 plant species as our food but 90 % of our food comes from only 20 species .Out of these 20 species, only 3 species namely wheat, rice and maize meet about 90% of our energy requirement. So emphasis is to be given on the vast domain of plant species, which can provide us with food and nutrition security. Unchecked population growth in the countries of Asia, Africa and Latin America has put tremendous pressure on land and a few domesticated principal food crops. These food crops have almost reached the yield plateau and further improvement of yield poses a great challenge. To ease the situation, we should learn to consume the large number of plant species recognized and used as food by rural and ethnic communities all over the globe. Genetic resources of a species often differ in nutritional qualities due to variation in biochemical properties.

b) Variety Development

Genetic resources are used in crop variety development through breeding programs. The genetic resources are source of valuable genes in any plant species. These valuable genes include the genes for resistance to diseases, pests and nematodes, genes for improved nutritional quality, wide adaptability, genes for resistance to abiotic stresses (like drought, heat, cold, flood, soil acidity and alkalinity), genes for plant architecture etc.

Modern high yielding varieties of different crops have been developed by tailoring the valuable genes from genetic resources to a suitable agronomic background of that crop species or through combination breeding. A few examples may be cited in this respect. Green revolution in India has been possible due to development of semi-dwarf wheat and rice varieties by Dr.M. S. Swaminathan in late 1960s. In India, transferring the gene for dwarfness from 'Norin 10' genetic resource of Japan developed semi-dwarf wheat variety with high yield potential. Transferring the "Dee-geo-woo-gen" gene from the genetic resource developed semi dwarf rice. The wild rice species, *Oryza nivara*, collected from Uttar Pradesh of India, contributed the gene for resistance to grassy stunt viral disease in rice and helped in the development of high yielding rice variety with resistance to grassy stunt disease. The first commercial production of hybrid rice in China was possible due to discovery of " wild abortive" (WA) type of cytoplasmic male sterility factor in a wild rice species *O. rufipogon*.

Further progress in hybrid rice technology was achieved through the discovery of genetic resource having photoperiod sensitive genic male sterility (PGMS) and temperature sensitive genic male sterility (TGMS) system. Both PGMS and TGMS brought about a change from the complex 3–line method to the simpler 2-line method of hybrid seed production in rice. PGMS and TGMS genetic resources exhibit genic male sterility separately under a specific photoperiod regime and temperature range. The first PGMS rice plant was found in a japonica variety called " Nong – keng 58" under natural condition in 1973. Many TGMS lines have been found in indica varieties of rice. These are just a few examples, out of many, which clearly reveal the importance of genetic resources in varietal development. In several crop species, the genes for resistance to economically important diseases and insect pests and also to abiotic stresses have been transferred either through conventional hybridization (backcrossing) or by using recombinant DNA technology.

c) Sustainable Agriculture

Modern agriculture is basically a production-oriented agriculture and so the emphasis is obviously on the use of high yielding varieties, chemical fertilizers, pesticides and irrigation. All these inputs of modern agriculture have increased crop yields and have been successful in providing food to the ever- increasing human population in the world. In spite of this great achievement, modern agriculture has also brought about some undesirable and harmful effects in the agro ecosystem. For example, high yielding varieties are the results of directional selection and possess a narrow genetic base. In other words, these high yielding varieties are genetically more uniform than the local landraces and as a result the former may face a sudden loss of total crop yield in the event of a disease and pest epiphytotic or a climatic aberration. Wide cultivation of high yielding varieties has caused genetic erosion through loss of genetic resources.

Indiscriminate use of chemical fertilizers and pesticides without the use of sufficient manures caused soil and air pollution, changed soil properties and makes it unsuitable for crop production. Even the apparently harmless irrigation is now known to cause several problems like salinity, alkalinity and iron toxicity etc. Realizing the problems associated with modern agriculture, the attitude towards agriculture has currently been changed to sustainable agriculture. Sustainable agriculture is a form of agriculture, which meets the current needs of agricultural production without comprising the prospects of meeting the needs of future generations. Sustainable agriculture ensures a continuous flow of goods and services without any degradation of the natural resources used in agriculture. Sustainable agriculture is also known as ecofarming or organic farming or natural farming or permaculture (Reddy *et al.*, 1995). Sustainable agriculture indicates minimal dependence on synthetic fertilizers, pesticides and antibiotics.

Conservation of genetic resources is highly emphasized in sustainable agriculture. Genetic resources are the reservoirs of genes for resistance to diseases, pests and nematodes and also for resistance to environmental stresses. Loss of genetic resources means loss of the valuable genes forever. Breeding program for developing the biotic and abiotic stress resistant crop varieties requires the genetic resources for gene transfer because these possess inherent ability to combat the stresses. Cultivation of the crop varieties with in-built resistance minimizes the need for pesticide application and thereby follows the pattern of sustainable agriculture. Many important rice genetic resources particularly *chengri, dumai* and *murali* groups of rice are now extinct in the entire North Eastern region and more particularly in Assam. This type of loss is indeed a great loss to our country's economy as well in view of their potential economic importance, which cannot be realized anymore since these rice groups are already extinct.

d) Medicinal Value

Plant species and their genetic resources provide us with medicines. In the past, nearly all medicines came from plants. In Ayurveda, we find the mention of about 2000 medicines of plant origin. In China nearly 5100 species are used in the traditional Chinese medicine. In addition, more than 3000 antibiotics are derived from microbes. The genetic resources within a species differ in their DNA make-up as well as in the biochemistry of gene products. Genetic resources belonging to different genera possess diverse medicinal properties. A few examples can be given to justify this statement.

A wild tree species, *Taxus brevifloria*, of pacific rain forest is now a priceless tree because it contains a compound called "taxol" which cures ovarian cancer. Some plants belonging to the genus *Taxus* have also been identified in Arunachal Pradesh, several plant species of high therapeutic value found in the forests of Arunachal Pradesh in the North Eastern Region. Two chemical compounds namely vincristine and vinblastine isolated from *Vinca rosea* are well known in medical science for their use in the treatment of leukaemia. A few biochemical compounds isolated from different plant species have been found to have strong anti-HIV activity and appear to be useful in the treatment of AIDS disease. These include *Calophyllum* species (Malaysia) and *Ancistrocladus* species. Another plant species, *Homolanthus hutants*, of pacific rain forest, used by the native people for the treatment of yellow fever, has been found to contain the compound "prostratin" having anti-HIV activity. There are quite a large number of plant species having therapeutic value, which have long been used in the traditional system of medicine of different countries.

Hotspots of Plant Genetic Resources

India is exceptionally rich in biodiversity and is one of the 12 mega-diversity centers of the world. India has been divided into 10 bio-geographic zones and 25 biotic provinces. All the major ecosystems are found in India. Four hotspots of biodiversity have been identified in our country. These hotspots are (1) The Eastern Ghats, (2) The Western Ghats, (3) The Andaman and Nicobar Islands and (4) The North Eastern Region. About 45,000 plant species grow in India, which represents 7% of world's flora. About 33% of Indian flora are endemic and are not found anywhere in the world. The hotspots of biodiversity are also the hotspots of plant genetic resources because maximum genetic variability of each plant species is observed in these hotspots. The North Eastern States located at the foothills of Eastern Himalayas are large reservoirs of plant genetic resources. 3165 plant species have been identified as endemic to the Eastern Himalayas including the North Eastern Region. The number of endemic species is not constant and continues to increase as more and more species are being discovered in the NER. Many areas in the North Eastern Region are yet to be explored and many plant species are likely to be discovered through exploration in this region.

Law of Parallel Variation

N.I. Vavilov, a Russian geneticist and plant breeder postulated the law of parallel variation. The law states that a character found in one plant species also occurs in other related species. The law of parallel variation is also known as "the law of homologous series in variation". The importance of the law lies in the fact that a character absent in a particular species but found in a related species is likely to be discovered in the particular species if an extensive exploration is made in the area of the diversity of the crop.

Creation of Genetic Resources

A question may arise in the minds of the reader as to how genetic resources are created in a species. Genetic resources of a species are, in fact, different lines or strains having different allelic combinations of the entire set of genes present in the species. Alleles are the alternative forms of a gene occupying a specific locus in the chromosome. One or more of the following biological phenomena creates genetic resources.

a) Mutation

Mutation is the sudden hereditary change in the gene or DNA molecule. Mutation creates a new allele or produces an already existing allele. At molecular level, a mutation is caused by the addition or deletion of a nitrogenous base (adenine, thymine, guanine and cytosine) or substitution of one base by the other. Mutation occurs randomly (i.e., at any gene, at any time) and is recurrent in nature. Most mutations are deleterious to the species but only a few mutations are favourable to the species. Mutation is responsible for the creation of multiple alleles of a single gene and these multiple alleles of different genes create the diverse genetic resources.

b) Hybridization

The mating of two plants or lines of dissimilar genotype is known as hybridization. Hybridization between two lines or strains or races belonging to the same species is called intervarietal hybridization or intraspecific hybridization, which can produce new genotypes or new genetic resources. The F1 hybrid is generally more vigorous than the parents with respect to various characters. Segregation in F2 and subsequent generations produces a vast range of genotypes, which show different phenotypes. Generally 7 or 8 generations of selfing from the F1 hybrid onwards results in the separation of the population into genotypically different lines due to continuous increase in homozygosity. The increased homozygosity leads tc random fixation of alleles in different lines. Therefore, the lines differ in their genotype and consequently in phenotype. The phenotypes appear as new genetic resources because of their new gene combinations, which are different from parents.

Sometimes two species belonging to the same genus are naturally crossable and this type of hybridization is known as interspecific hybridization. Naturally occurring interspecific hybridization has produced genetic resources in many plant species like wheat, tobacco, *Brassica*, strawberry, etc.

c) Polyploidy

Polyploidy plays an important role in the creation of genetic resources within a species. Any organism or individual having more than two basic sets of chromosomes is called a “polyploid” and such condition is known as "Polyploidy". In other words, a polyploid contains more than two genomes. A genome is a basic set of chromosomes, which do not undergo pairing and crossing over. The chromosomes in a genome are distinct. About one-third species of the flowering plants are polyploids. Polyploidy is of two types namely (i) autopolyploidy and (ii) allopolyploidy. An autopolyploid individual contains more than two identical genomes due to multiplication of the same genome. An allopolyploid individual contains more than two different genomes and is produced by combining complete genomes of two or more species.

Autopolyploidy appears to be more important in creating genetic resources with different ploidy levels than allopolyploidy. Autopolyploids may occur naturally due to chromosome doubling in some somatic tissues. Besides this, sometimes unreduced gametes are produced in low frequencies in some plant species due to action of certain specific genes. These unreduced gametes thus fuse together to form autopolyploids. Autopolyploidy has played a significant role in the creation of genetic resources with various ploidy levels in some plant species like banana, apple, water melon, potato, sweet potato, oat and alfalfa. Nearly 70% species of the grass family are natural polyploids.

(d) Artificial Means

Genetic resources of a species can be artificially created by hybridization or by induced mutagenesis and induced polyploidy. Several *in vitro* techniques are now available for producing different genotypes and thus enhancing genetic resources. These include the exploitation of somaclonal variation, protoclonal variation and gametoclonal variation occurring in plants. Artificial gene transfer through modern techniques of genetic engineering can also lead to the production of a wide range of genetic resources created artificially by mankind to meet their own needs.

Reasons for Loss of Genetic Resources in North East India

Genetic resources are being lost every day from the world. Even the species are getting extinct every day. It has been estimated that about 140 biological species including plants, animals and microbes are being lost everyday forever from the earth and this implies the loss of one species every 12 minutes. The North Eastern Region of India is also not free from the loss of genetic resources. The main reasons for the loss of genetic resources in North East India are briefly outlined below:

1. Shifting cultivation (jhuming) practised by the hill tribes in North East India.
2. Over-exploitation of some plant species by man for food, fibre and medicinal properties.
3. Rampant deforestation for fuel, timber and space in all the states in this region.
4. Habitat degradation of the natural flora due to human encroachment in forest land.
5. Cultivation of a few high yielding crop varieties by farmers over large areas by sidelining the low yielding land races and thereby causing great loss of genetic resources.
6. Unchecked population growth in some North Eastern States due to increased human birth rate and massive immigration from adjoining countries has put excess pressure on land and forest. This has also caused the loss of genetic resources in North East India.
7. Pollution of land, water and air from industrial wastes, automobile exhausts and indiscriminate use of fertilizers and pesticides.
8. Biological invasions by some notorious weed species like *Mikania, Parthenium, Lantana*, *Ipomoea* etc. causing loss of genetic resources in forests.
9. Lack of public sensibility towards genetic resources.

Means to Conserve Genetic Resources of North East India

The loss of genetic resources of any region is a great loss in terms of goods and services, which they offer to mankind. In North East India, the loss of plant genetic resources can be minimized, if not completely checked, by adopting the following means:

1. Restricting shifting cultivation in hilly areas by enacting stringent laws and implementing them.
2. Preventing over-exploitation of plant species by the people despite their great utility to human beings.
3. Reducing deforestation and at the same time enhancing afforestation measures.
4. Checking habitat destruction and restricting human encroachment of forest land.

5. Maintaining indigenous genetic resources in small areas of crop fields along with large-scale cultivation of high yielding varieties through farmers' participatory approach.
6. Checking human population growth for decreasing the burden on per unit of forest land and crop field.
7. Checking the pollution of land, water, and air through various measures namely optimum use of pesticides and chemical fertilizers, detoxification of industrial wastes and purification of automobiles exhausts prior to release to environment.
8. Preventing introduction of alien plant species and their genetic resources to the North Eastern Region of India.
9. Creating public awareness about the conservation of genetic resources through print and electronic media, seminars, workshop, group meet, face-to-face communication, adult education and formal education system etc.
10. Collecting the genetic resources of all plant species and conserving them in universities and research institutes.
11. *In situ* conservation of genetic resources in national parks and biosphere reserves and
12. Motivating bureaucrats about the conservation of genetic resources.

The Concept of Sustainable Management of Genetic Resources

The term "Sustainable Management" has been derived from two words "sustainability" and "management". Sustainability refers to a dynamic process which meets the needs of the present generation from a resource base without sacrificing the prospects for meeting the needs of future generations. Management means the act of managing a resource in a scientific way. In simple words, sustainability means a continuous flow of goods and services from a resource without causing any detriment to it.

Sustainable management of genetic resources implies that the needs of present generation can be met from the genetic resources without reducing the ability of the genetic resources to meet the needs of our future generations. Sustainable management concept is always associated with a long-term impact. Sustainable management of genetic resources involves a balanced management system of genetic resources and their ecosystems to provide continuous service to mankind over generations.

Sustainable management of genetic resources has several characteristics e.g.

a) It is an eco-friendly approach in the management of genetic resources because ecological balance is given importance in management strategy.

b) It does not involve any threat to genetic resources despite their current use in human welfare.

c) Sustainable management ensures that the rate of extraction from genetic resources does not exceed their rate of regeneration.

d) It promotes stable ecology.

e) It does not create any adverse impact on our environment due to use or exploitation of genetic resources.

f) It has a long-term impact in maintaining or improving the genetic resources on earth.

g) Sustainable management of genetic resources follows the trend of maintaining the genetic resources and, therefore, can be called nature - friendly.

h) Sustainable management of genetic resources offers opportunity for further research and development of genetic resources.

i) It also safeguards the existence of the total diversity of genes within a species.

It is easy to use the prefix "sustainable" before the term "management" but it is difficult to make the prefix really meaningful by getting its implications implemented in the management strategy. Sustainable management possibly arises as a middle path between two extreme schools of thought. At one extreme is the naturalist who discourages exploitation of genetic resources and want to keep nature as it is. At the other extreme is the technology driven groups who want to over exploit genetic resources at any cost without considering long-term impact of such over-exploitation on genetic resources. The concept of sustainable management is being increasingly adopted and implemented by the scientists and the policy makers throughout the world in natural resource management. The concept holds great promise in ensuring better life support systems to our offsprings in future.

Our life support systems are under stress and the loss of the biological resources of the country is increasing (Swaminathan, 1990). Good land for crop husbandry is shrinking, while biotic and abiotic threats to food security are expanding. The answer to overcome these threats lies in the sustainable management of plant genetic resources because genetic resources are the store house of valuable genes which can overcome the effects of biotic and abiotic stresses in plants. Sustainable management ensures the maintenance of plant genetic resources and therefore the valuable genes contained therein for potential use by our posterity. So Sustainable management and plant genetic resources are inter-linked in the sense that the former ensures the maintenance of the latter over time.

REFERENCES

Ahmed, T. (1986). Diversity, genetic associations and breeding potentiality of glutinous rice. Assam Agricultural University, Jorhat, India (Ph.D. thesis).

Bhattacharyya, R.K. (1987). Banana and pineapple germplasm situation in Assam. Summer institute on prospect of utilizing plant genetic resources of North East India, Assam Agricultural University, Jorhat, 15 June-4 July.

Bhattacharyya, R.K. (2001). Horticultural crops-fruits and vegetables. Agriculture in Assam, Assam Agricultural University, Jorhat, Assam: 104-113.

Bhattacharyya, S.C. and Dutta, S. (1957). Classification of citrus fruits of Assam. ICAR Monograph No. 20.

Das, G.R., Ahmed, T., Bhattacharyya, H.C. and Barthakur, B.C. (1981). Evaluation of germplasm: I. Characterisation and evaluation of *ahu* rice for important quantitative characters in relation to yield, J. Res., Assam Agricultural University, **2:**156-164.

Dutta, A.C. (1979). General description and economic plants. Botany for Degree Students (5th Ed.), Oxford University Press, Calcutta, India: 835-837.

Ganguli, P.M. (1950). Rice in Assam. Department of Agriculture, Assam State Government.

Garay, L. (1960). On the origin of Orchidaceae. Proc. 3rd World Orchid Conference: 172-195.

Gupta, S.R. (1999). Plant genetic resource-their maintenance and conservation: a need in pulses improvement program. Proceedings of the summer short course on "Recent Improved Methods for Increasing Pulses Production", IIPR, Kanpur: 23-35.

IRRI. (1974). Annual report for 1973. Los Banos, Philippines.

Maiti, S., Hedge, M.R. and Chattopadhyay, S.B. (1988). Handbook of annual oilseed crops. Oxford and IBH Publishing Co. Pvt. Ltd., New Delhi, India.

Neog, M. and Mohan, N.K. (1994). Minor and less-known fruits of Assam. Indian Hort., **39:**28-31.

Paroda, R.S. and Sharma, S.D. (1986). Collection of rice germplasm in India. Current status and future plans. Report presented at the AICRIP workshop, NDUAT, Faizabad.

Pathak, P.K., Pathak, A.K., Ahmed, T. and Roy, A.K. (1995). Status of rice germplasm in Assam, India. International Rice Commission Newsletter **44**:69-75.

Pathak, P.K. (2001). Major cereal crops of Assam. Agriculture in Assam, Assam Agricultural University, Jorhat, Assam: 53-74.

Reddy, T.Y. and Reddi, G.H.S. (1995). Sustainable agriculture. Principles of agronomy (2nd Ed.), Kalyani Publishers, Ludhiana, India: 502-509.

Sarma, B.K., Sharma, B.D. and Hazarika, G.N. (1991). Crop genetic resources of North East India: II, Pulses and Oilseeds – a review, Agric. Rev., **12:** 213-224.

Sharma, B.D., Vellanki, J.M.R., Hakim, K.I. and Singh, R.K. (1971). Primitive and current cultivars of rice in Assam- A rich source of valuable genes. Current Science, **40:**126-128.

Singh, B.D. (1986). Domestication, plant introduction and acclimatization. Plant Breeding Principles and Methods (2nd Ed.), Kalyani Publishers, Ludhiana, India: 15-38.

Singh, F. (1986). Orchids. Ornamental Horticulture in India, ICAR, New Delhi: 127-153.

Swaminathan, M.S. (1990). Sustainable food and nutrition security – challenges ahead. Agronomic Research Towards Sustainable Agriculture (Ed. K.N. Singh and R.P. Singh), Indian Soc. Agron. IARI, New Delhi: 1-6.

Wilson, E.O. (2001). The vanishing biodiversity. The Assam Tribune, Guwahati, Assam, April **21:**4.

Advancing Frontiers of Ecological Researches in India, 695-708, 2006
Edited by A.K. Kandya & Asha Gupta
Published by Bishen Singh Mahendra Pal Singh, Dehra Dun, India

Eco-Physiological Studies on Medicinal Plants IV. Experimental Studies on the Effect of Certain Growth Regulants on the Induction of Roots on Stem Cuttings of *Tylophora indica* (Brum. f.) Merr.

Sandhya Tyagi and Y.D. Tiagi
Department of Botany, College of Science
M.L. Sukhadia University, Udaipur (Raj.), India

Abstract: *Tylophora indica,* usually does not set seeds and fruits under the agroclimatic conditions of Udaipur, Jaipur and Jodhpur. The effect of different concentrations of IAA, NAA, IBA, GA_3, Kinetin and extracts of certain plants which can be profusely propagated by stem cuttings on the initiation of roots on stem cuttings of *Tylophora indica* was studied repeatedly year after years. It was found that some growth regulants at certain concentrations and the extracts of certain plants yield results far superior to the controls, in respect of initiation of roots on stem cuttings. In the following table, the maxima of the various treatments employed are given:-

It can, therefore be concluded that the number of roots induced on a stem cutting over a period of 60 days by certain high concentrations of IBA and the extract of the apical portion of the stem of *Ipomoea fistulosa* were rather spectacular, followed by that of IAA. The results of NAA and kinetin treatments were of medium category while GA_3 treatments were no better than the controls.

S.No.	Chemical	Concentration	Number of roots per cutting induced during 60 days
1.	IAA	1000	41
2.	NAA	100	24
3.	IBA	4000	42
4.	GA_3	100 & 500	3
5.	Kinetin	500	24
6.	Stem extract of *Ipomoea fistulosa*	Extract	48
7.	Control Distilled water	-	0 to 4

INTRODUCTION

The science of medicinal plants was at the peak of its glory in ancient India when people in the West lived in caves. Ayurveda, the science advocating the use of plants for the cure of diseases of man flourished during the Vedic period (2000 B.C. to 800 B.C.), "Vrikshayurveda", written by Parasara during the pre-christian era was the most authentic text-book for pre-medical students in ancient India (Core, 1955). The Atharveda contains a wealth of details regarding the subject of medicinal plants and the diseases against which they were applied. Thus, there appears to be a gap of some 4000 years in the study of the science of medicinal plants in India and Europe. Excepting some important contributions by the Greek philosophers in the early part of the post-christian era, the serious study of the medicinal plants in Europe was made only during the "Age of Herbals" which lasted from 1470 to 1670.

Notwithstanding the advances in the science of medicinal plants in ancient India, much of the earlier work has either been lost or lies in obscurity. The science of Ayurveda suffers from the non-availability of diagnostic descriptions of the medicinal plants and their correct identification and due to this lacuna passed into oblivion for quite sometime. It is only during the last few decades that after realising the importance of the uses and preservation of medicinal plants that Government of India constituted the **"Central Council for Research in Indian Medicine and Homeopathy"** in 1969.

The medicinal properties of *Tylophora indica* (Burm. f.) Merr, were known to the tribal people of the region where it occurred but no standard Sanskrit or Persian works on *Materia Medica* has any mention of it (Chopra, *et al.*, 1965) It was for the first time Sir William Roxburgh (1820-24) who brought its medicinal properties to the notice of allopathic medicine and during the early thirties, this plant species was recommended as a standard emetic in place of ipecacuanha in the Medical College at Madras. It is also now a standard drug in homeopathy against asthama (Verma *et al.*, 1976).

Earlier named as *Tylophora asthamatica* Wight and Arn. (Hooker, 1885), its correct name according to the International Rules of Botanical Nomenclature should be *Tylophora indica* (Burm. f.) Merr. Its naming as *T. asthamatica,* though incorrect shows that it must have had some use in the cure of asthama. Once widespread, throughout India, it is known by various varnacular names such as : Sanskrit- Anthrapachka; Hindi – Antamul, Jangli pikavan; Marathi-Antamul, Kharaki, Rasna, Pitakari, Pitmari; Bengali-Antamul; Malyalam-Vallippala; Telugu-Vettipala, Verripala; Karnataka-Adu Sopa Matta du (literally meaning "that which is not touched by goats); Tamil-Nayppalai, Peyppalai, Nachchuruppan etc. (Chopra, *et al.*, 1965).

Inquiries from practising Ayurvedic physicians shows that it is the roots, known as "antamul" which were used for the cure of biliousness. Since the plant usually does not set seeds and fruits in North India and propagates scantily by stem suckers, the use of roots by ayurvedic physicians seems to be main cause for the disappearance of this plant from wide stretches of this region. However, it was Prof. D.N. Shivpuri and his associates who for the first time popularized the use of its leaves as a powerful broncho-dilator, giving protracted protection to patients prone to bronchial asthama (Shivpuri, 1973, 1975; Shivpuri *et al.*, 1968, 1969, 1972; Mathew and Shivpuri, 1974; Agarwal *et al.*, 1973; Agarwal and Shivpuri, 1971; Mathew *et al.*, 1971). *Tylophora indica* does not set seeds and fruits under the agro-climatic conditions of Rajasthan and Delhi. In an earlier work, the artificial induction of fruit and seed setting was achieved. The present work was undertaken with the objective of experimentally inducing profuse rooting on stem cuttings.

MATERIAL AND METHODS

The experimental material of stem cuttings consisted of the third and the fourth node from the uppermost unfolded pair of leaves. It consisted of one complete internode between these nodes and portions of two internodes, one proximal and the other distal. The proximal internode was immersed for 48 hrs. in various concentrations of IAA, NAA, IBA, GA_3, kinetin and extracts of certain plants in which natural production of rooting is well known, as well as that, of *Helianthus annus* in which there is no natural rooting. 100 grams of the fresh weight of these plants taken from the apical region was thoroughly homogenised in a mixer blender for a total period of half an hour in 100 ml of distilled water. After homogenization, the volume of homogenate was raised to 1000 ml. The homogenate was filtered through Whatman's filter paper number 40. The filtrate was used as the plant extract.

After treatment of the proximal internode, the cuttings were washed in distilled water and then inserted in pots up to a depth of three centimeters. Each treatment was replicated thrice. Stem cuttings immersed in distilled water constituted the control. One kilogram of the garden soil was kept in each pot and 100 ml of tap water was added to each pot on alternate days after the initial drenching by 500 ml of tap water.

The cuttings were carefully taken out from the pots after 60 days. The adhering soil was removed by washing and kept on blotting sheets for the adhering water to drain off for a period of eight hours. The total number of roots appeared on a cutting and their fresh and dry weights were recorded.

OBSERVATIONS

These are given in Tables 1-6; fresh and dry weights were recorded in grams.

Table-1: Showing the Effect of IAA on the Induction of Roots on Stem cuttings of *Tylophora indica* during a period of 60 days.

(Values are mean of three replicates; all weights are in grams)

S.No.	Concentration (ppm)	Number of roots/ cutting	Fresh weight of roots	Dry weight of roots
1.	100	4.66	3.36	1.17
2.	250	6.33	4.66	1.57
3.	500	9.66	7.90	2.68
4.	750	20.66	13.35	4.62
5.	1000	37.66	20.77	7.55
6.	1500	29.66	15.64	5.63
7.	2000	11.00	8.87	3.18
8.	2500	1.00	9.36	0.33
9.	3000	0.00	0.00	0.00
	Control 00	1.33	1.06	0.42

DISCUSSION

Ever since man became conscious of agriculture, he has developed many good varieties of crops, vegetables, fruits and other useful plants by the mere process of selection, generation after generations. During the last one century or so, the technique of hybridization and mutation breeding have also been used to produce improved varieties of economically

Statistical Summary of the Data

Source	MSS	F	S.E.M.±	C.D. 5%	C.D. 1%
Number of Roots:					
Control *vs* Rest	504.166	154.667**			
Rest	507.119	155.593**	1.042	3.097	4.242
Error	3.259				
Fresh Weight					
Control *vs* Rest	186.892	92.430**			
Rest	135.794	67.158**	0.821	2.439	3.341
Error	2.021				
Dry Weight					
Control *vs* Rest	22.698	79.885**			
Rest	17.959	63.208**	0.307	0.914	1.252
Error	0.284				

Table-2: Showing the Effect of NAA on the Induction of Roots on Stem cuttings of *Tylophora indica* during a period of 60 days.

(Values are mean of three replicates; all weights are in grams)

S.No.	Concentration (ppm)	Number of roots/ cutting	Fresh weight of roots	Dry weight of roots
1.	100	22.66	10.43	3.30
2.	250	21.00	9.50	2.93
3.	500	15.00	8.20	3.16
4.	750	13.66	7.46	3.00
5.	1000	10.33	5.03	2.20
6.	1500	3.66	2.33	0.90
7.	2000	1.33	0.37	0.15
8.	2500	0.00	0.00	0.00
9.	3000	0.00	0.00	0.00
	Control 00	1.00	0.70	0.24

important plants. The importance of vegetative propagation in respect of these improved varieties does not need any emphasis (Barua and Barua, 1960; Hartmann *et al.*, 1990). Genetic uniformity of the improved stock can only be maintained by vegetative propagation. Moreover, in the case of *Tylophora indica*. vegetative propagation is the only method of propagation in all those areas where it does not set seeds. In the case of *Tylophora indica,* the different varieties (Chaudhary, 1976) differ statistically not only in respect of certain characters of microanatomy but

Statistical Summary of the Data

Source	MSS	F	S.E.M.±	C.D. 5%	C.D. 1%
Number of Roots					
Control *vs* Rest	348.595	363.751**			
Rest	195.317	203.809**	0.565	1.694	2.334
Error	0.958				
Fresh Weight					
Control *vs* Rest	79.158	396.919**			
Rest	42.338	212.294**	0.257	0.773	1.065
Error	0.199				
Dry Weight					
Control *vs* Rest	10.460	450.863**			
Rest	4.598	198.209**	0.087	0.263	0.363
Error	0.023				

Table-3: Showing the Effect of IBA on the Induction of Roots on Stem cuttings of *Tylophora indica* during a period of 60 days.

(Values are mean of three replicates; all weights are in grams)

S.No.	Concentration (ppm)	Number of roots/cutting	Fresh weight of roots	Dry weight of roots
1.	100	7.33	5.16	2.16
2.	250	6.66	4.60	1.96
3.	300	7.66	5.53	2.40
4.	500	14.00	9.96	4.16
5.	1000	20.66	13.66	6.63
6.	2000	30.33	21.46	8.66
7.	3000	37.66	26.43	11.30
8.	4000	41.33	29.00	12.30
9.	5000	14.00	9.73	4.13
10.	7500	1.00	0.73	0.29
	Control 00	3.33	2.43	0.94

also in respect of their crude alkaloid content (Tyagi, 1988). Since *Tylophora indica* is a cross-pollinated plant, the importance of induction of roots on stem cutting does not need any justification.

The subject of vegetative propagation by stem cuttings has been excellently reviewed by Adriance and Brison (1955), Hartmann *et al.*

Statistical Summary of The Data

Source	MSS	F	S.E.M. ±	C.D. 5%	C.D. 1%
Number of Roots:					
Control *vs* Rest	592.012	152.626**			
Rest	586.800	151.284**	1.137	3.335	3.335
Error	3.878				
Fresh Weight					
Control *vs* Rest	282.985	147.176**			
Rest	286.370	149.977**	0.800	2.348	2.367
Error	1.922				
Dry Weight					
Control *vs* Rest	53.659	124.686**			
Rest	51.499	119.668**	0.376	1.110	1.120
Error	0.430				

Table-4: Showing the Effect of GA_3 on the Induction of Roots on Stem cuttings of *Tylophora indica* during a period of 60 days.

(Values are mean of three replicates; all weights are in grams)

S.No.	Concentration (ppm)	Number of roots/cutting	Fresh weight of roots	Dry weight of roots
1.	100	2.33	1.83	0.75
2.	250	2.00	1.63	0.57
3.	500	2.33	1.93	0.63
4.	750	1.67	1.34	0.44
5.	1000	1.67	1.53	0.48
6.	1500	1.33	1.30	0.41
7.	2000	1.00	0.81	0.33
8.	2500	0.33	0.24	0.11
9.	3000	0.00	0.00	0.00
	Control 00	1.67	1.30	0.47

(1993) and Mahlstede and Haber (1957, 1966). It would indeed be marvellous if some one could discover a chemical which will induce rooting at will in stem cutting in all plant species. Though highly spectacular results have been obtained by treatment of stem cuttings of *Tylophora indica* with extract of apical portion of the plants of *Ipomoea fistulosa* and certain high concentrations of IBA, one would still hesitate to call them

Statistical Summary of the Data

Source	MSS	F	S.E.M.±	C.D. 5%	C.D. 1%
Number of Roots:					
Control *vs* Rest	0.018	0.071 NS			
Rest	0.404	5.418**	0.294	0.873	1.196
Error	0.259				
Fresh Weight					
Control *vs* Rest	0.002	0.014 NS			
Rest	0.914	4.801**	0.252	0.748	1.025
Error	0.198				
Dry Weight					
Control *vs* Rest	0.000017	0.0007 NS			
Rest	0.1136	5.326**	0.084	0.250	0.343
Error	0.021359				

Table-5: Showing the Effect of KINETIN on the Induction of Roots on Stem cuttings of *Tylophora indica* during a period of 60 days.

(Values are mean of three replicates; all weights are in grams)

S.No.	Concentration (ppm)	Number of roots/cutting	Fresh weight of roots	Dry weight of roots
1.	100	21.33	13.30	6.66
2.	500	23.33	13.83	6.93
3.	1000	8.33	10.13	5.13
4.	2000	6.66	6.36	3.20
5.	3000	7.66	6.93	3.40
6.	5000	1.66	0.66	0.36
	Control 00	15.33	9.80	4.86

as marvel chemical for induction of roots on stem cuttings. That the initiation of roots is due to chemical substances formed in the shoot apex and which then move down towards the base of the stem was long ago put forward by Duhamel (1758) and Sachs (1880, 1882, also Went, 1928, 1938) Later workers have also concluded likewise (Van-der-Lek, 1925; Sinha *et al.* 1962 Parihar; 1948, Bouillenne and Bouillenne, 1952; Pennock and Maldonago, 1963; Chadwick, 1930; Challenger *et al.,* 1964; Chandler, 1959; Chauhan and Maheshwari, 1970; Cheesman and Spencer, 1936; Clay, 1964; Cooper, 1944; Doran and Bailey, 1957; Gardner and Piper, 1943;

Statistical Summary of The Data

Source	MSS	F	S.E.M.±	C.D. 5%	C.D. 1%
Number of Roots:					
Control *vs* Rest	37.785	21.4459**			
Rest	228.900	129.9162**	0.766	2.324	3.226
Error	1.7619				
Fresh Weight					
Control *vs* Rest	4.089	26.1039**			
Rest	73.505	469.1865**	0.228	0.693	0.962
Error	0.156				
Dry Weight					
Control *vs* Rest	0.875	17.6124**			
Rest	18.431	371.0007**	0.128	0.390	0.541
Error	0.0497				

Table-6: Showing the Effect of certain plant extracts on the Induction of Roots on Stem cuttings of *Tylophora indica* during a period of 60 days.

(Values are mean of three replicates; all weights are in grams)

S.No.	Plant Extract	Number of roots/cutting	Fresh weight of roots	Dry weight of roots
1.	*Ipomoea fistulosa*	44.00	26.76	12.46
2.	Bryophyllum calycimum	39.66	26.33	11.73
3.	*Calotropis procera*	3.00	2.20	9.90
4.	Opuntia dillenii	39.33	26.06	11.70
5.	*Euphorbia caducifolia*	31.66	20.96	9.36
6.	*Lantana indica*	24.00	16.67	6.93
7.	*Helianthus annuus*	1.00	6.63	0.25
8.	*Croton gossypifolia*	23.66	15.63	6.63
	Control	1.66	7.46	0.52

Garner, 1958; Girouard and Hess, 1964, 1966; Grace, 1939; Hess, 1961, 1967; Migdon and Westwood, 1963). West (1928, 1929). Thimann and Went (1934) showed that auxins could induce root formation in stem cuttings in a number of species. Acidic root promoting substances have been reported in *Picea* and *Chamaecyparis* by Gangwar and Singh (1965). Similar results were obtained by Thimann and Koepfli (1935), Cooper (1936) and Thakurta and Dutt (1941) using IAA. Thimann (1935) and Zimmermann and Wilcoxon (1935) obtained similar results using NAA.

Statistical Summary of the Data

Source	MSS	F	S.E.M.±	C.D. 5%	C.D. 1%
Number of Roots:					
Control *vs* Rest	1562.782	124.4694**			
Rest	806.470	64.2321**	2.045	6.078	8.326
Error	12.555				
Fresh Weight					
Control *vs* Rest	237.909	12.0842**			
Rest	329.263	16.7383**	2.560	7.608	10.422
Error	19.671				
Dry Weight					
Control *vs* Rest	129.874	106.6585**			
Rest	68.919	56.5994**	0.637	1.893	2.593
Error	1.217				

However, later workers have found IBA to be more spectacular in this respect. Blommaert (1959) found that treatment of stem cuttings with 4000 ppm solutions of IBA induced successful rooting in not less than 80% stem cuttings of guava. Similar results have been obtained by treatment of stem cuttings with high concentrations of IBA in two varieties of guava by Jolicoeur (1962), in some varieties of pear by Floor (1963), in several species of tropical plants by Manohar (1966), in some soft wood cuttings of plum by Tarasenko and Vesitev (1966), in guava by Bhandary and Mukherjee (1969) and in *Ixora rosea* by Singh (1979) and in grape vines by Flervo and Kovalenko (1952). Calma and Richey (1930) showed that the amount of foliage on stem cuttings too had a influence on rooting in stem cuttings of *Coleus*. Thus, the role of IBA, IAA and NAA in respect of induction of rooting on stem cutting appears to be well established.

ACKNOWLEDGEMENT

We are grateful to Prof. N.C. Aery for help in the preparation of this paper and D.S.T., New Delhi for massive financial support in initial stages of the project under which this work was carried out.

REFERENCES

Adriance, G.W. and Brison, F.R. (1955). Propagation of horticultural plants. McGraw Hill Book Company, New York.

Agarwal, M.K. and Shivpuri, D.N. (1971). Preliminary studies on bronchial tolerance to histamine and methacholine chloride inhalation challenge before and after six days treatment with the leaves of *Tylophora indica.* Aspects Allergy Appl. Immunol. **5**:62-70.

Agarwal, M.K., Shivpuri, D.N., Gupta Manju and Ray, G.N. (1973). Studies on the effect of *Tylophora indica* on the immunological reaction in rats and *in vitro*. Aspects Allergy Appl. Immunol. **6**:107.

Barua, D.N. and Barua, S.C. (1960). Vegetative propagation of the tea plant in Northern India I. The choice of cuttings. J. Indian bot. Soc. **39**:480-489.

Bhandary, K.R. and Mukherjee, S.K. (1969). Relation between investigation and rooting of guava (*Psidium guajava* L.) stem cuttings. Curr. Sci. **38**:197-198.

Blommaert, K.L.J. (1959). Propagate guavas by means of leaf-bud cuttings. Fmg. S. Afr. **35**:29-32.

*Bouillenne, R. and Bouillenne, W.M. (1952). Experiments on root formation. Non-polar transport of rhizocaline in the epicotyls of *Impations balsamina* and *Phaseolus vulgaris* C.R. Rech. I.R.S.I.A. No. 6. Trav. Centre Rech. Hormones Veg. 95-11 (Hort. Abst. **22**:2081).

Calma, V.C. and Richey, H.W. (1930). Influence of amount of foliage on rooting of *Coleus* cuttings. Proc. Am. Soc. Hort. Sci. **27**:457-462.

Chaudhary, B.L. (1976). Microanatomy and varietal differentiation in *Tylophora indica* (Burm.f.) Merr. Geobios **3**:190-192.

Chadwick, L.C. (1930). The influence of the position of the basal cut on the rooting and arrangement of roots on deciduous softwood cuttings. Proc. Am. Soc. Hort. Sci. **27**: 487-494.

*Challenger, S., Jacey, H.J. and Howard, B.H. (1964). The demonstration of root promoting substances in apple and plum root stocks. Annu. Rep. E. Malling Res. Stn. p. 124.

Chandler, C. (1959). The propagation of *Larix* from softwood cuttings. Contr. Boyce. Thomp. Inst. **20**:231-238.

Chauhan, K.S. and Maheshwari, L.D. (1970). Effect of certain plant regulators, seasons and types of cutting on root initiation and vegetative growth in stem cuttings of peach variety Sharbati. Indian J. Hort. **27**:136-140.

*Cheesman, E.E. and Spencer, G.E.L. (1936). The vegetative propagation of Cacao VI. General notes on technique with cuttings. VII. Root system of cuttings. Fifth Ann. Rep. on Cacao Res. C.C.T.A., Trin. pp. 4-7 (Hort. Abst. 6 : 573).

Chopra, R.N., Badhwar, R.L. and Ghosh, S. (1965). Poisonous Plants of India. Vol. I. ICAR, New Delhi.

*Clay, D.W.T. (1964). Vegetative propagation of Kola (Colanitidavant) Schott and Endl. Trop. Agri. Trin. **41**:61-68.

Cooper, W.C. (1936). Transport of root forming hormones in woody cuttings. Plant Physiol. **11**:779-793.

Cooper, W.C. (1944). Vegetative propagation of *Derris* and *Lonchocarpus* with the aid of growth substances. Bot. Gaz. 1061.

Core, E.L. (1955). Plant Taxonomy. Prentice-Hall, Inc., Englewood Cliffs N.J., U.S.A.

*Doran, W.L. and Bailey, J.S. (1957). Vegetative propagation of beach plums. Nat. Hort. Mag. **36**:349-352. (Hort. Abst. 28 : 1106-1958).

*Flervo, A.F. and Kovalenko, E.I. (1952). Anatomical alteration in tissues of grape vine cuttings after treatment with d-naphthalene-acetic acid. Doklady Akad. Nauk. S.S.S.R. **85**:433-436. (Hort. Abst. 22 : 329).

*Floor, J. (1963). The propagation of pear varieties by cuttings. Meded. Dir. Tuinb. **26**:687-691 (Hort. Abst. 34 : 2208).

Gangwar, R.P. and Singh, S.N. (1965). Acidic root promoting growth inhibitor (s) found in *Picea* and *Chamaecyparis.* J. Amer. Soc. Hort. Sci. **97**:574-578.

Gardner, F.E. and Piper, R.B. (1943). Ease of propagation of some of the subtropical fruits by cuttings from young seedlings. Proc. Fla. St. Hort. Sci. LVI : 124-126.

Garner, R.J. (1958). Propagation of woody plants by cuttings. Recent research and its application to fruit tree rootstock. J. Roy. Hort. Soc. **83(8)**:335-343.

*Girouard, R.M. and Hess, C.E. (1964). The diffusion of root promoting substances from stems of *Hedera helix.* Combined Proc. Int. Plant Propagation Soc. **14**:162-166 (Hort. Abst. 36 : 5185).

Girouard, B.M. and Hess, C.E. (1966). Distribution and mobility of rooting 'Cofactors' in juvenile and adult growth phases of *Hedera helix*. Proc. 17th In. Hort. Congr. **1**:370.

Hartmann, H.T., Kester, D.E. and Davies (Jr.) F.T. (1993). Plant Propagation : Principles and Practices. Prentice – Hall of India, New Delhi.

Hess, C.E. (1961). The physiology of root initiation in easy and difficult to root cuttings. Hormolog. **3**:3-6.

Hess, C.E. (1967). Root initiation and juvenility. A possible implication of terpene compounds. Proc. XVII Int. Hort. Congr. **3**:433-451.

Higdon, R.J. and Westwood, M.N. (1963). Some factors affecting the rooting of hardwood pear cuttings. Proc. Amer. Soc. Hort. Sci. **83**:193-198.

Hooker, J.D. (1885). The Flora of British India. Vol. IV, Ashford, Kent.

*Jolicoeur, J.A. (1962). The rooting of guava cuttings treated with hormones under mist in Haiti. Proc. Caribb. Reg. Amer. Soc. Hort. Sci. 57-59 (Hort. Abst. **34**:7704).

Mahlstede, J.P. and Haber, E.S. (1957). Plant propagation. John Wiley and Sons, New York.

Mahlstede, J.P. and Haber, E.S. (1966). Plant Propagation. London, Sydney.

Manohar, M.S. (1966). Influence of source on the performance of softwood cuttings of guava. Allah. Fmr. **11**:191-194.

Mathew, K.S., Agarwal, M.K. and Shivpuri, D.N. (1971). Effect of six doses of alcoholic extract of *Tylophora indica* on bronchial sensitivity to inhalation challenges with specific allergens in 24 symptom free asthama patients. Aspects Allergy Appl. Immunol. **5**:57.

Mathew, K.K. and Shivpuri, D.N. (1974). Treatment of asthama with alkaloids of *Tylophora indica*- A double blind study. Aspects Allergy Appl. Immunol. **7**·166-179.

Parihar, N.S. (1948). Hormonal control of plant growth. M.Sc. Thesis, Allahabad University.

*Pennock, W. and Maldonago, G. (1963). The propagation of guavas from stem cuttings. J. Agric. Univ. P. Rico **47**:280-289 (Hort. Abst. 34 : 5779).

*Sachs, J. (1880 and 1882). Stoff und Form der Pflanzenorgane I and II. Arb. Bot. Inst. Wurzburg. **2**:452 and 689.

Shivpuri, D.N. (1973). A new approach in the treatment of asthama with *Tylophora indica.* Aspects Allergy Appl. Immunol. **6**:317-361.

Shivpuri, D.N. (1975). Studies on a new anti-asthamatic *Tylophora indica* in the treatment of bronchial asthama. Aspects of Allergy and Applied Immunology **8**:61-94.

Shivpuri, D.N. and Agarwal, M.K. (1973). Effect of *Tylophora indica* on bronchial tolerance to inhalation challenges with specific allergens. Ann. Allergy. **31**:87.

Shivpuri, D.N., Menon, M.P.S. and Parkash, D. (1968). Preliminary studies on leaves of *Tylophora indica* in the treatment of asthama and allergic rhinitis. Jour. Assoc. Physc., India **16**:9.

Shivpuri, D.N., Menon, M.P.S. and Parkash, D. (1969). Crossover double-blind study on leaves of *Tylophora indica* in the treatment of asthama and allergic rhinitis. J. Allergy **43**:145.

Shivpuri, D.N., Singhal, S.C. and Parkash, D. (1972). Treatment of asthama with an alcoholic extract of *tylophora indica*. A crossover double blind study. Ann Allergy **30**:407.

Singh, S.P. (1979). Note on auxins on root formation in *Ixora rosea* Sims. cuttings during winter under Intermittent Mist. Curr. Agri. **3**:225-227.

Sinha, R.B., Mallick, P.C. and Chhonkar, V.S. (1962). Effect of plant regulators on the rootage of guava cuttings. Indian J. Hort. **19**:120-127.

*Tarasenko, M.T. and Vasitev, B.S. (1966). The rooting of cuttings under artificial mist. Vestn. Sel'.-hoz. Nauki **11**:112-115 (Hort. Abst. 37 : 2257).

Taylor, G.G. and Odum, R.E. (1970). Some biochemical compounds associated with rooting of *Carya illinoensis* stem cuttings. J. Amer. Soc. Hort. Sci. **95**:146-151.

Thakurta, A.G. and Dutt, B.K. (1941). Vegetative propagation of mango from gootes (marcotte) and cuttings by treatment with high concentration auxin. Curr. Sci. **10**:297.

*Thimann, K.V. (1935). On an analysis of activity of two growth promoting substances on plant tissues. Proc. Kon. Ned. Akad. Wet. **38**:896-912.

*Thimann, K.V. and Behnke-Rogers, J. (1947). The use of auxins in the rooting of woody cuttings. Harvard Forest, Petersham, Mass.

Thimann, K.V. and Koepfil, J.B. (1935). Identity of the growth promoting and root forming substances of plants. Nature **135**:101-102.

*Thimann, K.V. and Went, F.W. (1934). On the chemical nature of the root forming hormone. Proc. Kon. Ned. Akad. Wet. **38**:896-912.

*Van der Lek, H.A.A. (1925). Over de Wortelvorming Van houtige stekken. Diss. Utrecht, pp. 230.

Varma, P.N., Vikramaditya and Singh, V.P. (1976). *Tylophora indica,* A new drug in homoeopathy. The Hahnemannian Gleanings **43**:103-124.

*Went, F.W. (1928). On a substance causing root formation. Proc. Kon. Ned. Akad. Wet. **32**:35-39.

Went, F.W. (1938). Specific factors other than auxin affecting growth and root formation. Plant Physiol. **13**:55.

Zimmerman, P.W. and Wilcoxon, F. (1935). Several chemical growth substances which cause initiation of roots and other responses in plants. Contrib. Boyce Thompson Inst. **7**:209-229.

Advancing Frontiers of Ecological Researches in India, 709-716, 2006
Edited by A.K. Kandya & Asha Gupta
Published by Bishen Singh Mahendra Pal Singh, Dehra Dun, India

Increasing Production Potential of Degraded Lands in Bundelkhand Through Agroforestry

S. R. Gupta and Vinod Kumar
Bipin Bihari (P.G.) College, Jhansi (U.P.) India

Abstract: The hostile climatic conditions, improper land use, coupled with low level of soil fertility, leads to low productivity of the region. The cropping intensity, in general, is low because of poor availability of water resources and problematic soils. Augmenting the productivity on sustainable basis is extremely important for development of the Bundelkhand region.

Under the above circumstances it can be seen that various alternate land use technology/agroforestry systems can increase the biomass production including food, fodder, fuel and timber, and also conserve and improve soil and water resources, thereby helping in sustainability of the resources poor farmers in the Bundelkhand region. Keeping in view of this, it is extremely important that in future developmental strategies and planning of the region, agroforestry component should be given more attention.

INTRODUCTION

Bundelkhand region of Uttar Pradesh is Sub-Zone I (Planning Commission, 1989) in the agro-climate region (Central Plateau and Hills Region). This area lies between 24°11′ - 26° 27′ N latitude and 78° 17′ - 80° 30′ E longitude comprises of Jhansi, Lalitpur, Hamirpur, Banda, and Jalaun Districts falling in the south-west corner of U.P. and separated from the mainland by the Yamuna river. The region is spread over 29.67 thousand sq. km. (2.967 m ha) which is about 11 per cent of the total geographical area of the state. The area is an undulating terrain dotted regularly with rock outcrops. As a whole, the region has very poor vegetative cover. It is faced with serious threat of soil erosion which has rendered extensive areas as westeland, some irreversible in the form of ravines. High surface run-off (34% of the total rainfall) promoted by barren land surface is the major cause of soil erosion and frequent drought.

Climate : The climate of the region is semi-arid and is characterized by long hot summer, highly erratic rainfall and a short mild winter. The annual rainfall ranges from 800 to 1150 mm with yearly mean of 936 mm especially for Jhansi. About 90 percent of the total precipitation is received from mid June to end of September with occasional showers during winter months of November to February. The distribution of rainfall is often erratic and even wet months (July and August) many a times experience long dry spells. The average annual temperatures of Bundelkhand are uniformly high and their mean monthly values vary considerably. Diurnal variation in temperature is quite high. May an June are the hottest months. The maximum recorded temperature on a particular day often touches 47°-49°C and minimum of 1°C recorded in the winter months. The frost are usually absent in this part of the country. The annual potential euapo-transpiration varies from 1400-1700 mm (Singh and Hazra, 1996).

Topography and Soils : The soils of the tracts are entirely different from those of the state as a whole, since these differ geographically from the rest in being non-alluvial nature. The general landscape of Bundelkhand is that of a bare undulating plain with irregular rocky hills covering to a level expanse of black soil towards the north reaching upto the river Yamuna. On the extreme south-east lie the Vindhyan ranges. The general slope of the region is from north to south and as one passes through the level plains on the south of the Yamuna to the rocky hills, the fertile black soil converges into coarse-grained, red soil inter-spersed with brownish to black patches.

The soils of this region have developed from Vinhyan rocky abounding in gneiss and granites of the Deccan trap with highly ferruginous beds and often soft limestone. The soils can be divided into two broad groups, *viz.*, red and black soils. Four soil associations have been recognized in the area which include (1) Bundelkhand-coarse-grained-reddish brown soils, (2) Bundelkhand-coarse-grained-grey to grayish brown soils, (3) Bundelkhand-clay-loam-black soils and (4) Bundelkhand-fine clayey-black soils. Locally these soils are termed as *Rakar, Parwa, Kabar* and *Mar*, respectively. The red soils (*Rakar and Parwa*) belong to the order Ultisols, Alfisols and Inceptisols and the black soils (*Kabar and Mar*) are Vertisols and grouped under the great groups Pellusterts, Chromusterts and Ustochrepts (Singh and Hazra, 1996).

The *Rakar* soils are residual, slightly acidic, coarse grained, shallow and excessively permeable soils occupying higher elevations, *Parwa* soils are alluvial, mildly alkaline, very deep soils with free $CaCO_3$ accumulation at lower depths. The black soils (*Kabar and Mar*) are medium to very deep soils confined to lowlying landscapes having fine texture and the remarkable property of shrinkage and swelling on drying and wetting.

Drainage : The river Yamuna flows along the northern boundary of the region. The other important rivers flowing from south towards north are the Sindh, the Betwa, the Ken and the Baghain which ultimately terminates into the Yamuna and several tributaries like Sahjad, Jamni, Pahuj and Dhasan and this contribute to the river system networking in the region. Most of these rivers forms districts/tehsil boundaries. Most of these are seasonal and become almost dry during the late winter and summer seasons. Stagnated water bodies both natural and created (tanks and lakes) supported by bunds also provide water for different purposes (Tyagi, 1997).

Vegetation : Bundelkhand is an ecologically degraded region. The natural vegetation comprises of tropical dry deciduous and thorn forest species and *Sehima-Dichanthium* grass cover. Formerly dense forests were present in the region particularly in the areas comprising of Vindhyan tableland and hill ranges spreading over the central and southern Bundelkhand. A thick forest cover existed along the rivers covering ravine lands in the north, but human interference has brought about destruction of vegetation. Most of the forest is now confined along the banks of rivers, Pahuj, Betwa and Yamuna. The important trees in these forests are Babul (*Acacia nilotica*), Dhak or Palas (*Butea monosperma*), Ber (*Zizyphus spp.*), Khair (*Acacia catechu*), Tendu (*Diospyrus melanoxylon*), Mahua (*Madhuca indica*), Semal (*Salmelia malabarica*), Kardhai (*Anogeissus pendula*), Teak (*Tectona grandis*) and Shisham (*Dalbergia sissoo*). Many species of shrubs and grasses represent the secondary vegetation succession. Among the most common grasses. Doob (*Cynodon dectylon*), Kans (*Saccharum spontaneum*), Lumpa (*Heteropogon contortus*), Musel (*Iseilema laxum*), Kail (*Dichanthium annulatum*), Guner (*Themeda quadrivolvis*) and Sen (*Sehima nervosum*). Extensive grasslands comes up during rainy season. After harvesting grasses hay is prepared for use in winter and summer (Srivastava, 1981).

Land Degradation : Out of total geographical area, nearly 50 per cent area suffers from varying degree of land degradation. Most of the land in the region has ridges to valley formation with slope grades ranging from 1 to 15 per cent. Excessive surface run-off leads to heavy soil erosion and formation of gullies and ravines.

Land Use Pattern : Table 1 shows that 81 per cent of the geographical area is arable. 12 per cent area devoted to fallow lands, while 11 per cent form cultivable wasteland. Net sown area (NSA) is 1885 thousand ha, which constitutes 63.5 per cent of the total geographical area. The forest

Table-1: Land use statistics of Bundelkhand region of U.P.

Land Category	Area (Thousand ha)	% Area of Total
Total Geographical Area	2967	
Cultivable Lands	2394	80.9
Fallow Lands	244	12.1
Forest	239	8.1
Pasture Lands	9	0.3
Usar Lands	128	4.3
Cultivable Wastelands	332	11.2
Net Sown Area (NSA)	1885	63.5
Irrigated Area (% of NSA)	26.6	-
Cropping Intensity (%)	111.0	-

(*Source* : Singh *et. al.*, 1993)

and pasture land occupy 8.0 and 0.3 per cent of the geographical area. Cropping intensity is 111 per cent with 26.6 per cent of the net sown area being irrigated. Canals and wells are the major sources of irrigation (84%). Tanks (8.5%), bandhis, check dams and bore wells (all together 7.5%) are other source of irrigation.

Farming Systems : There are two wide spread farming systems in the region, based on the nature of the soil, climatic characteristics and degree of assured irrigation available for farming.

Cropped Area and Cropping Systems : The dominant farming systems in the Bundelkhand is monocropping which is practiced on red and black soils during *kharif* and *rabi* seasons, respectively. The important crops of the region are sorghum, paddy, maize, millets, wheat and barley among cereals; pigeonpea, gram, lentil and mustard among oilseeds and sugarcane and potato among commercial crops of the total cropped area of 2394 thousand ha, *kharif* crops occupy 35 per cent whereas *rabi* crops occupy 65 per.cent.

In general, the productivity of all crops in this region is low, the average yield of chief cereals i.e. wheat, sorghum and paddy is far below the national average.

Bundelkhand region is, by and large, monocropped. Mixed cropping is a rule rather than exception. Sorghum is normally grown mixed with pigeonpea and sesame with black gram. Similarly, during *rabi* season mixed cropping of wheat/barley + gram and gram + linseed is a common practices.

Mixed Farming : Mixed farming (agriculture and animal husbandry) is an important enterprise of the Bundelkhand region. Almost all the farmers (95%) keep animals as a source of additional income. Cattles, buffaloes and goats are the principal livestock. On an average small farmers keeps 8 to 10 animals while large farmers keep 3 to 4 animals of good bread (Singh *et. al.*, 1993).

Constraints of the Region: The region has a number of natural constraints which badly affect agricultural as well as other spheres of development.

- Most of the rivers in Bundelkhand region flow only during the monsoon season and even the main rivers face scarcity of water during winter and summer seasons. It creates problem for irrigation as well as drinking water for animals and human beings.
- Climate conditions are also not much favourable for agricultural development. The rainfall pattern, which is very erratic and uncertain is responsible for creating water scarcity and crop failures. Due to heavy rainfall and uneven rocky topography, most of the rain water drains out side the region.
- The temperature starts rising very early which results in the early maturity of *rabi* crops. The temperature rises very high (47-49°C) during summer. Thus, there is considerable variation in the temperature of winter and summer seasons. It becomes difficult to take *jayad* crops due to extreme heat and paucity of irrigation resources.
- The region has acute problem of soil erosion due to thick depth of red soils of upland areas and of overgrazing on such lands. This leads to problems during rainy season when the water from heavy rainfall cuts out gullies in the naked, undulating areas resulting in heavy loss of soil. The soil thus collects into dams/tanks/ponds creating silting and reducing water storage capacity.
- The cropping pattern is affected by the soil type and the availability of irrigation resources. Only 20 per cent area is sown more than once. The crop varieties suited for dryland farming may be the only alternatives in such a region.

- Livestock is a major resource of the region but it is of poor breed having low milk yield. However, there is noticeable change from unproductive cattle to productive buffaloes. But poor feeding is again a problem in the livestock rearing. Most of the livestock depends upon the degraded grazing lands and crop residues which are nutritionally poor (Singh and Tyagi, 1996).

Role and relevance of Agroforestry : Degraded lands with rocky substratum, hostile climatic conditions, undulating terrain, sloppy land, vast area under barren and small hillocks, poor fertility of the soil, poor status of water resources, high fluctuations in ground water table, large area under culturable waste and fallow lands, dominance of monocropping and mixed farming in the farming situations, offer a wide scope for alternate land use/agroforestry in Bundelkhand.

An agroforestry system is more acceptable than exclusive tree farming, since intercropped annuals regulate income when the trees are too young to produce economic benefit. Mature trees bring about more stability in the system because of their innate ability to withstand destructive aberrations of rainfall. The perennial characters helps to make use of the non-seasonal rains which constitutes nearly one-fourth of the total precipitation received in the country.

The agroforestry systems suitable for sustainable development of degraded lands of Bundelkhand are agri-silviculture (crops + trees), agri-horticulture (crops + fruit trees), (agri-horti-silviculture (crops + fruit trees + MPTS), horti-silvipasture (fruit trees + pasture + animals) and silvipasture (trees + pasture + animals).

Agri-silviculture System : In the marginal and submarginal lands agrisilviculture system has been found to be the most recommended under semi-arid conditions. Studies at Jhansi revealed the possibilities of introducing over 800 fast growing multipurpose trees/ha like Ramkanthi (*Acacia cupressiformis*), casuarinas, subabul (*Leucaena leucocephala*) in the initial year (Deb Roy and Gill, 1991). In order to reduce the light competition the canopies are manipulated by pruning and removing 40 per cent plants in the fourth year without adversely affecting crop production. In the third/fourth year, the reduction in crop production was only 10-15 per cent for wheat and upto 20 per cent for gram. For alley cropping, *Leucaena leucocophala, Sesbania sesban* and perennial pigeonpea along with sorghum, pearl millet, ground nut and gram can be grown at the alley width of 7-8 m without much affecting the crop yield. The fuelwood production thus obtained from this system helps to divert the cowdung to the field for increasing biomass production and available fodder is utilized for livestock feeding.

Agri-horti-silviculture System : This system is suitable for the development of marginal land and wastelands. The 100 to 400 fruit trees like ber, anar, kinnow, guava and aonla can be grown successfully along with 200 to 400 MPTS like subabul in association with the crops without much affecting the crop production in the initial years. Subabul is being harvested annually from the second year onwards before sowing *rabi* crops which provide 2.5 to 3.0 t/ha of nutritious green fodder and 1.8 to 2.5 t/ha of dry fuelwood. The relative grain yield in respect of sorghum-wheat rotation was more than 85% of the normal yield in association with guava, kinnow and anar and more than 70% in association with ber in the fourth year with 200 fruit trees and another 200 trees of subabul (Gill and Deb Roy, 1991).

Agri-horticulture System : Under agri-horticulture system on marginal lands and wastelands, it has been observed that improved varieties of ber (Gola, Sev, Banarasi karaka and Local), aonla (Chakiya, Kanchan, Krishna and NA-7) and citrus showed very good promise in association with the sesamum, *urd* bean and pigeonpea. Under rainfed condition both ber and aonla produced a good quantity of fruits from the third/fourth year besides the usual production from dryland crops (Deb Roy, 1993).

Horti-silvipasture System : In the semi-arid rangelands of Bundelkhand tract, there are large number of jharberi shrubs (*Zizyphus nummularia*) available which could be successfully buded *in situ* with improved varieties of ber (Sev, Gola, Banarasi karaka, Umran). In association with ber, fast growing MPTS like Subabul, Anjan and good perennial grasses like *Cenchrus, Lasiurus, Chrysopogon* and *Stylosanthes* spp. could be introduced the system.

It has been observed from the work done at NRCAF, Jhansi that the buded jharberi plants could produce improved varieties of ber from the very second year from 2-4 kg/plant to 8-10 kg/plant in the fourth year giving an income of Rs. 1600 per ha in the second year and Rs. 9880 per ha in the fourth year. (Rai *et. al.*, 1993). This is a low input technology which needs no inputs except protection. Large scale application of this technology on the degraded lands of Bundelkhand region can help the farmers to improve their economy.

Silvipasture System : The silvipastoral systems ia an ideal combination of grasses/legumes and trees for optimizing land productivity, conserving plants, soils and nutrients and producing forage, timber and firewood on a sustainable basis. The gullies, ravinous and under utilized degraded forest lands could be sustainably developed through this system. NRCAF, Jhansi developed an improved silvipastoral system with woody component of *Albizia amara, Dichrostachys cineraria* and *Leucaena leucocephala* in association with *Chrysopogon fulvus* and a mixture of *Stylosanthes hamata* and *S. scabra* giving a forage production of 6 to 8 t/ha and introducing sheep and goats in the very second year of establishment, which showed encouraging results, that it is possible to maintain @ 12 sheep and goats/ha in this system even during summer months without giving any additional concentrates (Deb Roy, 1994).

REFERENCES

Deb Roy, R. (1993). Agroforestry in 2000 A.D. - Concept and role in meeting food, fodder and fuelwood. In : *Agroforestry in 2000 A.D. for the Semi-Arid Tropics* (eds. A.S. Gill, R. Deb Roy and A.K. Bisaria). NRCAF, Jhansi : 23-26.

Deb Roy, R. (1994). Role of agroforestry for sustainable agriculture and environment. In : *Souvenir on Integrated Wasteland Development and Drought Prone Area Development*, D.R.D.A., Jhansi : 25-34.

Deb Roy, R. and A.S. Gill. (1991). Tree growth and crop production under agri-silvicultural system. *Range Mgmt. & Agroforestry*, **12**(2) : 69-78.

Gill, A.S. and R. Deb Roy (1991). *Annual Report*, NRCAF, Jhansi.

Planning Commission. (1989). Agro-climatic Regional Planning - An Overview. Planning Commission, GOI, New Delhi, India : 144 p.

Singh, Daulat, Dhananjai Singh, A.K. Chauhan and R.A. Singh (1993). Report on rainfed farming study in Bundelkhand region of Uttar Pradesh, India World Bank - Winrock International : 75 p.

Singh, D.P. and C.R. Hazra (1996). Soil water resources and land degradation problem-Management of water and soil resources with special reference to Bundelkhand region. In : *Resources Conservation and Integrated Development of Bundelkhand* (eds. P.S. Tomar and A.K. Srivastava) IGFRI, Jhansi : 17-31.

Srivastava, P.C. (1981). Natural forests in Bundelkhand region. *Indian J. Range Mgmt.*, **2** (1 & 2) : 23-29.

Tyagi, R.K. (ed) (1996). *Grassland & Fodder Atlas of Bundelkhand.* IGFRI, Jhansi : pp. 178.

Rai, P., R. Deb Roy and C. K. Bajpai (1993). *Jharberi me budding dwara banjar bhumi ko udhyaniki charagah me badal kar adhik labh uthayen* (In Hindi). Extn. Bull. No. 2, NRCAF, Jhansi.

Advancing Frontiers of Ecological Researches in India, 717-736, 2006
Edited by A.K. Kandya & Asha Gupta
Published by Bishen Singh Mahendra Pal Singh, Dehra Dun, India

Vermiculture Technology–An Overview

B. Narsimha Rao[1] and Anjani Kumar Jha[2]
ICAR Research Complex for NEH Region,
Manipur Centre, Imphal-795 004

From the time human beings started cultivating land, for thousands of years, it remained sustainable. It is only in 20th century that modern technology was developed for cultivating the land and with it came the harmful effect of inorganic chemical inputs. A recent study in USA revealed that risk of cancer due to pesticides is 3 out of 1000 people. Besides, the pesticides and fertilizers persisting in the soil are harmful to the beneficial soil microorganisms and earthworms and thereby resulting in degradation of soil fertility. To restore the soil composition, farmers are now showing inclination to revert to organic farming.

The green revolution is the result of use of improved varieties, higher levels of fertilizers and pesticides. But now it has been realized that the increase in production was achieved at the cost of soil health and the sustainable production at higher levels is possible only by maintaining the fertility of the soil.

Therefore, the soil must be fed in such a way that the activities of beneficial soil organisms, necessary for recycling nutrients and producing humus, are not inhibited. Use of earthworms for degradation of organic wastes and production of vermicompost is becoming popular and is being commercialized.

The importance of organic manures in replenishing soil fertility is well documented. The organic matter content of tropical and subtropical soils is usually very low due to high temperature and intense microbial activity. Therefore, soil needs to be enriched periodically with organic manures. Organic manures were the main nutrient sources in India until the 1950s, when synthetic fertilizers began to gain popularity. Higher food grain production was concomitant with high synthetic nutrient consumption.

---

[1].Senior Scientist (Horticulture), [2] Scientist (Horticulture)

However, the rate of increase in productivity has drastically come down despite increasing consumption of synthetic inputs and spread in coverage of high yielding varieties. In recent years, there has been a renewed emphasis on integrated use of fertilizers and organic manures to sustain high yields and maintain the chemical, physical and microbial status of the soils.

Soil fauna play a prominent role in regulating soil processes and among them the termites and earthworms play a vital role. Vermiculture is the science of cultivating or raising earthworms. They are raised mainly for agricultural or fishing activities. In vermiculture biotechnology, the potentiality of earthworms is being exploited as natural biodegraders of non- toxic organic wastes for soil improvement and nutrient mobilization. Earthworms form a major component of the soil system and these organisms have been efficiently ploughing the land for millions of years and assisting in the recycling of organic nutrients or the efficient growth of the plants. The presence of earthworms in soils is generally indicative of the presence of other faunal representatives contributing to soil fertility. Therefore, it would be appropriate to call earthworms as indicators of soil health.

Earthworms population in organic matter rich soils act as natural bioreactors, harness beneficial soil micro flora, destroy soil pathogens and convert organic wastes into valuable products such as biofertilizers, vitamins, enzymes, antibiotics, growth hormones, proteinaceous worm biomass. Earthworm gut is a tiny fertilizer factory, where raw materials are entering from one end and the finished product vermicast comes out from the other end.

Earthworms and Plant growth

The effect of earthworms on plant growth may be due to several reasons, apart from the presence of macro- and micro-nutrients in their secretions and in vermicasts in considerable quantities. Certain metabolites produced by earthworms may also be responsible to stimulate plant growth. It is considered that earthworms release certain vitamins and similar ubstances into the soil. Though the NPK value of the vermicasts is always r than any chemical fertilizer. Several experiments have proved that sts can promote lush growth of plants. This may be due to plant moters like cytokinins and auxins present in casts.

They maintain stable body temperature through novel temperature regulation mechanism thus accelerating the rates of bioprocesses and preventing enzyme inactivation caused by high temperatures.

Gizzard is a novel colloidal mill in which the feed is ground in to particles smaller than two microns giving thereby an enhanced surface area for microbial processing. They have an inhouse supply of enzymes such as protease, lipase, amylase, cellulase and chitinase which biodegrades complex biomolecules into simple compounds utilizable by the symbiotic gut microflora.

Earthworms have a built-in oxygen plant, which can separate aerial oxygen by chemical absorption into blood hemoglobin. They promote growth of microorganisms in their gut by providing favourable conditions.

Vermicastings contain nutrients in a balanced proportion and are rich in vitamins, enzymes, antibiotics and growth hormones.

Earthworms and soil fertility

Earthworms modify soil physical and chemical and biological properties and it is believed that they enhance nutrient cycling by ingestion of soil and humus and by production of casts. The concentrations of exchangeable Ca, Na, Mg, K and available P and Mo are higher in earthworm casts than in surrounding soil.

Table-1: Composition of the available microbial elements in the cast of earthworms and in the upper layer of ploughed soil.

Particulars	Earthworm cast	Depth of soil layer	
		0-15 cm	20-40 cm
Less on ignition (%)	13.1	9.8	4.9
C:N ratio	14.7	13.8	13.8
NO_3 nitrogen (ppm)	21.9	4.7	1.7
Available P (ppm)	130.0	20.8	8.3
Available K (ppm)	335.0	32.0	27.0
Exchangeable Ca (ppm)	2.78	1.99	1.81
Total Ca (%)	1.19	0.88	0.81
Total Mg (%)	0.545	0.511	0.548
Exchangeable Mg (ppm)	49.2	162	69
pH	7.0	6.36	6.05

Table-2: Comparison of nutrient contents of various organic manures

Manure	Nutrient content (%)		
	N	P_2O_5	K_2O
FYM	0.80	0.41	0.74
Poultry manure	2.87	2.93	2.35
Vermicompost	**1.60**	**2.20**	**0.67**
Pressmud	1.15	2.40	1.98
Saw dust	0.25	0.20	-
Bio gas slurry	1.60	1.50	1.20
Coir pith	0.26	0.05	0.84

The worm cast or excretions of worms form the needed organic fertilizer. One worm weighs about 0.5 to 0.6 g and consumes wastes of the same weight as their body weight in a day. If one million worms exist in one acre, the cast they produce in that area is about 500 kg/day/acre i.e., about 200 tonnes per year. The worm cast contains all the nutrients in available form and in addition, a great deal of organic matter is provided to the soil. Therefore the activity of earthworms is recognized as beneficial for the improvement of soil physical conditions and plant growth. On the other hand, the burrowing activities of earthworms improve air and water penetration into soil. Vermicasts being granular with enhanced porosity and water absorption capacity absorb moisture particularly during night and hold it effectively for releasing it to micro-roots of the vegetation.

Vermicomposting

Vermiculture and vermicomposting are two interlinked and interdependent processes. Composting is the outcome of earthworm activities. We can multiply earthworms for various uses and can obtain vermicast or vermicompost. Earthworms feed upon degradable and decomposable organic matter like animal dung, plant waste etc. Such matters after ingestion are digested, assimilated in intestines and left over are excreted as casts which are popularly referred as vermicast.

Availability of Organic wastes

ith increasing human population and its demands there is increase on of various types of wastes, ranging from agricultural, domestic, nicipal sewage and thus resulting in various types of pollution problems. Leaving aside industrial non-recyclable waste, most decomposable organic wastes can be recycled with utilization of earthworms. This not only reduces the pollution but also reduces chemical fertilizer and other economic inputs besides creation of employment opportunities.

In India, the manurial resources and organic wastes are from human wastes, livestock, crop residues, plant wastes, aquatic weeds, green manure, urban rural waste, agro industries waste etc. It is estimated that annual production of organic wastes in India is about 3000 million tones in addition to 200 million tonnes of liquid and animal excreta. Besides this a huge quantity of garbage and industrial wastes are available.

Municipal wastes are largely generated by urban human population and are some of the most problematic, polluting and expensive to dispense. It is well known that higher living standard generates more wastes with lesser bio-degradable, while lower living standards mean less waste and lesser bio-degradable. In the developed cities like Mumbai, Kolkata, Delhi and Chennai daily production of Municipal Sewage Waste (MSW) per person is 600-800 g. It is estimated that in India about 25 million tones of MSW is generated annually. Unfortunately published figures on tonnage of production, recycling, expenditures involved and recovery are not available. These, with proper processing *viz.*, vermicomposting are expected to generate high economic returns.

Animal Dung

In India, annual production of cattle and buffalo dung is 2028 mt., which comprises of 80% of all recyclable organic wastes. It is estimated that manurial potential of N, P_20_5 and K_20 from bovine dung alone is to be 3442, 1307 and 2214 mt. annually.

Earthworm - Digestive and Reproductive system

Earthworm body is muscular tubular (elongated) with numerous external rings which more or less internally correspond to internal segments. Here we describe the external characters digestive system, reproductive system and cocoon production pertaining to vermiculture.

External characters

All earthworm species have long elongated tubular body, which is generally circular in section, with annuli (formed with superficial grooves). This represents typical vermiform body. However, close examination reveals that shape of body, if cut into section vary in different species, from nearly circular, squairish or trapezoidal. In some species like an arboreal *Perionyx* spp., body is dorsoventrally flattened, which is in most live forms of other species. Body colours also somewhat vary in different species due to pigments and range from typical rich brown to various hues of red, grey and purple. In some, there is some luminance too. In live earthworms, colour can help in differentiating habitat, *viz.,* burrowing forms have light pigmentation, while topsoil/surface dwellers have darker pigmentation.

Externally whole body has series of circular depressions or furrows called segments or metamers. These tally with internal segmentations too and their numbers, are well in relation to internal organs which are characteristic features in different species. Mouth lies at the anterior most tip with a bulbous lip called prostomium. Shape and placement of prostomium is also characteristic features in different species. This first metamere with prostomium is called buccal segment, which does not correspond to internal segmentation.

All along metameric segments or mid region are attached chitinous hook like structures called chaetae or setae. These are retracticle and provide hold in locomotion. Their number, placement, shape, arrangement pattern and descriptions in relation to characteristic features (the pores) are taxonomic characters.

There are numerous pores opening out externally. These have been named with functions and location on body and are of taxonomic importance. Most of these are not visible without use of magnifying glass. Various pores are: (1) Dorsal pores located within intersegmental furrows; (2) Nephridial pores, which are, minute and scattered irregularly over body surface.

Digestive System

The digestive system comprises a tubular alimentary canal extending from mouth to anus. On dissection various parts of the system can easily be differentiated, namely buccal cavity, pharynx, oesophagus, crop, gizzard, intestine and rectum.

Mouth is at extreme anterior end (1st segment) and opens into a short buccal cavity. At tip of mouth, a bulbous lip is attached and is called prostomium whose size, shape and location varies between species. So, it is a taxonomic character. In some species like common Indian earthworm, *Metaphire posthuma* buccal cavity is continuously protruded out to pick food particles. Union of prostomium and buccal cavity containing receptor cells is very sensitive as contain receptor cells. Due to these perhaps earthworms can discriminate and show preference to different kinds of food items.

Pharynx joins a tubular structure called oesophagus in which calciferous glands open. These secrete calcium and carbon dioxide. Calcium probably neutralises contents of alimentary canal making excreted cast alkaline, thereby helps in reduction of acidity of soil (*i.e.,* through cast and with use of Vermicompost). Posterior end of oesophagus is thin walled in some species and can be called as crop, which acts as storage organ. This joins gizzard.

Gizzard is a prominent muscular structure, which functions as masticatory or for grinding of food substrate. This is done with the help of grit particles within food and with internal cuticular epithelium or lining within gizzard, besides muscular contractions etc. Ingested food at this stage is grinded into fine particles. Some species of worms have two gizzards in sequence and some have upto ten gizzards that appear as a beaded muscular tube. These also comprise taxonomic characters. However, some aquatic species do not have crop and gizzard.

Gizzard follows tubular canal which has sphincters valves at both end and internally has epithelial folds having glandular and non-ciliated cells known popularly as stomach. In stomach proteolytic enzymes are secreted which digest proteins. In some species, namely common Indian earthworm (*Metaphire* spp.) and in *Eutyphoeus* spp. there are calciferous glands also. These, however, show structural variation between species and are important from research point of view. These were studied by few top zoologists so deserve mention here. Charles Darwin opined that functionally secretions of calciferous gland is to neutralize humic acid of decaying leaves that earthworm consume. Robertson disagreed and opined that the function is to excrete extra calcium. While top old era Indian Zoologist, namely S. Stephenson, Bani Prasad and K.N. Bahl gave structures of these calciferous glands between 1930-1950. Since then, however, not much has been studied despite curriculum studies of earthworm all over country and rather world over.

Stomach follows or joins a thin-walled tubular structure called intestine that ends at the anus. Functionally this is the main organ wherein digestion and absorption of ingested food takes place. Structurally in intestine, three regions can be differentiated. These are namely: (i) Pre-typhlosolar region, which has vascular internal folds. In the region after few segments, there are conical outgrowths called intestinal caeca. These, according to Chen and Puh, structurally, caecae have glandular epithelial cell and secrete amylatic enzyme, so these are digestive glands. (ii) Typhlosolar region is the middle part of intestine and structurally is differentiable with a medium longitudinal folds inside the intestine called typhlosole. This structure shows variation in content to its size, shape and blood supply. In common Indian earthworm, it is more or less rudimentary, while in common European earthworm, *Lumbicus* spp., typhlosole is very well developed and contain chlorgogen cells. Functionally this is to increase food absorption with increased surface area. Chlorgogen cells functionally, are believed to be either excretory or function as storage cells of nutrients in absence of liver in earthworms. (iii) Last part of intestine is known as rectum and contains faeces as rounded pellets, which are excreted through anus.

The excreted faeces are known as castings or casts. Collective mass of casts from earthworm culture or decomposable organic wastes is generally called "vermi-compost." However, technically cast and vermi-compost is differentiable, former being excretment from soil and later being casts from decomposed organic matter.

Physiology of digestion in earthworms in many respects is similar to function in higher animals. Depending upon feeding habits, different species have different enzymes for digestion. In most species, enzyme amylase is secreted that digests starch, thereby C/N ratio of substrate is reduced after passing through intestine, and the material is converted in readily assimilable form for plants. Other species may have enzymes like diastase, which converts starch into sugar. Some have glycogen hydrolysing ferment, lipase (for digesting fats), invertase (acting of cane-sugar) and oxidising ferment-catalase. In some earthworms evidence of presence of cellulase and chitinase have been reported, but are suggested to have been produced by symbiotic bacteria and protozoans within gut of worms. The absorbed food is carried through blood and coelomic fluid to different parts of the body.

Reproductive and cocoon production

Fertilization as a rule in earthworms is cross-fertilization. For this, exchange process copulation takes place which is typical. Two individuals lie opposed to each in head to tail position with spermathecal pores, closely addressed to each other pores. Each copulating individual move sequentially till all sperms are discharged into spermathecae. Separation of copulating individuals succeeds changes in clitellum, which gradually gets covered with mucilage. This follows release of ovum by individuals within mucilage cover and comprises the stage when fertilization takes place with simultaneous release of sperms. After fertilization, the cocoon or the egg capsule is formed from clitellar gland. It is three layerd and is believed to be made from some kind of chitinous material and also has proteins. This is actually formed in most species when individual worm retracts out from the albuminous mucilage ring over its clitellum.

Shapes of cocoon in general are ovoid but with careful observations, shapes can be defined in different species. Freshly laid cocoon are whitish or dull white, soft and jelly like. In most part of India *Metaphire posthuma* cocoons are laid from April–October with peaks between August-October. In general optimum condition, for production of cocoon and its survival for further progeny, are substrate temperature range between 15-25^{0}C with 20-40% moisture i.e. 60-100% air humidity, ionic conductivity below 3 m mhos/cm in darkened environment.

In each cocoon, number of eggs may range high. However, survival is only in one or two or three. Incubation period is 3-4 weeks in *Eisenia foetida* and 2-3 weeks in *Eudrilus eugeniae*. On an average, young ones grow to maturity in 2-13 months. In African species widely used for vermiculture, freshly emerged individuals have rapid growth and attain reproductive stage within 6 weeks. In such individuals cocoon laying continues for 3-6 months when gradually number decreases. Worms should then be harvested for other purposes.

Pre-requisite for successful Vermicomposting

In natural conditions, reproductive potential and other activities of earthworms depend upon some important factors, like environmental conditions and soil pH. Most earthworms prefer neutral soils. *Eisenia foetida*, common sewage worm, prefers alkaline soils and some prefers slightly acidic soils. Moisture conditions of soil as well as that of worm body influence its distribution, viz. vertical distribution to soil surface when availability of leaf litter is more. So better growth and soil conditioning is expected unless humus is available in lower reaches (as in pits). Under decreased moisture conditions, worms go in deeper layers and may undergo diapause. Temperature has also direct bearings to reproduction and distribution etc. as it affects various activities like metabolism, respiration, growth, reproduction and hatching of cocoons. Extremes of temperature range, limit tolerances and activities. Types of soil also determine the total biomass of worms to be supported. In general, light and medium loam soils have more worms than heavier clay and alluvial soils. The kind and amount of food influences their type, number, growth and fecundity.

Some experimental results revealed that earthworms produce more cocoons when feed on decaying organic matter or on nitrogen rich diets. Amongst vegetable matter, leaf litter is most preferred.

Temperature

Optimum temperature for earthworm is between 68° - 77°F (20-30 degrees C). However, they will tolerate a range of +40°F to +85°F. At 40° F the worms are less active and at 85° F they become stressed. If the worms freeze, they will die.

Moisture

Moisture should be about 40 percent. Too much moisture will close up the air spaces and induce formation of unpleasant odours. Moisture can be measured by squeezing a ball of composting matter in hand. If more than two or three drops come out, it is too moist. Shredded paper or leaves or dry cow dung manure or leaf litter can be mixed with substrate to absorb the excess moisture.

Light

The worms are sensitive to light so keep it dark. Although worms have no eyes, the sensory receptors in their skin are sensitive to light, particularly light of mixed wavelengths. When exposed, they will burry themselves in their bedding and will stay there until it is dark again.

Oxygen

Worms are aerobic digesters. They require oxygen just as we do. It diffuses through their skin and they exhale carbon dioxide by the same process. If their environment becomes anaerobic, they will seek more favourable conditions.

Vibration

Worms will try to leave their home if it is located on or near a source of vibration (e.g. next to heavy machinery, vibrators, refrigerators etc.).

pH:

pH 5.5 is preferred but can tolerate to wide a range of pH 4 to pH 9.

Step wise work at start of vermiculture, in particular for developing it eventually at commercial level, necessiate appreciation of some points at planning level *viz*. (1) site selection (2) availability of decomposable organic waste, its daily quantity and quality, alternate organic material, transportation and stocking etc. (3) marketable outlet and requirement with future scope (4) collection and study of know how on earthworms to be cultured (5) collection/procurement of suitable species (6) testing of suitability of species on performance in available organic waste (7) maintenance of seed culture for eventual large-scale culture.

In addition to these, numerous allied points like funding, availability of labour, quantity of water etc. are to be kept in mind. In terms of wider utilization of vermiculture, first step is selection of suitable species for culture and eventual vermicomposting.

Selection of species

The selection of suitable species for vermiculture is based on the requirement, either for vermicomposting or for fish baits. As yet demands in India for bait is meagre. For all culture feeds, care has to be taken that earthworm feed does not harbour parasites and pathogens. Earthworms are invertebrates assigned to Phyllum *Annelida*, Class *Chaetopoda* and Order *Oligochaeta*. *Oligochaeta* includes the major earthworms belonging to *Megascolicidae*, *Lumbricidae* and other families. More than half the earthworm species of the world belong to *Megascolicidae*. The genus *Pheritima* alone has a large number of species. Both Megascolicidae and Lumbricidae are valuable to agriculture and are therefore intimately limited to human welfare and development programmes.

The commonly used species are *Eisenia foetida, Perionyx excavatus, Lumbricus rubelus, Lumbricus terestris, Eudrilus spp., Lampito mauritii, Octochaetona serrata, Drawida willsii, Octochaetona surensis and Octochaectona thurstoni.*

Theoretical background of selection of suitable species is based on certain biological and ecological parameters. Technically these have been referred as habitat characteristics, distribution in soil or feed media and tropic functions. Based on these parameters, we find that there are three types of earthworms, which can be used in vermiculture. These are (1) Epiges, (2) Endoges and (3) Aneciques.

1. **Epiges**: These are small sized, with uniform body colouration, live on surface litter or dung, tolerate disturbance, have active gizzard; but have limited period of activity. These are phytophagous. Most of these species are good bio-degraders, so are good nutrient releasers, but not redistribute nutrients. These have short life cycle with high reproduction and regeneration rate.

2. **Endoges**: These are small or large sized worms with weak pigmentation, found in top soil layer of organic and mineralized matter. Burrow branching are horizontal and worms moderately tolerate to disturbance. These are geophagous. Life cycle is of intermediate duration, but is potential in soil improvements due to high efficiency in energy utilization from poor soils.

3. **Aneciques**: These are large sized worms with pigmentation only at anterior and posterior ends. These are largely nocturnal, deep burrowing and pull leaves or litter matter in to soil and are phytogeophagous. Tolerance to disturbance is poor and reproduction rate as evidenced from cocoon production is low. May have useful role in mixing nutrients as from deep burrow come up to surface for casting.

So far in whole world, only few species have been studied for vermiculture and majority remain unstudied.

Basic characteristics of suitable species

Farmers or gardeners need only those species, which can quickly convert organic waste to vermi-compost. Therefore, a worm should have following characteristics:

1. Worm species should be tolerant to diseases.
2. The culturing techniques should be simple enough to adopt.
3. Worm should be efficient converter of plant or animal bio-mass to body proteins, so that its growth rates are high.
4. It should have high consumption, digestion and assimilation rate (composting qualities).
5. The worms should have wider adaptability (tolerance) to environmental factors (capability to live in varying temperature conditions).
6. The worms should have feeding preference and adaptability for wide range of organic material (high and rich organic matter).
7. The worm should produce large number of cocoons that should not have long hatching time, so that multiplication and organic matter conversion is fast.
8. Growth rate, maturity from young one to adult stage should be fast.
9. Worm should have compatibility or tolerance with other worms (as with possibility of mixture of species) as it would add to productivity of biomass (worms) and conversion rate at different strata (layers) of organic matter, *i.e.,* faster composting. This feature i.e., composting with different layer feeders have so far received very little attention.
10. Worms on introduction in substrate, should have least inactivity period (vermistabilization period).
11. The worms should feed near the surface of organic matter.

Vermicompost materials

Almost all types of biologically degradable and decomposable organic wastes can be used in vermiculture and vermicomposting. However, nutrient constituents vary *viz.* poultry droppings are rich in calcium, phosphates and other salts. This has to be diluted by mixing with leaf litter, soil and cow dung. Commonly used composting materials are listed below:

Animal dung: Dung of almost all domestic animals like cattle, sheep, horse, goat and poultry droppings etc. In case of animal dungs other than cattle dung, various preliminary testing and precautions for pathogens and responses to earthworms are necessary.

Agricultural wastes: These comprise of all items discarded after harvesting and threshing of produce *e.g.* stem, leaves, husk, peels, vegetable waste, orchard leaf litter, processed food waste, sugarcane trash, press mud and processing waste etc.

Forestry wastes: These also comprise various types of plant products like wood shavings, peels, saw dust and pulp. All these materials besides various types of forest leaf litter can be used.

City leaf waste: As yet no data is available in tonnage of leaf litter from avenue or residential areas that is burnt. All this can be converted into vermicompost. This would also keep cities clean.

Waste paper and cotton cloth: These are decomposable organic wastes. These, if are not being recycled for other useful products, can be recycled with vermicomposting.

City refuge: Considerable portion of city refuge can be sorted and recycled or composted. For composting however, essential requirement is that citizens avoid mixing of toxicants. There may be sorting or separation right at generation or producers level.

Biogas slurry: After recovery of biogas, if not required for agricultural use *viz.*, in conventional composting, it can be used for vermicomposting.

Industrial waste: Some types of industrial wastes *viz.*, waste from food processing, distillery etc. can be used in vermiculture with some manipulations in regard to vermicompostable conditions.

Organic wastes that are decomposable are to be subjected to certain preliminary treatments to enhance vermicompostability. First step is proper collection, sorting or separation of compostable, non-compostable and non-biodegradables like plastics, stone, glass, ceramic and metals. Even in kitchen wastes, heavily spicing wastes should be got separated. Separated matter, if possible be spread in a layer up to one foot and be exposed to sun for a day. This removes foul smell and kills unwanted organisms. This is always not possible in present day city living where open space is a limiting factor. In such a situation, mixing of decaying organic waste produce be done with somewhat pretreated leaf litter in approximate ratio ranging from 10-40% of the waste to be vermicomposted. Leaf litter or agricultural wastes should be broken into pieces for enhancing decomposition. Vegetable produce infected with insect pests require some chemical treatment (4% neem insecticidal formulation) because their life stages continue in waste and hinder earthworm activities.

Precautions in addition of food material

Fresh feed material should be placed immediately after removing the compost on the worm bed, so that they again start endeavouring it.

Water logging leads to anaerobiosis and change in the pH of the medium. Moisture level should be only 30-40 per cent.

If too much food is added, the system can become overloaded and cause an odour. Anaerobic condition is harmful to the worms. Providing adequate aeration and moisture minimizes odours.

A mixture of cattle and sheep dung with gram and wheat bran and vegetable wastes forms the ideal feed for worms. Mixing of gram bran with dung mixture in 3:10 ratio hastens the growth of the worms. Addition of kitchen waste in the same proportion increases the population of worms.

Keep bedding moist, but not saturated. The vermicomposting process generates water, and too much will clog air spaces and reduce circulation. If air is limited, it will reduce the efficacy of the worms to process the waste.

The pH of the worm bed should not exceed 5.5 to 7. Egg shells are a good pH monitor, measuring the acidity of the bed. If the egg shells disappear too fast (within a week) it means the pH is too low.

Maintenance of vermicompost beds

Preparation of vermi-composting beds after necessary pre-tests and selection of suitable species is done. This is, in fact, one of the most important aspects of whole programme.

Production of Vermicompost

There are different methods of production of vermicompost.

1. Surface Bed Method

Vermi-composting can be done on ground. For this the ground is leveled. Over this, soil free of stones, glass or any form of chemical contaminants is plastered and a manageable sized platform is made. Depending upon the availability of raw materials and earthworms the beds of convenient length, width and height are to be prepared. However, the width and height should not exceed 3 and 2 feet, respectively. The layers in the bed are to be made in the following manner for effective composting. The whole bed is covered with a hessian cloth. If the beds are already under thatched roof shade, heaps can also be covered with broad leaves locally available. These, however, have to be periodically replaced as they are decomposed and eaten up by worms.

Layers in the bed (Top to Bottom)

Leaves/waste materials	25 - 30 cm
Partially digested dung	5 - 8 cm
Earthworms/cocoons	0.5 - 1 cm
Partially digested dung	8 - 10 cm
Bedding material	8 - 10 cm

Bedding material

The bedding needs to hold air spaces so that noxious odours do not form. Good bedding can be any combination of aged manures, i.e. shredded coconut husk, *coir* (coconut fibre), shredded paper products, sugarcane trash, decomposing leaves, straw, wood chips, peat moss, etc. The bedding material will also be eaten by the earthworms and should be discarded after six months and a fresh bed material should be laid.

2. Worm Bin Method (Containers)

The selected bin/box is to be cleaned for removing unwanted chemical or other material, if any present. At bottom a 2 to 3 inches thick layer of any biodegradable matter is laid. Over this layer, 2 to 3 inches thick layer of partly digested and powdered cow dung is put. The whole material in bedding is sufficiently moistened (up to 40% moisture) and then live earthworms are gently released over it. A bin/box of 1 m x 1 m x 0.5 m can hold around 1000 to 1500 worms which require about 30 to 40 kgs of whole organic matter. On top any of the earthworm feed matter is placed in 7 to 9 inches thick layer and watered. Feed layer should be replaced when previous layer disappears (converted into vermin-compost). Finally on top a moist gunny bag is spread and the vermi-compost should be periodically removed.

Pre treatment of wastes

Large or small quantities of plant waste to be vermicomposted, be heaped and mixed with small quantity of FYM (say ½ kg FYM/10-20 kg plant waste). This is only to give bacterial inoculum for enhancing decomposition process and eventual faster vermicomposting. These heaps should be sprinkled with little water (5 liters /20 kg waste) and covered with a Hessian cloth or left on ground. Whole waste heap need to be kept moist for 3-7 days and periodically it is upturned and mixed. In fact, only after 2-7 weeks heap comes in a stage that vermicomposting is fast. After such treatment waste is ready for vermicomposting.

Vermi-composting Efficiency

In general, a bed of 1x1x0.3m requires 30-40 kg of bedding and feed material. This can support 1000 to 1500 earthworms, which would multiply, and compost the matter from upper layers. So upper layers of organic matter should be periodically replenished. Many experts opine that from top layer, accumulating casts should be periodically removed as it reduces vermic activity or reproduction. Vermicompost is reported to contain 5 times more nitrogen, seven times more phosphorus, 11 times more potassium, 2 times more magnesium and calcium. These are in a form that is readily assimiable to plants.

It has already been discussed that decomposition of organic waste depends upon several factors. Earthworm activities like burrowing, feeding and defecating (casting) exhilarate process of decomposition. So larger the number of earthworms present in an appropriate medium and conditions, faster would be vermicomposting. Thus initially, *i.e.,* immediately after introduction of worms, first lot of vermicompost is ready within 60-70 days. Gradually with bacterial decomposition leading to breaking of larger masses and increase in numbers of worms, vermicompost is ready in 30-40 days only. According to available extrapolative estimates, 1kg of earthworms (i.e., nearly 1000 adult *Eugenia species* worms) would produce 10 kg casts in 60-70 days.

Collection of Vermicompost

When vermicompost and vermicastings are ready for collection, top layers appear somewhat dark brown, granular as if used dry tea leaves have been spread over the layer. Watering should then be stopped for 2-3 days and gently compost should be scrapped from top layers. This should then be left undisturbed for 6-24 hours. If there were any adult worms present these would move down or away from the heaped material. The above material is sieved to avoid any adult/cocoon being removed. After collection of vermicompost from top layer, feed material is again replenished and composting process is rescheduled. Collected vermicompost is air dried in shade before packing in convenient size.

Application of Vermicompost

Horticulture

- Quantity depends on age and size of the plant
- It is preferred to apply by mixing equal quantity of FYM
- Vermicompost and soil in the ratio of 1:2 is best for potted plants. Vermicompost is not phytotoxic
- Two applications of Vermicompost 150g and 500g per 5 kg soil of potted plants gave good results in most plants.

Agriculture

- Limited applicationdue to low level of vermicompost production in India
- At seedling stage mixed with FYM in equal proportion and applied
- Dosage 5 tones/ ha is advisable

Worm castings are very nutrient-rich, so use them sparingly. Although worm castings can be used in the same way as farmyard manure, they are so rich that they should be used in smaller amounts.

Potting mixture - Worm castings make an excellent addition to a home made or commercial potting mix. Mix one part of worm castings with three parts of a favourite potting mix. Equal amounts of worm castings, peat moss, perlite and sand or garden soil also make a good potting mix.

Transplants - When transplanting plants into the garden from nursery, place a handful of vermicompost into the hole before planting.

Amending planting beds - Spread 1/2 to 1 inch of castings and incorporate into the soil with a fork or spade.

Side dressing for houseplants or garden plants - Spread 1/2 to 1 inch of castings around established plants and mix into the soil with a trowel.

Caution

Worm castings are not recommended for seedbeds. Since compost beds do not get hot, it is common for the seeds of weeds to survive in the composting process and resprout. In addition, when worm castings are used in outdoor seedbeds, birds may disturb the sprouting seeds while hunting for worms and worm eggs in the castings.

Worm castings are also not recommended for mulching. Worm castings are too nutrient-rich to use as mulch on outdoor plants; they tend to dry out and the nutrients are wasted. Using worm castings to mulch houseplants can result in fruit fly problems.

The more you know about worms and what they need, the more you'll enjoy the vermicomposting experience!

REFERENCES

Bhatnagar, R.K. and R.K. Palta (2002). Earthworm vermiculture and vermicomposting. Kalyani Publishers, Ludhiana

Bhinday, M.R. (1995). Technology for earthworm rearing. Farm Digest, **5**(4) :19-20.

Palniappan, S. and K. Annadurai (1999). Application of vermiculture biotechnology. In – Organic farming : Theory and Practice. Scientific Publishers (India), Jodhpur, pp. 74 -92.

Renuka, B. and C. Ravishanker (2001). Effect of organic manures on growth and yield of tomato. South Indian Hort. 49 (Special) : 216-219 .

Robinson, Ch., P. Ineson, T. G Pierra and A. P. Rowland (1992). N mobilization by earthworms in limited peat soils under *Picea sitchensis*. J. Applied Ecology, **29**: 236-237.

Sharma, A. K. (2002). Vermicomposting. In – A Handbook of Organic Farming. Agrobios (India) Jodhpur, pp. 193-215.

Subba Rao, T. S. S. and C. Ravishanker (2001). Effect of organic manures on growth and yield of Brinjal. South Indian Hort. 49 (Special) : 288 -291.